solutions manual

to accompany

Paul A. Tipler

physics for scientists and engineers

Fourth Edition

Volume 1 Chapters 1–21

Frank J. Blatt
Professor Emeritus
Michigan State University

Solutions Manual, Volume 1, Chapters 1–21
by Frank J. Blatt
to accompany
Tipler: *Physics for Scientists and Engineers,* Fourth Edition

Printed in the United States of America

ISBN: 1-57259-514-0

Fifth Printing 2003

W. H. Freeman and Company
41 Madison Avenue
New York, New York 10010
http://www.whfreeman.com

contents

CHAPTER 1

Systems of Measurement

1* • Which of the following is *not* one of the fundamental physical quantities in the SI system?

(*a*) mass (*b*) length (*c*) force (*d*) time (*e*) All of the above are fundamental physical quantities.

(*c*) Force is *not* a fundamental physical quantity; see text.

---

2 • In doing a calculation, you end up with m/s in the numerator and m/s^2 in the denominator. What are your final units? (*a*) m^2/s^3 (*b*) 1/s (*c*) s^3/m^2 (*d*) s (*e*) m/s

(*d*) (m/s)/(m/s^2) = (m/s)(s^2/m) = s.

---

3 • Write the following using prefixes listed in Table 1-1 and the abbreviations listed on page EP-1; for example, 10,000 meters = 10 km. (*a*) 1,000,000 watts, (*b*) 0.002 gram, (*c*) 3×10^{-6} meter, (*d*) 30,000 seconds.

(*a*) 1 MW (*b*) 2 mg (*c*) 3 μm (*d*) 30 ks

---

4 • Write each of the following without using prefixes: (*a*) 40 μW, (*b*) 4 ns, (*c*) 3 MW, (*d*) 25 km.

(*a*) 0.000040 W (*b*) 0.000000004 s (*c*) 3,000,000 W (*d*) 25,000 m

---

5* • Write out the following (which are not SI units) without using any abbreviations. For example, 10^3 meters = 1 kilometer. (*a*) 10^{-12} boo, (*b*) 10^9 low, (*c*) 10^{-6} phone, (*d*) 10^{-18} boy, (*e*) 10^6 phone, (*f*) 10^{-9} goat (*g*) 10^{12} bull.

(*a*) 1 picoboo (*b*) 1 gigalow (*c*) 1 microphone (*d*) 1 attoboy (*e*) 1 megaphone (*f*) 1 nanogoat (*g*) 1 terabull

---

6 •• In the following equations, the distance x is in meters, the time t is in seconds, and the velocity v is in meters per second. What are the SI units of the constants C_1 and C_2? (*a*) $x = C_1 + C_2 t$ (*b*) $x = \frac{1}{2} C_1 t^2$ (*c*) $v^2 = 2C_1 x$ (*d*) $x = C_1 \cos C_2 t$ (*e*) $v = C_1 e^{-C_2 t}$

(*a*) C_1 and $C_2 t$ must be in meters	C_1 in m; C_2 in m/s
(*b*) $C_1 t^2$ must be in meters	C_1 in m/s^2
(*c*) v^2 is in m^2/s^2; so $C_1 x$ must be in m^2/s^2	C_1 in m/s^2
(*d*) cos $C_2 t$ is a number; $C_2 t$ is a number	C_1 in m; C_2 in s^{-1}
(*e*) v is in m/s. $C_2 t$ is a number	C_1 in m/s; C_2 in s^{-1}

---

7 •• If x is in feet, t is in seconds, and v is in feet per second, what are the units of the constants C_1 and C_2 in each part of Problem 6?

It is only necessary to change the unit of length in each case from m (meter) to ft (foot). Thus, for example, in part (*e*), we have C_1 in ft/s; C_2 in s^{-1}.

---

8 • From the original definition of the meter in terms of the distance from the equator to the North Pole, find in meters (*a*) the circumference of the earth, and (*b*) the radius of the earth. (*c*) Convert your answers for (*a*) and (*b*) from meters into miles.

(*a*) The Pole–Equator distance is one-fourth of the circumference — $C = 4 \times 10^7$ m

(*b*) Use $C = 2\pi R$ — $R = (4 \times 10^7\ \text{m})/2\pi = 6.37 \times 10^6$ m

(*c*) Use the conversion factor (1 mi)/(1.61 km) = 1 — $C = (4 \times 10^7\ \text{m})(1\ \text{mi}/1.61 \times 10^3\ \text{m}) = 2.46 \times 10^4$ mi

$R = (6.37 \times 10^6\ \text{m})(1\ \text{mi}/1.61 \times 10^3\ \text{m}) = 3.96 \times 10^3$ mi

---

9* • The speed of sound in air is 340 m/s. What is the speed of a supersonic plane that travels at twice the speed of sound? Give your answer in kilometers per hour and miles per hour.

1. Find the speed in m/s — v = 2(340 m/s) = 680 m/s = 0.680 km/s
2. Use [1 h/(60 min/h)(60 s/min)] = 1 — v = (0.680 km/s)/(1 h/3600 s/h) = 2450 km/h
3. Use (1 mi/1.61 km) = 1 — v = (2450 km/h)(1 mi/1.61 km) = 1520 mi/h

---

10 • A basketball player is 6 ft $10\frac{1}{2}$ in tall. What is his height in centimeters?

1. Express H in inches — $H = (6 \times 12 + 10.5)$ in = 82.5 in
2. Use (1 in/2.54 cm) = 1 — H = (82.5 in)/(1 in/2.54 cm) = 210 cm

---

11 • Complete the following: (*a*) 100 km/h = ___mi/h; (*b*) 60 cm = ___in; (*c*) 100 yd = ___m.

(*a*) (100 km/h)(1 mi/1.61 km) = 62.1 mi/h; (*b*) (60 cm)(1 in/2.54 cm) = 23.6 in;

(*c*) (100 yd)/(1 yd/0.9144 m) = 91.4 m.

---

12 • The main span of the Golden Gate Bridge is 4200 feet. Express this distance in kilometers.

(4200 ft)/(1 ft/0.3048 m)(1000 m/1 km) = 1.28 km.

13* • Find the conversion factor to convert from miles per hour into kilometers per hour.

Since 1 mi = 1.61 km, (v mi/h) = (v mi/h)(1.61 km/1 mi) = 1.61v km/h.

---

14 • Complete the following: (*a*) 1.296×10^5 km/h² = ___ km/h·s; (*b*) 1.296×10^5 km/h² = ___m/s²; (*c*) 60 mi/h = ___ft/s; (*d*) 60 mi/h = ___m/s.

(*a*) (1.296×10^5 km/h²)(1 h/3600 s) = 36.00 km/h·s; (*b*) (36.00 km/h·s)(1 h/3600 s)(1000 m/1 km) = 10 m/s²;

(*c*) (60 mi/h)(5280 ft/1 mi)(1 h/3600 s) = 88 ft/s; (*d*) (88 ft/s)(0.3048 m/1 ft) = 27 m/s.

---

15 • There are 1.057 quarts in a liter and 4 quarts in a gallon. (*a*) How many liters are there in a gallon? (*b*) A barrel equals 42 gallons. How many cubic meters are there in a barrel?

(*a*) 1 gal = (1 gal)(4 qt/1 gal)(1 L/1.057 qt) = 3.784 L;

(*b*) 1 ba = (1 ba)(42 ga/1 ba)(3.784 L/1 ga)(1 m³/1000 L) = 0.1589 m³.

16 • There are 640 acres in a square mile. How many square meters are there in one acre?

1 acre = (1 acre)(1 mi^2/640 acres)(1610 m/1 mi)^2 = 4050 m^2.

17* •• A right circular cylinder has a diameter of 6.8 in and a height of 2 ft. What is the volume of the cylinder in (*a*) cubic feet, (*b*) cubic meters, (*c*) liters?

(*a*) 1. Express the diameter in feet — $D = (6.8\text{ in})(1\text{ ft}/12\text{ in}) = 0.567\text{ ft}$

2. The volume of a cylinder is $V = (\pi D^2/4)H$ — $V = [\pi(0.567\text{ ft})^2/4](2\text{ ft}) = 0.505\text{ ft}^3$

(*b*) Use (1 ft/0.3048 m) = 1 — $V = (0.505\text{ ft}^3)(0.3048\text{ m}/1\text{ ft})^3 = 0.0143\text{ m}^3$

(*c*) Use (1000 L/1 m3) = 1 — $V = (0.0143\text{ m}^3)(1000\text{ L}/1\text{ m}^3) = 14.3\text{ L}$

18 •• In the following, x is in meters, t is in seconds, v is in meters per second, and the acceleration a is in meters per second squared. Find the SI units of each combination: (*a*) v^2/x; (*b*) $\sqrt{x/a}$; (*c*) $\frac{1}{2}at^2$.

(*a*) The units of v^2/x are $(m^2/s^2)/m = m/s^2$; (*b*) the units of $\sqrt{x/a}$ are $[m/(m/s^2)]^{1/2} = s$;

(*c*) the units of $\frac{1}{2}at^2$ are $(m/s^2)s^2 = m$.

19 • What are the dimensions of the constants in each part of Problem 6?

Referring to the results of Problem 6 we see that the dimensions are:

(*a*) $C_1 : L$, $C_2 : L/T$; (*b*) $C_1 : L/T^2$; (*c*) $C_1 : L/T^2$; (*d*) $C_1 : L$, $C_2 : 1/T$; (*e*) $C_1 : L/T$, $C_2 : 1/T$.

20 •• The law of radioactive decay is $N(t) = N_0e^{-\lambda t}$, where N_0 is the number of radioactive nuclei at $t = 0$, $N(t)$ is the number remaining at time t, and λ is a quantity known as the decay constant. What is the dimension of λ?

Since the exponent must be a number, the dimension of λ must be $1/T$.

21* •• The SI unit of force, the kilogram-meter per second squared (kg·m/s^2) is called the newton (N). Find the dimensions and the SI units of the constant G in Newton's law of gravitation $F = Gm_1m_2/r^2$.

1. Solve for G — $G = Fr^2/m_1m_2$
2. Replace the variables by their dimensions — $G = (ML/T^2)(L^2)/(M^2) = L^3/(MT^2)$
3. Use the SI units for L, M, and T — Units of G are $m^3/kg{\cdot}s^2$

22 •• An object on the end of a string moves in a circle. The force exerted by the string has units of ML/T^2 and depends on the mass of the object, its speed, and the radius of the circle. What combination of these variables gives the correct dimensions?

1. Write the relationship in terms of powers of the variables — $F = m^av^br^c$
2. Insert the appropriate dimensions — $MLT^{-2} = M^a(L/T)^bL^c$
3. Solve for the exponents a, b, and c — $a = 1;\ b = 2;\ b + c = 1,\ c = -1$
4. Write the corresponding expression — $F = mv^2/r$

23 •• Show that the product of mass, acceleration, and speed has the dimension of power.

1. Write the dimensions of the variables — $m : M,\ a : L/T^2,\ v : L/T,\ P : ML^2/T^3$
2. Find the dimensions of the product mav — $mav : M(L/T^2)(L/T) = ML^2/T^3 : P$

24 •• The momentum of an object is the product of its velocity and mass. Show that momentum has the dimension of force multiplied by time.

The dimension of mv is ML/T; the dimension of force times time is $(ML/T^2)(T) = ML/T$.

25* •• What combination of force and one other physical quantity has the dimension of power?

(ML/T^2)(dimension of Y) $= ML^2/T^3$; dimension of Y $= (ML^2/T^3)/(ML/T^2) = L/T$; Y = velocity.

26 •• When an object falls through air, there is a drag force that depends on the product of the surface area of the object and the square of its velocity, i.e., $F_{air} = CAv^2$, where C is a constant. Determine the dimension of C.

1. Solve for the constant C — $C = F_{air}/Av^2$
2. Replace the variables by their dimensions — $C = (ML/T^2)/[L^2(L/T)^2] = M/L^3$

27 •• Kepler's third law relates the period of a planet to its radius r, the constant G in Newton's law of gravitation ($F = Gm_1m_2/r^2$), and the mass of the sun M_S. What combination of these factors gives the correct dimensions for the period of a planet.

1. Write a general expression for the period in terms of the dimensions of the variables — $T = L^aG^bM^c$; note that $G = L^3/(MT^2)$ (see Problem 21); $T = L^a[L^3/(MT^2)]^bM^c = L^{a+3b}M^{c-b}T^{-2b}$
2. Solve for b, c, and a in that order — $-2b = 1$, $b = -½$; $c - b = 0$, $c = -½$; $a + 3b = 0$, $a = 3/2$
3. Write the result in terms of r, G, and M_S — $T = C(r^{3/2})/\sqrt{M_SG}$, where C is a numerical constant.

28 • The prefix giga means _____. (*a*) 10^3 (*b*) 10^6 (*c*) 10^9 (*d*) 10^{12} (*e*) 10^{15}

(*c*) 10^9; see Table 1-1.

29* • The prefix mega means _____. (*a*) 10^{-9} (*b*) 10^{-6} (*c*) 10^{-3} (*d*) 10^6 (*e*) 10^9

(*d*) 10^6; see Table 1-1.

30 • The prefix pico means _____. (*a*) 10^{-12} (*b*) 10^{-6} (*c*) 10^{-3} (*d*) 10^6 (*e*) 10^9

(*a*) 10^{-12}; see Table 1-1.

31 • The number 0.0005130 has _____ significant figures. (*a*) one (*b*) three (*c*) four (*d*) seven (*e*) eight

(*c*) four; the three zeros after the decimal point are not significant figures, but the last zero is significant.

32 • The number 23.0040 has _____ significant figures. (*a*) two (*b*) three (*c*) four (*d*) five (*e*) six

(*e*) six; all digits including the last zero are significant.

33* • Express as a decimal number without using powers of 10 notation: (*a*) 3×10^4 (*b*) 6.2×10^{-3} (*c*) 4×10^{-6} (*d*) 2.17×10^5

(*a*) $3 \times 10^4 = 30{,}000$; (*b*) $6.2 \times 10^{-3} = 0.0062$; (*c*) $4 \times 10^{-6} = 0.000004$; (*d*) $2.17 \times 10^5 = 217{,}000$.

34 • Write the following in scientific notation. (*a*) 3.1 GW = _____W. (*b*) 10 pm = _____m. (*c*) 2.3 fs = _____s. (*d*) 4 μs = _____s.

(*a*) 3.1 GW = 3.1×10^9 W; (*b*) 10 pm = 10^{-11} m; (*c*) 2.3 fs = 2.3×10^{-15} s; (*d*) 4 μs = 4×10^{-6} s.

35 • Calculate the following, round off to the correct number of significant figures, and express your result in scientific notation. (*a*) $(1.14)(9.99 \times 10^4)$ (*b*) $(2.78 \times 10^{-8}) - (5.31 \times 10^{-9})$ (*c*) $12\pi/(4.56 \times 10^{-3})$ (*d*) $27.6 + (5.99 \times 10^2)$

(*a*) The number of significant figures in each factor is three, so the result has three significant figures — $(1.14)(9.99 \times 10^4) = 11.4 \times 10^4 = 1.14 \times 10^5$

(*b*) We must first express both terms with the same power of ten. Since the first number has only two digits after the decimal point, the result can have only two digits after the decimal point. — $(2.78 \times 10^{-8}) - (5.39 \times 10^{-9}) = (2.78 - 0.539) \times 10^{-8}$
$(2.78 \times 10^{-8}) - (5.39 \times 10^{-9}) = 2.24 \times 10^{-8}$

(*c*) We assume here that 12 is an exact number. Hence the answer has three significant figures. — $12\pi/(4.56 \times 10^{-3}) = 8.27 \times 10^3$

(*d*) See (*b*) above — $27.6 + 599 = 627 = 6.27 \times 10^2$

---

36 • Calculate the following, round off to the correct number of significant figures, and express your result in scientific notation. (*a*) (200.9)(569.3) (*b*) $(0.000000513)(62.3 \times 10^7)$ (*c*) $28401 + (5.78 \times 10^4)$ (*d*) $63.25 / (4.17 \times 10^{-3})$

(*a*) Both factors have 4 significant figures — $(200.9)(569.3) = 1.144 \times 10^4$

(*b*) Express the first factor in scientific notation. Both factors have 3 significant figures. — $0.000000513 = 5.13 \times 10^{-7}$
$(5.13 \times 10^{-7})(62.3 \times 10^7) = 3.20 \times 10^2$

(*c*) Express both terms in scientific notation; the second term and result have only 3 significant figures — $28{,}401 = 2.8401 \times 10^4$
$2.8401 \times 10^4 + 5.78 \times 10^4 = 8.62 \times 10^4$

(*d*) The result has 3 significant figures — $63.25/(4.17 \times 10^{-3}) = 1.52 \times 10^4$

37* • A cell membrane has a thickness of about 7 nm. How many cell membranes would it take to make a stack 1 in high?

1. The number of membranes is the total thickness divided by the thickness per membrane — $N = (1 \text{ in})/(7 \times 10^{-9} \text{ m/membrane})$
2. Use all SI units — $N = (1 \text{ in})(2.54 \times 10^{-2} \text{ m/1 in})/(7 \times 10^{-9} \text{ m/membrane})$
3. Solve for *N*; give result to 1 significant figure — $N = 4 \times 10^6$ membranes

---

38 • Calculate the following, round off to the correct number of significant figures, and express your result in scientific notation. (*a*) $(2.00 \times 10^4)(6.10 \times 10^{-2})$ (*b*) $(3.141592)(4.00 \times 10^5)$ (*c*) $(2.32 \times 10^3)/(1.16 \times 10^8)$ (*d*) $(5.14 \times 10^3) + (2.78 \times 10^2)$; (*e*) $(1.99 \times 10^2) + (9.99 \times 10^{-5})$

(*a*) Both factors, and hence result, have 3 sig. figs. — $(2.00 \times 10^4)(6.10 \times 10^{-2}) = 1.22 \times 10^3$

(*b*) The 2nd factor, and hence the result, has 3 sig. figs. — $(3.141592)(4.00 \times 10^5) = 1.26 \times 10^6$

(*c*) See (*a*) — $(2.32 \times 10^3)/(1.16 \times 10^8) = 2.00 \times 10^{-5}$

(*d*) Write both terms using same power of 10; note that result has only 3 sig. figs. — $(5.14 \times 10^3) + (2.78 \times 10^2) = (5.14 \times 10^3) + (0.278 \times 10^3) = (5.14 + 0.278) \times 10^3 = 5.42 \times 10^3$

(*e*) Follow procedure as in (*d*) — $(1.99 \times 10^2) + (0.000000999 \times 10^2) = 1.99 \times 10^2$

39 • Perform the following calculations and round off the answers to the correct number of significant figures: (*a*) 3.141592654 × $(23.2)^2$ (*b*) 2 × 3.141592654 × 0.76 (*c*) $(4/3)\pi \times (1.1)^3$ (*d*) $(2.0)^5/3.141592654$

Note that 3.141592654 is π to 10 significant figures.

(*a*) The second factor and result have 3 sig.figs. $(23.2)^2\pi = 1.69 \times 10^3$

(*b*) We assume 2 is exact; result has 2 sig. figs. $2\pi \times 0.76 = 4.8$

(*c*) Again, (4/3) is assumed exact; 2 sig. figs. $(4/3)\pi \times (1.1)^3 = 5.6$

(*d*) 2.0 has 2 sig. figs., so result has 2 sig. figs. 10

---

40 •• The sun has a mass of 1.99×10^{30} kg and is composed mostly of hydrogen, with only a small fraction being heavier elements. The hydrogen atom has a mass of 1.67×10^{-27} kg. Estimate the number of hydrogen atoms in the sun.

1. Assume the sun is made up only of hydrogen $M_S = N_H M_H$, where N_H is number of H atoms
2. Solve for N_H $N_H = M_S/M_H = 1.99 \times 10^{30}/(1.67 \times 10^{-27}) = 1.19 \times 10^{57}$

41* • What are the advantages and disadvantages of using the length of your arm for a standard length?

The advantage is that the length measure is always with you. The disadvantage is that arm lengths are not uniform, so if you wish to purchase a board of "two arm lengths" it may be longer or shorter than you wish, or else you may have to physically go to the lumber yard to use your own arm as a measure of length.

---

42 • A certain clock is known to be consistently 10% fast compared with the standard cesium clock. A second clock varies in a random way by 1%. Which clock would make a more useful secondary standard for a laboratory? Why?

The first clock is a better secondary standard because one can make a precise correction for the discrepancy between it and the cesium standard.

---

43 • True or false:

(*a*) Two quantities to be added must have the same dimensions.

(*b*) Two quantities to be multiplied must have the same dimensions.

(*c*) All conversion factors have the value 1.

(*a*) True; you cannot add "apples to oranges" or a length (distance traveled) to a volume (liters of milk).

(*b*) False; the distance traveled is the product of speed (length/time) multiplied by the time of travel (time).

(*c*) True; see text.

---

44 • On many of the roads in Canada the speed limit is 100 km/h. What is the speed limit in miles per hour?

(100 km/h)(1 mi/1.61 km) = 62 mi/h.

45* • If one could count $1 per second, how many years would it take to count 1 billion dollars (1 billion = 10^9)?

It would take 10^9 seconds or (10^9 s)(1 h/3600 s)(1 day/24 h)(1 y/365 days) = (10^9 s)(1 y/3.154×10^7 s) = 31.7 y.

---

46 • Sometimes a conversion factor can be derived from the knowledge of a constant in two different systems. (*a*) The speed of light in vacuum is 186,000 mi/s = 3×10^8 m/s. Use this fact to find the number of kilometers in a mile. (*b*) The weight of 1 ft^3 of water is 62.4 lb. Use this and the fact that 1 cm^3 of water has a mass of 1 g to find the weight in pounds of a 1-kg mass.

(*a*) From the data, $(3.00 \times 10^8 \text{ m/s})/(1.86 \times 10^5 \text{ mi/s}) = 1 = 1.61 \times 10^3 \text{ m/mi} = 1.61 \text{ km/mi}$.

(*b*) 1. Find the volume of 1.00 kg of water — Volume of 1.00 kg = 10^3 g is 10^3 cm^3

2. Express 10^3 cm^3 in ft^3 — $(10 \text{ cm})^3[(1 \text{ in}/2.54 \text{ cm})(1 \text{ ft}/12 \text{ in})]^3 = 0.0353 \text{ ft}^3$

3. So $(1.00 \text{ kg}/0.0353 \text{ ft}^3) = (62.4 \text{ lb}/1 \text{ ft}^3)$ — $(62.4 \text{ lb}/1 \text{ ft}^3)/(1.00 \text{ kg}/0.0353 \text{ ft}^3) = 2.20 \text{ lb/kg}$

---

47 •• The mass of one uranium atom is 4.0×10^{-26} kg. How many uranium atoms are there in 8 g of pure uranium?

$N_U = (8.0 \text{ g})(1 \text{ U}/4.0 \times 10^{-26} \text{ kg})(10^{-3} \text{ kg}/1 \text{ g}) = 2.0 \times 10^{23}$.

---

48 •• During a thunderstorm, a total of 1.4 in of rain falls. How much water falls on one acre of land? (1 mi^2 = 640 acres.)

1. Express the area in ft^2 — $(1 \text{ acre})(1 \text{ mi}^2/640 \text{ acre})(5280 \text{ ft}/1 \text{ mi})^2 = 4.356 \times 10^4 \text{ ft}^2$

2. Find the total volume of water on 1 acre in cubic feet. — $V = (4.356 \times 10^4 \text{ ft}^2)(1.4 \text{ in})(1 \text{ ft}/12 \text{ in}) = 5.08 \times 10^3 \text{ ft}^3$

3. Use the fact that 1ft^3 of water weighs 62.4 lb — Weight of water $= (62.4 \text{ lb/ft}^3)(5.08 \times 10^3 \text{ ft}^3) = 3.17 \times 10^5 \text{ lb}$

49* •• The angle subtended by the moon's diameter at a point on the earth is about 0.524°. Use this and the fact that the moon is about 384 Mm away to find the diameter of the moon. (The angle subtended by the moon θ is approximately D/r_m, where D is the diameter of the moon and r_m is the distance to the moon.)

Note that $\theta \approx D/r_m$ for small values of θ only if θ is expressed in radians, and that radians are dimensionless.

1. Find θ in radians — $(0.524 \text{ deg})(2\pi \text{ rad}/360 \text{ deg}) = 0.00915 \text{ rad}$

2. Use $\theta = D/r_m$ and solve for D — $D = \theta r_m = (0.00915)(384 \text{ Mm}) = 3.51 \text{ Mm}$

---

50 •• The United States imports 6 million barrels of oil per day. This imported oil provides about one-fourth of our total energy. A barrel fills a drum that stands about 1 m high. (*a*) If the barrels are laid end to end, what is the length in kilometers of barrels of oil imported each day? (*b*) The largest tankers hold about a quarter-million barrels. How many tanker loads per year would supply our imported oil? (*c*) If oil costs $20 a barrel, how much do we spend for oil each year?

(*a*) $L = (6 \times 10^6 \text{ barrels})(1 \text{ m}/1 \text{ barrel}) = 6 \times 10^6 \text{ m} = 6 \times 10^3 \text{ km}$.

(*b*) Number of tankers per year $= (6 \times 10^6 \text{ barrels}/1 \text{ day})(365 \text{ day}/1 \text{ y})(1 \text{ tanker}/0.25 \times 10^6 \text{ barrels}) = 8.76 \times 10^3$ tankers.

(*c*) Cost per year $= (6 \times 10^6 \text{ barrels}/1 \text{ day})(365 \text{ days}/1 \text{ y})(\$20/\text{barrel}) = \$4.38 \times 10^{10} = \$43.8$ billion.

---

51 •• Every year the United States generates 160 million tons of municipal solid waste and a grand total of 10 billion tons of solid waste of all kinds. If one allows one cubic meter of volume per ton, how many square miles of area at an average height of 10 m is needed for landfill each year?

1. Find the total volume of landfill needed — $V = (10 \times 10^9 \text{ tons})(1 \text{ m}^3/1 \text{ ton}) = 10^{10} \text{ m}^3$

2. Find the surface area needed using $V = AH$ — $A = V/H = (10^{10} \text{ m}^3)/(10 \text{ m}) = 10^9 \text{ m}^2$

3. Express A in mi^2 — $A = (10^9 \text{ m}^2)(1 \text{ mi}/1610 \text{ m})^2 = 3.86 \times 10^2 \text{ mi}^2$

52 •• An iron nucleus has a radius of 5.4×10^{-15} m and a mass of 9.3×10^{-26} kg. (*a*) What is its mass per unit volume in kg/m^3? (*b*) If the earth had the same mass per unit volume, what would its radius be? (The mass of the earth is 5.98×10^{24} kg.)

(*a*) 1. Find the volume of an iron nucleus — $V = (4/3)\pi r^3 = (4/3)\pi(5.4 \times 10^{-15}\text{ m})^3 = 6.6 \times 10^{-43}\text{ m}^3$

2. Find the density, mass per unit volume — $\rho = (9.3 \times 10^{-26}\text{ kg})/(6.6 \times 10^{-43}\text{ m}^3) = 1.4 \times 10^{17}\text{ kg/m}^3$

(*b*) 1. Find the volume of the earth of density ρ — $V = (5.98 \times 10^{24}\text{ kg})/(1.4 \times 10^{17}\text{ kg/m}^3) = 4.2 \times 10^7\text{ m}^3$

2. Find the radius using $V = (4/3)\pi r^3$ — $r = (3V/4\pi)^{1/3} = [3 \times (4.2 \times 10^7\text{ m}^3)/(4\pi)]^{1/3} = 216\text{ m}$

53* •• Evaluate the following expressions. (*a*) $(5.6 \times 10^{-5})(0.0000075)/(2.4 \times 10^{-12})$; (*b*) $(14.2)(6.4 \times 10^7)(8.2 \times 10^{-9}) - 4.06$; (*c*) $(6.1 \times 10^{-6})^2(3.6 \times 10^4)^3/(3.6 \times 10^{-11})^{1/2}$; (*d*) $(0.000064)^{1/3}/[(12.8 \times 10^{-3})(490 \times 10^{-1})^{1/2}]$.

(*a*) $(5.6 \times 10^{-5})(7.5 \times 10^{-6})/(2.4 \times 10^{-12}) = 1.8 \times 10^2$ to two significant figures.

(*b*) $(14.2)(6.4 \times 10^7)(8.2 \times 10^{-9}) - 4.06 = 7.45 - 4.06 = 3.4$ to two significant figures.

(*c*) $(6.1 \times 10^{-6})^2(3.6 \times 10^4)^3/(3.6 \times 10^{-11})^{1/2} = 2.9 \times 10^8$ to two significant figures.

(*d*) $(6.4 \times 10^{-5})^{1/3}/[(12.8 \times 10^{-3})(49.0)^{1/2}] = 0.45$ to two significant figures.

54 •• The astronomical unit is defined in terms of the distance from the earth to the sun, namely 1.496×10^{11} m. The parsec is the radial length that one astronomical unit of arc length subtends at an angle of 1 s. The light-year is the distance that light travels in one year. (*a*) How many parsecs are there in one astronomical unit? (*b*) How many meters are in a parsec? (*c*) How many meters in a light-year? (*d*) How many astronomical units in a light-year? (*e*) How many light-years in a parsec?

Note: If S is the arc length and R the radius, then $\theta = S/R$, where θ is in radians.

(*a*) 1. Express 1 second in radian measure (see Problem 27) — $(1\text{ s})(1\text{ min}/60\text{ s})(1\text{ deg}/60\text{ min})(2\pi\text{ rad}/360\text{ deg}) = 4.85 \times 10^{-6}\text{ rad}$

2. Use $\theta = S/R$ and solve for R — $R = (1\text{ parsec})(4.85 \times 10^{-6}\text{ rad}) = 4.85 \times 10^{-6}\text{ parsec}$

(*b*) From $\theta = S/R$, $R = S/\theta$ — $R = (1.496 \times 10^{11}\text{ m})/(4.85 \times 10^{-6}\text{ rad}) = 3.08 \times 10^{16}\text{ m}$

(*c*) Speed of light $c = 3.00 \times 10^8$ m/s; $D = ct$ — $1\text{ l-y} = (3 \times 10^8\text{ m/s})(3.156 \times 10^7\text{ s/yr}) = 9.47 \times 10^{15}\text{ m}$

(*d*) Use definition of 1 AU and part (*c*) — $1\text{ l-y} = (9.47 \times 10^{15}\text{ m})(1\text{ AU}/1.496 \times 10^{11}\text{ m}) = 6.33 \times 10^4\text{ AU}$

(*e*) Use parts (*b*) and (*c*) — $1\text{ parsec} = (3.08 \times 10^{16}\text{ m})(1\text{ l-y}/9.47 \times 10^{15}\text{ m}) = 3.25\text{ light-years}$

55 •• If the average density of the universe is at least 6×10^{-27} kg/m^3, then the universe will eventually stop expanding and begin contracting. (*a*) How many electrons are needed in a cubic meter to produce the critical density? (*b*) How many protons per cubic meter would produce the critical density? ($m_e = 9.11 \times 10^{-31}$ kg; $m_p = 1.67 \times 10^{-27}$ kg)

(*a*) $N_e/\text{m}^3 = (6 \times 10^{-27}\text{ kg})(1\text{ electron}/9.11 \times 10^{-31}\text{ kg}) = 6.6 \times 10^3\text{ electrons/m}^3$.

(*b*) $N_p/\text{m}^3 = (N_e/\text{m}^3)(m_e/m_p) = (6.6 \times 10^3\text{ electrons/m}^3)(9.11 \times 10^{-31}\text{ kg/electron})/(1.67 \times 10^{-27}\text{ kg/proton}) = 3.6\text{ protons/m}^3$.

56 •• Observational estimates of the density of the universe yield an average of about 2×10^{-28} kg/m^3. (*a*) If a 100-kg football player had this mass uniformly spread out in a sphere to match the estimate for the average mass density of the universe, what would be the radius of the sphere? (*b*) Compare this radius with the earth–moon distance of 3.82×10^8 m.

(*a*) 1. Find the volume of a 100-kg mass — $V = M/\rho = (100\text{ kg})/(2\times10^{-28}\text{ kg/m}^3) = 5\times10^{29}\text{ m}^3$

2. Use $V = (4/3)\pi R^3$ and solve for R — $R = (3V/4\pi)^{1/3} = (1.5\times10^{30}\text{ m}^3/4\pi)^{1/3} = 5\times10^9$ m

(*b*) Divide R by the earth–moon distance — $R = (5\times10^9\text{ m})/(3.82\times10^8\text{ m}) = 13$ E–M distance

57* •• Beer and soft drinks are sold in aluminum cans. The mass of a typical can is about 0.018 kg. (*a*) Estimate the number of aluminum cans used in the United States in one year. (*b*) Estimate the total mass of aluminum in a year's consumption from these cans. (*c*) If aluminum returns \$1/kg at a recycling center, how much is a year's accumulation of aluminum cans worth?

(*a*) The population of the U.S. is about 3×10^8 persons. Assume 1 can per person per day. In one year the total number of cans used is $(3\times10^8$ persons)(1 can/person-day)(365 days/y) = 1×10^{11} cans/y.

(*b*) Total mass of aluminum per year = $(1\times10^{11}$ cans/y)$(1.8\times10^{-2}$ kg/can) = 2×10^9 kg/y.

(*c*) At \$1/kg this amounts to \$2 billion.

58 •• An aluminum rod is 8.00024 m long at 20.00°C. If the rod's temperature increases, it expands such that it lengthens by 0.0024% per degree temperature rise. Determine the rod's length at 28.00°C and at 31.45°C.

1. Find a relation between L_T, L_{20}, and T. Note that the <u>fractional</u> change is $0.0024\times\Delta T$. — $L_T = L_{20}[1 + (2.4\times10^{-5})(T - 20)]$, where T is in °C and L_{20} and L_T are the lengths at 20°C and temperature T

2. Solve for $L_{28.00}$; keep result to 6 sig. figs. — $L_{28.00} = (8.00024\text{ m})[1 + (2.4\times10^{-5})(8)] = 8.00178$ m

3. Solve for $L_{31.45}$; keep result to 6 sig.figs. — $L_{31.45} = (8.00024\text{ m})[(1 + (2.4\times10^{-5})(11.45)] = 8.0024$ m

59 ••• The table below gives experimental results for a measurement of the period of motion T of an object of mass m suspended on a spring versus the mass of the object. These data are consistent with a simple equation express ing T as a function of m of the form $T = Cm^n$, where C and n are constants and n is not necessarily an integer. (*a*) Find n and C. (There are several ways to do this. One is to guess the value of n and check by plotting T versus m^n on graph paper. If your guess is right, the plot will be a straight line. Another is to plot $\log T$ versus $\log m$. The slope of the straight line on this plot is n.) (*b*) Which data points deviate the most from a straight-line plot of T versus m^n?

Mass m, kg	0.10	0.20	0.40	0.50	0.75	1.00	1.50
Period T, s	0.56	0.83	1.05	1.28	1.55	1.75	2.22

(*a*) We will use a "judicious" guessing procedure. Note that as m increases from 0.1 kg to 1.0 kg, i.e., by a factor of 10, T increases from 0.56 to 1.75, i.e., by a factor of $3.13 \approx \sqrt{10}$. This suggests that we should try $T = Cm^{1/2}$.

A plot of T versus $m^{1/2}$ is shown.

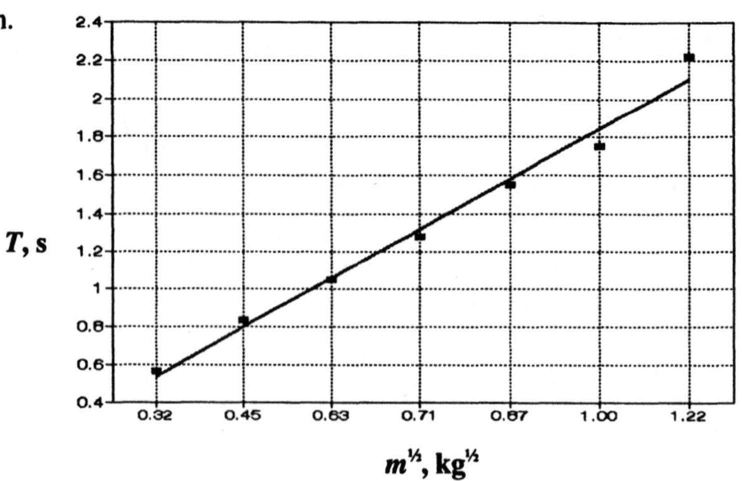

To find the constant C we use the point $m = 0.40$ kg, $T = 1.05$ s; $C = (1.05 \text{ s})/(0.40 \text{ kg})^{1/2} = 1.66 \text{ s/kg}^{1/2}$.

(*b*) From the plot we see that the data points $m = 1.00$ kg, $T = 1.75$ s, and $m = 1.50$ kg, $T = 2.22$ s deviate most from the straight-line plot.

---

60 ••• The table below gives the period T and orbit radius r for the motions of four satellites orbiting a dense, heavy asteroid. (*a*) These data can be fitted by the formula $T = Cr^n$. Find C and n. (*b*) A fifth satellite is discovered to have a period of 6.20 y. Find the radius for the orbit of this satellite, which fits the same formula.

Period T, y	0.44	1.61	3.88	7.89
Radius r, Gm	0.088	0.208	0.374	0.600

(*a*) We shall plot log T versus log r; the slope of the best-fit line determines the exponent n.

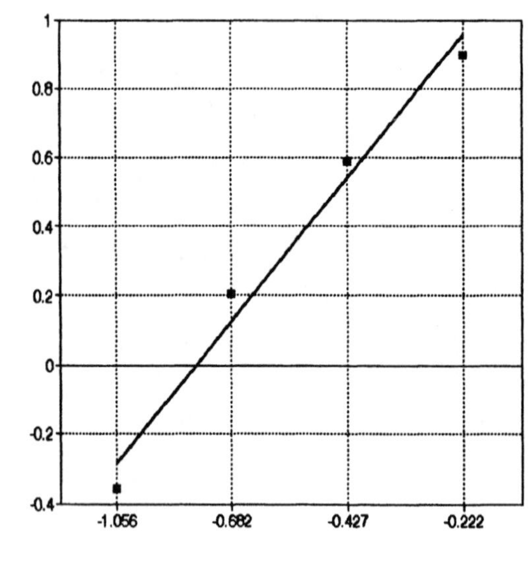

log *r*

(*a*) The slope of the best-fit straight line is 1.5; $n = 3/2$. To find C we use the point $r = 0.374$ Gm, $T = 3.88$ yr. This gives $C = (3.88 \text{ yr})/(0.374 \text{ Gm})^{3/2} = 17 \text{ yr/(Gm)}^{3/2}$.

(*b*) We use $r = (T/C)^{2/3}$; $r = [(6.20 \text{ yr})/(17 \text{ yr/(Gm)})]^{2/3} = 0.510$ Gm.

61* ••• The period T of a simple pendulum depends on the length L of the pendulum and the acceleration of gravity g (dimensions L/T^2). (*a*) Find a simple combination of L and g which has the dimensions of time. (*b*) Check the dependence of the period T on the length L by measuring the period (time for a complete swing back and forth) of a pendulum for two different values of L. (*c*) The correct formula relating T to L and g involves a constant which is a multiple of π, and cannot be obtained by the dimensional analysis of part (*a*). It can be found by experiment as in (*b*) if g is known. Using the value $g = 9.81$ m/s^2 and your experimental results from (*b*), find the formula relating T to L and g.

(*a*) 1. Write $T = CL^a g^b$ and express dimensionally — $T = L^a(L/T^2)^b = L^{a+b}T^{-2b}$.

2. Solve for a and b — $-2b = 1, b = -½; a + b = 0, a = ½$.

3. Write the expression for T — $T = C\sqrt{L/g}$

(*b*) Check by using pendulums of lengths 1 m and 0.5 m; the periods should be about 2 s and 1.4 s.

(*c*) Using $L = 1$ m, $T = 2$ s, $g = 9.81$ m/s^2, $C = (2.0\text{ s})/\sqrt{1.0\text{ m}/9.81\text{ m/s}^2} = 6.26 = 2\pi$. $T = 2\pi\sqrt{L/g}$.

62 ••• The weight of the earth's atmosphere pushes down on the surface of the earth with a force of 14.7 lbs for each square inch of earth's surface. What is the weight in pounds of the earth's atmosphere? (The radius of the earth is about 6370 km.)

1. Find the total surface area of the earth — $A = 4\pi r^2 = 4\pi(6.37 \times 10^6\text{ m})^2 = 5.1 \times 10^{14}\text{ m}^2$

2. Express A in square inches — $A = (5.1 \times 10^{14}\text{ m}^2)(10^4\text{ cm}^2/\text{m}^2)(1\text{ in}^2/2.54^2\text{ in}^2)$
$= 7.9 \times 10^{17}\text{ in}^2$

3. Multiply the area A by 14.7 lb/in^2 — $W = (14.7\text{ lbs/in}^2)(7.9 \times 10^{17}\text{ in}^2) = 1.16 \times 10^{19}$ lbs

63 ••• Each binary digit is termed a bit. A series of bits grouped together is called a word. An eight-bit word is called a byte. Suppose a computer hard disk has a capacity of 2 gigabytes. (*a*) How many bits can be stored on the disk? (*b*) Estimate the number of typical books that can be stored on the disk.

(*a*) Number of bits = N_{bytes}(number of bits/byte) — $N_{bits} = (2 \times 10^9\text{ bytes})(8\text{ bits/byte}) = 1.6 \times 10^{10}$ bits

(*b*) 1. Estimate the number of bits required for the alphabet — A bit is 0 or 1. So $2^5 = 32$ bits can be used to represent the alphabet of 26 letters. We need 4 bytes per letter.

2. Assume an average of 8 letters/word — Need 8×4 bytes/word = 32 bytes/word

3. Assume 10 words/line and 60 lines/page — 600 words/page = 600×32 bytes/page
$= 1.92 \times 10^4$ bytes/page

4. Assume book length of 300 pages — $(1.92 \times 10^4\text{ bytes/page})(300\text{ pages}) = 5.8 \times 10^6$ bytes

5. $N_{books} = (2 \times 10^9\text{ bytes/disk})/(N_{bytes/book})$ — $N_{books} = (2 \times 10^9\text{ bytes})/(5.8 \times 10^6\text{ bytes/book})$
$= 350$ books

CHAPTER 2

Motion in One Dimension

1* • What is the approximate average velocity of the race cars during the Indianapolis 500?

Since the cars go around a closed circuit and return nearly to the starting point, the displacement is nearly zero, and the average velocity is zero.

2 • Does the following statement make sense? "The average velocity of the car at 9 a.m. was 60 km/h."

No, it does not. Average velocity must refer to a finite time interval.

3 • Is it possible for the average velocity of an object to be zero during some interval even though its average velocity for the first half of the interval is not zero? Explain.

Yes, it is. In a round trip, *A* to *B* and back to *A*, the average velocity is zero; the average velocity between *A* and *B* is not zero.

4 • The diagram in Figure 2-21 tracks the path of an object moving in a straight line. At which point is the object farthest from its starting point? (*a*) *A* (*b*) *B* (*c*) *C* (*d*) *D* (*e*) *E*

(*b*) Starting point is at $x = 0$; point *B* is farthest from $x = 0$.

5* • (*a*) An electron in a television tube travels the 16-cm distance from the grid to the screen at an average speed of 4×10^7 m/s. How long does the trip take? (*b*) An electron in a current-carrying wire travels at an average speed of 4×10^{-5} m/s. How long does it take to travel 16 cm?

(*a*) From Equ. 2-3, $\Delta t = \Delta s/(\text{av. speed})$ $\Delta t = (0.16\text{ m})/(4 \times 10^7\text{ m/s}) = 4 \times 10^{-9}\text{ s} = 4\text{ ns}$

(*b*) Repeat as in (*a*) $\Delta t = (0.16\text{ m})/(4 \times 10^{-5}\text{ m/s}) = 4 \times 10^3\text{ s} = 4\text{ ks}$

6 • A runner runs 2.5 km in 9 min and then takes 30 min to walk back to the starting point. (*a*) What is the runner's average velocity for the first 9 min? (*b*) What is the average velocity for the time spent walking? (*c*) What is the average velocity for the whole trip? (*d*) What is the average speed for the whole trip?

Take the direction of running as the positive direction.

(*a*) Use Equ. 2-2 $v_{av} = (2.5\text{ km})/[(9\text{ min})(1\text{ h}/60\text{ min})] = 16.7\text{ km/h}$

(*b*) Use Equ. 2-2 $v_{av} = (-2.5\text{ km})/(0.5\text{ h}) = -5.0\text{ km/h}$

(*c*) $\Delta x = 0$ $v_{av} = 0$

(*d*) Total distance = 5.0 km; $\Delta t = 39$ min Av. speed = $(5.0\text{ km})/[(39\text{ min})(1\text{ h}/60\text{ min})]$ $= 7.7\text{ km/h}$

7 • A car travels in a straight line with an average velocity of 80 km/h for 2.5 h and then with an average velocity of 40 km/h for 1.5 h. (*a*) What is the total displacement for the 4-h trip? (*b*) What is the average velocity for the total trip?

(*a*) 1. Find the displacements of each segment; Use Equ. 2-2 — $S_1 = (80 \text{ km/h})(2.5 \text{ h}) = 200 \text{ km}$; $S_2 = (40 \text{ km/h})(1.5 \text{ h}) = 60 \text{ km}$

2. Add the two displacements — $S_{tot} = 260 \text{ km}$

(*b*) Use Equ. 2-2 — $v_{av} = (260 \text{ km})/(4 \text{ h}) = 65 \text{ km/h}$

8 • One busy air route across the Atlantic Ocean is about 5500 km. (*a*) How long does it take for a supersonic jet flying at 2.4 times the speed of sound to make the trip? Use 340 m/s for the speed of sound. (*b*) How long does it take a subsonic jet flying at 0.9 times the speed of sound to make the same trip? (*c*) Allowing 2 h at each end of the trip for ground travel, check-in, and baggage handling, what is your average speed door to door when traveling on the supersonic jet? (*d*) What is your average speed taking the subsonic jet?

(*a*) Find v, then use Equ. 2-2 — $v = 2 \times (0.340 \text{ km/s})(3600 \text{ s/h}) = 2448 \text{ km/h}$; $t = (5500 \text{ km})/(2448 \text{ km/h}) = 2.25 \text{ h}$

(*b*) Repeat as for part (*a*) — $v = 0.9 \times (0.340 \text{ km/s})(3600 \text{ s/h}) = 1102 \text{ km/h}$; $t = (5500 \text{ km})/(1102 \text{ km/h}) = 5.0 \text{ h}$

(*c*) Find the total time, then use Equ. 2-2 — $t_{tot} = (2.25 + 4)\text{h} = 6.25 \text{ h}$; $v_{av} = (5500 \text{ km})/(6.25 \text{ h}) = 880 \text{ km/h}$

(*d*) Repeat as for part (*c*) — $t_{tot} = 9 \text{ h}$; $v_{av} = (5500 \text{ km})/(9 \text{ h}) = 611 \text{ km/h}$

9* • As you drive down a desert highway at night, an alien spacecraft passes overhead, causing malfunctions in your speedometer, wristwatch, and short-term memory. When you return to your senses, you can't tell where you are, where you are going, or even how fast you are traveling. The passenger sleeping next to you never woke up during this incident. Although your pulse is racing, hers is steady at 55 beats per minute. (*a*) If she has 45 beats between the mile markers posted along the road, determine your speed. (*b*) If you want to travel at 120 km/h, how many heartbeats should there be between mile markers?

(*a*) Find the time between mile markers — $\Delta t = (45 \text{ beats/mile})/(55 \text{ beats/min}) = 0.818 \text{ min}$

$v = \Delta s/\Delta t$ — $v = 1 \text{ mi}/0.818 \text{ min} = 1.22 \text{ mi/min} = 73.3 \text{ mi/h}$

(*b*) $N = (\text{beats/min})(60 \text{ min/h})/(v \text{ mi/h})$ — $N = (55 \times 60 \text{ beats/h})/[(120 \text{ km/h})/(1.61 \text{ km/mi})] = 44.3$

10 • The speed of light, c, is 3×10^8 m/s. (*a*) How long does it take for light to travel from the sun to the earth, a distance of 1.5×10^{11} m? (*b*) How long does it take light to travel from the moon to the earth, a distance of 3.84×10^8 m? (*c*) A light-year is a unit of distance equal to that traveled by light in 1 year. Convert 1 light-year into kilometers and miles.

(*a*) Use Equ. 2-3 — $t = (1.5 \times 10^{11} \text{ m})/(3 \times 10^8 \text{ m/s}) = 500 \text{ s}$

(*b*) Use Equ. 2-3 — $t = (3.84 \times 10^8 \text{ m})/(3 \times 10^8 \text{ m/s}) = 1.28 \text{ s}$

(*c*) See Problem 1-54(*c*) — $1 \text{ l-y} = 9.48 \times 10^{15} \text{ m} = 9.48 \times 10^{12} \text{ km}$

Express km in miles — $= (9.48 \times 10^{12} \text{ km})(1 \text{ mi}/1.61 \text{ km}) = 5.9 \times 10^{12} \text{ mi.}$

---

11 • The nearest star, Proxima Centauri, is 4.1×10^{15} km away. From the vicinity of this star, Gregor places an order at Tony's Pizza in Hoboken, New Jersey, communicating via light signals. Tony's fastest delivery craft travels at $10^{-4}c$ (see Problem 10). (*a*) How long does it take Gregor's order to reach Tony's Pizza? (*b*) How long does Gregor wait between sending the signal and receiving the pizza? If Tony's has a 1000-years-or-it's-free delivery policy, does Gregor have to pay for the pizza?

(*a*) Use Equ. 2-3 — $t = (4.1 \times 10^{18}\text{ m})/(3 \times 10^{8}\text{ m/s}) = 1.37 \times 10^{10}\text{ s}$ $= 434\text{ y}$

(*b*) Traveling at $10^{-4}c$, time will be 10^4 times (*a*) — $t = 4.34 \times 10^{6}$ y; he need not pay for the pizza

---

12 • A car making a 100-km journey travels 40 km/h for the first 50 km. How fast must it go during the second 50 km to average 50 km/h?

1. Find the time for the total journey — $t_{\text{tot}} = (100\text{ km})/(50\text{ km/h}) = 2\text{ h}$
2. Find the time for the first 50 km — $t_1 = (50\text{ km})/(40\text{ km/h}) = 1.25\text{ h}$
3. Find the speed in the remaining 0.75 h — $v = (50\text{ km})/(0.75\text{ h}) = 66.7\text{ km/h}$

13* •• John can run 6.0 m/s. Marcia can run 15% faster than John. (*a*) By what distance does Marcia beat John in a 100-m race? (*b*) By what time does Marcia beat John in a 100-m race?

(*a*) 1. Find the running speed for Marcia — $v_M = 1.15(6.0\text{ m/s}) = 6.9\text{ m/s}$

2. Find the time for Marcia — $t_M = (100\text{ m}/6.9\text{ m/s}) = 14.5\text{ s}$

3. Find the distance covered by John in 14.5 s — $s_J = (6.0\text{ m/s})(14.5\text{ s}) = 87\text{ m}$; distance = 13 m

(*b*) Find the time required by John — $t_J = (100\text{ m})/(6\text{ m/s}) = 16.7\text{ s}$; time difference = 2.2 s

---

14 •• Figure 2-22 shows the position of a particle versus time. Find the average velocities for the time intervals *a*, *b*, *c*, and *d* indicated in the figure.

From the figure: $\Delta s_a = 0$, so $v_{\text{av}} = 0$; $\Delta s_b = 1\text{ m}$, $\Delta t_b = 3\text{ s}$, so $v_{\text{av}} = 0.33\text{ m/s}$; $\Delta s_c = -6\text{ m}$, $\Delta t_c = 3\text{ s}$, so $v_{\text{av}} = -2\text{ m/s}$; $\Delta s_d = 3\text{ m}$, $\Delta t_d = 3\text{ s}$, so $v_{\text{av}} = 1\text{ m/s}$.

---

15 •• It has been found that galaxies are moving away from the earth at a speed that is proportional to their distance from the earth. This discovery is known as Hubble's law. The speed of a galaxy at distance *r* from the earth is given by $v = Hr$, where *H* is the Hubble constant, equal to $1.58 \times 10^{-18}\text{ s}^{-1}$. What is the speed of a galaxy (*a*) 5×10^{22} m from earth and (*b*) 2×10^{25} m from earth? (*c*) If each of these galaxies has traveled with constant speed, how long ago were they both located at the same place as the earth?

(*a*) Use Hubble's law — $v_a = (5 \times 10^{22}\text{ m})(1.58 \times 10^{-18}\text{ s}^{-1}) = 7.9 \times 10^{4}\text{ m/s}$

(*b*) Use Hubble's law — $v_b = (2 \times 10^{25}\text{ m})(1.58 \times 10^{-18}\text{ s}^{-1}) = 3.16 \times 10^{7}\text{ m/s}$

(*c*) Use Equ. 2-3 for both galaxies — $t = r/v = r/rH = 1/H = 6.33 \times 10^{17}\text{ s} = 2 \times 10^{10}\text{ y}$

---

16 •• Cupid fires an arrow that strikes St. Valentine, producing the usual sound of harp music and bird chirping as Valentine swoons into a fog of love. If Cupid hears these telltale sounds exactly one second after firing the arrow, and the average speed of the arrow was 40 m/s, what was the distance separating them? Take 340 m/s for the speed of sound.

Let t_1 be the travel time of the arrow, and let t_2 be that of the sound. Both the sound and arrow travel a distance D.

1. Write expressions for D in terms of t_1 and t_2	$D = (40\text{ m/s})t_1 = (340\text{ m/s})t_2$	(1)
2. Write t_1 in terms of t_2	$t_1 = (340/40)t_2 = 8.5t_2$	(2)
3. The sum of t_1 and t_2 equals 1 s	$t_1 + t_2 = 1$ s	(3)
4. Solve (1) and (2) for t_1 and t_2	$9.5t_2 = 1$ s; $t_2 = 0.105$ s; $t_1 = 0.895$ s	(4)
5. Use (1) to find D	$D = 35.8$ m	

17* • If the instantaneous velocity does not change, will the average velocities for different intervals differ?

No, they will not. For constant velocity, the instantaneous and average velocities are equal.

18 • If $v_{av} = 0$ for some time interval Δt, must the instantaneous velocity v be zero at some point in the interval? support your answer by sketching a possible x-versus-t curve that has $\Delta x = 0$ for some interval Δt.

Yes, it must.

In the adjoining graph of x versus t, $\Delta x = 0$ in the interval between t = 0 and t = 4.0 s. Consequently, $v_{av} = \Delta x/\Delta t = 0$, although the instantaneous velocity is zero only at the point $t = 2$ s.

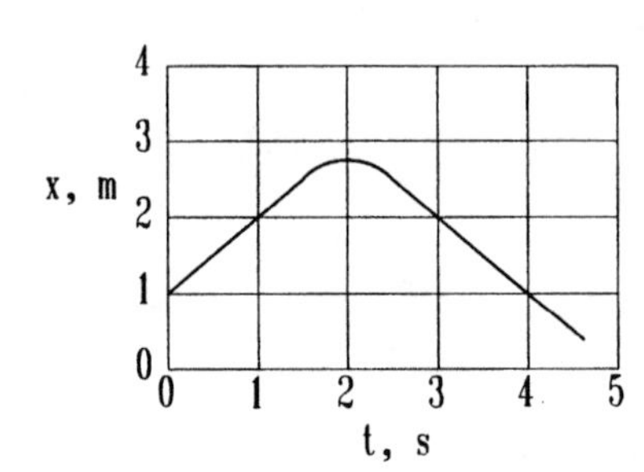

19 • An object moves along the x axis as shown in Figure 2-23. At which point or points is the magnitude of its velocity at a minimum? (*a*) A and E (*b*) B, D, and E (*c*) C only (*d*) E only (*e*) None of these is correct.

(*b*) At these points the slope of the position-versus-time curve is zero; therefore the velocity is zero.

20 • For each of the four graphs of x versus t in Figure 2-24, answer the following questions. (*a*) Is the velocity at time t_2 greater than, less than, or equal to the velocity at time t_1? (*b*) Is the speed at time t_2 greater than, less than, or equal to the speed at time t_1?

We shall use $\mathbf{v}$ to denote velocity and v to denote speed.

(*a*) curve *a*: $\mathbf{v}(t_2) < \mathbf{v}(t_1)$; curve *b*: $\mathbf{v}(t_2) = \mathbf{v}(t_1)$; curve *c*: $\mathbf{v}(t_2) > \mathbf{v}(t_1)$; curve *d*: $\mathbf{v}(t_2) < \mathbf{v}(t_1)$.

(*b*) curve *a*: $v(t_2) < v(t_1)$; curve *b*: $v(t_2) = v(t_1)$; curve *c*: $v(t_2) < v(t_1)$; curve *d*: $v(t_2) > v(t_1)$.

21* • Using the graph of x versus t in Figure 2-25, (*a*) find the average velocity between the times $t = 0$ and $t = 2$ s. (*b*) Find the instantaneous velocity at $t = 2$ s by measuring the slope of the tangent line indicated.

(*a*) Find Δx from graph; $v_{av} = \Delta x/\Delta t$	$\Delta x = 2$ m, $\Delta t = 2$ s; $v_{av} = 1$ m/s
(*b*) From graph, tangent passes through points $x = 0$, $t = 1$ s; $x = 4$ m, $t = 3$ s.	Slope of tangent line is (4 m)/(2 s) = 2 m/s $v(t = 2\text{ s}) = 2$ m/s

22 • Using the graph of x versus t in Figure 2-26, find (*a*) the average velocity for the time intervals $\Delta t = t_2 - 0.75$ s when t_2 is 1.75, 1.5, 1.25, and 1.0 s; (*b*) the instantaneous velocity at $t = 0.75$ s; (*c*) the approximate time when the instantaneous velocity is zero.

(*a*) From graph find $x(0.75)$, $x(1.75)$, $x(1.5)$, $x(1.25)$, $x(1.0)$	$x(0.75) = 4.0$ m, $x(1.75) = 5.9$ m, $x(1.5) = 6$ m, $x(1.25) = 5.6$ m, $x(1.0) = 5$ m

Find $v_{av} = \Delta x/\Delta t$ — 1.75 s, $v_{av} = 1.9$ m/s; 1.5 s, $v_{av} = 2.67$ m/s; 1.25 s, $v_{av} = 3.2$ m/s; 1.0 s, $v_{av} = 4$ m/s

(*b*) Draw tangent at $t = 0.75$ s and find slope — Slope = $v(0.75) = 4.2$ m/s

(*c*) Find the value of t where slope is zero — $t = 1.6$ s

---

23 •• The position of a certain particle depends on the time according to $x = (1\text{ m/s}^2)t^2 - (5\text{ m/s})t + 1$ m. (*a*) Find the displacement and average velocity for the interval $3\text{ s} \le t \le 4\text{ s}$. (*b*) Find the general formula for the displacement for the time interval from t to $t + \Delta t$. (*c*) Use the limiting process to obtain the instantaneous velocity for any time t.

(*a*) 1. Find $x(4)$ and $x(3)$ — $x(4) = (16 - 20 + 1)$ m $= -3$ m; $x(3) = (9 - 15 + 1)$ m $= -5$ m

2. Find Δx — $\Delta x = x(4) - x(3) = 2$ m

3. Use Equ. 2-2 — $v_{av} = \Delta x/\Delta t = (2\text{ m})/(1\text{ s}) = 2$ m/s

(*b*) 1. Find $x(t + \Delta t)$ — $x(t + \Delta t) = [(t^2 + 2t\Delta t + \Delta t^2) - 5(t + \Delta t) + 1]$ m

2. Find $x(t + \Delta t) - x(t) = \Delta x$ — $\Delta x = [(2t - 5)\Delta t + \Delta t^2]$ m

(*c*) From (*b*) find $\Delta x/\Delta t$ as $\Delta t \to 0$ — $v = \lim_{t\to 0}(\Delta x/\Delta t) = (2t - 5)$ m/s

---

24 •• The height of a certain projectile is related to time by $y = -5(t - 5)^2 + 125$, where y is in meters and t is in seconds. (*a*) Sketch y versus t for $0 \le t \le 10$ s. (*b*) Find the average velocity for each of the 1-s time intervals between integral time values from $0 \le t \le 10$ s. Sketch v_{av} versus t. (*c*) Find the instantaneous velocity as a function of time.

(*a*) The plot of y versus t is shown

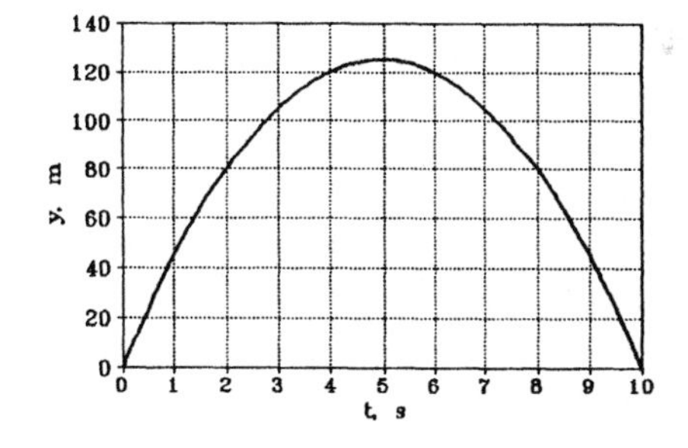

(*b*) Substitute $t = 0,1,2,\ldots,10$ into the expression for y. The table below lists the values. Then evaluate $\Delta y/\Delta t = \Delta y/(1\text{ s})$ to get v_{av}, shown in the table.

t	y, m	v_{av}, m/s
0	0	0
1	45	45
2	80	35
3	105	25
4	120	15
5	125	5
6	120	-5
7	105	-15
8	80	-25
9	45	-35
10	0	-45

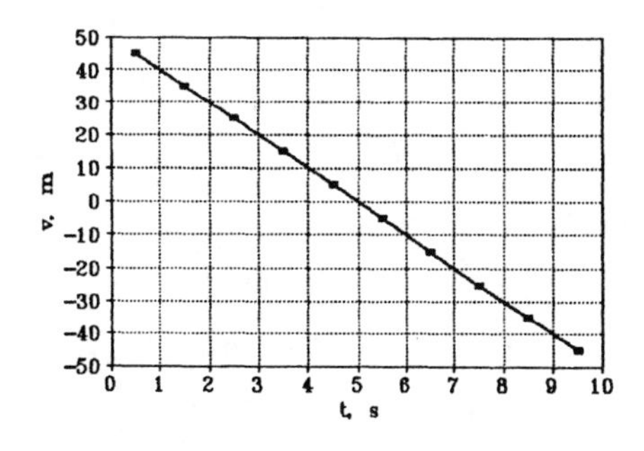

(*c*) To find the instantaneous velocity, take the derivative of the expression for $y(t)$;

$v(t) = dy/dt = (50 - 10t)$ m/s

25* •• The position of a body oscillating on a spring is given by $x = A \sin \omega t$, where A and ω are constants with values $A = 5$ cm and $\omega = 0.175\ \text{s}^{-1}$. (*a*) Sketch x versus t for $0 \le t \le 36$ s. (*b*) Measure the slope of your graph at $t = 0$ to find the velocity at this time. (*c*) Calculate the average velocity for a series of intervals beginning at $t = 0$ and ending at $t = 6, 3, 2, 1, 0.5$, and 0.25 s. (*d*) Compute dx/dt and find the velocity at time $t = 0$.

(*a*) The plot of x versus t is shown

(*b*) The slope of the dotted line (tangent at $t = 0$) is 0.875; $v(0) = 0.875$ cm/s

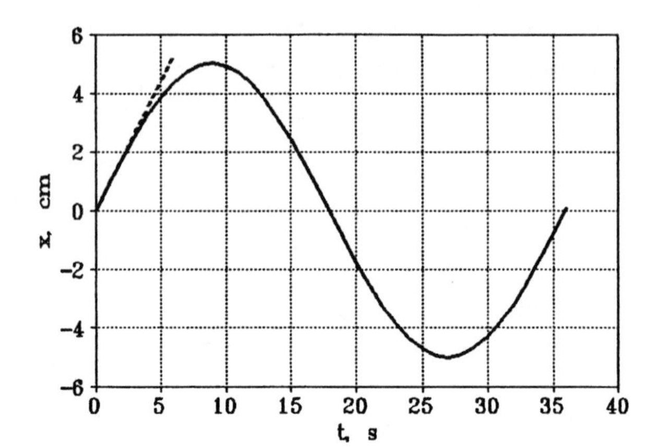

(*c*)

t	x	$\Delta x/\Delta t$
6	4.34	0.723
3	2.51	0.835
2	1.71	0.857
1	0.174	0.871
0.5	0.437	0.874
0.25	0.219	0.875

(*d*) $dx/dt = A\omega \cos \omega t$; at $t = 0$ $\cos \omega t = 1$; dx/dt at $t = 0$ is $A\omega = 0.875$ cm/s

26 • To avoid falling too fast during a landing, an airplane must maintain a minimum airspeed (the speed of the plane relative to the air). However, the slower the ground speed (speed relative to the ground) during a landing, the safer the landing. Is it safer for an airplane to land with the wind or against the wind?

It is safer to land against the wind.

27 •• Two cars are traveling along a straight road. Car A maintains a constant speed of 80 km/h; car B maintains a constant speed of 110 km/h. At $t = 0$, car B is 45 km behind car A. How far will car A travel from $t = 0$ before it is overtaken by car B?

1. Find the velocity of car B relative to car A — $v_{rel} = v_B - v_A = (110 - 80)$ km/h $= 30$ km/h
2. Find the time before overtaking — $t = s/v_{rel} = (45\text{ km})/(30\text{ km/h}) = 1.5$ h
3. Find the distance traveled by car A in 1.5 h — $d = (1.5\text{ h})(80\text{ km/h}) = 120$ km

28 •• A car traveling at constant speed of 20 m/s passes an intersection at time $t = 0$, and 5 s later another car traveling 30 m/s passes the same intersection in the same direction. (*a*) Sketch the position functions $x_1(t)$ and $x_2(t)$ for the two cars. (*b*) Determine when the second car will overtake the first. (*c*) How far from the intersection will the two cars be when they pull even?

(*a*) The plot of $x_1(t)$ and $x_2(t)$ are shown

(*b*) From the plot, the second will overtake the first at $t = 15$ s

(*c*) The two cars will be 300 m from the intersection

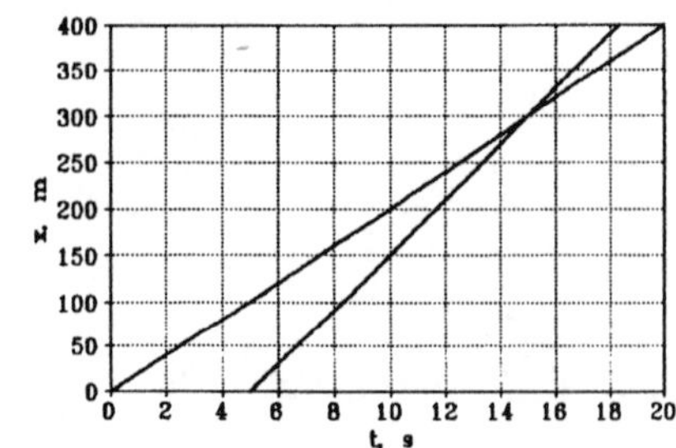

29* •• Margaret has just enough gas in her speedboat to get to the marina, an upstream journey that takes 4.0 hours. Finding it closed for the season, she spends the next 8.0 hours floating back downstream to her shack. The entire trip took 12.0 h; how long would it have taken if she had bought gas at the marina?

Let D = distance to marina, v_W = velocity of stream, v_{rel} = velocity of boat under power relative to stream.

1. Express the times of travel with gas in terms of D, v_W, and v_{rel} — $t_1 = D/(v_{rel} - v_W) = 4$ h; $t_2 = D/(v_{rel} + v_W)$
2. Express the time required to drift distance D — $t_3 = D/v_W = 8$ h; $v_W = D/(8\text{ h})$
3. From $t_1 = 4$ h, find v_{rel} — $4(v_{rel} - D/8) = D$; $v_{rel} = 1.5D/(4\text{ h})$
4. Solve for t_2 — $t_2 = D/[(1.5D/4\text{ h}) + (D/8\text{ h}) = 2$ h
5. Add t_1 and t_2 — $t_{tot} = t_1 + t_2 = 6$ h

---

30 •• Joe and Sally tend to argue when they travel. Just as they reached the moving sidewalk at the airport, their struggle for itinerary-making powers peaked. Though they stepped on the moving belt at the same time, Joe chose to stand and ride, while Sally opted to keep walking. Sally reached the end in 1 min, while Joe took 2 min. How long would it have taken Sally if she had walked twice as fast?

Let v_B = velocity of belt, v_S velocity of Sally when walking normally, and D the length of the belt.

1. Write D in terms of v_B; Joe's speed on the belt — $D = (2\text{ min})(v_B)$; $v_B = D/(2\text{ min})$
2. Write D in terms of $v_B + v_S$; Sally's speed on the belt — $D = (1\text{ min})(v_B + v_S) = (1\text{ min})[(D/2\text{ min}) + v_S]$
3. Solve for v_S — $v_S = D/(2\text{ min})$
4. Write t_2 = time for fast walk — $t_2 = D/(v_B + 2v_S) = D/[(D/2\text{ min}) + (D/1\text{ min}) = 40$ s

---

31 • Walk across the room in such a way that, after getting started, your velocity is negative but your acceleration is positive. (*a*) Describe how you did it. (*b*) Sketch a graph of v versus t for your motion.

(*a*) Start walking with velocity in the negative direction. Gradually slow the speed of walking, until the other end of the room is reached.

(*b*) A sketch of v versus t is shown

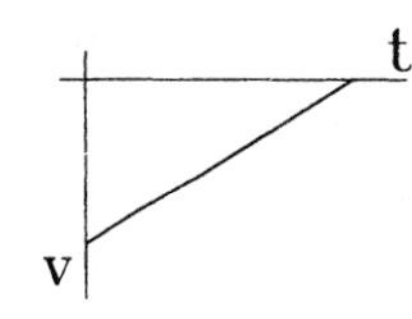

---

32 • Give an example of a motion for which both the acceleration and the velocity are negative.

Let up be the positive direction. A falling object then has a negative velocity and a negative acceleration.

33* • Is it possible for a body to have zero velocity and nonzero acceleration?

Yes, it is. An object tossed up has a constant downward acceleration; at its maximum height, its instantaneous velocity is zero.

---

34 • True or false: (*a*) If the acceleration is zero, the body cannot be moving. (*b*) If the acceleration is zero, the x-versus-t curve must be straight line.

(*a*) False (*b*) True

---

35 •• State whether the acceleration is positive, negative, or zero for each of the functions $x(t)$ in Figure 2-27.

(*a*) $a = 0$, constant velocity. (*b*) $a > 0$, v changes from negative to positive. (*c*) $a < 0$. (*d*) $a = 0$.

36 •• Answer the following question for each of the graphs in Figure 2-28: (*a*) At what times are the accelerations of the objects positive, negative, and zero? (*b*) At what times are the accelerations constant? (*c*) At what times are the instantaneous velocities zero?

Curve 1: v versus t. Hence, $a < 0$ for $t < 3$ s and $t > 7$ s; $a = 0$ for $t = 3$ s and 6 s $\le t \le 7$ s; $a > 0$ for 3 s $\le t \le 6$ s.
(*b*) a is constant for $t < 2.7$ s, 3.2 s $\le t \le 5.8$ s, 6 s $\le t \le 7$ s, 7.1 s $\le t$. (*c*) $v = 0$ at $t = 8.6$ s.

Curve 2: x versus t. Here, $a < 0$ for $0 \le t \le 3$ s and $t > 7$ s; $a = 0$ for 3 s $\le t \le 5$ s; $a > 0$ for 5 s $\le t \le 7$ s.
(*b*) It is difficult to tell where a is constant from the curve; if the curved segments are parabolas, then a is constant everywhere. (*c*) $v = 0$ at $t = 2$ s, 5.8 s, and 7.8 s.

37* • A BMW M3 sports car can accelerate in third gear from 48.3 km/h (30 mi/h) to 80.5 km/h (50 mi/h) in 3.7 s. (*a*) What is the average acceleration of this car in m/s^2? (*b*) If the car continued at this acceleration for another second, how fast would it be moving?

(*a*) 1. Use Equ. 2-8	$a_{av} = [(80.5 - 48.3)\text{ km/h}]/(3.7\text{ s}) = 8.7\text{ km/h·s}$
2. Convert to m/s^2	$(8.7 \times 10^3\text{ m/h·s})(1\text{ h}/3600\text{ s}) = 2.42\text{ m/s}^2$
(*b*) In 1 s, its speed increases by 8.7 km/h	$v = (80.5 + 8.7)\text{ km/h} = 89.2\text{ km/h}$

38 • At $t = 5$ s, an object at $x = 3$ m is traveling at 5 m/s. At $t = 8$ s, it is at $x = 9$ m and its velocity is -1 m/s. Find the average acceleration for this interval.

Use Equ. 2-8, $a_{av} = \Delta v/\Delta t$	$a_{av} = [(-1\text{ m/s}) - (5\text{ m/s})]/[(8\text{ s}) - (5\text{ s})] = -2\text{ m/s}^2$

39 •• A particle moves with velocity $v = 8t - 7$, where v is in meters per second and t is in seconds. (*a*) Find the average acceleration for the one-second intervals beginning at $t = 3$ s and $t = 4$ s. (*b*) Sketch v versus t. What is the instantaneous acceleration at any time?

(*a*) 1. Determine v at $t = 3$ s, $t = 4$ s, $t = 5$ s	$v(3) = 17$ m/s; $v(4) = 25$ m/s; $v(5) = 33$ m/s
2. Find a_{av} for the two 1-s intervals	$a_{av}(3 - 4) = 8/1\text{ m/s}^2 = 8\text{ m/s}^2$; $a_{av}(4 - 5) = 8\text{ m/s}^2$

(*b*) v versus t is shown in the adjacent plot
Use Equ. 2-10: $a = dv/dt = 8\text{ m/s}^2$.

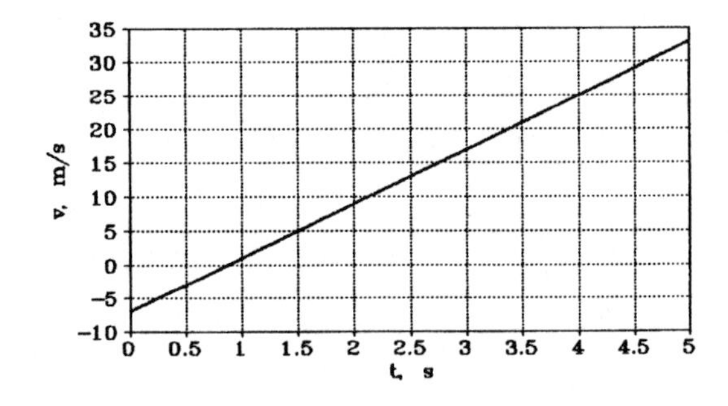

40 •• The position of an object is related to time by $x = At^2 - Bt + C$, where $A = 8\text{ m/s}^2$, $B = 6$ m/s, and $C = 4$ m. Find the instantaneous velocity and acceleration as functions of time.

1. Use Equs. 2-5 and 2-10	$v = dx/dt = 2At - B$; $a = dv/dt = 2A$
2. Substitute numerical values for A and B	$v = (16t - 6)$ m/s; $a = 16\text{ m/s}^2$

41* • Identical twin brothers standing on a bridge each throw a rock straight down into the water below. They throw rocks at exactly the same time, but one hits the water before the other. How can this occur if the rocks have the same acceleration?

The initial downward velocities of the two rocks are not the same.

---

42 • A ball is thrown straight up. What is the velocity of the ball at the top of its flight? What is its acceleration at that point?

At the top of its flight, its velocity is instantaneously zero; its acceleration is 9.81 m/s^2 downward.

---

43 • An object thrown straight up falls back to the ground T seconds later. Its maximum height is H meters. Its average velocity during those T seconds is (*a*) H/T, (*b*) 0, (*c*) $H/2T$, (*d*) $2H/T$.

(*b*) The displacement is zero.

---

44 • For an object thrown straight up, which of the following is true while it is in the air? (*a*) The acceleration is always opposite to the velocity. (*b*) The acceleration is always directed downward. (*c*) The acceleration is always in the direction of motion. (*d*) The acceleration is zero at the top of the trajectory.

(*b*) The acceleration is $-g$, always directed downward.

45* • An object projected up with initial velocity v attains a height H. Another object projected up with initial velocity $2v$ will attain a height of (*a*) $4H$, (*b*) $3H$, (*c*) $2H$, (*d*) H.

(*a*) $4H$; from Equ. 2-15, with $a = -g$ and $v = 0$ at top of trajectory, $H = v_0^2/2g$, v_0 is initial velocity. So $H \propto v_0^2$.

---

46 • A ball is thrown upward. While it is in the air, its acceleration is (*a*) decreasing, (*b*) constant, (*c*) zero, (*d*) increasing.

(*b*) constant; downward with value g.

---

47 • At $t = 0$, object A is dropped from the roof of a building. At the same instant, object B is dropped from a window 10 m below the roof. During their descent to the ground the distance between the two objects (*a*) is proportional to t. (*b*) is proportional to t^2. (*c*) decreases. (*d*) remains 10 m throughout.

(*d*) Both move with the same acceleration and velocity at all times; the distance between them remains that at $t = 0$.

---

48 •• A Porsche accelerates uniformly from 80.5 km/h (50 mi/h) at $t = 0$ to 113 km/h (70 mi/h) at $t = 9$ s. Which graph in Figure 2-29 best describes the motion of the car?

(*c*) $v(0) > 0$ and $v(t)$ increases at a uniform rate.

49* •• An object is dropped from rest. If the time during which it falls is doubled, the distance it falls will (*a*) double, (*b*) decrease by one-half, (*c*) increase by a factor of four, (*d*) decrease by a factor of four, (*e*) remain the same.

(*c*) increase by a factor of 4; see Equ. (2-14): $\Delta x \propto t^2$ if $v_0 = 0$.

---

50 •• A ball is thrown upward with an initial velocity v_0. Its velocity halfway to its highest point is (*a*) $0.5v_0$, (*b*) $0.25v_0$, (*c*) v_0, (*d*) $0.707v_0$, (*e*) cannot be determined from the information given.

(*d*) $0.707v_0$; $H = v_0^2/2g$, $v^2 = v_0^2 - 2g(H/2) = \frac{1}{2}v_0^2$; $v = 0.707v_0$.

---

51 • A car starting at $x = 50$ m accelerates from rest at a constant rate of 8 m/s^2. (*a*) How fast is it going after 10 s? (*b*) How far has it gone after 10 s? (*c*) What is its average velocity for the interval $0 \le t \le 10$ s?

(*a*) Use Equ. 2-12 $\quad v = 0 + (8\ m/s^2)(10\ s) = 80\ m/s$

(*b*) Use Equ. 2-14 — $\Delta x = \frac{1}{2}(8\ \text{m/s}^2)(10\ \text{s})^2 = 400\ \text{m}$

(*c*) Use Equ. 2-2 — $v_{av} = (400\ \text{m})/(10\ \text{s}) = 40\ \text{m/s}$

---

52 • An object with an initial velocity of 5 m/s has a constant acceleration of 2 m/s^2. When its speed is 15 m/s, how far has it traveled?

Use Equ. 2-15 — $\Delta x = [(15^2 - 5^2)\ \text{m}^2/\text{s}^2]/[2(2\ \text{m/s}^2)] = 50\ \text{m}$

53* • An object with constant acceleration has velocity $v = 10$ m/s when it is at $x = 6$ m and $v = 15$ m/s when it is at $x = 10$ m. What is its acceleration?

Use Equ. 2-15 — $a = [(15^2 - 10^2)\ \text{m}^2/\text{s}^2]/[2(4\ \text{m})] = 15.6\ \text{m/s}^2$

---

54 • An object has constant acceleration $a = 4\ \text{m/s}^2$. At $t = 0$, its velocity is 1 m/s and it is at $x = 7$ m. How fast is it moving when it is at $x = 8$ m? What is t at that point?

1. Use Equ. 2-15 — $v^2 = 1\ \text{m}^2/\text{s}^2 + 2(4\ \text{m/s}^2)(1\ \text{m}) = 9\ \text{m}^2/\text{s}^2$; $v = 3$ m/s
2. Use Equ. 2-12 — $t = [(3 - 1)\ \text{m/s}]/(4\ \text{m/s}^2) = 0.5\ \text{s}$

---

55 • If a rifle fires a bullet straight up with a muzzle speed of 300 m/s, how high will the bullet rise? (Ignore air resistance.)

At top, $v = 0$; use Equ. 2-15 — $H = v_0^2/2g = 4.59 \times 10^3\ \text{m}$

---

56 • A test of the prototype of a new automobile shows that the minimum distance for a controlled stop from 98 km/h to zero is 50 m. Find the acceleration, assuming it to be constant, and express your answer as a fraction of the free-fall acceleration due to gravity. How long does the car take to stop?

1. Use Equ. 2-15 — $a = -(98\ \text{km/h})^2/(100\ \text{m}) = -9.6 \times 10^7\ \text{m/h}^2 = -7.41\ \text{m/s}^2$
2. Divide a by $g = -9.81\ \text{m/s}^2$ — $a = 0.755g$
3. Use Equ. 2-12 — $t = (98\ \text{km/h})/(9.6 \times 10^7\ \text{m/h}^2) = 1.02 \times 10^{-3}\ \text{h} = 3.67\ \text{s}$

57* •• A ball is thrown upward with an initial velocity of 20 m/s. (*a*) How long is the ball in the air? (*b*) What is the greatest height reached by the ball? (*c*) When is the ball 15 m above the ground?

(*a*) 1. Take upward as positive; use Equ. 2-14 — $\Delta x = 0 = (20\ \text{m/s})t - \frac{1}{2}(9.81\ \text{m/s}^2)t^2$

2. Solve for t — $t = 0$; $t = 4.08$ s; $t = 4.08$ s is the proper result

(*b*) See Problem 55 — $H = (400\ \text{m}^2/\text{s}^2)/[2(9.81\ \text{m/s}^2)] = 20.4\ \text{m}$

(*c*) 1. Use Equ. 2-14 — $15\ \text{m} = (20\ \text{m/s})t - \frac{1}{2}(9.81\ \text{m/s}^2)t^2$

2. Use quadratic formula $t = \dfrac{-b \pm \sqrt{b^2 - 4ac}}{2a}$ — $t = 0.991$ s, $t = 3.09$ s; both are acceptable solutions

---

58 •• A particle moves with constant acceleration of 3 m/s^2. At $t = 4$ s, it is at $x = 100$ m; at $t = 6$ s, it has a velocity $v = 15$ m/s. Find its position at $t = 6$ s.

1. Find $v(4\ \text{s}) = v_0$ using Equ. (2-12) — $v_0 = (15 - 2 \times 3)\ \text{m/s} = 9\ \text{m/s}$
2. Use Equ. 2-14; elapsed time = 2 s — $x = [100\ \text{m} + (9\ \text{m/s})(2\ \text{s}) + \frac{1}{2}(3\ \text{m/s}^2)(2\ \text{s})^2] = 124\ \text{m}$

---

59 •• A bullet traveling at 350 m/s strikes a telephone pole and penetrates a distance of 12 cm before stopping. (*a*) Estimate the average acceleration by assuming it to be constant. (*b*) How long did it take for the bullet to stop?

(*a*) Use Equ. 2-15 $a = -(350 \text{ m/s})^2/[2(0.12 \text{ m})] = -5.1 \times 10^5 \text{ m/s}^2$

(*b*) Use Equ. 2-12 $t = (350 \text{ m/s})/(5.1 \times 10^5 \text{ m/s}^2) = 0.686 \text{ ms}$

---

60 •• A plane landing on an aircraft carrier has just 70 m to stop. If its initial speed is 60 m/s, (*a*) what is the acceleration of the plane during landing, assuming it to be constant? (*b*) How long does it take for the plane to stop?

(*a*) Use Equ. 2-15 $a = -(60 \text{ m/s})^2/[2(70 \text{ m})] = -25.7 \text{ m/s}^2$

(*b*) Use Equ. 2-12 $t = (60 \text{ m/s})/(25.7 \text{ m/s}^2) = 2.33 \text{ s}$

61* •• An automobile accelerates from rest at 2 m/s² for 20 s. The speed is then held constant for 20 s, after which there is an acceleration of −3 m/s² until the automobile stops. What is the total distance traveled?

1. Determine the distance traveled during first 20 s and the speed at the end of first 20 s — $v(20) = at = (2 \text{ m/s}^2)(20 \text{ s}) = 40 \text{ m/s}$; $\Delta x_1 = v_{av}t = (20 \text{ m/s})(20 \text{ s}) = 400 \text{ m}$
2. Find Δx_2 = distance covered $20 \text{ s} \le t \le 40 \text{ s}$ — $\Delta x_2 = (40 \text{ m/s})(20 \text{ s}) = 800 \text{ m}$
3. Find Δx_3 = distance during deceleration — $\Delta x_3 = (40 \text{ m/s})^2/[2(3 \text{ m/s}^2)] = 267 \text{ m}$
4. Find total distance — $x = \Delta x_1 + \Delta x_2 + \Delta x_3 = 1467 \text{ m}$

---

62 •• In the Blackhawk landslide in California, a mass of rock and mud fell 460 m down a mountain and then traveled 8 km across a level plain on a cushion of compressed air. Assume that the mud dropped with the free-fall acceleration due to gravity and then slid horizontally with constant deceleration. (*a*) How long did the mud take to drop the 460 m? (*b*) How fast was it traveling when it reached the bottom? (*c*) How long did the mud take to slide the 8 km horizontally?

(*a*) Use Equ. 2-14 $t^2 = 2(460 \text{ m})/(9.81 \text{ m/s}^2) = 93.8 \text{ s}^2$; $t = 9.68 \text{ s}$

(*b*) Use Equ. 2-12 $v = (9.81 \text{ m/s}^2)(9.68 \text{ s}) = 95.0 \text{ m/s}$

(*c*) Use Equ. 2-13 $t = 2(8000 \text{ m})/(95.0 \text{ m/s}) = 168 \text{ s}$

---

63 •• A load of bricks is being lifted by a crane at a steady velocity of 5 m/s when one brick falls off 6 m above the ground. (*a*) Sketch *x(t)* to show the motion of the free brick. (*b*) What is the greatest height the brick reaches above the ground? (*c*) How long does it take to reach the ground? (*d*) What is its speed just before it hits the ground?

(*a*) Use Equ. 2-14; $x_0 = 6 \text{ m}$, $v_0 = 5 \text{ m/s}$, $a = -9.81 \text{ m/s}^2$

(*b*) Use Equ. 2-15; find Δx and add to x_0;

$\Delta x = (25/2 \times 9.81) \text{ m} = 1.27 \text{ m}$;

$x = 7.27 \text{ m}$ (see also graph)

(*c*) Use Equ. 2-14; then solve quadratic equ. for t

$0 = 6 + 5t - \frac{1}{2} \times 9.81t^2$; $t = 1.73 \text{ s}$, $t = -0.708 \text{ s}$

Second (negative solution) is non-physical; $t = 1.73 \text{ s}$

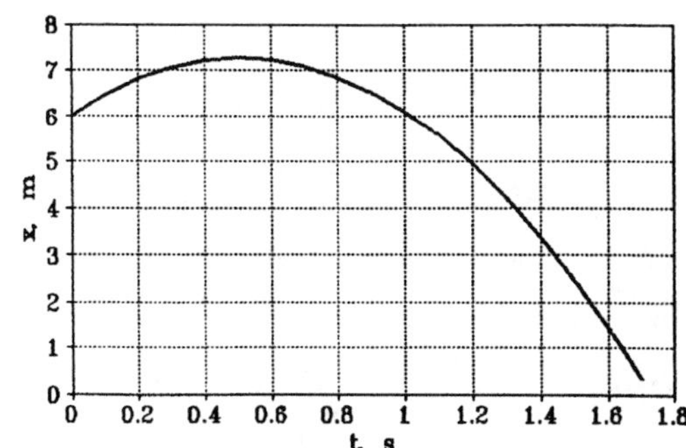

(*d*) Use $v^2 = 2gH$, with $H = 7.27$ m

$v^2 = 2 \times 9.81 \times 7.27\ \text{m}^2/\text{s}^2$; $v = 11.9$ m/s

---

64 •• An egg with a mass of 50 g rolls off a table at a height of 1.2 m and splatters on the floor. Estimate the average acceleration of the egg while it is in contact with the floor.

We shall take 5 cm as the diameter of an egg, and consider the time of contact while crashing, during which the "egg" travels 5 cm. Just before contact with the floor, $v = \sqrt{2gH} \approx 5$ m/s, so the time of contact $\approx$ (0.05/2.5) s = 20 ms. In that time the egg decelerates to $v = 0$. So the magnitude of a_{av} = (5/0.02) m/s² = 250 m/s².

65* •• To win publicity for her new CD release, Sharika, the punk queen, jumps out of an airplane without a parachute. She expects a stack of loose hay to break her fall. If she reaches a speed of 120 km/h prior to impact, and if a 35 g deceleration is the greatest deceleration she can withstand, how high must the stack of hay be in order for her to survive? Assume uniform acceleration while she is in contact with the hay.

1. Use Equ. 2-15 $$H = \frac{(1.20\times10^5\ \text{m/h})^2}{(3600\ \text{s/1 h})^2[2\times35\times(9.81\ \text{m/s}^2)]} = 1.62\ \text{m}$$

---

66 •• A bolt comes loose from underneath an elevator that is moving upward at a speed of 6 m/s. The bolt reaches the bottom of the elevator shaft in 3 s. (*a*) How high up was the elevator when the bolt came loose? (*b*) What is the speed of the bolt when it hits the bottom of the shaft?

We use up as the positive direction; $x = 0$ at bottom of shaft.

(*a*) Use Equ. 2-14; $v_0 = 6$ m/s, $t = 3$ s, $x(3\ \text{s}) = 0$ — $x_0 = [0 - (6 \times 3) - ½(-9.81)(9)]$ m = 26.1 m

(*b*) Use Equ. 2-12 — $v = [6 - 9.81 \times 3]$ m/s = -23.4 m/s; $|v| = 23.4$ m

---

67 •• An object is dropped from a height of 120 m. Find the distance it falls during its final second in the air.

1. Find final (impact) speed, v_f — $v_f = \sqrt{2(9.81)(120)}$ m/s = 48.5 m/s
2. Find v_{f-1}, speed 1 s prior to impact — v_{f-1} = (48.5 − 9.81) m/s = 38.7 m/s
3. Find the average speed during this 1 s — v_{av} = ½(48.5 + 38.7) m/s = 43.6 m/s
4. Find the distance traversed, $s = v_{av}\Delta t$ — s = (43.6 × 1) m = 43.6 m

---

68 •• An object is dropped from a height H. During the final second of its fall, it traverses a distance of 38 m. What was H?

Let v_f be the final speed before impact, v_{f-1} the speed 1 s before impact.

1. Find the average speed in last second — v_{av} (1 s) = 38 m = ½($v_f + v_{f-1}$)(1 s); $v_f + v_{f-1}$ = 76 m/s
2. Express v_{f-1} in terms of v_f and solve for v_f — $v_f - v_{f-1}$ = 9.81 m/s; v_f = 42.9 m/s
3. Use v_f to determine H — $H = v_f^2/2g$ = 93.8 m

69* • A stone is thrown vertically from a cliff 200 m tall. During the last half second of its flight the stone travels a distance of 45 m. Find the initial velocity of the stone.

We take <u>down</u> as the positive direction. Let v_1 be the velocity ½ s before impact, v_f velocity at impact.

1. Find v_{av} during last ½ s — v_{av} = ½($v_1 + v_f$) = (45/0.5) m/s; $v_1 + v_f$ = 180 m/s
2. Write v_f in terms of v_1 and g — $v_f = v_1 + gt$; $v_f = v_1$ + (9.81 × 0.5) m/s
3. Solve for v_f — v_f = 92.5 m/s

4. Use Equ. 2-15 to find v_0 — $v_0 = \sqrt{92.5^2 - 2\times 9.81\times 200}$ m/s $= \pm 68$ m/s; the stone may be thrown either up or down

---

70 •• An object in free-fall from a height H traverses $0.4H$ during the first second of its descent. Determine the average speed of the object during free-fall.

1. Find Δx for first second and H — $\Delta x = \frac{1}{2}gt^2 = 4.9$ m $= 0.4H$; $H = 12.3$ m
2. Find v_f — $v_f = [2(9.81)(12.3)]^{1/2}$ m/s $= 15.5$ m/s
3. Determine v_{av}, $v_0 = 0$ — $v_{av} = \frac{1}{2}v_f = 7.77$ m/s

---

71 •• A bus accelerates at 1.5 m/s² from rest for 12 s. It then travels at constant speed for 25 s, after which it slows to a stop with an acceleration of − 1.5 m/s². (*a*) How far did the bus travel? (*b*) What was its average velocity?

(*a*) 1. Find $v(12)$ and $v_{av}(0-12)$ — $v(12) = (1.5 \times 12)$ m/s $= 18$ m/s; $v_{av}(0-12) = 9$ m/s

2. Find $x(12) = v_{av}(0-12)\Delta t$ — $x(12) = (9 \times 12)$ m $= 108$ m

3. Find $\Delta x(12-37) = v(12) \times \Delta t$ — $\Delta x(12-37) = (18 \times 25)$ m $= 450$ m

4. |deceleration| = acceleration; $\Delta x_{dec} = \Delta x_{acc}$ — $\Delta x_{dec} = 108$ m and $\Delta t_{dec} = 12$ s

5. Add dislacements to find x_{tot} — Total distance = 666 m

(*b*) $v_{av} = x_{tot}/t_{tot}$ — $v_{av} = (666\text{ m})/(49\text{ s}) = 13.6$ m/s

---

72 •• A basketball is dropped from a height of 3 m and rebounds from the floor to a height of 2 m. (*a*) What is the velocity of the ball just as it reaches the floor? (*b*) What is its velocity just as it leaves the floor? (*c*) Estimate the magnitude and direction of its average acceleration during this interval.

(*a*) $v = \sqrt{2gH}$ — $v = -(2 \times 9.81 \times 3)^{1/2}$ m/s $= -7.67$ m/s

(*b*) Use the above expression — $v = +(2 \times 9.81 \times 2)^{1/2}$ m/s $= 6.26$ m/s

(*c*) A reasonable estimate for Δt is 0.05 s — $a_{av} = (6.26 + 7.67)/0.05$ m/s² $= 279$ m/s² upward

---

73* •• A rocket is fired vertically with an upward acceleration of 20 m/s². After 25 s, the engine shuts off and the rocket continues as a free particle until it reaches the ground. Calculate (*a*) the highest point the rocket reaches, (*b*) the total time the rocket is in the air, (*c*) the speed of the rocket just before it hits the ground.

Take up as the positive direction.

(*a*) 1. Find x_1 and v_1 at $t = 25$ s; use Equ. 2-14 — $x_1 = \frac{1}{2}(20)(25)^2$ m $= 6250$ m; $v_1 = 20 \times 25$ m/s $= 500$ m/s

2. Find x_2 = distance above x_1 when $v = 0$ — $x_2 = (500)^2/(2 \times 9.81)$ m $= 1.274 \times 10^4$ m

3. Total height $= x_1 + x_2$ — $H = 1.90 \times 10^4$ m $= 19.0$ km

(*b*) 1. Find time, t_2, for part (*a*) — $t_2 = x_2/v_{av} = (1.274 \times 10^4/250)$ s $= 51$ s

2. Find time, t_3, to drop 19.0 km — $t_3 = [2(1.90 \times 10^4)/9.81]^{1/2}$ s $= 62.2$ s

3. Total time, $T = 25\text{ s} + t_2 + t_3$ — $T = (25 + 51 + 62.2)$ s $= 138$ s $= 2$ min 18 s

(*c*) Use Equ. 2-12 — $v_f = (9.81)(62.2)$ m/s $= 610$ m/s

---

74 •• A flowerpot falls from the ledge of an apartment building. A person in an apartment below, coincidentally holding a stopwatch, notices that it takes 0.2 s for the pot to fall past his window, which is 4 m high. How far above the top of the window is the ledge from which the pot fell?

Take down as positive, $x = 0$ at ledge. Let t = time when pot is at top of window, $t + \Delta t$ when pot is at bottom of window.

1. Express $\Delta x = 4$ m in terms of t, Δt, and g — $4\text{ m} = \tfrac{1}{2}g[(t+\Delta t)^2 - t^2] = \tfrac{1}{2}g(\Delta t)^2 + gt\Delta t$
2. Solve for t, with $\Delta t = 0.2$ s, $g = 9.81\text{ m/s}^2$ — $t = 1.94$ s
3. Find distance pot falls from rest in 1.94 s — $H = \tfrac{1}{2}(9.81)(1.94)^2 = 18.4$ m

---

75 •• Sharika arrives home late from a gig, only to find herself locked out. Her roommate and bass player Chico is practicing so loudly that he can't hear Sharika's pounding on the door downstairs. One of the band's props is a small trampoline, which Sharika places under Chico's window. She bounces progrssively higher trying to get Chico's attention. Propelling herself furiously upward, shw miscalculates on the last bounce and flies past the window and out of sight. Chico sees her face for 0.2 s as she moves a distance of 2.4 m from the bottom to the top of the window. (*a*) How long until she reappears? (*b*) What is her greatest height above the top of the window? (Treat Sharika as a point-particle punk.)

Take up as positive; $x = 0$ at bottom of window, v_0 = upward velocity at top of window.

(*a*) 1. Use Equ. 2-14 and solve for v_0 — $2.4\text{ m} = v_0(0.2\text{ s}) + \tfrac{1}{2}(9.81)(0.2)^2\text{ m};\ v_0 = 11.0\text{ m/s}$

2. Find the time to reach $v = 0$ — $t = (11.0/9.81)\text{ s} = 1.12$ s

3. Time when her face reappears is $2t$ — Time till her face reappears is 2.24 s

(*b*) Find H using Equ. 2-15 — $H = v_0^2/2g = 6.17$ m

---

76 •• In a classroom demonstration, a glider moves along an inclined air track with constant acceleration a. It is projected from the start of the track ($x = 0$) with an initial velocity v_0. At time $t = 8$ s, it is at $x = 100$ cm and is moving along the track at velocity $v = -15$ cm/s. Find the initial speed v_0 and the acceleration a.

1. Use Equ. (2-13) and solve for v_0 — $v_0 = 2x/t - v = [(200/8) - (-15)]\text{ m/s} = 40\text{ cm/s}$
2. Use the definition of $a = \Delta v/\Delta t$ — $a = (-55/8)\text{ cm/s}^2 = -6.88\text{ cm/s}^2$

77* •• A rock dropped from a cliff falls one-third of its total distance to the ground in the last second of its fall. How high is the cliff?

1. Write the final speed in terms of H and g — $v_f^2 = 2gH$
2. Write the average speed in the last second — $v_{av} = \tfrac{1}{2}[v_f + (v_f - g)] = (v_f - g/2)$
3. Set the distance in last second = $H/3$; solve for v_f — $(v_f - g/2) = H/3;\ v_f = H/3 + g/2$
4. Set $v_f^2 = 2gH$ and solve for H — $2gH = H^2/9 + gH/3 + g^2/4;\ H = 14.85g = 145.7$ m

---

78 ••• A typical automobile has a maximum deceleration of about 7 m/s²; the typical reaction time to engage the brakes is 0.50 s. A school board sets the speed limit in a school zone to meet the condition that all cars should be able to stop in a distance of 4 m. (*a*) What maximum speed should be allowed for a typical automobile? (*b*) What fraction of the 4 m is due to the reaction time?

(*a*) 1. Write the total distance as the sum of distance traveled in 0.50 s plus that during deceleration — $\Delta x_1 = 0.50v_0;\ \Delta x_2 = -v_0^2/2a\ \ (a = -7\text{ m/s}^2)$; $4\text{ m} = \Delta x_1 + \Delta x_2 = (0.5v_0 + v_0^2/14)\text{ m}$

2. Solve quadratic equ. keeping positive result — $v_0 = 4.76\text{ m/s} = 4.76\text{ m/s} = 10.6\text{ mi/h}$

(*b*) Find reaction time distance — $\Delta x_1 = 2.38\text{ m} = (2.38/4)\Delta x_{tot} = 59.5\%\text{ of }\Delta x_{tot}$

---

79 •• Two trains face each other on adjacent tracks. They are initially at rest 40 m apart. The train on the left accelerates rightward at 1.4 m/s². The train on the right accelerates leftward at 2.2 m/s². How far does the train on the left travel before the two trains pass?

Take $x = 0$ as position of train on left at $t = 0$.

1. Write positions of each train as a function of time	$x_L = 0.7t^2$ m; $x_R = 40 - 1.1t^2$
2. Set $x_L = x_R$ and solve for t	$0.7t^2 = 40 - 1.1t^2$; $t = 4.71$ s
3. Find distance traveled , i.e., x_L	$x_L = 15.6$ m

---

80 •• Two stones are dropped from the edge of a 60-m cliff, the second stone 1.6 s after the first. How far below the cliff is the second stone when the separation between the two stones is 36 m?

1. Write expressions for x_1 and x_2	$x_1 = \frac{1}{2}gt^2$; $x_2 = \frac{1}{2}g(t - 1.6\text{ s})^2$
2. Set $x_1 - x_2 = 36$ m	$36\text{ m} = \frac{1}{2}g[(3.2\text{ s})t - 2.56\text{ s}^2]$
3. Solve for t, then find x_2	$t = 3.09$ s; $x_2 = 10.9$ m

81* •• A motorcycle policeman hidden at an intersection observes a car that ignores a stop sign, crosses the intersection, and continues on at constant speed. The policeman takes off in pursuit 2.0 s after the car has passed the stop sign, accelerates at 6.2 m/s² until his speed is 110 km/h, and then continues at this speed until he catches the car. At that instant, the car is 1.4 km from the intersection. How fast was the car traveling?

1. Find the time of travel of policeman: t_1 is the time of acceleration, t_2 the time of travel at 110 km/h; and d_1 and d_2 the corresponding distances	$t_1 = (1.10 \times 10^3\text{ m/h})/[(3600\text{ s/1 h})(6.2\text{ m/s}^2)] = 4.93$ s $d_1 = \frac{1}{2}(30.5\text{ m/s})(4.93\text{ s}) = 75.3$ m; $d_2 = (1400 - 75.3)$ m $= 1325$ m; $t_2 = d_2/(30.6\text{ m/s}) = 43.3$ s
2. The time of travel of car is 2.0 s + t_1 + t_2	$t_C = (2.0 + 43.3 + 4.93)\text{ s} = 50.2$ s
3. Find the speed of the car	$v_C = (1400/50.2)\text{ m/s} = 27.9\text{ m/s} = 100.4$ km/h

---

82 •• At $t = 0$, a stone is dropped from a cliff above a lake; 1.6 seconds later another stone is thrown downward from the same point with an initial speed of 32 m/s. Both stones hit the water at the same instant. Find the height of the cliff.

1. Write the distances dropped and equate	$\frac{1}{2}gt^2 = (32\text{ m/s})(t - 1.6\text{ s}) + \frac{1}{2}g(t - 1.6\text{ s})^2$
2. Solve for t	$t = 2.37$ s
3. Find H; Use Equ. 2-14	$H = \frac{1}{2}gt^2 = 27.6$ m

---

83 •• A passenger train is traveling at 29 m/s when the engineer sees a freight train 360 m ahead traveling on the same track in the same direction. The freight train is moving at a speed of 6 m/s. If the reaction time of the engineer is 0.4 s, what must be the deceleration of the passenger train if a collision is to be avoided? If your answer is the maximum deceleration of the passenger train but the engineer's reaction time is 0.8 s, what is the relative speed of the two trains at the instant of collision and how far will the passenger train have traveled in the time between the sighting of the freight train and the collision?

Let $x = 0$ be the location of the passenger train engine at moment of sighting of freight train's end; let $t = 0$ be the instant the passenger train decelerates.

1. Write expressions for x_P and x_F, positions of the passenger train engine and the freight train's end	$x_P = [(29\text{ m/s})(t + 0.4\text{ s}) - \frac{1}{2}at^2]$; $x_P = 11.6 + 29t - 4.91t^2$ $x_F = (360\text{ m}) + (6\text{ m/s})(t + 0.4\text{ s})$; $x_F = 362.4 + 6t$

2. Set $x_F = x_P$ and obtain an equation for t	$\frac{1}{2}at^2 - (23\text{ m/s})t + 351\text{ m} = 0$
3. If equation has a real solution collision occurs	If $(23^2 - 702a) > 0$, collision
4. Solve for a so there is no real solution for t	$a \geq 0.754\text{ m/s}^2$
5. Repeat, with $a = 0.754\text{ m/s}^2$ and 0.8 s reaction time	quadratic equation is $0.377t^2 - 23t + 342 = 0$
6. Solve for t, keeping only smaller value	$t = 25.5$ s; ($t = 35.5$ s: trains have already collided)
7. Find x_P at that instant	$x_P = (29\text{ m/s})(26.3\text{ s}) - (0.377\text{ m/s}^2)(25.5\text{ s})^2 = 518\text{ m}$
8. Find the speeds of the two trains	$v_P = (29\text{ m/s}) - (0.754\text{ m/s}^2)(25.5\text{ s}) = 9.77\text{ m/s}$; $v_F = 6\text{ m/s}$; $v_{rel} = 3.77\text{ m/s}$

The adjacent figure shows the location of the trains. The solid straight line is for the freight train; the solid and dashed curved lines are for the passenger train, with reaction times of 0.4 s and 0.8 s, respectively.

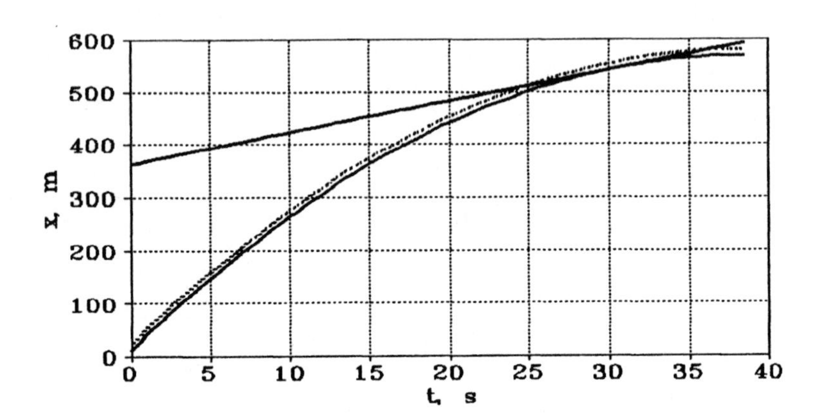

84 •• After being forced out of farming, Lou has given up on trying to find work locally and is about to "ride the rails" to look for a job. Running at his maximum speed of 8 m/s, he is a distance d from the train when it begins to accelerate from rest at 1.0 m/s². (*a*) If d = 30 m and Lou keeps running, will he be able to jump into the train? (*b*) Sketch the position function $x(t)$ for the train, with $x = 0$ at $t = 0$. On the same graph, sketch $x(t)$ for Lou for various distances d, including $d = 30$ m and the critical separation distance d_c, the distance at which he just catches the train. (*c*) For the situation $d = d_c$, what is the speed of the train when Lou catches it? What is the train's average speed for the time interval between $t = 0$ and the moment Lou catches the train? What is the exact value of d_c?

(*a*), (*b*) and (*c*) For the train, $x_T(t) = \frac{1}{2}(1.0\text{ m/s}^2)t^2$; for Lou, $x_L(t) = (8\text{ m/s})t - d$

(*a*) Yes, he can jump on the train. (see graph)

(*c*) When Lou just manages to catch the train, his speed and that of the train are equal. The speed of the train is $v_T(t) = 8\text{ m/s} = at = t$ m/s.

Hence the critical time is $t_C = 8$ s. At that time, Lou has run 64 m and the train has traveled 32 m. Consequently, $d_C = (64 - 32)\text{ m} = 32\text{ m}$. $v_{av} = (32/8)\text{ m/s} = 4\text{ m/s}$.

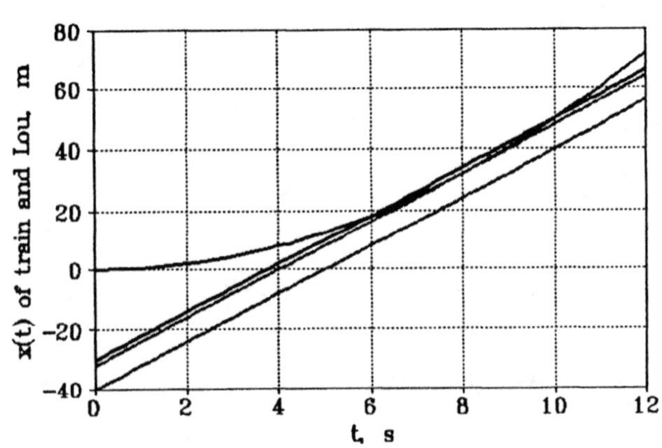

The three straight lines on the graph correspond to
$d = 30$ m, $d = d_C = 32$ m, and $d = 40$ m.

85* •• A train pulls away from a station with a constant acceleration of 0.4 m/s². A passenger arrives at the track 6.0 s after the end of the train has passed the very same point. What is the slowest constant speed at which she can run and catch the train? Sketch curves for the motion of passenger and the train as functions of time.

1. As in Problem 84, the critical conditions are $v_T = v_P$ and $x_T = x_P$ — $v_T = 0.4t$ m/s; $x_T = 0.2t^2$ m; $x_P = v_P(t - 6\text{ s})$; $0.2t^2 = 0.4t(t - 6)$
2. Solve for t — $0.2t^2 = 2.4t$; $t = 12$ s
3. Find v_T at $t = 12$ s, which is also v_{PC} — $v_{PC} = 4.8$ m/s

The positions of the train and passenger as functions of time are shown in the adjoining figure.

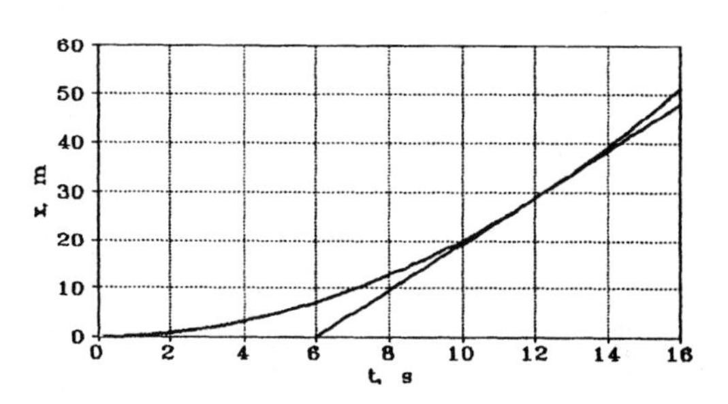

---

86 ••• Lou applies for a job as a perfume salesman. He tries to convince the boss to try his daring, aggressive promotional gimmick: dousing prospective customers as they wait at bus stops. A hard ball is to be thrown straight upward with an initial speed of 24 m/s. A thin-skinned ball filled with perfume is then thrown straight upward along the same path with a speed of 14 m/s. The balls are to collide when the perfume ball is at the high point of its trajectory, so that it breaks open and everyone gets a free sample. If $t = 0$ when the first ball is thrown, find the time when the perfume ball should be thrown.

1. Find Δt for the perfume ball to reach the top — $v = 0 = (14 - 9.81\Delta t)$ m/s; $\Delta t = 1.427$ s
2. Find height at $\Delta t = 1.427$ s — $H = (14\text{ m/s})^2/2g = 9.99$ m
3. Find the time for the hard ball to be at H on its downward motion — $9.99 = 24t - \frac{1}{2}(9.81)t^2$; solve for t; $t = 4.43$ s ($t = 0.46$ s corresponds to upward motion)
4. Time delay is $t - \Delta t$ — Throw the perfume ball at $t = (4.43 - 1.43)$ s $= 3$ s

---

87 ••• Ball A is dropped from the top of a building at the same instant that ball B is thrown vertically upward from the ground. When the balls collide, they are moving in opposite directions, and the speed of A is twice the speed of B. At what fraction of the height of the building does the collision occur?

Take $x = 0$ at ground, upward positive.

1. Describe the conditions at collision — $x_A = x_B$; $v_A = -2v_B$ (Note: A is moving down)
2. Write expressions for x_A and x_B — $x_A = H - \frac{1}{2}gt^2$; $x_B = v_0t - \frac{1}{2}gt^2$
3. Write expressions for v_A and v_B — $v_A = -gt$; $v_B = v_0 - gt$
4. Set $v_A = -2v_B$ and solve for t — $t = 2v_0/3g$
5. Use this result in part 2 with $x_A = x_B$ — $H = 2v_0^2/3g$
6. Write x_A (see part 2) in terms of v_0 — $x_A = 4v_0^2/9g = 2H/3$

88 ••• Solve Problem 87 if the collision occurs when the balls are moving in the same direction and the speed of A is 4 times that of B.

Proceed as in Problem 87 except that now the condition on velocities is $v_A = 4v_B$. One obtains for time $t = 4v_0/3g$, and for velocities $v_A = -(4/3)v_0$ and $v_B = -(1/3)v_0$. Now $H = 4v_0^2/3g$, giving $x_A = H/3$.

89* ••• The Sprint missile, designed to destroy incoming ballistic missiles, can accelerate at $100g$. If an ICBM is detected at an altitude of 100 km moving straight down at a constant speed of 3×10^4 km/h and the Sprint missile is launched to intercept it, at what time and altitude will the interception take place? (*Note*: You can neglect the acceleration due to gravity in this problem. Why?)

1. Neglect g; see below. Find x_{ICBM}, x_S — $x_S = \frac{1}{2}at^2$; $x_{ICBM} = H - vt$
2. Express v in m/s — $v = (3 \times 10^7 \text{ m/h})(1 \text{ h}/3600 \text{ s}) = 8.33 \times 10^3$ m/s
3. Set $x_{ICBM} = x_S$ with $H = 10^5$ m and find t — $\frac{1}{2} \times 981t^2 + 8.33 \times 10^3 t - 10^5 = 0$; $t = 8.12$ s
4. Find x_{ICBM} — $x_{ICBM} = [981 \times 8.12^2/2]$ m $= 3.24 \times 10^4$ m $= 32.4$ km

Note: In 8 s, the change in velocity Δv of the ICBM due to g is less than 80 m/s , i.e., less than 1% of v; also, g = 1% of a_S. So the result is good to about 1%. Also, if $a_S = 100g$ is taken to be the possible horizontal acceleration, then both objects suffer the same downward acceleration due to gravity, and this contribution cancels.

90 ••• When a car traveling at speed v_1 rounds a corner, the driver sees another car traveling at a slower speed v_2 a distance d ahead. (*a*) If the maximum acceleration the driver's brakes can provide is a, show that the distance d must be greater than $(v_1 - v_2)^2/(2a)$ if a collision is to be avoided. (*b*) Evaluate this distance for $v_1 = 90$ km/h, $v_2 = 45$ km/h, and $a = 6$ m/s^2. (*c*) Estimate or measure your reaction time and calculate the effect it would have on the distance found in part (*b*).

Note: This problem is similar to Problem 83. Again, the critical conditions are $x_1 = x_2$ and $v_1 = v_2$.

(*a*) 1. Write expressions for x_1, x_2, v_1, and v_2 — $x_1 = v_1t - \frac{1}{2}at^2$; $x_2 = d + v_2t$; $v_1(t) = v_1 - at$; $v_1 = v_2$

2. Set $v_1(t) = v_2$ — $t = (v_1 - v_2)/a$

3. Set $x_1 = x_2$ and solve for d — $d = (v_1 - v_2)^2/(2a)$

(*b*) Convert km/h to m/s and substitute to find d — $d = [(25 \text{ m/s}) - (12.5 \text{ m/s})]^2/[2(6 \text{ m/s}^2)] = 13.0$ m

(*c*) Assume reaction time of 0.8 s. In that time, the distance between cars diminishes by 10 m. — $d' = (13.0 + 10)$ m $= 23$ m

91 • The velocity of a particle is given by $v = 6t + 3$, where t is in seconds and v is in meters per second. (*a*) Sketch $v(t)$ versus t, and find the area under the curve for the interval $t = 0$ to $t = 5$ s. (*b*) Find the position function $x(t)$. Use it to calculate the displacement during the interval $t = 0$ to $t = 5$ s.

(*a*) The graph is shown. The area under the straight line is 90 m.

(*b*) $x(t) = \int_0^t (6t+3)dt = 3t^2 + 3t$; $\Delta x = x(5) - x(0) = 90$ m.

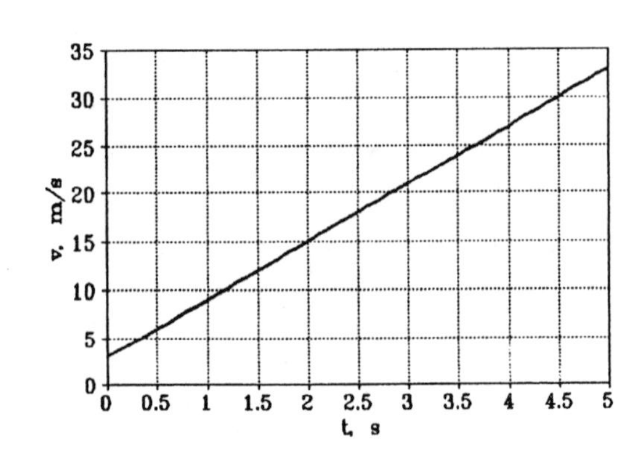

92 • Figure 2-30 shows the velocity of a particle versus time. (*a*) What is the magnitude in meters of the area of the rectangle indicated? (*b*) Find the approximate displacement of the particle for the one-second intervals beginning at $t = 1$ s and $t = 2$ s. (*c*) What is the approximate average velocity for the interval $1\text{ s} \le t \le 3\text{ s}$?

(*a*) Find the area of the rectangle — $A = (1\text{ m/s})(1\text{ s}) = 1\text{ m}$

(*b*) 1. Find approx. area under curve — $\Delta x(1 - 2) \approx 1\text{ m}$

2. Find approx. area under curve — $\Delta x(2 - 3) \approx 3\text{ m}$

(*c*) $v_{av} = \Delta x/\Delta t$ — $v_{av} \approx (4/2)\text{ m/s} = 2\text{ m/s}$

93* •• The velocity of a particle is given by $v = 7t^2 - 5$, where t is in seconds and v is in meters per second. Find the general position function $x(t)$.

$x(t) = \int (7t^2 - 5)\,dt = (7/3)t^3 - 5t + C.$

94 •• The equation of the curve shown in Figure 2-30 is $v = 0.5t^2$ m/s. Find the displacement of the particle for the interval $1\text{ s} \le t \le 3\text{ s}$ by integration, and compare this answer with your answer for Problem 92. Is the average velocity equal to the mean of the initial and final velocities for this case?

1. $\Delta x = \int_1^3 0.5t^2\,dt = (1/6)t^3\Big]_1^3 = 4.33\text{ m}.$

2. $a = dv/dt = 1.0t\text{ m/s}^2$; a is not constant, therefore $v_{av} \neq v_{mean}$.

95 ••• Figure 2-31 shows the acceleration of a particle versus time. (*a*) What is the magnitude of the area of the rectangle indicated? (*b*) The particle starts from rest at $t = 0$. Find the velocity at $t = 1$ s, 2 s, and 3 s by counting the rectangles under the curve. (*c*) Sketch the curve $v(t)$ versus t from your results for part (*b*), and estimate how far the particle travels in the interval $t = 0$ to $t = 3$ s.

(*a*) Find the area of the rectangle — $A = (0.5\text{ m/s}^2)(0.5\text{ s}) = 0.25\text{ m/s}$

(*b*) $v = 0$ at $t = 0$; count squares and multiply by 0.25 — $v(1) = 0.9$ m/s, $v(2) = 3$ m/s, $v(3) = 6$ m/s

(*c*) The curve of $v(t)$ versus t is shown. By counting squares under the curve, we find the distance traveled is approximately 6.5 m

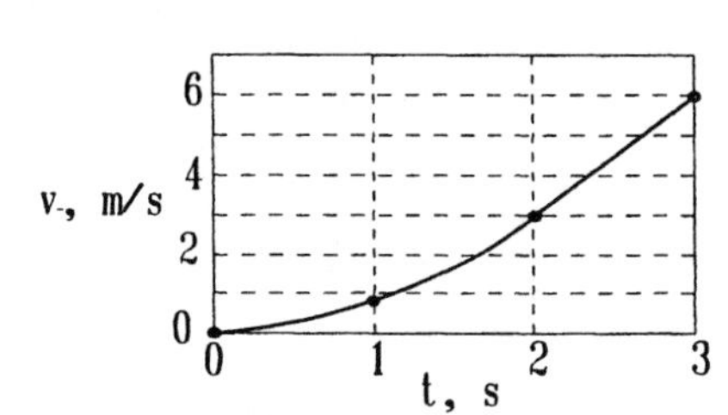

96 •• Figure 2-32 is a graph of v versus t for a particle moving along a straight line. The position of the particle at time $t = 0$ is $x_0 = 5$ m. (*a*) Find x for various times t by counting squares, and sketch x versus t. (*b*) Sketch the acceleration a versus t.

(*a*) Count squares to $t = 1$ s, 2 s, 3 s, 4 s, 5 s, 6 s, 7 s, 8 s, 9 s, 10 s; add $x(0) = 5$ m

(*b*) $a = dv/dt$ = slope of v versus t curve. Take slopes at various times.

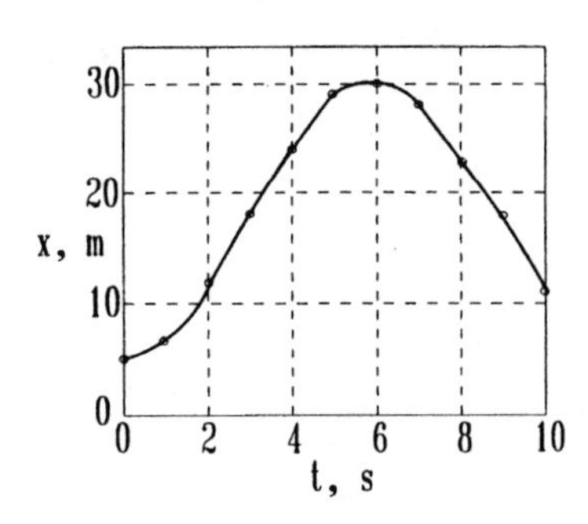

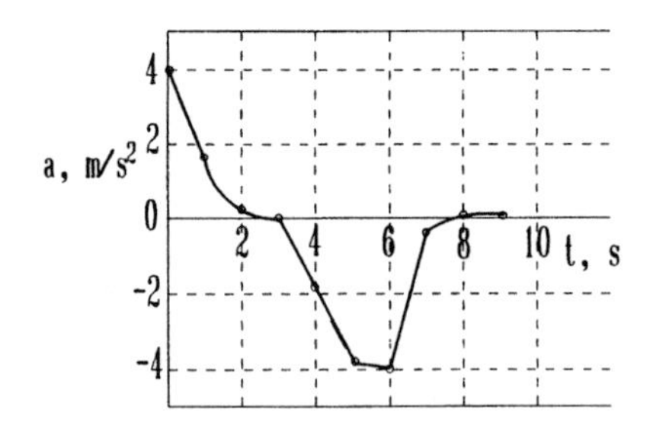

97* ••• Figure 2-33 shows a plot of x versus t for a body moving along a straight line. Sketch rough graphs of v versus t and a versus t for this motion.

Note: The curve of x versus t appears to be a sine curve; hence $v(t) = dx/dt$ is a cosine curve and $a(t) = dv/dt$ is a negative sine curve.

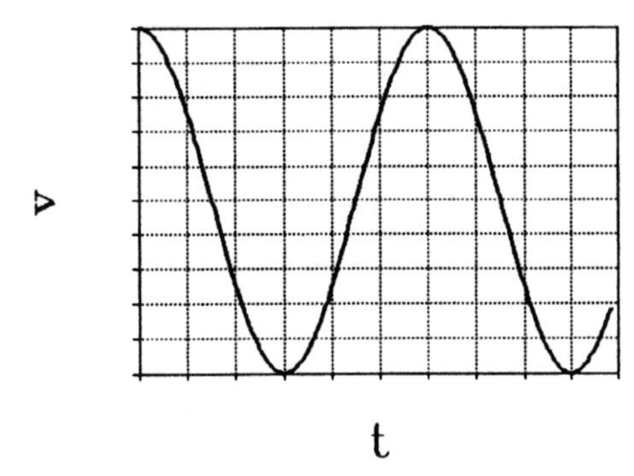

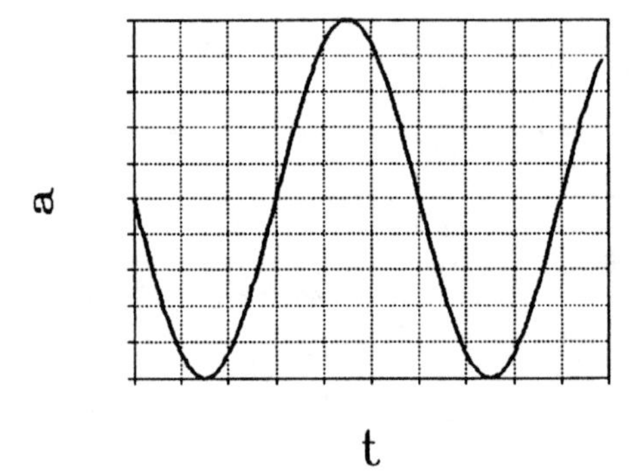

98 • True or false: (*a*) The equation $\Delta x = v_0 t + \frac{1}{2}at^2$ is valid for all particle motion in one dimension. (*b*) If the velocity at a given instant is zero, the acceleration at that instant must also be zero. (*c*) The equation $\Delta x = v_{av}\Delta t$ holds for all motion in one dimension.

(*a*) False; valid only for constant a. (*b*) False (*c*) True; by definition.

99 • If an object is moving at constant acceleration in a straight line, its instantaneous velocity halfway through any time interval is (*a*) greater than its average velocity. (*b*) less than its average velocity. (*c*) equal to its average velocity. (*d*) half its average velocity. (*e*) twice its average velocity.

(*c*) is correct, by definition of v_{av} for constant acceleration.

100 • On a graph showing position on the vertical axis and time on the horizontal axis, a straight line with a negative slope represents (*a*) a constant positive acceleration. (*b*) a constant negative acceleration. (*c*) zero velocity. (*d*) a constant positive velocity. (*e*) a constant negative velocity.

(*e*) The slope represents the velocity; negative slope corresponds to a negative velocity

101* •• On a graph showing position on the vertical axis and time on the horizontal axis, a parabola that opens upward represents (*a*) a positive acceleration. (*b*) a negative acceleration. (*c*) no acceleration. (*d*) a positive followed by a negative acceleration. (*e*) a negative followed by a positive acceleration.

(*a*) it represents a positive acceleration; the slope—velocity—is increasing.

102 •• On a graph showing velocity on the vertical axis and time on the horizontal axis, zero acceleration is represented by (*a*) a straight line with positive slope. (*b*) a straight line with negative slope. (*c*) a straight line with zero slope. (*d*) either (*a*), (*b*), or (*c*). (*e*) none of the above.

(*c*) Zero acceleration means constant velocity.

---

103 •• On a graph showing velocity on the vertical axis and time on the horizontal axis, constant acceleration is represented by (*a*) a straight line with positive slope. (*b*) a straight line with negative slope. (*c*) a straight line with zero slope. (*d*) either (*a*), (*b*), or (*c*). (*e*) none of the above.

(*d*) any line with constant slope, including zero slope.

---

104 •• Which graph of v versus t in Figure 2-34 best describes the motion of a particle with positive velocity and negative acceleration?

(*e*) $v > 0$ and the slope of $v(t)$ is negative.

105* •• Which graph of v versus t in Figure 2-34 best describes the motion of a particle with negative velocity and negative acceleration?

(*d*) $v \leq 0$ and the slope of $v(t)$ is negative.

---

106 •• A graph of the motion of an object is plotted with the velocity on the vertical axis and time on the horizontal axis. The graph is a straight line. Which of these quantities *cannot* be determined from this graph? (*a*) The displacement from time $t = 0$ (*b*) The initial velocity at $t = 0$ (*c*) The acceleration of the object (*d*) The average velocity of the object (*e*) None of the above.

(*e*) All of the quantities can be determined.

---

107 •• Figure 2-35 shows the position of a car plotted as a function of time. At which times t_0 to t_7 is the velocity (*a*) negative? (*b*) positive? (*c*) zero? At which times is the acceleration (*d*) negative? (*e*) positive? (*f*) zero?

Velocity: (*a*) Negative at t_0 and t_1. (*b*) Positive at t_3, t_6, and t_7. (*c*) Zero at t_2, t_4, and t_5.

Acceleration: (*d*) Negative at t_4. (*e*) Positive at t_2 and t_6. (*f*) Zero at t_0, t_1, t_3, t_5, and t_7

---

108 •• Sketch v-versus-t curves for each of the following conditions: (*a*) Acceleration is zero and constant while velocity is not zero. (*b*) Acceleration is constant but not zero. (*c*) Velocity and acceleration are both positive. (*d*) Velocity and acceleration are both negative. (*e*) Velocity is positive and acceleration is negative. (*f*) Velocity is negative and acceleration is positive. (*g*) Velocity is zero at a point but the acceleration is not zero.

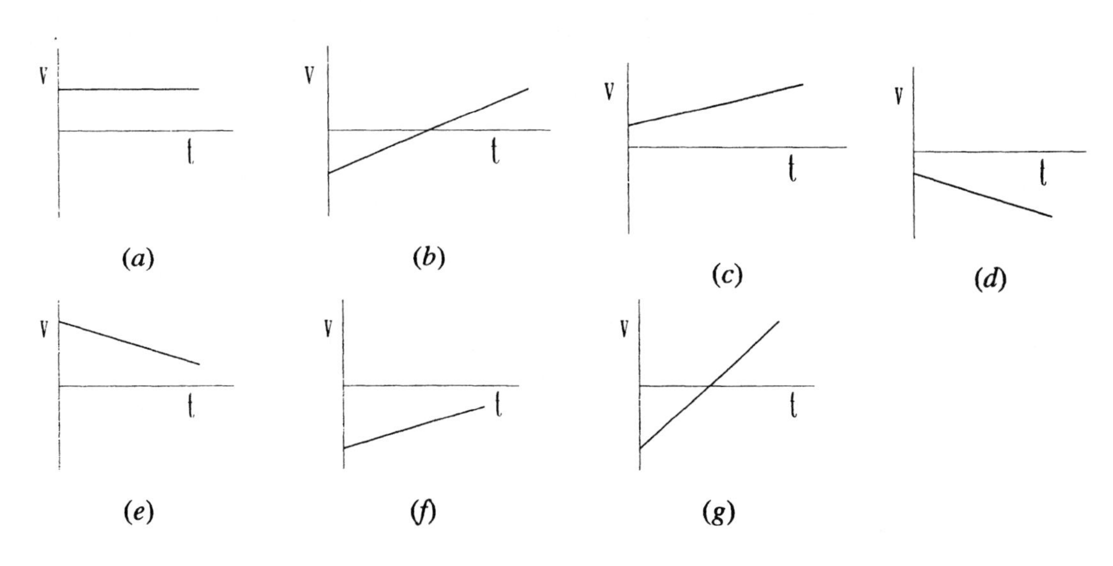

109* •• Figure 2-36 shows nine graphs of position, velocity, and acceleration for objects in linear motion. Indicate the graphs that meet the following conditions: (*a*) Velocity is constant (*b*) Velocity has reversed its direction (*c*) Acceleration is constant (*d*) Acceleration is not constant. Which graphs of velocity and acceleration are mutually consistent?

(*a*) *a*, *f*, *i*; (*b*) *c*, *d*; (*c*) *a*, *d*, *e*, *f*, *h*, *i*; (*d*) *b*, *c*, *g*. The graphs *d* and *h*, and *f* and *i* are mutually consistent.

110 • Two cars are being driven at the same speed v, one behind the other, with a distance d between them. The first driver jams on her brakes and decelerates at a rate $a = 6$ m/s^2. The second driver sees her brake lights and reacts, decelerating at the same rate starting 0.5 s later. (*a*) What is the minimum distance d such that the two cars do not collide? (*b*) Express this answer in meters for $v = 100$ km/h (62 mi/h).

(*a*) Since $v_1 = v_2 = v$, $d = 0.5v$.

(*b*) Converting 100 km/ to m/s we obtain $v = 27.8$ m/s. Hence $d = 13.9$ m.

111 • The velocity of a particle in meters per second is given by $v = 7 - 4t$, where t is in seconds. (*a*) Sketch $v(t)$ versus t, and find the area between the curve and the t axis from $t = 2$ s to $t = 6$ s. (*b*) Find the position $x(t)$ by integration, and use it to find the displacement during the interval $t = 2$ s to $t = 6$ s. (*c*) What is the average velocity for this interval?

(*a*) The sketch is shown here. The area between the curve and the t axis between $t = 2$ s and $t = 6$ s is about 36 m.

(*b*) $x(t) = \int(7 - 4t)\,dt = (7t - 2t^2 + C)$ m;

$\Delta x = x(6) - x(2) = -36$ m

(*c*) Note that this is an instance of constant acceleration. Hence,

$v_{av} = [v(6) + v(2)]/2 = -9$ m/s; note: $v_{av}\Delta t = \Delta x$.

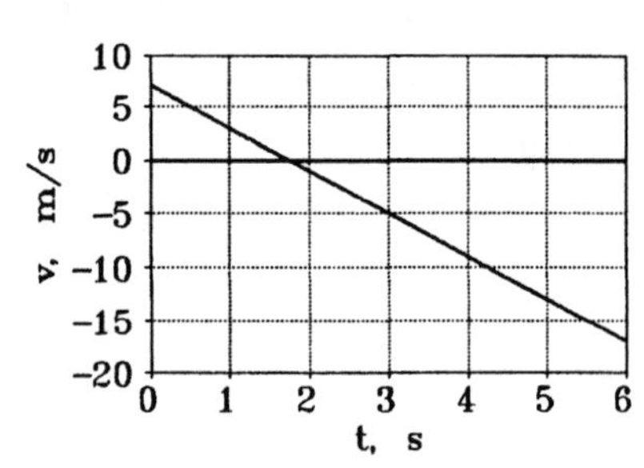

112 •• Estimate how high a ball or small rock can be thrown if it is thrown straight up.

1. Estimate the initial velocity — $v_0 \approx 100$ km/h ≈ 28 m/s
2. Use $H = v_0^2/2g$ to find H — $H \approx [(28)^2/19.6]$ m $= 40$ m

113* •• The cheetah can run as fast as $v_1 = 100$ km/h, the falcon can fly as fast as $v_2 = 250$ km/h, and the sailfish can swim as fast as $v_3 = 120$ km/h. The three of them run a relay with each covering a distance L at maximum speed. What is the average speed v of this triathlon team?

1. Express the time required for each animal — $t_1 = L/v_1$; $t_2 = L/v_2$; $t_3 = L/v_3$
2. Write the total time, Δt — $\Delta t = L(1/v_1 + 1/v_2 + 1/v_3)$
3. Find $v = \Delta x/\Delta t = 3L/\Delta t$; use values for v_1 etc. — $v = [3/(0.01 + 0.004 + 0.00833)]$ m/s $= 134$ m/s

114 •• In 1997, the men's world record for the 50-m freestyle was held by Tom Jager of the United States, who covered $d = 50$ m in $t = 21.81$ s. Suppose Jager started from rest at constant acceleration a, and reached his maximum speed in 2.00 s, which he then kept constant until the finish line. Find Jager's acceleration a.

1. Write d in terms of a, v_0, and t — 50 m $= [(\frac{1}{2}a \times 2^2) + (2a \times 19.81)]$ m
2. Solve for a — $a = 1.20$ m/s^2

115 •• The click beetle can project itself vertically with an acceleration of about $a = 400\ g$ (an order of magnitude more than a human could stand). The beetle jumps by "unfolding" its legs, which are about $d = 0.6$ cm long. How high can the click beetle jump? How long is the beetle in the air? (Assume constant acceleration while in contact with the ground, and neglect air resistance.)

1. Find the time of contact with the ground — $\frac{1}{2}(400g)t_1^2 = 6 \times 10^{-3}$ m; $t_1 = 1.75 \times 10^{-3}$ s
2. Find the velocity at take-off — $v = (400g)(1.75 \times 10^{-3}\text{ s}) = 6.86$ m/s
3. Find height H — $H = v^2/2g = 2.4$ m
4. Find the time to return to $x = 0$ — $T = 2v/g = 1.4$ s

116 •• The one-dimensional motion of a particle is plotted in Figure 2-37. (*a*) What is the acceleration in the intervals AB, BC, and CE? (*b*) How far is the particle from its starting point after 10 s? (*c*) Sketch the displacement of the particle as a function of time; label the instants A, B, C, D, and E on your figure. (*d*) At what time is the particle traveling most slowly?

(*a*) $a = \Delta v/\Delta t$; determine a for the three intervals — AB: $a = (10/3)$ m/s^2 = 3.33 m/s^2; BC: $a = 0$; CD: $a = (-30/4)$ m/s^2 = -7.5 m/s^2

(*b*) Find the area under v versus t — $\Delta x = [5 \times 3 + \frac{1}{2}(10 \times 3) + 15 \times 3]$ m = 75 m

(*c*) The displacement, x, versus t is shown in the figure

(*d*) At D, $t = 8$ s, $v = 0$

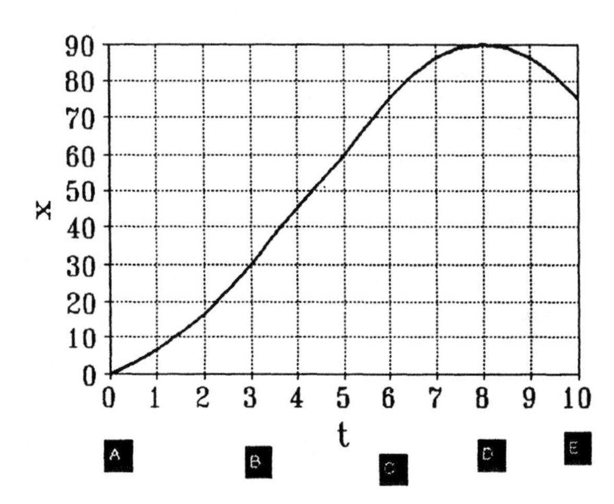

117* •• Consider the velocity graph in Figure 2-38. Assuming $x = 0$ at $t = 0$, write correct algebraic expressions for $x(t)$, $v(t)$, and $a(t)$ with appropriate numerical values inserted for all constants.

1. Write $v(t)$; note that a is constant and < 0 — $v(t) = (50 - 10t)$ m/s; $a = -10$ m/s^2
2. $x(t) = \int v(t)dt$ — $x(t) = 50t - 5t^2$

118 •• Starting at one station, a subway train accelerates from rest at a constant rate of 1.0 m/s^2 for half the distance to the next station, then slows down at the same rate for the second half of the journey. The total distance between stations is 900 m. (*a*) Sketch a graph of the velocity v as a function of time over the full journey. (*b*) Sketch a graph of the distance covered as a function of time over the full journey. Place appropriate numerical values on both axes.

(*a*) The graph for $v(t)$ is shown below

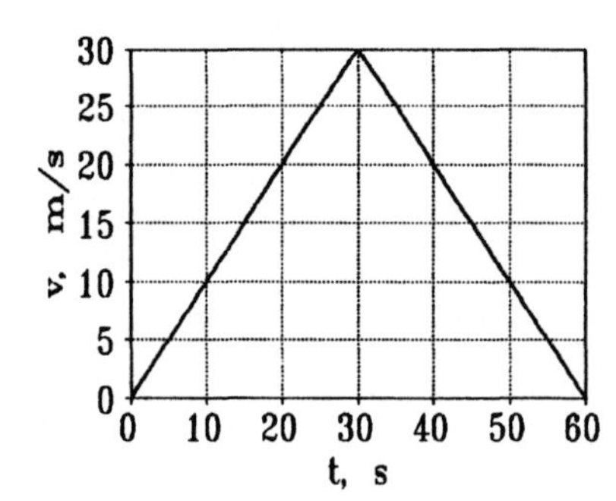

(*b*) The graph for $x(t)$ is shown below

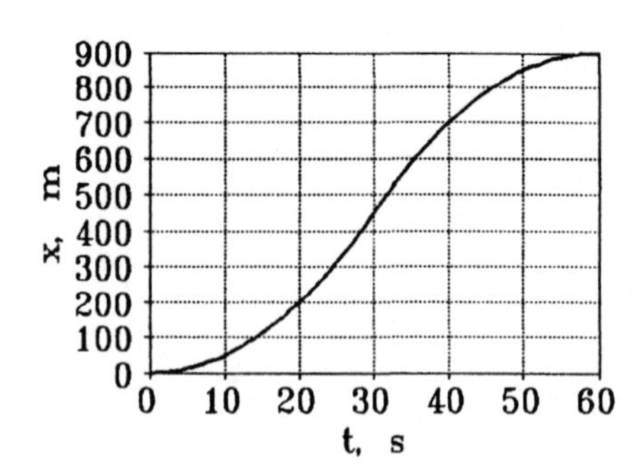

---

119 •• The acceleration of a certain rocket is given by $a = Ct$, where C is a constant. (*a*) Find the general position function $x(t)$. (*b*) Find the position and velocity at $t = 5$ s if $x = 0$ and $v = 0$ at $t = 0$ and $C = 3$ m/s^2.

(*a*) 1. Integrate $a(t)$ to find $v(t)$ — $v(t) = \int a(t)\,dt = C\int t\,dt = \frac{1}{2}Ct^2 + A$; A is a constant

2. Integrate $v(t)$ to find $x(t)$ — $x(t) = \int v(t)\,dt = \int(\frac{1}{2}Ct^2 + A)\,dt = Ct^3/6 + At + B$

(*b*) Find constants A, B from boundary conditions — $v(0) = 0$: $A = 0$; $x(0) = 0$: $B = 0$

Evaluate $v(5)$ and $x(5)$ with $A = B = 0$ — $v(5) = 37.5$ m/s; $x(5) = 62.5$ m

---

120 •• A physics professor demonstrates his new "anti-gravity parachute" by exiting from a helicopter at an altitude of 1500 m with zero initial velocity. For 8 s he falls freely. Then he switches on the "parachute" and falls with constant upward acceleration of 15 m/s^2 until his downward speed reaches 5 m/s, whereupon he adjust his controls to maintain that speed until he reaches the ground. (*a*) On a single graph, sketch his acceleration and velocity as functions of time. (Take upward to be positive.) (*b*) What is his speed at the end of the first 8 s? (*c*) For how long does he maintain the constant upward acceleration of 15 m/s^2? (*d*) How far does he travel during the upward acceleration in part (c)? (*e*) How many seconds are required for the entire trip from the helicopter to the ground? (*f*) What is his average velocity for the entire trip?

Note: we shall do part (*a*) last.

(*b*) Use Equ. 2-12 — $v(8) = [-(9.81)(8)]$ m/s $= -78.5$ m/s

(c) Use Equ. 2-12 — $t_2 = \Delta v/a = (73.5/15)$ s $= 4.9$ s

(*d*) Use Equ. 2-13 — $\Delta x_2 = \frac{1}{2}[(-78.5 - 5)(4.9)]$ m $= 205$ m

(*e*) 1. Find the remaining distance — $\Delta x_3 = [1500 - \frac{1}{2}(78.5)(8) - 205]$ m $= 981$ m

2. Find the total time — $t_{tot} = [(981/5) + 4.9 + 8]$ s $= 209$ s

(*a*) The sketch of $a(t)$ and $v(t)$ is shown.

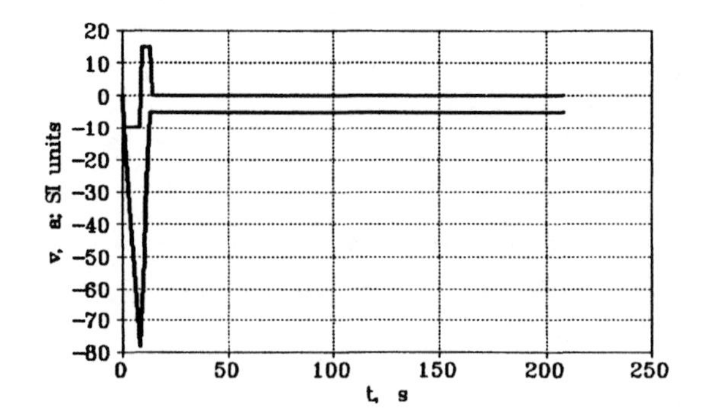

121* •• Without telling Sally, Joe made travel arrangements that include a stopover in Toronto to visit Joe's old buddy. Sally doesn't like Joe's buddy and wants to change their tickets. She hops on a courtesy motor scooter and begins accelerating at 0.9 m/s^2 toward the ticket counter to make arrangements. As she begins moving, Joe is 40 m behind her, running at constant speed of 9 m/s. (*a*) How long does it take for Joe to catch up with her? (*b*) What is the time interval during which Joe remains ahead of Sally?

(*a*) 1. Write expressions for x_S and x_J	$x_S = 0.45t^2$ m; $x_J = (-40 + 9t)$ m
2. Set $x_S = x_J$ to obtain equation for t	$0.45t^2 - 9t + 40 = 0$
3. Solve for t; keep smallest answer	$t = 6.67$ s, $t = 13.33$ s; $t = 6.67$ s
(*b*) Sally will catch Joe at 13.33 s	$\Delta t = 6.67$ s

122 •• A speeder races past at 125 km/h. A patrol car pursues from rest with constant acceleration of 8 km/h·s until it reaches its maximum speed of 190 km/h, which it maintains until it catches up with the speeder. (*a*) How long until the patrol car catches the speeder if it starts moving just as the speeder passes? (*b*) How far does each car travel? (*c*) Sketch $x(t)$ for each car.

(*a*) 1. Write expressions for x_S and x_P; convert all quantities to m/s and m/s^2	$x_S = 34.7t$ m; $t_{acc} = 52.8/2.22$ s $= 23.8$ s; $x_{acc} =[\frac{1}{2}(52.8)(23.8)]$ m; $x_P = [628 + 52.8 \times (t - 23.8)]$ m
2. Set $x_S = x_P$ and solve for t	$18.1t$ m $= 628$ m, $t = 34.7$ s
(*b*) The distance traveled is x_S	$x_S = 1204$ m

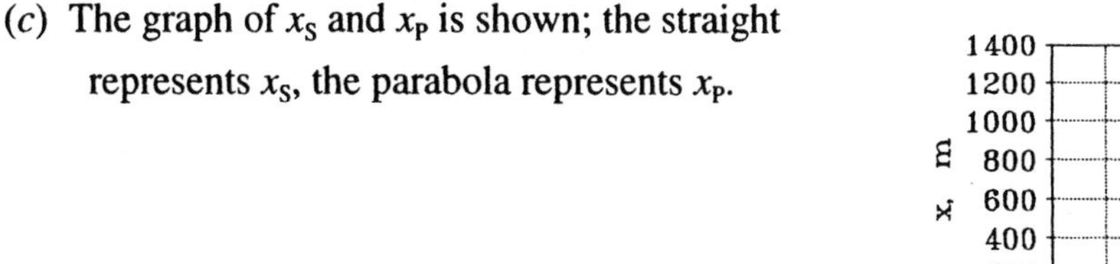

(*c*) The graph of x_S and x_P is shown; the straight represents x_S, the parabola represents x_P.

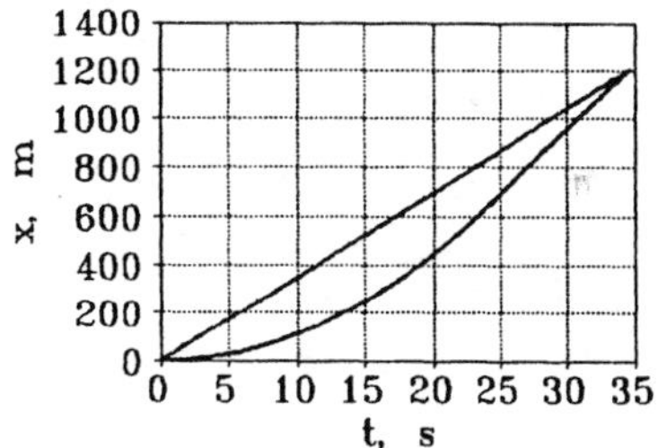

123 •• When the patrol car in Problem 122 (traveling at 190 km/h), pulls within 100 m behind the speeder (traveling at 125 km/h), the speeder sees the police car and slams on his brakes, locking the wheels. (*a*) Assuming that each car can brake at 6 m/s^2 and that the driver of the police car brakes instantly as she sees the brake lights of the speeder (reaction time = 0 s), show that the cars collide. (*b*) At what time after the speeder applies his brakes do the two cars collide? (*c*) Discuss how reaction time affects this problem.

(*a*) $\Delta x = v^2/2a$ (Equ. 2-15) Note that $\Delta x_S + 100 \text{ m} < \Delta x_P$	$\Delta x_S = (34.7^2/12)$ m$=100.3$ m; $\Delta x_P = (52.8^2/12)$ m $= 232.3$ m. The cars collide.
(*b*) 1. Write x_S and x_P as functions of t; set $t = 0$ at brake time, $x = 0$ at patrol car at $t = 0$	$x_S = (100 + 34.7t - 3t^2)$ m; $x_P = (52.8t - 3t^2)$ m
2. Set $x_S = x_P$ and solve for t	$t = 5.52$ s

(*c*) If you take reaction time into account, the collision will occur sooner and be more severe.

124 •• The speed of a good base runner is 9.5 m/s. The distance between bases is 26 m, and the pitcher is about 18.5 m from home plate. If a runner on first base edges 2 m off the base and takes off for second the instant the ball leaves the pitcher's hand, what is the likelihood that the runner will steal second base safely?

We will neglect the time of acceleration of the runner; we will assume that the speed of a fast ball is 28 m/s, and use that also for the speed of the ball thrown by the catcher. We will take 0.6 s as the reaction time of the catcher. The time taken by the runner is $24/9.5 = 2.53$ s. The time of flight of the ball is $3 \times 18.5/28 = 1.98$ s; add 0.6 s as the reaction time to get 2.58 s. It will be a very close call!

125* •• Repeat Problem 124, but with the runner attempting to steal third base, starting from second base with a lead of 3 m.

We use the same speeds and reaction times as before. The distance the runner travels is now 23 m. The distance the ball travels is now $(18.5 + 26) = 44.5$ m so the running time is $23/9.5 = 2.42$ s. The time of flight of the ball is $44.5/28 = 1.59$ s. Add to this a reaction time of 0.6 s to get 2.19 s. A good umpire will call him out!

---

126 •• Urgently needing the cash prize, Lou enters the Rest-to-Rest auto competition, in which each contestant's car begins and ends at rest, covering a distance L in as short a time as possible. The intention is to demonstrate mechanical and driving skills, and to consume the largest amount of fossil fuels in the shortest time possible. The course is designed so that maximum speeds of the cars are never reached. If Lou's car has a maximum acceleration of a, and a maximum deceleration of $2a$, then at what fraction of L should Lou move his foot from the gas pedal to the brake? What fraction of the time for the trip has elapsed at that point?

Let t_1 be the time when the brake is applied, L_1 the distance traveled from $t = 0$ to $t = t_1$. Do (*b*) first.

(*b*) 1. Write the expressions for x, v for $0 \le t \le t_1$ — $v = at;\ x = \tfrac{1}{2}at^2$

2. Write expressions for x, v for $t_1 \le t \le t_{fin}$ — $v = at_1 - 2a(t - t_1);\ x = \tfrac{1}{2}at^2 + at_1(t - t_1) - \tfrac{1}{2}(2a)(t - t_1)^2$

3. At $t = t_{fin}$ $v = 0$; find t_1 in terms of t_{fin} from 2 — $t_1 = (2/3)t_{fin}$

(*a*) 1. Set $v = 0$ at $x = L$ to find $t = t_{fin}$ — $t_{fin} = \sqrt{3L/a}$

2. Find $x(t_1) = L_1$ — $L_1 = \tfrac{1}{2}at_1^2 = \tfrac{1}{2}a(4/9)(3L/a) = (2/3)L$

---

127 ••• The acceleration of a badminton birdie falling under the influence of gravity and a resistive force, such as air resistance, is given by $a = dv/dt = g - bv$, where g is the free-fall acceleration due to gravity and b is a constant that depends on the mass and shape of the birdie and on the properties of the medium. Suppose the birdie begins with zero velocity at time $t = 0$. (*a*) Discuss qualitatively how the speed v varies with time from your knowledge of the rate of change dv/dt given by this equation. What is the velocity when the acceleration is zero? This is called the *terminal velocity*. (*b*) Sketch the solution $v(t)$ versus t without solving the equation. This can be done as follows: at $t = 0$, v is zero and the slope is g. Sketch a straight-line segment, neglecting any change in slope for a short time interval. At the end of the interval, the velocity is not zero, so the slope is less than g. Sketch another straight-line segment with a smaller slope. Continue until the slope is zero and the velocity equals the terminal velocity.

(*a*) Initially $v = 0$ and increases as gt, but as v becomes finite, the acceleration diminishes to $g - bv$. Ultimately, the acceleration approaches zero and v remains constant. To find this terminal velocity set $dv/dt = 0$ and solve for $v = v_{term} = g/b$.

(b) A curve showing the general behavior of v as a function of time is shown below.

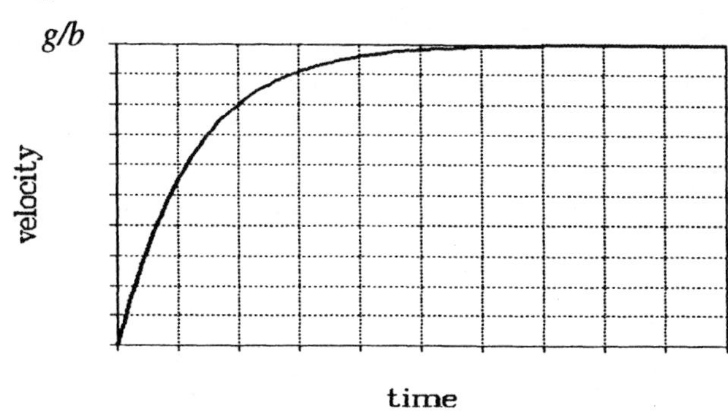

---

128 ••• Suppose acceleration is a function of x, where $a(x) = (2\text{ s}^{-2})x$. ($a$) If the velocity at $x = 1$ m is zero, what is the velocity at $x = 3$ m? (b) How long does it take to travel from $x = 1$ m to $x = 3$ m?

(a) Since $a = dv/dt$ we shall have to integrate to find v. Write dv/dt as $(dv/dx)(dx/dt) = v(dv/dx)$. Hence, the problem statement can be rewritten $v(dv/dx) = 2x$, or $v\,dv = 2x\,dx$. Integrate: $\int v\,dv = \int x\,dx$ which gives the relationship $v^2 - v_0^2 = 2x^2 - 2x_0^2$. Now set $v_0 = 0$, $x_0 = 1$ and find v; $v(x) = \sqrt{2(x^2 - 1)}$ and $v(3) = 4$ m/s.

(b) We have the expression for $v(x)$, given the initial conditions. So we can now integrate the differential equation $dx/dt = v(x)$ or $\int dt = \int dx/v(x)$, with limits $t = 0$ to t, and $x = 1$ to $x = 3$. From standard integral tables we obtain

$$t - t_0 = \frac{1}{\sqrt{2}} \ln\left(\frac{x + \sqrt{x^2 - 1}}{x_0 + \sqrt{x_0^2 - 1}}\right)$$

setting $t_0 = 0$, $x_0 = 1$, and $x = 3$ we find $t(3) = 1.25$ s.

129* ••• Suppose that a particle moves in a straight line such that, at any time t, its position, velocity, and acceleration all have the same numerical value. Give the position x as a function of time.

We are given that $v = 1 \times x = 1 \times a$. So $dx/dt = x$, and integrating we obtain $t - t_0 = \ln(x/x_0)$ or $x(t) = x_0 e^{t - t_0}$. The velocity and acceleration are obtained from $v = dx/dt$ and $a = dv/dt$: $v(t) = x_0 e^{t - t_0} = a(t)$.

---

130 ••• An object moving in a straight line doubles its velocity each second for the first 10 s. Let the initial speed be 2 m/s. (a) Sketch a smooth function $v(t)$ that gives the velocity. (b) What is the average velocity over the first 10 s?

(a) According to the statement $v(t) = 2^t v_0$.

A plot of this function is shown.

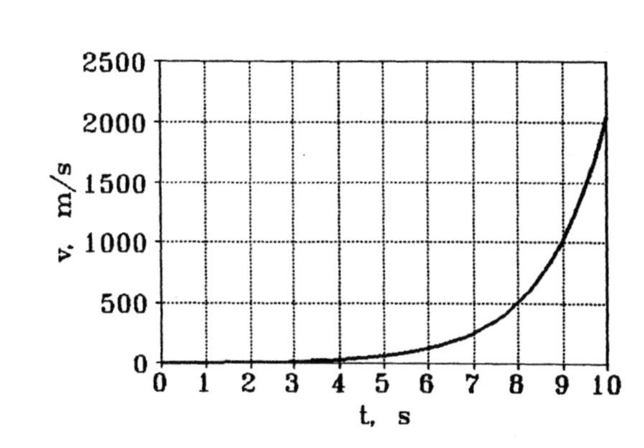

(b) The average velocity is given by $v_{av} = \Delta x/\Delta t$. We need to find Δx. We do so by integrating $v(t) = dx/dt$ from $t = 0$ to $t = 10$ s.

$$\int_0^{\Delta x} dx = \int_0^{10} 2^t v_0\,dt\ ;\ \Delta x = (v_0/\ln 2)(2^{10} - 1);$$

Setting $v_0 = 2$ m/s, $\Delta x = 2952$ m and $v_{av} = 295.2$ m/s

131 ••• In a dream, you find that you can run at superhuman speeds, but there is a resistant force that reduces your speed by one-half for each second that passes. Assume that the laws of physics still hold in your dreamworld, and that your initial speed is 1000 m/s. (*a*) Sketch a smooth function $v(t)$ that gives your velocity. (*b*) What is your average velocity over the first 10 s?

Note: This problem is the same as Problem 130 except that now $v(t) = 2^{-t} v_0$. We can follow the procedure of the preceding problem.

(*a*) A sketch of $v(t)$ is shown

(*b*) We now find $\Delta x = (v_0/\ln 2)(1 - 2^{-10}) = 1441$ m,

and $v_{av} = 144.1$ m/s

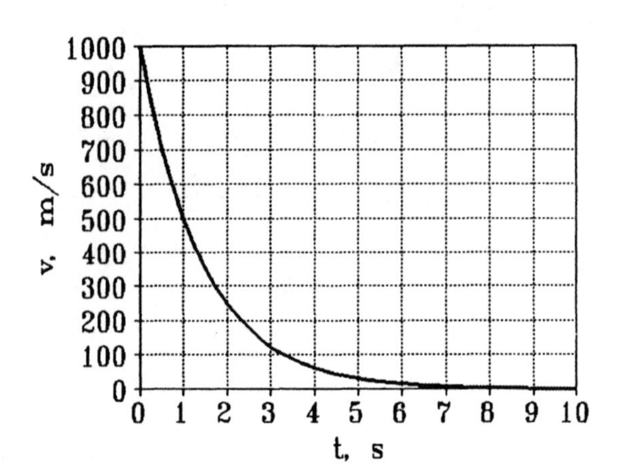

CHAPTER 3

Motion in Two and Three Dimensions

1* • Can the magnitude of the displacement of a particle be less than the distance traveled by the particle along its path? Can its magnitude be more than the distance traveled? Explain.

The magnitude of the displacement can be less but never more than the distance traveled. If the path is a semi-circle of radius R, the distance traveled is $2\pi R$, the magnitude of the displacement is $2R$. The maximum magni tude of the displacement occurs when the path is a straight line; then the two quantities are equal.

---

2 • Give an example in which the distance traveled is a significant amount yet the corresponding displacement is zero.

The path is a complete circle; Displacement = 0; distance traveled $= 2\pi R$.

---

3 • The magnitude of the displacement of a particle is ___ the distance the object has traveled. (*a*) larger than (*b*) smaller than (*c*) either larger or smaller (*d*) the same as (*e*) smaller than or equal to

(*e*) See Problem 1.

---

4 • A bear walks northeast for 12 m and then east for 12 m. Show each displacement graphically, and find the resultant displacement vector.

The two vector displacements are shown. The total displacement is 22.2 m at an angle of 22.5° with respect to the horizontal.

12

12

R

5* • (*a*) A man walks along a circular arc from the position $x = 5$ m, $y = 0$ to a final position $x = 0$, $y = 5$ m. What is his displacement? (*b*) A second man walks from the same initial position along the x axis to the origin and then along the y axis to $y = 5$ m and $x = 0$. What is his displacement?

(*a*), (*b*) Since the initial and final positions are the same, the displacements are equal. $\boldsymbol{D} = (-5\,\boldsymbol{i} + 5\,\boldsymbol{j})$ m.

---

6 • A circle of radius 8 m has its center on the y axis at $y = 8$ m. You start at the origin and walk along the circle at a steady speed, returning to the origin exactly 1 min after you started. (*a*) Find the magnitude and direction of your displacement from the origin 15, 30, 45, and 60 s after you start. (*b*) Find the magnitude and direction of your displacement for each of the four successive 15-s intervals of your walk. (*c*) How is your displacement for the first 15 s related to that for the second 15 s? (*d*) How is your displacement for the second 15-s interval related to that for the last 15-s interval?

The adjacent figure shows the path around the circle. Since the radius of the circle is 8 m, the length of each vector is $(8 \times \sqrt{2})$ m = 11.3 m.

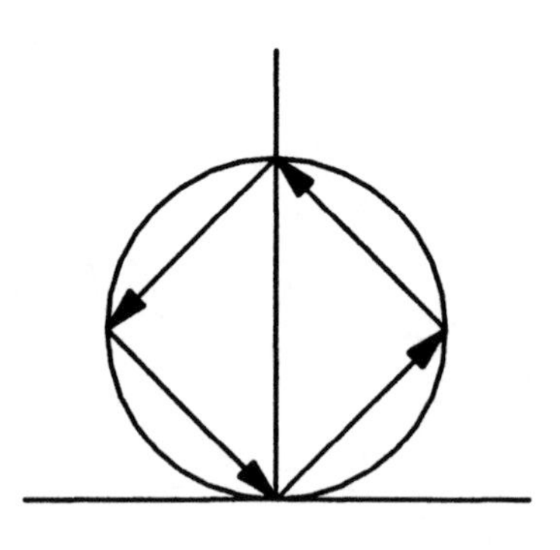

(*a*) 1. Find the coordinates for each point — $\boldsymbol{r}(0) = \boldsymbol{r}(60) = 0$; $\boldsymbol{r}(15) = 8\,\boldsymbol{i} + 8\,\boldsymbol{j}$; $\boldsymbol{r}(30) = 8\,\boldsymbol{j}$; $\boldsymbol{r}(45) = -8\,\boldsymbol{i} + 8\,\boldsymbol{j}$

2. Find $r(t - 0)$ and $\theta(t - 0)$ — $r(15 - 0) = 11.3$ m, $\theta = 45°$; $r(30 - 0) = 8$ m, $\theta = 180°$; $r(45 - 0) = 11.3$ m, $\theta = 135°$; $r(60 - 0) = 0$

(*b*) The displacement magnitudes are equal; the angles rotate by 90° — displacement magnitudes = 11.3 m; $\theta_1 = 45°$; $\theta_2 = 135°$; $\theta_3 = 225°$; $\theta_4 = 315°$

(*c*) See the figure — The second is the first rotated 90°

(*d*) See the figure — The two are oppositely directed.

---

7 • For the two vectors ***A*** and ***B*** of Figure 3-31, find the following graphically: (*a*) ***A*** + ***B***, (*b*) ***A*** − ***B***, (*c*) 2***A*** + ***B***, (*d*) ***B*** − ***A***, (*e*) 2***B*** − ***A***

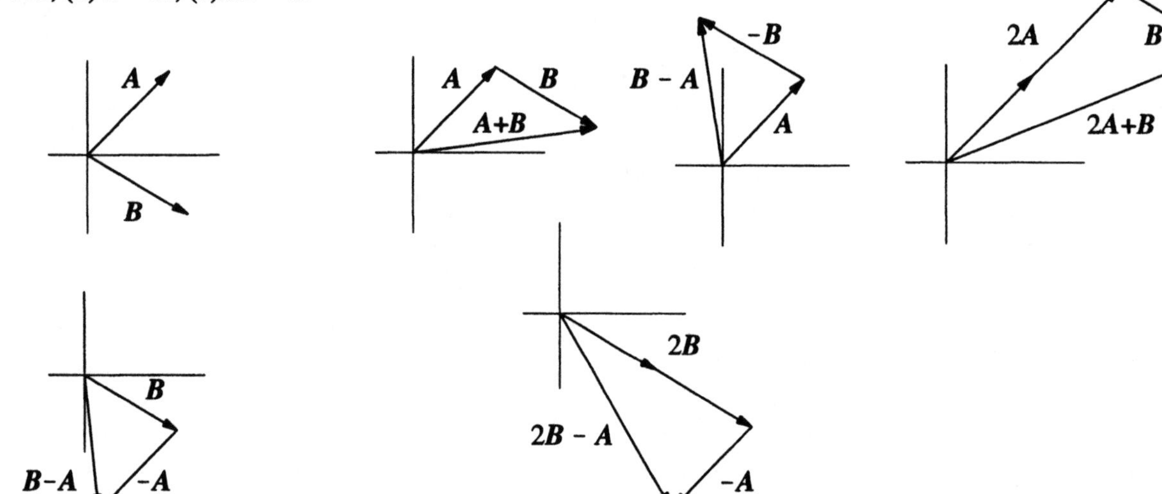

---

8 • A scout walks 2.4 km due east from camp, then turns left and walks 2.4 km along the arc of a circle centered at the campsite, and finally walks 1.5 km directly toward the camp. (*a*) How far is the scout from camp at the end of his walk? (*b*) In what direction is the scout's position relative to the campsite? (*c*) What is the ratio of the final magnitude of the displacement to the total distance walked?

The figure shows the path. Part A is the walk of 2.4 km to the east; part B is the walk of 2.4 km along the arc of radius 2.4 km; part C is the walk of 1.5 km toward the starting point.

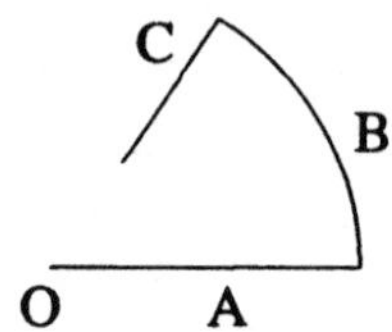

He walks to, then along the circle of radius 2.4 km, then 1.5 km along radius toward origin (campsite). Distance from campsite, $d = (2.4 - 1.5)$ km $= 0.9$ km

(*b*) Use the definition of the radian — He is 1 rad north of east of campsite

(*c*) Find total distance, D — $D = (2.4 + 2.4 + 1.5)$ km $= 6.3$ km; $d/D = 1/7$

9* • Can a component of a vector have a magnitude greater than the magnitude of the vector? Under what circumstances can a component of a vector have a magnitude equal to the magnitude of the vector?

No; $A_x = A\cos\theta \le A$. $A_x = A$ for $\theta = 0$, i.e., if the vector is along the component direction.

---

10 • Can a vector be equal to zero and still have one or more components not equal to zero?

No; see definition of component.

---

11 • Are the components of $\boldsymbol{C} = \boldsymbol{A} + \boldsymbol{B}$ necessarily larger than the corresponding components of either $\boldsymbol{A}$ or $\boldsymbol{B}$?

No; consider the case when $\boldsymbol{B} = -\boldsymbol{A}$, then $\boldsymbol{C} = 0$, and the components of $\boldsymbol{A}$ and $\boldsymbol{B}$ are larger than 0.

---

12 • The components of a vector are $A_x = -10$ m and $A_y = 6$ m. What angle does this vector make with the positive x axis? (*a*) 31° (*b*) −31° (*c*) 180° − 31° (*d*) 180° + 31° (*e*) 90° − 31°.

$\theta = \tan^{-1}(6/-10) = (180 - 31)°$ or $-31°$. Vector in second quadrant; therefore (*c*).

13* • A velocity vector has an x component of +5.5 m/s and a y component of −3.5 m/s. Which diagram in Figure 3-32 gives the direction of the vector?

(*b*) The vector is in the fourth quadrant; $\theta = \tan^{-1}(-3.5/5.5) = -32.5°$.

---

14 • Three vectors $\boldsymbol{A}$, $\boldsymbol{B}$, and $\boldsymbol{C}$ have the following x and y components: $A_x = 6$, $A_y = -3$; $B_x = -3$, $B_y = 4$; $C_x = 2$, $C_y = 5$. The magnitude of $\boldsymbol{A} + \boldsymbol{B} + \boldsymbol{C}$ is ____. (*a*) 3.3 (*b*) 5.0 (*c*) 11 (*d*) 7.8 (*e*) 14

Add x and y components and use Equ. 3-5. The magnitude is (*d*) 7.8.

---

15 • Find the rectangular components of the following vectors $\boldsymbol{A}$, which lie in the xy plane, and make an angle θ with the x axis (Figure 3-33) if (*a*) $A = 10$ m, $\theta = 30°$; (*b*) $A = 5$ m, $\theta = 45°$; (*c*) $A = 7$ km, $\theta = 60°$; (*d*) $A = 5$ km, $\theta = 90°$; (*e*) $A = 15$ km/s, $\theta = 150°$; (*f*) $A = 10$ m/s, $\theta = 240°$; and (*g*) $A = 8$ m/s^2, $\theta = 270°$.

Apply Equs. 3-2 and 3-3 — (*a*) $A_x = 8.66$ m, $A_y = 5$ m (*b*) $A_x = 3.54$ m, $A_y = 3.54$ m (*c*) $A_x = 3.5$ km, $A_y = 6.06$ km (*d*) $A_x = 0$ km, $A_y = 5$ km (*e*) $A_x = -13$ km/s, $A_y = 7.5$ km/s (*f*) $A_x = -5$ m/s, $A_y = -8.66$ m/s (*g*) $A_x = 0$, $A_y = -8$ m/s^2

---

16 • Vector $\boldsymbol{A}$ has a magnitude of 8 m at an angle of 37° with the x axis; vector $\boldsymbol{B} = 3$ m $\boldsymbol{i} - 5$ m $\boldsymbol{j}$; vector $\boldsymbol{C} = -6$ m $\boldsymbol{i} + 3$ m $\boldsymbol{j}$. Find the following vectors: (*a*) $\boldsymbol{D} = \boldsymbol{A} + \boldsymbol{C}$; (*b*) $\boldsymbol{E} = \boldsymbol{B} - \boldsymbol{A}$; (*c*) $\boldsymbol{F} = \boldsymbol{A} - 2\boldsymbol{B} + 3\boldsymbol{C}$; (*d*) A vector $\boldsymbol{G}$ such that $\boldsymbol{G} - \boldsymbol{B} = \boldsymbol{A} + 2\boldsymbol{C} + 3\boldsymbol{G}$.

Write $\boldsymbol{A}$ in component form using Equs. 3-2 and 3-3 — $\boldsymbol{A} = 6.4$ m $\boldsymbol{i} + 4.8$ m $\boldsymbol{j}$

(*a*), (*b*), (*c*) Add (or subtract) x and y components — $\boldsymbol{D} = 0.4$ m $\boldsymbol{i} + 7.8$ m $\boldsymbol{j}$, $\boldsymbol{E} = -3.4$ m $\boldsymbol{i} - 9.8$ m $\boldsymbol{j}$; $\boldsymbol{F} = -17.6$ m $\boldsymbol{i} + 23.8$ m $\boldsymbol{j}$

(*d*) 1. Solve for $\boldsymbol{G}$ — $\boldsymbol{G} = -\frac{1}{2}(\boldsymbol{A} + \boldsymbol{B} + 2\boldsymbol{C})$

2. Add components — $\boldsymbol{G} = -1.3$ m $\boldsymbol{i} + 2.9$ m $\boldsymbol{j}$

17* • Find the magnitude and direction of the following vectors: (*a*) $\boldsymbol{A} = 5\,\boldsymbol{i} + 3\,\boldsymbol{j}$; (*b*) $\boldsymbol{B} = 10\,\boldsymbol{i} - 7\,\boldsymbol{j}$; (*c*) $\boldsymbol{C} = -2\,\boldsymbol{i} - 3\,\boldsymbol{j} + 4\,\boldsymbol{k}$.

(*a*), and (*b*) Use Equs. 3-4 and 3-5 — $A = 5.83$, $\theta = 31°$; $B = 12.2$, $\theta = -35°$

(*c*) $C = \sqrt{C_x^2 + C_y^2 + C_z^2}$; the angle between $\boldsymbol{C}$ and the *z* axis is $\theta = \cos^{-1}(C_z/C)$; the angle with the *x* axis is $\phi = \tan^{-1}(C_y/C_z)$ — $C = 5.39$; $\theta = 42°$, $\phi = 236°$

---

18 • Find the magnitude and direction of $\boldsymbol{A}$, $\boldsymbol{B}$, and $\boldsymbol{A} + \boldsymbol{B}$ for (*a*) $\boldsymbol{A} = -4\,\boldsymbol{i} - 7\,\boldsymbol{j}$, $\boldsymbol{B} = 3\,\boldsymbol{i} - 2\boldsymbol{j}$, and (*b*) $\boldsymbol{A} = 1\,\boldsymbol{i} - 4\boldsymbol{j}$, $\boldsymbol{B} = 2\,\boldsymbol{i} + 6\,\boldsymbol{j}$.

Let $\boldsymbol{C} = \boldsymbol{A} + \boldsymbol{B}$

(*a*) 1. Use Equs. 3-4 and 3-5 — $A = 8.06$, $\theta = 240°$, $B = 3.61$, $\theta = -33.7°$

2. Use Equs. 3-6, 3-4, and 3-5 — $C = 9.06$, $\theta = 264°$

(*b*) Follow the same steps as in (*a*) — $A = 4.12$, $\theta = -76°$, $B = 6.32$, $\theta = 71.6°$, $C = 3.61$, $\theta = 33.7°$

---

19 • Describe the following vectors using the unit vectors $\boldsymbol{i}$ and $\boldsymbol{j}$: (*a*) a velocity of 10 m/s at an angle of elevation of 60°; (*b*) a vector $\boldsymbol{A}$ of magnitude $A = 5$ m and $\theta = 225°$; (*c*) a displacement from the origin to the point $x = 14$ m, $y = -6$ m.

(*a*), (*b*), and (*c*) Use Equ. 3-7 — $\boldsymbol{v} = 5$ m/s $\boldsymbol{i}$ + 8.66 m/s $\boldsymbol{j}$; A = −3.54 m $\boldsymbol{i}$ − 3.54 m $\boldsymbol{j}$; $\boldsymbol{r} = 14$ m $\boldsymbol{i}$ − 6 m $\boldsymbol{j}$

---

20 • For the vector $\boldsymbol{A} = 3\,\boldsymbol{i} + 4\,\boldsymbol{j}$, find any three other vectors $\boldsymbol{B}$ that also lie in the *xy* plane and have the property that $A = B$ but $\boldsymbol{A} \neq \boldsymbol{B}$. Write these vectors in terms of their components and show them graphically.

1. Determine the magnitude of A — $A = (3^2 + 4^2)^{1/2} = 5$
2. Write three vectors of magnitude 5 — $\boldsymbol{B} = 5\,i$; $\boldsymbol{B} = -5\,i$; $\boldsymbol{B} = 5\,j$

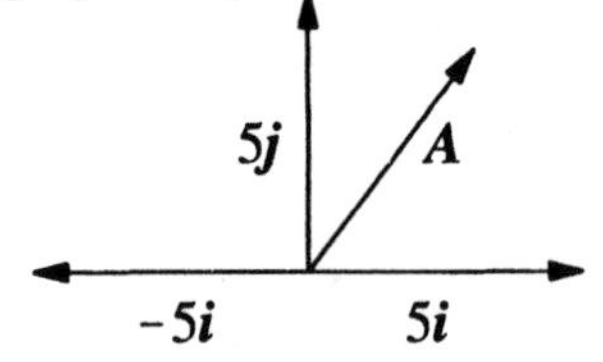

21* • If $\boldsymbol{A} = 5\,\boldsymbol{i} - 4\,\boldsymbol{j}$ and $\boldsymbol{B} = -7.5\,\boldsymbol{i} + 6\,\boldsymbol{j}$, write an equation relating $\boldsymbol{A}$ to $\boldsymbol{B}$.

Note $B_x = -1.5A_x$ and $B_y = -1.5A_y$. Consequently, $\boldsymbol{B} = -1.5\boldsymbol{A}$.

---

22 • The faces of a cube of side 3 m are parallel to the coordinate planes with one corner at the origin. A fly begins at the origin and walks along three edges until it is at the far corner. Write the displacement vector of the fly using the unit vectors $\boldsymbol{i}$, $\boldsymbol{j}$, and $\boldsymbol{k}$, and find the magntiude of this displacement.

The displacement vector is $\boldsymbol{r} = 3$ m $(\boldsymbol{i} + \boldsymbol{j} + \boldsymbol{k})$; the magnitude of $\boldsymbol{r}$ is $r = 3\sqrt{3}$ m = 5.2 m.

---

23 • For an arbitrary motion of a given particle, does the direction of the velocity vector have any particular relation to the direction of the position vector?

No; the position and velocity vectors are unrelated.

---

24 • Give examples in which the directions of the velocity and position vectors are (*a*) opposite, (*b*) the same, and (*c*) mutually perpendicular.

(*a*) A car moving at constant velocity toward the origin. (*b*) A car moving at constant velocity away from the origin. (*c*) A particle moving in a circle centered at the origin.

25* • How is it possible for a particle moving at constant speed to be accelerating? Can a particle with constant velocity be accelerating at the same time?

A particle moving at constant speed in a circular path is accelerating (the direction of the velocity vector is changing). If a particle is moving at constant velocity, it is not accelerating.

---

26 • If an object is moving toward the west, in what direction is its acceleration? (*a*) North (*b*) East (*c*) West (*d*) South (*e*) May be any direction.

(*e*) The instantaneous velocity and acceleration are unrelated.

---

27 •• Consider the path of a particle as it moves in space. (*a*) How is the velocity vector related geometrically to the path of the particle? (*b*) Sketch a curved path and draw the velocity vector for the particle for several positions along the path.

(*a*) The velocity vector is always tangent to the path

(*b*) The sketch is shown

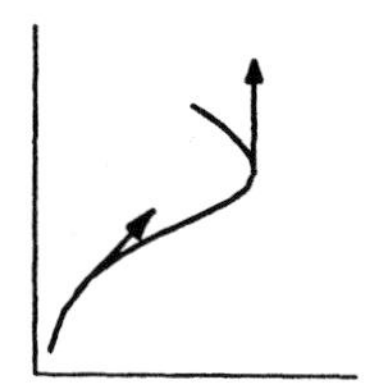

---

28 •• A dart is thrown straight up. After it leaves the player's hand, it steadily loses speed as it gains altitude until it lodges in the ceiling of the game room. Draw the dart's velocity vector at times t_1 and t_2, where $\Delta t = t_2 - t_1$ is small. From your drawing find the direction of the change in velocity $\Delta v = v_2 - v_1$, and thus the direction of the acceleration vector.

The sketch is shown.

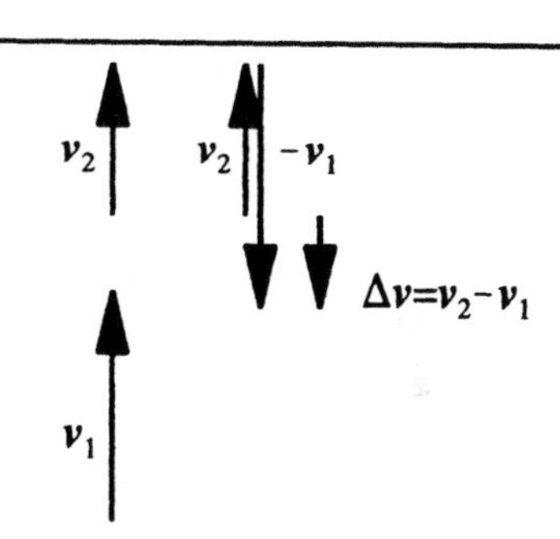

29* •• As a bungee jumper approaches the lowest point in her drop, she loses speed as she continues to move downward. Draw the velocity vectors of the jumper at times t_1 and t_2, where $\Delta t = t_2 - t_1$ is small. From your drawing find the direction of the change in velocity $\Delta v = v_2 - v_1$, and thus the direction of the acceleration vector.

The sketch is shown.

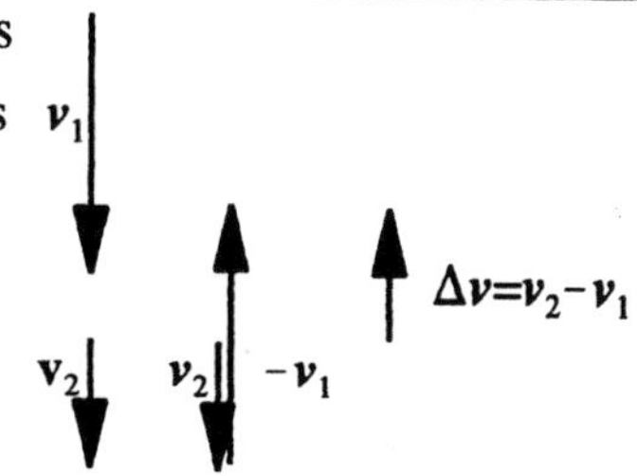

---

30 •• After reaching the lowest point in her jump at time t_{low}, the bungee jumper in the previous problem then moves upward, gaining speed for a short time until gravity again dominates her motion. Draw her velocity vectors at times t_1 and t_2, where $\Delta t = t_2 - t_l$ is small and $t_l < t_{low} < t_2$. From your drawing find the direction of the change in velocity $\Delta v = v_2 - v_1$, and thus the direction of the acceleration vector.

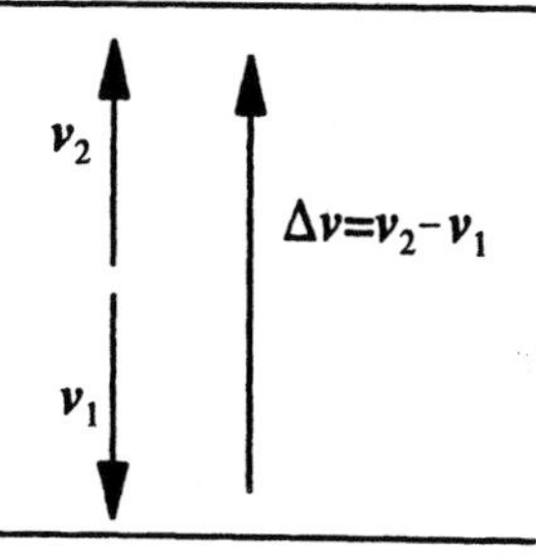

---

31 • A stationary radar operator determines that a ship is 10 km south of him. An hour later the same ship is 20 km southeast. If the ship moved at constant speed and always in the same direction, what was its velocity during this time?

Take the origin to be the location of the radar.

1. Determine the position vectors — $\mathbf{r}_1 = -10\ \text{km}\,\mathbf{j}$; $\mathbf{r}_2 = 14.1\ \text{km}\,\mathbf{i} - 14.1\ \text{km}\,\mathbf{j}$
2. Find the displacement vector, $\Delta\mathbf{r}$ — $\Delta\mathbf{r} = \mathbf{r}_2 - \mathbf{r}_1 = 14.1\ \text{km}\,\mathbf{i} - 4.1\ \text{km}\,\mathbf{j}$
3. $\mathbf{v} = \Delta\mathbf{r}/\Delta t$ — $\mathbf{v} = 14.1\ \text{km/h}\,\mathbf{i} - 4.1\ \text{km/h}\,\mathbf{j}$

---

32 • A particle's position coordinates (x, y) are (2 m, 3 m) at $t = 0$; (6 m, 7 m) at $t = 2$ s; and (13 m, 14 m) at $t = 5$ s. (*a*) find v_{av} from $t = 0$ to $t = 2$ s. (*b*) Find v_{av} from $t = 0$ to $t = 5$ s.

(*a*) 1. Find the magnitude of the displacement vector — $\Delta r = (4^2 + 4^2)^{1/2}$ m $= 5.66$ m

2. $v_{av} = \Delta r/\Delta t$ — $v_{av} = (5.66/2)$ m/s $= 2.83$ m/s

(*b*) Repeat as in (*a*) — $\Delta r = (11^2 + 11^2)^{1/2}$ m $= 15.6$ m; $v_{av} = 3.1$ m/s

---

33* • A particle moving at 4.0 m/s in the positive x direction is given an acceleration of 3.0 m/s^2 in the positive y direction for 2.0 s. The final speed of the particle is ____. (*a*) −2.0 m/s (*b*) 7.2 m/s (*c*) 6.0 m/s (*d*) 10 m/s (*e*) None of the above

1. Find the final velocity vector — $\mathbf{v} = v_x\mathbf{i} + a_y t\mathbf{j} = 4.0\ \text{m/s}\,\mathbf{i} + 6.0\ \text{m/s}\,\mathbf{j}$
2. Find the magnitude of the velocity vector — $v = 7.2$ m/s (*b*)

---

34 • A ball is thrown directly upward. Consider the 2-s time interval $\Delta t = t_2 - t_1$, where t_1 is 1 s before the ball reaches its highest point and t_2 is 1 s after it reaches its highest point. For the time interval Δt, find (*a*) the change in speed, (*b*) the change in velocity, and (*c*) the average acceleration.

Take up as the positive direction. Then $a = -g = -9.81$ m/s^2.

(*a*) $y(t_1) = y(t_2)$, and $v(t_2) = -v(t_1)$ — Change in speed $= 0$

(*b*) 1. Find $v(t_2)$; note $v = 0$ at highest point — $v(t_2) = a \times (1\text{ s}) = -9.81$ m/s

2. Find Δv — $\Delta v = v(t_2) - v(t_1) = -19.6$ m/s

(*c*) Constant acceleration: $a_{av} = a$ — $a_{av} = -9.81$ m/s^2

---

35 • Initially, a particle is moving due west with a speed of 40 m/s; 5 s later it is moving north with a speed of 30 m/s. (*a*) What was the change in the magnitude of the particle's velocity during this time? (*b*) What was the change in the direction of the velocity? (*c*) What are the magnitude and direction of $\Delta\mathbf{v}$ for this interval? (*d*) What are the magnitude and direction of $\mathbf{a}_{av}$ for this interval?

(*a*) The magnitudes of $\mathbf{v}_1$ and $\mathbf{v}_2$ are 40 m/s and 30 m/s — $\Delta v = v_2 - v_1 = -10$ m/s

(*b*) The change is from west to north — Change in direction is by 90°

(*c*) $\mathbf{v}_1 = -40\ \text{m/s}\,\mathbf{i}$, $\mathbf{v}_2 = 30\ \text{m/s}\,\mathbf{j}$ — $\Delta\mathbf{v} = \mathbf{v}_2 - \mathbf{v}_1 = 30\ \text{m/s}\,\mathbf{j} + 40\ \text{m/s}\,\mathbf{i}$

(*d*) Use Equ. 3-15 — $\mathbf{a}_{av} = 8\ \text{m/s}^2\,\mathbf{i} + 6\ \text{m/s}^2\,\mathbf{j}$; $\mathbf{a} = 10$ m/s^2 at 37° with x axis

---

36 • At $t = 0$, a particle located at the origin has a velocity of 40 m/s at $\theta = 45°$. At $t = 3$ s, the particle is at $x =$ 100 m and $y = 80$ m with a velocity of 30 m/s at $\theta = 50°$. Calculate (*a*) the average velocity and (*b*) the average acceleration of the particle during this interval.

(*a*) Find $\Delta\boldsymbol{r}$ and use Equ. 3-11	$\Delta\boldsymbol{r} = 100\text{ m }\boldsymbol{i} + 80\text{ m}\boldsymbol{j}$; $\boldsymbol{v}_{av} = 33.3\text{ m/s }\boldsymbol{i} + 26.7\text{ m/s}\boldsymbol{j}$
(*b*) 1. Find $\boldsymbol{v}_1$ and $\boldsymbol{v}_2$, using Equs. 3-2 and 3-3	$\boldsymbol{v}_1 = 28.3\text{ m/s }(\boldsymbol{i} + \boldsymbol{j})$; $\boldsymbol{v}_2 = 19.3\text{ m/s }\boldsymbol{i} + 23.0\text{ m/s}\boldsymbol{j}$
2. Use Equ. 3-15	$\boldsymbol{a}_{av} = -3\text{ m/s}^2\,\boldsymbol{i} - 1.77\text{ m/s}^2\boldsymbol{j}$

37* •• A particle moves in the *xy* plane with constant acceleration. At time zero, the particle is at $x = 4$ m, $y = 3$ m, and has velocity $\boldsymbol{v} = 2\text{ m/s }\boldsymbol{i} - 9\text{ m/s}\boldsymbol{j}$. The acceleration is given by the vector $\boldsymbol{a} = 4\text{ m/s}^2\,\boldsymbol{i} + 3\text{ m/s}^2\boldsymbol{j}$. (*a*) Find the velocity vector at $t = 2$ s. (*b*) Find the position vector at $t = 4$ s. Give the magnitude and direction of the position vector.

(*a*) $\boldsymbol{v}(t) = \boldsymbol{v}_0 + \boldsymbol{a}t$	$\boldsymbol{v}(2) = 10\text{ m/s }\boldsymbol{i} - 3\text{ m/s}\boldsymbol{j}$
(*b*) $\boldsymbol{r}(t) = \boldsymbol{r}_0 + \boldsymbol{v}_0t + ½\boldsymbol{a}t^2$	$\boldsymbol{r}(4) = 44\text{ m }\boldsymbol{i} - 9\text{ m}\boldsymbol{j}$; $r = 44.9$ m, $\theta = -11.6°$

---

38 •• A particle has a position vector given by $\boldsymbol{r} = 30t\,\boldsymbol{i} + (40t - 5t^2)\boldsymbol{j}$, where r is in meters and t in seconds. Find the instantaneous-velocity and instantaneous-acceleration vectors as functions of time t.

1. $\boldsymbol{v}(t) = d\boldsymbol{r}/dt$	$\boldsymbol{v}(t) = [30\,\boldsymbol{i} + (40 - 10t)\,\boldsymbol{j}]$ m/s
2. $\boldsymbol{a}(t) = d\boldsymbol{v}/dt$	$\boldsymbol{a}(t) = -10\boldsymbol{j}$ m/s^2

---

39 •• A particle has a constant acceleration of $\boldsymbol{a} = (6\,\boldsymbol{i} + 4\,\boldsymbol{j})$ m/s^2. At time $t = 0$, the velocity is zero and the position vector is $\boldsymbol{r}_0 = 10\text{ m }\boldsymbol{i}$. (*a*) Find the velocity and position vectors at any time t. (*b*) Find the equation of the particle's path in the *xy* plane, and sketch the path.

(*a*) 1. Constant $\boldsymbol{a}$; therefore $\boldsymbol{v} = \boldsymbol{v}_0 + \boldsymbol{a}t$	$\boldsymbol{v} = [\,6t\,\boldsymbol{i} + 4t\,\boldsymbol{j}]$ m/s
2. $\boldsymbol{r} = \boldsymbol{r}_0 + \boldsymbol{v}_0t + ½\boldsymbol{a}t^2$; $\boldsymbol{r}_0 = 10\text{ m }\boldsymbol{i}$	$\boldsymbol{r} = [(10 + 3t^2)\,\boldsymbol{i} + 2t^2\boldsymbol{j}]$ m

(*b*) The *x* and *y* components of the path are $(10 + 3\,t^2)$ and $2t^2$, respectively; $t^2 = y/2$, so $x = 10 + (3/2)y$. The path in the *xy* plane is a straight line as shown.

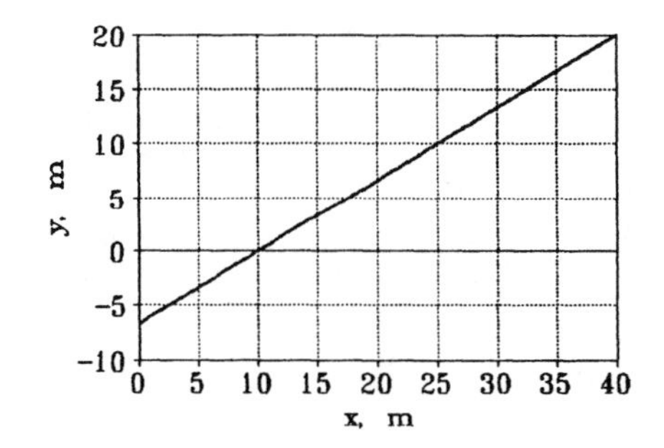

---

40 ••• Mary and Robert decide to rendezvous on Lake Michigan. Mary departs in her boat from Petoskey at 9:00 a.m. and travels due north at 8 mi/h. Robert leaves from his home on the shore of Beaver Island, 26 miles 30° west of north of Petoskey, at 10:00 a.m. and travels at a constant speed of 6 mi/h. In what direction should Robert be heading to intercept Mary, and where and when will they meet?

We will use the following conventions: Take the origin of coordinates at Petoskey; let $t = 0$ at 9:00 a.m.; let $\boldsymbol{i}$ be a unit vector pointing east, $\boldsymbol{j}$ a unit vector pointing north. Let θ be the angle between the velocity vector of

Robert's boat and the easterly direction. All quantities will be in miles and hours. "M" and "R" denote Mary and Robert, respectively.

1. Write the position vectors of M and R	$\boldsymbol{r}_M = 8t\boldsymbol{j}$
	$\boldsymbol{r}_R = [-26 \sin 30° + 6(t-1) \cos \theta] \boldsymbol{i} +$
	$[26 \cos 30° + 6(t-1) \sin \theta] \boldsymbol{j}$
2. Equate $\boldsymbol{i}$ and $\boldsymbol{j}$ components of $\boldsymbol{r}_M$ and $\boldsymbol{r}_R$	$-13 + (6t \cos \theta) - (6 \cos \theta) = 0;$
	$22.5 + (6t \sin \theta) - (6 \sin \theta) = 8t$
3. Solve the first equation for $t - 1$ and t	$t - 1 = 13/(6 \cos \theta); t = 13/(6 \cos \theta) + 1$
4. Substitute $(t - 1)$ and t into the other equation	$13 \tan \theta = 104/(6 \cos \theta) - 14.5;$
	$(78 \tan \theta + 87) \cos \theta = 104$

5. The transcendental equation must be solved by trial and error. Note that if $\theta = 45°$, the left hand side (LHS) is $165\sqrt{2} = 116.7$; if $\theta = 0°$, the LHS = 87. So the correct value of θ must be between 0° and 45°.

θ	LHS
30°	114
20°	108
15°	104

6. Solve for t from (3)	$t = 3.24 = 3$ hours 15 minutes
7. Find distance traveled due north by Mary	$8t = 26$ miles, due north of Petoskey

---

41* • A river is 0.76 km wide. The banks are straight and parallel (Figure 3-34). The current is 5.0 km/h and is parallel to the banks. A boat has a maximum speed of 3 km/h in still water. The pilot of the boat wishes to go on a straight line from *A* to *B*, where *AB* is perpendicular to the banks. The pilot should (*a*) head directly across the river. (*b*) head 68° upstream from the line *AB*. (*c*) head 22° upstream from the line *AB*. (*d*) give up—the trip from *A* to *B* is not possible with this boat. (*e*) None of the above.

(*d*) give up; since the speed of the stream is greater than that of the boat in still water, it will always drift down stream.

---

42 •• A plane flies at a speed of 250 km/h relative to still air. There is a wind blowing at 80 km/h in the northeast direction at exactly 45° to the east of north. (*a*) In what direction should the plane head so as to fly due north? (*b*) What is the speed of the plane relative to the ground?

Let θ be the angle between north and the direction of the plane's heading. All velocities in km/h.

1. $\boldsymbol{v} = \boldsymbol{v}_{air} + \boldsymbol{v}_{wind}$; east-west component of $\boldsymbol{v} = 0$	$250 \sin \theta = 80 \sin 45° = 56.6$
2. Solve for θ	$\theta = 13.1°$
3. Add the north components of $\boldsymbol{v}_{air}$ and $\boldsymbol{v}_{wind}$	$v = 250 \cos 13.1° + 56.6 = 300$ km/h

---

43 •• A swimmer heads directly across a river, swimming at 1.6 m/s relative to still water. She arrives at a point 40 m downstream from the point directly across the river, which is 80 m wide. (*a*) What is the speed of the river current? (*b*) What is the swimmer's speed relative to the shore? (*c*) In what direction should the swimmer head so as to arrive at the point directly opposite her starting point?

(*a*) 1. Velocity component across river is 1.6 m/s	time of swim, $t = (80/1.6)$ s = 50 s
2. In 50 s she has drifted 40 m downstream	$v_{water} = (40/50)$ m/s = 0.8 m/s

(*b*) Use Equ. 3-14 — $v = (1.6^2 + 0.8^2)^{1/2}$ m/s = 1.79 m/s

(*c*) Set "downstream" components = 0 — $1.6 \sin \theta = 0.8$; $\theta = 30°$ upstream

---

44 •• A small plane departs from point *A* heading for an airport at point *B* 520 km due north. The airspeed of the plane is 240 km/h and there is a steady wind of 50 km/h blowing northwest to southeast. Determine the proper heading for the plane and the time of flight.

Follow the procedure of Problem 42.

1. Set east-west component of $\boldsymbol{v}$ equal to zero — $(240 \sin\theta)$ km/h = $(50 \sin 45°)$ km/h = 35.35 km/h
2. Solve for θ, angle west of north — $\theta = 8.47°$
3. Find $\boldsymbol{v} = \boldsymbol{v}_{\text{air}} + \boldsymbol{v}_{\text{wind}}$ — $v = (240 \cos 8.47° - 50 \cos 45°)$ km/h = 202 km/h
4. Find time of travel — $T = d/v = (520/202)$ h = 2.57 h = 2 h 34 min

---

45* •• Two boat landings are 2.0 km apart on the same bank of a stream that flows at 1.4 km/h. A motorboat makes the round trip between the two landings in 50 minutes. What is the speed of the boat relative to the water?

1. Find the speeds upstream and downstream — $v_{\text{up}} = v - 1.4$ km/h; $v_{\text{down}} = v + 1.4$ km/h
2. Express total time in terms of v_{up} and v_{down} — $T = (2 \text{ km})/v_{\text{up}} + (2 \text{ km})/v_{\text{down}} = 5/6$ h
3. Obtain equation for v and solve — $v^2 - 4.8\, v - 1.96 = 0$; $v = -0.378, 5.18$
4. Select physically acceptable solution — $v = 5.18$ km/h

---

46 •• A model airplane competition has the following rules: Each plane must fly to a point 1 km from the start and then back again. The winner is the plane with the shortest round-trip time. The contestants are free to launch their planes in any direction, so long as the plane travels exactly 1 km out and then returns. On the day of the race, a steady wind blows from the north at 5 m/s. Your plane can maintain an airspeed (speed relative to the air) of 15 m/s, and you know that starting, stopping, and turning times will be negligible. The question: Should you plan to fly into the wind and against the wind on your round-trip, or across the wind flying east and west? Make a reasoned choice by working out the following round-trip times: (1) The plane goes 1 km due north and then back; (2) the plane goes to a point 1 km due east of the start, and then back.

1. For case 1 find v for each section of trip — $v_{\text{N}} = (15 - 5)$ m/s = 10 m/s; $v_{\text{S}} = (15 + 5)$ m/s = 20 m/s
2. For case 1 determine time of round trip — $T_{\text{NS}} = (10^3/10 + 10^3/20)$ s = 150 s
3. For case 2 find v for each section of trip — See Fig. 3-17; $v_{\text{WE}} = v_{\text{EW}} = \sqrt{15^2 - 5^2}$ m/s = 14.14 m/s
4. Determine round-trip time T_{EW} and compare with T_{NS} — $T_{\text{EW}} = (2 \times 10^3/14.14)$ s = 141 s; this is shorter than T_{NS}

---

47 •• Car A is traveling east at 20 m/s. As car A crosses the intersection shown in Figure 3-35, car B starts from rest 40 m north of the intersection and moves south with a constant acceleration of 2 m/s^2. (*a*) What is the position of B relative to A 6 s after A crosses the intersection? (*b*) What is the velocity of B relative to A for $t = 6$ s? (*c*) What is the acceleration of B relative to A for $t = 6$ s?

Note: Position of B relative to A is the vector from A to B, i.e. $\boldsymbol{r}_{\text{AB}} = \boldsymbol{r}_{\text{B}} - \boldsymbol{r}_{\text{A}}$; similarly $\boldsymbol{v}_{\text{AB}} = \boldsymbol{v}_{\text{B}} - \boldsymbol{v}_{\text{A}}$ and $\boldsymbol{a}_{\text{AB}} = \boldsymbol{a}_{\text{B}} - \boldsymbol{a}_{\text{A}}$. As before, we take $\boldsymbol{i}$ east and $\boldsymbol{j}$ north. Origin at intersection. All quantities in SI units.

(*a*) 1. Find $\boldsymbol{r}_{\text{B}}$ and $\boldsymbol{r}_{\text{A}}$ — $\boldsymbol{r}_{\text{B}} = (40 - t^2)\boldsymbol{j}$; $\boldsymbol{r}_{\text{A}} = 20t\,\boldsymbol{i}$

2. Find $\boldsymbol{r}_{\text{AB}}$ at $t = 6$ s — $\boldsymbol{r}_{\text{AB}} = -120 \text{ m}\,\boldsymbol{i} + 4 \text{ m}\,\boldsymbol{j}$

(*b*) 1. Find $\boldsymbol{v}_A$ and $\boldsymbol{v}_B$ — $\boldsymbol{v}_A = 20\,\boldsymbol{i};\ \boldsymbol{v}_B = -2t\boldsymbol{j}$

2. Find $\boldsymbol{v}_{AB}$ at $t = 6$ s — $\boldsymbol{v}_{AB} = -20$ m/s $\boldsymbol{i}$ − 12 m/s $\boldsymbol{j}$

(*c*) Since $\boldsymbol{a}_A = 0$, $\boldsymbol{a}_{AB} = \boldsymbol{a}_B$ — $\boldsymbol{a}_{AB} = -2\ \text{m/s}^2\,\boldsymbol{j}$

---

48 •• Bernie is showing Margaret his new boat and its autonavigation feature, of which he is particularly proud. "That island is 1 km east and 3 km north of this dock. So I just punch in the numbers like this, and we get ourselves a refreshment and enjoy the scenery." Forty-five minutes later, they find themselves due east of the island. "OK, something went wrong. I'll just reverse the instructions, and we'll go back to the dock and try again." But 45 min later, the boat is 6 km east of the original position at the dock. " Did you allow for the cur rent?" asks Margaret. "For the what?" (*a*) What is the velocity of the current in the waterway where Bernie and Margaret are boating? (*b*) What is the velocity of the boat, relative to the water, for the first 45 min? (*c*) What is the velocity of the boat relative to the island for the first 45 min?

Let $\boldsymbol{v}_{WB}$ be the velocity of the boat relative to the water, $\boldsymbol{v}_W$ the velocity of the water (current), $\boldsymbol{v}_B$ the velocity of the boat relative to the land. Take $\boldsymbol{i}$ and $\boldsymbol{j}$ as in preceding problem. Assume constant $\boldsymbol{v}_W$.

(*a*) Boat has drifted east 6 km in 1.5 h — $\boldsymbol{v}_W = (6/1.5)\,\boldsymbol{i}$ km/h $= 4\,\boldsymbol{i}$ km/h

(*b*) Displacement on water in 0.75 h is $(1\,\boldsymbol{i} + 3\boldsymbol{j})$ km — $\boldsymbol{v}_{WB} = (1.33\,\boldsymbol{i} + 4\boldsymbol{j})$ km/h

(*c*) $\boldsymbol{v}_B = \boldsymbol{v}_{WB} + \boldsymbol{v}_W$ — $\boldsymbol{v}_B = (5.33\,\boldsymbol{i} + 4\boldsymbol{j})$ km/h

---

49* ••• Airports A and B are on the same meridian, with B 624 km south of A. Plane P departs airport A for B at the same time that an identical plane, Q, departs airport B for A. A steady 60 km/h wind is blowing from the south 30° east of north. Plane Q arrives at airport A 1 h before plane P arrives at airport B. Determine the airspeeds of the two planes (assuming that they are the same) and the heading of each plane.

Let $\boldsymbol{v}_{AP}$ and $\boldsymbol{v}_{AQ}$ be the velocity of P and Q relative to air; $\boldsymbol{v}_A$ is the velocity of the air (wind); $\boldsymbol{v}_P$ and $\boldsymbol{v}_Q$ the veloci ties of planes P and Q relative to the ground. Chose $\boldsymbol{i}$ and $\boldsymbol{j}$ as before.

1. Set east-west component of $\boldsymbol{v}_P$ and $\boldsymbol{v}_Q$ equal to zero. — $v_{AP}\sin\theta_P = (-60\ \text{km/h})\sin 30° = -30\ \text{km/h} = v_{AQ}\sin\theta_Q$;

   Let θ be the angle relative to north direction. — $\theta_P = \theta_Q = \theta$

2. Find north-south components of $\boldsymbol{v}_P$ and $\boldsymbol{v}_Q$; these are the ground speeds of P and Q — $v_P = v_{AP}\cos\theta - (60\ \text{km/h})\cos 30° = v_{AP}\cos\theta - 52$ km/h; $v_Q = v_{AQ}\cos\theta + 52$ km/h

3. Write the time difference in terms of $v_{AQ} = v_{AP} = v$ — $1\ \text{h} = (624\ \text{km})[1/(v\cos\theta - 52\ \text{km/h}) + 1/(v\cos\theta + 52\ \text{km/h})]$

4. Solve for $v\cos\theta$ — $v\cos\theta = 260$ km/h

5. Note that $v\sin\theta = -30$ km/h — $\theta = \tan^{-1}(30/260) = -6.58°$, i.e., 6.58° west of north

6. Solve for v — $v = v_{AP} = v_{AQ} = (260\ \text{km/h})/\cos 6.58° = 261.7$ km/h

---

50 • What is the acceleration of a projectile at the top of its flight?

The acceleration is the acceleration due to gravity, g.

---

51 • True or false: When a bullet is fired horizontally, it takes the same amount of time to reach the ground as a bullet dropped from rest from the same height.

True

52 • A golfer drives her ball from the tee a distance of 240 yards down the fairway in a high arcing shot. When the ball is at the highest point of its flight, (*a*) its velocity and acceleration are both zero. (*b*) its velocity is zero but its acceleration is nonzero. (*c*) its velocity is nonzero but its acceleration is zero. (*d*) its velocity and accelera tion are both nonzero. (*e*) Insufficient information is given to answer correctly.

(*d*) its velocity is horizontal and finite; its acceleration is g.

53* • A projectile was fired at 35° above the horizontal. At the highest point in its trajectory, its speed was 200 m/s. The initial velocity had a horizontal component of (*a*) 0. (*b*) (200 cos 35°) m/s. (*c*) (200 sin 35°) m/s. (*d*) (200/cos 35°) m/s. (*e*) 200 m/s.

(*e*) At the highest point, the speed is the horizontal component of the initial velocity.

54 • Figure 3-36 represents the parabolic trajectory of a ball going from *A* to *E*. What is the diretion of the acceleration at point *B*? (*a*) Up and to the right (*b*) Down and to the left (*c*) Straight up (*d*) Straight down (*e*) The acceleration of the ball is zero.

(*d*) Throughout the flight the acceleration is g directed down.

55 • Referring to Figure 3-36, (*a*) at which point(s) is the speed the greatest? (*b*) At which point(s) is the speed the lowest? (*c*) At which two points is the speed the same? Is the velocity the same at those points?

(*a*) The speed is greatest at A and E. (*b*) The speed is least at point C. (*c*) The speed is the same at A and E; the velocities are not equal at these points; the vertical components are opposite.

56 • A bullet is fired horizontally with an initial velocity of 245 m/s. The gun is 1.5 m above ground. How long is the bullet in the air?

For vertical motion $y = \frac{1}{2}gt^2$; solve for t — $t = (2 \times 1.5/9.81)^{1/2}$ s $= 0.553$ s

57* • A pitcher throws a fast ball at 140 km/h toward home plate, which is 18.4 m away. Neglecting air resistance (not a good idea if you are the batter), find how far the ball drops because of gravity by the time it reaches home plate.

1. Find time of flight — $t = [18.4$ m/ (140/3.6) m/s$] = 0.473$ s
2. Find distance dropped in 0.473 s — $d = \frac{1}{2}gt^2 = 1.1$ m

58 • A projectile is launched with speed v_0 at an angle of θ_0 with the horizontal. Find an expression for the maximum height it reaches above its starting point in terms of v_0, θ_0, and g.

1. Write v_{0y} — $v_{0y} = v_0 \sin \theta_0$
2. Use Equ. 2-15 to find H, setting $v_y = 0$ — $H = (v_0^2 \sin^2 \theta_0)/2g$

59 • A projectile is fired with an initial velocity of 30 m/s at 60° above horizontal. At the projectile's highest point, what is its velocity? Its acceleration?

At the highest point, the velocity is $v_{0x} = (30 \cos 60°)$ m/s $= 15$ m/s. Its acceleration is 9.81 m/s² downward.

60 •• A projectile is fired with initial speed v at an angle 30° above the horizontal from a height of 40 m above the ground. The projectile strikes the ground with a speed of $1.2v$. Find v.

We will use a general approach: find the velocity on impact for a projectile launched at an arbitrary angle θ with initial speed v_0 from a height h. We take our origin at the launch site.

1. Use Equ. 2-15 — $v_y^2 = v_{0y}^2 - 2gh = (v_0 \sin \theta)^2 + 2gh$

2. Use Equ. 3-2 to find $v_x = v_{0x}$ — $v_x = v_0 \cos\theta$
3. Write the expression for v — $v = [v_x^2 + v_y^2]^{1/2} = [v_0^2(\sin^2\theta + \cos^2\theta) + 2gh]^{1/2}$
4. Note that $\sin^2\theta + \cos^2\theta = 1$ — $v = (v_0^2 + 2gh)^{1/2}$; NOTE: v is independent of θ
5. Set $v = 1.2v_0$, $h = 40$ m and solve for v_0 — $1.44v_0^2 = v_0^2 + 785\ \mathrm{m^2/s^2}$; $v_0 = 42.2$ m/s

---

61* •• If the tree in Example 3-11 is 50 m away and the monkey hangs from a branch 10 m above the muzzle position, what is the minimum initial speed of the dart if it is to hit the monkey before hitting the ground? We shall assume that the muzzle of the dart gun is 1.2 m above ground.

1. Find the time for monkey to fall to ground — $t = (2h/g)^{1/2} = (22.4/9.81)^{1/2} = 1.51$ s
2. Projection angle $\theta = \tan^{-1}(10/50)$ — $\theta = 11.3°$
3. $v_x = (50/1.51)$ m/s; $v_0 = v_x/\cos 11.3°$ — $v_0 = [50/(1.51 \cos 11.3°)]$ m/s $= 33.8$ m/s

---

62 •• A projectile is fired with an initial speed of 53 m/s. Find the angle of projection such that the maximum height of the projectile is equal to its horizontal range.

Let v_0 be the initial speed of projectile. Then maximum height, $H = (v_0 \sin\theta)^2/2g$, and range, $R = (2v_0^2/g)\sin\theta\cos\theta$; Set $H = R$: $\frac{1}{2}\sin^2\theta = 2\sin\theta\cos\theta$; $\tan\theta = 4$ and $\theta = 76°$. (Independent of v_0)

---

63. A ball thrown into the air lands 40 m away 2.44 s later. Find the direction and magnitude of the initial velocity.

1. Find v_{0y} using Equ. 2-12 — $v_{0y} = \frac{1}{2}g(2.44\ \mathrm{s}) = 12$ m/s
2. Find v_{0x} using Equ. 2-14 — $v_{0x} = (40/2.44)$ m/s $= 16.4$ m/s
3. Find v and θ — $\theta = \tan^{-1}(12/16.4) = 36.2°$; $v = 16.4/\cos 36.2° = 20.3$ m/s

---

64 •• Show that if an object is thrown with speed v_0 and an angle θ above the horizontal, its speed at some height h is independent of θ.

$v_y^2 = v_0^2\sin^2\theta - gh$; $v_x^2 = v_0^2\cos^2\theta$, $v^2 = v_0^2(\sin^2\theta + \cos^2\theta) - 2gh = v_0^2 - 2gh$; Q.E.D.

65* •• At half its maximum height, the speed of a projectile is three-fourth its initial speed. What is the angle of the initial velocity vector with respect to the horizontal?

1. Find v_{0y}^2 in terms of H and g — $v_{oy}^2 = 2gH$
2. Find v_y^2 at $H/2$; use Equ. 2-15 — $v_y^2 = 2gH - 2gH/2 = gH$
3. Find v^2 at $H/2$ — $v^2 = v_y^2 + v_x^2 = gH + v_x^2 = (9/16)[2gH + v_x^2]$
4. Solve for v_x^2 in terms of v_{0y}^2 — $7v_x^2 = v_{0y}^2$; $v_{0y} = \sqrt{7}\,v_x$
5. $\theta = \tan^{-1}(v_{0y}/v_x)$ — $\theta = \tan^{-1}(\sqrt{7}) = 69.3°$

---

66 ••• Wally and Luke advertise their circus act as "The Human Burrs—Trapeze Artists for the New Millennium." Their specialty involves wearing padded Velcro suits that cause them to stick together when they make contact in midair. While working on their act, Wally is shot from a cannon with a speed of 20 m/s at an angle of 30° above the horizontal. At the same moment, Luke drops from a platform having (x,y) coordinates of (8 m, 16 m), if the cannon is taken to sit at the origin. (*a*) Will they make contact? (*b*) What is the minimum distance separating Wally and Luke during their flight paths? (*c*) At what time does this minimum separation occur? (*d*) Give the coordinates of each daredevil at that time.

(*a*) See Example 3-11; $\theta = \tan^{-1}(16/8) = 63.4°$ — But here $\theta = 30°$, so no contact

(*c*) Write x and y components of Luke and Wally — $x_L = 8$ m, $y_L = (16 - \frac{1}{2}gt^2)$ m; $x_W = (20t\cos 30°)$ m

	$y_W = (20t \sin 30° - ½gt^2)$ m
Find Δx and Δy, and $D^2 = \Delta x^2 + \Delta y^2$	$D^2 = (400t^2 - 597t + 320)$ m^2
Set $d(D^2)/dt = 0$, solve for t	$800t - 597 = 0$; $t = 0.746$ s
(*b*) With $t = 0.746$ s, find D	$D = 9.86$ m
(*d*) Use $t = 0.746$ s in exprssions for x_L, y_L, x_W, y_W	$x_L = 8$ m, $y_L = 13.27$ m; $x_W = 12.9$ m, $y_W = 4.73$ m

---

67 • A cargo plane is flying horizontally at an altitude of 12 km with a speed of 900 km/h when a battle tank falls out of the rear loading ramp. (*a*) How long does it take the tank to hit the grounf? (*b*) How far horizontally is the tank from where it fell off when it hits the ground? (*c*) How far is the tank from the aircraft when the tank hits the ground, assuming that the plane continues to fly with constant velocity?

(*a*) Use Equ. 2-14 with $v_{0y} = 0$	$t = (2 \times 12 \times 10^3/9.81)^{½}$ s $= 49.5$ s
(*b*) $v_x = 900$ km/h $= 250$ m/s, constant	$\Delta x = (250 \times 49.5)$ m $= 12.4$ km
(*c*) Distance to aircraft = vertical distance	$\Delta y = 12.0$ km

---

68 • A cannon barrel is elevated at an angle of 45°. It fires a ball with a speed of 300 m/s. (*a*) What height does the ball reach? (*b*) How long is the ball in the air? (*c*) What is the horizontal range of the cannon?

(*a*) 1. Find v_{0y}	$v_{0y} = (300 \sin 45°)$ m/s $= 212$ m/s
2. Find H, use Equ. 2-15	$H = v_{0y}{}^2/2g = 2293$ m
(*b*) Find t; use Equ. 2-12	$t = 2(v_{0y}/g) = 43.2$ s
(*c*) Use Equ. 3-22	$R = (9 \times 10^4/9.81)$ m $= 9.17$ km

69* •• A stone thrown horizontally from the top of a 24-m tower hits the ground at a point 18 m from the base of the tower. (*a*) Find the speed at which the stone was thrown. (*b*) Find the speed of the stone just before it hits the ground.

(*a*) 1. Find the time to fall 24 m	$t = \sqrt{2H/g} = 2.21$ s
2. Find $v_x = v_0$	$v_x = v_0 = \Delta x/t = (18/2.21)$ m $= 8.14$ m/s
(*b*) 1. Find v_y at 2.21 s	$v_y = gt = 21.7$ m/s
2. Find v_f	$v_f = (v_x^2 + v_y^2)^{½} = 23.2$ m/s

---

70 •• A projectile is fired into the air from the top of a 200-m cliff above a valley (Figure 3-37). Its initial velocity is 60 m/s at 60° above the horizontal. Where does the projectile land?

1. Find the time of flight; use Equ. 2-14	-200 m $= [(60 \sin 60°)t - ½ \times 9.81t^2]$ m
2. Solve the quadratic equation for t	$t = 13.6$ s
3. Find Δx	$\Delta x = [(60 \cos 60°)13.6]$ m $= 408$ m

---

71 •• The range of a projectile fired horizontally from a cliff is equal to the height of the cliff. What is the direction of the velocity vector when the projectile strikes the ground?

1. Set $\Delta x = \Delta y$	$\Delta x = v_x t = \Delta y = ½gt^2$; $v_x = ½gt$
2. Find v_y	$v_y = -gt = -2v_x$
3. Find θ, angle of v with horizontal	$\theta = \tan^{-1}(v_y/v_x) = \tan^{-1}(-2) = -63.4°$

72 •• Find the range of the projectile of Problem 60.

Note: The projectile is fired with an initial speed of 42.2 m/s at an angle of 30° above the horizontal from an initial height of 40 m. This problem is identical to Problem 70 with different numerical values. Following the same procedure one obtains $\Delta x = 209$ m.

73* •• Compute $dR/d\theta$ from Equation 3-22 and show that setting $dR/d\theta = 0$ gives $\theta = 45°$ for the maximum range.

$d[(v_0^2 \sin 2\theta)/g]/d\theta = (v_0^2/g)d(\sin 2\theta)/d\theta = (2v_0^2/g)\cos 2\theta$; set equal to 0; $2\theta = 90°$, $\theta = 45°$.

74 •• A rock is thrown from the top of a 20-m building at an angle of 53° above the horizontal. If the horizontal range of the throw is equal to the height of the building, with what speed was the rock thrown? What is the velocity of the rock just before it strikes the ground?

1. Write v_x and v_y	$v_x = v_0 \cos 53° = 0.6v_0$, $v_y = v_0 \sin 53° - gt = 0.8v_0 - gt$
2. Write expressions for x and y	$20\text{ m} = 0.6v_0 t$; $t = (20/0.6v_0)$; $-20\text{ m} = 0.8v_0 t - \frac{1}{2}gt^2$
3. Substitute t into the expression for y	$-20\text{ m} = (0.8 \times 20/0.6)\text{ m} - \frac{1}{2}g(20/0.6v_0)^2$
4. Solve for v_0	$v_0^2 = 116.9\text{ m}^2/\text{s}^2$; $v_0 = 10.8$ m/s
5. Find t at impact	$t = [20/(0.6 \times 10.8)]\text{ s} = 3.086$ s
6. Find v_y at impact; write $\boldsymbol{v}$	$v_y = v_{0y} - gt = -21.6$ m/s; $\boldsymbol{v} = (6.48\,\boldsymbol{i} - 21.6\,\boldsymbol{j})$ m/s

75 •• A stone is thrown horizontally from the top of an incline that makes an angle ϕ with the horizontal. If the stone's initial speed is v, how far down the incline will it land?

1. Write expressions for x, y, and y/x	$x = vt$, $y = \frac{1}{2}gt^2$, $y/x = \tan\phi = \frac{1}{2}gt/v$
2. Solve for $x = vt$	$x = \frac{1}{2}gt^2/\tan\phi = L\cos\phi$
3. Find L	$L = \frac{1}{2}gt^2/(\tan\phi\cos\phi) = 2v^2\tan\phi/(g\cos\phi)$

76 •• A flock of seagulls has decided to mount an organized response to the human overpopulation of their favorite beach. One tactic popular among the innovative radicals is bombing the sunbathers with clams. The gull dives with a speed of 16 m/s, at an angle of 40° below the horizontal. He releases a projectile when his vertical distance above his traget, a sunbather's bronzed tummy, is 8.5 m, and scores a bull's-eye. (*a*) Where is the sunbather in relation to the gull at the instant of release? (*b*) How long is the projectile in the air? (*c*) What is the velocity of the projectile upon impact?

(*b*) 1. Determine v_{0x} and v_{0y}	$v_{0x} = 16\cos 40°\text{ m/s} = 12.3$ m/s; $v_{0y} = -16\sin 40°$ m/s $= -10.3$ m/s
2. Use Equ. 2-14 to find time of flight	$-8.5 = -10.3t - \frac{1}{2}gt^2$; solve for t, $t = 0.633$ s
(*a*) Find x and y	$x = (12.3 \times 0.633)\text{ m} = 8.15$ m, $y = -8.5$ m
(*c*) 1. Find v_y at t = 0.633 s; $v_x = v_{0x}$	$v_y = (-10.3 - 9.81 \times 0.633)\text{ m/s} = -16.5$ m/s; $v_x = 12.3$ m/s
2. Write $\boldsymbol{v}$	$\boldsymbol{v} = (12.3\,\boldsymbol{i} - 16.5\,\boldsymbol{j})$ m/s

77* ••• A girl throws a ball at a vertical wall 4 m away (Figure 3-38). The ball is 2 m above ground when it leaves the girl's hand with an initial velocity of $\boldsymbol{v}_0 = (10\,\boldsymbol{i} + 10\,\boldsymbol{j})$ m/s. When the ball hits the wall, the horizontal component of its velocity is reversed; the vertical component remains unchanged. Where does the ball hit the ground?

Note: The wall acts like a mirror. We shall determine the range, neglecting the wall. Then consider the mirror-like reflection.

1. Use Equ. 2-14 to find t, time of flight — $0 = (2 + 10t - ½gt^2)$ m; $t = 2.22$ s
2. Find the distance Δx — $\Delta x = 10t$ m $= 22.2$ m
3. Consider wall reflection; x from wall $= \Delta x - 4$ m — The ball will land 18.2 m from the wall

---

78 • A boy uses a slingshot to project a pebble at a shoulder-height target 40 m away. He finds that to hit the target he must aim 4.85 m above the target. Determine the velocity of the pebble on leaving the slingshot and the time of flight.

1. Find angle of $\mathbf{v}_0$ with the horizontal — $\theta = \tan^{-1}(4.85/40) = 6.91°$
2. Use Equ. 3-22 to find $\mathbf{v}_0$ — $v_0 = [40 \times 9.81/\sin 13.82°]^{½} = 40.5$ m/s at 6.91°
3. $t = d/v_x = d/(v_0 \cos \theta)$ — $t = 40/(40.5 \times \cos 6.91°) = 0.995$ s

---

79 •• The distance from the pitcher's mound to home plate is 18.4 m. The mound is 0.2 m above the level of the field. A pitcher throws a fast ball with an initial speed of 37.5 m/s. At the moment the ball leaves the pitcher's hand, it is 2.3 m above the mound. What should the angle between $\mathbf{v}$ and the horizontal be so that the ball crosses th plate 0.7 m above ground?

1. Assume $\mathbf{v} = 37.5\,\mathbf{i}$ m/s; find distance it drops — $t = (18.4/37.5)$ s $= 0.49$ s; $\Delta y = -½gt^2 = -1.18$ m
2. Find where it would be at plate — $y = (2.5 - 1.18)$ m $= 1.32$ m above ground
3. Must drop additonal 0.62 m; throw downward. Find v_y (assume v_x unchanged) — $v_y = (-0.62/0.49)$ m/s $= -1.265$ m/s
4. Find angle with horizontal — $\theta = \tan^{-1}(-1.265/37.5) = -1.93°$

One can readily show that $(v_x^2 + v_y^2)^{½} = 37.5$ m/s to within 1%; so the assumption that v and t are unchanged by throwing the ball downward at an angle of 1.93° is justified.

---

80 •• Suppose the puck in Example 3-12 is struck in such a way that it just clears the Plexiglas wall when it is at its highest point. Find v_{0y}, the time t to reach the wall, and v_{0x}, v_0, and θ_0 for this case.

Assume the same initial conditions, i.e., $h = 2.80$ m, $x_1 = 12.0$ m

1. Use Equ. 2-15 to find v_{0y} — $v_{0y} = (5.60g)^{½} = 7.41$ m/s
2. Use Equ. 2-12 to find T — $T = (7.41/9.81)$ s $= 0.756$ s
3. Use Equ. 2-14 to find v_{0x} — $v_{0x} = (12.0/0.756)$ m/s $= 15.9$ m/s
4. Use Equ. 3-5 to find v_0 — $v_0 = (15.9^2 + 7.41^2)^{½}$ m/s $= 17.5$ m/s
5. Find θ — $\theta = \tan^{-1}(7.41/15.9) = 25.0°$

81* •• The coach throws a baseball to a player with an initial speed of 20 m/s at an angle of 45° with the horizontal. At the moment the ball is thrown, the player is 50 m from the coach. At what speed and in what direction must the player run to catch the ball at the same height at which it was released?

1. Use Equ. 3-22 to find the range of ball — $R = (400/9.81)$ m $= 40.8$ m
2. Find the distance the runner must cover — $\Delta x = 50 - 40.8 = 9.2$ m
3. Find the time of flight — $\Delta t = [40.8/(20 \cos 45°)] = 2.88$ s
4. Find the speed of runner toward coach — $v = \Delta x/\Delta t = (9.2/2.88)$ m/s $= 3.19$ m/s

82 •• Carlos is on his trail bike, approaching a creek bed that is 7 m wide. A ramp with an incline of 10° has been built for daring people who try to jump the creek. Carlos is traveling at his bike's maximum speed, 40 km/h. (*a*) Should Carlos attempt to jump or emphatically hit the brakes? (*b*) What is the minimum speed a bike must have to make this jump? (Assume equal elevations on either side of creek.)

(*a*) Use Equ. 3-22 to find R with $v_0 = 40$ km/h $= 11.1$ m/s — $R = [(11.1^2 \times \sin 20°)/9.81] = 4.3$ m; apply brakes!

(*b*) Use Equ. 3-22 to find v_{0min} with $R = 7$ m, $\theta = 10°$ — $v_{0min} = [(7 \times 9.81)/\sin 20°]^{1/2}$ m/s $= 14.17$ m/s $= 51$ km/h

---

83 •• It's the bottom of the ninth with two outs and the winning run is on base. You hit a knee-high fastball that just clears the leaping third baseman's glove. He is standing 28 m from you and his glove reaches 3.2 m above the ground. The flight time to that point is 0.64 s. Assume the ball's initial height was 0.6 m. Find (*a*) the initial speed and direction of the ball; (*b*) the time at which the ball reaches its maximum height; (*c*) the maximum height of the ball.

(*a*) 1. Determine v_{0x} — $v_{0x} = (28/0.64)$ m/s $= 43.75$ m/s

2. Determine distance ball drops in 0.64 s — $\Delta y = \frac{1}{2}gt^2 = 2.0$ m

3. Find v_{0y} to make up the total of 4.6 m — $v_{0y} = (4.6/0.64)$ m/s $= 7.19$ m/s

4. Find v_0 and θ_0 — $v_0 = (43.75^2 + 7.19^2)^{1/2}$ m/s $= 44.3$ m/s; $\theta_0 = \tan^{-1}(7.19/43.75) = 9.33°$ above horizontal

(*b*) Use Equ. 2-12 to find t_{max} — $t_{max} = (7.19/9.81)$ s $= 0.733$ s

(*c*) Use Equ. 2-15 and add knee-height of 0.6 m — $h = [7.19^2/(2 \times 9.81) + 0.6]$ m $= 3.23$ m

---

84 •• Noobus is a death-defying squirrel with miraculous jumping abilities. Running to the edge of a flat rooftop, she leaps horizontally with a speed of 6 m/s. If she just clears the 3-m gap between the houses and lands on the neighbor's roof, what is her speed upon landing?

1. Find the flight time — $t = (3/6)$ s $= 0.5$ s

2. Find v_y on landing — $v_y = -gt = -4.9$ m/s

3. Write $\mathbf{v}$ — $\mathbf{v} = (6\,\mathbf{i} - 4.9\,\mathbf{j})$ m/s

---

85* ••• If a bullet that leaves the muzzle of a gun at 250 m/s is to hit a target 100 m away at the level of the muzzle, the gun must be aimed at a point above the target. How far above the target is that point?

1. Find θ_0 using Equ. 3-22 — $\sin 2\theta_0 = 981/250^2 = 0.0157$; $\theta_0 = 0.45°$ (or 89.55°)

2. $y = R \tan \theta_0$ — $y = 0.785$ m (disregard $\theta_0 = 89.55°$ as unrealistic)

---

86 ••• A baseball just clears a 3-m wall that is 120 m from home plate. If the ball leaves the bat at 45° and 1.2 m above the ground, what must its initial speed be?

1. Use Equ. 2-14 and set $x(t) = 120$, $y(t) = 3$ — $v_{0x}t = 120$ m; $(1.2 + v_{0y}t - \frac{1}{2}gt^2)$ m $= 3.0$ m

2. Use $v_{0x} = v_{0y} = v_0 \cos 45° = v_0/\sqrt{2}$ — $(1.2 + 120 - \frac{1}{2}gt^2)$ m $= 3.0$ m; $t = 4.91$ s;

3. Solve for v_0 — $v_0 = [(120 \times \sqrt{2})/4.91]$ m/s $= 34.6$ m/s

87 ••• A baseball is struck by a bat, and 3 s later is caught 30 m away. (*a*) If the baseball was 1 m above ground when it was struck and caught, what was the greatest height it reached above ground? (*b*) What were the hori zontal and vertical components of its velocity when it left was struck? (*c*) What was its speed when it was caught? (*d*) At what angle with the horizontal did it leave the bat?

(*b*) Find v_{0x}, v_{0y} using Equ.2-14	$v_{0x} = (30/3)$ m/s $= 10$ m/s; $v_{0y} = (9.81 \times 3/2)$ m/s $= 14.7$ m/s
(*a*) Use Equ. 2-15	$H = (14.7^2/19.62)$ m $= 11.0$ m
(*d*) Find θ_0	$\theta_0 = \tan^{-1}(14.7/10) = 55.8°$
(*c*) It was caught 1 m above ground; $v_f = v_0$	$v_f = v_{0x}/\cos\theta_0 = 17.8$ m/s

---

88 ••• A baseball player hits a baseball that drops into the stands 22 m above the playing field. The ball lands with a velocity of 50 m/s at an angle of 35° below the horizontal. (*a*) If the batter contacted the ball 1.2 m above the playing field, what was the velocity of the ball upon leaving the bat? (*b*) What was the horizontal distance traveled by the ball? (*c*) How long was the ball in the air?

(*a*) 1. Find v_{0x}, and v_{0y} using Equ. 2-15	$v_{0x} = (50$ m/s$) \cos 35° = 41.0$ m/s;
	$v_{oy} = [(50 \times \sin 35°)^2 + 19.62 \times (22 - 1.2)]^{½}$ m/s $= 35.1$ m/s
2. Write $\boldsymbol{v}$	$\boldsymbol{v} = (41.0\,\boldsymbol{i} + 35.1\,\boldsymbol{j})$ m/s
(*c*) Find t using Equ. 2-12	$t = [(35.1 + 50 \times \sin 35°)/9.81]$ s $= 6.5$ s
(*b*) Find Δx	$\Delta x = v_{0x}t = 266$ m

89* • True or false: (*a*) The magnitude of the sum of two vectors must be greater than the magnitude of either vector. (*b*) If the speed is constant, the acceleration must be zero. (*c*) If the acceleration is zero, the speed must be constant.

(*a*) False (*b*) False (*c*) True

---

90 • The initial and final velocities of an object are as shown in Figure 3-39. Indicate the direction of the average acceleration.

$\boldsymbol{a} = \Delta\boldsymbol{v}/\Delta t = (\boldsymbol{v}_f - \boldsymbol{v}_i)/\Delta t$

The direction of $\boldsymbol{a}$ is that of $\boldsymbol{v}_f - \boldsymbol{v}_i$ as shown.

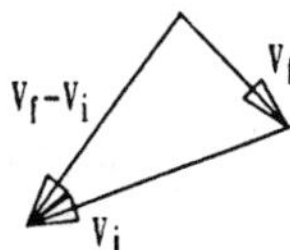

---

91 • The velocities of objects A and B are shown in Figure 3-40. Draw a vector that represents the velocity of B relative to A.

Velocity of B relative to A is $\boldsymbol{v}_{BA} = \boldsymbol{v}_B - \boldsymbol{v}_A$, as shown in the adjoining figure.

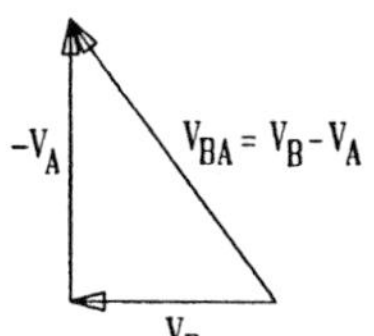

---

92 •• A vector $A(t)$ has a constant magnitude but is changing direction in a uniform way. Draw the vectors $A(t + \Delta t)$ and $A(t)$ for a small time interval Δt, and find the difference $\Delta A = A(t + \Delta t) - A(t)$ graphically. How is the direction of ΔA related to A for small time intervals?

The vectors $A(t)$ and $A(t + \Delta t)$ are of equal length but point in slightly different directions. ΔA is shown in the second diagram. Note that ΔA is nearly perpendicular to $A(t)$. For very small time intervals, ΔA and $A(t)$ are perpendicular to one another.

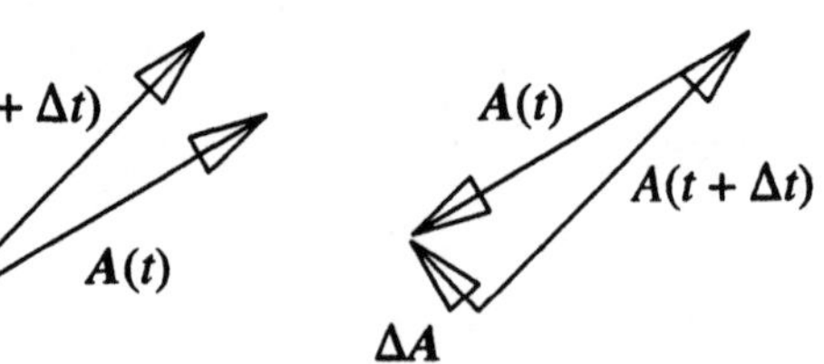

93* •• The automobile path shown in Figure 3-41 is made up of straight lines and arcs of circles. The automobile starts from rest at point A. After it reaches point B, it travels at constant speed until it reaches point E. It comes to rest at point F. (*a*) At the middle of each segment (AB, BC, CD, DE, and EF), what is the direction of the velocity vector? (*b*) At which of these points does the automobile have an acceleration? In those cases, what is the direction of the acceleration? (*c*) How do the magnitudes of the acceleration compare for segments BC and DE?

Let y direction be north, x direction be east.

(*a*) AB, $\boldsymbol{v}$ directed north; BC, $\boldsymbol{v}$ directed northeast; CD, $\boldsymbol{v}$ directed east; DE, $\boldsymbol{v}$ directed southeast; EF, $\boldsymbol{v}$ directed south.

(*b*) AB, $\boldsymbol{a}$ to north; BC, $\boldsymbol{a}$ to southeast; CD, $\boldsymbol{a} = 0$; DE, $\boldsymbol{a}$ to southwest; EF, $\boldsymbol{a}$ to north.

(*c*) The magnitudes are equal.

---

94 • The displacement vectors $\boldsymbol{A}$ and $\boldsymbol{B}$ in Figure 3-42 both have a magnitude of 2 m. (*a*) Find their x and y components. (*b*) Find the components, magnitude, and direction of the sum $\boldsymbol{A} + \boldsymbol{B}$. (*c*) Find the components, magnitude, and direction of the difference $\boldsymbol{A} - \boldsymbol{B}$.

(*a*) $A_x = 2\cos 45° = 1.41$ m, $A_y = 2\sin 45° = 1.41$ m; $B_x = 2\cos -30° = 1.73$ m, $B_y = 2\sin -30° = -1$ m.

(*b*) $C_x = A_x + B_x = 3.14$ m, $C_y = A_y + B_y = 0.41$ m; $C = (C_x^2 + C_y^2)^{1/2} = 3.17$ m; $\theta_C = \tan^{-1}(C_y/C_x) = 7.44°$.

(*c*) $D_x = A_x - B_x = -0.32$ m, $D_y = A_y - B_y = 2.41$ m; $D = (D_x^2 + D_y^2)^{1/2} = 2.43$ m; $\theta_D = \tan^{-1}(D_y/D_x) = 97.5°$.

---

95 • A plane is inclined at an angle of 30° from the horizontal. Choose the x axis pointing down the slope of the plane and the y axis perpendicular to the plane. Find the x and y components of the acceleration of gravity, which has the magnitude 9.81 m/s² and points vertically down.

The coordinate axes, properly rotated, are shown as the vector $\boldsymbol{g}$.
Note that $g_x = g\sin 30° = 4.9$ m/s², $g_y = -g\cos 30° = 8.5$ m/s².

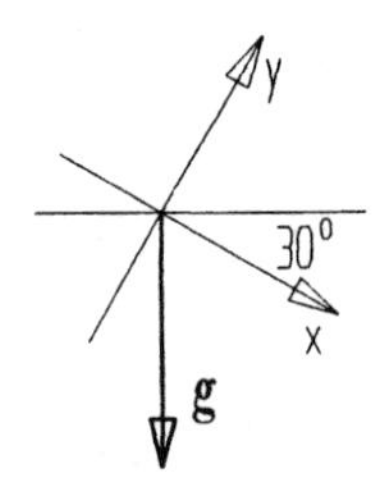

---

96 • Two vectors $\boldsymbol{A}$ and $\boldsymbol{B}$ lie in the xy plane. Under what conditions does the ratio A/B equal A_x/B_x?

$A_x/B_x = A/B$ if $\boldsymbol{A}$ and $\boldsymbol{B}$ are parallel ($\theta_A = \theta_B$) or if $\theta_A = -\theta_B$.

97* • The position vector of a particle is given by $\boldsymbol{r} = 5t\,\boldsymbol{i} + 10t\,\boldsymbol{j}$, where t is in seconds and $\boldsymbol{r}$ is in meters. (*a*) Draw the path of the particle in the xy plane. (*b*) Find $\boldsymbol{v}$ in component form and then find its magnitude.

(*b*) $\boldsymbol{v} = d\boldsymbol{r}/dt$ $\boldsymbol{v} = (5\,\boldsymbol{i} + 10\,\boldsymbol{j})$ m/s; $v = \sqrt{125}$ m/s $= 11.2$ m/s

(*a*) The path is shown in the figure

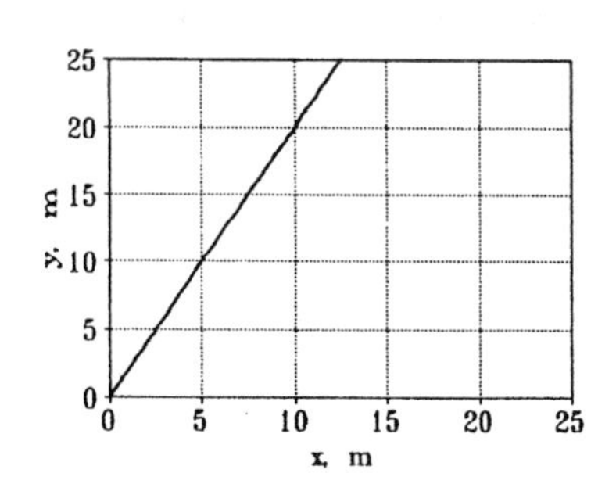

---

98 • Off the coast of Chile, a spotter plane sees a school of tuna swimming at a steady 5 km/h northwest (Figure 3-43). The pilot informs a fishing trawler located 100 km due south of the fish. The trawler sails at full steam along the best straight-line course and intercepts the tuna after 4 hours. How fast did the trawler move?

1. Find the distance covered $D = [(100 + 4 \times 5/\sqrt{2})^2 + (4 \times 5/\sqrt{2})^2]^{1/2}$ km = 115 km
2. Find speed of trawler $v = D/t = (115/4)$ km/h = 28.74 km/h

---

99 •• A workman on the roof of a house drops her hammer, which slides down the roof at constant speed of 4 m/s. The roof makes an angle of 30° with the horizontal, and its lowest point is 10 m from the ground. What is the horizontal distance traveled by the hammer after it leaves the roof of the house and before it hits the ground?

1. Use Equ. 2-14 to find *t*, time from roof edge to ground $0 = 10 - (4 \sin 30°)t - \frac{1}{2}gt^2$; $t = 1.24$ s
2. Find Δx $\Delta x = (4 \cos 30°)(1.24)$ m = 4.29 m

---

100 •• A freight train is moving at a constant speed of 10 m/s. A man standing on a flatcar throws a ball into the air and catches it as it falls. Relative to the flatcar, the initial velocity of the ball is 15 m/s straight up. (*a*) What are the magnitude and direction of the initial velocity of the ball as seen by a second man standing next to the track? (*b*) How long is the ball in the air according to the man on the train? According to the man on the ground? (*c*) What horizontal distance has the ball traveled by the time it is caught according to the man on the train? Accord ing to the man on the ground? (*d*) What is the minimum speed of the ball during its flight according to the man on the train? According to the man on the ground? (*e*) What is the acceleration of the ball according to the man on the train? According to the man on the ground?

Let $\boldsymbol{v}_{BG}$, $\boldsymbol{v}_{BT}$, and $\boldsymbol{v}_{TG}$ be the velocities of the ball relative to the ground, of the ball relative to the train, and of the train relative to the ground.

(*a*) $\boldsymbol{v}_{BG} = \boldsymbol{v}_{BT} + \boldsymbol{v}_{TG} = (10\,\boldsymbol{i} + 15\,\boldsymbol{j})$ m/s; $v_{BG} = 18.0$ m/s, $\theta = \tan^{-1}1.5 = 56.3°$.

(*b*) $t = 2v_{BT}/g = 3.06$ s.

(*c*) $\Delta x_{BT} = 0$; $\Delta x_{BG} = v_{TG}t = 30.6$ m.

(*d*) $v_{min,BT} = 0$; $v_{min,BG} = 10$ m/s.

(*e*) $\boldsymbol{a} = -g\,\boldsymbol{j}$ for both men.

101*•• Estimate how far you can throw a ball if you throw it (*a*) horizontally while standing on level ground; (*b*) at $\theta = 45°$ while standing on level ground; (*c*) horizontally from the top of a building 12 m high; (*d*) at $\theta = 45°$ from the top of a building 12 m high.

The answer will, of course, depend on the initial speed. Assume an initial speed of 90 km/h = 25 m/s.

(*a*) Assume the throwing arm is 1.5 m off the ground. Time to drop 1.5 m under gravity is $t = (3/9.81)^{1/2}$ s $= 0.55$ s. In that time the ball travels a distance $\Delta x = (25 \text{ m/s})(0.55 \text{ s}) = 14$ m.

(*b*) Neglect the initial height of the arm. Use Equ. 3-22 to get range $R = v^2/g = 64$ m.

(*c*) Now the distance to drop is 13.5 m and $t = 1.66$ s. $\Delta x = 25 \times 1.66 \text{ m} = 41$ m.

(*d*) Now we cannot neglect the difference between initial and final elevations. So we use Equ. 2-14 to determine t. We find $t = 4.25$ s and $\Delta x = (25/\sqrt{2})(4.25) \text{ m} = 75$ m.

---

102 •• A stunt motorcyclist wants to jump over 10 cars parked side by side below a horizontal launching ramp, as shown in Figure 3-44. With what minimum horizontal speed v_0 must the cyclist leave the ramp in order to clear the top of the last car?

1. Find the time to drop 5.0 m — $t = (10/9.81)^{1/2}$ s $= 1.01$ s
2. $v_0 = \Delta x/t$ — $v_0 = 24/1.01 \text{ m/s} = 23.8 \text{ m/s} = 85.6$ km/h

---

103 •• In 1978, Geoff Capes of Great Britain threw a heavy brick a horizontal distance of 44.5 m. Find the velocity of the brick at the highest point of its flight.

As a rough guess, we will assume that the throwing arm was not much above ground level and use $\theta = 45°$ in Equ. 3-22.

Then $v_0 = (44.5 \times 9.81)^{1/2}$ m/s $= 21$ m/s and $v_{0x} = (21 \times 0.707) \text{ m/s} = 14.8$ m/s; this is the horizontal velocity of the brick at its highest point.

---

104 •• In 1940, Emanuel Zacchini flew about 53 m as a human cannonball, a record that remains unbroken. His initial velocity was 24.2 m/s at an angle θ. Find θ, and the maximum height h Emanuel achieved during the record flight.

1. Use Equ. 3-22 to determine θ — $\sin 2\theta = Rg/v_0^2 = (53 \times 9.81/24.2^2) = 0.888$; $\theta = 31.3°$
2. Use Equ. 2-15 with $v = 0$, $v_0 = v_{0y}$ $= (24.2 \sin 31.3°)$ m/s — $h = [(24.2 \sin 31.3°)^2/(2 \times 9.81)] \text{ m} = 8.06$ m

---

105* •• A particle moves in the xy plane with constant acceleration. At $t = 0$ the particle is at $\boldsymbol{r}_1 = 4 \text{ m } \boldsymbol{i} + 3 \text{ m} \boldsymbol{j}$, with velocity $\boldsymbol{v}_1$. At $t = 2$ s the particle has moved to $\boldsymbol{r}_2 = 10 \text{ m } \boldsymbol{i} - 2 \text{ m} \boldsymbol{j}$, and its velocity has changed to $\boldsymbol{v}_2 = (5\,\boldsymbol{i} - 6\boldsymbol{j})$ m/s. (*a*) Find $\boldsymbol{v}_1$. (*b*) What is the acceleration of the particle? (*c*) What is the velocity of the particle as a function of time? (*d*) What is the position vector of the particle as a function of time?

Note: For constant acceleration: $\boldsymbol{v}_{av} = \frac{1}{2}(\boldsymbol{v}_1 + \boldsymbol{v}_2)$, $\boldsymbol{a} = (\boldsymbol{v}_2 - \boldsymbol{v}_1)/\Delta t$, $\boldsymbol{v}_{av} = (\boldsymbol{r}_2 - \boldsymbol{r}_1)/\Delta t$; $\boldsymbol{v}(t) = \boldsymbol{v}_1 + \boldsymbol{a}t$; $\boldsymbol{r}(t) = \boldsymbol{r}_1 + \boldsymbol{v}_1 t + \frac{1}{2}\boldsymbol{a}t^2$.

(*a*) 1. Use the third of the above expressions — $\boldsymbol{v}_{av} = [(10 - 4) \text{ m } \boldsymbol{i} + (-2 - 3)\text{m} \boldsymbol{j}]/2$ s $= 3 \text{ m/s } \boldsymbol{i} - 2.5 \text{ m/s } \boldsymbol{j}$

2. Use the first of the above to find $\boldsymbol{v}_1$ — $\boldsymbol{v}_1 = 2\boldsymbol{v}_{av} - \boldsymbol{v}_2$; $\boldsymbol{v}_1 = (6 - 5) \text{ m/s } \boldsymbol{i} + (-5 + 6) \text{ m/s } \boldsymbol{j}$ $= (\boldsymbol{i} + \boldsymbol{j})$ m/s

(*b*) Use the second of the above relations — $\boldsymbol{a} = [(5\,\boldsymbol{i} - 6\,\boldsymbol{j} - \boldsymbol{i} - \boldsymbol{j})/2] \text{ m/s}^2 = (2\,\boldsymbol{i} - 3.5\,\boldsymbol{j}) \text{ m/s}^2$

(*c*) Use the fourth relation above — $\boldsymbol{v}(t) = [\boldsymbol{i} + \boldsymbol{j} + (2\,\boldsymbol{i} - 3.5\,\boldsymbol{j})t]$ m/s

(*d*) Use the last of the above relations — $\boldsymbol{r}(t) = [4\,\boldsymbol{i} + 3\,\boldsymbol{j} + (\boldsymbol{i} + \boldsymbol{j})t + \frac{1}{2}(2\,\boldsymbol{i} - 3.5\,\boldsymbol{j})t^2]$ m

---

106 •• A small steel ball is projected horizontally off the top landing of a long rectangular staircase (Figure 3-45). The initial speed of the ball is 3 m/s. Each step is 0.18 m high and 0.3 m wide. Which step does the ball strike first?

Note: This problem is essentially identical to Problem 75, and we can use that result. We need only determine the angle of the line joining the tops of the steps with the horizontal.

1. From Problem 75, $x = (2v^2/g)\tan\phi/\cos\phi$ — $\phi = \tan^{-1}(0.18/0.3) = 31°$; $x = 1.28$ m
2. Find number of steps with $x < 1.28$ m — number of steps with $x < 1.28$ m is 4; fifth step is hit

---

107 •• As a car travels down a highway at 25 m/s, a passenger flips out a can at a 45° angle of elevation in a plane perpendicular to the motion of the car. The initial speed of the can relative to the car is 10 m/s. The can is released at a height 1.2 m above the road. (*a*) Write the initial velocity of the can (relative to the road) in terms of unit vectors ***i***, ***j***, and ***k***. (*b*) Where does the can land?

Let $\mathbf{v}_{CA}$, denote the velocity of the can relative to the car. Assume the car is moving with velocity $\mathbf{v}_A = 25\,\mathbf{i}$ m/s. Take ***i***, ***j*** to be in the plane of the road and origin at point where can is tossed.

(*a*) 1. Write $\mathbf{v}_{CA}$ — $\mathbf{v}_{CA} = (10\sin 45°\,\mathbf{j} + 10\cos 45°\,\mathbf{k})$ m/s

2. Use (3-14) to find $\mathbf{v}_C$ — $\mathbf{v}_C = (25\,\mathbf{i} + 7.07\,\mathbf{j} + 7.07\,\mathbf{k})$ m/s

(*b*) 1. Find time of flight; use (2-14) — $1.2 + 7.07t - \tfrac{1}{2}gt^2 = 0$; $t = 1.59$ s

2. Find Δx and Δy, i.e., $\mathbf{r}$ — $\mathbf{r} = [(25 \times 1.59)\,\mathbf{i} + (7.07 \times 1.59)\,\mathbf{j}]$ m $= (39.9\,\mathbf{i} + 11.3\,\mathbf{j})$ m

---

108 •• Suppose you can throw a ball a distance x_0 when standing on level ground. How far can you throw it from a building of height $h = x_0$ if you throw it at (*a*) 0°? (*b*) 30°? (*c*) 45°?

We will solve the problem for a general angle θ first, and then substitute $\theta = 0°$, 30°, and 45°.

1. Find v_0 in terms of x_0, assuming $\theta_0 = 45°$; use Equ. 3-22 — $v_0 = \sqrt{x_0 g}$
2. Use Equ. 3-20b and solve for t — $t = \dfrac{\sqrt{x_0 g}\sin\theta + \sqrt{x_0 g\sin^2\theta + 2gx_0}}{g}$
3. Write $x = v_{0x}t = [(x_0 g)^{1/2}\cos\theta]t$; simplify — $x = x_0\cos\theta(\sin\theta + \sqrt{\sin^2\theta + 2})$
4. Substitute the values $\theta = 0°$, 30°, 45° — $x(0°) = 1.41x_0$; $x(30°) = 1.73x_0$; $x(45°) = 1.62x_0$

109*•• A baseball hit toward center field will land 72 m away unless caught first. At the moment the ball is hit, the center fielder is 98 m away. He uses 0.5 s to judge the flight of the ball, then races to catch it. The ball's speed as it leaves the bat is 35 m/s. Can the center fielder catch the ball before it hits the ground?

1. Find the angle θ when ball leaves bat from Equ. 3-22 — $\sin 2\theta = (72 \times 9.81/35^2) = 0.5766$; $2\theta = 35.2°$ or $125.2°$; (*a*) $\theta = 17.6°$ (line drive); (*b*) $\theta = 62.6°$ (high fly)

(*a*) 1. Find time of flight (line drive); use Equ. 2-12 — $t = 2(35\sin 17.6°)/9.81$ s $= 2.16$ s

2. Time to reach ball = 2.16 s − 0.5 s = 1.66 s — $v_{min} = (26/1.66)$ m/s $= 15.7$ m/s ; faster than the world record. The ball cannot be caught.

(*b*) 1. Find time of flight (high fly); use Equ. 2-12 — $t = 2(35 \sin 62.6°)/9.81$ s = 6.34 s

2. Find v_{min} for fly ball — $v_{min} = 4.45$ m/s; the ball can be caught

---

110 ••• Darlene is a stunt motorcyclist in a traveling circus. For the climax of her show, she takes off from the ramp at angle θ, clears a fiery ditch of width x, and lands on an elevated platform (height H) on the other side (Figure 3-46). Darlene notices, however, that night after night, the circus owner keeps raising the height of the platform and the flames to make the jump more spectacular. She is beginning to worry about how far this trend can be taken before she becomes a spectacular casualty, so she decides that it is time for some calculations. (*a*) For a given angle θ and distance x, what is the upper limit H_{max} such that the bike can make the jump? (*b*) For H less than H_{max}, what is the minimum takeoff speed necessary for a successful jump? (Neglect the size of the bike.)

1. Use (3-20a) and (3-20b) — $x = (v \cos \theta)t$; $y = (v \sin \theta)t - \frac{1}{2}gt^2$; $t = x/(v \cos \theta)$

2. Write y as function of x — $y(x) = x \tan \theta - \frac{1}{2}gx^2/(v^2 \cos^2 \theta)$

(*a*) Find $y_{max} = H_{max}$ — H_{max} when $v = \infty$; $H_{max} = x \tan \theta$

(*b*) Solve 2. for v, setting $y(x) = H$ — $v_{min} = \dfrac{x}{\cos \theta}\sqrt{\dfrac{g}{2(x \tan \theta - H)}}$

---

111 ••• A small boat is headed for a harbor 32 km northwest of its current position when it is suddenly engulfed in heavy fog. The captain maintains a compass bearing of northwest and a speed of 10 km/h relative to the water. Three hours later, the fog lifts and the captain notes that he is now exactly 4.0 km south of the harbor. (*a*) What was the average velocity of the current during those three hours? (*b*) In what direction should the boat have been heading to reach its destination along a straight course? (*c*) What would its travel time have been if it had followed a straight course?

We take the origin at the position of the boat when fogged in. Take x and y direction east and north, respectively. Let $\mathbf{v}$ be the velocity of the current, and θ the angle of $\mathbf{v}$ with the x (east) direction.

(*a*) 1. Write the coordinates of the boat at $t = 3$ h — $r_x = (32 \cos 135°)$ km $= [(10 \times 3) \cos 135° + 3v \cos \theta]$ km; $r_y = (32 \sin 135° - 4)$ km $= [(10 \times 3)\sin 135° + 3v \sin \theta]$ km

2. Solve for θ — $3v \cos \theta = -1.41$ km, $3v \sin \theta = -2.586$ km; $\theta = \tan^{-1}(1.83) = 61.3°$ or $241.3°$

3. Since boat has drifted south, use $\theta = 241.3°$ — $v = (-1.41/3\cos 241.3°) = 0.98$ km/h

(*b*) 1. Let ϕ be the angle between east and the proper heading; to travel northwest $v_{Bx} = -v_{By}$, $\mathbf{v}_B$ is the velocity relative to land. — $10 \cos \phi + 0.98 \cos 241.3° = 10 \sin \phi + 0.98 \sin 241.3°$; $\sin \phi + \cos \phi = 0.1325$; $\sin^2 \phi + \cos^2 \phi + 2 \sin \phi \cos \phi = 0.01756 = 1 + \sin 2\phi$

2. Solve for ϕ — $\phi = 129.6°$ or $140.4°$

3. Since current pushes south, boat must head more northerly; use $\phi = 129.6°$ from east. — Using $\phi = 129.6°$, the correct heading is 39.6 west of north.

(*c*) 1. Find v_B — $v_{Bx} = -6.84$ km/h; $v_B = v_{Bx}/\cos 45° = 9.64$ km/h

2. Find the time to travel 32 km — $t = (32/9.64)$ h = 3.32 h = 3 h 19 min

---

112 ••• Galileo showed that, if air resistance is neglected, the ranges for projectiles whose angles of projection exceed or fall short of 45° by the same amount are equal. Prove Galileo's result.

1. Use Equ. 3-22 and set $\theta = 45° \pm \phi$ — $R = (v_0^2/g)\sin(90° \pm 2\phi) = (v_0^2/g)\cos(\pm 2\phi)$
2. Note that $\cos(-\phi) = \cos(+\phi)$ — $R(45° + \phi) = R(45° - \phi)$; Q.E.D.

113* ••• Two balls are thrown with equal speeds from the top of a cliff of height H. One ball is thrown upward and an angle α above the horizontal. The other ball is thrown downward at an angle β below the horizontal. Show that each ball strikes the ground with the same speed, and find that speed in terms of H and the initial speed v_0.

1. Note that from Equ. 2-15, $v_y^2 = v_{0y}^2 + 2gH$, regardless of direction (up or down).
2. $v^2 = v_x^2 + v_y^2$, and since $v_x = v_{0x}$, $v^2 = v_{0x}^2 + v_{0y}^2 + 2gH = v_0^2 + 2gH$, for any angle, positive or negative.
3. $v = (v_0^2 + 2gH)^{½}$.

CHAPTER 4

Newton's Laws

Note: For all problems we shall take the upward direction as positive unless otherwise stated.

1* •• How can you tell if a particular reference frame is an inertial reference frame?

If Newton's first law is obeyed, the reference frame is an inertial reference frame.

---

2 •• Suppose you find that an object in a particular frame has an acceleration $\boldsymbol{a}$ when there are no forces acting on it. How can you use this information to find an inertial reference frame?

If the object has an acceleration $\boldsymbol{a}$ in the absence of external forces, its reference frame has an acceleration $-\boldsymbol{a}$. An inertial reference frame is then one that accelerates with acceleration $-\boldsymbol{a}$ relative to the reference frame in which the object has the acceleration $\boldsymbol{a}$ in the absence of external forces.

---

3 • If an object has no acceleration in an inertial reference frame, can you conclude that no forces are acting on it?

No; one can only conlcude that the *net* force is zero.

---

4 • If only a single force acts on an object, must the object accelerate in an inertial reference frame? Can it ever have zero velocity?

Yes, it must accelerate. It can have zero velocity at some instant.

5* • If an object is acted upon by a single known force, can you tell in which direction the object will move using no other information?

No; only its acceleration is known (assuming one knows the mass).

---

6 • An object is observed to be moving at constant velocity in an inertial reference frame. It follows that (*a*) no forces act on the object. (*b*) a constant force acts on the object in the direction of motion. (*c*) the net force acting on the object is zero. (*d*) the net force acting on the object is equal and opposite to its weight.

(*c*) See Problem 4-3.

---

7 • A body moves with constant speed in a straight line in an inertial reference frame. Which of the following statements must be true? (*a*) No force acts on the body. (*b*) A single constant force acts on the body in the direction of motion. (*c*) A single constant force acts on the body in the direction opposite to the motion. (*d*) A net force of zero acts on the body. (*e*) A constant net force acts on the body in the direction of motion.

(*d*) $\boldsymbol{a} = 0$, therefore $\boldsymbol{F}_{net} = 0$.

---

8 •• Figure 4-23 shows the position x versus time t of a particle moving in one dimension. During what time intervals is there a net force acting on the particle? Give the direction (+ or −) of the net force during these time intervals.

Between $t = 0$ and $t = 2$ s, and between $t = 8$ s and $t = 10$ s, the velocity is constant and $\boldsymbol{a} = 0$ and $\boldsymbol{F}_{net} = 0$. Between $t = 2$ s and $t = 5$ s, $\boldsymbol{a}$ is negative and $\boldsymbol{F}_{net}$ is negative; between $t = 5$ s and $t = 8$ s, $\boldsymbol{a}$ is positive and so is $\boldsymbol{F}_{net}$.

9* • A particle of mass m is traveling at an initial speed $v_0 = 25.0$ m/s. It is brought to rest in a distance of 62.5 m when a net force of 15.0 N acts on it. What is m? (*a*) 37.5 kg (*b*) 3.00 kg (*c*) 1.50 kg (*d*) 6.00 kg (*e*) 3.75 kg

$a = v^2/2s = F/m$; $m = v^2/2Fs$ — $m = 3.00$ kg; (*b*)

---

10 • (*a*) An object experiences an acceleration of 3 m/s² when a certain force F_0 acts on it. What is its acceleration when the force is doubled? (*b*) A second object experiences an acceleration of 9 m/s² under the influence of the force F_0. What is the ratio of the masses of the two objects? (*c*) If the two objects are tied together, what acceleration will the force F_0 produce?

(*a*) Use $\boldsymbol{F} = m\boldsymbol{a}$ — $a = 2 \times (3 \text{ m/s}^2) = 6 \text{ m/s}^2$

(*b*) $m_2a_2 = m_1a_1 = F_0$; $m_2/m_1 = a_1/a_2$ — $m_2/m_1 = 3/9 = 1/3$

(*c*) $m_1 + m_2 = (4/3)m_1$; $a = (3/4)a_1$ — $a = (3/4)(3 \text{ m/s}^2) = 2.25 \text{ m/s}^2$

---

11 • A tugboat tows a ship with a constant force F_1. The increase in the ship's speed in a 10-s interval is 4 km/h. When a second tugboat applies a second constant force F_2 in the same direction, the speed increases by 16 km/h in a 10-s interval. How do the magnitudes of the two forces compare? (Neglect water resistance.)

$F_1 = ma_1$; $F_1 + F_2 = 4ma_1 = 4F_1$. Consequently, $F_2 = 3F_1$.

---

12 • A force F_0 causes an acceleration of 3 m/s² when it acts on an object of mass m sliding on a frictionless surface. Find the acceleration of the same object in the circumstances shown in Figure 4-24*a* and *b*.

(*a*) $F = (F_x^2 + F_y^2)^{1/2} = \sqrt{2}F_0$ — $a = \sqrt{2}(3 \text{ m/s}^2) = 4.24 \text{ m/s}^2$

(*b*) $F_x = 2F_0 + F_0 \cos 45°$; $F_y = F_0 \sin 45°$ — $F_x = 2.707F_0$; $F_y = 0.707F_0$; $F = 2.8F_0$; $a = 8.39 \text{ m/s}^2$

13* • A force $\boldsymbol{F} = 6 \text{ N}\,\boldsymbol{i} - 3 \text{ N}\,\boldsymbol{j}$ acts on an object of mass 1.5 kg. Find the acceleration $\boldsymbol{a}$. What is the magnitude a?

$\boldsymbol{a} = \boldsymbol{F}/m = (4\boldsymbol{i} - 2\boldsymbol{j}) \text{ m/s}^2$; $a = (16 + 4)^{1/2} \text{ m/s}^2 = 4.46 \text{ m/s}^2$.

---

14 • A single force of 12 N acts on a particle of mass m. The particle starts from rest and travels in a straight line a distance of 18 m in 6 s. Find m.

$a = 2s/t^2 = F/m$; $m = Ft^2/2s$ — $m = [(12 \times 36)/(2 \times 18)]$ kg $= 12$ kg

---

15 • To drag a 75-kg log along the ground at constant velocity, you have to pull on it with a horizontal force of 250 N. (*a*) What is the resistive force exerted by the ground? (*b*) What force must you exert if you want to give the log an acceleration of 2 m/s²?

(*a*) Since $a = 0$, $F_{net} = 0$; F_{res}, resistive force, = 250 N.

(*b*) $F_{net} = (75 \text{ kg})(2 \text{ m/s}^2) = 150 \text{ N} = F_{appl} - F_{res}$; $F_{appl} = (250 + 150) \text{ N} = 400 \text{ N}$.

---

16 • Figure 4-25 shows a plot of v_x versus t for an object of mass 8 kg moving in a straight line. Make a plot of the net force acting on the object as a function of time.

Between $t = 0$ and $t = 3$ s, $a = 1 \text{ m/s}^2$ and $F = 8$ N;

between $t = 5$ s and 6.5 s, $a = -2.75$ m/s^2 and $F = -22$ N;
between $t = 7.5$ s and 8.5 s, $a = 2$ m/s^2 and $F = 16$ N;
between $t = 3$ s and 5 s, F decreases from 8 N to -22 N;
between $t = 6.5$ s and 7.5 s, F increases from -22 N to 16 N.

A sketch of F versus t is shown.

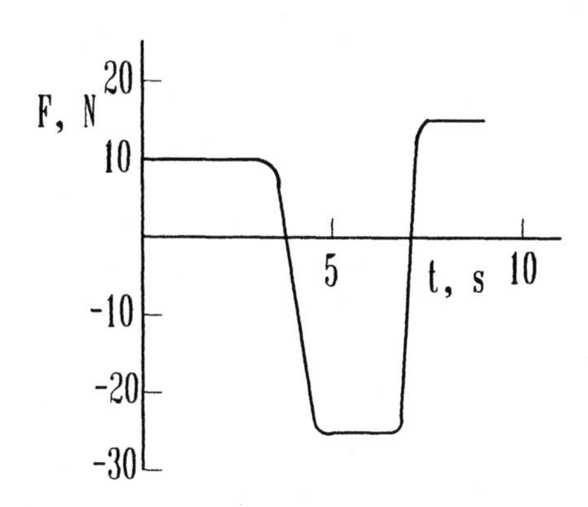

17* • A 4-kg object is subjected to two forces, $\boldsymbol{F}_1 = 2\text{ N}\,\boldsymbol{i} - 3\text{ N}\,\boldsymbol{j}$ and $\boldsymbol{F}_2 = 4\text{ n}\,\boldsymbol{i} - 11\text{ N}\,\boldsymbol{j}$. The object is at rest at the origin at time $t = 0$. (*a*) What is the object's acceleration? (*b*) What is its velocity at time $t = 3$ s? (*c*) Where is the object at time $t = 3$ s?

(*a*) Find the net force, $\boldsymbol{F} = \boldsymbol{F}_1 + \boldsymbol{F}_2$ — $\boldsymbol{F} = (2\boldsymbol{i} + 4\boldsymbol{i})\text{ N} - (3\boldsymbol{j} + 11\boldsymbol{j})\text{ N} = (6\boldsymbol{i} - 14\boldsymbol{j})\text{ N}$

$\boldsymbol{a} = \boldsymbol{F}/m$ — $\boldsymbol{a} = (1.5\boldsymbol{i} - 3.5\boldsymbol{j})\text{ m/s}^2$

(*b*) $\boldsymbol{v} = \boldsymbol{a}t$ — $\boldsymbol{v} = (4.5\boldsymbol{i} - 10.5\boldsymbol{j})\text{ m/s}$

(*c*) $\boldsymbol{r} = \boldsymbol{v}_{av}t$; $\boldsymbol{v}_{av} = \frac{1}{2}\boldsymbol{v}(3\text{ s})$ — $\boldsymbol{r} = (6.75\boldsymbol{i} - 15.75\boldsymbol{j})\text{ m}$

---

18 • Suppose an object was sent far out in space, away from galaxies, stars, or other bodies. How would its mass change? Its weight?

Its mass does not change. Its weight would be zero.

---

19 • How would an astronaut in apparent weightlessness be aware of her mass?

To accelerate, she must exert a force proportional to her mass.

---

20 • Under what circumstances would your apparent weight be greater than your true weight?

When in a reference frame that is accelerating upward.

21* • On the moon, the acceleration due to gravity is only about 1/6 of that on earth. An astronaut whose weight on earth is 600 N travels to the lunar surface. His mass as measured on the moon will be (*a*) 600 kg. (*b*) 100 kg. (*c*) 61.2 kg. (*d*) 9.81 kg. (*e*) 360 kg.

(*c*) The mass is (600 N)/(9.81 m/s^2) = 61.2 kg, and remains the same on the moon.

---

22 • Find the weight of a 54-kg girl in (*a*) newtons and (*b*) pounds.

(*a*) $w = mg$ — $w = 54 \times 9.81\text{ N} = 530\text{ N} = (530/4.45)\text{ lb} = 119\text{ lb}$

---

23 • Find the mass of a 165-lb man in kilograms.

165 lb = 165 × 4.45 N = 734 N; $m = w/g$ = (734 N)/(9.81 m/s^2) = 74.8 kg.

---

24 • After watching a space documentary, Lou speculates that there is money to be made by combining the phenomenon of weightlessness in space with the widespread longing for weight loss in the general population. Researching the matter, he learns that the gravitational force on a mass m at a height h above the earth's surface is given by $F = mgR_E^2/(R_E + h)^2$, where R_E is the radius of the earth (about 6370 km) and g is the acceleration due to gravity at the earth's surface. (*a*) Using this expression, find the weight in newtons and pounds of an 83-kg person at the earth's surface. (*b*) If this person were weight-conscious and rich, and Lou managed to sell the person a trip to a height of 400 km above the earth's surface, how much weight would the person lose? (*c*) What is the person's mass at this altitude?

(*a*) $w = mg$ $w = (83 \times 9.81)$ N = 814 N = (814/4.45) lb = 183 lb

(*b*) Find change in acceleration of gravity, Δg $\Delta g = g_E[1 - R_E^2/(R_E + h)^2] = g_E(1 - 6370^2/6770^2)$ $= 0.115g_E$

Weight loss is 11.5% $\Delta w = (814 \times 0.115)$ N = 93 N = 21 lb

(*c*) Mass is unchanged $m = 83$ kg

25* •• Caught without a map again, Hayley lands her spacecraft on an unknown planet. Visibility is poor, but she finds someone on a local communications channel and asks for directions to Earth. "You are already on Earth," is the reply, "Wait there and I'll be right over." Hayley is suspicious, however, so she drops a lead ball of mass 76.5 g from the top of her ship, 18 m above the surface of the planet. It takes 2.5 s to reach the ground. (*a*) If Hayley's mass is 68.5 kg, what is her weight on this planet? (*b*) Is she on Earth?

(*a*) Use $s = \frac{1}{2}at^2$ to find accel. of gravity, g' $g' = (2 \times 18/2.5^2)\ \text{m/s}^2 = 5.76\ \text{m/s}^2$

$w = mg'$ $w = (68.5 \times 5.76)$ N = 395 N

(*b*) Evidently, she is not on Earth.

---

26 • True or false: (*a*) Action-reaction forces never act on the same object. (*b*) Action equals reaction only if the objects are not accelerating.

(*a*) True (*b*) False

---

27 • An 80-kg man on ice skates pushes a 40-kg boy also on skates with a force of 100 N. The force exerted by the boy on the man is (*a*) 200 N. (*b*) 100 N. (*c*) 50 N. (*d*) 40 N.

(*b*) 100 N; the reaction force = action force, in magnitude.

---

28 • A boy holds a bird in his hand. The reaction force to the force exerted on the bird by the boy's hand is (*a*) the force of the earth on the bird. (*b*) the force of the bird on the earth. (*c*) the force of the hand on the bird. (*d*) the force of the bird on the hand. (*e*) the force of the earth on the hand.

(*d*) the force exerted by the bird on the hand.

The reaction force to the weight of the bird is (*a*) the force of the earth on the bird. (*b*) the force of the bird on the earth. (*c*) the force of the hand on the bird. (*d*) the force of the bird on the hand. (*e*) the force of the earth on the hand.

(*b*) the force exerted by the bird on the earth.

29* • A baseball player hits a ball with a bat. If the force with which the bat hits the ball is considered the action force, what is the reaction force? (*a*) The force the bat exerts on the batter's hands. (*b*) The force on the ball exerted by the glove of the person who catches it. (*c*) The force the ball exerts on the bat. (*d*) The force the pitcher exerts on the ball while throwing it. (*e*) Friction, as the ball rolls to a stop.

(*c*)

---

30 •• Dean reads in his physics book that when two people pull on the end of a rope in a tug-of-war, the forces exerted by each on the other are equal and opposite, according to Newton's third law. Misunderstanding the law tragically, Dean runs out to challenge Hugo the Large, convinced that the laws of physics guarantee a tie. Hugo lumbers over, picks up the rope, pulls Dean off his feet, and then drags him through a puddle, across the road, and up the steps of the physics building. Use a force diagram to show Dean that, in spite of Newton's third law, it is possible for one side to win a tug-of-war.

In the figure, (1) is Dean. The two persons pull directly on each otherss hands with forces F_{12} and F_{21} of equal magnitude (action-reaction). However, the forces that their feet exert on the ground, and the corresponding reaction forces of the ground on their feet (friction force) need not be the same. So (1), Dean, will be pulled to the right.

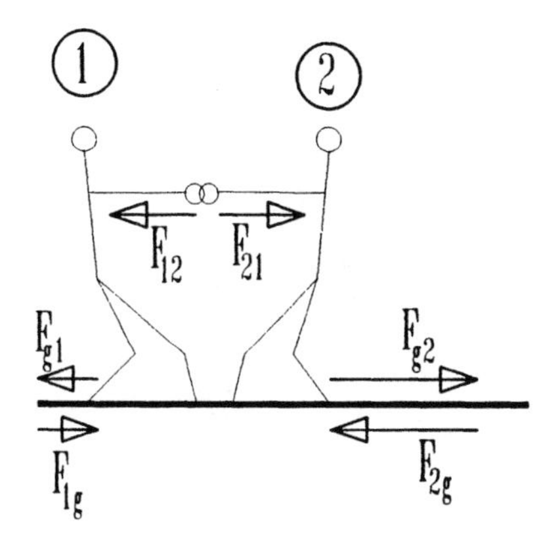

31 • A 2.5-kg object hangs at rest from a string attached to the ceiling. (*a*) Draw a diagram showing all forces acting on the object and indicate each reaction force. (*b*) Do the same for each force acting on the string.

(*a*) The forces acting on the 2.5-kg mass are its weight, *W*, and the tension T_1 in the string. The reaction forces are *W*' and T_1' as shown.

(*b*) The forces acting on the string are T_1, the force that the mass exerts on the string, which is the same as T_1' of part (*a*), and T_2, the force that the ceiling exerts on the string. The reaction forces are shown as T_1' and T_2'.

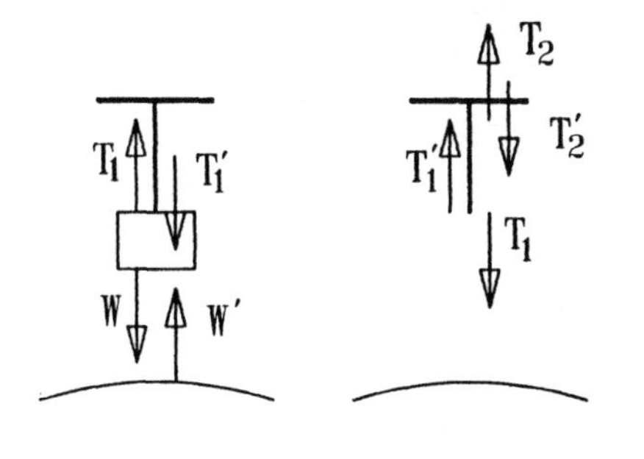

32 • A 9-kg box rests on a 12-kg box that rests on a horizontal table. (*a*) Draw a diagram showing all forces acting on the 9-kg box and indicate each reaction force. (*b*) Do the same for all forces acting on the 12-kg box.

(*a*) The forces acting on the 9-kg box, m_1, are its weight, W_1, and the normal reaction force of the 12-kg box, m_2, on m_1, F_{n1}. The reaction forces are F_{n1}' and W_1'. (*b*)The forces acting on m_2 are F_{n1}' (equal to W_1), its weight, W_2, and the normal reation force of the table on m_2, F_{n2}. The reaction forces are F_{n1}, F_{n2}', and W_2'.

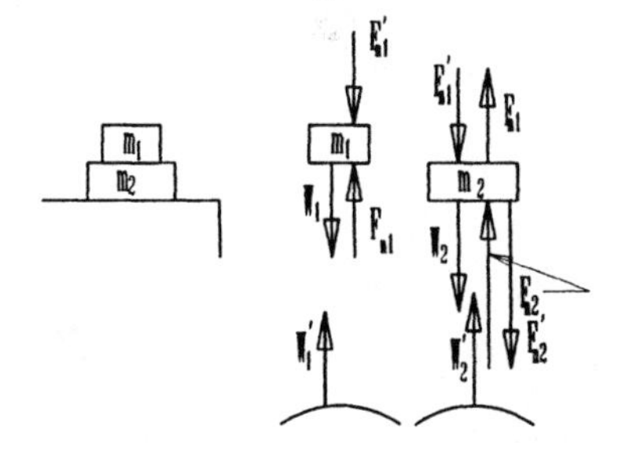

33* • A vertical spring of force constant 600 N/m has one end attached to the ceiling and the other to a 12-kg block resting on a horizontal surface so that the spring exerts an upward force on the block. The spring is stretched by 10 cm. (*a*) What force does the spring exert on the block? (*b*) What is the force that the surface exerts on the block?

Draw a free body diagram and show the forces acting on the 12-kg block. $\boldsymbol{F}_k$ is the force exerted by the spring; $\boldsymbol{W} = m\boldsymbol{g}$ is the weight of the block; $\boldsymbol{F}_n$ is the normal force exerted by the horizontal surface.

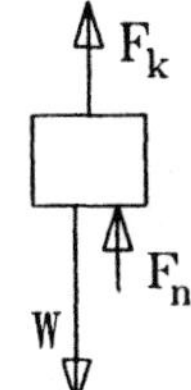

(*a*) $F_k = -kx$, where x is the extension of the spring $F_k = (600 \times 0.1)$ N = 60 N

(*b*) $\Sigma F = 0$; solve for F_n $F_n = (12 \times 9.81 - 60)$ N = 57.7 N

---

34 • A 6-kg box on a frictionless horizontal surface is attached to a horizontal spring with a force constant of 800 N/m. If the spring is stretched 4 cm from its equilibrium length, what is the acceleration of the box?

$a = F/m$; $F = F_k = -kx$ $a = (800 \times 0.04/6.00)\ \text{m/s}^2 = 5.33\ \text{m/s}^2$

---

35 •• The acceleration a versus spring length L observed when a 0.5-kg mass is pulled along a frictionless table by a single spring is shown in the following table:

L, cm	4	5	6	7	8	9	10	11	12	13	14
a, m/s^2	0	2.0	3.8	5.6	7.4	9.2	11.2	12.8	14.0	14.6	14.6

(*a*) Make a plot of the force exerted by the spring versus length L. (*b*) If the spring is extended to 12.5 cm, what force does it exert? (*c*) How much does the spring extend when the mass is suspended from it at rest near sea level, where $g = 9.81$ N/kg?

(*a*) $F = ma$; since $m = 0.5$ kg, the values of F are ½a. The plot of F versus L is shown.

(*b*) From the graph one finds that for $L = 12.5$ cm, $F = 7.15$ N

(*c*) $F = mg = 4.905$ N; for $F = 4.905$ N, $L = 9.3$ cm. Extension, $x = L - L_0 = (9.3 - 4.0)$ cm = 5.3 cm.

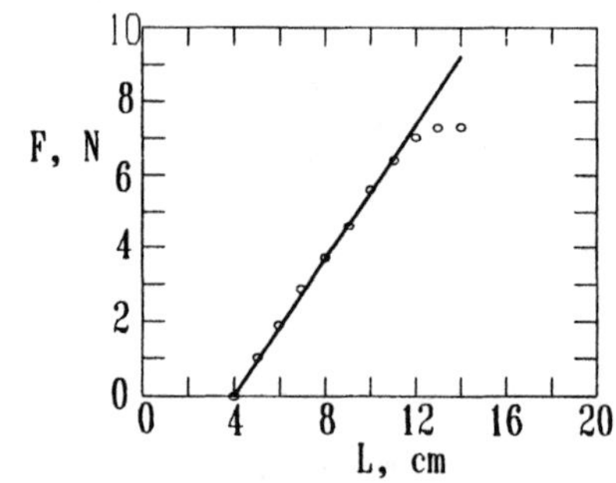

---

36 • A picture is supported by two wires as in Example 4-9. Do you expect the tension in the wire that is more nearly vertical to be greater than or less than the tension in the other wire?

The tension in the more nearly vertical wire is greater. (Consider a picture supported by one wire, and then pulled slightly to one side by a second wire.)

37* • A clothesline is stretched taut between two poles. Then a wet towel is hung at the center of the line. Can the line remain horizontal? Explain.

No; the tension in the line must have a vertical component to support the weight of the towel.

---

38 • Which of the free-body diagrams in Figure 4-26 represents a block sliding down a frictionless inclined surface?

(*c*) Two forces act on the block, its weight, acting vertically downward, and the normal reaction force of the surface on the block. The magnitude of the normal force is less than that of the weight, since it supports only a portion of the weight.

---

39 • A lamp with a mass $m = 42.6$ kg is hanging from wires as shown in Figure 4-27. The tension T_1 in the vertical handle is (*a*) 209 N. (*b*) 418 N. (*c*) 570 N. (*d*) 360 N. (*e*) 730 N.

(*b*) T_1 supports the full weight $mg = 418$ N.

---

40 • A 40.0-kg object supported by a vertical rope is initially at rest. The object is then accelerated upward. The tension in the rope needed to give the object an upward speed of 3.50 m/s in 0.700 s is (*a*) 590 N. (*b*) 390 N. (*c*) 200 N. (*d*) 980 N. (*e*) 720 N.

(*a*) $a = (3.5/0.7)$ m/s^2 = 5 m/s^2. $F_{net} = (40 \times 5)$ N = 200 N = $T - (40 \times 9.81)$ N; T = 592 N.

41* • A hovering helicopter of mass m_h is lowering a truck of mass m_t. If the truck's downward speed is increasing at the rate $0.1g$, what is the tension in the supporting cable? (*a*) $1.1m_tg$ (*b*) m_tg (*c*) $0.9m_tg$ (*d*) $1.1(m_h + m_t)g$ (*e*) $0.9(m_h + m_t)g$

(*c*) similar to Problem 4-40.

---

42 • A 10-kg object on a frictionless table is subjected to two horizontal forces, $\boldsymbol{F}_1$ and $\boldsymbol{F}_2$, with magnitudes F_1 = 20 N and F_2 = 30 N, as shown in Figure 4-28. (*a*) Find the acceleration $\boldsymbol{a}$ of the object. (*b*) A third force $\boldsymbol{F}_3$ is applied so that the object is in static equilibrium. Find $\boldsymbol{F}_3$.

(*a*) Find the components of $\boldsymbol{F}_1$ and $\boldsymbol{F}_2$; add components to find $\boldsymbol{F}_{tot}$ — $\boldsymbol{F}_1 = 20\boldsymbol{j}$ N; $\boldsymbol{F}_2 = (30\cos 30°\,\boldsymbol{i} - 30\sin 30°\,\boldsymbol{j})$ N = $(26\boldsymbol{i} - 15\boldsymbol{j})$ N; $\boldsymbol{F}_{tot} = (26\boldsymbol{i} + 5\boldsymbol{j})$ N

(*b*) $\boldsymbol{a} = \boldsymbol{F}_{tot}/m$ — $\boldsymbol{a} = (2.6\,\boldsymbol{i} + 0.5\,\boldsymbol{j})$ m/s^2

(*c*) $\Sigma F = 0 = \boldsymbol{F}_3 + \boldsymbol{F}_{tot}$ — $\boldsymbol{F}_3 = -\boldsymbol{F}_{tot} = -(26\boldsymbol{i} + 5\boldsymbol{j})$ N

---

43 • A vertical force $\boldsymbol{T}$ is exerted on a 5-kg body near the surface of the earth, as shown in Figure 4-29. Find the acceleration of the body if (*a*) T = 5 N, (*b*) T = 10 N, and (*c*) T = 100 N.

(*a*), (*b*), and (*c*) $a = (T - mg)/m = T/m - g$, directed up. — (*a*) -8.81 m/s^2; (*b*) -7.81 m/s^2; (*c*) 10.19 m/s^2.

---

44 •• To compensate for a distinct lack of personality, Herbert relies on the Grand Entrance technique when he attends parties. His latest plan for appearing at a pool party is to arrive by helicopter and then slide down a nylon rope as the helicopter hovers above poolside. However, as the helicopter approaches its destination, the pilot tells Herbert that the rope will break if the tension exceeds 300 N. Herbert, whose mass is 61.2 kg, realizes that the rope won't hold him unless he slides down with an appropriate acceleration. What must his acceleration be if the rope is not to break and ruin the whole effect?

$ma = T - mg$; $a = T/m - g$ — $a = -4.91$ m/s^2, i.e., 4.91 m/s^2 directed downward

---

45* •• A student has to escape from his girlfriend's dormitory through a window that is 15.0 m above the ground. He has a 24-m rope, but it will break when the tension exceeds 360 N, and the student weighs 600 N. The student will be injured if he hits the ground with a speed greater than 8 m/s. (*a*) Show that he cannot safely slide down the rope. (*b*) Find a strategy using the rope that will permit the student to reach the ground safely.

(*a*) $a = T/m - g$; $v = (2as)^{1/2}$ — $a = -0.4g$; $v = -(2 \times 0.4 \times 9.81 \times 15)^{1/2}$ m/s = -10.8 m/s

(*b*) Double the rope and drop last 3 m — Now $v = -(2 \times 9.81 \times 3)^{1/2} = -7.7$ m/s

---

46 •• A rifle bullet of mass 9 g starts from rest and exits from the 0.6-m barrel at 1200 m/s. Find the force exerted on the bullet, assuming it to be constant, while the bullet is in the barrel.

$a = v^2/2s$; $F = ma = mv^2/2s$ — $F = [(1200^2 \times 0.009)/(2 \times 0.6)]$ N = 1.08×10^4 N

---

47 •• A 2-kg picture is hung by two wires of equal length. Each makes an angle of θ with the horizontal, as shown in Figure 4-30. (*a*) Find the general equation for the tension T, given θ and weight w for the picture. For what angle θ is T the least? The greatest? (*b*) If $\theta = 30°$, what is the tension in the wires?

(a) $w = 2T \sin \theta$; $T = w/(2 \sin \theta)$. T least for $\theta = 90°$; T greatest as $\theta \to 0°$.

(b) Use result of (a); $T = (2 \times 9.81)/(2 \times \sin 30°) = 19.6$ N.

---

48 •• A bullet of mass 1.8×10^{-3} kg moving at 500 m/s impacts with a large fixed block of wood and travels 6 cm before coming to rest. Assuming that the deceleration of the bullet is constant, find the force exerted by the wood on the bullet.

Proceed as in Problem 4-46; $F = mv^2/2s$ — $F = [(1.8 \times 10^{-3})(500)^2/(2 \times 0.06)]$ N $= 3.75 \times 10^3$ N

49* •• A 1000-kg load is being moved by a crane. Find the tension in the cable that supports the load as (a) it is accelerated upward at 2 m/s^2, (b) it is lifted at constant speed, and (c) it moves upward with speed decreasing by 2 m/s each second.

(a), (b), and (c) $T = m(a - g)$; $g = -9.81$ m/s^2 — (a) $T = 11810$ N; (b) $T = 9810$ N; (c) $T = 7810$ N

---

50 •• A horse-drawn coach is decelerating at 3.0 m/s^2 while moving in a straight line. A lamp of mass 0.844 kg is hanging from the ceiling of the coach on a string 0.6 m long. The angle that the string makes with the vertical is (a) 8.5° toward the front of the coach. (b) 17° toward the front of the coach. (c) 17° toward the back of the coach. (d) 2.5° toward the front of the coach. (e) 0° or straight down.

$\theta = \tan^{-1}(a/g)$; $\theta = 17°$ toward front of coach. (b) is correct.

---

51 •• For the systems in equilibrium in Figure 4-31, find the unknown tensions and masses.

In each case, consider the junction of the three strings.

(a) Set $\Sigma F_x = 0$, $\Sigma F_y = 0$ — 30 N $= T_1 \cos 60°$; $T_1 \sin 60° = T_2$

Solve for T_1 and T_2 — $T_1 = 60$ N, $T_2 = 52$ N

$M = T_2/g$ — $M = 5.3$ kg

(b) Proceed as in part (a) — $T_1 \sin 60° = 80 \cos 60°$ N; $T_1 \cos 60° + T_2 = 80 \sin 60°$ N

$T_1 = (80 \text{ N})/\tan 60° = 46.2$ N; $T_2 = 46.2$ N

$M = (46.2/9.81)$ kg $= 4.71$ kg

(c) Proceed as in part (a); note that $Mg = T_1$ — $T_1 \sin 30° = T_3 \sin 30°$; $T_1 = T_3$;

$2T_1 \cos 30° = T_2 = 6 \times 9.81$ N

$T_1 = T_3 = 34$ N; $M = (34/9.81)$ kg $= 3.46$ kg

---

52 •• Your car is stuck in a mudhole. You are alone, but you have a long, strong rope. Having studied physics, you tie the rope tautly to a telephone pole and pull on it sideways, as shown in Figure 4-32. (a) Find the force exerted by the rope on the car when the angle θ is 3° and you are pulling with a force of 400 N but the car does not move. (b) How strong must the rope be if it takes a force of 600 N to move the car when $\theta = 4°$?

(a) $\Sigma F_y = 0$; solve for T — $2T \sin 3° = 400$ N; $T = 3.82$ kN

(b) Proceed as in (a) — $T = (600/2 \sin 4°)$ N $= 4.30$ kN

53* • A box slides down a frictionless inclined plane. Draw a diagram showing the forces acting on the box. For each force in your diagram, indicate the reaction force.

The forces acting on the box are its weight, $\boldsymbol{W}$, and the normal reaction force of the inclined plane on the box, $\boldsymbol{F}_n$. The reaction forces are indicated with primes.

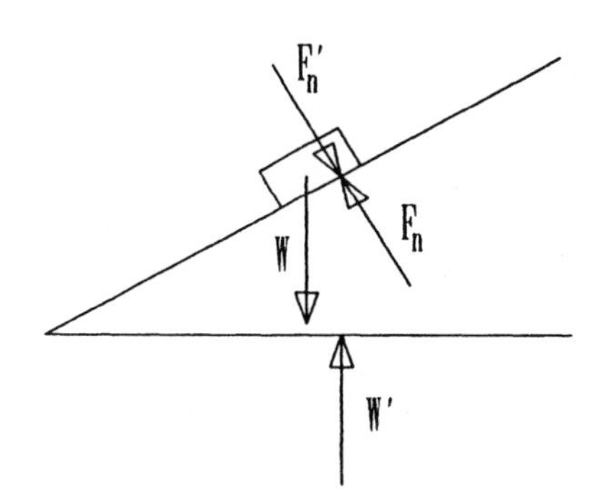

---

54 • The system shown in Figure 4-33 is in equilibrium. It follows that the mass m is (*a*) 3.5 kg. (*b*) 3.5 sin 40° kg. (*c*) 3.5 tan 40° kg. (*d*) none of the above.

(*d*) M must be greater than 3.5 kg.

---

55 • In Figure 4-34, the objects are attached to spring balances calibrated in newtons. Give the readings of the balances in each case, assuming that the strings are massless and the incline is frictionless.

(*a*) and (*b*) $F = 98.1$ N. (*c*) F(per spring balance) = 49 N. (*d*) $F = (98.1 \sin 30°)$ N = 49 N.

---

56 •• A box is held in position by a cable along a frictionless incline (Figure 4-35). (*a*) If $\theta = 60°$ and $m = 50$ kg, find the tension in the cable and the normal force exerted by the incline. (*b*) Find the tension as a function of θ and m, and check your result for $\theta = 0°$ and $\theta = 90°$.

We shall use a coordinate system with x pointing to the right and parallel to the inclined plane, y along $\boldsymbol{F}_n$.

(*b*) $\Sigma F_x = 0$; $\Sigma F_y = 0$ — $T = mg \sin \theta$, $F_n = mg \cos \theta$. For $\theta = 0°$, $T = 0$, $F_n = mg$, and for $\theta = 90°$, $T = mg$ and $F_n = 0$, as should be

(*a*) Use results of (*b*) with $m = 50$ kg, $\theta = 60°$ — $T = 508$ N, $F_n = 245$ N

57* •• A horizontal force of 100 N pushes a 12-kg block up a frictionless incline that makes an angle of 25° with the horizontal. (*a*) What is the normal force that the incline exerts on the block? (*b*) What is the acceleration of the block?

Draw a free-body diagram for the box. Let x point to the right and up along the plane with y in the direction of $\boldsymbol{F}_n$.

(*a*) Write $\Sigma F_y = 0$ and solve for F_n — $F_n - mg \cos 25° - (100 \text{ N})\sin 25° = 0$; $F_n = 149$ N

(*b*) Write $\Sigma F_x = ma$ and solve for a — $(100 \text{ N})\cos 25° - mg \sin 25° = ma$; $a = 3.41 \text{ m/s}^2$

---

58 •• A 65-kg boy weighs himself by standing on a scale mounted on a skateboard that is rolling down an incline, as shown in Figure 4-36. Assume there is no friction so that the force exerted by the incline on the skateboard is perpendicular to the incline. What is the reading on the scale if $\theta = 30°$?

The force on the scale is the normal reaction force; $F_n = 65 \times 9.81 \times \cos 30°$ N = 552 N.

59 • An object is suspended from the ceiling of an elevator that is descending at a constant speed of 9.81 m/s. The tension in the string holding the object is (*a*) equal to the weight of the object. (*b*) less than the weight of the object but not zero. (*c*) greater than the weight of the object. (*d*) zero.

(*a*) The acceleration is zero, so $T = mg$.

---

60 • What effect does the velocity of an elevator have on the apparent weight of a person in the elevator?

None.

61* • Suppose you are standing on a scale in a descending elevator as it comes to a stop on the ground floor. Will the scale's report of your weight be high, low, or correct?

It will be high because the acceleration is upward.

---

62 • A person of weight w is in an elevator going up when the cable suddenly breaks. What is the person's apparent weight immediately after the cable breaks? (*a*) w (*b*) Greater than w (*c*) Less than w (*d*) $9.8w$ (*e*) Zero

(*e*) zero; the floor of the elevator exerts no force on the person.

---

63 • A person in an elevator is holding a 10-kg block by a cord rated to withstand a tension of 150 N. When the elevator starts up, the cord breaks. What was the minimum acceleration of the elevator?

$150 \text{ N} = (10 \text{ kg})(a - g) = (10 \text{ kg})(a + 9.81 \text{ m/s}^2)$; $a = 5.19 \text{ m/s}^2$.

---

64 • A 60-kg girl weighs herself by standing on a scale in an elevator. What does the scale read when (*a*) the elevator is descending at a constant rate of 10 m/s; (*b*) the elevator is descending at 10 m/s and gaining speed at a rate of 2 m/s^2; (*c*) the elevator is ascending at 10 m/s but its speed is decreasing by 2 m/s each second?

(*a*) $a = 0$; $w = mg = 589$ N. (*b*) $a = -2 \text{ m/s}^2$; $w = (60 \times 7.81) \text{ N} = 469$ N. (*c*) Again, $a = -2 \text{ m/s}^2$ and $w = 469$ N.

65* •• A 2-kg block hangs from a spring balance calibrated in newtons that is attached to the ceiling of an elevator (Figure 4-37). What does the balance read when (*a*) the elevator is moving up with a constant velocity of 30 m/s; (*b*) the elevator is moving down with a constant velocity of 30 m/s; (*c*) the elevator is ascending at 20 m/s and gaining speed at a rate of 10 m/s^2? From $t = 0$ to $t = 2$ s, the elevator moves up at 10 m/s. Its velocity is then reduced uniformly to zero in the next 2 s, so that it is at rest at $t = 4$ s. Describe the reading of the balance during the interval $0 < t < 4$ s.

(*a*) $a = 0$; $F = mg = 19.6$ N. (*b*) $F = 19.6$ N. (*c*) $a = 10 \text{ m/s}^2$; $F = (2 \text{ kg})[(10 + 9.81) \text{ m/s}^2] = 39.6$ N.

For $0 \le t \le 2$ s, $F = 19.6$ N; for $2 \text{ s} \le t \le 4$ s, $a = -5 \text{ m/s}^2$, so $F = (2 \text{ kg})(4.81 \text{ m/s}^2) = 9.62$ N.

---

66 •• A man stands on a scale in an elevator that has an upward acceleration a. The scale reads 960 N. When he picks up a 20-kg box, the scale reads 1200 N. Find the mass of the man, his weight, and the acceleration a.

1. A mass of 20 kg has an apparent weight of 240 N $240/20 = 9.81 \text{ m/s}^2 + a$; $a = 2.19 \text{ m/s}^2$
2. Finds the mass and weight of the man $m = (960/12.0) \text{ kg} = 80$ kg; $w = (80 \times 9.81) \text{ N} = 785$ N

---

67 • Two boxes of mass m_1 and m_2 connected together by a massless string are accelerated uniformly on a frictionless surface, as shown in Figure 4-38. The ratio of the tensions T_1/T_2 is given by (*a*) m_1/m_2. (*b*) m_2/m_1. (*c*) $(m_1+m_2)/m_2$. (*d*) $m_1/(m_1 + m_2)$ (*e*) $m_2/(m_1 + m_2)$

$T_2 = (m_1 + m_2)a$; $T_1 = m_1 a$; $T_1/T_2 = m_1/(m_1 + m_2)$. (*d*) is correct.

---

68 • A box of mass $m_2 = 3.5$ kg rests on a frictionless horizontal shelf and is attached by strings to boxes of masses $m_1 = 1.5$ kg and $m_3 = 2.5$ kg, which hang freely, as shown in Figure 4-39. Both pulleys are frictionless and massless. The system is initially held at rest. After it is released, find (*a*) the acceleration of each of the boxes and (*b*) the tension in each string.

Evidently, m_1 will accelerate up, m_2 to the right, and m_3 down, with the same acceleration a.

1. Draw free-body diagrams for each box

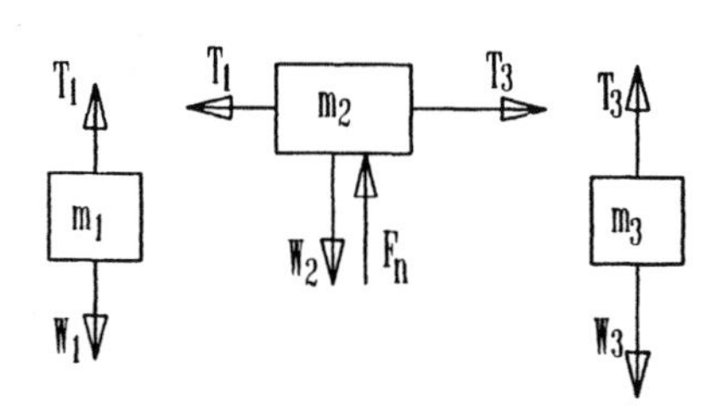

(*a*) Write $\Sigma F = m_i a$ for each box — $T_1 - 1.5g = 1.5a$; $T_3 - T_1 = 3.5a$; $2.5g - T_3 = 2.5a$

Add the three equations to find a — $a = (9.81/7.5)\ \text{m/s}^2 = 1.31\ \text{m/s}^2$

(*b*) Use the three equations to find the tensions — $T_1 = (1.5 \times 11.12)\ \text{N} = 16.7\ \text{N}$, $T_3 = (2.5 \times 8.5)\ \text{N} = 21.3\ \text{N}$

69* •• Two blocks are in contact on a frictionless, horizontal surface. The blocks are accelerated by a horizontal force $\boldsymbol{F}$ applied to one of them (Figure 4-40). Find the acceleration and the contact force for (*a*) general values of F, m_1, and m_2, and (*b*) for $F = 3.2$ N, $m_1 = 2$ kg, and $m_2 = 6$ kg.

(*a*) $a = F/(m_1 + m_2)$ and contact force $F_c = m_2 a$. So $F_c = Fm_2/(m_1 + m_2)$.

(*b*) Substitute numerical values in above expressions. $a = (3.2/8)\ \text{m/s}^2 = 0.4\ \text{m/s}^2$; $F_c = (3.2 \times 6/8)\ \text{N} = 2.4\ \text{N}$.

---

70 • Repeat the previous problem, but with the two blocks interchanged.

(*a*) Interchange subscripts 1 and 2. $a = F/(m_1 + m_2)$; $F_c = F[m_1/(m_1 + m_2)]$.

(*b*) $a = 0.4\ \text{m/s}^2$; $F_c = (3.2 \times 2/8)\ \text{N} = 0.8\ \text{N}$.

---

71 •• Two 100-kg boxes are dragged along a frictionless surface with a constant acceleration of 1.6 m/s², as shown in Figure 4-41. Each rope has a mass of 1 kg. Find the force F and the tension in the ropes at points A, B, and C.

1. Total mass accelerated is 202 kg. $F = ma$ — $F = 323.2$ N
2. Apply $F = ma$ at points A, B, and C — $F_A = 160$ N; $F_B = 161.2$ N; $F_C = 321.6$ N

---

72 •• Two objects are connected by a massless string, as shown in Figure 4-42. The incline and pulley are frictionless. Find the acceleration of the objects and the tension in the string for (*a*) general values of θ, m_1, and m_2, and (*b*) $\theta = 30°$ and $m_1 = m_2 = 5$ kg.

(*a*) Draw a free-body diagram for each of the two masses.

Take the x axis for the first diagram to the right to be along the inclined plane.

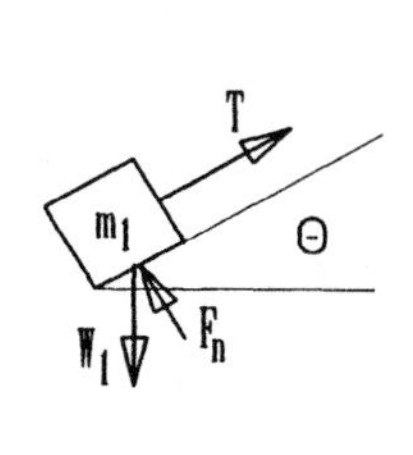

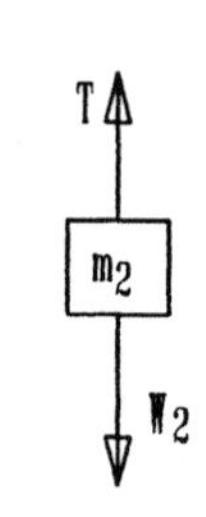

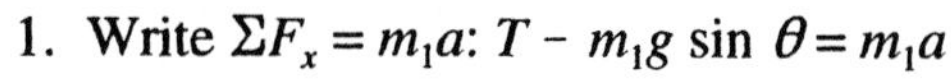

1. Write $\Sigma F_x = m_1 a$: $T - m_1 g \sin\theta = m_1 a$
2. Write $\Sigma F = m_2 a$: $m_2 g - T = m_2 a$
3. Add the two equations and solve for a. $a = g[(m_2 - m_1 \sin\theta)/(m_1 + m_2)]$
4. Use expression for a to find T. $T = g[m_1 m_2/(m_1 + m_2)](1 + \sin\theta)]$

(*b*) Substitute numerical values into expressions for a and T: $T = 36.8$ N; $a = 2.45\ \text{m/s}^2$

73* •• Two climbers on an icy (frictionless) slope, tied together by a 30-m rope, are in the predicament shown in Figure 4-43. At time $t = 0$, the speed of each is zero, but the top climber, Paul (mass 52 kg), has taken one step too many and his friend Jay (mass 74 kg) has dropped his pick. (*a*) Find the tension in the rope as Paul falls and his speed just before he hits the ground. (*b*) If Paul unhooks his rope after hitting the ground, find Jay's speed as he hits the ground.

(*a*) Use results of Problem 4-72 to find a and T — $a = (9.81)[(52 - 74 \sin 40°)/126]$ m/s² $= 0.345$ m/s²

$T = (9.81)(52 \times 74/126)(1 + \sin 40°) = 492$ N

$v = (2as)^{½}$ — $v = (2 \times 0.345 \times 20)^{½}$ m/s $= 3.71$ m/s

(*b*) 1. Find the distance Jay slides — $s_J = (25\text{ m})/\sin 40° = 33.9$ m

2. Find the acceleration of Jay — $a_J = g \sin 40° = 6.31$ m/s²

3. Find $v_J = (2a_Js_J)^{½}$ — $v_J = (2 \times 33.9 \times 6.31)^{½}$ m/s $= 20.7$ m/s

---

74 • The northwest face of Half Dome, a large rock in Yosemite National Park, makes an angle of $\theta = 7.0°$ with the vertical. Suppose a rock climber lying horizontal on the top is trying to support her unfortunate friend of equal mass who is hanging from a rope over the edge, as shown in Figure 4-44. If friction is negligible (the top is icy!), at what acceleration will they slide down before the top partner manages to grab someone's hand and stop?

1. Write the $\Sigma F = ma$ for each mass — $ma = T$; $mg \cos 7° - T = ma$
2. Solve for a — $a = (g \cos 7°)/2 = 4.87$ m/s²

---

75 • In a stage production of Peter Pan, the 50-kg actor playing Peter has to fly in vertically, and to be in time with the music, he must be lowered a distance of 3.2 m in 2.2 s. Backstage, a smooth surface sloped at 50° supports a counterweight of mass m, as shown in Figure 4-45. Show the calculations that the stage manager must perform to find (*a*) the mass of the counterweight that must be used, and (*b*) the tension in the wire.

We can use the results of Problem 4-72.

(*a*) Find the acceleration $a = 2s/t^2$ — $a = (2 \times 3.2)/2.2^2$ m/s² $= 1.32$ m/s²

$a = g[(m_2 - m_1 \sin \theta)/(m_1 + m_2)$; solve for m_1 — $m_1 = m_2(g - a)/(a + g \sin \theta) = 48$ kg

(*b*) $T = m_1m_2g(1 + \sin \theta)/(m_1 + m_2)$ — $T = [(50 \times 48 \times 9.81 \times (1 + \sin 50°)/98]$ N $= 424$ N

---

76 •• An 8-kg block and a 10-kg block connected by a rope that passes over a frictionless peg slide on frictionless inclines, as shown in Figure 4-46. (*a*) Find the acceleration of the blocks and the tension in the rope. (*b*) The two blocks are replaced by two others of mass m_1 and m_2 such that there is no acceleration. Find whatever information you can about the mass of these two new blocks.

(*a*) Draw the free-body diagrams for the two boxes. We shall use coordinate systems with the x axis pointing toward the right and parallel to the inclined planes.

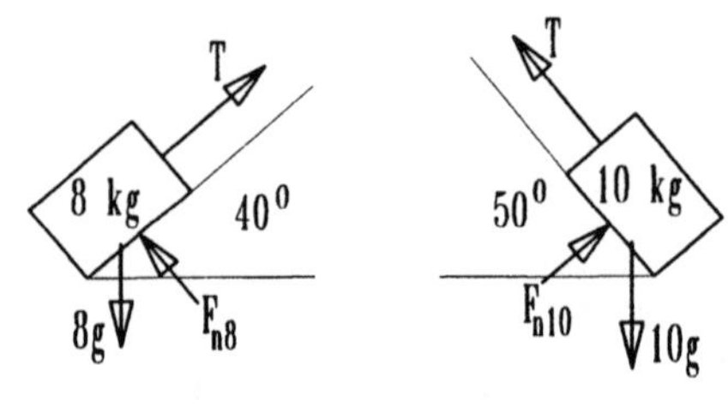

Write $\Sigma F_x = ma$ for the two boxes — $T - 8g \sin 40° = 8a$; $10g \sin 50° - T = 10a$

Solve for a — $a = (10g \sin 50° - 8g \sin 40°)/18$ m/s² $= 1.37$ m/s²

Solve for T using $a = 1.37$ m/s^2 $T = 61.4$ N

(*b*) Set $a = 0$ and find ratio m_1/m_2 $0 = m_1 \sin 40° - m_2 \sin 50°$; $m_1/m_2 = 1.19$

77* •• A heavy rope of length 5 m and mass 4 kg lies on a frictionless horizontal table. One end is attached to a 6-kg block. At the other end of the rope, a constant horizontal force of 100 N is applied. (*a*) What is the accelera tion of the system? (*b*) Give the tension in the rope as a function of position along the rope.

(*a*) $F = (m_1 + m_2)a$ $a = (100/10)$ m/s^2 = 10 m/s^2

(*b*) T at 6-kg mass is 60 N. x is distance from 6-kg. $T(x) = [60 + (40/5)x]$ N $= 60 + 8x$ N

---

78 •• A 60-kg housepainter stands on a 15-kg aluminum platform. The platform is attached to a rope that passes through an overhead pulley, which allows the painter to raise herself and the platform (Figure 4-47). (*a*) To accelerate herself and the platform at a rate of 0.8 m/s^2, with what force must she pull on the rope? (*b*) When her speed reaches 1 m/s, she pulls in such a way that she and the platform go up at a constant speed. What force is she exerting on the rope? (Ignore the mass of the rope.)

(*a*) $\Sigma F = ma$; $2T - mg = ma$ $T = \frac{1}{2}m(a + g) = [75 \times 10.61/2]$ N = 398 N

(*b*) Set $a = 0$ and solve for T $T = (75 \times 9.81/2)$ N = 368 N

---

79 ••• Figure 4-48 shows a 20-kg block sliding on a 10-kg block. All surfaces are frictionless. Find the acceleration of each block and the tension in the string that connects the blocks.

1. Draw a free-body diagram for each block

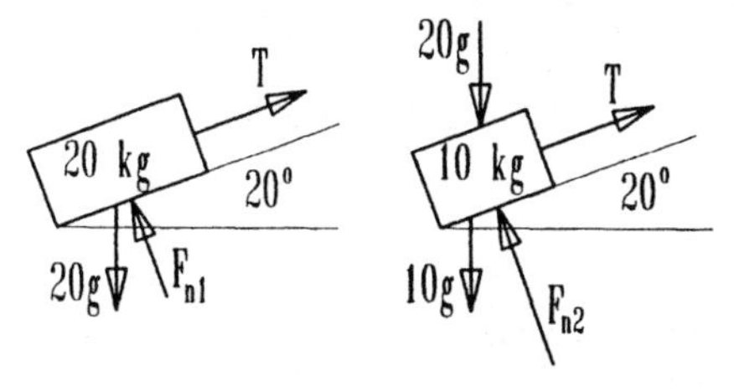

2. Write $\Sigma F_x = ma$ for each block $T - 20g \sin 20° = 20a_{20}$; $T - 10g \sin 20° = 10a_{10}$; $a_{20} = -a_{10}$

3. Solve for a_{20} and a_{10}. $a_{10} = [(10g \sin 20°)/30]$ m/s^2 = 1.12 m/s^2; $a_{20} = -1.12$ m/s^2

4. Solve for T $T = [11.2 + 98.1 \sin 20°]$ N = 44.8 N

---

80 ••• A 20-kg block with a pulley attached slides along a frictionless ledge. It is connected by a massless string to a 5-kg block via the arrangement shown in Figure 4-49. Find the acceleration of each block and the tension in the connecting string.

1. Draw a free-body diagram for each block. Note that the distance the 5-kg block moves in a time Δt is twice the distance the 20-kg block moves. Thus, if we designate by a the acceleration of the 20-kg block, that of the 5-kg block is $2a$.

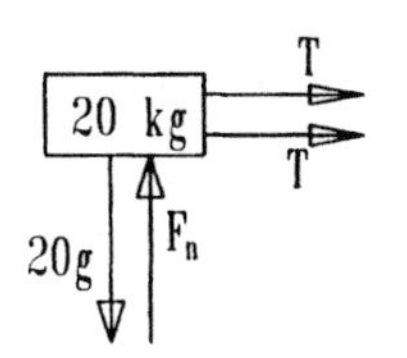

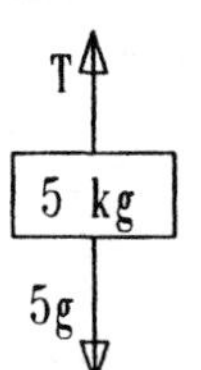

2. Write $\Sigma F = ma$ for each block $2T = 20a$; $5g - T = 5 \times 2a$

3. Solve for T and a $T = 24.5$ N; $a = a_{20} = 2.45$ m/s^2, $a_5 = 4.9$ m/s^2

81* •• The apparatus in Figure 4-50 is called an *Atwood's machine* and is used to measure acceleration due to gravity g by measuring the acceleration of the two blocks. Assuming a massless, frictionless pulley amd a massless string, show that the magnitude of the acceleration of either body and the tension in the string are

$$a = \frac{m_1 - m_2}{m_1 + m_2} g \quad \text{and} \quad T = \frac{2m_1 m_2}{m_1 + m_2} g$$

This is identical to Problem 4-72 with $\theta = 90°$. Setting $\sin\theta = 1$ in the expressions for a and T one obtains the above.

82 •• If one of the masses of the Atwood's machine of Figure 4-50 is 1.2 kg, what should the other mass be so that the displacement of either mass during the first second following release is 0.3 m?

1. Find the acceleration $a = 2s/t^2$ — $a = 0.6$ m/s^2
2. Solve for m_1 — $m_1 = m_2[(g + a)/(g - a)] = m_2(10.41/9.21)$
3. Find the second mass for m_2 or $m_1 = 1.2$ kg — $m_1 = 1.356$ kg, or $m_2 = 1.06$ kg

83 •• A small pebble of mass m rests on the block of mass m_2 of the Atwood's machine in Figure 4-50. Find the force exerted by the pebble on m_2.

Since $m_2 < m_1$, m_2 accelerates up. Hence, the force on a small mass m exerted by m_2 is:

$F = m(g + a) = mg[1 + (m_1 - m_2)/(m_1 + m_2)] = 2m_1mg/(m_1 + m_2)$.

84 •• Find the force exerted by the Atwood's machine on the hanger to which the pulley is attached, as shown in Figure 4-50, while the blocks accelerate. Neglect the mass of the pulley. Check your answer by considering appropriate variations for m_1 and/or m_2.

$F = 2T = 4m_1m_2g/(m_1 + m_2)$. If $m_1 = m_2 = m$, $F = 4m^2g/2m = 2mg$ as expected; if either m_1 or $m_2 = 0$, $F = 0$.

85* ••• The acceleration of gravity g can be determined by measuring the time t it takes for a mass m_2 in an Atwood's machine to fall a distance L, starting from rest. (*a*) Find an expression for g in terms of m_1, m_2, L, and t. (*b*) Show that if there is a small error in the time measurement dt, it will lead to an error in the determination of g by an amount dg given by $dg/g = -2\,dt/t$. If $L = 3$ m and m_1 is 1 kg, find the value of m_2 such that g can be measured with an accuracy of ±5% with a time measurement that is accurate to 0.1 s. Assume that the only significant uncertainty in the measurement is the time of fall.

(*a*) Use $a = 2L/t^2$; from Problem 4-81, $g = a[(m_1 + m_2)/(m_1 - m_2)] = (2L/t^2)[(m_1 + m_2)/(m_1 - m_2)]$.

(*b*) Differentiate with respect to t. $dg/dt = -2g/t$ or $dg/g = -2dt/t$.

With $dg/g = \pm0.05$, $dt/t = \pm0.025$ and $t = (0.1\text{ s})/0.025 = 4$ s. Now find $a = 2L/t^2 = 0.375$ m/s^2 and solve for m_2 with $m_1 = 1$ kg, using the result of Problem 4-82. $m_2 = 0.926$ kg or 1.08 kg.

86 • True or false: (*a*) If there are no forces acting on an object, it will not accelerate. (*b*) If an object is not accelerating, there must be no forces acting on it. (*c*) The motion of an object is always in the direction of the resultant force. (*d*) The mass of an object depends on its location.

(*a*) True (*b*) False (*c*) False (*d*) False

87 • A skydiver of weight w is descending near the surface of the earth. What is the magnitude of the force exerted by her body *on the earth*? (*a*) w (*b*) Greater than w (*c*) Less than w (*d*) $9.8w$ (*e*) 0 (*f*) It depends on the air resistance.

(*a*) It is the reaction force of w.

---

88 • The net force on a moving object is suddenly reduced to zero. As a consequence, the object (*a*) stops abruptly.

(*b*) stops during a short time interval. (*c*) changes direction. (*d*) continues at constant velocity. (*e*) changes velocity in an unknown manner.

(*d*) See Newton's first law.

89* • A force of 12 N is applied to an object of mass m. The object moves in a straight line, with its speed increasing by 8 m/s every 2 s. Find m.

1. Find a — $a = (8/2)\text{ m/s}^2 = 4\text{ m/s}^2$
2. $F = ma$; $m = F/a$ — $m = (12/4)\text{ kg} = 3\text{ kg}$

---

90 • A certain force F_1 gives an object an acceleration of $6 \times 10^6\text{ m/s}^2$. Another force F_2 gives the same object an acceleration of $15 \times 10^6\text{ m/s}^2$. What is the acceleration of the object if (*a*) the two forces act together on the object in the same direction; (*b*) the two forces act in opposite directions on the object; (*c*) the two forces act on the object at 90° to each other?

(*a*) $F_1 = ma_1$; $F_2 = ma_2$; $F_1 + F_2 = m(a_1 + a_2) = ma$ — $a = 21 \times 10^6\text{ m/s}^2$

(*b*) $F_1 - F_2 = m(a_1 - a_2) = ma$ — $a = 9 \times 10^6\text{ m/s}^2$

(*c*) $F = (F_1^2 + F_2^2)^{1/2}$; $a = (a_1^2 + a_2^2)^{1/2}$ — $a = 16.2 \times 10^6\text{ m/s}^2$

---

91 • A certain force applied to a particle of mass m_1 gives it an acceleration of 20 m/s². The same force applied to a particle of mass m_2 gives it an acceleration of 50 m/s². If the two particles are tied together and the same force is applied to the pair, find the acceleration.

$m_1 = F/a_1$; $m_2 = F/a_2$; $m_1 + m_2 = F(1/a_1 + 1/a_2)$ — $a = F/(m_1 + m_2) = (1/20 + 1/50)^{-1}\text{ m/s}^2 = 14.3\text{ m/s}^2$

---

92 • A 6-kg object is pulled along a frictionless horizontal surface by a horizontal force of 10 N. (*a*) If the object is at rest at $t = 0$, how fast is it moving after 3 s? (*b*) How far does it travel during these 3 s?

(*a*) $v = at = Ft/m$ — $v = (10 \times 3/6)\text{ m/s} = 5\text{ m/s}$

(*b*) $s = v_{av}t = \frac{1}{2}vt$ — $s = (\frac{1}{2} \times 5 \times 3)\text{ m} = 7.5\text{ m}$

93* • If you weigh 125 lb on the earth, what would your weight be in pounds on the moon, where the free-fall acceleration due to gravity is 5.33 ft/s²?

$g_E = 32\text{ ft/s}^2$; $w_M = w_E(g_M/g_E)$ — $w_M = (125 \times 5.33/32)\text{ lb} = 20.8\text{ lb}$

---

94 • A redheaded woodpecker hits the bark of a tree extremely hard—the speed of its head reaches approximately $v = 3.5$ m/s before impact. If the mass of the bird's head is 0.060 kg, and the average force acting on the head during impact is $F = 6.0$ N, find (*a*) the acceleration of its head (assuming it is constant); (*b*) the depth of penetration into the bark; (*c*) the time t it takes the woodpecker's head to stop.

(*a*) $a = F/m$ $a = (6/0.06)\ \mathrm{m/s^2} = 100\ \mathrm{m/s^2}$

(*b*) $s = v^2/2a$ $s = (3.5^2/200)\ \mathrm{m} = 6.13\ \mathrm{cm}$

(*c*) $t = v/a$ $t = (3.5/100)\ \mathrm{s} = 35\ \mathrm{ms}$

---

95 •• A simple accelerometer can be made by suspending a small object from a string attached to a fixed point on an accelerating object—to the ceiling of a passenger car, for example. When there is an acceleration, the object will deflect and the string will make some angle with the vertical. (*a*) How is the direction in which the suspended object deflects related to the direction of the acceleration? (*b*) Show that the acceleration *a* is related to the angle θ that the string makes by $a = g \tan \theta$. (*c*) Suppose the accelerometer is attached to the ceiling of an automobile that brakes to rest from 50 km/h in a distance of 60 m. What angle will the accelerometer make? Will the object swing forward or backward?

(*a*) The object will swing backward; see figure.

(*b*) $T_x = ma;\ T_y = mg.\ T_x/T_y = \tan \theta = a/g.\ a = g \tan \theta.$

(*c*) The object will swing forward

Find $a = v^2/2s$ $a = (13.9^2/120)\ \mathrm{m/s^2} = 1.6\ \mathrm{m/s^2}$

Find $\theta = \tan^{-1}(a/g)$ $\theta = \tan^{-1}(1.6/9.81) = 9.3°$

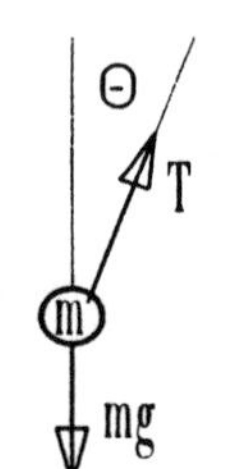

---

96 •• The mast of a sloop is supported at bow and stern by stainless steel wires, the forestay and backstay, anchored 10 m apart (Figure 4-51). The 12-m long mast weighs 800 N and stands vertically on the deck of the sloop. The mast is positioned 3.6 m behind where the forestay is attached. The tension in the forestay is 500 N. Find the tension in the backstay and the force that the mast exerts on the deck of the sloop.

Take as the "free-body" the top of the mast.

1. Find the angles that the forestay and backstay make with the vertical $\theta_F = \tan^{-1}(3.6/12) = 16.7°;\ \theta_B = \tan^{-1}(6.4/12) = 28.1°$

2. $\Sigma F_x = 0$ $500 \sin 16.7° = T_B \sin 28.1°;\ T_B = 305\ \mathrm{N}$

3. Find ΣF_y $F_y = -(500\cos 16.7° + 305 \cos 28.1°)\ \mathrm{N} = 748\ \mathrm{N}$

4. Set $\Sigma F_y = 0$ at deck; $F_n = (800 + 748)\ \mathrm{N}$ $F_n = 1548\ \mathrm{N}$

97* •• A box of mass m_1 is pulled along a smooth horizontal surface by a force ***F*** exerted at the end of a rope that has a much smaller mass m_2, as shown in Figure 4-52. (*a*) Find the acceleration of the rope and block, assuming them to be one object. (*b*) What is the net force acting on the rope? (*c*) Find the tension in the rope at the point where it is attached to the block. (*d*) The diagram, with the rope perfectly horizontal along its length, is not quite accurate. Correct the diagram, and state how this correction affects your solution.

(*a*) $a = F/(m_1 + m_2)$. (*b*) $F_{net} = m_2 a = Fm_2/(m_1 + m_2)$. (*c*) $T = m_1 a = Fm_1/(m_1 + m_2)$.

(*d*) The rope sags and ***F*** has a vertical component and its horizontal component is less than *F*. Consequently, *a* will be somewhat smaller.

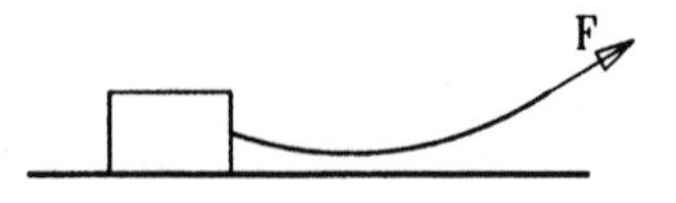

---

98 •• Joe and Sal are in a rollerbladers' club that is building a ramp to reach new levels of extremeness. The ramp is to be a simple incline, so that after coasting horizontally, a skater will ride up the slope at some angle θ. Sal suggests making the slope as steep as possible to maximize the height that will be reached. Joe whips out a pencil and paper to prove to Sal that, if the surfaces are smooth, the height reached is independent of the angle of the slope. Sal acknowledges that even though Joe is being smug and obnoxious, his argument is sound. Show Joe's proof.

The acceleration along the length of the slope is $g \sin \theta$. The distance traveled along the slope is $s = v^2/(2g \sin \theta)$ and the height reached is $h = s \sin \theta = v^2/2g$, independent of θ.

---

99 •• A car traveling 90 km/h crashes into the rear end of an unoccupied stalled vehicle. Fortunately, the driver is wearing a seat belt. Using reasonable values for the mass of the driver and the stopping distance, estimate the force (assuming it to be constant) exerted on the driver by the seat belt.

Assume a stopping distance of 25 m. Then the acceleration is 50 m/s^2. Assume the mass of the driver is 80 kg. Then F = 4000 N.

---

100 •• A 2-kg body rests on a frictionless wedge that has an inclination of 60° and an acceleration a to the right such that the mass remains stationary relative to the wedge (Figure 4-53). (*a*) Find a. (*b*) What would happen if the wedge were given a greater acceleration?

(*a*) Since $a_y = 0$, $\Sigma F_y = 0$ — $F_n \cos 60° = mg$

Find $a = a_x = F_{nx}/m$ — $a = F_n \sin 60°/m$; $a = g \cot 60° = 17$ m/s^2

(*b*) For $a > 17$ m/s^2 block will move up the plane.

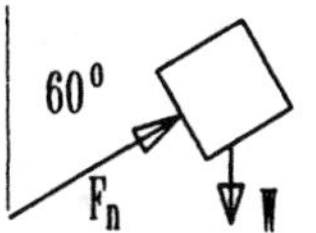

101* •• The masses attached to each side of an Atwood's machine consist of a stack of five washers each of mass m, as shown in Figure 4-54. The tension in the string is T_0. When one of the washers is removed from the left side, the remaining washers accelerate and the tension decreases by 0.3 N. (*a*) Find m. (*b*) Find the new tension and the acceleration of each mass when a second washer is removed from the left side.

(*a*) Use the result of Problem 4-81 — $T_0 = 5mg$; $T_0 - T = 5mg - (2 \times 4m \times 5m)g/9m = 0.3$ N

Solve for m — $m = (0.3 \text{ N}) \times 9/5g = 0.055$ kg = 55 g

(*b*) Use the results of Problem 4-81 — $T = (2 \times 3 \times 5 \times 9.81m/8)$ N = 2.02 N; $a = (2/8)g$ = 2.45 m/s^2

---

102 •• Consider the Atwood's machine in Figure 4-54. When N washers are transferred from the left side to the right side, the right side drops 47.1 cm in 0.40 s. Find N.

1. Find the acceleration $a = 2s/t^2$ — $a = (2 \times 0.471/0.16)$ m/s^2 = 5.89 m/s^2 = $0.6g$

2. Use the result of Problem 4-81 — $0.6 = 2N/10$; $N = 3$

---

103 •• Blocks of mass m and $2m$ are connected by a string (Figure 4-55). (*a*) If the forces are constant, find the tension in the connecting string. (*b*) If the forces vary with time as $F_1 = Ct$ and $F_2 = 2Ct$, where C is a constant and t is time, find the time t_0 at which the tension in the string is T_0.

(*a*) Find the acceleration a — $a = (F_2 - F_1)/3m$

Find the force acting on m, and solve for T — $T - F_1 = ma = (F_2 - F_1)/3$; $T = (F_2 + 2F_1)/3$

(*b*) Write the expression for T_0 and solve for t_0 — $T_0 = 4Ct_0/3$; $t_0 = 3T_0/4C$

---

104 ••• Find the normal force and the tangential force exerted by the road on the wheels of your bicycle (*a*) as you climb an 8% grade at constant speed, (*b*) as you descend the 8% grade at constant speed. (An 8% grade means that the angle of inclination θ is given by $\tan \theta = 0.08$.)

Assume a total mass of 80 kg. Then $mg = 785$ N. The angle $\theta = \tan^{-1}(0.08) = 4.57°$. The total normal and tangential forces on the bicycle are $F_n = mg \cos \theta$ and $F_t = mg \sin \theta$. Thus, $F_n = 782$ N, $F_t = 62.6$ N. Since there is no acceleration, these forces are the same going up and going down.The forces on each wheel are $F_n = 391$ N, $F_t = 31.3$ N.

105* ••• The pulley in an Atwood's machine is given an upward acceleration $\boldsymbol{a}$, as shown in Figure 4-56. Find the acceleration of each mass and the tension in the string that connects them.

A constant upward acceleration has the same effect as an increase in the acceleration of gravity from g to $g + a$. Thus, the tension in the string is given by the expression of Problem 4-81 with g replaced by $(g + a)$: $T = 2m_1m_2(g + a)/(m_1 + m_2)$. To find the acceleration of the mass m_2 consider the forces acting on m_2; they are the tension T and the weight m_2g. Thus, $a_2 = (T - m_2g)/m_2$. Substituting the expression just derived for T and simplifying, one obtains $a_2 = [(m_1 - m_2)g + 2m_1a]/(m_1 + m_2)$. To check these results consider the some limiting cases: 1. $a = 0$; 2. $m_1 = m_2 = $ m; 3. $a = -g$ (free fall of the system).

1. Setting $a = 0$, T and a_2 reduce to the expression given in Problem 4-81, as they should.
2. Setting $m_1 = m_2 = m$, the tension in the string is $T = m(g + a)$, as expected, and $a_2 = a$, as expected.
3. Setting $a = -g$, as in free fall, $T = 0$, as expected, and $a_2 = -g$, as expected.

The expression for $a_1$is the same as for a_2 with all subscripts interchanged, i.e., $a_1 = [(m_2 - m_1)g + 2m_2a]/(m_1 + m_2)$.

---

106 ••• The pulley in an Atwood's machine has a mass m_P. A force $\boldsymbol{F}$ is exerted on the pulley, as shown in Figure 4-57. Find the acceleration of each mass and the tension in the string that connects them.

Consider the pulley of mass m_p as a free body, and list the forces acting on it; they are F, directed up, the tension T in each string, directed down, and m_pg, directed down. The acceleration of the pulley is therefore $a_p = (F - 2T - m_pg)/m_p$. The problem now is the same as Problem 4-105, with a replaced by a_p. Substituting the expression for a_p into the expression for T of Problem 4-105 and simplifying, one obtains $T = 2m_1m_2F/[m_p(m_1 + m_2) + 4m_1m_2]$.

To find the acceleration of the mass m_2, we again write $a_2 = (T - m_2g)/m_2$ and use the above expression for T to obtain $a_2 = \{(2m_1F)/[m_p(m_1 + m_2) + 4m_1m_2]\} - g$. To check, again consider special cases: 1. $F = 0$; 2. $m_1 = m_2 = m$ and $F = (2m + m_p)g$:

1. If $F = 0$, the system is in free fall. $T = 0$ and $a_2 = -g$, as expected.
2. If $m_1 = m_2 = m$ and $F = (2m + m_p)g$, the system is at rest. $a_2 = 0$ and $T = mg$, as expected.

CHAPTER 5

Applications of Newton's Laws

1* • Various objects lie on the floor of a truck moving along a horizontal road. If the truck accelerates, what force acts on the objects to cause them to accelerate?

Force of friction between the objects and the floor of the truck.

2 • Any object resting on the floor of a truck will slide if the truck's acceleration is too great. How does the critical acceleration at which a light object slips compare with that at which a much heavier object slips?

They are the same.

3 • True or false: (*a*) The force of static friction always equals $\mu_s F_n$. (*b*) The force of friction always opposes the motion of an object. (*c*) The force of friction always opposes sliding. (*d*) The force of kinetic friction always equals $\mu_k F_n$.

(*a*) False (*b*) True (*c*) True (*d*) True

4 • A block of mass *m* rests on a plane inclined at an angle θ with the horizontal. It follows that the coefficient of static friction between the block and the plane is (*a*) $\mu_s \geq 1$. (*b*) $\mu_s = \tan\theta$. (*c*) $\mu_s \leq \tan\theta$. (*d*) $\mu_s \geq \tan\theta$.

(*d*)

5* • A block of mass *m* is at rest on a plane inclined at angle of 30° with the horizontal, as in Figure 5-38. Which of the following statements about the force of static friction is true? (*a*) $f_s > mg$ (*b*) $f_s > mg \cos 30°$ (*c*) $f_s = mg \cos 30°$ (*d*) $f_s = mg \sin 30°$ (*e*) None of these statements is true.

(*d*) f_s must equal in magnitude the component of the weight along the plane.

6 • A block of mass *m* slides at constant speed down a plane inclined at an angle θ with the horizontal. It follows that (*a*) $\mu_k = mg \sin\theta$. (*b*) $\mu_k = \tan\theta$. (*c*) $\mu_k = 1 - \cos\theta$. (*d*) $\mu_k = \cos\theta - \sin\theta$.

(*a*) Acceleration = 0, therefore $f_k = mg \sin\theta$. With $F_n = mg \cos\theta$, it follows that $\mu_k = \tan\theta$

7 • A block of wood is pulled by a horizontal string across a horizontal surface at constant velocity with a force of 20 N. The coefficient of kinetic friction between the surfaces is 0.3. The force of friction is (*a*) impossible to determine without knowing the mass of the block. (*b*) impossible to determine without knowing the speed of the block. (*c*) 0.3 N. (*d*) 6 N. (*e*) 20 N.

(*e*) The net force is zero.

8 • A 20-N block rests on a horizontal surface. The coefficients of static and kinetic friction between the surface and the block are $\mu_s = 0.8$ and $\mu_k = 0.6$. A horizontal string is attached to the block and a constant tension T is maintained in the string. What is the force of friction acting on the block if (*a*) $T = 15$ N, or (*b*) $T = 20$ N.

(*a*) If $\mu_s mg > 15$, then $f = f_s = 15$ N — $0.8 \times (20 \text{ N}) = 16 \text{ N}; f = f_s = 15$ N

(*b*) $T > f_{s,max}; f = f_k = \mu_k mg$ — $f = f_k = 0.6 \times (20 \text{ N}) = 12$ N

9* • A block of mass m is pulled at constant velocity across a horizontal surface by a string as in Figure 5-39. The magnitude of the frictional force is (*a*) $\mu_k mg$. (*b*) $T \cos \theta$. (*c*) $\mu_k(T - mg)$. (*d*) $\mu_k T \sin \theta$. (*e*) $\mu_k(mg + T \sin \theta)$.

(*b*) The net force is zero.

---

10 • A tired worker pushes with a force of 500 N on a 100-kg crate resting on a thick pile carpet. The coefficients of static and kinetic friction are 0.6 and 0.4, respectively. Find the frictional force exerted by the surface.

1. Draw the free-body diagram

2. Apply $\Sigma F = ma$ — $F_n - (100 \times 9.81) \text{ N} = 0; F_n = 981$ N

3. $f_{s,max} = \mu_s F_n$ — $f_{s,max} = 589 \text{ N} > 500$ N

4. Since 500 N $< f_{s,max}$ the box does not move — $F = f_s = 500$ N

---

11 • A box weighing 600 N is pushed along a horizontal floor at constant velocity with a force of 250 N parallel to the floor. What is the coefficient of kinetic friction between the box and the floor?

Draw the free-body diagram.

1. Apply $\Sigma F = ma$; $a_x = a_y = 0$ — $250 \text{ N} = f_k$; $F_n = 600$ N

2. $f_k = \mu_k F_n$; solve for μ_k — $\mu_k = (250/600) = 0.417$

---

12 • The coefficient of static friction between the tires of a car and a horizontal road is $\mu_s = 0.6$. If the net force on the car is the force of static friction exerted by the road, (*a*) what is the maximum acceleration of the car when it is braked? (*b*) What is the least distance in which the car can stop if it is initially traveling at 30 m/s?

Draw the free-body diagram.

(*a*) 1. Apply $\Sigma F = ma$; $a_y = 0$ — $f_{s,max} = ma_{max}$; $F_n = mg$

2. Use $f_{s,max} = \mu_s F_n$ and solve for a_{max} — $a_{max} = \mu_s g = (0.6 \times 9.81) \text{ m/s}^2 = 5.89 \text{ m/s}^2$.

(*b*) Use $v^2 = v_0^2 + 2as$; solve for s with $v = 0$ — $s = [30^2/(2 \times 5.89)]$ m $= 76.5$ m

13* • The force that accelerates a car along a flat road is the frictional force exerted by the road on the car's tires. (*a*) Explain why the acceleration can be greater when the wheels do not spin. (*b*) If a car is to accelerate from 0 to 90 km/h in 12 s at constant acceleration, what is the minimum coefficient of friction needed between the road and tires? Assume that half the weight of the car is supported by the drive wheels.

(*a*) $\mu_s > \mu_k$; therefore f is greater if the wheels do not spin.

(*b*) 1. Draw the free-body diagram; the normal force on each pair of wheels is $\frac{1}{2}mg$.

2. Apply $\Sigma F = ma$ — $f_s = ma = \mu_s F_n$; $F_n = \frac{1}{2}mg$

3. Solve for a — $a = \frac{1}{2}\mu_s g = (25\ \text{m/s}^2)/(12\ \text{s}) = 2.08\ \text{m/s}^2$

4. Find μ_s — $\mu_s = (2 \times 2.08/9.81) = 0.425$

---

14 • On the current tour of the rock band Dead Wait, the show opens with a dark stage. Suddenly there is the sound of a large automobile accident. Lead singer Sharika comes sliding to the front of the stage on her knees. Her initial speed is 3 m/s. After sliding 2 m, she comes to rest in a dry ice fog as flash pots explode on either side. What is the coefficient of kinetic friction between Sharika and the stage?

Draw the free-body diagram

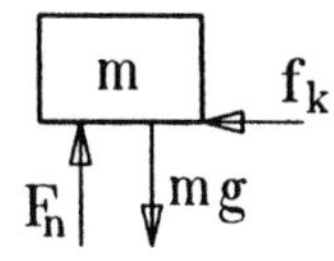

1. Use $v^2 = v_0^2 + 2as$; solve for a — $a = (9/4)\ \text{m/s}^2$

2. $f_k = ma = \mu_k F_n$; $F_n = mg$ — $\mu_k = a/g = [9/(4 \times 9.81)] = 0.23$

---

15 • A 5-kg block is held at rest against a vertical wall by a horizontal force of 100 N. (*a*) What is the frictional force exerted by the wall on the block? (*b*) What is the minimum horizontal force needed to prevent the block from falling if the coefficient of friction between the wall and the block is $\mu_s = 0.40$?

(*a*) 1. Draw the free-body diagram

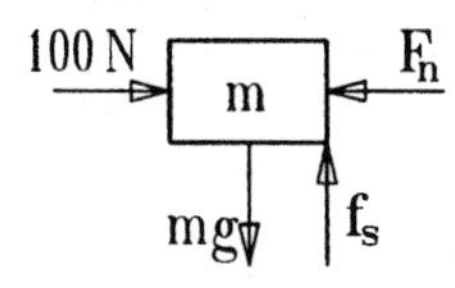

2. Apply $\Sigma F = ma$ — $F_n = 100\ \text{N}$; $f_s = mg = 49.05\ \text{N}$

(*b*) $f_s = \mu_s F_n$; solve for F_n — $F_n = (49.05\ \text{N})/0.4 = 123\ \text{N}$

---

16 • On a snowy day with the temperature near the freezing point, the coefficient of static friction between a car's tires and an icy road is 0.08. What is the maximum incline that this four-wheel-drive vehicle can climb with zero acceleration?

1. Draw the free-body diagram

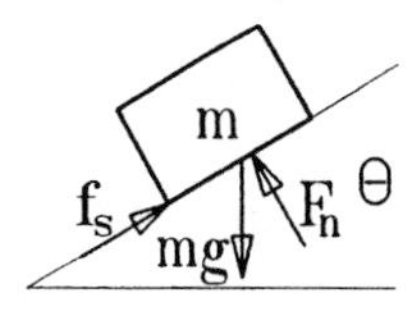

2. Apply $\Sigma F = ma$ — $f_s - mg \sin\theta = 0$; $f_s = mg \sin\theta$

$F_n - mg \cos\theta = 0$; $F_n = mg \sin\theta$

3. $f_s = \mu_s F_n$	$mg \sin\theta = \mu_s mg \cos\theta$
4. Solve for μ_s and find θ	$\mu_s = \tan\theta = 0.08$; $\theta = \tan^{-1}(0.08) = 4.57°$

17* • A 50-kg box that is resting on a level floor must be moved. The coefficient of static friction between the box and the floor is 0.6. One way to move the box is to push down on it at an angle θ with the horizontal. Another method is to pull up on the box at an angle θ with the horizontal. (*a*) Explain why one method is better than the other. (*b*) Calculate the force necessary to move the box by each method if $\theta = 30°$ and compare the answers with the result when $\theta = 0°$.

The free-body diagram for both cases, $\theta > 0$ and $\theta < 0$, is shown.

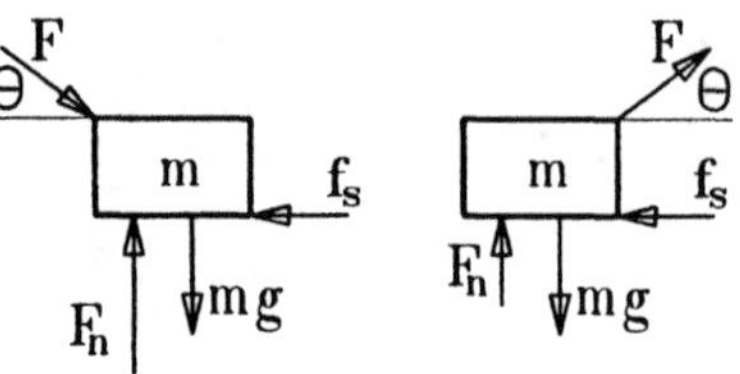

(*a*) $\theta > 0$ is preferable; it reduces F_n and therefore f_s.

(*b*) 1. Use $\Sigma F = ma$ to determine F_n	$F \sin\theta + F_n - mg = 0$. $F_n = mg - F\sin\theta$
2. $f_{s,max} = \mu_s F_n$	$f_{s,max} = \mu_s(mg - F\sin\theta)$
3. To move the box, $F_x = F\cos\theta \ge f_{s,max}$	$F = \mu_s(mg - F\sin\theta)/\cos\theta$; $F = \dfrac{\mu_s mg}{\cos\theta + \mu_s \sin\theta}$
4. Find F for $m = 50$ kg, $\mu_s = 0.6$, and $\theta = 30°$, $\theta = -30°$, and $\theta = 0°$	$F(30°) = 252$ N, $F(-30°) = 520$ N, $F(0°) = 294$ N

18 • A 3-kg box resting on a horizontal shelf is attached to a 2-kg box by a light string as in Figure 5-40. (*a*) What is the minimum coefficient of static friction such that the objects remain at rest? (*b*) If the coefficient of static friction is less than that found in part (*a*), and the coefficient of kinetic friction between the box and the shelf is 0.3, find the time for the 2-kg mass to fall 2 m to the floor if the system starts from rest.

1. Draw a free-body diagram for each object. In the absence of friction, m_1 will move to the right, m_2 will move down. The friction force is indicated by f without subscript; it is f_s for (*a*), f_k for (*b*).

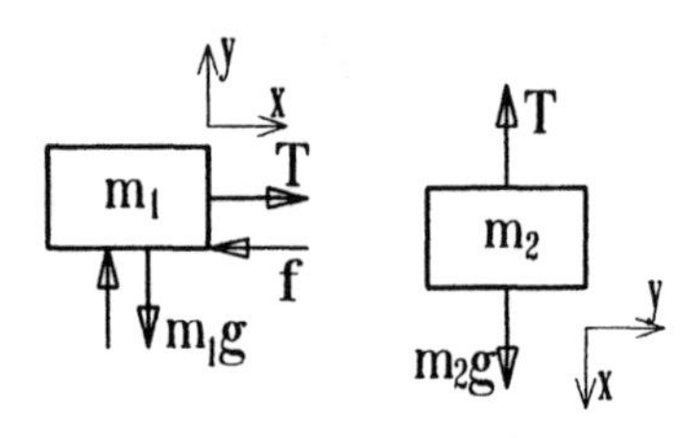

(*a*) 1. Apply $\Sigma F = ma$ for each mass. Note that $a = 0$	$F_n - m_1 g = 0$; $T - f = 0$; $m_2 g - T = 0$; $f = f_s = T = m_2 g$ $F_n = m_1 g$
2. $f_s = f_{s,max} = \mu_s F_n$; solve for μ_s	$m_2 g = \mu_s m_1 g$; $\mu_s = m_2/m_1 = 2/3 = 0.667$
(*b*) If $\mu_s < 0.667$, the system will accelerate.	
1. Apply $\Sigma F = ma$; $a_y = 0$; $a = a_x$	$F_n = m_1 g$; $T - f_k = m_1 a$; $m_2 g - T = m_2 a$; $f_k = \mu_k m_1 g$
2. Solve for a	$a = (m_2 - \mu_k m_1)g/(m_1 + m_2)$
3. Find a for $m_1 = 3$ kg, $m_2 = 2$ kg, $\mu_k = 0.3$	$a = 2.16$ m/s^2
4. Use $s = \frac{1}{2}at^2$; solve for and find t	$t = (2s/a)^{1/2} = (2 \times 2/2.16)^{1/2}$ s $= 1.36$ s

19 •• A block on a horizontal plane is given an initial velocity v. It comes to rest after a displacement d. The coefficient of kinetic friction between the block and the plane is given by (*a*) $\mu_k = v^2 d/2g$. (*b*) $\mu_k = v^2/2dg$.

(*c*) $\mu_k = v^2g/d^2$. (*d*) none of the above.

(*b*) $v^2 = 2ad$; $a = f_k/m = \mu_k mg/m = \mu_k g$; solve for μ_k.

---

20 •• A block of mass $m_1 = 250$ g is at rest on a plane that makes an angle $\theta = 30°$ above the horizontal (Figure 5-41). The coefficient of kinetic friction between the block and the plane is $\mu_k = 0.100$. The block is attached to a second block of mass $m_2 = 200$ g that hangs freely by a string that passes over a frictionless and massless pulley. When the second block has fallen 30.0 cm, its speed is (*a*) 83 cm/s. (*b*) 48 cm/s. (*c*) 160 cm/s. (*d*) 59 cm/s. (*e*) 72 cm/s.

1. Draw free-body diagrams for each object. Note that both objects have the same acceleration.

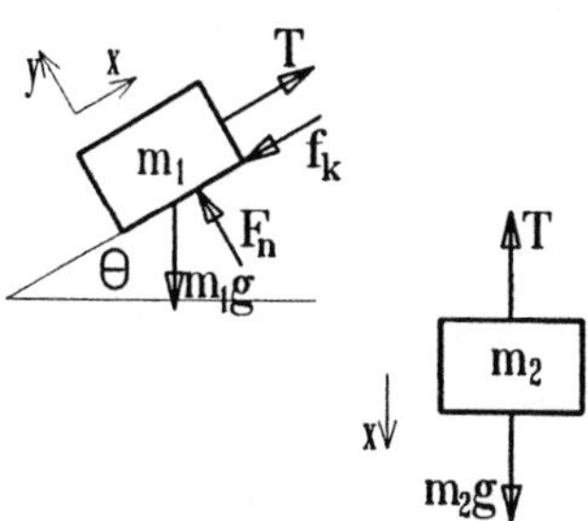

2. Apply $\Sigma F = ma$ to each object and use $f_k = \mu_k F_n$ — $T - m_1g\sin\theta - f_k = m_1a$; $m_1g\cos\theta = F_n$; $m_2g - T = m_2a$; $T = m_2g - m_2a$; $f_k = \mu_k m_1 g\cos\theta$

3. Solve for a — $a = \dfrac{[m_2 - m_1(\sin\theta + \mu_k\cos\theta)]g}{m_1 + m_2}$; $a = 1.16$ m/s^2

4. Find $v = (2ad)^{1/2}$ — $v = (2 \times 1.16 \times 0.3)^{1/2} = 0.83$ m; (*a*) is correct.

21* •• Returning to Figure 5-41, this time $m_1 = 4$ kg. The coefficient of static friction between the block and the incline is 0.4. (*a*) Find the range of possible values for m_2 for which the system will be in static equilibrium. (*b*) What is the frictional force on the 4-kg block if $m_2 = 1$ kg?

(*a*) 1. Use the result of Problem 5-20; set $a = 0$. Note that f_s may point up or down the plane. — $0 = m_2 - m_1(\sin\theta \pm \mu_s\cos\theta)$

2. Solve for m_2 with $m_1 = 4$ kg, $\mu_s = 0.4$ — $m_2 = 3.39$ kg, 0.614 kg.

$m_{2,\text{max}} = 3.39$ kg, $m_{2,\text{min}} = 0.614$ kg

(*b*) 1. Apply $\Sigma F = ma$; set $\boldsymbol{a} = 0$ — $m_2g + f_s - m_1g\sin\theta = 0$

2. Solve for and find f_s — $f_s = [(4.0 \times 0.5 - 1.0) \times 9.81]$ N $= 9.81$ N

---

22 •• Returning once again to Figure 5-41, this time $m_1 = 4$ kg, $m_2 = 5$ kg, and the coefficient of kinetic friction between the inclined plane and the 4-kg block is $\mu_k = 0.24$. Find the acceleration of the masses and the tension in the cord.

1. Use the result of Problem 5-20; substitute numerical values. — $a = 2.36$ m/s^2

2. Use $T = m_2g - m_2a$ and Problem 5-20 to obtain an expression for T and substitute numerical values. — $T = \dfrac{m_1m_2g(1 + \sin\theta + \mu_s\cos\theta)}{m_1 + m_2}$; $T = 37.2$ N

23 •• The coefficient of static friction between the bed of a truck and a box resting on it is 0.30. The truck is traveling at 80 km/h along a horizontal road. What is the least distance in which the truck can stop if the box is not to slide?

1. Draw the free-body diagram.

2. Apply $\Sigma F = ma$ to find a_{max} — $f_{s,max} = \mu_s mg = ma_{max}$; $a_{ax} = \mu_s g$

3. Use $v^2 = v_0^2 + 2ax$; solve for $x = d_{min}$ — $d_{min} = (v_0^2/2a_{max})^{1/2} = (v_0^2/2\mu_s g)^{1/2}$; $d_{min} = 83.9$ m

---

24 •• A 4.5-kg mass is given an initial velocity of 14 m/s up an incline that makes an angle of 37° with the horizontal. When its displacement is 8.0 m, its upward velocity has diminished to 5.2 m/s. Find (*a*) the coefficient of kinetic friction between the mass and the plane, (*b*) the displacement of the mass from its starting point at the time when it momentarily comes to rest, and (*c*) the speed of the block when it again reaches its initial position.

Draw the free-body diagram

(*a*) 1. Apply $\Sigma F = ma$ — $F_n = mg \cos \theta$; $ma_x = -mg \cos \theta - f_k$

2. Replace $f_k = \mu_k F_n$ and solve for a_x — $a_x = -(\sin \theta + \mu_k \cos \theta)g$

3. Use $a_x = (v^2 - v_0^2)/2s$ and solve for μ_k — $\mu_k = (v_0^2 - v^2)/(2gs\cos \theta) - \tan \theta$; $\mu_k = 0.594$

(*b*) Set $v^2 = 0$ and solve for s. — $s = v_0^2/2g(\sin \theta + \mu_k \cos \theta)$; $s = 9.28$ m

(*c*) 1. Note that now f_k points upward; write a_x — $a_x = (\mu_k \cos \theta - \sin \theta)g$

2. $v = (2as)^{1/2}$; note that s and a are negative and solve for and evaluate v — $v = v_0\sqrt{\dfrac{\sin \theta - \mu_k \cos \theta}{\sin \theta + \mu_k \cos \theta}}$; $v = 4.82$ m/s

---

25* •• An automobile is going up a grade of 15° at a speed of 30 m/s. The coefficient of static friction between the tires and the road is 0.7. (*a*) What minimum distance does it take to stop the car? (*b*) What minimum distance would it take if the car were going down the grade?

The free-body diagram is shown for part (*a*).
For part (*b*), f_s points upward along the plane.

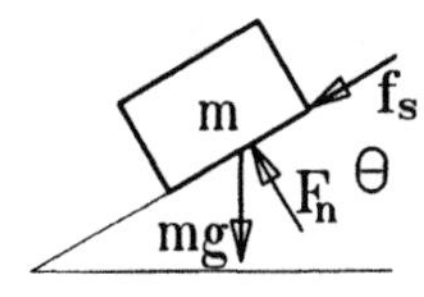

(*a*) We can use the result of Problem 5-24(*a*) and (*b*), replacing μ_k by μ_s — $s = v_0^2/2g(\sin \theta + \mu_s \cos \theta)$; $s = 49.1$ m

(*b*) Replace $\sin \theta$ by $-\sin \theta$ — $s = v_0^2/2g(\mu_s \cos \theta - \sin \theta)$; $s = 110$ m

---

26 •• A block of mass m slides with initial speed v_0 on a horizontal surface. If the coefficient of kinetic friction between the block and the surface is μ_k, find the distance d that the block moves before coming to rest.

The problem is essentially identical to Problem 5-14, with $s = d$ the unknown. $a = -\mu_k g$ and $d = -v_0^2/2a = v_0^2/2\mu_k g$.

---

27 •• A rear-wheel-drive car supports 40% of its weight on its two drive wheels and has a coefficient of static friction of 0.7. (*a*) What is the vehicle's maximum acceleration? (*b*) What is the shortest possible time in which this car can achieve a speed of 100 km/h? (Assume that the engine has unlimited power.)

Draw the free-body diagram.

$f_{s,max} = 0.4\mu_s mg$.

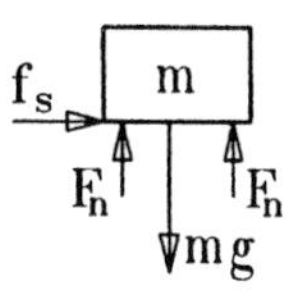

(*a*) Apply $\Sigma F = ma$	$0.4\mu_s mg = ma$; $a = 0.4\mu_s g$; $a = 2.75$ m/s^2
(*b*) $v = at$; solve for and find t	$t = v/a$; $t = (27.8/2.75)$ s $= 10.1$ s

---

28 •• Lou bets an innocent stranger that he can place a 2-kg block against the side of a cart, as in Figure 5-42, and that the block will not fall to the ground, even though Lou will use no hooks, ropes, fasteners, magnets, glue, or adhesives of any kind. When the stranger accepts the bet, Lou begins to push the cart in the direction shown. The coefficient of static friction between the block and the cart is 0.6. (*a*) Find the minimum acceleration for which Lou will win the bet. (*b*) What is the magnitude of the frictional force in this case? (*c*) Find the force of friction on the block if a is twice the minimum needed for the block not to fall. (*d*) Show that, for a block of any mass, the block will not fall if the acceleration is $a \geq g/\mu_s$, where μ_s is the coefficient of static friction.

(*a*) 1. The normal force acting on the block is the force exerted by the cart.	
2. Apply $\Sigma F = ma$	$F_n = F = ma$
(*b*) $f_s = f_{s,max}$	$f_{s,max} = \mu_s ma = mg$; $a_{min} = g/\mu_s = 16.4$ m/s^2
(*c*) f_s is again mg	$f_s = mg = 19.6$ N
(*d*) Since g/μ_s is a_{min}, block will not fall if $a \geq g/\mu_s$	$f_s = 19.6$ N

29* •• Two blocks attached by a string slide down a 20° incline. The lower block has a mass of $m_1 = 0.25$ kg and a coefficient of kinetic friction $\mu_k = 0.2$. For the upper block, $m_2 = 0.8$ kg and $\mu_k = 0.3$. Find (*a*) the acceleration of the blocks and (*b*) the tension in the string.

Since the coefficient of friction for the lower block is the smaller, the string will be under tension.

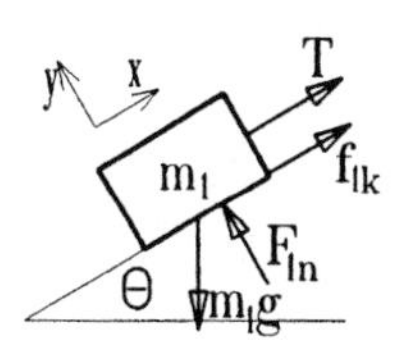

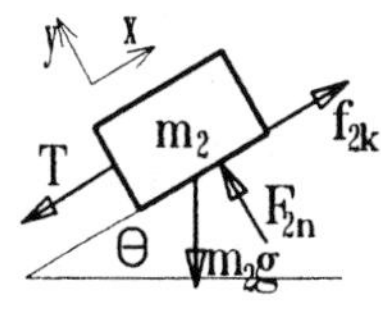

2. Apply $\Sigma F = ma$ to each block

$T + f_{1k} - m_1 g\sin\theta = m_1 a \qquad -T + f_{2k} - m_2 g\sin\theta = m_2 a$

$F_{1n} - m_1 g\cos\theta = 0 \qquad F_{2n} - m_2 g\cos\theta = 0$

3. Add the first pair of equations; use $f_k = \mu_k F_n$

$(m_1\mu_{1k} + m_2\mu_{2k})g\cos\theta - (m_1 + m_2)g\sin\theta = (m_1 + m_2)a$

4. Solve for a

$$a = \frac{(m_1\mu_{1k} + m_2\mu_{2k})\cos\theta - (m_1 + m_2)\sin\theta}{m_1 + m_2}g$$

5. Solve for T

$$T = \frac{m_1 m_2(\mu_{2k} - \mu_{1k})g\cos\theta}{m_1 + m_2}$$

6. Substitute numerical values for the masses, friction coefficients, and θ to find a and T. — $a = -0.809$ m/s^2 (i.e., down the plane); $T = 0.176$ N

---

30 •• Two blocks attached by a string are at rest on an inclined surface. The lower block has a mass of $m_1 = 0.2$ kg and a coefficient of static friction $\mu_s = 0.4$. The upper block has a mass $m_2 = 0.1$ kg and $\mu_s = 0.6$. (*a*) At what angle θ_c do the blocks begin to slide? (*b*) What is the tension in the string just before sliding begins?

(*a*) 1. Referring to Problem 5-29, replace μ_k by μ_s and set $a = 0$. — $(m_1\mu_{1s} + m_2\mu_{2s})\cos\theta - (m_1 + m_2)\sin\theta = 0$

2. Solve for and evaluate $\theta = \theta_c$ — $\theta_c = \tan^{-1}[(m_1\mu_{1s} + m_2\mu_{2s})/(m_1 + m_2)$; $\theta_c = 25°$

(*b*) 1. Since $\tan^{-1}(0.4) = 21.8°$, lower block would slide if $T = 0$. Set $a = 0$ and solve for T — $m_1g\sin\theta - \mu_{1s}m_1g\cos\theta - T = 0$; $T = m_1g\sin\theta - \mu_{1s}m_1g\cos\theta$

2. Evaluate T — $T = 0.118$ N

---

31 •• Two blocks connected by a massless, rigid rod slide on a surface inclined at an angle of 20°. The lower block has a mass $m_1 = 1.2$ kg, and the upper block's mass is $m_2 = 0.75$ kg. (*a*) If the coefficients of kinetic friction are $\mu_k = 0.3$ for the lower block and $\mu_k = 0.2$ for the upper block, what is the acceleration of the blocks? (*b*) Determine the force transmitted by the rod.

(*a*), (*b*) We can use the results of Problem 5-29 and evaluate a and T. — $a = -0.944$ m/s^2 (downward acceleration); $T = -0.425$ N (rod under compression)

---

32 •• A block of mass m rests on a horizontal surface (Figure 5-43). The box is pulled by a massless rope with a force $\boldsymbol{F}$ at an angle θ. The coefficient of static friction is 0.6. The minimum value of the force needed to move the block depends on the angle θ. (*a*) Discuss qualitatively how you would expect this force to depend on θ. (*b*) Compute the force for the angles $\theta = 0°$, 10°, 20°, 30°, 40°, 50°, and 60°, and make a plot of F versus θ for $mg = 400$ N. From your plot, at what angle is it most efficient to apply the force to move the block?

(*a*) F will decrease with increasing θ for small values of θ since the normal component diminishes; it will reach a minimum and then increase as the tangential component of F decreases.

(*b*) The expression for F is given in Problem 5-17

$$F = \frac{\mu_s mg}{\cos\theta + \mu_s \sin\theta}$$

θ (degrees)	0	10	20	30	40	50	60
F (N)	240	220	210	206	208	218	235

A plot of F versus θ is shown here.

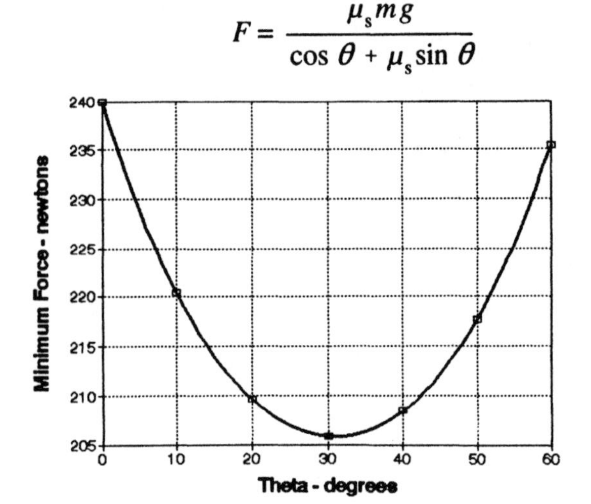

From the graph it appears that F is a minimum at $\theta = 30°$.

33* •• Answer the same questions as in Problem 32, only this time with a force $\boldsymbol{F}$ that pushes down on the block in Figure 5-44 at an angle θ with the horizontal.

(*a*) As in Problem 5-17, replace θ by $-\theta$ in the expression for F. One expects that F will increase with increasing magnitude of the angle since the normal component increases and tangential component decreases.

(*b*)

θ (degrees)	0	-10	-20	-30	-40	-50	-60
F (N)	240	272	327	424	631	1310	diverged

A plot of F versus the magnitude of θ is shown

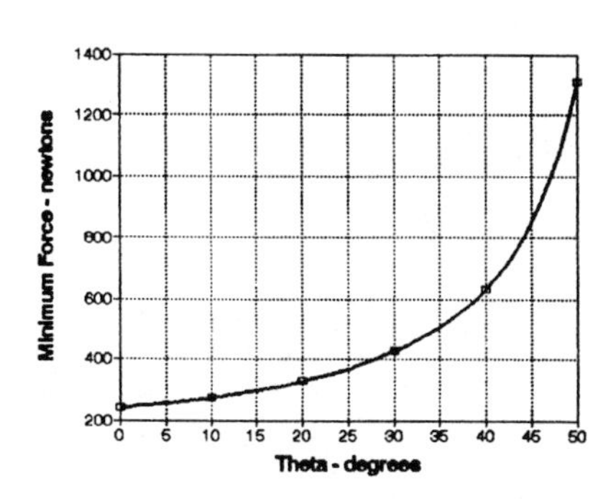

34 •• A 100-kg mass is pulled along a frictionless surface by a horizontal force **F** such that its acceleration is 6 m/s² (Figure 5-45). A 20-kg mass slides along the top of the 100-kg mass and has an acceleration of 4 m/s². (It thus slides backward relative to the 100-kg mass.) (*a*) What is the frictional force exerted by the 100-kg mass on the 20-kg mass? (*b*) What is the net force acting on the 100-kg mass? What is the force F? (*c*) After the 20-kg mass falls off the 100-kg mass, what is the acceleration of the 100-kg mass? (Assume that the force F does not change.)

(*a*) 1. Draw the free-body diagram for the masses.

2. Apply $\Sigma\boldsymbol{F} = m\boldsymbol{a}$. Note that by Newton's third law, the normal reaction force, F_{n1}, and the friction force acts on both masses but in opposite directions.

$f_k = m_1a_1$ (1)

$F_{n1} - m_1g = 0$ (2)

$F - f_k = m_2a_2 = F_{net}$ (3)

3. Evaluate f_k from (1) — $f_k = (20 \times 4)$ N $= 80$ N

(*b*) Evaluate F and F_{net} from (3) — $F_{net} = (100 \times 6)$ N $= 600$ N; $F = 680$ N

(*c*) Use $F = ma$ — $a = (680/100)$ m/s² $= 6.80$ m/s²

35 •• A 60-kg block slides along the top of a 100-kg block with an acceleration of 3 m/s² when a horizontal force $\boldsymbol{F}$ of 320 N is applied, as in Figure 5-46. The 100-kg block sits on a horizontal frictionless surface, but there is friction between the two blocks. (*a*) Find the coefficient of kinetic friction between the blocks. (*b*) Find the acceleration of the 100-kg block during the time that the 60-kg block remains in contact.

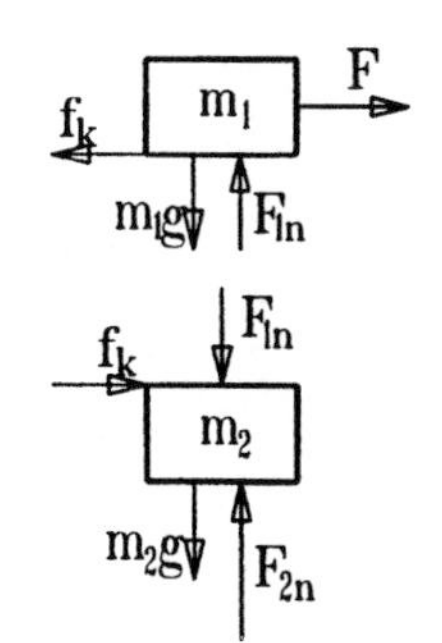

(*a*) 1. The solution is similar to that of the previous problem except that now the force F acts on the upper mass m_1. The corresponding equations are listed.

$F - f_k = m_1a_1$ (1)

$F_n - m_1g = 0$ (2)

$f_k = m_2a_2$ (3)

$f_k = \mu_kF_n = \mu_km_1g$ (4)

2. Replace f_k in (1) by (4) and solve for μ_k — $\mu_k = (F - m_1a_1)/m_1g$; $\mu_k = 0.238$

(*b*) From (3) and (4) $a_2 = \mu_km_1g/m_2$ — $a_2 = 1.4$ m/s²

36 •• The coefficient of static friction between a rubber tire and the road surface is 0.85. What is the maximum acceleration of a 1000-kg four-wheel-drive truck if the road makes an angle of 12° with the horizontal and the truck is (*a*) climbing, and (*b*) descending?

(*a*) 1. Draw the free-body diagram

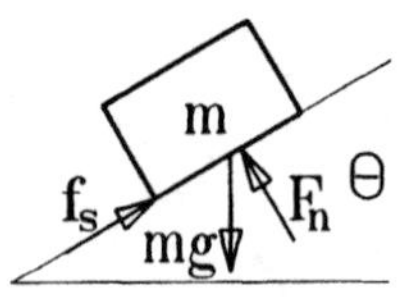

2. Apply $\Sigma F = ma$ — $\mu_s mg \cos\theta - mg \sin\theta = ma_{max}$

Solve for and find a_{max} — $a_{max} = g(\mu_s \cos\theta - \sin\theta)$; $a_{max} = 6.12\ \text{m/s}^2$

(*b*) Replace θ by $-\theta$. — $a_{max} = g(\mu_s \cos\theta + \sin\theta)$; $a_{nax} = 10.2\ \text{m/s}^2$

37* •• A 2-kg block sits on a 4-kg block that is on a frictionless table (Figure 5-47). The coefficients of friction between the blocks are $\mu_s = 0.3$ and $\mu_k = 0.2$. (*a*) What is the maximum force *F* that can be applied to the 4-kg block if the 2-kg block is not to slide? (*b*) If *F* is half this value, find the acceleration of each block and the force of friction acting on each block. (*c*) If *F* is twice the value found in (*a*), find the acceleration of each block.

(*a*) 1. Draw the free-body diagram

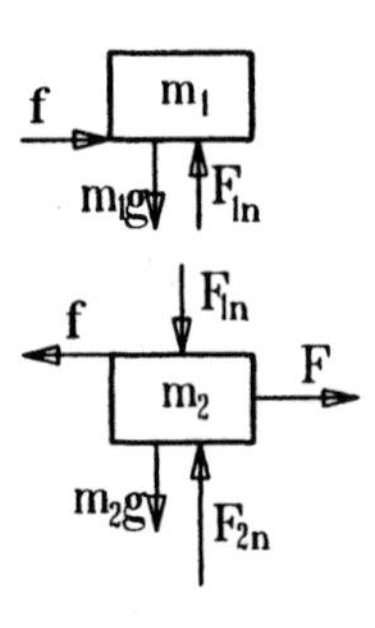

2. Apply $\Sigma F = ma$ — $f_{s,max} = m_1 a_{max}$; $F_{n1} = m_1 g$; $F_{max} - f_{s,max} = m_2 a_{max}$

3. Use $f_{s,max} = \mu_s F_{n1}$ and solve for a_{max} and F_{max} — $a_{max} = \mu_s g$; $F_{max} = (m_1+m_2)g\mu_s$;

4. Evaluate a_{max} and F_{max} — $a_{max} = 2.94\ \text{m/s}^2$, $F_{max} = 17.7\ \text{N}$

(*b*) 1. The blocks move as a unit. The force on m_1 is $m_1 a = f_s$. — $a = F/(m_1 + m_2)$; $a = 2.95\ \text{m/s}^2$; $f_s = (2.95 \times 2)\ \text{N} = 5.9\ \text{N}$

(*c*) 1. If $F = 2F_{max}$ then m_1 slips on m_2. — $f = f_k = \mu_k m_1 g$

2. Apply $\Sigma F = ma$ — $m_1 a_1 = f_k = \mu_k m_1 g$; $m_2 a_2 = F - \mu_k m_1 g$

3. Solve for and evaluate a_1 and a_2 for $F = 35.4\ \text{N}$ — $a_1 = \mu_k g$; $a_2 = (F - \mu_k g m_1)/m_2$; $a_1 = 1.96\ \text{m/s}^2$, $a_2 = 7.87\ \text{m/s}^2$

38 •• In Figure 5-48, the mass $m_2 = 10$ kg slides on a frictionless table. The coefficients of static and kinetic friction between m_2 and $m_1 = 5$ kg are $\mu_s = 0.6$ and $\mu_k = 0.4$. (*a*) What is the maximum acceleration of m_1? (*b*) What is the maximum value of m_3 if m_1 moves with m_2 without slipping? (*c*) If $m_3 = 30$ kg, find the acceleration of each body and the tension in the string.

The free-body diagrams for m_1 and m_2 are identical to those of the previous problem. Now the force F arises from the tension T in the string supporting m_3, as shown.

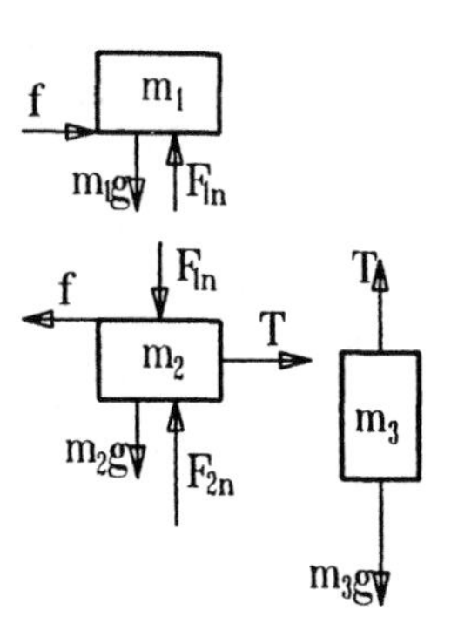

(*a*) See Problem 5-37 — $a_{max} = \mu_s g$; $a_{max} = 5.89\ \text{m/s}^2$

(*b*) 1. Apply $\Sigma F = ma$ — $T = (m_1 + m_2)a_{max}$; $m_3 g - T = m_3 a_{max}$

2. Solve for and evaluate m_3 — $m_3 = \mu_s(m_1 + m_2)/(1 - \mu_s)$; $m_3 = 22.5$ kg

(*c*) 1. For $m_3 = 30$ kg, m_1 will slide on m_2. Follow the procedure of Problem 5-37(*c*). Note that $a_3 = a_2$ — $m_1 a_1 = f_k = \mu_k m_1 g$; $m_2 a_2 = T - \mu_k m_1 g$; (1)

$m_3 a_3 = m_3 g - T = m_3 a_2$ (2)

2. Add the equations involving T to find a_2 — $a_2 = (m_3 - m_1\mu_k)g/(m_2 + m_3)$; $a_2 = a_3 = 6.87\ \text{m/s}^2$

3. Evaluate a_1 and T using equation (1) — $a_1 = (0.4 \times 9.81)\ \text{m/s}^2 = 3.92\ \text{m/s}^2$; $T = 88.3$ N

---

39 ••• A box of mass m rests on a horizontal table. The coefficient of static friction is μ_s. A force $\boldsymbol{F}$ is applied at an angle θ as shown in Problem 5-32. (*a*) Find the force F needed to move the box as a function of the angle θ. (*b*) At the angle θ for which this force is a minimum, the slope $dF/d\theta$ of the curve F versus θ is zero. Compute $dF/d\theta$ and show that this derivative is zero at the angle θ that obeys $\tan\theta = \mu_s$. Compare this general result with that obtained in Problem 5-32.

The expression for F was obtained previously: $F = \mu_s mg/(\cos\theta + \mu_s \sin\theta)$. F is a minimum when the denominator is a maximum. Differentiate $(\cos\theta + \mu_s\sin\theta)$ and set to 0. $(d/d\theta)(\cos\theta + \mu_s\sin\theta) = -\sin\theta + \mu_s\cos\theta = 0$. Solve for θ: $\theta = \tan^{-1}\mu_s$. For $\mu_s = 0.6$, $\theta = 31°$, in agreement with the result of Problem 5-32.

---

40 ••• A 10-kg block rests on a 5-kg bracket like the one shown in Figure 5-49. The 5-kg bracket sits on a frictionless surface. The coefficients of friction between the 10-kg block and the bracket on which it rests are $\mu_s = 0.40$ and $\mu_k = 0.30$. (*a*) What is the maximum force F that can be applied if the 10-kg block is not to slide on the bracket? (*b*) What is the corresponding acceleration of the 5-kg bracket?

(*a*), (*b*) 1. Draw the free-body diagrams for the two objects. The net force acting on m_2 in the direction of motion is $f_s - F$. $a_{2,max} = f_{s,max} = \mu_s F_{n2}$ and since m_2 does not move relative to m_1, this is also the acceleration of m_1.

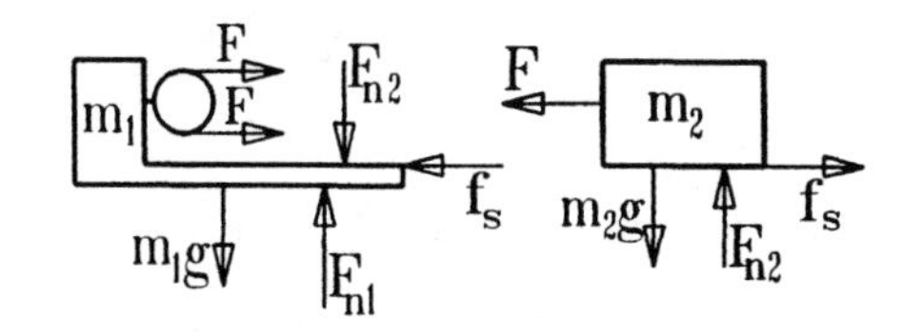

2. Apply $\Sigma F = ma$ — $F_{n2} = m_2 g$; $2F - f_s = m_1 a$; $f_s = m_2 a$

3. Use $f_s = \mu_s F_{n2}$ and solve for $a = a_{max}$ — $a_{max} = \mu_s m_2 g/(m_1 + 2m_2)$; $a_{max} = 1.57\ \text{m/s}^2$

4. Solve for $F = F_{max}$ — $F_{max} = m_2(\mu_s g - a_{max})$; $F_{max} = 23.5$ N

41* ••• Lou has set up a kiddie ride at the Winter Ice Fair. He builds a right-angle triangular wedge, which he intends to push along the ice with a child sitting on the hypotenuse. If he pushes too hard, the kid will slide up and over the top, and Lou could be looking at a lawsuit. If he doesn't push hard enough, the kid will slide down the wedge, and the parents will want their money back. If the angle of inclination of the wedge is 40°, what are the minimum and maximum values for the acceleration that Lou must achieve? Use m for the child's mass, and μ_s for the coefficient of static friction between the child and the wedge.

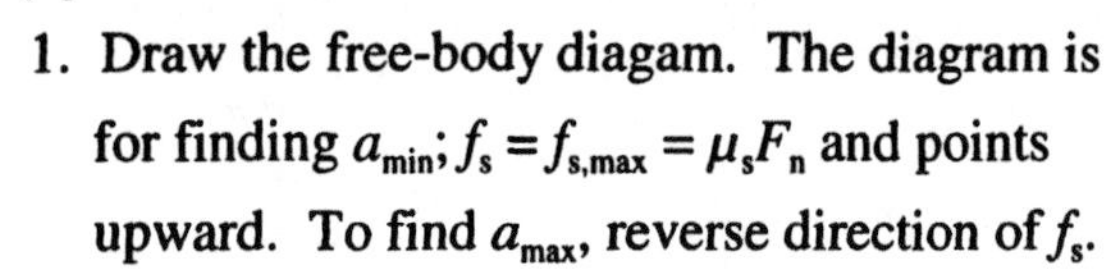

1. Draw the free-body diagam. The diagram is for finding a_{min}; $f_s = f_{s,max} = \mu_s F_n$ and points upward. To find a_{max}, reverse direction of f_s.

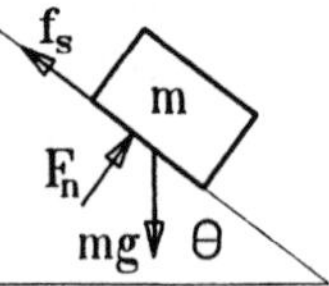

2. Apply $\Sigma F = ma$ — $F_n \sin\theta - \mu_s F_n \cos\theta = ma$; $F_n \cos\theta + \mu_s F_n \sin\theta - mg = 0$
3. Use the second equation to solve for F_n — $F_n = mg/(\cos\theta + \mu_s \sin\theta)$
4. Substitute F_n into first equation and solve for $a = a_{min}$ — $$a_{min} = g\frac{\sin\theta - \mu_s\cos\theta}{\cos\theta + \mu_s\sin\theta}$$
5. Reverse the direction of f_s and follow the same procedure to find a_{max}. — $$a_{max} = g\frac{\sin\theta + \mu_s\cos\theta}{\cos\theta - \mu_s\sin\theta}$$

---

42 ••• A block of mass 0.5 kg rests on the inclined surface of a wedge of mass 2 kg, as in Figure 5-50. The wedge is acted on by a horizontal force **F** and slides on a frictionless surface. (*a*) If the coefficient of static friction between the wedge and the block is $\mu_s = 0.8$, and the angle of the incline is 35°, find the maximum and minimum values of F for which the block does not slip. (*b*) Repeat part (*a*) with $\mu_s = 0.4$.

(*a*) Use results of Problem 5-41 for a_{min} and a_{max}. Then set $F = m_{tot}a$. Substitute numerical values. — $a_{min} = -0.627\ \text{m/s}^2$, $F_{min} = -1.57$ N; (accelerate backward); $a_{max} = 33.5\ \text{m/s}^2$, $F_{max} = 83.7$ N

(*b*) Repeat (*a*) with $\mu_s = 0.4$ — $F_{min} = 6.49$ N; $F_{max} = 37.5$ N

---

43 • True or false: An object cannot move in a circle unless there is a net force acting on it.

True; it requires centripetal force.

---

44 • An object moves in a circle counterclockwise with constant speed (Figure 5-51). Which figure shows the correct velocity and acceleration vectors?

(*c*)

45* • A particle is traveling in a vertical circle at constant speed. One can conclude that the ________ is constant.

(*a*) velocity (*b*) acceleration (*c*) net force (*d*) apparent weight (*e*) none of the above

(*e*)

---

46 • An object travels with constant speed v in a circular path of radius r. (*a*) If v is doubled, how is the acceleration a affected? (*b*) If r is doubled, how is a affected? (*c*) Why is it impossible for an object to travel around a perfectly sharp angular turn?

(*a*) $a \propto v^2$; a is quadrupled. (*b*) $a \propto 1/r$; a is halved. (*c*) Would require an infinite centripetal force ($r = 0$).

47 • A boy whirls a ball on a string in a horizontal circle of radius 0.8 m. How many revolutions per minute must the ball make if the magnitude of its centripetal acceleration is to be the same as the free-fall acceleration due to gravity g?

$a_c = \omega^2 r = g$; $\omega = \sqrt{g/r} = \sqrt{9.81/0.8}$ rad/s $= 3.5$ rad/s $= 33.4$ rpm

---

48 • A 0.20-kg stone attached to a 0.8-m long string is rotated in a horizontal plane. The string makes an angle of 20° with the horizontal. Determine the speed of the stone.

1. Draw the free-body diagram. Note that $a_y = 0$ and $a_x = v^2/r$. The radius of the circle is $r = L\cos\theta$, where L is the length of the string.

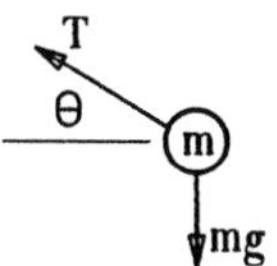

2. Apply $\Sigma F = ma$ — $T\sin\theta = mg$; $T\cos\theta = mv^2/(L\cos\theta)$
3. Solve for and evaluate v. — $v = \sqrt{Lg\cos\theta\cot\theta}$; $v = 4.5$ m/s

---

49* • A 0.75-kg stone attached to a string is whirled in a horizontal circle of radius 35 cm as in the conical pendulum of Example 5-10. The string makes an angle of 30° with the vertical. (*a*) Find the speed of the stone. (*b*) Find the tension in the string.

This problem is identical to Problem 5-48; since the angle θ is with respect to the vertical, the expressions for v and T must be changed accordingly.

(*a*), (*b*) Write v and T in terms of θ and r — $v = \sqrt{rg\tan\theta}$; $T = mg/\cos\theta$

Evaluate v and T — $v = 1.41$ m/s; $T = 8.5$ N

---

50 •• A stone with a mass $m = 95$ g is being whirled in a horizontal circle at the end of a string that is 85 cm long. The length of required time for the stone to make one complete revolution is 1.22 s. The angle that the string makes with the horizontal is _____. (*a*) 52° (*b*) 46° (*c*) 26° (*d*) 23° (*e*) 3°

$v^2/Lg = \cos\theta\cot\theta$ (see Problem 5-48); substitute $v = r\omega = \omega L\cos\theta$ and obtain $\sin\theta = g/L\omega^2$.

$\omega = (2\pi/1.22)$ rad/s.

$\theta = \sin^{-1}[(9.81 \times 1.22^2)/(0.85 \times 4\pi^2)] = 25.8°$; (*c*) is correct.

---

51 •• A pilot of mass 50 kg comes out of a vertical dive in a circular arc such that her upward acceleration is 8.5g. (*a*) What is the magnitude of the force exerted by the airplane seat on the pilot at the bottom of the arc? (*b*) If the speed of the plane is 345 km/h, what is the radius of the circular arc?

(*a*) 1. Draw the free-body diagram

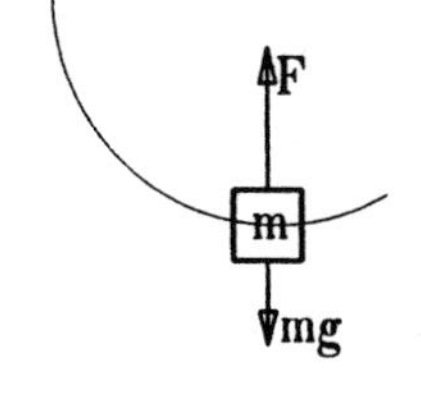

2. Apply $\Sigma F = ma$ — $F - mg = ma$
3. Solve for and evaluate F — $F = 9.5mg = 4660$ N

(*b*) $r = v^2/a_c$; evaluate for $a_c = 8.5g$, $v = 95.8$ m/s — $r = 110$ m

---

52 •• A 65-kg airplane pilot pulls out of a dive by following the arc of a circle whose radius is 300 m. At the bottom of the circle, where her speed is 180 km/h, (*a*) what are the direction and magnitude of her acceleration? (*b*) What is the net force acting on her at the bottom of the circle? (*c*) What is the force exerted on the pilot by the airplane seat?

(*a*) 1. See Problem 5-51 for the free-body diagram.

2. $a = a_c = v^2/r$ — $a = (50^2/300)$ m/s^2 = 8.33 m/s^2, directed up

(*b*) $F_{net} = ma$ — $F_{net} = (65 \times 8.33)$ N = 542 N, directed up

(*c*) $F = mg + F_{net}$ — $F = (542 + 65 \times 9.81)$ N = 1179 N, directed up

---

53* •• Mass m_1 moves with speed v in a circular path of radius R on a frictionless horizontal table (Figure 5-52). It is attached to a string that passes through a frictionless hole in the center of the table. A second mass m_2 is attached to the other end of the string. Derive an expression for R in terms of m_1, m_2, and v.

1. Draw the free-body diagrams for the two masses

2. Apply $\Sigma F = ma$ — $T = m_1v^2/R$; $T - m_2g = 0$

3. Solve for R — $R = (m_1/m_2)v^2/g$

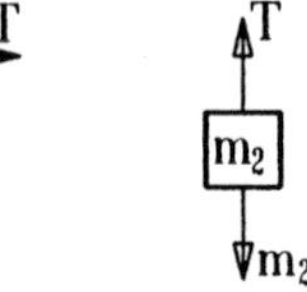

---

54 •• In Figure 5-53, particles are shown traveling counterclockwise in circles of radius 5 m. The acceleration vectors are indicated at three specific times. Find the values of v and dv/dt for each of these times.

(*a*) The acceleration is radial; $a = a_c = v^2/r$; $v = (a_cr)^{1/2}$ =$(20 \times 5)^{1/2}$ m/s = 10 m/s. $dv/dt = 0$

(*b*) $\boldsymbol{a}$ has radial and tangential components. $a_r = a_c = (30 \cos 30°)$ m/s^2; $v = (a_cr)^{1/2}$; for $r = 5$ m, $v = 11.4$ m/s. The tangential acceleration is $a \sin 30° = 15$ m/s^2 = dv/dt.

(*c*) Here $a_c = (50 \cos 45°)$ m/s^2 = v^2/r. For $r = 5$ m, $v = 13.3$ m/s. The tangential acceleration is directed opposite to $\boldsymbol{v}$ and its magnitude is $(50 \sin 45°)$ m/s^2 = 35.4 m/s^2. Hence, $dv/dt = -35.4$ m/s^2.

---

55 •• A block of mass m_1 is attached to a cord of length L_1, which is fixed at one end. The block moves in a horizontal circle on a frictionless table. A second block of mass m_2 is attached to the first by a cord of length L_2 and also moves in a circle, as shown in Figure 5-54. If the period of the motion is T, find the tension in each cord.

1. Draw the free-body diagrams for the two blocks
Note that there is no vertical motion.

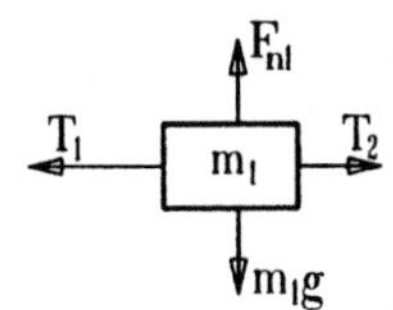

2. Apply $\Sigma F = ma$ to each mass — $T_2 = m_2(L_1 + L_2)\omega^2 = m_2(L_1 + L_2)(2\pi/T)^2$; $T_1 - T_2 = m_1L_1(2\pi/T)^2$

3. Solve for T_1 — $T_1 = [m_1L_1 + m_2(L_1 + L_2)](2\pi/T)^2$

---

56 •• A particle moves with constant speed in a circle of radius 4 cm. It takes 8 s to make a complete trip. Draw the path of the particle to scale, and indicate the particle's position at 1-s intervals. Draw displacement vectors for each interval. These vectors also indicate the directions for the average-velocity vectors for each interval. Find graphically the change in the average velocity $\Delta\boldsymbol{v}$ for two consecutive 1-s intervals. Compare $\Delta\boldsymbol{v}/\Delta t$, measured in this way, with the instantaneous acceleration computed from $a = v^2/r$.

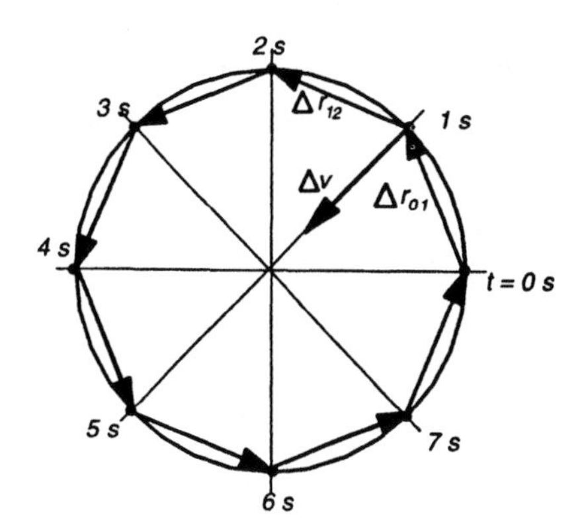

The path of the particle and its position at 1 s intervals are shown. The displacement vectors are also shown. The velocity vectors for the average velocities in the first and second intervals are along $\Delta \boldsymbol{r}_{01}$ and $\Delta \boldsymbol{r}_{12}$, respectively, and are shown in the lower diagram. $\Delta \boldsymbol{v}$ points toward the center of the circle. Since the angle between $\boldsymbol{v}_1$ and $\boldsymbol{v}_2$ is 45°, $\Delta v = 2v_1 \sin 22.5°$. Also $\Delta r = 2r \sin 22.5° = 3.06$ cm. Thus $v_{av} = 3.06$ cm/s and $\Delta v = 2.34$ cm/s and $\Delta v/\Delta t = 2.34$ cm/s^2. The instantaneous speed is $2\pi r/T = \pi$ cm/s and the instantaneous acceleration is then $v^2/r = (3.14^2/4)$ cm/s^2 = 2.47 cm/s^2.

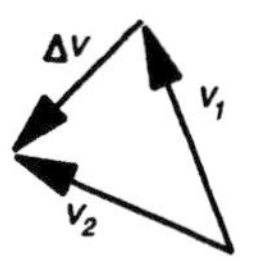

57* •• A man swings his child in a circle of radius 0.75 m, as shown in the photo. If the mass of the child is 25 kg and the child makes one revolution in 1.5 s, what are the magnitude and direction of the force that must be exerted by the man on the child? (Assume the child to be a point particle.)

1. See Problem 5-49. In this problem T stands for the period.
2. $\tan \theta = v^2/rg = r\omega^2/g = 4\pi^2 r/gT^2$ (see Problem 5-49) — $\tan \theta = (4\pi^2 \times 0.75)/(9.81 \times 1.5^2) = 1.34;\ \theta = 53.3°$
3. $F = mg/\cos \theta$ — $F = (25 \times 9.81/\cos 53.3°) = 410$ N

---

58 •• The string of a conical pendulum is 50 cm long and the mass of the bob is 0.25 kg. Find the angle between the string and the horizontal when the tension in the string is six times the weight of the bob. Under those conditions, what is the period of the pendulum?

1. See Problem 5-48 for the free-body diagram and the relevant equations.
2. $\sin \theta = mg/T$; solve for and evaluate θ — $\theta = \sin^{-1}(1/6) = 9.6°$
3. $T = 2\pi r/v = 2\pi r/\sqrt{Lg \cos \theta \cot \theta}$; evaluate T — $T = 2\pi\sqrt{L \sin \theta/g} = 0.58$ s

---

59 •• Frustrated with his inability to make a living through honest channels, Lou sets up a deceptive weight-loss scam. The trick is to make insecure customers believe they can "think those extra pounds away" if they will only take a ride in a van that Lou claims to be "specially equipped to enhance mental-mass fluidity." The customer sits on a platform scale in the back of the van, and Lou drives off at constant speed of 14 m/s. Lou then asks the customer to "think heavy" as he drives through the bottom of a dip in the road having a radius of curvature of 80 m. Sure enough, the scale's reading increases, until Lou says, "Now think light," and drives over the crest of a hill having a radius of curvature of 100 m. If the scale reads 800 N when the van is on level ground, what is the range of readings for the trip described here?

1. Draw the free-body diagrams for each case.

Passing through the dip, a_c is upward; driving over the crest, a_c is downward. The apparent weights are F_1 and F_2, respectively.

2. Apply $\Sigma F = ma$ — $F_1 - mg = mv^2/r_1$ — $F_2 - mg = -mv^2/r_2$

3. Evaluate F_1 and F_2. — $F_1 = 1000$ N; $F_2 = 640$ N

---

60 •• A 100-g disk sits on a horizontally rotating turntable. The turntable makes one revolution each second. The disk is located 10 cm from the axis of rotation of the turntable. (*a*) What is the frictional force acting on the disk? (*b*) The disk will slide off the turntable if it is located at a radius larger than 16 cm from the axis of rotation. What is the coefficient of static friction?

(*a*) 1. Draw the free-body diagram.

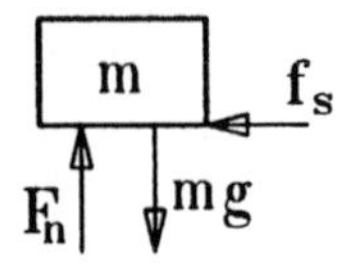

2. Apply $\Sigma F = ma$ — $F_n = mg; f_s = mr\omega^2 = mr(2\pi/T)^2$

3. Evaluate f_s — $f_s = 0.395$ N

(*b*) For $r = 0.16$ m, $f_s = \mu_s F_n$. Find μ_s — $\mu_s = 4\pi^2 r/gT^2 = 0.644$

61* •• A tether ball of mass 0.25 kg is attached to a vertical pole by a cord 1.2 m long. Assume the cord attaches to the center of the ball. If the cord makes an angle of 20° with the vertical, then (*a*) what is the tension in the cord? (*b*) What is the speed of the ball?

This problem is identical to Problem 5-48, except that the angle θ is now with respect to the vertical. Consequently, the relevant equations are: $T\cos\theta = mg$ and $v = \sqrt{Lg\sin\theta\tan\theta}$. Substituting the appropriate numerical values one obtains (*a*) $T = 2.61$ N, (*b*) $v = 1.21$ m/s

---

62 •• An object on the equator has an acceleration toward the center of the earth because of the earth's rotation and an acceleration toward the sun because of the earth's motion along its orbit. Calculate the magnitudes of both accelerations, and express them as fractions of the free-fall acceleration due to gravity g.

1. Evaluate ω_R (rotation) and ω_O (orbital motion) — $\omega_R = 2\pi/(24 \times 60 \times 60)$ rad/s $= 7.27 \times 10^{-5}$ rad/s; $\omega_O = \omega_R/365 = 19.9 \times 10^{-8}$ rad/s

2. Evaluate $R_e\omega_R^2$ and $R_o\omega_O^2$ (see Appendix B on page AP-3) — $a_R = 3.37 \times 10^{-2}$ m/s$^2 = 3.44 \times 10^{-3}g$; $a_O = 5.95 \times 10^{-3}$ m/s$^2 = 6.1 \times 10^{-4}g$

---

63 •• A small bead with a mass of 100 g slides along a semicircular wire with a radius of 10 cm that rotates about a vertical axis at a rate of 2 revolutions per second, as shown in Figure 5-55. Find the values of θ for which the bead will remain stationary relative to the rotating wire.

The semicircular wire of radius 10 cm limits the motion of the bead in the same manner as would a 10-cm string attached to the bead and fixed at the center of the semicircle. Consequently, we can use the expression for T derived in Problem 5-58, $T = 2\pi\sqrt{L\sin\theta/g}$, where here θ is the angle with respect to the horizontal. Thus, we shall use $T = 2\pi\sqrt{L\cos\theta/g}$. Solving for θ we obtain $\theta = \cos^{-1}(T^2 g/4\pi^2 L) = 51.6°$.

---

64 ••• Consider a bead of mass m that is free to move on a thin, circular wire of radius r. The bead is given an initial speed v_0, and there is a coefficient of kinetic friction μ_k. The experiment is performed in a spacecraft drifting in space. Find the speed of the bead at any subsequent time t.

1. Draw the free-body diagram. Note that the acceleration of the bead has two components, the radial component perpendicular to v, and a tangential component due to friction directed opposite to v.

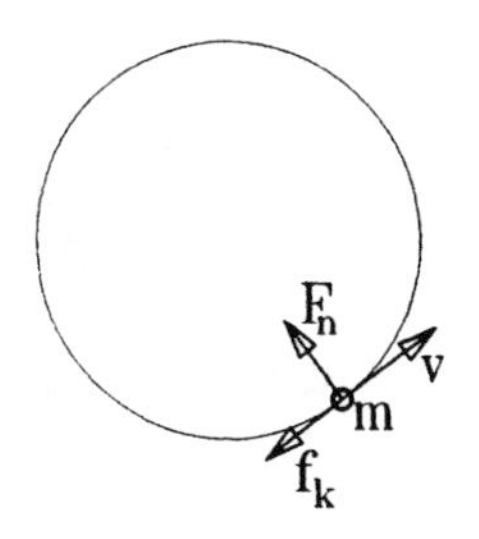

2. Apply $\Sigma F = ma$ — $F_n = mv^2/r;\ f_k = \mu_k mv^2/r = -m(dv/dt)$
3. Rewrite the differential equation — $dv/v^2 = -(\mu_k/r)dt$
4. Integrate the differential equation; the limits on v and t are v_0 and v, and 0 and t, respectively. — $-\left(\frac{1}{v} - \frac{1}{v_0}\right) = -\left(\frac{\mu_k}{r}\right)t;\ v = v_o\left(\frac{1}{1 + (\mu_k v_0/r)t}\right)$

65* ••• Revisiting the previous problem, (*a*) find the centripetal acceleration of the bead. (*b*) Find the tangential acceleration of the bead. (*c*) What is the magnitude of the resultant acceleration?

(*a*) Use the result of Problem 5-64 — $a_c = v^2/r = \frac{v_0^2}{r}\left(\frac{1}{1 + (\mu_k v_0/r)t}\right)^2$

(*b*) $a_t = -\mu_k v^2/r$ — $a_t = -\mu_k a_c$

(*c*) $a = (a_c^2 + a_t^2)^{1/2}$ — $a = a_c(1 + \mu_k^2)^{1/2}$, where a_c is given above.

---

66 • A block is sliding on a frictionless surface along a loop-the-loop, as shown in Figure 5-56. The block is moving fast enough that it never loses contact with the track. Match the points along the track to the appropriate free-body diagrams (Figure 5-57).

A : 3; B : 4; C : 5; D : 2.

---

67 • A person rides a loop-the-loop at an amusement park. The cart circles the track at constant speed. At the top of the loop, the normal force exerted by the seat equals the person's weight, mg. At the bottom of the loop, the force exerted by the seat will be _____. (*a*) 0 (*b*) mg (*c*) $2mg$ (*d*) $3mg$ (*e*) greater than mg, but the exact value cannot be calculated from the information given

At the top, $mv^2/r = 2mg$. At the bottom, $F_n = mg + 2mg = 3mg$. (*d*) is correct.

---

68 • The radius of curvature of a loop-the-loop roller coaster is 12.0 m. At the top of the loop, the force that the seat exerts on a passenger of mass m is $0.4mg$. Find the speed of the roller coaster at the top of the loop.

$mv^2/r = 1.4mg;\ v = (1.4rg)^{1/2} = 12.8$ m/s

69* • Realizing that he has left the gas stove on, Aaron races for his car to drive home. He lives at the other end of a long, unbanked curve in the highway, and he knows that when he is traveling alone in his car at 40 km/h, he can just make it around the curve without skidding. He yells at his friends, "Get in the car! With greater mass, I can take the curve at higher speed!" Carl says, "No, that will make you skid at even lower speed." Bonita says, "The mass does not matter. Just get going!" Who is right?

Bonita is right.

---

70 • A car speeds along the curved exit ramp of a freeway. The radius of the curve is 80 m. A 70-kg passenger holds the arm rest of a car door with a 220-N force to keep from sliding across the front seat of the car.

(Assume the exit ramp is not banked and ignore friction with the car seat.) What is the car's speed? (*a*) 16 m/s (*b*) 57 m/s (*c*) 18 m/s (*d*) 50 m/s (*e*) 28 m/s

$mv^2/r = F;\ v = (Fr/m)^{½};\ v = 15.9$ m/s. (*a*) is correct.

---

71 ••• Suppose you ride a bicycle on a horizontal surface in a circle with a radius of 20 m. The resultant force exerted by the road on the bicycle (normal force plus frictional force) makes an angle of 15° with the vertical. (*a*) What is your speed? (*b*) If the frictional force is half its maximum value, what is the coefficient of static friction?

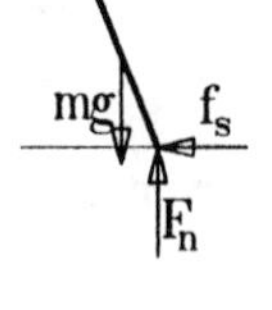

(*a*) 1. Draw the free-body diagram

2. Apply $\Sigma F = ma$ — $F_n = mg;\ f_s = mv^2/r$

3. $F_n + f_s$ makes an angle of 15° with vertical — $v^2/rg = \tan 15°$

4. Solve for v — $v = (20 \times 9.81 \times \tan 15°)^{½} = 7.25$ m/s

(*b*) $\mu_s = f_{s,max}/F_n = 2f_s/F_n$ — $\mu_s = 2v^2/rg = 2\tan 15° = 0.536$

---

72 • A 750-kg car travels at 90 km/h around a curve with a radius of 160 m. What should the banking angle of the curve be so that the only force between the pavement and tires of the car is the normal reaction force?

1. See Example 5-12. — $\theta = \tan^{-1}(v^2/rg) = 21.7°$.

73* •• A curve of radius 150 m is banked at an angle of 10°. An 800-kg car negotiates the curve at 85 km/h without skidding. Find (*a*) the normal force on the tires exerted by the pavement, (*b*) the frictional force exerted by the pavement on the tires of the car, and (*c*) the minimum coefficient of static friction between the pavement and tires.

(*a*), (*b*) 1. Draw the free-body diagram

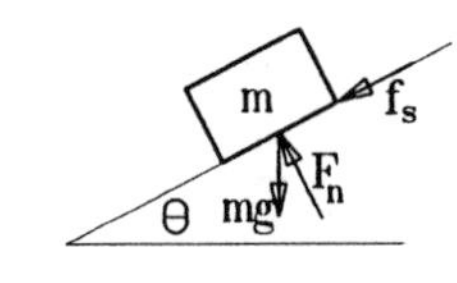

2. Apply $\Sigma F = ma$ — $F_n\sin\theta + f_s\cos\theta = mv^2/r$ (1)

$F_n\cos\theta - f_s\sin\theta = mg$ (2)

3. Multiply (1) by $\sin\theta$, (2) by $\cos\theta$ and add — $F_n = (mv^2/r)\sin\theta + mg\cos\theta$

4. Evaluate F_n and use (2) to evaluate f_s — $F_n = 8245$ N; $f_s = 1565$ N

(*c*) $\mu_{s,min} = f_s/F_n$ — $\mu_{s,min} = 0.19$

---

74 •• On another occasion, the car in the previous problem negotiates the curve at 38 km/h. Find (*a*) the normal force exerted on the tires by the pavement, and (*b*) the frictional force exerted on the tires by the pavement.

Proceed as in the previous problem. One obtains (*a*) $F_n = 7832$ N, and (*b*) $f_s = -766$ N (f_s points up along the plane.)

---

75 ••• A civil engineer is asked to design a curved section of roadway that meets the following conditions: With ice on the road, when the coefficient of static friction between the road and rubber is 0.08, a car at rest must not slide into the ditch and a car traveling less than 60 km/h must not skid to the outside of the curve. What is the minimum radius of curvature of the curve and at what angle should the road by banked?

1. The free-body diagram for the car at rest is that of Problem 5-73; for the car at 60 km/h, reverse f_s. In each case we require that $f_s = f_{s,max} = \mu_s F_n$.
2. Apply $\Sigma F = ma$ for $v = 0$ — $F_n(\cos\theta + \mu_s \sin\theta) = mg$ (1); $F_n(\mu_s\cos\theta - \sin\theta) = 0$ (2)
3. Solve (2) for θ and evaluate — $\theta = \tan^{-1}(\mu_s)$; $\theta = \tan^{-1}(0.08) = 4.57°$
4. Apply $\Sigma F = ma$ for $v \neq 0$ — $F_n(\cos\theta - \mu_s \sin\theta) = mg$ (1*a*); $F_n(\mu_s\cos\theta + \sin\theta) = mv^2/r$ (2*a*)
5. Substitute numerical values into (1*a*) and (2*a*) — $0.9904F_n = mg$; $0.1595F_n = mv^2/r$
6. Evaluate *r* for *v* = 16.67 m/s — $r = 176$ m

---

76 ••• A curve of radius 30 m is banked so that a 950-kg car traveling 40 km/h can round it even if the road is so icy that the coefficient of static friction is approximately zero. Find the range of speeds at which a car can travel around this curve without skidding if the coefficient of static friction between the road and the tires is 0.3.

This problem is similar to the preceding problem, and we shall use the free-body diagram of Problem 5-73.

1. Determine the banking angle — $\theta = \tan^{-1}(v^2/rg) = 22.8°$
2. Apply $\Sigma F = ma$ for $v = v_{min}$ (diagram $v = 0$ of Problem 5-75) — $F_n(\cos\theta + \mu_s\sin\theta) = mg$; $F_n(\mu_s\cos\theta - \sin\theta) = mv_{min}^2/r$
3. Evaluate for $\theta = 22.8°$, $\mu_s = 0.3$. — $1.038F_n = mg$; $0.1102F_n = mv_{min}^2/r$; $v_{min}^2 = 0.106rg$
4. Evaluate v_{mn} — $v_{min} = 5.59$ m/s = 20.1 km/h
5. Repeat steps 3, 4 using (1*a*) and (2*a*) of Problem 5-75 — $v_{max} = 15.57$ m/s = 56.1 km/h

---

77* • How would you expect the value of *b* for air resistance to depend on the density of air?

The constant *b* should increase with density as more air molecules collide with the object as it falls.

---

78 • True or false: The terminal speed of an object depends on its shape.

True.

---

79 • As a skydiver falls through the air, her terminal speed (*a*) depends on her mass. (*b*) depends on her orientation as she falls. (*c*) equals her weight. (d) depends on the density of the air. (*e*) depends on all of the above.

(*a*), (*b*), and (*d*)

---

80 • What are the dimensions and SI units of the constant *b* in the retarding force bv^n if (*a*) $n = 1$, and (*b*) $n = 2$?

(*a*) For $n = 1$, [b] = [F]/[v] = $[ML/T^2]/[L/T]$ = [M/T], kg/s; (*b*) for $n = 2$, [b] = $[ML/T^2]/[L^2/T^2]$ = [M/L], kg/m

---

81* • A small pollution particle settles toward the earth in still air with a terminal speed of 0.3 mm/s. The particle has a mass of 10^{-10} g and a retarding force of the form *bv*. What is the value of *b*?

When $v = v_t$, $bv = mg$, $b = mg/v_t$ — $b = (10^{-13} \times 9.81/3 \times 10^{-4})$ kg/s $= 3.27 \times 10^{-9}$ kg/s

---

82 • A Ping-Pong ball has a mass of 2.3 g and a terminal speed of 9 m/s. The retarding force is of the form bv^2. What is the value of *b*?

For $v = v_t$, $bv_t^2 = mg$, $b = mg/v_t^2$ — $b = [(2.3 \times 10^{-3} \times 9.81)/(9^2)]$ kg/m $= 2.79 \times 10^{-4}$ kg/m

---

83 • A sky diver of mass 60 kg can slow herself to a constant speed of 90 km/h by adjusting her form. (*a*) What is the magnitude of the upward drag force on the sky diver? (*b*) If the drag force is equal to bv^2, what is the value of b?

(*a*) Since $a = 0$, $F_d = mg = 589$ N. (*b*) (See Problem 5-82) $b = mg/v_t^2 = (589/25^2)$ kg/m $= 0.942$ kg/m

---

84 • Newton showed that the air resistance on a falling object with circular cross section should be approximately $\frac{1}{2}\rho\pi r^2v^2$, where $\rho = 1.2$ kg/m^3, the density of air. Find the terminal speed of a 56-kg sky diver, assuming that his cross-sectional area is equivalent to a disk of radius 0.30 m.

For $v = v_t$, $F_d = mg = \frac{1}{2}\pi\rho r^2v^2$; $v_t = \sqrt{(2mg)/(\rho\pi r^2)} = 56.9$ m/s

---

85* •• An 800-kg car rolls down a very long 6° grade. The drag force for motion of the car has the form $F_d = 100\text{ N} + (1.2\text{ N}\cdot\text{s}^2/\text{m}^2)v^2$. What is the terminal velocity of the car rolling down this grade?

1. Draw the free-body diagram. Note that the car moves at constant velocity, i.e., $a = 0$.

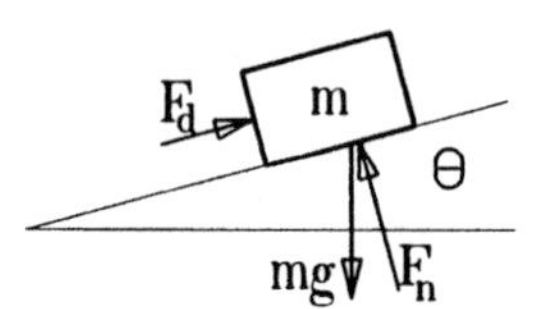

2. Apply $\Sigma F = ma$ — $F_d = mg\sin\theta$
3. Use F_d as given, $m = 800$ kg, and $\theta = 6°$ — $(100 + 1.2v^2)$ N $= 820$ N
4. Evaluate $v = v_t$ — $v_t = 24.5$ m/s $= 88.2$ km/h

---

86 •• While claims of hailstones the size of golf balls may be a slight exaggeration, hailstones are often substantially larger than raindrops. Estimate the terminal velocity of a raindrop and a large hailstone. (See Problem 5-84.)

1. Estimate the radius of a raindrop and a hailstone — Raindrop, $r_r \simeq 0.5$ mm; hailstone, $r_h \simeq 1$ cm
2. Evaluate b_r and b_h using $b = \frac{1}{2}\pi\rho r^2$ — $b_r = 4.7\times10^{-7}$ kg/m; $b_h = 1.9\times10^{-4}$ kg/m
3. Find m_r and m_h using $m = 4\pi r^3\rho/3$ — $\rho_r = 10^3$ kg/m^3, $m_r = 5.2\times10^{-7}$ kg; $\rho_h = 920$ kg/m^3, $m_h = 3.8\times10^{-3}$ kg
4. Find $v_{t,r}$ and $v_{t,h}$ using $v_t = (mg/b)^{1/2}$ — $v_{t,r} \simeq 3$ m/s; $v_{t,h} \simeq 14$ m/s

---

87 •• (*a*) A parachute creates enough air resistance to keep the downward speed of an 80-kg sky diver to a constant 6 m/s. Assuming the force of air resistance is given by $f = bv^2$, calculate b for this case. (*b*) A sky diver free-falls until his speed is 60 m/s before opening his parachute. If the parachute opens instantaneously, calculate the initial upward force exerted by the chute on the sky diver moving at 60 m/s. Explain why it is important that the parachute takes a few seconds to open.

(*a*) For $v = v_t$, $f = mg = bv_t^2$; solve for and find b — $b = mg/v_t^2 = (80\times9.81/36)$ kg/m $= 21.8$ kg/m

(*b*) Find f — $f = 78.48$ kN, corresponds to $a = 100g$

This initial acceleration would cause internal damage

---

88 ••• An object falls under the influence of gravity and a drag force $F_d = -bv$. (*a*) By applying Newton's second law, show that the acceleration of the object can be written $a = dv/dt = g - (b/m)v$. (*b*) Rearrange this equation to obtain $dv/(v - v_t) = -(g/v_t)dt$, where $v_t = mg/b$. (*c*) Integrate this equation to obtain the exact solution $v = \frac{mg}{b}\left(1 - e^{-bt/m}\right) = v_t\left(1 - e^{-gt/v_t}\right)$. (*d*) Plot v versus t for $v_t = 60$ m/s.

(*a*) From Newton's second law, $ma = F_{\text{net}} = mg - bv$. Divide both sides by m and replace a by dv/dt to obtain the result.

(*b*) Multiply both sides by $-dt/[g - (b/m)v]$; the multiply by (b/m) and replace gm/b by v_t.

(*c*) Integrate over v between the limits of 0 and v; integrate over t between the limits of 0 and t.

$\int_0^v \frac{dv}{v - v_t} = \ln\left(\frac{v_t - v}{v_t}\right) = \ln(1 - v/v_t) = -\frac{g}{v_t}\int_0^t dt = -\frac{gt}{v_t}$; take antilogs of both sides to obtain $1 - v/v_t = e^{-gt/v_t}$. Solving for v gives $v = v_t(1 - e^{-gt/v_t})$.

(*d*) The graph of v versus t for $v_t = 60$ m/s is shown.

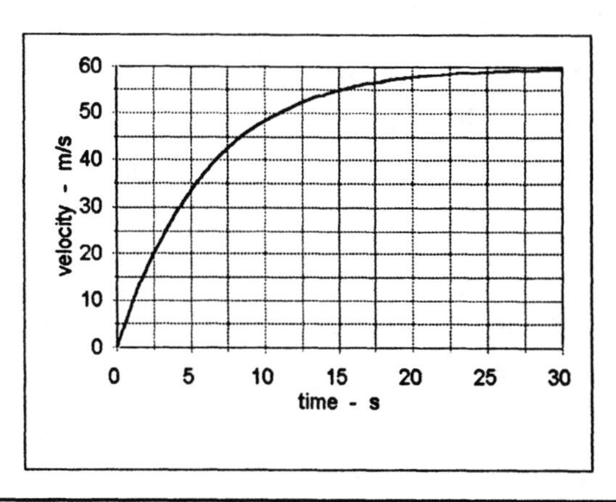

89* ••• Small spherical particles experience a viscous drag force given by Stokes' law: $F_d = 6\pi\eta rv$, where r is the radius of the particle, v is its speed, and η is the viscosity of the fluid medium. (*a*) Estimate the terminal speed of a spherical pollution particle of radius 10^{-5} m and density 2000 kg/m^3. (*b*) Assuming that the air is still and η is 1.8×10^{-5} N·s/m^2, estimate the time it takes for such a particle to fall from a height of 100 m.

Assume a spherical particle. Also, neglect the time required to attain terminal velocity; we will later confirm that this assumption is justified.

(*a*) Using Stokes's law and $m = (4/3)\pi r^3\rho$ solve for v_t — $v_t = (2r^2\rho g)/(9\eta) = 2.42$ cm/s

(*b*) Find the time to fall 100 m at 2.42 cm/s — $t = (10^4 \text{ cm})/(2.42 \text{ cm/s}) = 4.13 \times 10^3 \text{ s} = 1.15 \text{ h}$

Find time, t', to reach v_t (see Problem 5-88) — $t' \approx 5v_t/g = 12$ ms <<< 1.15 h; neglect of t' is justified

90 ••• An air sample containing pollution particles of the size and density given in Problem 5-89 is captured in a test tube 8.0 cm long. The test tube is then places in a centrifuge with the midpoint of the test tube 12 cm from the center of the centrifuge. The centrifuge spins at 800 revolutions per minute. Estimate the time required for nearly all of the pollution particles to sediment at the end of the test tube and compare this to the time required for a pollution particle to fall 8 cm under the action of gravity and subject to the viscous drag of air.

1. The effective acceleration is $a_c = r\omega^2$ — $r\omega^2 = [0.12 \times (2\pi \times 800/60)^2]$ m/s^2 = 840 m/s^2 >> g
2. Find $v_t = (2r^2\rho a_c)/(9\eta)$ (See Problem 5-89) — $v_t = (2.42 \text{ cm/s})(840/9.81) \approx 200$ cm/s
3. Find time to move 8 cm — $t = (8/200)$ s = 40 ms
4. Find time to fall under g, i.e., at 2.42 cm/s — $t_g = (8/2.42)$ s $\approx$ 3 s

91 • The mass of the moon is about 1% that of the earth. The centripetal force that keeps the moon in its orbit around the earth (*a*) is much smaller than the gravitational force exerted by the moon on the earth. (*b*) depends on the phase of the moon. (*c*) is much greater than the gravitational force exerted by the moon on the earth. (*d*) is the same as the gravitational force exerted by the moon on the earth. (*e*) I cannot answer; we haven't studied Newton's law of gravity yet.

(*d*) by Newton's third law.

---

92 • True or false: Centripetal force is one of the four fundamental forces.

False

93* • On an icy winter day, the coefficient of friction between the tires of a car and a roadway might be reduced to one-half its value on a dry day. As a result, the maximum speed at which a curve of radius R can be safely negotiated is (*a*) the same as on a dry day. (*b*) reduced to 70% of its value on a dry day. (*c*) reduced to 50% of its value on a dry day. (*d*) reduced to 37% of its value on a dry day. (*e*) reduced by an unknown amount depending on the car's mass.

(*b*) $v_{max} = (\mu_s gR)^{½}$; therefore $v'_{max} = v_{max}/\sqrt{2}$.

---

94 • A 4.5-kg block slides down an inclined plane that makes an angle of 28° with the horizontal. Starting from rest, the block slides a distance of 2.4 m in 5.2 s. Find the coefficient of kinetic friction between the block and plane.

1. Draw the free-body diagram

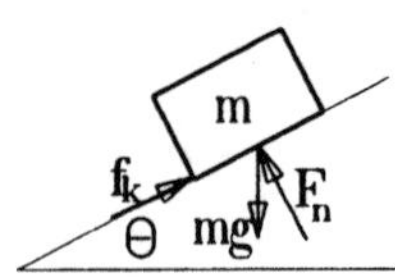

2. $s = ½at^2$; solve for and find a	$a = 2s/t^2 = 0.1775\ \text{m/s}^2$
3. Apply $\Sigma F = ma$	$mg \sin\theta - f_k = ma$; $F_n = mg\cos\theta$
4. Set $f_k = \mu_k F_n$ and solve for μ_k	$\mu_k = (g\sin\theta - a)/(g\cos\theta)$
5. Find μ_k for $a = 0.1775\ \text{m/s}^2$, $\theta = 28°$	$\mu_k = 0.51$

---

95 • A model airplane of mass 0.4 kg is attached to horizontal string and flies in a horizontal circle of radius 5.7 m. (The weight of the plane is balanced by the upward "lift" force of the air on the wings of the plane.) The plane makes 1.2 revolutions ever 4 s. (*a*) Find the speed v of the plane. (*b*) Find the tension in the string.

(*a*) $v = r\omega = 2\pi r/T$	$v = [2\pi \times 5.7/(4/1.2)]\ \text{m/s} = 10.7\ \text{m/s}$
(*b*) F (tension) $= ma = ma_c = mv^2/r$	$F = (0.4 \times 10.7^2/5.7)\ \text{N} = 8.0\ \text{N}$

---

96 •• Show with a force diagram how a motorcycle can travel in a circle on the inside vertical wall of a hollow cylinder. Assume reasonable parameters (coefficient of friction, radius of the circle, mass of the motorcycle, or whatever is required), and calculate the minimum speed needed.

We shall take the following values for the numerical calculation: $R = 6.0$ m, $\mu_s = 0.8$.

1. The appropriate free-body diagram is shown. The normal reaction force F_n provides the centripetal force, and the force of static friction, $\mu_s F_n$ keeps the cycle from sliding down the wall.

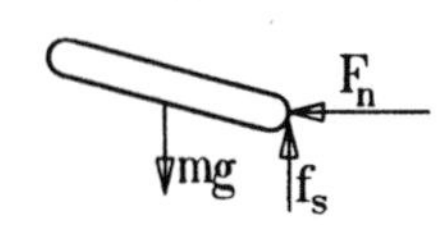

2. Apply $\Sigma F = ma$ $\mu_s F_n = mg$; $F_n = mv^2/R$

3. Solve for and evaluate $v = v_{min}$ $v_{min} = \sqrt{Rg/\mu_s} = 8.6$ m/s $= 31$ km/h

---

97* •• An 800-N box rests on a plane inclined at 30° to the horizontal. A physics student finds that she can prevent the box from sliding if she pushes with a force of at least 200 N parallel to the surface. (*a*) What is the coefficient of static friction between the box and the surface? (*b*) What is the greatest force that can be applied to the box parallel to the incline before the box slides up the incline?

(*a*) 1. Draw the free-body diagram.

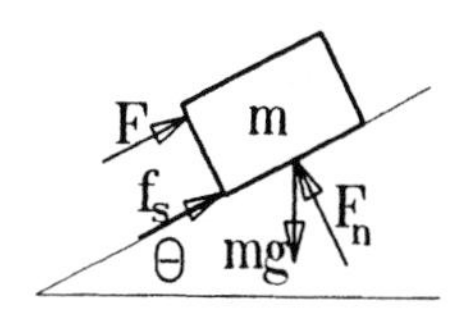

2. Apply $\Sigma F = ma$ $F + f_s - mg \sin\theta = 0$; $F_n = mg\cos\theta$

3. Use $f_{s,max} = \mu_s F_n$ and solve for and find μ_s $\mu_s = \tan\theta - F/(mg\cos\theta)$; $\mu_s = 0.289$

(*b*) 1. Find $f_{s,max}$ from part (*a*) $f_{s,max} = mg\sin\theta - F = 400\text{ N} - 200\text{ N} = 200\text{ N}$

2. Reverse the direction of $f_{s,max}$ and evaluate F $F = mg\sin\theta + f_{s,max} = 400\text{ N} + 200\text{ N} = 600\text{ N}$

---

98 • The position of a particle is given by the vector $\boldsymbol{r} = -10\text{ m}\cos\omega t\,\boldsymbol{i} + 10\text{ m}\sin\omega t\,\boldsymbol{j}$, where $\omega = 2\text{ s}^{-1}$. (*a*) Show that the path of the particle is a circle. (*b*) What is the radius of the circle? (*c*) Does the particle move clockwise or counterclockwise around the circle? (*d*) What is the speed of the particle? (*e*) What is the time for one complete revolution?

(*a*), (*b*) We need to show that r is constant. $r = \sqrt{r_x^2 + r_y^2} = \sqrt{100(\cos^2\omega t + \sin^2\omega t)}\text{ m} = 10\text{ m}$.

(*c*) Note that at $t = 0$, $x = -10$ m, $y = 0$; at $t = \Delta t$ (Δt small), the particle is at $x \approx -10$ m, $y = \Delta y$, where Δy is positive. It follows that the motion is clockwise.

(*d*) $\boldsymbol{v} = d\boldsymbol{r}/dt = (10\omega\sin\omega t)\,\boldsymbol{i}\text{ m} + (10\omega\cos\omega t)\,\boldsymbol{j}\text{ m}$; $v = \sqrt{v_x^2 + v_y^2} = 10\omega = 20$ m/s

(*e*) $T = 2\pi/\omega = \pi$ s

---

99 •• A crate of books is to be put on a truck with the help of some planks sloping up at 30°. The mass of the crate is 100 kg, and the coefficient of sliding friction between it and the plank is 0.5. You and your friends push *horizontally* with a force $\boldsymbol{F}$. Once the crate has started to move, how large must F be in order to keep the crate moving at constant speed?

1. Draw the free-body diagram

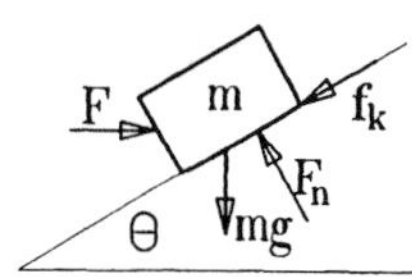

2. Note that $a = 0$. Apply $\Sigma F = ma$ $F\cos\theta - \mu_k F_n - mg\sin\theta = 0$; $F_n - F\sin\theta - mg\cos\theta = 0$

3. Solve for F and evaluate

$$F = \frac{mg(\sin\theta + \mu_k\cos\theta)}{\cos\theta - \mu_k\sin\theta};\ F = 1486\text{ N}$$

---

100 •• Brother Bernard is a very large dog with a taste for tobogganing. Ernie gives him a ride down Idiots' Hill—so named because it is steep slope that levels out at the bottom for 10 m, and then drops into a river. When they reach the level ground at the bottom, their speed is 40 km/h, and Ernie, sitting in front, starts to dig in his heels to make the toboggan stop. He knows, however, that if he brakes too hard, he will be mashed by Brother Bernard. If the coefficient of static friction between the dog and the toboggan is 0.8, what is the minimum stopping distance that will keep Brother Bernard off Ernie's back?

1. Draw the free-body diagram

m, f_s, F_n, mg

2. Apply $\Sigma F = ma$ — $F_n = mg$; $f_{s,max} = \mu_s F_n = ma_{max}$; $a_{max} = \mu_s g$
3. $s_{min} = v^2/2a_{max} = v^2/2\mu_s g$ — $s_{min} = 7.86$ m

---

101* •• An object with a mass of 5.5 kg is allowed to slide from rest down an inclined plane. The plane makes an angle of 30° with the horizontal and is 72 m long. The coefficient of kinetic friction between the plane and the object is 0.35. The speed of the object at the bottom of the plane is (*a*) 5.3 m/s. (*b*) 15 m/s. (*c*) 24 m/s. (*d*) 17 m/s. (*e*) 11 m/s.

1. Draw the free-body diagram

y, x, m, f_k, F_n, θ, mg

2. Apply $\Sigma F = ma$ — $mg \sin\theta - \mu_k F_n = ma$; $F_n - mg\cos\theta = 0$
3. Solve for a — $a = g(\sin\theta - \mu_k \cos\theta)$
4. Use $v^2 = 2as$ — $v = \sqrt{2(g\sin\theta - \mu_k g\cos\theta)s} = 16.7$ m/s
5. (*d*) is correct

---

102 •• A brick slides down an inclined plank at constant speed when the plank is inclined at an angle θ_0. If the angle is increased to θ_1, the block accelerates down the plank with acceleration a. The coefficient of kinetic friction is the same in both cases. Given θ_0 and θ_1, calculate a.

The free-body diagram is the same as for the preceding problem. We now have $mg\sin\theta_0 = f_k = \mu_k F_n$, and $mg\cos\theta_0 = F_n$. Solving for μ_k we obtain $\mu_k = \tan\theta_0$. With $\theta = \theta_1$, $mg\sin\theta_1 - \mu_k mg\cos\theta_1 = ma$, and using the result $\mu_k = \tan\theta_0$ one finds $a = g(\sin\theta_1 - \tan\theta_0\cos\theta_1)$.

---

103 •• One morning, Lou was in a particularly deep and peaceful slumber. Unfortunately, he had spent the night in the back of a dump truck, and Barry, the driver, was keen to go off to work and start dumping things. Rather than risk a ruckus with Lou, Barry simply raised the back of the truck, and when it reached an angle of 30°, Lou slid down the 4-m incline in 2 s, plopped onto a pile of sand, rolled over, and continued to sleep. Calculate the coefficients of static and kinetic friction between Lou and the truck.

1. Use Equ. 5-4 — $\mu_s = \tan 30° = 0.577$
2. Apply $\Sigma F = ma$ and use $s = \frac{1}{2}at^2$ — $F_n = mg\cos\theta$; $mg\sin\theta - \mu_k mg\cos\theta = ma$; $a = 2s/t^2$
3. Solve for and evaluate μ_k — $\mu_k = \tan\theta - 2s/(t^2 g\cos\theta)$; $\mu_k = 0.342$

---

104 •• In a carnival ride, the passenger sits on a seat in a compartment that rotates with constant speed in a vertical circle of radius $r = 5$ m. The heads of the seated passengers always point toward the axis of rotation. (*a*) If the carnival ride completes one full circle in 2 s, find the acceleration of the passenger. (*b*) Find the slowest rate of rotation (in other word, the longest time T to complete one full circle) if the seat belt is to exert no force on the passenger at the top of the ride.

(*a*) $a = r\omega^2 = 4\pi^2 r/T^2$ $a = 5\pi^2$ m/s^2

(*b*) In this case, $a = a_c = g$; solve for and find T $T = 2\pi(r/g)^{1/2}$; $T = 4.49$ s

---

105* •• A flat-topped toy cart moves on frictionless wheels, pulled by a rope under tension T. The mass of the cart is m_1. A load of mass m_2 rests on top of the cart with a coefficient of static friction μ_s. The cart is pulled up a ramp that is inclined at an angle θ above the horizontal. The rope is parallel to the ramp. What is the maximum tension T that can be applied without making the load slip?

1. Draw the free-body diagrams for the two objects.

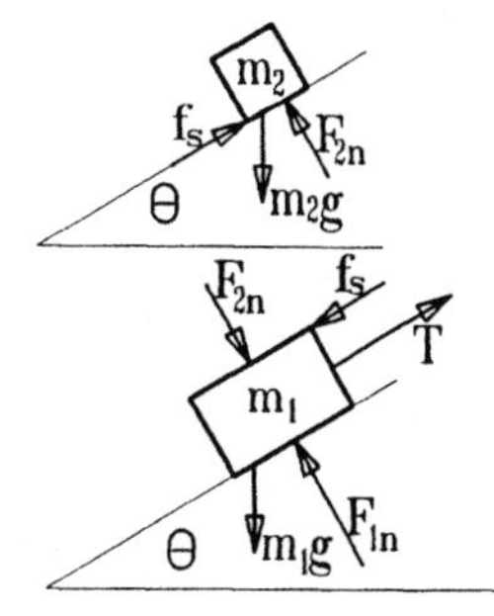

1. m_2 is accelerated by f_s. Apply $\Sigma F = ma$ — $F_{n2} = m_2 g \cos\theta$; $\mu_s m_2 g \cos\theta - m_2 g \sin\theta = m_2 a_{max}$

2. Solve for a_{max} — $a_{max} = g(\mu_s \cos\theta - \sin\theta)$

3. The masses move as single unit. Apply $\Sigma F = ma$ — $T - (m_1 + m_2)g\sin\theta = (m_1 + m_2)g(\mu_s\cos\theta - \sin\theta)$

4. Solve for T — $T = (m_1 + m_2)g\mu_s\cos\theta$

---

106 •• A sled weighing 200 N rests on a 15° incline, held in place by static friction (Figure 5-58). The coefficient of static friction is 0.5. (*a*) What is the magnitude of the normal force on the sled? (*b*) What is the magnitude of the static friction on the sled? (*c*) The sled is now pulled up the incline at constant speed by a child. The child weighs 500 N and pulls on the rope with a constant force of 100 N. The rope makes an angle of 30° with the incline and has negligible weight. What is the magnitude of the kinetic friction force on the sled? (*d*) What is the coefficient of kinetic friction between the sled and the incline? (*e*) What is the magnitude of the force exerted on the child by the incline?

Draw the free-body diagrams for the sled (m_1) and the child (m_2).

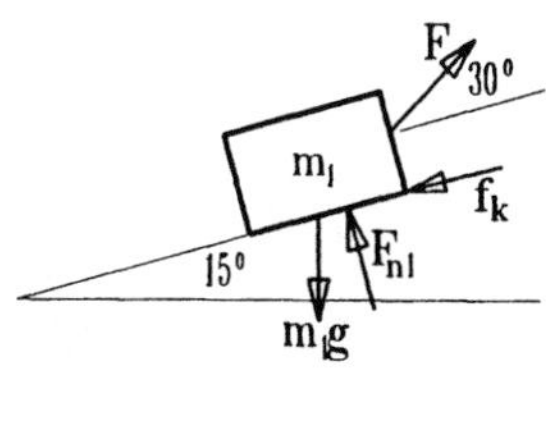

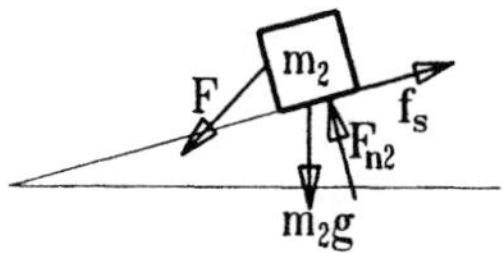

(*a*) $F_n = m_1 g\cos\theta$ — $F_n = 200\cos 15° = 193$ N

(*b*) Apply $\Sigma F = ma$ — $f_s = 200\sin 15° = 51.8$ N

(*c*) Determine if the sled moves — $100\cos 30° - 200\sin 15° - f_{s,max} = F_{net}$;

$F_{n1} = 200\cos 15° - 100\sin 30° = 143$ N; $f_{s,max} = 0.5 \times F_{n1}$;

The sled does not move! f_k is undetermined. — $f_{s,max} = 71.5$ N. $F_{net} = -36.7 \text{ N} < 0$

(*d*) μ_k undetermined.

(*e*) Apply $\Sigma \boldsymbol{F} = m\boldsymbol{a}$ — $F_{n2} = (500\cos 15° + 100\sin 30°)$ N $= 533$ N;

Note that the child is stationary — $F_x = (500\sin 30° + 100\cos 30°)$ N $= 216$ N

$F = (F_{n2}^2 + F_x^2)^{1/2}$ — $F = 575$ N

---

107 •• A child slides down a slide inclined at 30° in time t_1. The coefficient of kinetic friction between her and the slide is μ_k. She finds that if she sits on a small cart with frictionless wheels, she slides down the same slide in time $\frac{1}{2}t_1$. Find μ_k.

1. Apply $\Sigma \boldsymbol{F} = m\boldsymbol{a}$ for case with and without friction — $a_1 = g(\sin 30° - \mu_k\cos 30°)$; $a_2 = g\sin 30°$
2. $s = \frac{1}{2}a_1t_1^2 = \frac{1}{2}a_2t_2^2 = a_2t_1^2/8$; $a_2/a_1 = 4$; solve for μ_k — $\mu_k = (3/4)\tan 30° = 0.433$

---

108 •• The position of a particle of mass $m = 0.8$ kg as a function of time is $\boldsymbol{r} = x\boldsymbol{i} + y\boldsymbol{j} = R\sin\omega t\,\boldsymbol{i} + R\cos\omega t\,\boldsymbol{j}$, where $R = 4.0$ m, and $\omega = 2\pi\text{ s}^{-1}$. (*a*) Show that this path of the particle is a circle of radius R with its center at the origin. (*b*) Compute the velocity vector. Show that $v_x/v_y = -y/x$. (*c*) Compute the acceleration vector and show that it is in the radial direction and has the magnitude v^2/r. (*d*) Find the magnitude and direction of the net force acting on the particle.

(*a*) See Problem 5-98. $r = R = 4$ m.

(*b*) See Problem 5-98. $\mathbf{v} = (\omega R\cos\omega t)\,\mathbf{i} - (\omega R\sin\omega t)\,\mathbf{j} = [(8\pi\cos\omega t)\,\mathbf{i} - (8\pi\sin\omega t)\,\mathbf{j}]$ m/s;

$v_x/v_y = -\cot\omega t = -y/x$.

(*c*) $\boldsymbol{a} = d\boldsymbol{v}/dt = [(-16\pi^2\sin\omega t)\,\boldsymbol{i} - (16\pi^2\cos\omega t)\boldsymbol{j}]$ m/s²; note that $\boldsymbol{a} = -4\pi^2\boldsymbol{r}$, i.e., in the radial direction toward origin. The magnitude of $\boldsymbol{a}$ is $16\pi^2$ m/s² $= [(8\pi)^2/4]$ m/s² $= v^2/r$.

(*d*) $F = ma = 12.8\pi^2$ N; the direction of $\boldsymbol{F}$ is that of $\boldsymbol{a}$, i.e., toward the center of the circle.

109* •• In an amusement-park ride, riders stand with their backs against the wall of a spinning vertical cylinder. The floor falls away and the riders are held up by friction. If the radius of the cylinder is 4 m, find the minimum number of revolutions per minute necessary to prevent the riders from dropping when the coefficient of static friction between a rider and the wall is 0.4.

1. Apply $\Sigma \boldsymbol{F} = m\boldsymbol{a}$ — $F_n = mr\omega^2$; $f_{s,max} = \mu_s F_n = mg$
2. Solve for and evaluate ω — $\omega = (g/\mu_s r)^{1/2}$; $\omega = 2.476$ rad/s $= 23.6$ rpm

---

110 •• Some bootleggers race from the police down a road that has a sharp, level curve with a radius of 30 m. As they go around the curve, the bootleggers squirt oil on the road behind them, reducing the coefficient of static friction from 0.7 to 0.2. When taking this curve, what is the maximum safe speed of (*a*) the bootleggers' car, and (*b*) the police car?

(*a*), (*b*) Apply $\Sigma \boldsymbol{F} = m\boldsymbol{a}$ — $F_n = mg$; $f_{s,max} = \mu_s mg = mv_{max}^2/r$

Solve for v_{max} — $v_{max} = (\mu_s gr)^{1/2}$

(*a*) Evaluate v_{max} for $\mu_s = 0.7$ — $v_{max} = 14.35$ m/s $= 51.7$ km/h

(*b*) Evaluate v_{max} for $\mu_s = 0.2$ — $v_{max} = 7.67$ m/s $= 27.6$ km/h

111 •• A mass m_1 on a horizontal shelf is attached by a thin string that passes over a frictionless peg to a 2.5-kg mass m_2 that hangs over the side of the shelf 1.5 m above the ground (Figure 5-59). The system is released from rest at $t = 0$ and the 2.5-kg mass strikes the ground at $t = 0.82$ s. The system is now placed in its initial position and a 1.2-kg mass is placed on top of the block of mass m_1. Released from rest, the 2.5-kg mass now strikes the ground 1.3 seconds later. Determine the mass m_1 and the coefficient of kinetic friction between m_1 and the shelf.

1. Draw the free-body diagrams.

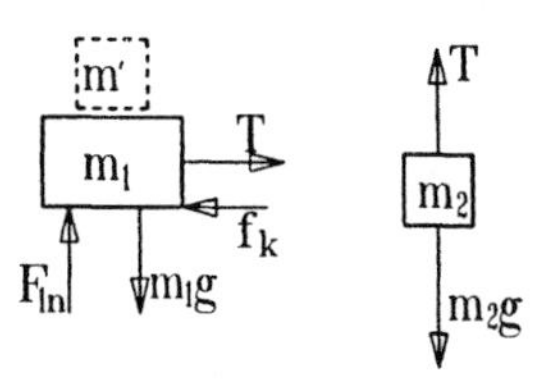

2. Use $s = \frac{1}{2}at^2$ to find the acceleration a_1 of the first run. — $a_1 = (3\text{ m})/(0.82\text{ s})^2$; $a_1 = 4.46\text{ m/s}^2$.
3. Apply $\Sigma F = ma$ to m_2 — $T_1 - m_2g = -m_2a_1$;
4. Evaluate T_1 — $T_1 = (2.5\text{ kg})[(9.81 - 4.46)\text{ m/s}^2]$; $T_1 = 13.375$ N
5. Apply $\Sigma F = ma$ to m_1. — $13.375\text{ N} - \mu_k m_1 g = m_1(4.46\text{ m/s}^2)$. (1)
6. Repeat part 2 for the second run to find a_2. — $a_2 = (3\text{ m})/1.3\text{ s})^2 = 1.775\text{ m/s}^2$
7. Repeat parts 3 and 4 to find T_2. — $T_2 = (2.5\text{ kg})[(9.81 - 1.775)\text{ m/s}^2]$; $T_2 = 20.1$ N
8. Apply $\Sigma F = ma$ to $m_1 + 1.2$ kg. — $20.1\text{ N} - \mu_k(m_1 + 1.2\text{ kg})g = (1.775\text{ m/s}^2)(m_1 + 1.2\text{ kg})$
9. Simplify the preceding result. — $17.97\text{ N} - \mu_k m_1 g - (1.2\text{ kg})\mu_k g = (1.775\text{ m/s}^2)m_1$ (2)
10. Solve (1) for μ_k — $\mu_k = [(13.375\text{ N}) - (4.46\text{ m/s}^2)m_1]/m_1 g$ (3)
11. Substitute (3) into (2), simplify to obtain a quadratic equation for m_1. — $2.685m_1^2 + 9.947m_1 - 16.05 = 0$
12. Use the standard solution for m_1. — $$m_1 = \frac{-9.947 \pm \sqrt{9.947^2 + 4(2.655)(16.05)}}{2(2.685)}\text{ kg}$$
13. m_1 must be positive, only one solution applies. — $m_1 = (-1.85 \pm 3.07)$ kg; $m_1 = 1.22$ kg.
14. Substitute $m_1 = 1.22$ kg into (3) and evaluate μ_k — $\mu_k = 0.672$

112 ••• (*a*) Show that a point on the surface of the earth at latitude θ has an acceleration relative to a reference frame not rotating with the earth with a magnitude of $3.37\cos\theta\text{ cm/s}^2$. What is the direction of this acceleration? (*b*) Discuss the effect of this acceleration on the apparent weight of an object near the surface of the earth. (*c*) The free-fall acceleration of an object at sea level measured *relative to the earth's surface* is 9.78 m/s^2 at the equator and 9.81m/s^2 at latitude $\theta = 45°$. What are the values of the gravitational field g at these points?

(*a*) $R = 6.37 \times 10^8$ cm is the radius of the earth. At a latitude of θ the distance from the surface of the earth to the axis of rotation is $r = R\cos\theta$. The rotational speed of the earth is $\omega = 2\pi/86400\text{ rad/s} = 7.27 \times 10^{-5}\text{ rad/s}$. The acceleration is $r\omega^2 = (6.37 \times 10^8\cos\theta)(7.27 \times 10^{-5})^2\text{ cm/s}^2 = 3.37\cos\theta\text{ cm/s}^2$. The acceleration is directed toward the axis of rotation.

(*b*) Since the force of gravity supplies the required centripetal force, the acceleration of gravity at the surface of the earth is reduced in magnitude. Consequently, the apparent weight is slightly reduced. This effect is greatest at the equator.

(*c*) The free-body diagram for a mass m at $\theta = 45°$ is shown. We also show the acceleration at $\theta = 45°$.

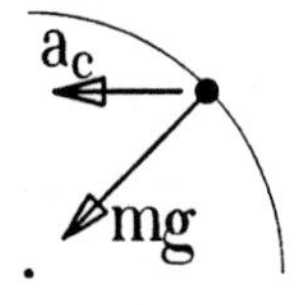

1. At $\theta = 0°$, $g_{\text{eff}} = g - a_c$ | $g = g_{\text{eff}} + a_c = (978 + 3.37)\ \text{cm/s}^2 = 981.4\ \text{cm/s}^2$
2. At $\theta = 45°$, $\boldsymbol{g}$ and $\boldsymbol{a}_c$ are not colinear (see diagram). Use law of cosines. | $g_{\text{eff}}^2 = g^2 + a_c^2 - 2ga_c\cos 45° = g^2 + a_c^2 - 1.41ga_c$

   $g^2 - 4.75g - 962350 = 0$

   Find solution of the quadratic equation. | $g = 983.4\ \text{cm/s}^2$

CHAPTER 6

Work and Energy

1* • True or false: (*a*) Only the net force acting on an object can do work. (*b*) No work is done on a particle that remains at rest. (*c*) A force that is always perpendicular to the velocity of a particle never does work on the particle.

(*a*) False (*b*) True (*c*) True

2 • A heavy box is to be moved from the top of one table to the top of another table of the same height on the other side of the room. Is work required to do this?

No; there is no change in potential or kinetic energy.

3 • To get out of bed in the morning, do you have to do work?

Yes, if your center of mass is higher when standing than when lying in bed. Otherwise, no.

4 • By what factor does the kinetic energy of a car change when its speed is doubled?

Note that $K \propto v^2$; $K' = 4K$.

5* • An object moves in a circle at constant speed. Does the force that accounts for its acceleration do work on it? Explain.

No; $dW = \mathbf{F}\cdot d\mathbf{r}$ and here $\mathbf{F}$ is perpendicular to $d\mathbf{r}$.

6 • An object initially has kinetic energy K. The object then moves in the opposite direction with three times its initial speed. What is the kinetic energy now? (*a*) K (*b*) $3K$ (*c*) $-3K$ (*d*) $9K$ (*e*) $-9K$.

(*d*) $K \propto v^2$.

7 • A 15-g bullet has a speed of 1.2 km/s. (*a*) What is its kinetic energy in joules? (*b*) What is its kinetic energy if its speed is halved? (*c*) What is its kinetic energy if its speed is doubled?

(*a*) $K = ½mv_2$ $\quad$ $K = ½(0.015)(1.2 \times 10^3)^2$ J $= 10.8$ kJ

(*b*) K is reduced by a factor of 4 $\quad$ $K = (10.8$ kJ$)/4 = 2.7$ kJ

(*c*) K is increased by a factor of 4 $\quad$ $K = 4(10.8$ kJ$) = 43.2$ kJ

8 • Find the kinetic energy in joules of (*a*) a 0.145-kg baseball moving with a speed of 45 m/s and (*b*) a 60-kg jogger running at a steady pace of 9 min/mi.

(*a*) $K = ½mv^2$ $\quad$ $K = ½(0.145)(45)^2$ J $= 147$ J

(*b*) $K = ½mv^2$; (1/9)mi/min=0.179km/min=2.98m/s $\quad$ $K = ½(60$ kg$))(2.98$ m/s$)^2 = 267$ J

9* • A 6-kg box is raised from rest a distance of 3 m by a vertical force of 80 N. Find (*a*) the work done by the force, (*b*) the work done by gravity, and (*c*) the final kinetic energy of the box.

(*a*) $W = Fy$ — $W = 3 \times 80\text{ J} = 240\text{ J}$

(*b*) $W_g = -mgy$ — $W_g = -(6)(9.81)(3)\text{ J} = -177\text{ J}$

(*c*) $K = W + W_g$ — $K = 63\text{ J}$

10 • A constant force of 80 N acts on a box of mass 5.0 kg that is moving in the direction of the applied force with a speed of 20 m/s. Three seconds later the box is moving with a speed of 68 m/s. Determine the work done by this force.

$W = K_f - K_i = \frac{1}{2}m(v_f^2 - v_i^2)$ — $W = \frac{1}{2}(5)[(68)^2 - (20)^2]\text{ J} = 10.56\text{ kJ}$

11 •• You run a race with your girlfriend. At first you each have the same kinetic energy, but you find that she is beating you. When you increase your speed by 25%, you are running at the same speed she is. If your mass is 85 kg, what is her mass?

$\frac{1}{2}m_1v^2 = \frac{1}{2}m_2(1.25v)^2$ — $m_1 = 85$ kg; solve for $m_2 = (85/1.56)\text{ kg} = 54.4\text{ kg}$

12 • How does the work required to stretch a spring 2 cm from its natural length compare with that required to stretch it 1 cm from its natural length?

$W = \frac{1}{2}kx^2$; $W \propto x^2$; therefore work is increased by factor of 4.

13* •• A 3-kg particle is moving with a speed of 2 m/s when it is at $x = 0$. It is subjected to a single force F_x that varies with position as shown in Figure 6-30. (*a*) What is the kinetic energy of the particle when it is at $x = 0$? (*b*) How much work is done by the force as the particle moves from $x = 0$ to $x = 4$ m? (*c*) What is the speed of the particle when it is at $x = 4$ m?

(*a*) $K(0) = \frac{1}{2}mv^2$ — $K(0) = \frac{1}{2}(3)(4)\text{ J} = 6\text{ J}$

(*b*) W = area under curve — $W = 2 \times 6\text{ J} = 12\text{ J}$

(*c*) $K(4) = K(0) + W = \frac{1}{2}mv^2$ — $\frac{1}{2}(3)v^2 = 18$ J; $v = 3.46$ m/s

14 •• A 4-kg particle is initially at rest at $x = 0$. It is subjected to a single force F_x that varies with position as shown in Figure 6-31. Find the work done by the force as the particle moves (*a*) from $x = 0$ to $x = 3$ m, and (*b*) from $x = 3$ m to $x = 6$ m. Find the kinetic energy of the particle when it is at (*c*) $x = 3$ m and (*d*) $x = 6$ m.

(*a*) W = area under curve from $x = 0$ to $x = 3$ — $W(0-3) = 7.5$ J

(*b*) See above but for $x = 3$ to $x = 6$ — $W(3-6) = -3$ J

(*c*) Since $K(0) = 0$, $K(3) = W(0-3)$, $K(6) = W(0-6)$ — $K(3) = 7.5$ J; $K(6) = 4.5$ J

15 •• A force F_x acts on a particle. The force is related to the position of the particle by the formula $F_x = Cx^3$, where C is a constant. Find the work done by this force on the particle when the particle moves from $x = 1.5$ m to $x = 3$ m.

$W = \int F\,dx$ — $W = C\int_{1.5}^{3} x^3\,dx = \frac{C}{4}(3^4 - 1.5^4) = 19C\text{ J}$

16 •• Lou's latest invention, aimed at urban dog owners, is the X-R-Leash. It is made of a rubber-like material that exerts a force $F_x = -kx - ax^2$ when it is stretched a distance x, where k and a are constants. The ad claims,

"You'll never go back to your old dog leash after you've had the thrill of an X-R-Leash experience. And you'll see a new look of respect in the eyes of your proud pooch." Find the work done on a dog by the leash if the person remains stationary and the dog bounds off, stretching the X-R-Leash from $x = 0$ to $x = x_0$.

$W = \int F dx$ $\qquad$ $W = \int_0^{x_0}(-kx - ax^2)dx = -\frac{1}{2}kx_0^2 - \frac{1}{3}ax_0^3$

17* •• A 3-kg object is moving with a speed of 2.40 m/s in the x direction when it passes the origin. It is acted on by a single force F_x that varies with x as shown in Figure 6-32. (*a*) What is the work done by the force from $x = 0$ to $x = 2$ m? (*b*) What is the kinetic energy of the object at $x = 2$ m? (*c*) What is the speed of the object at $x = 2$ m? (*d*) What is the work done on the object from $x = 0$ to $x = 4$ m? (*e*) What is the speed of the object at $x = 4$ m?

(*a*) W = area under curve; count squares, each square = 0.125 J $\qquad$ Between $x = 0$ and $x = 2$ there are about 22 squares, corresponding to $W = 2.75$ J

(*b*) $K = K_i + W$ $\qquad$ $K = ½(3)(2.4)^2$ J + 2.75 J = 11.4 J

(*c*) $v = (2K/m)^{½}$ $\qquad$ $v = (2 \times 11.4/3)^{½} = 2.76$ m/s

(*d*) As in (*a*) count squares $\qquad$ Net number of squares = 28; $W = 3.5$ J

(*e*) Proceed as in (*b*) and (*c*) $\qquad$ $K = (8.64 + 3.5)$ J = 12.14 J; $v = 2.84$ m/s

---

18 •• Near Margaret's cabin is a 20-m water tower that attracts many birds during the summer months. During a hot spell last year, the tower went dry, and Margaret had to have her water hauled in. She got lonesome without the birds visiting, so she decided to carry some water up the tower to attract them back. Her bucket has a mass of 10 kg and holds 30 kg of water when it is full. However, the bucket has a hole, and as Margaret climbed at a constant speed, water leaked out at a constant rate. Several birds took advantage of the shower below, but when she got to the top, only 10 kg of water remained for the birdbath. (*a*) Write an expression for the mass of the bucket plus water as a function of the height y climbed. (*b*) Find the work done by Margaret on the bucket.

(*a*) Evidently, she loses 1 kg of water per meter $\qquad$ $M(y) = (40 - y)$ kg

(*b*) $W = \int M(y) g dy$ $\qquad$ $W = 9.81\int_0^{20}(40 - y)dy = 9.81[40 \times 20 - \frac{1}{2}(20)^2] = 5886$ J

---

19 • Suppose there is a net force acting on a particle but it does no work. Can the particle be moving in a straight line?

No; force is perpendicular to motion, therefore will cause departure from straight line motion.

---

20 • A 6-kg block slides down a frictionless incline making an angle of 60° with the horizontal. (*a*) List all the forces acting on the block, and find the work done by each force when the block slides 2 m (measured along the incline). (*b*) What is the total work done on the block? (*c*) What is the speed of the block after it has slid 1.5 m if it starts from rest? (*d*) What is its speed after 1.5 m if it starts with an initial speed of 2 m/s?

(*a*) Forces acting are $F_g = 6g$ down, $F_N = 6g \cos 60°$ normal to plane $\qquad$ $F_{net} = 6g \sin 60°$ down along plane. $W = 12 \times 9.81 \times 0.866 = 102$ J = work done by F_g; F_N does no work.

(*b*) $W_{tot} = W_g$ $\qquad$ $W_{tot} = 102$ J

(*c*) $v = (2K/m)^{½} = (2mg\Delta y/m)^{½} = (2g\Delta y)^{½}$ $\qquad$ $\Delta y = (1.5 \sin 60°)$ m; $v = (2 \times 9.81 \times 1.5 \times 0.866)^{½} = 5.05$ m/s

(*d*) $a = g \sin 60°$; use Equ. 2-15 $\qquad$ $v = (4 + 2 \times 9.81 \times 0.866 \times 1.5)^{½}$ m/s = 5.43 m/s

---

21* • An 85-kg cart is deposited on a 1.5-m platform after being rolled up an incline formed by a plank of length L that has been laid from the lower level to the top of the platform. (Assume the rolling is equivalent to sliding without friction.) (*a*) Find the force parallel to the incline needed to push the cart up without acceleration for $L = 3$, 4, and 5 m. (*b*) Calculate directly from Equation 6-15 the work needed to push the cart up the incline for each value of L. (*c*) Since the work found in (*b*) is the same for each value of L, what advantage, if any, is there in choosing one length over another?

(*a*) $F = mg \sin\theta = mg(1.5/L)$; $mg = 834$ N — $L = 3$ m, $F = 417$ N; $L = 4$ m, $F = 313$ N; $L = 5$ m, $F = 250$ N

(*b*) $W = F\cdot s$ — For $L = 3$ m, $W = 3 \times 417$ J $= 1.25$ kJ; same for other L

(*c*) Choosing longer length means one can exert a smaller force

---

22 • A 2-kg object attached to a horizontal string moves with a speed of 2.5 m/s in a circle of radius 3 m on a frictionless horizontal surface. (*a*) Find the tension in the string. (*b*) List the forces acting on the object, and find the work done by each force during one revolution.

(*a*) $T = mv^2/r$ — $T = 2(2.5)^2/3$ N $= 4.17$ N

(*b*) F_g, F_N, and T — None of the forces does any work.

---

23 • What is the angle between the vectors $\boldsymbol{A}$ and $\boldsymbol{B}$ if $\boldsymbol{A}\cdot\boldsymbol{B} = -AB$?

$\boldsymbol{A}\cdot\boldsymbol{B} = AB\cos\theta$; $\cos\theta = -1$; $\theta = 180°$.

---

24 • Two vectors $\boldsymbol{A}$ and $\boldsymbol{B}$ have magnitudes of 6 m and make an angle of 60° with each other. Find $\boldsymbol{A}\cdot\boldsymbol{B}$.

$\boldsymbol{A}\cdot\boldsymbol{B} = (36\cos 60°)\ \text{m}^2 = 18\ \text{m}^2$.

25* • Find $\boldsymbol{A}\cdot\boldsymbol{B}$ for the following vectors: (*a*) $\boldsymbol{A} = 3\,\boldsymbol{i} - 6\,\boldsymbol{j}$, $\boldsymbol{B} = -4\,\boldsymbol{i} + 2\boldsymbol{j}$; (*b*) $\boldsymbol{A} = 5\,\boldsymbol{i} + 5\boldsymbol{j}$, $\boldsymbol{B} = 2\,\boldsymbol{i} - 4\boldsymbol{j}$; and (*c*) $\boldsymbol{A} = 6\,\boldsymbol{i} + 4\boldsymbol{j}$, $\boldsymbol{B} = 4\,\boldsymbol{i} - 6\boldsymbol{j}$.

(*a*), (*b*), (*c*) Use Equ. 6-12 — (*a*) $\boldsymbol{A}\cdot\boldsymbol{B} = -24$; (*b*) $\boldsymbol{A}\cdot\boldsymbol{B} = -10$; (*c*) $\boldsymbol{A}\cdot\boldsymbol{B} = 0$.

---

26 • Find the angles between the vectors $\boldsymbol{A}$ and $\boldsymbol{B}$ in Problem 25.

(*a*), (*b*), (*c*) $\theta = \cos^{-1}(\boldsymbol{A}\cdot\boldsymbol{B}/AB)$ — (*a*) $AB = 30, \theta = 143°$; (*b*) $AB = 31.62, \theta = 108°$; (*c*) $\theta = 90°$

---

27 • A 2-kg object is given a displacement $\Delta\boldsymbol{s} = (3\,\boldsymbol{i} + 3\,\boldsymbol{j} - 2\,\boldsymbol{k})$ m along a straight line. During the displacement, a constant force $\boldsymbol{F} = (2\,\boldsymbol{i} - 1\,\boldsymbol{j} + 1\,\boldsymbol{k})$ N acts on the object. (*a*) Find the work done by $\boldsymbol{F}$ for this displacement. (*b*) Find the component of $\boldsymbol{F}$ in the direction of the displacement.

(*a*) $W = \boldsymbol{F}\cdot\Delta\boldsymbol{s}$ — $W = (6 - 3 - 2)$ J $= 1$ J

(*b*) $W = F\cdot s\cos\theta = (F\cos\theta)\Delta s$ — $F\cos\theta = (1\ \text{J})/(22\ \text{m}^2)^{1/2} = 0.213$ N

---

28 •• (*a*) Find the unit vector that is parallel to the vector $\boldsymbol{A} = A_x\,\boldsymbol{i} + A_y\boldsymbol{j} + A_z\,\boldsymbol{k}$. (*b*) Find the component of the vector $\boldsymbol{A} = 2\,\boldsymbol{i} - \boldsymbol{j} - \boldsymbol{k}$ in the direction of the vector $\boldsymbol{B} = 3\,\boldsymbol{i} + 4\boldsymbol{j}$.

(*a*) The vector of unit length and parallel to $\boldsymbol{A}$ is $\boldsymbol{u}_A = \boldsymbol{A}/A$.

(*b*) 1. Find the unit vector parallel to $\boldsymbol{B}$ — $\boldsymbol{u}_B = (3/5)\,\boldsymbol{i} + (4/5)\boldsymbol{j} = 0.6\,\boldsymbol{i} + 0.8\boldsymbol{j}$

2. The component of $\boldsymbol{A}$ along $\boldsymbol{B}$ is $\boldsymbol{A}\cdot\boldsymbol{u}_B$ — $\boldsymbol{A}\cdot\boldsymbol{u}_B = 1.2 + 0.8 = 2.0$

29* •• When a particle moves in a circle with constant speed, the magnitudes of its position vector and velocity vector are constant. (*a*) Differentiate $\mathbf{r}\cdot\mathbf{r} = r^2$ = constant with respect to time to show that $\mathbf{v}\cdot\mathbf{r} = 0$ and therefore $\mathbf{v}\perp\mathbf{r}$. (*b*) Differentiate $\mathbf{v}\cdot\mathbf{v} = v^2$ = constant with respect to time and show that $\mathbf{a}\cdot\mathbf{v} = 0$ and therefore $\mathbf{a}\perp\mathbf{v}$. What do the results of (*a*) and (*b*) imply about the direction of $\mathbf{a}$? (*c*) Differentiate $\mathbf{v}\cdot\mathbf{r} = 0$ with respect to time and show that $\mathbf{a}\cdot\mathbf{r} + v^2 = 0$ and therefore $a_r = -v^2/r$.

(*a*) $(d/dt)(\mathbf{r}\cdot\mathbf{r}) = \mathbf{r}\cdot(d\mathbf{r}/dt) + (d\mathbf{r}/dt)\cdot\mathbf{r} = 2\mathbf{v}\cdot\mathbf{r} = 0$. Therefore $\mathbf{v}\perp\mathbf{r}$.

(*b*) $(d/dt)(\mathbf{v}\cdot\mathbf{v}) = 2\mathbf{a}\cdot\mathbf{v} = 0$. Therefore $\mathbf{a}\perp\mathbf{v}$.

(*c*) The above implies that the component of $\mathbf{a}$ in the plane formed by $\mathbf{r}$ and $\mathbf{v}$ is colinear with $\mathbf{r}$.

(*d*) $(d/dt)(\mathbf{v}\cdot\mathbf{r}) = \mathbf{v}\cdot(d\mathbf{r}/dt) + \mathbf{r}\cdot(d\mathbf{v}/dt) = v^2 + \mathbf{r}\cdot\mathbf{a} = 0$. Therefore, $a_r = -v^2/r$.

30 •• Vectors $\mathbf{A}$, $\mathbf{B}$, and $\mathbf{C}$ for a triangle as shown in Figure 6-33. The angle between $\mathbf{A}$ and $\mathbf{B}$ is θ, and the vectors are related by $\mathbf{C} = \mathbf{A} - \mathbf{B}$. Compute $\mathbf{C}\cdot\mathbf{C}$ in terms of A, B, and θ, and derive the law of cosines, $C^2 = A^2 + B^2 - 2AB\cos\theta$.

1. Write $\mathbf{C}\cdot\mathbf{C}$ in terms of $\mathbf{A}$ and $\mathbf{B}$	$\mathbf{C}\cdot\mathbf{C} = (\mathbf{A} - \mathbf{B})\cdot(\mathbf{A} - \mathbf{B}) = \mathbf{A}\cdot\mathbf{A} + \mathbf{B}\cdot\mathbf{B} - 2\mathbf{A}\cdot\mathbf{B}$
2. Write result in terms of magnitudes and θ	$C^2 = A^2 + B^2 - 2AB\cos\theta$

31 • The dimension of power is (*a*) $[M][L]^2[T]^2$. (*b*) $[M][L]^2/[T]$. (*c*) $[M][L]^2/[T]^2$. (*d*) $[M][L]^2/[T]^3$.

(*d*) $P = \mathbf{F}\cdot\mathbf{v} = [M][L/T^2][L/T] = [M][L]^2/[T]^3$

32 • True or false: A kilowatt-hour is a unit of power.

False; it is a unit of energy.

33* • The engine of a car operates at constant power. The ratio of acceleration of the car at a speed of 60 km/h to that at 30 km/h (neglecting air resistance) is (*a*) $\frac{1}{2}$. (*b*) $1/\sqrt{2}$. (*c*) $\sqrt{2}$. (*d*) 2.

(*a*) mav = constant; av = constant and $a \propto 1/v$.

34 •• A car starts from rest and travels at constant acceleration. Which of the following statements are true? (*a*) The power delivered by the engine is constant. (*b*) The power delivered by the engine increases as the car gains speed. (*c*) The power delivered by the engine decreases as the car gains speed. (*d*) both (*b*) and (*c*) are correct.

(*b*) since v increases and a is constant, av must increase.

35 •• Force *A* does 5 J of work in 10 s. Force *B* does 3 J of work in 5 s. Which force delivers greater power?

P_A = (5/10) W = 0.5 W; P_B = (3/5) W = 0.6 W; $P_B > P_A$.

36 • A 5-kg box is lifted by a force equal to the weight of the box. The box moves upward at a constant velocity of 2 m/s. (*a*) What is the power input of the force? (*b*) How much work is done by the force in 4 s?

(*a*) Find *F*; use Equ. 6-17	$F = mg$; $P = Fv = mgv = (5 \times 9.81 \times 2)$ W = 98.1 W
(*b*) $W = Pt$	$W = (98.1 \times 4)$ J = 392.4 J

37* • Fluffy has just caught a mouse, and decides that the only decent thing to do is to bring it to the bedroom so that his human roommate can admire it when she wakes up. A constant horizontal force of 3 N is enough to drag the mouse across the rug at constant speed v. If Fluffy's force does work at the rate of 6 W, (*a*) what is her speed, v? (*b*) How much work does Fluffy do in 4 s?

(*a*) Use Equ. 6-17	v = (6/3) m/s = 2 m/s
(*b*) $W = Pt$	$W = (6 \times 4)$ J = 24 J

38 • A single force of 5 N acts in the x direction on an 8-kg object. (*a*) If the object starts from rest at $x = 0$ at time $t = 0$, find its velocity v as a function of time. (*b*) Write an expression for the power input as a function of time. (*c*) What is the power input of the force at time $t = 3$ s?

(*a*) $v = at$; $a = F/m$ — $v = (5/8)t$ m/s $= 0.625t$ m/s

(*b*) $P = Fv$ — $P = (5 \times 0.625t)$ W $= 3.125t$ W

(*c*) Substitute $t = 3$ s — $P(3) = (3.125 \times 3)$ W $= 9.375$ W

39 • Find the power input of a force $\boldsymbol{F}$ acting on a particle that moves with a velocity $\boldsymbol{v}$ for (*a*) $\boldsymbol{F} = 4\text{ N}\,\boldsymbol{i} + 3\text{ N}\,\boldsymbol{k}$, $\boldsymbol{v} = 6\text{ m/s}\,\boldsymbol{i}$; (*b*) $\boldsymbol{F} = 6\text{ N}\,\boldsymbol{i} - 5\text{ N}\,\boldsymbol{j}$, $\boldsymbol{v} = -5\text{ m/s}\,\boldsymbol{i} + 4\text{ m/s}\,\boldsymbol{j}$; and (*c*) $\boldsymbol{F} = 3\text{ N}\,\boldsymbol{i} + 6\text{ N}\,\boldsymbol{j}$, $\boldsymbol{v} = 2\text{ m/s}\,\boldsymbol{i} + 3\text{ m/s}\,\boldsymbol{j}$.

(*a*), (*b*), (*c*) Use $P = \boldsymbol{F}\cdot\boldsymbol{v}$ — (*a*) $P = 24$ W; (*b*) $P = -50$ W; (*c*) $P = 24$ W

40 •• A particle of mass m moves from rest at $t = 0$ under the influence of a single force of magnitude F. Show that the power delivered by the force at time t is $P = F^2t/m$.

$P = \boldsymbol{F}\cdot\boldsymbol{v}$; $\boldsymbol{v} = \boldsymbol{a}t = (\boldsymbol{F}/m)t$; $P = (\boldsymbol{F}\cdot\boldsymbol{F})(t/m) = F^2t/m$.

41* •• At a speed of 20 km/h, a 1200-kg car accelerates at 3 m/s^2 using 20 kW of power. How much power must be expended to accelerate the car at 2 m/s^2 at a speed of 40 km/h?

$P = Fv = mav$ — $P = (1200 \times 2 \times 40/3.6)$ W $= 26.7$ kW

42 •• A car manufacturer claims that his car can accelerate from rest to 100 km/h in 8 s. The car's mass is 800 kg. (*a*) Assuming that this performance is achieved at constant power, determine the power developed by the car's engine. (*b*) What is the car's speed after 4 s? (Neglect friction and air resistance.)

(*a*) $K = \frac{1}{2}mv^2 = Pt$; solve for $P = mv^2/2t$ — $P = [800 \times (100/3.6)^2/16] = 38.6$ kW

(*b*) $K(4) = \frac{1}{2}K(8)$; $K \propto v^2$ — $v(4) = v(8)/\sqrt{2} = 70.7$ km/h

43 •• Show that the position of the truck in Example 6-11 is related to its speed by $x = (m/3P)v^3$.

From Example 6-11, we have $x = (8P/9m)^{1/2}t^{3/2}$; from $v = (2P/m)^{1/2}t^{1/2}$, $t^{3/2} = (m/2P)^{3/2}v^3$. Substituting this into the expression for x and simplifying, we obtain $x = (m/3P)v^3$.

44 •• A 700-kg car accelerates from rest under constant power. At the end of 8.0 s, its speed is 90 km/h and it is located 133 m from its starting point. If the car continues to accelerate using the same power, what will its speed be at the end of 10 s, and how far will the car be from the starting point?

From Example 6-11, $v \propto t^{1/2}$ and $x \propto t^{3/2}$ — $v(10) = v(8)(10/8)^{1/2} = 100.6$ km/h; $x(10) = x(8)(10/8)^{3/2} = 186$ m

45* •• A 4.0-kg object initially at rest at $x = 0$ is accelerated at constant power of 8.0 W. At $t = 9.0$ s, it is at $x = 36.0$ m. Find its speed at $t = 6.0$ s and its position at that instant.

$v = (2Pt/m)^{1/2}$; $x(6) = x(9)(6/9)^{3/2}$ — $v = (2 \times 8 \times 6/4)^{1/2} = 4.9$ m/s; $x(6) = (36)(6/9)^{3/2} = 19.6$ m

46 •• A 700-kg car accelerates from rest under constant power at $t = 0$. At $t = 9$ s it is 117.7 m from its starting point and its acceleration is then 1.09 m/s^2. Find the power expended by the car's engine, neglecting frictional losses.

From Example 6-11, $P = 9mx^2/8t^3$ — $P = [9 \times 700 \times (117.7)^2/8 \times 9^3]$ W $= 15$ kW

47 • Two knowledge seekers decide to ascend a mountain. Sal chooses a short, steep trail, while Joe, who weighs the same as Sal, goes up via a long, gently sloped trail. At the top, they get into an argument about who gained more potential energy. Which of the following is true? (*a*) Sal gains more gravitational potential energy than Joe. (*b*) Sal gains less gravitational potential energy than Joe. (*c*) Sal gains the same gravitational potential energy as Joe. (*d*) To compare energies, we must know the height of the mountain. (*e*) To compare energies, we must know the length of the two trails.

(*c*)

---

48 • The gravitational potential energy of an object changes by −6 J. It follows that the work done by the gravitational force on this object is (*a*) − 6 J and the elevation of the object is increased. (*b*) −6 J and the elevation of the object is decreased. (*c*) +6 J and the elevation of the object is increased. (*d*) +6 J and the elevation of the object is decreased.

(*d*)

49* • A woman runs up a flight of stairs. The gain in her gravitational potential energy is U. If she runs up the same stairs with twice the speed, what will be her gain in potential energy? (*a*) U (*b*) $2U$ (*c*) $U/2$ (*d*) $4U$ (*e*) $U/4$

(*a*) The change in U does not depend on speed, only on difference in elevation.

---

50 • Which of the following statements is true? (*a*) The kinetic and potential energies of an object must always be positive quantities. (*b*) The kinetic and potential energies of an object must always be negative quantities. (*c*) Kinetic energy can be negative, but potential energy cannot. (*d*) Potential energy can be negative, but kinetic energy cannot. (*e*) None of the preceding statements is true.

(*d*) U can be negative, depending on choice of zero; $K = \frac{1}{2}mv^2 \geq 0$ since m and v^2 are both positive or zero.

---

51 • A block slides a certain distance down an incline. The work done by gravity is W. What is the work done by gravity if this block slides the same distance up the incline? (*a*) W (*b*) Zero (*c*) $-W$ (*d*) Gravity can't do work; some other force does work. (*e*) Cannot be determined unless given the distance traveled.

(*c*) Force opposite to displacement, hence $W < 0$.

---

52 • True or false: (*a*) Only conservative forces can do work. (*b*) If only conservative forces act, the kinetic energy of a particle does not change. (*c*) The work done by a conservative force equals the decrease in potential energy associated with that force.

(*a*) False (*b*) False (for example, object in free fall) (*c*) True, by definition of U

53* • When you climb a mountain, is the work done on you by gravity different if you take a short, steep trail instead of a long, gentle trail? If not, why do you find one trail easier?

No; a steeper trail requires more effort per step, but fewer steps.

---

54 • Which of the following forces are conservative and which are nonconservative? (*a*) the frictional force exerted on a sliding box (*b*) the force exerted by a linear spring that obeys Hooke's law (*c*) the force of gravity (*d*) the wind resistance on a moving car

(*a*) nonconservative (*b*) conservative (*c*) conservative (*d*) nonconservative

---

55 • An 80-kg man climbs up a 6-m high flight of stairs. What is the increase in gravitational potential energy?

$\Delta U = mgh$ $\qquad$ $\Delta U = (80 \times 6 \times 9.81)\text{ J} = 4.71\text{ kJ}$

---

56 • One of the highlights of Sharika's concert is her daredevil swan dive into the audience from a height of 2 m above the crowd's outstretched hands. If her mass is 60 kg, and the time of her dive is defined as $t = 0$, (*a*) what is her initial potential energy relative to $U = 0$ at the position of the crowd's hands? (*b*) From Newton's laws, find the distance she has fallen, and her speed at $t = 0.20$ s. (*c*) Find her potential and kinetic energy at $t = 0.40$ s. (*d*) Find her kinetic energy and speed just as she reaches the hands of the crowd in the mosh pit.

(*a*) $U = mgh$ — $U = (60 \times 2 \times 9.81)$ J $= 1.18$ kJ

(*b*) $v = gt$; $y = \frac{1}{2}gt^2$ — $v = (9.81 \times 0.2)$ m/s $= 1.96$ m/s; $y = \frac{1}{2}(9.81 \times 0.04)$ m $= 0.196$ m

(*c*) $y = \frac{1}{2}gt^2$; $\Delta U = -mgy$, and $K = -\Delta U$ — $y = 0.785$ m; $\Delta U = -0.462$ kJ, $U = (1.18 - 0.462)$ kJ $= 0.718$ kJ; $K = 0.462$ kJ

(*d*) $K = -\Delta U = \frac{1}{2}mv^2$; $v = (2K/m)^{1/2}$ — $K = 1.18$ kJ; $v = (2 \times 1180/60)^{1/2}$ m/s $= 6.27$ m/s

---

57* • Water flows over Victoria Falls, which is 128 m high, at an average rate of 1.4×10^6 kg/s. If half the potential energy of this water were converted to electric energy, how much power would be produced by these falls?

$P = -\frac{1}{2}(dU/dt) = -\frac{1}{2}gh(dm/dt)$ — $P = \frac{1}{2}(9.81 \times 128 \times 1.4 \times 10^6)$ W $= 879$ MW

---

58 • A 2-kg box slides down a long, frictionless incline of angle 30°. It starts from rest at time $t = 0$ at the top of the incline at a height of 20 m above ground. (*a*) What is the original potential energy of the box relative to the ground? (*b*) From Newton's laws, find the distance the box travels in 1 s and its speed at $t = 1$ s. (*c*) Find the potential energy and the kinetic energy of the box at $t = 1$ s. (*d*) Find the kinetic energy and the speed of the box just as it reaches the bottom of the incline.

(*a*) $U_i = mgh$ — $U_i = (2 \times 9.81 \times 20)$ J $= 392$ J

(*b*) $a = g \sin 30°$; $x = \frac{1}{2}at^2$; $v = at$ — $x = \frac{1}{2}(9.81 \times 0.5 \times 1)$ m $= 2.45$ m; $v = 4.91$ m/s

(*c*) $K = \frac{1}{2}mv^2$; $U = U_i - K$ — $K = \frac{1}{2}(2 \times 4.91^2)$ J $= 24.1$ J; $U = (392 - 24.1)$ J $= 368$ J

(*d*) $K = U_i$; $v = (2K/m)^{1/2}$ — $K = 392$ J; $v = (2 \times 392/2)^{1/2}$ m/s $= 19.8$ m/s

---

59 • A force $F_x = 6$ N is constant. (*a*) Find the potential energy function U associated with this force for an arbitrary reference position at x_0 at which $U = 0$. (*b*) Find U such that $U = 0$ at $x = 4$ m. (*c*) Find U such that $U = 14$ J at $x = 6$ m.

(*a*) From Equ. 6-21*a*, $U(x) - U(x_0) = -\int_{x_0}^{x} F\,dx$ — $U(x) = U(x_0) - \int_{x_0}^{x} 6\,dx = -6(x - x_0)$ with $U(x_0) = 0$

(*b*) $U(x_0)$ and x_0 arbitrary — $U(x) = 24 - 6x$

(*c*) See (*b*) above — $U(x) = 50 - 6x$

---

60 • A spring has a force constant of $k = 10^4$ N/m. How far must it be stretched for its potential energy to be (*a*) 50 J and (*b*) 100 J?

(*a*), (*b*) $U = \frac{1}{2}kx^2$ — (*a*) $x = (2 \times 50/10^4)^{1/2}$ m $= 0.1$ m; (*b*) $x = (0.1 \text{ m})\sqrt{2} = 0.141$ m

---

61* •• A simple Atwood's machine uses two masses, m_1 and m_2 (Figure 6-34). Starting from rest, the speed of the two masses is 4.0 m/s at the end of 3.0 s. At that instant, the kinetic energy of the system is 80 J and each mass has moved a distance of 6.0 m. Determine the values of m_1 and m_2.

1. $K = \frac{1}{2}(m_1 + m_2)v^2$; solve for $m_1 + m_2$ — $m_1 + m_2 = (2 \times 80/4^2)$ kg $= 10$ kg

2. $a = \dfrac{m_1 - m_2}{m_1 + m_2}g = v(t)/t$; solve for $m_1 - m_2$ — $a = 1.33$ m/s^2; $m_1 - m_2 = 1.33(10/9.81)$ kg $= 1.36$ kg

3. Solve for m_1 and m_2 — $m_1 = 5.68$ kg, $m_2 = 4.32$ kg

---

62 •• A straight rod of negligible mass is mounted on a frictionless pivot as in Figure 6-35. Masses m_1 and m_2 are suspended at distances l_1 and l_2. (*a*) Write an expression for the gravitational potential energy of the masses as a function of the angle θ made by the rod and the horizontal. (*b*) For what angle θ is the potential energy a minimum? Is the statement "systems tend to move toward a configuration of minimum potential energy" consistent with your result? (*c*) Show that if $m_1l_1 = m_2l_2$, the potential energy is the same for all values of θ. (When this holds, the system will balance at any angle θ. This result is known as *Archimedes' law of the lever.*)

(*a*) Set $U = 0$ for $\theta = 0$; — $U(\theta) = m_2gl_2 \sin\theta - m_1gl_1 \sin\theta = (m_2l_2 - m_1l_1)g \sin\theta$

(*b*) Note that $\sin\theta$ is maximum for $\theta = +\pi/2$; $\sin\theta$ is minimum for $\theta = -\pi/2$ — If $m_2l_2 > m_1l_1 > 0$, $\theta = -\pi/2$ for U_{min}; if $m_2l_2 < m_1l_1$, $\theta = \pi/2$ for U_{min}

This is consistent with the tendency for systems to move toward minimum potential energy.

(*c*) If $m_1l_1 = m_2l_2$, then $(m_2l_2 - m_1l_1) = 0$ and U is independent of θ.

---

63 •• Figure 6-36 shows a plot of a potential-energy function U versus x. (*a*) At each point indicated, state whether the force F_x is positive, negative, or zero. (*b*) At which point does the force have the greatest magnitude? (*c*) Identify any equilibrium points, and state whether the equilibrium is stable or unstable.

(*a*) Use $F_x = -dU/dx$ — A and E: $F_x < 0$; B, D and E: $F_x = 0$; C: $F_x > 0$

(*b*) Find the point where slope is steepest — At point A $|F_x|$ is greatest

(*c*) Stable: $d^2U/dx^2 > 0$; unstable: $d^2U/dx^2 < 0$ — B: unstable equilibrium; D: stable equilibrium

---

64 • (*a*) Find the force F_x associated with the potential-energy function $U = Ax^4$, where A is a constant. (*b*) At what point(s) is the force zero?

(*a*) $F_x = -dU/dx$ — $F_x = -4Ax^3$

(*b*) Set $F_x = 0$, solve for x — $x = 0$

65* •• A potential-energy function is given by $U = C/x$, where C is a positive constant. (*a*) Find the force F_x as a function of x. (*b*) Is this force directed toward the origin or away from it? (*c*) Does the potential energy increase or decrease as x increases? (*d*) Answer parts (*b*) and (*c*) where C is a negative constant.

(*a*), (*b*), (*c*) $F_x = -dU/dx$ — (*a*) $F_x = C/x^2$; (*b*) directed away from the origin ($-x$ direction); (*c*) $U(x)$ increases with increasing x

(*d*) Repeat above with $C < 0$ — Directed toward the origin; U increases as x increases

---

66 •• On the potential-energy curve for U versus y shown in Figure 6-37, the segments AB and CD are straight lines. Sketch a plot of the force F_y versus y.

Note that between A and B, $F = +2$ N; between C and D, $F = -2$ N; between B and C, F changes gradually from +2 N to −2 N.

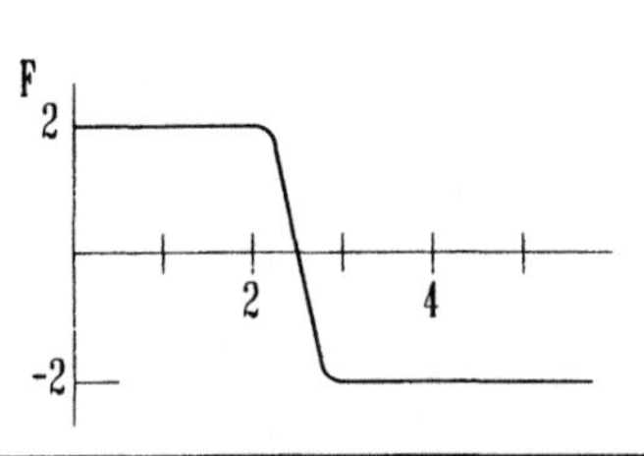

67 •• The force acting on an object is given by $F_x = a/x^2$. Determine the potential energy of the object as a function of x.

$U(x) = -\int F(x)\,dx = -\int a/x^2\,dx = a/x + U_0.$

68 •• The potential energy of an object is given by $U(x) = 3x^2 - 2x^3$, where U is in joules and x is in meters. (*a*) Determine the force acting on this object. (*b*) At what positions is this object in equilibrium? (*c*) Which of these equilibrium positions are stable and which are unstable?

(*a*) $F = -dU/dx$ — $F = 6x^2 - 6x = 6x(x - 1)$

(*b*) At equilibrium, $F = 0$ — Equilibrium at $x = 0$, $x = 1$

(*c*) $d^2U/dx^2 > 0$: stable equilibrium — $d^2U/dx^2 = 6 - 12x$; stable at $x = 0$, unstable at $x = 1$

69* •• During a Dead Wait concert, Sharika and Chico, each of mass M, are attached to the ends of a light rope that is hung over two frictionless pulleys, as shown in Figure 6-38. A large gong of mass m is attached to the middle of the rope, between the pulleys, and Sharika and Chico beat it madly in lieu of the usual guitar solo. (*a*) Find the potential energy of the system as a function of the distance y to the center of the gong. (*b*) Find the value of y for which the potential energy function of the system is a minimum. (*c*) Find the equilibrium distance y_0 using the potential energy function. (*d*) Check your answer by applying Newton's laws to the gong.

(*a*) Set $U = 0$ for $y = 0$; note that each M is raised by $h = (d^2 + y^2)^{1/2} - d$ as m drops a distance y — $U(y) = 2Mg[(d^2 + y^2)^{1/2} - d] - mgy$

(*b*) Find dU/dy and set equal to zero — $dU/dy = 2Mgy/(d^2 + y^2)^{1/2} - mg$; $dU/dy = 0$ for $y = d(m/2M)/\sqrt{1 - (m/2M)^2}$

(*c*) Set net force on $m = 0$ — $F_{net} = mg - 2Mg \sin\theta$; $\sin\theta = y/(y^2 + d^2)^{1/2}$

Solve for y — $y = d(m/2M)/\sqrt{1 - (m/2M)^2}$

70 ••• The potential energy of an object is given by $U(x) = 8x^2 - x^4$, where U is in joules and x is in meters. (*a*) Determine the force acting on this object. (*b*) At what positions is this object in equilibrium? (*c*) Which of these equilibrium positions are stable and which are unstable?

(*a*), (*b*), (*c*) Proceed as in Problem 6-68 — $F = 4x^3 - 16x = 4x(x + 2)(x - 2)$

$F = 0$ at $x = -2, 0, +2$; these are equilibrium points.

$d^2U/dx^2 = 16 - 12x^2 > 0$ at $x = 0$, < 0 at $x \pm 2$

$x = 0$ stable, $x = \pm 2$ unstable equilibrium

71 ••• The force acting on an object is given by $F(x) = x^3 - 4x$. Locate the positions of unstable and stable equilibrium and show that at these points $U(x)$ is a local maximum or minimum, respectively.

1. $F(x) = x(x^2 - 4)$; $F(x) = 0$ at $x = 0, \pm 2$. — At $x = 0$, a small increase in x makes $F(x)$ negative, i.e., a restoring force. Hence at $x = 0$, stable equilibrium.

At $x = +2$, an increase in x increases $F(x)$; thus, unstable equilibrium. Similarly, at $x = -2$, equilibrium is unstable.

2. $U(x) = -x^4/4 + x^2/2$ — $U(x)$ has local min. at $x = 0$, local max. at $x = \pm 2$

---

72 ••• The potential energy of a 4-kg object is given by $U = 3x^2 - x^3$ for $x \leq 3$ m, and $U = 0$ for $x \geq 3$ m, where U is in joules and x is in meters. (*a*) At what positions is this object in equilibrium? (*b*) Sketch a plot of U versus x. (*c*) Discuss the stability of the equilibrium for the values of x found in (*a*). (*d*) If the total energy of the particle is 12 J, what is its speed at $x = 2$ m?

(*a*) Proceed as in Problem 6-68 — $dU/dx = 3x(2 - x) = 0$ for $x = 0$, 2 m; also for $x \geq 3$ m

(*b*) $U(x)$ is shown on the right

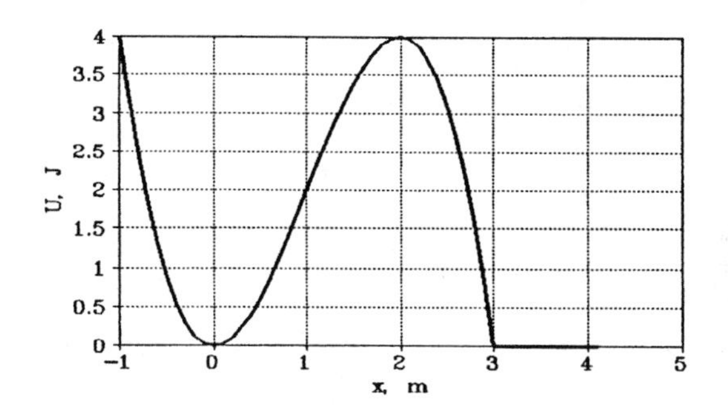

(*c*) Note that $U(x)$ is a minimum at $x = 0$ m; a maximum at $x = 2$ m; and constant for $x \geq 3$ m — Equilibrium stable at $x = 0$ m; unstable at $x = 2$ m; neutral at $x \geq 3$ m

(*d*) $K = E - U$; $v = (2K/m)^{1/2}$ — $v = [2(12 - 4)/4]^{1/2}$ m/s $= 2$ m/s

---

73* ••• A force is given by $F_x = Ax^{-3}$, where $A = 8$ N·m³. (*a*) For positive values of x, does the potential energy associated with this force increase or decrease with increasing x? (You can determine the answer to this question by imagining what happens to a particle that is placed at rest at some point x and is then released.) (*b*) Find the potential-energy function U associated with this force such that U approaches zero as x approaches infinity. (*c*) Sketch U versus x.

(*a*) Apply Equ. 6-21a — $U(x) = \frac{1}{2}A/x^2 + U_0$; for $x > 0$, U decreases as x increases

(*b*) Let $x \to \infty$ and set $U(x) \to 0$ — $0 = 0 + U_0$; $U_0 = 0$; $U(x) = \frac{1}{2}(8\text{ N·m}^3)/x^2$ J $= 4/x^2$ J

(*c*) The plot of $U(x)$ is shown

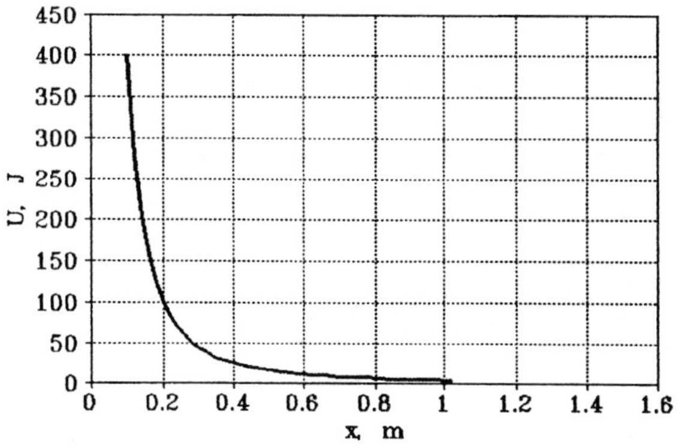

74 • True or false: (*a*) The gravitional force cannot do work because it acts at a distance. (*b*) Work is the area under the force-versus-time curve.

(*a*) False (*b*) False

75 • Negative work by an applied force implies that (*a*) the kinetic energy of the object increases. (*b*) the applied force is variable. (*c*) the applied force is perpendicular to the displacement. (*d*) the applied force has a component that is opposite to the displacement. (*e*) nothing; there is no such thing as negative work.

(*d*) See Equ. 6-1.

76 •• A movie crew is in the Badlands when their car overheats. After they stop to let it cool down, an argument breaks out. They agree that they must go easy on the engine, but they disagree about when the engine works the hardest, and therefore about how they should drive for the rest of the trip. Carolyn claims that the work done by the car in accelerating from 0 to 20 km/h is less than that required to accelerate from 20 km/h to 30 km/h, meaning they should drive more slowly. Ted says no, the work done between 0 and 20 km/h is more than the work done between 20 km/h and 30 km/h. Ernie says it all depends on the mass of the car, and Bloop says it all depends on how long you take to change from one speed to another. Who is right?

Note that $W(0-20) = \frac{1}{2}m \times 400$; $W(20-30) = \frac{1}{2}m(900 - 400) = \frac{1}{2}m \times 500$; Carolyn is right.

77* • Figure 6-39 shows two pulleys arranged to help lift a heavy load. A rope runs around two massless, frictionless pulleys and the weight w hangs from one pulley. You exert a force of magnitude F on the free end of the cord. (*a*) If the weight is to move up a distance h, through what distance must the force move? (*b*) How much work is done by the ropes on the weight? (*c*) How much work do you do? (This is an example of a simple machine in which a small force F_1 moves through a large distance x_1 in order to exert a large force F_2 (= w) through a smaller distance $x_2 = h$.)

(*a*) If w moves a distance h, F moves a distance $2h$. (*b*) $W = wh$. (*c*) $W = F \times 2h$; note that the tension, F, in the string is $\frac{1}{2}w$. Thus $F \times 2h = wh$.

78 • In February 1995, a total of 54.3 billion kW-h of electrical energy was generated by nuclear power plants in the United States. At the same time, the population of the United States was about 255 million people. If the average American has a mass of 60 kg, and if the entire energy output of all nuclear power plants was diverted to supplying energy for a single giant elevator, estimate the height h at which the entire population of the country could be lifted by the elevator. In your calculations, assume that 25% of the energy goes into lifting the people; assume also that g is constant over the entire height h.

$(NMgh) = 0.25E$; $h = 0.25E/NMg$

$$h = \frac{(0.25\times54.3\times10^{12}\times3600)\ \text{W}}{(2.55\times10^{8}\times60\times9.81)\ \text{N}} = 3.26\times10^{5}\ \text{m} = 326\ \text{km}$$

79 • One of the most powerful cranes in the world, operating in Switzerland, can slowly raise a load of M = 6000 tonne to a height of h = 12.0 m (1 tonne = 1000 kg). (*a*) How much work is done by the crane? (*b*) If it takes 1.00 min to lift the load at constant velocity to this height, find the power developed by the crane.

(*a*) $W = mgh$ $W = (6 \times 10^6 \times 9.81 \times 12)\ \text{J} = 7.06 \times 10^8\ \text{J} = 706\ \text{MJ}$

(*b*) $P = W/t$ $P = (706/60)\ \text{MW} = 11.8\ \text{MW}$

80 • In Australia, there used to be a ski lift of length 5.6 km. It took about 60 min for the gondola to travel all the way up. If there were 12 gondolas going up at one, each of mass 550 kg, and the angle of ascent was 30°, estimate the power P of the engine needed to operate the ski lift.

$P = NM\Delta h/\Delta t$ $P = (12 \times 550 \times 5.6 \times 10^3 \times \sin 30°)/60$ W = 50.4 kW

81* • A 2.4-kg object attached to a horizontal string moves with constant speed in a circle of radius R on a frictionless horizontal surface. The kinetic energy of the object is 90 J and the tension in the string is 360 N. Find R.

$\frac{1}{2}mv^2 = K$; $mv^2/R = T$; thus, $R = 2K/T$ $R = (2 \times 90/360)$ m = 0.5 m

---

82 • How high must an 800-kg Ford Escort be lifted to gain an amount of potential energy equal to the kinetic energy it has when it is moving at 100 km/h?

$\frac{1}{2}mv^2 = mgh$; $h = v^2/2g$ $h = [(100/3.6)^2/(2 \times 9.81)]$ m = 39.3 m

---

83 • The movie crew arrives in the Badlands ready to shoot a scene. The script calls for a car to crash into a vertical rock face at 100 km/h. Unfortunately, the car won't start, and there is no mechanic in sight. They are about to skulk back to the studio to face the producer's wrath when the cameraman get an idea. They use a crane to lift the car by its rear end and then drop it, filming at an angle that makes the car appear to be traveling horizontally. How high should the 800-kg car be lifted so that it reaches a speed of 100 km/h in the fall?

From Problem 6-82, we see that $h = 39.3$ m.

---

84 •• The force acting on a particle that is moving along the x axis is given by $F_x = -ax^2$, where a is a constant. Calculate the potential-energy function U relative to $U = 0$ at $x = 0$, and sketch a graph of U versus x.

Apply Equ. (6-21a) $U(x) = \int F_x\, dx = ax^3/3 + U_0$

Find U_0 by setting $U(0) = 0$ $U_0 = 0$. $U(x) = ax^3/3$

The plot of $U(x)$ is shown

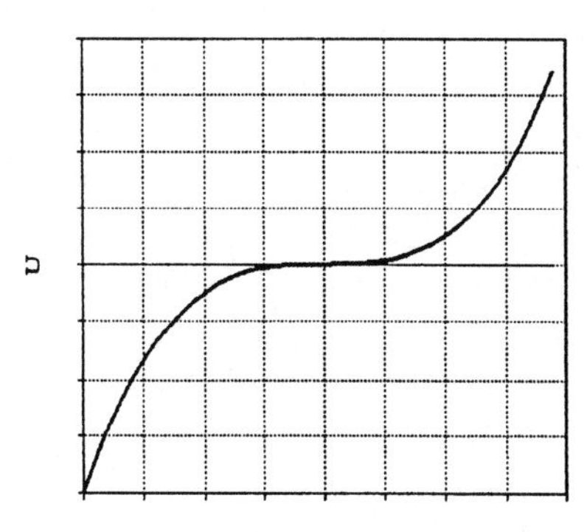

85* •• Water from behind a dam flows through a large turbine at a rate of 1.5×10^6 kg/min. The turbine is located 50 m below the surface of the reservoir, and the water leaves the turbine with a speed of 5 m/s. (a) Neglecting any energy dissipation, what is the power output of the turbine? (b) How many U.S. citizens would be supplied with energy by this dam if each citizen uses 3×10^{11} J of energy per year?

(a) $E_{init}/kg = gh = E_{fin}/kg = \frac{1}{2}v_f^2 + W_{el}/kg$;

$P_{el} = MW_{el}/\Delta t$ $P_{el} = [1.5 \times 10^6(9.81 \times 50 - \frac{1}{2} \times 5^2)/(60)]$ W = 12 MW (*b*) 3×10^{11} J/y = 9.51 kW $\qquad N = 12 \times 10^6/9.51 \times 10^3 = 1261$

---

86 •• A force acts on a cart of mass *m* in such a way that the speed *v* of the cart increases with distance *x* as $v = Cx$, where *C* is a constant. (*a*) Find the force acting on the cart as a function of position. (*b*) What is the work done by the force in moving the cart from $x = 0$ to $x = x_1$?

(*a*) No change in *U*; thus $E = \int F\,dx$ or $F = dE/dx = d/dx(\tfrac{1}{2}mC^2x^2) = mC^2x$. (*b*) $W = E = \tfrac{1}{2}mC^2x^2$.

---

87 •• A force $\boldsymbol{F} = (2\text{ N/m}^2)x^2\,\boldsymbol{i}$ is applied to a particle. Find the work done on the particle as it moves a total distance of 5 m (*a*) parallel to the *y* axis from point (2 m, 2 m) to point (2 m, 7 m) and (*b*) in a straight line from (2 m, 2 m) to (5 m, 6 m).

(*a*) Since ***F*** is along *x* direction and displacement is along *y* direction $W = \int \boldsymbol{F}\cdot d\boldsymbol{s} = 0$.

(*b*) Evaluate $\int \boldsymbol{F}\cdot d\boldsymbol{s}$ $\qquad W = \int_2^5 2x^2\,dx = \frac{2}{3}x^3\Big|_2^5 = 78\text{ J}$

---

88 •• A particle of mass *m* moves along the *x* axis. Its position varies with time according to $x = 2t^3 - 4t^2$, where *x* is in meters and *t* is in seconds. (*a*) Find the velocity and acceleration of the particle at any time *t*; (*b*) the power delivered to the particle at any time *t*; and (*c*) the work done by the force from $t = 0$ to $t = t_1$.

(*a*) $v = dx/dt$; $a = dv/dt = d^2x/dt^2$ $\qquad v = (6t^2 - 8t)$ m/s; $a = (12t - 8)$ m/s²

(*b*) $P = Fv = mav$ $\qquad P = 8mt(9t^2 - 18t + 8)$ W

(*c*) $W = \tfrac{1}{2}m\{[v(t_1)]^2 - [v(0)]^2\}$ $\qquad W = 2mt_1^2(3t_1 - 4)^2$

Note: *W* can also be obtained by integrating *P*(*t*) *dt*.

89* •• A 3-kg particle starts from rest at $x = 0$ and moves under the influence of a single force $F_x = 6 + 4x - 3x^2$, where F_x is in newtons and *x* is in meters. (*a*) Find the work done by the force as the particle moves from $x = 0$ to $x = 3$ m. (*b*) Find the power delivered to the particle when it is at $x = 3$ m.

(*a*) $W = \int F_x\,dx$ $\qquad W = \int_0^3 F(x)\,dx = (6x + 2x^2 - x^3)\Big|_0^3 = 9\text{ J}$

(*b*) Find *v* from $v = (2K/m)^{1/2}$, $K = W$ $\qquad v = (18/3)^{1/2}$ m/s = 2.45 m/s

Use $P = Fv$ $\qquad P = (-9\text{ N})(2.45\text{ m/s}) = -22$ W

---

90 •• The initial kinetic energy imparted to a 20-g bullet is 1200 J. Neglecting air resistance, find the range of this projectile when it is fired at an angle such that the range equals the maximum height attained.

Use Equ. 3-22 and the result of Problem 3-62 $\qquad \theta = 76°$;

$$R = \frac{v_0^2}{g}\sin 2\theta = \frac{2 \times 1200}{0.02 \times 9.81}\sin 152°\text{ m} = 5.74\text{ km}$$

---

91 •• A force F_x acting on a particle is shown as a function of *x* in Figure 6-40. (*a*) From the graph, calculate the work done by the force when the particle moves from $x = 0$ to the following values of *x*: −4, −3, −2, −1, 0, 1, 2, 3, and 4 m. (*b*) Plot the potential energy *U* versus *x* for the range of values of *x* from −4 m to +4 m, assuming that $U = 0$ at $x = 0$.

(*a*) Work = area under force-versus-displacement curve. Note that for negative displacements, *F* is positive, so *W* for $x < 0$ is negative. $W(0,1) = 1$ J; $W(1,2) = -1$ J; and for $x > 2$ m, *W* is negative.

Below is a tabulation of W for -4 m $< x < 4$ m.

x, m	-4	-3	-2	-1	0	1	2	3	4
W, J	-11	-10	-7	-3	0	1	0	-2	-3

(*b*) From Equ. 6-21*a*, $\Delta U = -W$ ($U = 0$ at $x = 0$ where $W = 0$).
The plot of $U = \Delta U = -W$ is shown.

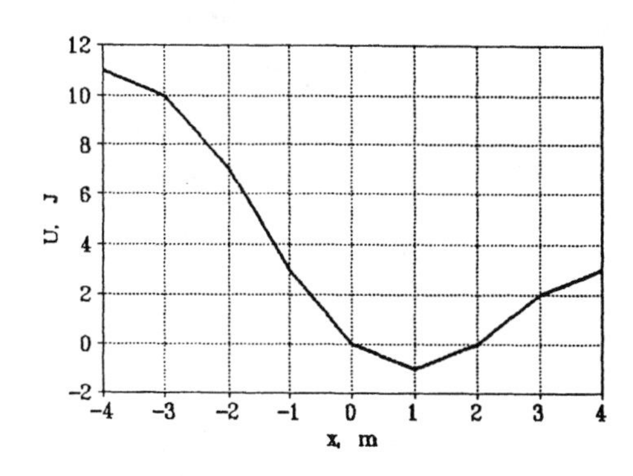

---

92 •• Repeat Problem 91 for the force F_x shown in Figure 6-41.

(*a*) Proceed as in Problem 6-91. The tabulation of x and W is below.

x, m	-4	-3	-2	-1	0	1	2	3	4
W, J	6	4	2	½	0	½	3/2	5/2	3

(*b*) As before, $U = -W$. See the plot.

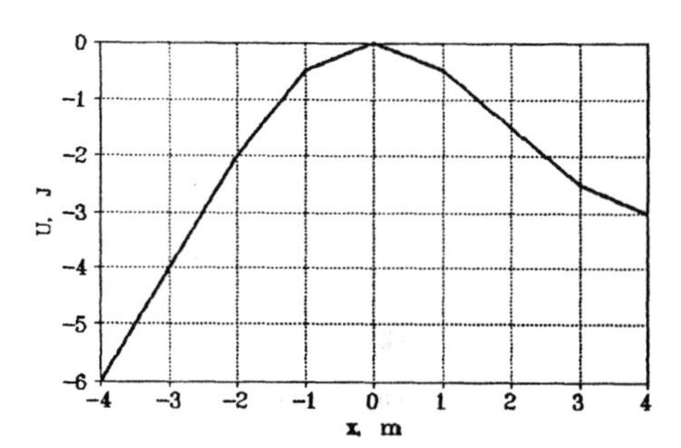

93* •• A rope of length L and mass per unit length of μ lies coiled on the floor. (*a*) What force F is required to hold one end of the rope a distance $y < L$ above the floor as shown in Figure 6-42? (*b*) Find the work required to lift one end of the rope from the floor to a height $l < L$ by integrating $F\,dy$ from $y = 0$ to $y = l$.

(*a*) The mass supported is μy; $F = \mu yg$. (*b*) $W = \int_0^l \mu g y\,dy = \mu g \frac{l^2}{2}$.

---

94 ••• A box of mass M is at the bottom of a frictionless inclined plane (Figure 6-43). The box is attached to a string that pulls with a constant tension T. (*a*) Find the work done by the tension T when the box has moved a distance x along the plane. (*b*) Find the speed of the box as a function of x and θ. (*c*) Determine the power produced by the tension in the string as a function of x and θ.

(*a*) By definition, $W = Tx$. (*b*) $F_{\text{net}} = T - Mg\sin\theta$; $a = (T/M) - g\sin\theta$; from Equ. 2-15,

$v(x) = \sqrt{2(\frac{T}{M} - g\sin\theta)x}$. (*c*) $P = Tv = T[2(T/M - g\sin\theta)x]^{1/2}$.

---

95 ••• A force in the xy plane is given by $\boldsymbol{F} = (F_0/r)(y\boldsymbol{i} - x\boldsymbol{j})$, where F_0 is a constant and $r = \sqrt{x^2 + y^2}$. (*a*) Show that the magnitude of this force is F_0 and that its direction is perpendicular to $\boldsymbol{r} = x\,\boldsymbol{i} + y\,\boldsymbol{j}$. (*b*) Find the work done by this force on a particle that moves in a circle of radius 5 m centered at the origin. Is this force conservative?

(*a*) $|\boldsymbol{F}| = (F_x^2 + F_y^2)^{1/2}$; If $\boldsymbol{F} \perp \boldsymbol{r}$, $\boldsymbol{F}\cdot\boldsymbol{r} = 0$

$F_x^2 + F_y^2 = (F_0^2/r^2)(y^2 + x^2) = F_0^2$;
$\boldsymbol{F}\cdot\boldsymbol{r} = (F_0/r)(y\boldsymbol{i} - x\boldsymbol{j})\cdot(x\boldsymbol{i} + y\boldsymbol{j}) = (F_0/r)(yx - xy) = 0$

(*b*) Since $\boldsymbol{F} \perp \boldsymbol{r}$, $\boldsymbol{F}$ is tangential to circle and constant. At $x = 5$ m, $y = 0$, $\boldsymbol{F}$ points in the $-\boldsymbol{j}$ direction. If $d\boldsymbol{s}$ is in $-\boldsymbol{j}$ direction (clockwise rotation), $dW > 0$.

$W = F_0 \times (2\pi r) = 10\pi F_0$ (clockwise rotation);
$W = -10\pi F_0$ (counterclockwise rotation).
Not conservative; $W \neq 0$ for complete circuit.

96 ••• A theoretical formula for the potential energy associated with the nuclear force between two protons, two neutrons, or a neutron and a proton is the *Yukawa potential*: $U = -U_0(a/x)e^{-x/a}$, where U_0 and a are constants. (*a*) Sketch U versus x using $U_0 = 4$ pJ (a picojoule, pJ, is 1×10^{-12} J) and $a = 2.5$ fm (a femtometer, fm, is 1×10^{-15} m). (*b*) Find the force F_x. (*c*) Compare the magnitude of the force at the separation $x = 2a$ to that at $x = a$. (*d*) Compare the magnitude of the force at the separation $x = 5a$ to that at $x = a$.

(*a*) A plot of $U(x)$ is shown

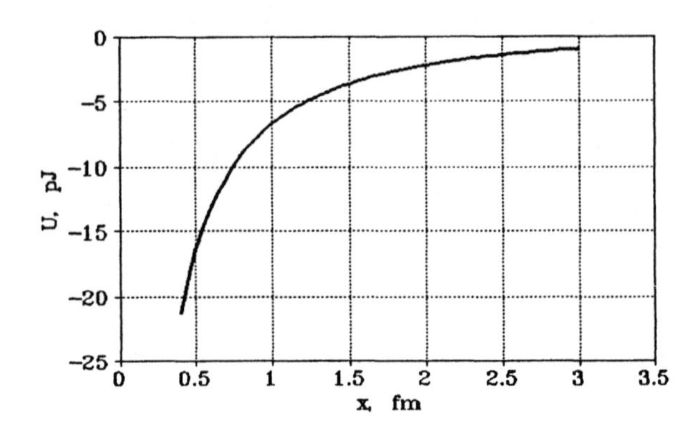

(*b*) $F_x = -(dU/dx)$ $\qquad F(x) = -U_0 e^{-x/a}[(a/x^2) + (1/x)]$

(*c*) $F(2a) = -U_0(3/4a)e^{-2}$; $F(a) = -U_0(2/a)e^{-1}$ $\qquad F(2a)/F(a) = (3/8)e^{-1} = 0.138$

(*d*) $F(5a) = -U_0(6/25a)e^{-5}$ $\qquad F(5a)/F(a) = (6/50)e^{-4} = 0.0022$

CHAPTER 7

Conservation of Energy

1* •• What are the advantages and disadvantages of using the conservation of mechanical energy rather than Newton's laws to solve problems?

It is generally simpler, involving only scalars, but some details cannot be obtained, e.g., trajectories.

2 •• Two objects of unequal mass are connected by a massless cord passing over a frictionless peg. After the objects are released from rest, which of the following statements are true? (U = gravitational potential energy, K = kinetic energy of the system.) (*a*) $\Delta U < 0$ and $\Delta K > 0$ (*b*) $\Delta U = 0$ and $\Delta K > 0$ (*c*) $\Delta U < 0$ and $\Delta K = 0$ (*d*) $\Delta U = 0$ and $\Delta K = 0$ (*e*) $\Delta U > 0$ and $\Delta K < 0$

(*a*)

3 •• Two stones are thrown with the same initial speed at the same instant from the roof of a building. One stone is thrown at an angle of 30° above the horizontal, the other is thrown horizontally. (Neglect air resistance.) Which statement is true? (*a*) The stones strike the ground at the same time and with equal speeds. (*b*) The stones strike the ground at the same time with different speeds. (*c*)The stones strike the ground at different times with equal speeds. (*d*) The stones strike the ground at different times with different speeds.

(*c*) Their kinetic energies are equal.

4 • A block of mass m is pushed up against a spring, compressing it a distance x, and the block is then released. The spring projects the block along a frictionless horizontal surface, giving the block a speed v. The same spring projects a second block of mass $4m$, giving it a speed of $3v$. What distance was the spring compressed in the second case?

$K_1 = \frac{1}{2}mv^2 = \frac{1}{2}kx^2$; $mv^2 = kx_1^2$; $kx_2^2 = (4m)(3v)^2 = 36mv^2 = 36kx_1^2$; $x_2 = 6x_1$.

5* • A woman on a bicycle traveling at 10 m/s on a horizontal road stops pedaling as she starts up a hill inclined at 3.0° to the horizontal. Ignoring friction forces, how far up the hill will she travel before stopping? (*a*) 5.1 m (*b*) 30 m (*c*) 97 m (*d*) 10.2 m (*e*) The answer depends on the mass of the woman.

(*c*) $h = v^2/2g = 50/9.81$ m = 5.1 m; $d = (5.1/\sin 3.0°)$ m = 97.4 m.

6 • A pendulum of length L with a bob of mass m is pulled aside until the bob is a distance $L/4$ above its equilibrium position. The bob is then released. Find the speed of the bob as it passes the equilibrium position.

$\frac{1}{2}mv^2 = mg\Delta h$; $\Delta h = L/4$; $v = (gL/2)^{1/2}$.

---

7 • When she hosts a garden party, Julie likes to launch bagels to her guests with a spring device that she has devised. She places one of her 200-g bagels against a horizontal spring mounted on her gazebo. The force constant of the spring is 300 N/m, and she compresses it 9 cm. (*a*) Find the work done by Julie and the spring when Julie launches a bagel. (*b*) If the released bagel leaves the spring at the spring's equilibrium position, find the speed of the bagel at that point. (*c*) If the bagel launcher is 2.2 m above the grass, what is Julie's horizontal range firing 200-g bagels?

(*a*) $W = \frac{1}{2}kx^2$	Work by Julie = 1.215 J = work by spring
(*b*) $\frac{1}{2}mv^2 = W$; solve for v	$v = 3.49$ m/s
(*c*) 1. Find time of flight using $y = \frac{1}{2}gt^2$	$t = (2.2 \times 2/9.81)^{1/2}$ s = 0.67 s
2. $R = vt$	$R = (3.49 \times 0.67)$ m = 2.34 m

---

8 • A 3-kg block slides along a frictionless horizontal surface with a speed of 7 m/s (Figure 7-17). After sliding a distance of 2 m, the block makes a smooth transition to a frictionless ramp inclined at an angle of 40° to the horizontal. How far up the ramp does the block slide before coming monetarily to rest?

Use Equ. 7-6; $mgs \sin \theta = \frac{1}{2}mv^2$, solve for s — $s = v^2/(2g \sin \theta) = [49/(2 \times 9.81 \times \sin 40°)] = 3.89$ m

---

9* • The 3-kg object in Figure 7-18 is released from rest at a height of 5 m on a curved frictionless ramp. At the foot of the ramp is a spring of force constant $k = 400$ N/m. The object slides down the ramp and into the spring, compressing it a distance x before coming momentarily to rest. (*a*) Find x. (*b*) What happens to the object after it comes to rest?

(*a*) $U_i = mgh$; $U_f = \frac{1}{2}kx^2$; $K_i = K_f = 0$; use Equ. 7-6 — $x = \sqrt{2mgh/k} = 0.858$ m

(*b*) The spring will accelerate the mass and it will then retrace its path, rising to a height of 5 m.

---

10 • A vertical spring compressed a distance x sits on a concrete floor. When a block of mass m_1 is placed on the spring and the spring is released, the block is projected upward to a height h. If a block of mass $m_2 = 2m_1$ is placed on the spring and the spring is again compressed a distance x and released, to what height will the block rise? (*a*) $h/4$ (*b*) $h/2$ (*c*) $h/\sqrt{2}$ (*d*) h

(*b*)

---

11 • If the spring in Problem 10 is compressed an amount $2x$ when the block of mass m_2 is placed on it, to what height will the block rise when the spring is released? (*a*) $2h$ (*b*) $\sqrt{2}h$ (*c*) h (*d*) $h/\sqrt{2}$

(*a*)

---

12 • A 15-g ball is shot from a spring gun whose spring has a force constant of 600 N/m. The spring can be compressed 5 cm. How high will the ball go if the gun is aimed vertically?

We take $U = 0$ at uncompressed level of the spring.

1. $E_i = -mgx + \frac{1}{2}kx^2 = mgh$	$h = \frac{1}{2}kx^2/mg - x$
2. Find h for $x = 0.05$ m, $m = 0.015$ kg, $k = 600$ N/m	$h = 5.05$ m

---

13* • A stone is projected horizontally with a speed of 20 m/s from a bridge 16 m above the surface of the water. What is the speed of the stone as it strikes the water?

Take $U = 0$ at surface of water.

$E_i = \frac{1}{2}mv_i^2 + mgh = E_f = \frac{1}{2}mv_f^2$ — $v_f = (v_i^2 + 2gh)^{1/2} = 26.7$ m/s

14 • At a dock, a crane lifts a 4000-kg container 30 m, swings it out over the deck of a freighter, and lowers the container into the hold of the freighter, which is 8 m below the level of the dock. How much work is done by the crane? (Neglect friction losses.)

$W = \Delta U = mg\Delta h$; Note that $K_i = K_f = 0$ $\quad$ $W = (4000 \times 9.81 \times -8)$ J = -314 kJ

15 • A 16-kg child on a playground swing moves with a speed of 3.4 m/s when the 6-m-long swing is at its lowest point. What is the angle that the swing makes with the vertical when the child is at the highest point?

1. $U_i = 0$, $K_i = \frac{1}{2}mv^2$; $U_f = mgh$; $K_f = 0$ $\quad$ $h = v^2/2g$
2. Express h in terms of L and θ $\quad$ $h = L(1 - \cos\theta)$
3. Solve for and evaluate θ $\quad$ $\theta = \cos^{-1}(1 - v^2/2gL) = 25.6°$

16 •• In 1983, Jacqueline De Creed, driving a 1967 Ford Mustang, made a jump of 71 m, taking off from a ramp inclined at 30° with the horizontal. If the mass of the car and driver was about 900 kg, find the kinetic energy K and potential energy U of De Creed's vehicle at the top point of her flight.

1. Use $R = v_0^2 \sin 2\theta/g$; find v_0 $\quad$ $v_0 = (71 \times 9.81/\sin 60°)^{1/2}$ m/s = 28.36 m/s
2. At top, $v = v_x = v_0 \cos\theta$, $K = \frac{1}{2}mv_x^2$ $\quad$ $K = \frac{1}{2} \times 900 \times (28.36 \cos 30°)^2$ J = 271.4 kJ
3. At top, $U = K_i - 271$ kJ $\quad$ $U = (\frac{1}{2} \times 900 \times 28.36^2 - 271.4)$ kJ = 90.5 kJ

17* •• The system shown in Figure 7-19 is initially at rest when the lower string is cut. Find the speed of the objects when they are at the same height.

1. $E_i = E_f$; $\Delta U + \Delta K = 0$ $\quad$ $\Delta U = (2 \times 0.5 - 3 \times 0.5) \times 9.81$ J = -4.91 J
2. $K = \frac{1}{2}mv^2 = -\Delta U$, $m = 5$ kg; find v $\quad$ $v = (2 \times 4.91/5)^{1/2}$ m/s = 1.40 m/s

18 •• While traveling in the far north, one of your companions gets snow blindness, and you have to lead him along by the elbow. Looking back, you see your other companion, Sandy, fall and slide along the frictionless surface of the frozen river valley shown in Figure 7-20. If point Q is 4.5 m higher than point P, and your hapless companion fell at point P with a velocity v_0 down the slope, describe his motion to your snow-blind friend if *(a)* $v_0 = 2$ m/s and *(b)* $v_0 = 5$ m/s. *(c)* What is the minimum initial speed required for the fall to carry your partner past point Q?

(c) To rise h m above initial height, $v^2 \geq 2gh$ $\quad$ $v_{min} = (2 \times 9.81 \times 4.5)^{1/2}$ m/s = 9.4 m/s

(a) and *(b)* $v < v_{min}$ $\quad$ Sandy will slide back and forth in valley. For *(a)*, height above initial position is 0.2 m, for *(b)* 1.27 m.

$h = v_i^2/2g$; Find h for $v_i = 2$ m/s, 5 m/s

19 •• A block rests on an inclined plane as shown in Figure 7-21. A spring to which it is attached via a pulley is being pulled downward with gradually increasing force. The value of μ_s is known. Find the potential energy U of the spring at the moment when the block begins to move.

1. When mass begins to move, $kx = f_s + mg\sin\theta = \mu_s mg\cos\theta + mg\sin\theta = mg(\sin\theta + \mu_s\cos\theta)$.
2. $U = \frac{1}{2}kx^2 = [mg(\sin\theta + \mu_s\cos\theta)]^2/2k$.

20 •• Sandy is sliding helplessly across the frictionless ice with her climbing rope trailing behind (Figure 7-22). Racing after her, you get hold of her rope just as she goes over the edge of a cliff. You manage to grab a tree branch in time to keep from going over yourself. Let $U = 0$ for the position of Sandy dangling in midair at the other end of the rope. Snap! The branch to which you are clinging breaks. (*a*) Write an expression for the total mechanical energy of this two-body system after Sandy has fallen a distance *y*. (*b*) There is another tree branch 2 m closer to the cliff edge than the first. What is your speed as you reach it?

(*a*) Let m_1 be your mass and m_2 that of Sandy. Then $K_1 = K_2 = 0$, and $U_1 = m_1gy$, $U_2 = 0$. Thus $E = m_1gy$.

(*b*) $\Delta U + \Delta K = 0$; $\Delta U = -m_2g\Delta y$; both masses move with speed v; $\frac{1}{2}(m_1 + m_2)v^2 = m_2g(2\text{ m})$; solving for v we find $v = \sqrt{(4\text{ m})m_1g/(m_1 + m_2)}$.

21* •• A 2.4-kg block is dropped from a height of 5.0 m onto a spring of spring constant 3955 N/m. When the block is momentarily at rest, the spring has compressed by 25 cm. Find the speed of the block when the compression of the spring is 15.0 cm.

1. Let ΔU_g be the change in gravitational potential energy U; use Equ. 7-6 — $\Delta U_g = -mg(h + x) = \frac{1}{2}mv^2 + \frac{1}{2}kx^2$
2. Find v at $x = 0.15$ m — $v = \sqrt{2g(h + x) - kx^2/2} = \pm 8.0$ m/s

---

22 •• Red is a girl of mass *m* who is taking a picnic lunch to her grandmother. She ties a rope of length *R* to a tree branch over a creek and starts to swing from rest at point *A*, which is a distance *R*/2 lower than the branch (Figure 7-23). What is the minimum breaking tension for the rope if it is not to break and drop Red in the creek?

At bottom, $T = m(g + v^2/R)$; $mv^2 = 2mg\Delta h = mgR$; $T = mg + mg = 2mg$.

---

23 •• A ball at the end of a string moves in a vertical circle with constant energy *E*. What is the difference between the tension at the bottom of the circle and the tension at the top?

Since E = constant, $\frac{1}{2}m(v_B^2 - v_T^2) = 2mgR$, where R is the radius of the circle. The tension at top is $T_T = mv_T^2/R - mg$ and at bottom it is $T_B = mv_B^2/R + mg$. But $(m/R)(v_B^2 - v_T^2) = 4mg$. Thus, $T_B - T_T = 6mg$.

---

24 •• A roller coaster car of mass 1500 kg starts a distance $H = 23$ m above the bottom of a loop 15 m in diameter (Figure 7-24). If friction is negligible, the downward force of the rails on the car when it is upside down at the top of the loop is __. (*a*) 4.6×10^4 N (*b*) 3.1×10^4 N (*c*) 1.7×10^4 N (*d*) 980 N (*e*) 1.6×10^3 N

(*c*) $\frac{1}{2}mv^2 = mg\Delta h = mg(H - 2R)$; $F = mv^2/R - mg = (2mg/R)(H - 2R) - mg = 16.7$ kN.

25* •• A stone is thrown upward at an angle of 53° above the horizontal. Its maximum height during the trajectory is 24 m. What was the stone's initial speed?

1. $\frac{1}{2}mv_y^2 = mgh$; solve for and evaluate v_y — $v_y = (2gh)^{1/2} = 21.7$ m/s
2. $v = v_y/\sin 53°$ — $v = 27.2$ m/s

---

26 •• A baseball of mass 0.17 kg is thrown from the roof of a building 12 m above the ground. Its initial velocity is 30 m/s at an angle of 40° above the horizontal. (*a*) What is the maximum height of the ball? (*b*) What is the work done by gravity as the ball moves from the roof to its maximum height? (*c*) What is the speed of the ball as it strikes the ground?

(*a*) $\frac{1}{2}mv_y^2 = mg\Delta h$; $v_y = v \sin\theta$, evaluate Δh — $\Delta h = (30 \times \sin 40°)^2/(2 \times 9.81)$ m $= 19$ m; $h = 31$ m

(*b*) $W_g = -mg\Delta h$ — $W_g = -(0.17 \times 9.81 \times 19)$ J $= -31.6$ J

(*c*) $v^2 = v_0^2 + 2g(12\text{ m})$; solve for and evaluate v — $v = 33.7$ m/s

---

27 •• An 80-cm-long pendulum with a 0.6-kg bob is released from rest at an initial angle θ_0 with the vertical. At the bottom of the swing, the speed of the bob is 2.8 m/s. (*a*) What was the initial angle of the pendulum? (*b*) What angle does the pendulum make with the vertical when the speed of the bob is 1.4 m/s?

(*a*) 1. Set $U = 0$ at $\theta = 0$; find U for $\theta = \theta_0$ — $U(\theta_0) = mgL(1 - \cos\theta_0)$

2. Use $\Delta K + \Delta U = 0$ — $\frac{1}{2}mv^2 = mgL(1 - \cos\theta_0)$

3. Solve for and find θ_0 — $\theta_0 = \cos^{-1}(1 - v^2/2gL) = 60°$

(*b*) 1. $\frac{1}{2}mv^2 = mg\Delta h$ — $\Delta h = 1.4^2/(2 \times 9.81)\text{ m} = 0.1\text{ m}$

2. Express Δh in terms of L, θ_0 and θ — $\Delta h = L(\cos\theta - \cos\theta_0)$

3. Solve for and evaluate θ — $\theta = \cos^{-1}(\cos\theta_0 + \Delta h/L) = \cos^{-1}(0.5 + 0.125) = 51.3°$

---

28 •• The Royal Gorge bridge over the Arkansas River is about $L = 310$ m high. A bungee jumper of mass 60 kg has an elastic cord of length $d = 50$ m attached to her feet. Assume that the cord acts like a spring of force constant k. The jumper leaps, barely touches the water, and after numerous ups and downs comes to rest at a height h above the water. (*a*) Find h. (*b*) Find the maximum speed of the jumper.

(*a*) 1. $U = 0$, $K = 0$ at top; write U at bottom, find k — $U = \frac{1}{2}k(310 - 50)^2 - (60 \times 9.81 \times 310)\text{ J}$; $k = 5.4$ N/m

2. At h, $kx = mg$; find x — $x = (60 \times 9.81/5.4)\text{ m} = 109\text{ m}$

3. $h = L - d - x$ — $h = 151$ m

(*b*) 1. Write expression for E — $E = 0 = \frac{1}{2}mv^2 - mg(50 + x) + \frac{1}{2}kx^2$

2. v is maximum when $K = \frac{1}{2}mv^2$ is max — $dK/dx = 0 = mg - kx$; $x = 109$ m

3. Evaluate v for $x = 109$ m — $v = [2 \times 9.81 \times 159 - 5.4 \times 109^2/60]^{1/2}\text{ m/s} = 45.3\text{ m/s}$

29* •• A pendulum consists of a 2-kg bob attached to a light string of length 3 m. The bob is struck horizontally so that it has an initial horizontal velocity of 4.5 m/s. For the point at which the string makes an angle of 30° with the vertical, what is (*a*) the speed? (*b*) the potential energy? (*c*) the tension in the string? (*d*) What is the angle of the string with the vertical when the bob reaches its greatest height?

(*a*) 1. Find h at $\theta = 30°$ — $h = L(1 - \cos\theta)$; $h = 0.402$ m

2. $v^2 = v_0^2 - 2gh$ — $v = (4.5^2 - 2 \times 0.402 \times 9.81)^{1/2}\text{ m/s} = 3.52\text{ m/s}$

(*b*) $U = mgh$ — $U = (2 \times 9.81 \times 0.402)\text{ J} = 7.89\text{ J}$

(*c*) $T = mg\cos\theta + mv^2/L$ — $T = (2 \times 9.81 \times 0.866 + 2 \times 3.52^2/3)\text{ N} = 25.3\text{ N}$

(*d*) $\theta_0 = \cos^{-1}(1 - v^2/2gL)$ (see Problem 7-27) — $\theta_0 = \cos^{-1}(1 - 4.5^2/2 \times 9.81 \times 3) = 49°$

---

30 •• Lou is trying to kill mice by swinging a clock of mass m attached to one end of a light (massless) stick 1.4 m in length hanging on a nail in the wall (Figure 7-25). The clock end of the stick is free to rotate around its other end in a vertical circle. Lou raises the clock until the stick is horizontal, and when mice peek their heads out from the hole to their den, he gives it an initial downward velocity v. The clock misses a mouse and continues on its circular path with just enough energy to complete the circle and bonk Lou on the back of his head, to the sound of cheering mice. (*a*) What was the value of v? (*b*) What was the clock's speed at the bottom of its swing?

(*a*) $\Delta U = mgR$; $\Delta K = -\frac{1}{2}mv^2$; $\Delta U + \Delta K = 0$ — $v = \sqrt{2gR} = \sqrt{2 \times 9.81 \times 1.4}\text{ m/s} = 5.24\text{ m/s}$

(*b*) $\Delta U' = -2mgR$, relative to top — $\frac{1}{2}mv_B^2 = 2mgR$; $v_B = 2\sqrt{gR} = 5.24 \times \sqrt{2}\text{ m/s} = 7.41\text{ m/s}$

---

31 •• A pendulum consists of a string of length L and a bob of mass m. The string is brought to a horizontal position and the bob is given the minimum initial speed enabling the pendulum to make a full turn in the vertical plane. (*a*) What is the maximum kinetic energy K of the bob? (*b*) What is the tension in the string when the kinetic energy is maximum?

(*a*) 1. At top, $mv_T^2/L = mg$ — $v_T^2 = Lg$

2. $\frac{1}{2}mv_B^2 = K_{max} = \frac{1}{2}mv_T^2 + 2mgL$ — $K_{max} = (5/2)mgL$

(*b*) $T_T = 0$; $T_B - T_T = 6mg$ (see Problem 7-23) — $T_B = 6mg$

---

32 •• A child whose weight is 360 N swings out over a pool of water using a rope attached to the branch of a tree at the edge of the pool. The branch is 12 m above ground level and the surface of the pool is 1.8 m below ground level. The child holds on to the rope at a point 10.6 m from the branch and moves back until the angle between the rope and the vertical is 23°. When the rope is in the vertical position, the child lets go and drops into the pool. Find the speed of the child at the surface of the pool.

1. $\Delta U = -[mgL(1 - \cos\theta) + mg\Delta h]$, where $\Delta h = 3.2$ m — $\Delta U = -[360 \times 10.6 \times (1 - \cos 23°) + 360 \times 3.2]$ J $= -1455$ J

2. $\frac{1}{2}mv^2 = -\Delta U$; $m = 360/9.81$ kg; solve for v — $v = (2 \times 1455 \times 9.81/360)^{1/2}$ m/s $= 8.9$ m/s

---

33* •• Walking by a pond, you find a rope attached to a tree limb 5.2 m off the ground. You decide to use the rope to swing out over the pond. The rope is a bit frayed but supports your weight. You estimate that the rope might break if the tension is 80 N greater than your weight. You grab the rope at a point 4.6 m from the limb and move back to swing out over the pond. (*a*) What is the maximum safe initial angle between the rope and the vertical so that it will not break during the swing? (*b*) If you begin at this maximum angle, and the surface of the pond is 1.2 m below the level of the ground, with what speed will you enter the water if you let go of the rope when the rope is vertical?

Here we must supply "your weight." We shall take your weight to be 650 N (about 145 lb).

(*a*) 1. $T_m = mg + mv^2/L = mg + 80$ N; $L = 4.6$ m — $mv^2 = 4.6 \times 80$ J

2. $\frac{1}{2}mv^2 = mgL(1 - \cos\theta)$ — $\cos\theta = 1 - mv^2/2mgL = 0.938$; $\theta = 20.2°$

(*b*) See Problem 7-32: $v^2 = 2g[L(1 - \cos\theta) + 1.8\text{ m}]$ — $v = 6.4$ m/s

---

34 •• A pendulum of length L has a bob of mass m attached to a light string, which is attached to a spring of force constant k. With the pendulum in the position shown in Figure 7-26, the spring is at its unstretched length. If the bob is now pulled aside so that the string makes a *small* angle θ with the vertical, what is the speed of the bob after release as it passes through the equilibrium position?

1. Write ΔU — $\Delta U = -[mgL(1 - \cos\theta) + \frac{1}{2}k(L\sin\theta)^2]$

2. For $\theta \ll 1$, $\sin\theta \simeq \theta$, $\cos\theta \simeq 1 - \frac{1}{2}\theta^2$ — $\Delta U = -[\frac{1}{2}mgL\theta^2 + \frac{1}{2}kL^2\theta^2]$

3. Set $\Delta K = -\Delta U = \frac{1}{2}mv^2$, and solve for v — $v^2 = \theta^2(gL + kL^2/m)$; $v = L\theta\sqrt{\frac{g}{L} + \frac{k}{m}}$

---

35 ••• A pendulum is suspended from the ceiling and attached to a spring fixed at the opposite end directly below the pendu-lum support (Figure 7-27). The mass of the pendulum bob is m, the length of the pendulum is L, and the spring constant is k. The unstretched length of the spring is $L/2$ and the distance between the bottom of the spring and the ceiling is $1.5L$. The pendulum is pulled aside so that it makes a small angle θ with the vertical and then released from rest. Obtain an expression for the speed of the pendulum bob when $\theta = 0$.

1. Write ΔU; note that for spring, $x = \frac{1}{2}L(\sec 2\theta - 1)$ — $\Delta U = -[mgL(1 - \cos\theta) + \frac{1}{2}k(\frac{1}{2}L)^2(\sec 2\theta - 1)^2]$

2. For $\theta << 1$, $\cos\theta \approx 1 - \frac{1}{2}\theta^2$, $\sec\theta \approx 1 + \frac{1}{2}\theta^2$ $\Delta U = -[\frac{1}{2}mgL\theta^2 + \frac{1}{2}kL^2\theta^4]$

3. $\frac{1}{2}mv^2 = -\Delta U$; solve for v $v^2 = mgL\theta^2 + kL^2\theta^4/m$; $v = L\theta\sqrt{\frac{g}{L} + \frac{k}{m}\theta^2}$

---

36 • True or false: (*a*) The total energy of a system cannot change. (*b*) When you jump into the air, the floor does work on you, increasing your potential energy.

(*a*) False (*b*) False

37* • A man stands on roller skates next to a rigid wall. To get started, he pushes off against the wall. Discuss the energy changes pertinent to this situation.

The kinetic energy of the man increases at the expense of metabolic (chemical) energy.

---

38 • Discuss the energy changes involved when a car starts from rest and accelerates such that the car's wheels do not slip. What external force accelerates the car? Does this force do work?

Kinetic energy increases, chemical energy decreases, and thermal energy increases. The force acting on the car is the static friction force of the road surface on the wheels. This force does no work.

---

39 • A body falling through the atmosphere (air resistance is present) gains 20 J of kinetic energy. The amount of gravitational potential energy that is lost is (*a*) 20 J. (*b*) more than 20 J. (*c*) less than 20 J. (*d*) impossible to tell without knowing the mass of the body. (*e*) impossible to tell without knowing how far the body falls.

(*b*) $\Delta U = -(\Delta K + \text{friction energy})$.

---

40 • Assume that you can expend energy at a constant rate of 250 W. Estimate how fast you can run up four flights of stairs, with each flight 3.5 m high.

Take $mg = 600$ N; $mgh = 8400\text{ J} = Pt = (250\text{ W})t$; $t = 37$ s.

41* • A 70-kg skater pushes off the wall of a skating rink, acquiring a speed of 4 m/s. (*a*) How much work is done on the skater? (*b*) What is the change in mechanical energy of the skater? (*c*) Discuss the conservation of energy as applied to the skater.

(*a*) No work is done *on* the skater. The displacement of the force exerted by the wall is zero.

(*b*) $\Delta K = \frac{1}{2}mv^2 = 560$ J.

(*c*) The increase in kinetic energy is at the loss of metabolic (chemical) energy.

---

42 • In a volcanic eruption, 4 km^3 of mountain with a density of 1600 kg/m^3 was lifted an average height of 500 m. (*a*) How much energy in joules was released in this eruption? (*b*) The energy released by thermonuclear bombs is measured in megatons of TNT, where 1 megaton of TNT = 4.2×10^{15} J. Convert your answer for (*a*) to megatons of TNT.

(*a*) $E = mgh = g\rho Vh$; evaluate E $E = 3.14 \times 10^{16}$ J

(*b*) Convert to Mton TNT $3.14 \times 10^{16}\text{ J} = (3.14 \times 10^{16}/4.2 \times 10^{15})$
$= 7.47$ Mton TNT

---

43 •• An 80-kg physics student climbs a 120-m hill. (*a*) What is the increase in the gravitational potential energy of the student? (*b*) Where does this energy come from? (*c*) The student's body is 20% efficient; that is, for every 20 J that are converted to mechanical energy, 100 J of internal energy are expended, with 80 J going into thermal energy. How much chemical energy is expended by the student during the climb?

(*a*) $\Delta U = mgh$ $\Delta U = (80 \times 9.81 \times 120)\text{ J} = 94.2$ kJ (*b*) Energy comes from metabolic energy

(*c*) At 20% efficiency, $E = 5 \times \Delta U$ $\qquad E = (5 \times 94.2)$ kJ = 471 kJ

---

44 •• In 1993, Carl Fentham of Great Britain raised a full keg of beer (mass 62 kg) to a height of about 2 m 676 times in 6 h. Assuming that work was done only as the keg was going up, estimate how many such kegs of beer he would have to drink to reimburse his energy expenditure. (1 liter of beer is approximately 1 kg and provides about 1.5 MJ of energy; in your calculations, neglect the mass of the empty keg.)

1. Determine the work done. $\qquad W = mgNh = (62 \times 9.81 \times 676 \times 2)$ J = 822 kJ

2. Number of kegs = $W/(E$ per keg) $\qquad$ Number of kegs = $(8.22 \times 10^5/62 \times 1.5 \times 10^6) = 0.00884$

45* • Discuss the energy considerations when you pull a box along a rough road.

Metabolic (chemical) energy is converted to thermal energy released through friction.

---

46 • A 2000-kg car moving at an initial speed of 25 m/s along a horizontal road skids to a stop in 60 m. (*a*) Find the energy dissipated by friction. (*b*) Find the coefficient of kinetic friction between the tires and the road.

We shall use W_f to denote the energy dissipated by friction.

(*a*) $K_i - K_f = W_f$; $K_f = 0$ $\qquad W_f = K_i = ½(2000)(25)^2$ J = 625 kJ

(*b*) Find μ_k using $W_f = f_k\Delta s$ and $f_k = \mu_k mg$ $\qquad \mu_k = W_f/mg\Delta s = [6.25 \times 10^5/(2000 \times 9.81 \times 60)] = 0.531$

---

47 • An 8-kg sled is initially at rest on a horizontal road. The coefficient of kinetic friction between the sled and the road is 0.4. The sled is pulled a distance of 3 m by a force of 40 N applied to the sled at an angle of 30° to the horizontal. (*a*) Find the work done by the applied force. (*b*) Find the energy dissipated by friction. (*c*) Find the change in the kinetic energy of the sled. (*d*) Find the speed of the sled after it has traveled 3 m.

(*a*) Use $W = \mathbf{F}\cdot\mathbf{s} = Fs\cos\theta$ $\qquad W = (40 \times 3 \times \cos 30°)$ J = 104 J

(*b*) $W_f = \mu_k F_n s$; $F_n = mg - F\sin\theta$ $\qquad W_f = [0.4 \times 3(8 \times 9.81 - 40\sin 30°)]$ J = 70.2 J

(*c*) $\Delta K = W - W_f$ $\qquad \Delta K = 33.8$ J

(*d*) Since $K_i = 0$, $K_f = \Delta K = ½mv^2$ $\qquad v = (2\Delta K/m)^{½} = 2.91$ m/s

---

48 • Returning to Problem 8, suppose the surfaces described are not frictionless and that the coefficient of kinetic friction between the block and the surfaces is 0.30. Find (*a*) the speed of the block when it reaches the ramp, and (*b*) the distance that the block slides up the ramp before coming momentarily to rest. (Neglect the energy dissipated along the transition curve.)

(*a*) 1. $\Delta U = 0$; $K_f = K_i - W_f$; $W_f = \mu_k F_n s = \mu_k mgs$ $\qquad ½mv_f^2 = ½mv_i^2 - \mu_k mgs$

2. Solve for and evaluate v_f $\qquad v_f = \sqrt{v_i^2 - 2\mu_k gs} = 6.10$ m/s

(*b*) 1. Now $F_n = mg\cos\theta$, $W_f = \mu_k F_n L$; at end $K = 0$.

So, $K_f = mgL\sin\theta + \mu_k mgL\cos\theta$, solve for L. $\qquad L = K_f/[mg(\sin\theta + \mu_k\cos\theta)]$

2. Evaluate L for $K_f = ½ \times 3 \times 6.10^2$ J = 55.8 J $\qquad L = 2.17$ m

---

49* • The 2-kg block in Figure 7-28 slides down a frictionless curved ramp, starting from rest at a height of 3 m. The block then slides 9 m on a rough horizontal surface before coming to rest. (*a*) What is the speed of the block at the bottom of the ramp? (*b*) What is the energy dissipated by friction? (*c*) What is the coefficient of friction between the block and the horizontal surface?

(*a*) Use $\Delta K + \Delta U = 0$; $\Delta U = -mg\Delta h$; $\Delta h = 3$ m $\quad$ $K = mg\Delta h = \frac{1}{2}mv^2$; $v = \sqrt{2g\Delta h} = 7.67$ m/s

(*b*) At end, $K = 0$; so $W_f = K = mg\Delta h$ $\quad$ $W_f = 58.9$ J

(*c*) $W_f = f_k s = \mu_k mgs$; solve for and find μ_k $\quad$ $\mu_k = W_f/mgs = mg\Delta h/mgs = \Delta h/s = 1/3$

---

50 •• A 20-kg girl slides down a playground slide that is 3.2 m high. When she reaches the bottom of the slide, her speed is 1.3 m/s. (*a*) How much energy was dissipated by friction? (*b*) If the slide is inclined at 20°, what is the coefficient of friction between the girl and the slide?

(*a*) Use $\Delta U + \Delta K + W_f = 0$; $\Delta U = -mg\Delta h$; $\Delta h = 3.2$ m $\quad$ $W_f = mg\Delta h - \frac{1}{2}mv^2 = 611$ J

(*b*) $W_f = \mu_k F_n L = \mu_k(mg\cos\theta)(\Delta h/\sin\theta)$. Solve for μ_k. $\quad$ $\mu_k = W_f\tan\theta/(mg\Delta h)$; $\mu_k = 0.354$

---

51 •• In Figure 7-29, the coefficient of kinetic friction between the 4-kg block and the shelf is 0.35. (*a*) Find the energy dissipated by friction when the 2-kg block falls a distance *y*. (*b*) Find the total mechanical energy *E* of the two-block system after the 2-kg block falls a distance *y*, assuming that $E = 0$ initially. (*c*) Use your result for (*b*) to find the speed of either block after the 2-kg block falls 2 m.

(*a*) Use $W_f = f_k s = \mu_k m_1 gs$, where $s = y$ $\quad$ $W_f = (0.35 \times 4 \times 9.81 \times y)$ J $= 13.73y$ J

(*b*) $E_{mech} = \Delta E = -W_f$ $\quad$ $E = -13.73y$ J

(*c*) 1. Use $E_{mech} = \Delta U + \Delta K$; $\Delta K = E_{mech} - \Delta U$ $\quad$ $\frac{1}{2}(m_1 + m_2)v^2 = m_2gy - W_f$

2. Solve for and evaluate v $\quad$ $v = \sqrt{2(m_2gy - W_f)/(m_1 + m_2)}$; $v = 1.98$ m/s

---

52 •• Nils Lied, an Australian meteorologist, once played golf on the ice in Antarctica and drove a ball a horizontal distance of 2400 m. For a rough estimate, let us assume that the ball took off at $\theta = 45°$, flew a horizontal distance of 200 m without air resistance, and then slid on the ice without bouncing, its velocity being equal to the horizontal component of the initial velocity. Estimate the coefficient of kinetic friction μ_k between the ice and ball.

1. Use $R = (v^2 \sin 2\theta)/g$ to determine v and v_x $\quad$ $v = (Rg)^{1/2} = 44.3$ m/s; $v_x = (44.3 \sin 45°)$ m/s $= 31.3$ m/s
2. Find acceleration using $v_x^2 = 2as$ $\quad$ $a = (31.3^2/2 \times 2200)$ m/s^2 $= 0.223$ m/s^2
3. $a = f_k/m = \mu_k mg/m$; solve for and evaluate μ_k $\quad$ $\mu_k = a/g = (0.223/9.81) = 0.0227$

53* •• A particle of mass *m* moves in a horizontal circle of radius *r* on a rough table. It is attached to a horizontal string fixed at the center of the circle. The speed of the particle is initially v_0. After completing one full trip around the circle, the speed of the particle is $\frac{1}{2}v_0$. (*a*) Find the energy dissipated by friction during that one revolution in terms of *m*, v_0, and *r*. (*b*) What is the coefficient of kinetic friction? (*c*) How many more revolutions will the particle make before coming to rest?

(*a*) Here, $W_f = K_i - K_f$ since *U* is constant. Thus, $W_f = \frac{1}{2}m(v_0^2 - v_0^2/4) = (3/8)mv_0^2$.

(*b*) Distance traveled in one revolution is $2\pi r$; so $W_f = \mu_k mg(2\pi r)$ and $\mu_k = (3v_0^2)/(16\pi gr)$.

(*c*) Since in one revolution it lost $(3/4)K_i$, it will only require another 1/3 revolution to lose the remaining $K_i/4$.

---

54 •• In 1987, British skier Graham Wilkie achieved a speed of $v = 211$ km/h going downhill. Assuming that he reached the maximum speed at the end of the hill and then continued on the horizontal surface, find the maximum distance *d* he *could have* covered on the horizontal surface. Take the coefficient of kinetic friction μ_k

to be constant throughout the run; neglect air resistance. Assume the hill is 225 m high with a constant slope of 30° with the horizontal.

1. Use $K_f = -\Delta U - W_f$, and $W_f = f_k$ — $\frac{1}{2}mv^2 = mg\Delta h - \mu_k F_n L$
   $= mg\Delta h - \mu_k(mg\cos\theta)(\Delta h/\sin\theta)$
2. Solve for and evaluate μ_k — $\mu_k = [1 - v^2/(2g\Delta h)](\tan\theta)$; $\mu_k = 0.128$
3. Use $d = v^2/2a$, and $a = \mu_k g$ to find d — $d = 1368$ m

55 •• During a move, Kate and Lou have to push Kate's 80-kg stove up a rough loading ramp, pitched at an angle of 10°, to get it into a truck. They push it along the horizontal floor to pick up speed and give it one last push at the bottom of the ramp, hoping for the best. Unfortunately, the stove stops short and then slides down the ramp, sending them leaping to the side. (*a*) If the stove has a speed of 3 m/s at the bottom of the ramp, and a speed of 0.8 m/s when it is 2-m up the ramp, what is the maximum height reached by the stove? (*b*) What is the stove's speed when it passes the 2 m spot again? (*c*) What is the energy dissipated by friction during the complete round trip back to the bottom of the ramp?

(*a*) 1. Use conservation of energy — $mgL\sin\theta + \mu_k mgL\cos\theta + \frac{1}{2}mv^2 = \frac{1}{2}mv_0^2$ (1)
2. Solve for and evaluate μ_k — $\mu_k = [\frac{1}{2}(v_0^2 - v^2) - gL\sin\theta]/(gL\cos\theta) = 0.04$ (2)
3. In Equ. (1) set $v = 0$ and solve for L — $L = v_0^2/[2g(\sin\theta + \mu_k\cos\theta)] = 2.15$ m

(*b*) Use conservation of energy — $v = [2 \times 0.15 \times 9.81(\sin\theta - \mu_k\cos\theta)]^{1/2} = 0.63$ m/s

(*c*) $W_f = f_k s$ — $W_f = (0.04 \times 4.3 \times 80 \times 9.81 \times \cos 10°)$ J $= 133$ J

56 •• A 2.4-kg box has an initial velocity of 3.8 m/s upward along a rough plane inclined at 37° to the horizontal. The coefficient of kinetic friction between the box and plane is 0.30. How far up the incline does the box travel? What is its speed when it passes its starting point on its way down the incline?

1. We use $\Delta K + \Delta U = -W_f$; write an expression for W_f — $W_f = \mu_k F_n L = \mu_k(mg\cos\theta)(h/\sin\theta) = \mu_k mgh\cot\theta$
2. Use $\Delta U = mgh$; $\Delta K = -\frac{1}{2}mv^2$; solve for h and L — $h = v^2/[2g(1 + \mu_k\cot\theta)]$; $L = h/\sin\theta$
3. Evaluate h and L — $h = 0.526$ m, $L = 0.875$ m
4. On return, $\Delta K = \frac{1}{2}mv_f^2$, $\Delta U = -mgh$; W_f as before — $v_f^2 = 2gh(1 - \mu_k\cot\theta)$; $v_f = 2.49$ m/s

57* ••• A block of mass m rests on a rough plane inclined at θ with the horizontal (Figure 7-30). The block is attached to a spring of constant k near the top of the plane. The coefficients of static and kinetic friction between the block and plane are μ_s and μ_k, respectively. The spring is slowly pulled upward along the plane until the block starts to move. (*a*) Obtain an expression for the extension d of the spring the instant the block moves. (*b*) Determine the value of μ_k such that the block comes to rest just as the spring is in its unstressed condition, i.e., neither extended nor compressed.

(*a*) The force exerted by the spring must equal the sum $mg\sin\theta + f_s$; hence, $d = (mg/k)(\sin\theta + \mu_s\cos\theta)$.

(*b*) Take $U_{grav} = 0$ at initial position of m. To meet the condition, $U_{grav} - W_f = \frac{1}{2}kd^2$, where $U_{grav} = mgd\sin\theta$, and $W_f = \mu_k mgd\cos\theta$. Solving for μ_k one finds $\mu_k = \tan\theta - \frac{1}{2}(1 + \mu_s\cot\theta)$.

58 • How much rest mass is consumed in the core of a nuclear-fueled electric generating plant in producing (*a*) one joule of thermal energy? (*b*) enough energy to keep a 100-W light bulb burning for 10 years?

(*a*) Use Equ. 7-15 — $m = E/c^2 = (1/9 \times 10^{16})$ kg $= 1.11 \times 10^{-17}$ kg (*b*) 1. Find the energy required $E = Pt = (100 \times 10 \times 365 \times 24 \times 60 \times 60)$ J $= 3.16 \times 10^{10}$ J

2. Use the result of part (*a*) $m = (1.11 \times 10^{-17})(3.16 \times 10^{10})$ kg = 0.351 mg

---

59 • (*a*) Calculate the rest energy in 1 g of dirt. (*b*) If you could convert this energy into electrical energy and sell it for 10 cents per kilowatt-hour, how much money would you get? (*c*) If you could power a 100-W light bulb with this energy, for how long could you keep the bulb lit?

(*a*) Use Equ. 7-15 $E = (10^{-3} \times 9 \times 10^{16})$ J $= 9 \times 10^{13}$ J

(*b*) 1. Convert to kWh 1 kWh $= 10^3 \times 3600$ J $= 3.6 \times 10^6$ J; $E = 2.5 \times 10^7$ kWh

2. Determine the price of E $(2.5 \times 10^7$ kWh)($ 0.10/1 kWh) = $ 2.5×10^6

(*c*) Use $E = Pt$ $t = 9 \times 10^{11}$ s = 28,400 y

---

60 • A muon has a rest energy of 105.7 MeV. Calculate its rest mass in kilograms.

1 MeV $= 1.6 \times 10^{-13}$ J; $m_\mu = E/c^2 = (105.7 \times 1.6 \times 10^{-13}/9 \times 10^{16})$ kg $= 1.88 \times 10^{-28}$ kg .

61* • For the fusion reaction in Example 7-14, calculate the number of reactions per second that are necessary to generate 1 kW of power.

From Example 7-14, energy per reaction is 17.59 MeV $= 28.1 \times 10^{-13}$ J; to generate 1000 J/s then requires $(1000/28.1 \times 10^{-13})$ reactions $= 3.56 \times 10^{14}$ reactions.

---

62 • How much energy is needed to remove one neutron from ^{4}He, leaving ^{3}He plus a neutron? (The rest energy of ^{3}He is 2808.41 MeV.)

Find Δmc^2 in MeV from Table 7-1 $E = (939.573 + 2808.41 - 3727.409)$ MeV = 20.574 MeV

---

63 • A free neutron at rest decays into a proton plus an electron: $n \rightarrow p + e$. Use Table 7-1 to calculate the energy released in this reaction.

Find Δmc^2 in MeV from Table 7-1 $E = (939.573 - 938.280 - 0.511)$ MeV = 0.782 MeV

---

64 •• In one nuclear fusion reaction, two ^{2}H nuclei combine to produce ^{4}He. (*a*) How much energy is released in this reaction? (*b*) How many such reactions must take place per second to produce 1 kW of power?

(*a*) $E = (2 \times 1875.628 - 3727.409)$ MeV = 23.847 MeV $= 3.816 \times 10^{-12}$ J.

(*b*) To obtain 1000 W requires $(1000/3.816 \times 10^{-12})$ reaction/s $= 2.62 \times 10^{14}$ reactions/s.

65* •• A large nuclear power plant produces 3000 MW of power by nuclear fission, which converts matter into energy. (*a*) How many kilograms of matter does the plant consume in one year? (*b*) In a coal-burning power plant, each kilogram of coal releases 31 MJ of energy when burned. How many kilograms of coal are needed each year for a 3000-MW plant?

(*a*) 1. Find the energy produced per year $E = (3 \times 10^9 \times 3.16 \times 10^7)$ J $= 9.48 \times 10^{16}$ J

2. Find $m = E/c^2$ $m = (9.48 \times 10^{16}/9 \times 10^{16})$ kg = 1.05 kg

(*b*) Find m_{coal} $m_{coal} = (9.48 \times 10^{16}/3.1 \times 10^7)$ kg $= 3.06 \times 10^9$ kg

---

66 •• A block of mass m, starting from rest, is pulled by a string up a frictionless inclined plane that makes an angle θ with the horizontal. The tension in the string is T and the string is parallel to the plane. After traveling a distance L, the speed of the block is v. The work done by the tension T is (*a*) $mgL \sin \theta$. (*b*) $mgL \cos \theta + \frac{1}{2}mv^2$. (*c*) $mgL \sin \theta + \frac{1}{2}mv^2$. (*d*) $mgL \cos \theta$. (*e*) $TL \cos \theta$.

(*c*) $W = \Delta U + \Delta K$.

---

67 •• A block of mass m slides with constant velocity v down a plane inclined at θ with the horizontal. During the time interval Δt, what is the magnitude of the energy dissipated by friction? (*a*) $mgv\Delta t \tan \theta$ (*b*) $mgv\Delta t \sin \theta$ (*c*) $\frac{1}{2}mv^3\Delta t$ (*d*) The answer cannot be determined without knowing the coefficient of kinetic friction.

(*b*) $W_f = -\Delta U$.

---

68 •• Assume that on applying the brakes a constant frictional force acts on the wheels of a car. If that is so, it follows that (*a*) the distance the car travels before coming to rest is proportional to the speed of the car before the brakes are applied. (*b*) the car's kinetic energy diminishes at a constant rate. (*c*) the kinetic energy of the car is inversely proportional to the time that has elapsed since the application of the brakes. (*d*) none of the above apply.

(*d*)

69* • Our bodies convert internal chemical energy into work and heat at the rate of about 100 W, which is called our metabolic rate. (*a*) How much internal chemical energy do we use in 24 h? (*b*) The energy comes from the food that we eat and is usually measured in kilocalories, where 1 kcal = 4.184 kJ. How many kilocalories of food energy must we ingest per day if our metabolic rate is 100 W?

(*a*) $E = Pt$; $E = (100 \times 24 \times 60 \times 60)$ J = 8.64 MJ.

(*b*) $E = (8.64/4.184)$ Mcal = 2.065 Mcal = 2065 kcal.

---

70 • A 3.5-kg box rests on a horizontal frictionless surface in contact with a spring of spring constant 6800 N/m. The spring is fixed at its other end and is initially at its uncompressed length. A constant horizontal force of 70 N is applied to the box so that the spring compresses. Determine the distance the spring is compressed when the box is momentarily at rest.

Since $\Delta K = 0$, $W = Fs = \frac{1}{2}ks^2$; solve for and evaluate s $\quad s = 2F/k = (140/6800)$ m = 0.0206 m = 2.06 cm

---

71 • The average energy per unit time per unit area that reaches the upper atmosphere of the earth from the sun, called the solar constant, is 1.35 kW/m^2. Because of absorption and reflection by the atmosphere, about 1 kW/m^2 reaches the surface of the earth on a clear day. How much energy is collected in 8 h of daylight by a solar panel 1 m by 2 m on a rotating mount that is always perpendicular to the sun's rays?

$E = (10^3 \times 2 \times 8 \times 60 \times 60)$ J = 57.6 MJ.

---

72 • When the jet-powered car *Spirit of America* went out of control during a test drive at Bonneville Salt Flats, Utah, it left skid marks about 9.5 km long. (*a*) If the car was moving initially at a speed of v = 708 km/h, estimate the coefficient of kinetic friction μ_k. (*b*) What was the kinetic energy K of the car at time t = 60 s after the brakes were applied? Take the mass of the car to be 1250 kg.

(*a*) 1. Use $W_f + \Delta K = 0$ $\quad \mu_k mgs = \frac{1}{2}mv^2$

2. Solve for and evaluate μ_k $\quad \mu_k = v^2/2gs$; $\mu_k = [196.7^2/(2 \times 9.81 \times 9.5 \times 10^3)] = 0.208$

(*b*) 1. Use $v = v_0 - \mu_k gt$ $\quad v = (196.7 - 0.208 \times 9.81 \times 60)$ m/s = 74.3 m/s

2. $K = \frac{1}{2}mv^2$; evaluate K $\quad K = \frac{1}{2}(1250 \times 74.3^2)$ J = 3.45 MJ

73* •• A T-bar tow is required to pull 80 skiers up a 600-m slope inclined at 15° above horizontal at a speed of 2.5 m/s. The coefficient of kinetic friction is 0.06. Find the motor power required if the mass of the average skier is 75 kg.

1. Find the force required — $F = m_{\rm tot}g \sin\theta + \mu_k g m_{\rm tot} \cos\theta$
2. Use $P = Fv$; evaluate P — $P = (600 \times 75 \times 9.81 \times 2.5)(\sin 15° + 0.06 \cos 15°)$W $= 350$ kW

---

74 •• A 2-kg box is projected with an initial speed of 3 m/s up a rough plane inclined at 60° above horizontal. The coefficient of kinetic friction is 0.3. (*a*) List all the forces acting on the box. (*b*) How far up the plane does the box slide before it stops momentarily? (*c*) What is the energy dissipated by friction as the box slides up the plane? (*d*) What is the speed of the box when it again reaches its initial position?

(*a*) See Problem 5-24 — $mg,\ f_k = \mu_k mg \cos\theta$

(*b*) $a = -(\sin\theta + \mu_k \cos\theta)g$; find s for $v = 0$ — $s = -v_0^2/2a = v_0^2/[2g(\sin\theta + \mu_k\cos\theta)] = 0.451$ m

(*c*) $W_f = f_k s = \mu_k mgs \cos\theta$ — $W_f = (0.3 \times 2 \times 9.81 \times 0.451 \times 0.5)$ J $= 1.33$ J

(*d*) 1. Use $\frac{1}{2}mv^2 = -\Delta U - W_f$; $\Delta U = -mg\Delta h$ — $\frac{1}{2}mv^2 = mgs \sin\theta - \mu_k mgs \cos\theta$

2. Solve for and evaluate v — $v = \sqrt{2gs(\sin\theta - \mu_k\cos\theta)} = 2.52$ m/s

---

75 •• A 1200-kg elevator driven by an electric motor can safely carry a maximum load of 800 kg. What is the power provided by the motor when the elevator ascends with a full load at a speed of 2.3 m/s?

1. Find the tension in the cable for $a = 0$ — $F = mg$
2. Use $P = Fv$ — $P = mgv = (2000 \times 9.81 \times 2.3)$ W $= 45.1$ kW

---

76 •• To reduce the power requirement of elevator motors, elevators are counterbalanced with weights connected to the elevator by a cable that runs over a pulley at the top of the elevator shaft. If the elevator in Problem 75 is counterbalanced with a mass of 1500 kg, what is the power provided by the motor when the elevator ascends fully loaded at a speed of 2.3 m/s? How much power is provided by the motor when the elevator ascends without a load at 2.3 m/s?

Note that the counterweight does negative work of $m_{\rm cw}gv$ per second. Hence $P_{\rm motor} = (500 \times 9.81 \times 2.3)$ W $=$ 11.3 kW. Without a load, $P_{\rm motor} = [(1200 - 1500) \times 9.81 \times 2.3]$ W $= -6.77$ kW.

77* •• The spring constant of a toy dart gun is 5000 N/m. To cock the gun the spring is compressed 3 cm. The 7-g dart, fired straight upward, reaches a maximum height of 24 m. Determine the energy dissipated by air friction during the dart's ascent. Estimate the speed of the projectile when it returns to its starting point.

1. Use conservation of energy — $\frac{1}{2}kx^2 = mgh + W_f$; $W_f = \frac{1}{2}kx^2 - mgh$
2. Evaluate W_f — $W_f = 0.602$ J
3. We assume that W_f same on descent (not true) — $\frac{1}{2}mv^2 = mgh - W_f$; $v = \sqrt{2gh - W_f/m} = 17.3$ m/s

---

78 •• A 0.050-kg dart is fired vertically from a spring gun that has a spring constant of 4000 N/m. Prior to release, the spring was compressed by 10.7 cm. When the dart is 6.8 m above the gun, its upward speed is 28 m/s. Determine the maximum height reached by the dart.

1. Neglect air friction: $\frac{1}{2}mv^2 = mgh_{\max} = \frac{1}{2}kx^2$ — $h_{\max} = [4000 \times 0.107^2/(2 \times 0.05 \times 9.81)]$ m $= 46.7$ m
2. To see if air friction negligible, find v at 6.8 m — $\frac{1}{2}mv^2 = \frac{1}{2}kx^2 - mgh = 19.6$ J; $v = 28$ m/s

79 •• In a volcanic eruption, a 2-kg piece of porous volcanic rock is thrown vertically upward with an initial speed of 40 m/s. It travels upward a distance of 50 m before it begins to fall back to the earth. (*a*) What is the initial kinetic energy of the rock? (*b*) What is the increase in thermal energy due to air friction during ascent? (*c*) If the increase in thermal energy due to air friction on the way down is 70% of that on the way up, what is the speed of the rock when it returns to its initial position?

(*a*) $K_i = \frac{1}{2}mv_i^2$ — $K_i = 1600$ J

(*b*) $W_f = K_i - \Delta U$ — $W_f = (1600 - 2 \times 9.81 \times 50)$ J $= 619$ J

(*c*) $K_f = mgh - 0.7 \times 619$ J; $v_f = (2K_f/m)^{1/2}$ — $v_f = [(981 - 0.7 \times 619)/2]^{1/2}$ m/s $= 16$ m/s

80 •• A block of mass m starts from rest at a height h and slides down a frictionless plane inclined at θ with the horizontal as shown in Figure 7-31. After sliding a distance L, the block strikes a spring of force constant k. Find the compression of the spring when the block is momentarily at rest.

Using energy conservation we have $\frac{1}{2}kx^2 - mg(L+x)\sin\theta = 0$; $\frac{1}{2}kx^2 - mgx\sin\theta - mgL\sin\theta = 0$. Solve the quadratic equation for x. $x = (mg/k)\sin\theta + [(mg/k)^2\sin^2\theta + 2(mgL/k)\sin\theta]^{1/2}$. The negative sign on the square root leads to a non-physical solution.

81* •• A car of mass 1500 kg traveling at 24 m/s is at the foot of a hill that rises 120 m in 2.0 km. At the top of the hill, the speed of the car is 10 m/s. Find the average power delivered by the car's engine, neglecting any frictional losses.

1. Find increase in mechanical energy; $\Delta E = \Delta K + \Delta U$ — $\Delta E = [750(24^2 - 10^2) + 1500 \times 9.81 \times 120]$ J $= 2.12$ MJ

2. Assume constant a; then $v_{av} = 17$ m/s — $\Delta t = (2000/17)$ s $= 118$ s; $P_{av} = \Delta E/\Delta t = 18$ kW

82 •• In a new ski jump event, a loop is installed at the end of the ramp as shown in Figure 7-32. This problem addresses the physical requirement such an event would place on the skiers. Neglect friction. (*a*) Along the track of the ramp and the loop, where do the legs of the skier have to support the maximum weight? (*b*) If the loop has a radius R, where should the starting gate (indicated by h) be placed so that the maximum force on the skier's legs is 4 times the skier's body weight? (*c*) With the starting gate at the height found in part (*b*), will the skier be able to make it completely around the loop? Why or why not? (*d*) What is the minimum height h such that the skier can make it around the loop? What is the maximum force on the skier's legs for this height?

(*a*) Maximum effective weight occurs at bottom of loop where v is a maximum

(*b*) 1. At bottom, $F = mg + mv^2/R = 4mg$ — $v^2 = 3gR$

2. Use $\Delta K + \Delta U = 0$ — $\frac{1}{2}mv^2 = mgh$; $v^2 = 2gh = 3gR$; $h = 1.5R$

(*c*) 1. To maintain contact at top, $v_T^2/R \geq g$ — $v_{T,min}^2 = Rg$

2. Use $\Delta E = 0$; $K_{T,min} = mg(h - 2R)$ — $\frac{1}{2}v_{T,min}^2 = gh - 2gR = \frac{1}{2}gR$; so $h_{min} = 2.5R$

3. Skier will not make it around loop

(*d*) 1. See part (*c*) — $h_{min} = 2.5R$

2. $F = mg + mv^2/R$ — $mv^2 = 2mgh$; $v^2 = 5Rg$; $F = 6mg$

(see also Problem 7-23)

83 •• A mass m is suspended from the ceiling by a spring and is free to move vertically in the y direction as indicated in Figure 7-33. We are given that the potential energy as a function of position is $U = \frac{1}{2}ky^2 - mgy$. (*a*) Sketch U as a function of y. What value of y corresponds to the *unstretched* condition of the spring, y_0? (*b*) From the given expression for U, find the net downward force acting on m at any position y. (*c*) The mass is released from rest at $y = 0$; if there is no friction, what is the maximum value y_{max} that will be reached by the mass? Indicate $y_{max.}$ on your sketch. (*d*) Now consider the effect of friction. The mass ultimately settles down into an equilibrium position y_{es}. Find this point on your sketch. (*e*) Find the amount of thermal energy produced by friction from the start of the operation to the final equilibrium.

(*a*) A plot of $U(y)$ is shown. Since k and m are not specified, we have plotted $U(y)$ for $k = 2$, $mg = 1$. The spring is unstretched when $y = y_0 = 0$.

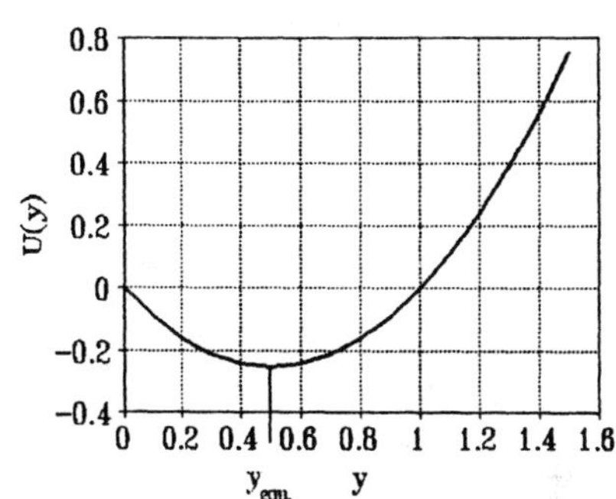

(*b*) $F = -dU/dy$ $\quad$ $F = mg - ky$

(*c*) $\Delta E = 0$ and $\Delta K = 0$, so $U = 0$; solve for y $\quad$ $y = 2mg/k = y_{max}$

(*d*) At equilibrium $F = 0$ $\quad$ $y_{eq} = mg/k$

(*e*) $W_f = U_{in} - U_{fin}$ $\quad$ $W_f = 0 - (\frac{1}{2}ky_{eq}^2 - mgy_{eq}) = \frac{1}{2}(m^2g^2/k)$

84 •• A spring-loaded gun is cocked by compressing a short, strong spring by a distance d. The gun fires a signal flare of mass m directly upward. The flare has speed v_0 as it leaves the spring and is observed to rise to a maximum height h above the point where it leaves the spring. After it leaves the spring, effects of drag force by the air on the packet are significant. (Express answers in terms of m, v_0, d, h, and g, the acceleration due to gravity.) (*a*) How much work is done on the spring in the course of the compression? (*b*) What is the value of the spring constant k? (*c*) How much mechanical energy is converted to thermal energy because of the drag force of the air on the flare between the time of firing and the time at which maximum elevation is reached?

(*a*) The work done on the spring is equal to the initial K of m, i.e., $\frac{1}{2}mv_0^2$.

(*b*) Since $\frac{1}{2}kd^2 = \frac{1}{2}mv_0^2$, $k = mv_0^2/d^2$.

(*c*) $W_f = E_{in,mech} - E_{fin,mech} = \frac{1}{2}mv_0^2 - mgh$.

85* •• A roller-coaster car having a total mass (including passengers) of 500 kg travels freely along the winding frictionless track in Figure 7-34. Points A, E, and G are horizontal straight sections, all at the same height of 10 m above ground. Point C is at a height of 10 m above ground on a section sloped at an angle of 30°. Point B is at the top of a hill, while point D is at ground level at the bottom of a valley. The radius of curvature at each of these points is 20 m. Point F is at the middle of a banked horizontal curve of radius of curvature of 30 m, and at the same height of 10 m above the ground as points A, E, and G. At point A the speed of the car is 12 m/s. (*a*) If the car is just barely able to make it over the hill at point B, what is the height of that point above ground? (*b*) If the car is just barely able to make it over the hill at point B, what is the magnitude of the total force exerted on the car by the track at that point? (*c*) What is the acceleration of the car at point C? (*d*) What are the magnitude and direction of the total force exerted on the car by the track at point D? (*e*) What are the magnitude and direction of the total force exerted on the car by the track at point F? (*f*) At point G, a constant braking force is applied to the car, bring the car to a halt in a distance of 25 m. What is the braking force?

Take $U = 0$ at A, E, and G.

(*a*) Use Equ. 7-6; $U_i = 0$, $K_f = 0$ — $mg\Delta h = \frac{1}{2}mv_A^2$; $\Delta h = 7.34$ m; $h = 17.34$ m

(*b*) $F = F_n = mg$ — $F = (500 \times 9.81)$ N $= 4905$ N

(*c*) $a = g \sin\theta$ — $a = (9.81 \sin 30°)$ m/s^2 $= 4.905$ m/s^2

(*d*) 1. At D, $F = mg + mv^2/R$ directed up; find v — $v_A = 0$, $v_D^2 = 2(17.34\text{ m})g$ m^2/s^2; $mv_D^2/R = 867g$ N

2. Find F — $F = (867 + 500)g$ N $= 13.41$ kN

(*e*) F has two components, vertical and horizontal — $F_v = mg = 4905$ N; $F_h = mv^2/R = 2400$ N

Find F and θ (angle with vertical) — $\theta = \tan^{-1}(2.4/4.905) = 26°$;

$F = (2400\text{ N})/\sin 26° = 5461$ N

(*f*) Use $\Delta K + W_f$; $\Delta K = -\frac{1}{2}mv_A^2$; $W_f = F_{brake}d$ — $F_{brake} = \frac{1}{2}mv_A^2/d = 1440$ N

---

86 • An elevator (mass $M = 2000$ kg) is moving downward at $v_0 = 1.5$ m/s. A braking system prevents the downward speed from increasing. (*a*) At what rate (in J/s) is the braking system converting mechanical energy to thermal energy? (*b*) While the elevator is moving downward at $v_0 = 1.5$ m/s, the braking system fails and the elevator is in free fall for a distance $d = 5$ m before hitting the top of a large safety spring with a force constant $k = 1.5 \times 10^4$ N/m. After the elevator cage hits the top of the spring, we want to know the distance Δy that the spring is compressed before the cage is brought to rest. Write an algebraic expression for the value of Δy in terms of the known quantities M, v_0, g, k, and d, and substitute the given values to find Δy.

(*a*) Use $P = Fv = Mgv$ — $P = (2000 \times 9.81 \times 1.5)$ J/s $= 29.43$ kJ/s

(*b*) 1. $\Delta E = 0$ — $\frac{1}{2}Mv_0^2 + Mg(d + \Delta y) = \frac{1}{2}k\Delta y^2$

2. Simplify to the quadratic equation for Δy — $\Delta y^2 - (2Mg/k)\Delta y - (M/k)(2gd - v_0^2) = 0$

3. Solve for and evaluate Δy — $\Delta y = \dfrac{Mg}{k} \pm \sqrt{\dfrac{M^2g^2}{k^2} + \dfrac{M}{k}(v_0^2 + 2gd)} = 5.19$ m

---

87 • To measure the force of friction on a moving car, engineers turn off the engine and allow the car to coast down hills of known steepness. The engineers collect the following data: 1. On a 2.87° hill, the car can coast at a steady 20 m/s. 2. On a 5.74° hill, the steady coasting speed is 30 m/s. The total mass of the car is 1000 kg. (*a*) What is the force of friction at 20 m/s (F_{20}) and at 30 m/s (F_{30})? (*b*) How much useful power must the engine deliver to drive the car on a level road at steady speeds of 20 m/s (P_{20}) and 30 m/s (P_{30})? (*c*) At full throttle, the engine delivers 40 kW. What is the angle of the steepest incline up which the car can maintain a steady 20 m/s? (*d*) Assume that the engine delivers the same total useful work from each liter of gas, no matter what speed. At 20 m/s on a level road, the car goes 12.7 km/L. How many kilometers per liter does it get if it goes 30 m/s instead?

(*a*) $F_{net} = 0 = mg\sin\theta - F_{frict}$ — $F_{20} = (1000 \times 9.81 \times \sin 2.87°)$ N $= 491$ N; $F_{30} = 981$ N

(*b*) On a level road, the engine must overcome friction loss — $P_{20} = F_{20}v = (491 \times 20)$ W $= 9.82$ kW; $P_{30} = 29.4$ kW (*c*) $a = 0$; $F_{net} = 0 = mg \sin \theta + F_{20} - P/v$ — (9810 N)sin θ = (20,000/20 - 491) N; $\theta = 8.85°$

(*d*) $F_{20} \times$ (12.7 km/liter) = $W = F_{30} \times$ (N km/liter) — N = 6.36 km/liter

---

88 •• A 50,000-kg barge is pulled along a canal at a constant speed of 3 km/h by a heavy tractor. The towrope makes an angle of 18° with the velocity vector of the barge. The tension in the towrope is 1200 N. If the towrope breaks, how far will the barge move before coming to rest? Assume that the drag force between the barge and water is independent of velocity.

1. Find the drag force — F_f = (1200 N)cos 18° = 1141 N
2. Set $F_f d = \frac{1}{2}mv^2$; find d — $d = [\frac{1}{2} \times 5 \times 10^4 \times (3/3.6)^2/1141]$ m = 15.2 m

---

89* •• A 2-kg block is released 4 m from a massless spring with a force constant k = 100 N/m that is fixed along a frictionless plane inclined at 30°, as shown in Figure 7-35. (*a*) Find the maximum compression of the spring. (*b*) If the plane is rough rather than frictionless, and the coefficient of kinetic friction between the plane and the block is 0.2, find the maximum compression. (*c*) For the rough plane, how far up the incline will the block travel after leaving the spring?

(*a*) 1. Use $\Delta E = 0$; let x (along plane) = 0 at equilibrium position of spring — $mgL \sin \theta + mgx \sin \theta = \frac{1}{2}kx^2$; for $\theta = 30°$, k = 100 N/m, m = 2 kg, L = 4 m, $50x^2 - 9.81x - 39.24 = 0$

2. Solve quadratic equation for x — x = 0.989 m

(*b*) 1. Express energy loss due to friction, W_f — $W_f = \mu_k mg \cos \theta(L + x)$

2. Apply work-energy theorem — $mg \sin \theta(L + x) - \mu_k mg \cos \theta(L + x) = \frac{1}{2}kx^2$

3. Obtain quadratic equation for x and solve — $50x^2 - 6.41x - 25.65 = 0$; x = 0.783 m

(*c*) 1. Now $E_{in} = \frac{1}{2}kx^2 = E_{mech} + W_f$ — $\frac{1}{2}kx^2 - mg \sin \theta(x + L') - \mu_k mg \cos \theta(x + L') = 0$

2. Solve for L' with x = 0.783 m — L' = 1.54 m

---

90 •• A train with a total mass of 2×10^6 kg rises 707 m in a travel distance of 62 km at an average speed of 15.0 km/h. If the frictional force is 0.8% of the weight, find (*a*) the kinetic energy of the train, (*b*) the total change in its potential energy, (*c*) the energy dissipated by kinetic friction, and (*d*) the power output of the train's engines.

(*a*) $K = \frac{1}{2}mv^2$ — $K = \frac{1}{2}[2 \times 10^6 \times (15/3.6)^2]$ J = 1.74×10^7 J

(*b*) $\Delta U = mgh$ — $\Delta U = (2 \times 10^6 \times 9.81 \times 707)$ J = 1.387×10^{10} J

(*c*) $W_f = F_f d$ — $W_f = (8 \times 10^{-3} \times 2 \times 10^6 \times 9.81 \times 62 \times 10^3)$ = 9.73×10^9 J

(*d*) 1. Find total work done by engine — $W = \Delta U + W_f = 2.36 \times 10^{10}$ J; note $\Delta K = 0$

2. Find time interval — t = (62/15) h = (62 × 3600/15) s = 1.49×10^4 s

3. $P = W/t$ — $P = (2.36 \times 10^{10}/1.49 \times 10^4)$ W = 1.58 MW

---

91 •• While driving, one expects to spend more energy accelerating than driving at a constant speed. (*a*) Neglecting friction, calculate the energy required to give a 1200-kg car a speed of 50 km/h. (*b*) If friction results in a retarding force of 300 N at a speed of 50 km/h, what is the energy needed to move the car a distance of

300 m at a constant speed of 50 km/h? (*c*) Assuming that the energy losses due to friction in part (*a*) are 75% of those found in part (*b*), estimate the ratio of the energy consumption for the two cases considered.

(*a*) $E = K = \frac{1}{2}mv^2$ — $E = \frac{1}{2}[1200 \times (50/3.6)^2]$ J = 115.7 kJ

(*b*) $E = W_f = F_f d$ — $E = (300 \times 300)$ J $= 9 \times 10^4$ J = 90 kJ

(*c*) $W_f = 67.5$ kJ; $E' = K + W_f$ — $E' = 183.2$ kJ; $E'/E = 1.58$

---

92 •• In one model of jogging, the energy expended is assumed to go into accelerating and decelerating the legs. If the mass of the leg is m and the running speed is v, the energy needed to accelerate the leg from rest to v is $\frac{1}{2}m^2$, and the same energy is needed to decelerate the leg back to rest for the next stride. Thus, the energy required for each stride is mv^2. Assume that the mass of a man's leg is 10 kg and that he runs at a speed of 3 m/s with 1 m between one footfall and the next. Therefore, the energy he must provide to his legs in each second is $3 \times mv^2$. Calculate the rate of the man's energy expenditure using this model and assuming that his muscles have an efficiency of 25%.

Since P = energy per second, $P = 3mv^2 = 270$ W; given an efficiency of 25%, the energy expended per second = 1.08 kW.

93* •• On July 31, 1994, Sergey Bubka pole-vaulted over a height of 6.14 m. If his body was momentarily at rest at the top of the leap, and all the energy required to raise his body derived from his kinetic energy just prior to planting his pole, how fast was he moving just before takeoff? Neglect the mass of the pole. If he could maintain that speed for a 100-m sprint, how fast would he cover that distance? Since the world record for the 100-m dash is just over 9.8 s, what do you conclude about world-class pole-vaulters?

1. $\Delta E = 0$; $\frac{1}{2}mv^2 = mgh$ — $v = (2gh)^{1/2}$; $v = 10.96$ m/s

2. Find t for 100-m sprint at 10.96 m/s — $t = 100/10.96$ s = 9.12 s

A pole vaulter uses additional metabolic energy to raise himself on pole.

---

94 •• A 5-kg block is held against a spring of force constant 20 N/cm, compressing it 3 cm. The block is released and the spring extends, pushing the block along a rough horizontal surface. The coefficient of friction between the surface and the block is 0.2. (*a*) Find the work done on the block by the spring as it extends from its compressed position to its equilibrium position. (*b*) Find the energy dissipated by friction while the block moves the 3 cm to the equilibrium position of the spring. (*c*) What is the speed of the block when the spring is at its equilibrium position? (*d*) If the block is not attached to the spring, how far will it slide along the rough surface before coming to rest?

(*a*) $W_{spring} = \frac{1}{2}kx^2$ — $W_{spring} = (10 \times 3^2) = 90$ N·cm = 0.9 J

(*b*) $W_f = \mu_k mgx$ — $W_f = (0.2 \times 5 \times 9.81 \times 0.03)$ J = 0.294 J

(*c*) $\frac{1}{2}mv^2 = W_{spring} - W_f$; evaluate v — $v = [2(0.9 - 0.294)/5]$ J = 0.242 m/s

(*d*) Use $f_k d = \mu_k mgd = K = (0.9 - 0.294)$ J; find d — $d = [0.606/(0.2 \times 5 \times 9.81)]$ m = 0.0618 m = 6.18 cm

---

95 •• A pendulum of length L has a bob of mass m. It is released from some angle θ_1. The string hits a peg at a distance x directly below the pivot as in Figure 7-36, effectively shortening the length of the pendulum. Find the maximum angle θ_2 between the string and the vertical when the bob is to the right of the peg.

Since $K_i = K_f = 0$, $U_i = U_f$. $U_i = mgL(1 - \cos\theta_1)$; $U_f = mg(L - x)(1 - \cos\theta_2)$; solving for θ_2 one finds:

$$\theta_2 = \cos^{-1}\left[1 - \frac{L}{L-x}(1 - \cos\theta_1)\right]$$

96 ••• A block of mass m is dropped onto the top of a vertical spring whose force constant is k. If the block is released from a height h above the top of the spring, (*a*) what is the maximum kinetic energy of the block? (*b*) What is the maximum compression of the spring? (*c*) At what compression is the block's kinetic energy half its maximum value?

(*a*) 1. Write an expression for K — $K = mgh + mgx - \frac{1}{2}kx^2$

2. Set $dK/dx = 0$ and solve for x — $mg - kx = 0$; $x = mg/k$

3. Find K at $x = mg/k$ — $K_{max} = mgh + m^2g^2/2k$

(*b*) Set $K = 0$ and solve quadratic equation in x for $x = x_{max}$ — $$x_{max} = \frac{mg}{k} + \sqrt{\frac{m^2g^2}{k^2} + \frac{2mgh}{k}}$$

(*c*) Repeat (*b*) with $K = \frac{1}{2}K_{max}$. — $mgh/2 + m^2g^2/4k = mgh + mgx - \frac{1}{2}kx^2$

Solve the quadratic equation in x — $$x = \frac{mg}{k} + \frac{1}{2}\sqrt{\frac{2m^2g^2}{k^2} + \frac{4mgh}{k}}$$

97* ••• The bob of a pendulum of length L is pulled aside so the string makes an angle θ_0 with the vertical, and the bob is then released. In Example 7-2, the conservation of energy was used to obtain the speed of the bob at the bottom of its swing. In this problem, you are to obtain the same result using Newton's second law. (*a*) Show that the tangential component of Newton's second law gives $dv/dt = -g\sin\theta$, where v is the speed and θ is the angle made by the string and the vertical. (*b*) Show that v can be written $v = L\,d\theta/dt$. (*c*) Use this result and the chain rule for derivatives to obtain $dv/dt = (dv/d\theta)(d\theta/dt) = (dv/d\theta)(v/L)$. (*d*) Combine the results of (*a*) and (*c*) to obtain $v\,dv = -gL\sin\theta\,d\theta$. (*e*) Integrate the left side of the equation in part (*d*) from $v = 0$ to the final speed v and the right side from $\theta = \theta_0$ to $\theta = 0$, and show that the result is equivalent to $v = \sqrt{2gh}$, where h is the original height of the bob above the bottom.

(*a*) $F_{tan} = -mg\sin\theta$; therefore $a = dv/dt = -g\sin\theta$

(*b*) For circular motion, $v = r\omega = L\,d\theta/dt$; $d\theta/dt = v/L$

(*c*) $dv/dt = (dv/d\theta)(d\theta/dt) = (v/L)(dv/d\theta)$

(*d*) $dv/d\theta = (L/v)(dv/dt) = -(L/v)\,g\sin\theta$; $v\,dv = -gL\sin\theta\,d\theta$

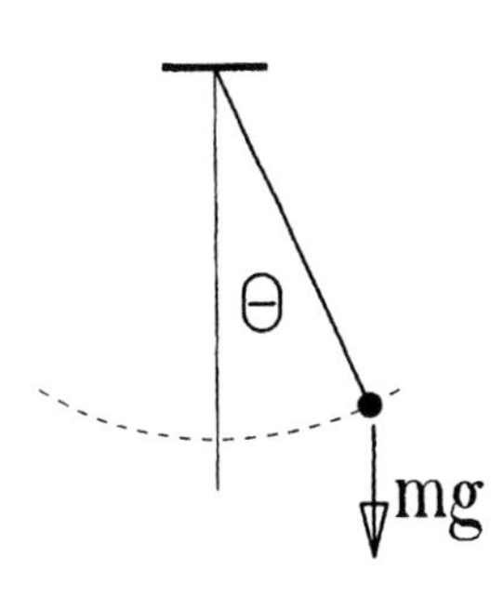

(*e*) $\int_0^v v\,dv = \int_{\theta_0}^0 -gL\sin\theta\,d\theta$; $\frac{1}{2}v^2 = gL(1 - \cos\theta_0)$

Note that $L(1 - \cos\theta_0) = h$; consequently, $v = \sqrt{2gh}$.

CHAPTER 8

Systems of Particles and Conservation of Momentum

1* • Give an example of a three-dimensional object that has no mass at its center of mass.

A hollow sphere.

---

2 • Three point masses of 2 kg each are located on the x axis at the origin, $x = 0.20$ m, and $x = 0.50$ m. Find the center of mass of the system.

Use Equ. 8-4; note that $y_{cm} = 0$ — $x_{cm} = [(2 \times 0 + 2 \times 0.2 + 2 \times 0.5)/6]$ m $= 0.233$ m

---

3 • A 24-kg child is 20 m from an 86-kg adult. Where is the center of mass of this system?

Take the origin at the position of the child.

Use Equ. 8-4 — $x_{cm} = (86 \times 20/110)$ m $= 15.6$ m

---

4 • Three objects of 2 kg each are located in the xy plane at points (10 cm, 0), (0, 10 cm), and (10 cm, 10 cm). Find the location of the center of mass.

Use Equ. 8-4 — $x_{cm} = [(10 \times 2 + 10 \times 2)/6]$ cm $= 6.67$ cm;

$y_{cm} = [(10 \times 2 + 10 \times 2)/6]$ cm $= 6.67$ cm

5* • Find the center of mass x_{cm} of the three masses in Figure 8-46.

Use Equ. 8-4 — $x_{cm} = [(1 \times 1 + 2 \times 2 + 8 \times 4)/11]$ m $= 3.36$ m

---

6 • Alley Oop's club-ax consists of a symmetrical 8-kg stone attached to the end of a uniform 2.5-kg stick that is 98 cm long . The dimensions of the club-ax are shown in Figure 8-47. How far is the center of mass from the handle end of the club-ax?

1. Locate CM of stick and of stone — By symmetry, x_{cm}(stick) $= 0.49$ m; x_{cm}(stone) $= 0.89$ m
2. Use Equ. 8-4 — $x_{cm} = [(2.5 \times 0.49 + 8 \times 0.89)/10.5]$ m $= 0.795$ m

---

7 • Three balls A, `B, and C, with masses of 3 kg, 1 kg, and 1 kg, respectively, are connected by massless rods. The balls are located as in Figure 8-48. What are the coordinates of the center of mass?

Use Equ. 8-4 — $x_{cm} = [(3 \times 2 + 1 \times 1 + 1 \times 3)/5]$ m $= 2$ m

$y_{cm} = [(3 \times 2 + 1 \times 1 + 1 \times 0)/5]$ m $= 1.4$ m

8 • By symmetry, locate the center of mass of an equilateral triangle of side length a with one vertex on the y axis and the others at $(-a/2, 0)$ and $(+a/2, 0)$.

1. Draw the triangle; assume vertex at $y > 0$
2. Locate the intersection of the bisectors; see sketch

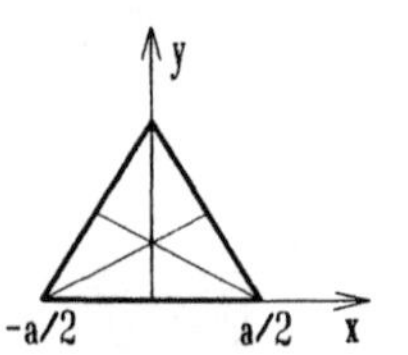

3. Give the coordinates of CM $\quad x_{cm} = 0;\ y_{cm} = ½a \tan 30° = 0.289a$

9* •• The uniform sheet of plywood in Figure 8-49 has a mass of 20 kg. Find its center of mass. We shall consider this as two sheets, a square sheet of 3 m side length and mass m_1 and a rectangular sheet 1m × 2m with a mass of $-m_2$. Let coordinate origin be at lower left hand corner of the sheet. Let σ be the surface density of the sheet.

1. Find $x_{cm}(m_1)$, $y_{cm}(m_1)$ and $x_{cm}(m_2)$, $y_{cm}(m_2)$ $\quad$ By symmetry, $x_{cm}(m_1) = 1.5$ m, $y_{cm}(m_1) = 1.5$ m and $x_{cm}(m_2) = 1.5$ m, $y_{cm}(m_2) = 2.0$ m
2. Determine m_1 and m_2 $\quad m_1 = 9\sigma\,\text{kg},\ m_2 = 2\sigma\,\text{kg}$
3. Use Equ. 8-4 $\quad x_{cm} = (9\sigma \times 1.5 - 2\sigma \times 1.5)/7\sigma = 1.5$ m

$y_{cm} = (9\sigma \times 1.5 - 2\sigma \times 2.0)/7\sigma = 1.36$ m

10 •• Show that the center of mass of a uniform semicircular disk of radius R is at a point $(4/3\pi)R$ from the center of the circle.

1. The semicircular disk is shown; we also show here the surface element dA
2. Use Equ. 8-5 to find y_{cm}; $x_{cm} = 0$ by symmetry. $y_{cm} = (1/M)\int y\sigma\, dA$
3. $y = r \sin\theta$, $dA = r\, d\theta dr$, and $M = \pi R^2\sigma/2$; make the appropriate substitutions
4. $y_{cm} = \dfrac{\sigma}{M}\displaystyle\int_0^R\int_0^\pi r^2 \sin\theta\, d\theta dr = \dfrac{2\sigma}{M}\int_0^R r^2\, dr = \dfrac{2\sigma}{3M}R^3 = \dfrac{4}{3\pi}R$

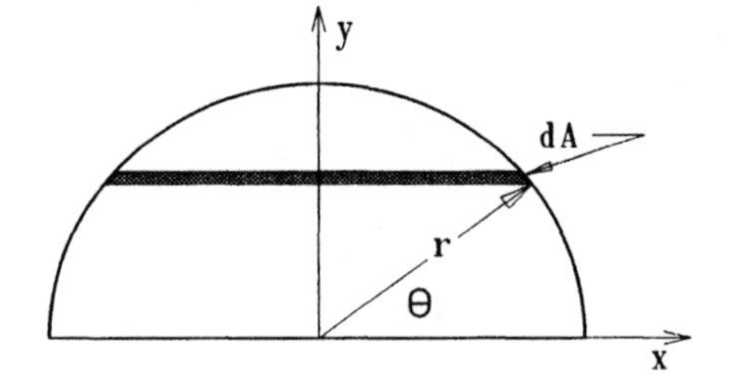

11 •• A baseball bat of length L has a peculiar linear density (mass per unit length) given by $\lambda = \lambda_0(1 + x^2/L^2)$. Find the x coordinate of the center of mass in terms of L.

Use Equ. 8-5. $x_{cm} = \int x\, dm/\int dm$. Here $M = \int dm = \int \lambda\, dx = \int_0^L \lambda_0(1 + x^2/L^2)dx = 4\lambda_0 L/3$ and $\int_0^L x dm = \int_0^L x\lambda_0(1 + x^2/L^2)dx = 3\lambda_0 L^2/4$. We find that $x_{cm} = (9/16)L$.

12 ••• Find the center of mass of a homogeneous solid hemisphere of radius R and mass M.

The volume element for a sphere is $dV = r^2 \sin\theta\, d\theta\, d\phi\, dr$, where θ is the polar angle and ϕ the azimuthal angle. Let the base of the hemisphere be the xy plane and ρ be the mass density. Then $z = r\cos\theta$. Now use Equ. 8-5:

$$z_{cm} = \frac{\rho}{M}\int_0^R\int_0^{\pi/2}\int_0^{2\pi} r^3 \sin\theta\cos\theta\, d\theta d\phi dr = \frac{\pi\rho R^4}{2M}\left[½\sin^2\theta\right]_0^{\pi/2} = \frac{\pi\rho R^4}{4M};\quad M = \frac{1}{2}\left(\frac{4}{3}\pi\rho R^3\right);\quad z_{cm} = \frac{3}{8}R.$$

13* ••• Find the center of mass of a thin hemispherical shell.

The element of area on the shell is $dA = 2\pi R^2 \sin\theta\, d\theta$, where R is the radius of the hemisphere. Let σ be the surface mass density. Then $M = ½(4\pi R^2\sigma) = 2\pi R^2\sigma$. Use the same coordinates as in Problem 8-12, and apply Equ. 8-5:

$$z_{cm} = \frac{\int z\sigma dA}{M} = \frac{2\pi R^3\sigma}{2\pi R^2\sigma}\int_0^{\pi/2}\sin\theta\cos\theta\, d\theta = R/2.$$

14 ••• A sheet of metal is cut in the shape of a parabola. The edge of the sheet is given by the expression $y = ax^2$, and y ranges from $y = 0$ to $y = b$. Find the center of mass in terms of a and b.

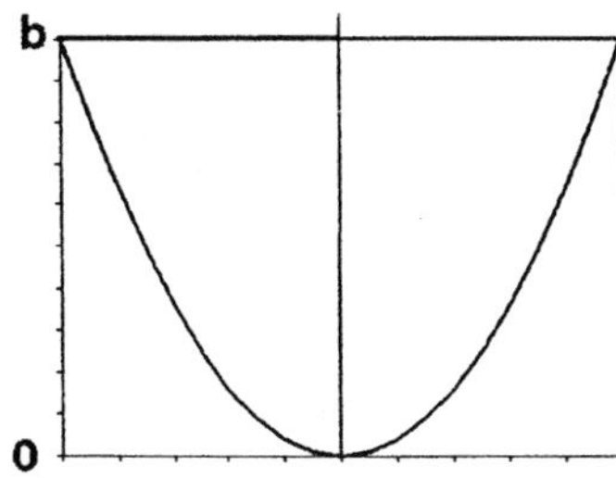

The parabolic sheet is shown here. Note that by symmetry, $x_{cm} = 0$. The element of area is $dA = xdy$, and $x = y^{1/2}/\sqrt{a}$.

$$y_{cm} = \frac{\int_0^b xydy}{\int_0^b xdy} = \frac{\int_0^b y^{3/2}dy}{\int_0^b y^{1/2}dy} = \frac{\frac{2}{5}b^{5/2}}{\frac{2}{3}b^{3/2}} = \frac{3}{5}b.$$

15 • On the night before your physics exam, you hear a banging on your door, and in walks Kelly. She says, "There's a big problem here. According to Newtonian physics, only external forces can cause the center of mass of a system to accelerate. But a car accelerates because of its own engine, so obviously Newton was wrong." She crosses her arms in a way that suggests that she is not going anywhere until she gets a satisfactory explanation. How can you explain Kelly's error to her in order to rescue Newton and get back to your studying?

The external force here is the force of static friction between the tires and the road.

16 • Two pucks of mass m_1 and m_2 lie unconnected on a frictionless table. A horizontal force F_1 is exerted on m_1 only. What is the magnitude of the acceleration of the center of mass of the pucks? (*a*) F_1/m_1 (*b*) $F_1/(m_1 + m_2)$ (*c*) F_1/m_2 (*d*) $(m_1 + m_2)F_1/m_1m_2$

(*b*) by application of Equ. 8-10.

17* • The two pucks in Problem 16 are lying on a frictionless table and connected by a spring of force constant k. A horizontal force F_1 is again exerted only on m_1 along the spring away from m_2. What is the magnitude of the acceleration of the center of mass? (*a*) F_1/m_1 (*b*) $F_1/(m_1 + m_2)$ (*c*) $(F_1 + k\Delta x)/(m_1 + m_2)$, where Δx is the amount the spring is stretched. (*d*) $(m_1 + m_2)F_1/m_1m_2$

(*b*) by application of Equ. 8-10; the spring force is an internal force.

18 • Two 3-kg masses have velocities $\boldsymbol{v}_1 = 2$ m/s $\boldsymbol{i} + 3$ m/s $\boldsymbol{j}$ and $\boldsymbol{v}_2 = 4$ m/s $\boldsymbol{i} - 6$ m/s $\boldsymbol{j}$. Find the velocity of the center of mass for the system.

Use Equ. 8-13; $\boldsymbol{v}_{cm} = (\Sigma m_i\boldsymbol{v}_i)/M = (\boldsymbol{v}_1 + \boldsymbol{v}_2)/2$. $\boldsymbol{v}_{cm} = 3$ m/s $\boldsymbol{i} - 1.5$ m/s $\boldsymbol{j}$

19 • A 1500-kg car is moving westward with a speed of 20 m/s, and a 3000-kg truck is traveling east with a speed of 16 m/s. Find the velocity of the center of mass of the system.

Let east be the positive $\boldsymbol{i}$ direction. $\boldsymbol{v}_{cm} = (\Sigma m_i\boldsymbol{v}_i)/M$ $\boldsymbol{v}_{cm} = [(3000 \times 16 - 1500 \times 20)/4500]$ m/s $\boldsymbol{i} = 4$ m/s $\boldsymbol{i}$

20 • A force $\boldsymbol{F} = 12$ N $\boldsymbol{i}$ is applied to the 3-kg ball in Problem 7. What is the acceleration of the center of mass?

Use Equ. 8-10 $\boldsymbol{a}_{cm} = (12/5)$ m/s^2 $\boldsymbol{i} = 2.4$ m/s^2 $\boldsymbol{i}$

21* •• A block of mass m is attached to a string and suspended inside a hollow box of mass M. The box rests on a scale that measures the system's weight. (*a*) If the string breaks, does the reading on the scale change? Explain your reasoning. (*b*) Assume that the string breaks and the mass m falls with constant acceleration g. Find the acceleration of the center of mass, giving both direction and magnitude. (*c*) Using the result from (*b*), determine the reading on the scale while m is in free fall.

(*a*) Yes; initially the scale reads $(M + m)g$; while m is in free fall, the reading is Mg.

(*b*) $a_{cm} = mg/(M + m)$, directed downward.

(*c*) $F_{net} = (M + m)g - (M + m)a_{cm} = Mg$.

---

22 •• A vertical spring of force constant k is attached at the bottom to a platform of mass m_p, and at the top to a massless cup, as in Figure 8-50. The platform rests on a scale. A ball of mass m_b is placed in the cup. What is the reading on the scale when (*a*) the spring is compressed an amount $d = m_bg/k$? (*b*) the ball comes to rest momentarily with the spring compressed? (*c*) the ball again comes to rest in its original position?

(*a*) The force exerted by the spring on $m_p = m_bg$ Scale reading $= m_pg + m_bg$

(*b*) When the ball is at rest, $m_bgd = \frac{1}{2}kd^2$; Scale reading $= kd + m_pg = (2m_b + m_p)g$

$d = 2m_bg/k$

(*c*) In this case, the spring exerts no force on m_b Scale reading $= m_pg$

---

23 •• In the Atwood's machine in Figure 8-51, the string passes over a fixed, frictionless cylinder of mass m_c. (*a*) Find the acceleration of the center of mass of the two-block-and-cylinder system. (*b*) Use Newton's second law for systems to find the force F exerted by the support. (*c*) Find the tension in the string connecting the blocks and show that $F = m_cg + 2T$.

(*a*) 1. Use Equ. 8-8 and results of Problem 4-81 $a_{cm} = (m_1a - m_2a)/(m_1 + m_2 + m_c)$;

$a = (m_1 - m_2)/(m_1 + m_2)$

2. Simplify the expression for a_{cm} $a_{cm} = [(m_1 - m_2)^2g]/[(m_1 + m_2)(m_1 + m_2 + m_c)]$

(*b*) $F = Mg - Ma_{cm}$; $M = m_1 + m_2 + m_c$ $F = Mg - (m_1 - m_2)^2g/(m_1 + m_2)$

$= [4m_1m_2/(m_1 + m_2) + m_c]g$

(*c*) From Problem 4-81, $T = 2m_1m_2g/(m_1 + m_2)$ $F = 2T + m_c g = [4m_1m_2/(m_1 + m_2) + m_c]g$; Q.E.D.

---

24 •• Repeat Problems 22*a* and 22*b* with the ball dropped into the cup from a height h above the cup.

(*a*) There is no change here. Scale reading $= (m_p + m_b)g$

(*b*) See Prob. 7-96; $x_{max} = \dfrac{m_bg}{k}\left(1 + \sqrt{1 + \dfrac{2kh}{m_bg}}\right)$ $F = m_pg + m_bg\left(1 + \sqrt{1 + \dfrac{2kh}{m_bg}}\right)$

25* • True or false: (*a*) The momentum of a heavy object is greater than that of a light object moving at the same speed. (*b*) The momentum of a system may be conserved even when mechanical energy is not. (*c*) The velocity of the center of mass of a system equals the total momentum of the system divided by its total mass.

(*a*) True (for magnitude) (*b*) True (inelastic collision) (*c*) True

---

26 • How is the recoil of a rifle related to momentum conservation?

$\mathbf{p}_{recoil} + \mathbf{p}_{gun} = 0$; $\mathbf{p}_{recoil} = -\mathbf{p}_{gun}$.

---

27 • A man is stranded in the middle of an ice rink that is perfectly frictionless. How can he get to the edge?

If he throws something forward, he will move backward.

---

28 • A girl jumps from a boat to a dock. Why does she have to jump with more energy than she would need if she were jumping the same distance from one dock to another?

She must give the boat a recoil momentum, and $E_{boat} = p_{boat}^2/2m_{boat}$.

29* •• Much early research in rocket motion was done by Robert Goddard, physics professor at Clark College in Worcester, Mass. A quotation from a 1921 editorial in the *New York Times* illustrates the public acceptance of his work: "That Professor Goddard with his 'chair' at Clark College and the countenance of the Smithonian Institution does not know the relation between action and reaction, and the need to have something better than a vacuum against which to react—to say that would be absurd. Of course, he only seems to lack the knowledge ladled out daily in high schools." The belief that a rocket needs something to push against was a prevalent misconception before rockets in space were commonplace. Explain why that belief is wrong.

Conservation of momentum does not require the presence of a medium such as air.

---

30 •• Liz, Jay, and Tara discover that sinister chemicals are leaking at a steady rate from a hole in the bottom of a railway car. To collect evidence of a potential environmental mishap, they videotape the car as it rolls without friction at an initial speed v_0. Tara claims that careful analysis of the videotape will show that the car's speed is increasing, because it is losing mass as it drains. The increase in speed will help to prove that the leak is occurring. Liz says no, that with a loss of mass, the car's speed will be decreasing. Jay says the speed will remain the same. (*a*) Who is right? (*b*) What forces are exerted on the system of the car plus chemical cargo?

(*a*) Jay is right.

(*b*) No net external force acts on the car plus cargo.

---

31 • A girl of mass 55 kg jumps off the bow of a 75-kg canoe that is initially at rest. If her velocity is 2.5 m/s to the right, what is the velocity of the canoe after she jumps?

$F_{ext} = 0$; $\Sigma m_i v_i = 0$ $\qquad$ $v_c = (55 \times 2.5/75)$ m/s = 1.83 m/s, opposite to girl's $\mathbf{v}$

---

32 • Two masses of 5 kg and 10 kg are connected by a compressed spring and rest on a frictionless table. After the spring is released, the smaller mass has a velocity of 8 m/s to the left. What is the velocity of the larger mass?

$F_{ext} = 0$; $\Sigma m_i v_i = 0$ $\qquad$ $v_{10} = (5 \times 8/10)$ m/s = 4 m/s to right

---

33* • Figure 8-52 shows the behavior of a projectile just after it has broken up into three pieces. What was the speed of the projectile the instant before it broke up? (*a*) v_3 (*b*) $v_3/3$ (*c*) $v_3/4$ (*d*) $4v_3$ (*e*) $(v_1 + v_2 + v_3)/4$

(*c*) Use $\mathbf{p}_i = \mathbf{p}_f = m\mathbf{v}_3 = 4m\mathbf{v}_i$.

---

34 • A shell of mass *m* and speed *v* explodes into two identical fragments. If the shell was moving horizontally with respect to the earth, and one of the fragments is subsequently moving vertically with the speed *v*, find the velocity $\mathbf{v}'$ of the other fragment.

Use $\mathbf{p}_i = \mathbf{p}_f$; $\mathbf{p}_i = mv\,\mathbf{i} = \frac{1}{2}mv\,\mathbf{j} + \frac{1}{2}m\mathbf{v}'$ $\qquad$ $\mathbf{v}' = 2v\,\mathbf{i} - v\,\mathbf{j}$

---

35 •• In a circus act, Marcello (mass 70 kg) is shot from a cannon with a muzzle velocity of 24.00 m/s at an angle of 30° above horizontal. His partner, Tina (mass 50 kg), stands on an elevated platform located at the top of his trajectory. He grabs her as he flies by and the two fly off together. They land in a net at the same elevation as the cannon a horizontal distance *x* away. Find *x*.

1. Find $\mathbf{p}_i$ at contact; $v_{xi} = (24 \cos 30°)$ m/s = 20.8 m/s $\qquad$ (70×20.8) kg·m/s = $(120 \times v_{xf})$ kg·m/s; $v_{xf} = 12.1$ m/s

2. Find the time of flight to the platform — $t = (24 \sin 30°)/g$ s = 1.22 s

3. Find $x = v_{xi}t + v_{xf}t$ — $x = (20.8 + 12.1)1.22$ m= 40.1 m

---

36 •• A block and a handgun loaded with one bullet are firmly affixed to opposite ends of a massless cart that rests on a level frictionless air table (Figure 8-53). The mass of the handgun is m_g, the mass of the block is m_{bk}, and the mass of the bullet is m_{bt}. The gun is aimed so that when fired, the bullet will go into the block. When the bullet leaves the barrel of the handgun, it has a velocity v_b as measured by an observer at rest with the table. Take the fall of the bullet to be negligible and its penetration into the block to be small. (*a*) What is the velocity of the cart immediately after the bullet leaves the gun barrel? (*b*) What is the velocity of the cart immediately after the bullet comes to rest in the block? (*c*) How far has the block moved from its initial position at the moment when the bullet comes to rest in the block?

(*a*) Use $\mathbf{p}_b = -\mathbf{p}_{cart}$; $\mathbf{v}_{cart} = -[m_b/(m_g + m_{bk})]\mathbf{v}_b$.

(*b*) $F_{ext} = 0$; $\mathbf{p}_i = \mathbf{p}_f = 0$; $\mathbf{v}_{cart} = 0$.

(*c*) $v_{rel} = v_b + v_{cart} = v_b[(m_g+m_b+m_{bk})/(m_g+m_{bk})]$; time of flight = L/v_{rel}; $d = v_{cart}t = L[m_b/(m_g+m_b+m_{bk})]$.

37* •• A small object of mass m slides down a wedge of mass $2m$ and exits smoothly onto a frictionless table. The wedge is initially at rest on the table. If the object is initially at rest at a height h above the table, find the velocity of the wedge when the object leaves it.

$p_x = 0 = mv_x - 2mV$; $V = ½v_x$; $v_x = \sqrt{2gh}$; $V = \sqrt{gh/2}$, directed opposite to that of the mass m.

---

38 • Describe how a basketball is moving when (*a*) its total kinetic energy is just the energy of motion of its center of mass, and (*b*) its total kinetic energy is the energy of its motion relative to its center of mass.

(*a*) It has a speed v but is not spinning about its axis.

(*b*) It has no velocity but is spinning about its center of mass.

---

39 • Two bowling balls are moving with the same velocity, but one just slides down the alley, whereas the other rolls down the alley. Which ball has more energy?

The rolling ball has more energy, $½mv_{cm}^2 + K_{rel}$.

---

40 • A 3-kg block is traveling to the right at 5 m/s, and a second 3-kg block is traveling to the left at 2 m/s. (*a*) Find the total kinetic energy of the two blocks in this reference frame. (*b*) Find the velocity of the center of mass of the two-body system. (*c*) Find the velocities of the two blocks relative to the center of mass. (*d*) Find the kinetic energy of the motion of the blocks relative to the center of mass. (*e*) Show that your answer for part (*a*) is greater than your answer for part (*d*) by an amount equal to the kinetic energy of the center of mass.

(*a*) $K = K_1 + K_2$ — $K = ½[3 \times 25 + 3 \times 4]$ J = 43.5 J

(*b*) Use Equ. 8-13 — $\mathbf{v}_{cm} = (3 \times 5 - 3 \times 2)/6\, \mathbf{i}$ m/s = 1.5 $\mathbf{i}$ m/s

(*c*) $\mathbf{v}_{rel} = \mathbf{v} - \mathbf{v}_{cm}$ — $\mathbf{v}_{1,rel} = 3.5\, \mathbf{i}$ m/s; $\mathbf{v}_{2,rel} = -3.5\, \mathbf{i}$ m/s

(*d*) $K_{rel} = K_{1,rel} + K_{2,rel}$ — $K_{rel} = ½(3 \times 3.5^2 + 3 \times 3.5^2)$ J = 36.75 J

(*e*) Find K_{cm} — $K_{cm} = ½(6 \times 1.5^2)$ J = 6.75 J = $K - K_{rel}$

41* • Repeat Problem 40 with the second, 3-kg block replaced by a block having a mass of 5 kg and moving to the right at 3 m/s.

We follow the same procedures as in the preceding problem.

(*a*) $K = ½(3 \times 5^2 + 5 \times 3^2)$ J = 60 J; (*b*) $\boldsymbol{v}_{cm} = (3 \times 5 + 5 \times 3)/8\ \boldsymbol{i}$ m/s = 3.75 $\boldsymbol{i}$ m/s; (*c*) $v_{1,rel} = 1.25$ m/s, $v_{2,rel} = -0.75$ m/s; (*d*) $K_{rel} = ½(3 \times 1.25^2 + 5 \times 0.75^2)$ J = 3.75 J; (*e*) $K_{cm} = ½(8 \times 3.75^2) = 56.25$ J = $K - K_{rel}$.

---

42 • Explain why a safety net can save the life of a circus performer.

It reduces the force acting on the performer by increasing Δt. Note that $\Delta p = F\Delta t$ is constant.

---

43 • How might you estimate the collision time of a baseball and bat?

Assume that the ball travels at 80 mi/h = 35 m/s. The ball stops in a distance of about 1 cm. So the distance traveled is about 2 cm at an average speed of about 18 m/s. The collision time is 0.02/18 = 1 ms.

---

44 • Why does a wine glass survive a fall onto a carpet but not onto a concrete floor?

The average force on the glass is less when falling on a carpet because Δt is longer.

45* • A soccer ball of mass 0.43 kg leaves the foot of the kicker with an initial speed of 25 m/s. (*a*) What is the impulse imparted to the ball by the kicker? (*b*) If the foot of the kicker is in contact with the ball for 0.008 s, what is the average force exerted by the foot on the ball?

(*a*) Use Equ. 8-19 — $I = mv = 10.75$ N·s

(*b*) Use Equ. 8-20 — $F_{av} = 10.75/0.008$ N = 1344 N

---

46 • A 0.3-kg brick is dropped from a height of 8 m. It hits the ground and comes to rest. (*a*) What is the impulse exerted by the ground on the brick? (*b*) If it takes 0.0013 s from the time the brick first touches the ground until it comes to rest, what is the average force exerted by the ground on the brick?

(*a*) $I = \Delta p$ — $I = mv = 0.3 \times (2 \times 9.81 \times 8)^{½}$ N·s = 3.76 N·s

(*b*) $F_{av} = I/\Delta t$ — $F_{av} = (3.76/0.0013)$ N = 2891 N

---

47 • A meteorite of mass 30.8 tonne (1 tonne = 1000 kg) is exhibited in the Hayden Planetarium in New York. Suppose the kinetic energy of the meteorite as it hit the ground was 617 MJ. Find the impulse I experienced by the meteorite up to the time its kinetic energy was halved (which took about $t = 3.0$ s). Find also the average force F exerted on the meteorite during this time interval.

1. $I = \Delta p$; find p_i and p_f; use Equ. 8-23 — $p_i = (2 \times 30.8 \times 10^3 \times 617 \times 10^6)^{½}$ kg·m/s $= 6.165 \times 10^6$ kg·m/s

2. $p_f^2 = p_i^2/2$ — $p_f = p_i/\sqrt{2}$; $\Delta p = p_i(1 - 1/\sqrt{2}) = I = 1.81$ MN·s

3. $F_{av} = I/\Delta t$ — $F_{av} = 0.602$ MN

---

48 •• When a 0.15-kg baseball is hit, its velocity changes from +20 m/s to −20 m/s. (*a*) What is the magnitude of the impulse delivered by the bat to the ball? (*b*) If the baseball is in contact with the bat for 1.3 ms, what is the average force exerted by the bat on the ball?

(*a*) $I = \Delta p$ — $I = 0.15 \times 40$ N·s = 6 N·s

(*b*) $F_{av} = I/\Delta t$ — $F_{av} = (6/1.3 \times 10^{-3})$ N = 4.62 kN

49* •• A 300-g handball moving with a speed of 5.0 m/s strikes the wall at an angle of 40° and then bounces off with the same speed at the same angle. It is in contact with the wall for 2 ms. What is the average force exerted by the ball on the wall?

1. Find Δv; $v_{xi} = v_0 \cos 40°$, $v_{xf} = -v_0 \cos 40°$ — $\Delta v = 2 \times 5.0 \times \cos 40°$ m/s = 7.66 m/s

2. $F_{av} = m\Delta v/\Delta t$ — $F_{av} = 0.3 \times 7.66/2 \times 10^{-3}$ N = 1.15 kN

---

50 •• A 2000-kg car traveling at 90 km/h crashes into a concrete wall that does not give at all. (*a*) Estimate the time of collision, assuming that the center of the car travels halfway to the wall with constant deceleration. (Use any reasonable length for the car.) (*b*) Estimate the average force exerted by the wall on the car.

(*a*) Assume a car length of 6 m; then $\Delta t = (3/2v_{av})$ s = 1.5/(25/2) s = 0.12 s.

(*b*) $F_{av} = \Delta p/\Delta t$ = (2000 × 25/0.12) N = 417 kN.

---

51 •• You throw a 150-g ball to a height of 40 m. (*a*) Use a reasonable value for the distance the ball moves while it is in your hand to calculate the average force exerted by your hand and the time the ball is in your hand while you throw it. (*b*) Is it reasonable to neglect the weight of the ball while it is being thrown?

(*a*) Take d = 0.7 m; find v on leaving hand and v_{av}	$v = (2gh)^{1/2} = (2 \times 9.81 \times 40)^{1/2}$ = 28 m/s; v_{av} = 14 m/s
$\Delta t = d/v_{av}$	Δt = 0.7/14 s = 0.05 s
$F_{av} = \Delta p/\Delta t$	F_{av} = (0.15 × 28/0.05) N = 84 N
(*b*) $w/F_{av} = mg/F_{av}$ = 0.0175	Yes; w is less than 2% of the average force

---

52 •• A handball of mass 300 g is thrown straight against a wall with a speed of 8 m/s. It rebounds with the same speed. (*a*) What impulse is delivered to the wall? (*b*) If the ball is in contact with the wall for 0.003 s, what average force is exerted on the wall by the ball? (*c*) The ball is caught by a player who brings it to rest. In the process, her hand moves back 0.5 m. What is the impulse received by the player? (*d*) What average force was exerted on the player by the ball?

(*a*) See Problem 8-48	I = 0.3 × 16 N·s = 4.8 N·s
(*b*) See Problem 8-48	F_{av} = 4.8/0.003 N = 1.6 kN
(*c*) $I = m\Delta v$	I = 0.3 × 8 N·s = 2.4 N·s
(*d*) $\Delta t = d/v_{av}$; $F_{av} = I/\Delta t = Iv_{av}/d$	F_{av} = (2.4 × 4/0.5) N = 19.2 N

53* ••• The great limestone caverns were formed by dripping water. (*a*) If water droplets of 0.03 mL fall from a height of 5 m at a rate of 10 per minute, what is the average force exerted on the limestone floor by the droplets of water? (*b*) Compare this force to the weight of a water droplet.

(*a*) 1. Find the mass of the droplet	$m = (3 \times 10^{-5}$ L)(1.0 kg/L) = 3×10^{-5} kg
2. Find v at impact	$v = (2 \times 9.81 \times 5)^{1/2}$ m/s = 9.9 m/s
3. Find $F_{av} = Nm\Delta v/\Delta t$; $N/\Delta t = (10/60)$ s^{-1}	$F_{av} = (3 \times 10^{-5} \times 9.9/6)$ N = 4.95×10^{-5} N
(*b*) w/F_{av} = 3 × 9.81/4.95 = 6	w is about six times the average force due to 10 drops

---

54 ••• A favorite game at picnics is the egg toss. Two people toss a raw egg back and forth as they move farther apart. If the force required to break the egg's shell is about 5 N and the mass of the egg is 50 g, estimate the maximum separation distance for the egg throwers. Make whatever assumptions seem reasonable.

Assume θ = 45° for the maximum range; then $R = v_0^2/g$. Assume the movement of the hand is d = 0.7 m. With F_{max} = 5 N, we have 5 × 0.7 J = ½ × 0.05 × 9.81 × R J. So R = 14.3 m.

---

55 • True or false: (*a*) In any perfectly inelastic collision, all the kinetic energy of the bodies is lost. (*b*) In a head-on elastic collision, the relative speed of recession after the collision equals the relative speed of approach before the collision. (*c*) Kinetic energy is conserved in an elastic collision.

(*a*) False (*b*) True (*c*) True

---

56 •• Under what conditions can all the initial kinetic energy of colliding bodies be lost in a collision?

This occurs in a perfectly inelastic collision in which the velocity of the center of mass is zero.

57* •• Consider a perfectly inelastic collision of two objects of equal mass. (*a*) Is the loss of kinetic energy greater if the two objects have oppositely directed velocities of equal magnitude $v/2$, or if one of the two objects is initially at rest and the other has an initial velocity of v? (*b*) In which situation is the percentage loss in kinetic energy the greatest?

(*a*) Case 1: $K_i = 2(\frac{1}{2}mv^2/4)$, $K_f = 0$; $\Delta K = mv^2/4$. Case 2: $K_i = \frac{1}{2}mv^2$, $K_f = \frac{1}{2}[2\mathrm{m} \times (v/2)^2] = mv^2/4$; $\Delta K = mv^2/4$. The energy losses are the same.

(*b*) The percentage loss is greatest (infinite) in case 1.

---

58 •• A mass m_1 traveling with a speed v makes a head-on elastic collision with a stationary mass m_2. In which scenario will the energy imparted to m_2 be greatest? (*a*) $m_2 << m_1$ (*b*) $m_2 = m_1$ (*c*) $m_2 >> m_1$ (*d*) none of the above

(*b*) All of the energy is imparted to m_2.

---

59 • Joe and Sal decide that little Ronny is well-behaved enough to sit at the table with the family for Thanksgiving dinner. They are wrong. Ronny throws a 150-g handful of mashed potatoes horizontally with a speed of 5 m/s. It strikes a 1.2-kg gravy boat that is initially at rest on the frictionless table. If the potatoes stick to the gravy boat, what is the speed of the combined system as it slides down the table toward Grandpa?

Use $p_i = p_f$; $v_f = v_i m_1/(m_1 + m_2)$ $v_f = (5 \times 0.15/1.35)$ m/s $= 0.556$ m/s

---

60 • A 2000-kg car traveling to the right at 30 m/s is chasing a second car of the same mass that is traveling to the right at 10 m/s. (*a*) If the two cars collide and stick together, what is their speed just after the collision? (*b*) What fraction of the initial kinetic energy of the cars is lost during this collision? Where does it go?

(*a*) Use $p_i = m_1v_1 + m_2v_2 = (m_1 + m_2)v$; $m_1=m_2=m$ $v_f = m(v_1 + v_2)/2m = 20$ m/s

(*b*) $K_i = \frac{1}{2}m(v_1^2 + v_2^2)$; $K_f = mv_f^2$ $\Delta K/K_i = (v_1^2 + v_2^2 - 2v_f^2)/(v_1^2 + v_2^2) = 0.2$

The energy goes into heat, sound, and the deformation of metal.

61* • An 85-kg running back moving at 7 m/s makes a perfectly inelastic collision with a 105-kg linebacker who is initially at rest. What is the speed of the players just after their collision?

Use $p_i = p_f$; $v_f = v_i m_1/(m_1 + m_2)$ $v_f = 85 \times 7/190 = 3.13$ m/s

---

62 • A 5.0-kg object with a speed of 4.0 m/s collides head-on with a 10-kg object moving toward it with a speed of 3.0 m/s. The 10-kg object stops dead after the collision. (*a*) What is the final speed of the 5-kg object? (*b*) Is the collision elastic?

(*a*) $p_i = p_f$ $v_{f,5} = (10 \times 3.0 - 5 \times 4.0)/5$ m/s $= 2$ m/s

(*b*) If $K_i = K_f$, the collision is elastic $K_i = (40 + 45)$ J; $K_f = 10$ J; inelastic collision

---

63 • A ball of mass m moves with speed v to the right towards a much heavier bat that is moving to the left with speed v. Find the speed of the ball after it makes an elastic collision with the bat.

Use Equ. 8-29 and the fact that $v_{i,bat} \approx v_{f,bat}$; take left as the positive direction $v_{f,ball} = v_{i,ball} + 2v_{i,bat}$

$v_{f,ball} = 3v$

64 •• During the Great Muffin Wars of '98, students from rival residences became familiar with the characteristics of various muffins. Mushy Pumpkin Surprise, for example, was good for temporarily blinding an attacker, while Mrs. O'Brien's Bran Muffins, having the density of lacrosse balls, were used more sparingly, and mainly as a deterrent. According to the rules, all muffins must have a mass of 0.3 kg. During one of the more memorable battles, a muffin moving to the right at 5 m/s collides with a muffin moving to the left at 2 m/s. Find the final velocities if (*a*) it is a perfectly inelastic collision of two pumpkin muffins and (*b*) it is an elastic collision of two bran muffins.

(*a*) Use $p_i = p_f$; $v_f = (v_{1i}m_1 + v_{2i}m_2)/(m_1 + m_2)$ — $v_f = m(5 - 2)/2m$ m/s = 1.5 m/s, to the right

(*b*) 1. Find v_{cm} and transform to CM system — $v_{cm} = 1.5$ m/s; $u_{1i} = 3.5$ m/s, $u_{2i} = -3.5$ m/s

2. Use $u_f = -u_i$ and transform back — $u_{1f} = -3.5$ m/s, $u_{2f} = 3.5$ m/s; $v_{1f} = -2$ m/s, $v_{2f} = 5$ m/s

65* •• Repeat Problem 64 with a second (illegal) muffin having a mass of 0.5 kg and moving to the right at 3 m/s.

(*a*) Use $p_i = p_f$; $v_f = (v_{1i}m_1 + v_{2i}m_2)/(m_1 + m_2)$ — $v_f = [(0.3 \times 5 + 0.5 \times 3)/0.8]$ m/s = 3.75 m/s

(*b*) 1. Transform to CM system; $u = v - v_{cm}$ — $v_{cm} = 3.75$ m/s; $u_{1i} = 1.25$ m/s, $u_{2i} = -0.75$ m/s

2. Use $u_f = -u_i$; transform back to lab system — $u_{1f} = -1.25$ m/s, $u_{2f} = 0.75$ m/s; $v_{1f} = 2.5$ m/s, $v_{2f} = 4.5$ m/s

66 •• A proton of mass m undergoes a head-on elastic collision with a stationary carbon nucleus of mass $12m$. The speed of the proton is 300 m/s. (*a*) Find the velocity of the center of mass of the system. (*b*) Find the velocity of the proton after the collision.

(*a*) Use Equ. 8-13 — $v_{cm} = 300m/13m$ m/s = 23.1 m/s

(*b*) Use Equ. 8-30*a* — $v_{pf} = [(-11/13)300]$ m/s = -254 m/s

67 •• A 3-kg block moving at 4 m/s makes an elastic collision with a stationary block of mass 2 kg. Use conservation of momentum and the fact that the relative velocity of recession equals the relative velocity of approach to find the velocity of each block after the collision. Check your answer by calculating the initial and final kinetic energies of each block.

Equs. 8-30*a* and 8-30*b* are a direct consequence of momentum conservation and Equ. 8-29. (see Example 8-17)

1. Use Equs. 8-30*a* and 8-30*b* — $v_{3f} = (1 \times 4/5)$ m/s = 0.8 m/s; $v_{2f} = (6 \times 4/5)$ m/s = 4.8 m/s

2. Evaluate K_i and K_f — $K_i = 1.5 \times 16$ J = 24 J; $K_f = (1.5 \times 0.8^2 + 1 \times 4.8^2)$ J = 24 J

68 •• Night after night, Lucy is tormented by nocturnal wailing from the house next door. One day she seizes a revolver, stalks to the neighbor's window with a crazed look in her eye, takes aim, and fires a 10-g bullet into her target: a 1.2 kg saxophone that rests on a frictionless surface. The bullet passes right through and emerges on the other side with a speed of 100 m/s, and the saxophone is given a speed of 4 m/s. Find the initial speed of the bullet, and the amount of energy dissipated in its trip through the saxophone.

1. Use $p_i = p_f$ — $(0.01v_i)$ kg·m/s = $(0.01 \times 100 + 1.2 \times 4)$ kg·m/s; $v_i = 580$ m/s

2. Find $\Delta K = K_i - K_f$ — $\Delta K = [0.005 \times 580^2 - 0.6 \times 4^2 - 0.005 \times 100^2]$ J = 1622 J

69* •• A block of mass $m_1 = 2$ kg slides along a frictionless table with a speed of 10 m/s. Directly in front of it, and moving in the same direction with a speed of 3 m/s, is a block of mass $m_2 = 5$ kg. A massless spring with spring constant $k = 1120$ N/m is attached to the second block as in Figure 8-54. (*a*) Before m_1 runs into the spring, what is the velocity of the center of mass of the system? (*b*) After the collision, the spring is compressed by a maximum amount Δx. What is the value of Δx? (*c*) The blocks will eventually separate again. What are the final velocities of the two blocks measured in the reference frame of the table?

(*a*) Use Equ. 8-13 — $v_{cm} = (2 \times 10 + 5 \times 3)/7$ m/s $= 5$ m/s

(*b*) 1. At max. compression, $u_1 = u_2 = 0$; $K = K_{cm}$ — $K_{cm} = \frac{1}{2} \times 7 \times 5^2$ J $= 87.5$ J

2. Use conservation of energy: $\frac{1}{2}k\Delta x^2 = K_i - K_{cm}$ — $K_i = (100 + 22.5)$ J $= 122.5$ J; $\frac{1}{2}k\Delta x^2 = 35$ J

3. Solve for and evaluate Δx — $\Delta x = (2 \times 35/1120)^{1/2}$ m $= 0.25$ m $= 25$ cm

(*c*) 1. Collision is elastic; find u_{1i} and u_{2i}; u_{1f} and u_{2f} — $u_{1i} = 5$ m/s, $u_{2i} = -2$ m/s; $u_{1f} = -5$ m/s, $u_{2f} = 2$ m/s

2. Transform to reference frame of table — $v_{1f} = 0$ m/s, $v_{2f} = 7$ m/s

70 •• A bullet of mass m is fired vertically from below into a block of wood of mass M that is initially at rest, supported by a thin sheet of paper. The bullet blasts through the block, which rises to a height H above its initial position before falling back down. The bullet continues rising to a height h. (*a*) Express the upward velocity of the bullet and the block immediately after the bullet exits the block in terms of h and H. (*b*) Use conservation of momentum to express the speed of the bullet before it enters the block of wood in terms of given parameters. (*c*) Obtain expressions for the mechanical energies of the system before and after the inelastic collision. (*d*) Express the energy dissipated in the block of wood in terms of m, h, M, and H.

(*a*) $v_m = \sqrt{2gh}$; $v_M = \sqrt{2gH}$.

(*b*) $mv_{mi} = mv_m + Mv_M$; $v_{mi} = \sqrt{2gh} + (M/m)\sqrt{2gH}$.

(*c*) $E_i = \frac{1}{2}mv_{mi}^2 = mg[h + 2(M/m)\sqrt{hH} + (M/m)^2 H]$; $E_f = g(mh + MH)$.

(*d*) $W_f = E_i - E_f = gMH[2\sqrt{h/H} + (M/m) - 1]$.

71 •• A proton of mass m is moving with initial speed v_0 toward an α particle of mass $4m$, which is initially at rest. Because both particles carry positive electrical charge, they repel each other. Find the speed v' of the α particle (*a*) when the distance between the two particles is least, and (*b*) when the two particles are far apart.

(*a*) This problem is similar to Problem 8-69; here the electrostatic repulsion takes the place of the spring. When the distance between the two particles is least, both move at the same speed, namely v_{cm}.

Find v_{cm} — $v_{cm} = v_\alpha' = v_0/5 = 0.2v_0$

(*b*) Use Equ. 8-30*b* — $v_{\alpha f} = (2/5)v_0 = 0.4v_0$

72 •• A 16-g bullet is fired into the bob of a ballistic pendulum of mass 1.5 kg. When the bob is at its maximum height, the strings make an angle of 60° with the vertical. The length of the pendulum is 2.3 m. Find the speed of the bullet.

1. Use $p_i = p_f$ — $0.016v_1 = 1.516v_2$

2. From energy conservation, $v_2 = \sqrt{2gh}$ — $v_2 = [2 \times 9.81 \times 2.3 \times (1 - \cos 60°)]^{1/2}$

3. Evaluate v_2 and v_1, the speed of the bullet — $v_2 = 4.75$ m/s, $v_1 = 450$ m/s

---

73* •• A bullet of mass m_1 is fired with a speed v into the bob of a ballistic pendulum of mass m_2. The bob is attached to a very light rod of length L that is pivoted at the other end. The bullet is stopped in the bob. Find the minimum v such that the bob will swing through a complete circle.

1. Find v_i of bob + bullet to make complete circle — $\frac{1}{2}(m_1+m_2)v_i^2 = 2gL(m_1+m_2)$; $v_i = 2\sqrt{gL}$
2. Use $p_i = p_f$ to find v — $v = 2[(m_1 + m_2)/m_1]\sqrt{gL}$

---

74 •• A bullet of mass m_1 is fired with a speed v into the bob of a ballistic pendulum of mass m_2. Find the maximum height h attained by the bob if the bullet passes through the bob and emerges with a speed $v/2$.

1. Use $p_i = p_f$ to find v_2 — $m_1v = \frac{1}{2}m_1v + m_2v_2$; $v_2 = \frac{1}{2}(m_1/m_2)v$
2. Use conservation of energy to find h — $\frac{1}{2}m_2v_2^2 = m_2gh$; $h = (m_1v/2m_2)^2/2g = (v^2/8g)(m_1/m_2)^2$

---

75 •• A 3-kg bomb slides along a frictionless horizontal plane in the x direction at 6 m/s. It explodes into two pieces, one of mass 2 kg and the other of mass 1 kg. The 1-kg piece moves along the horizontal plane in the y direction at 4 m/s. (*a*) Find the velocity of the 2-kg piece. (*b*) What is the velocity of the center of mass after the explosion?

We shall do part (*b*) first and use the result to solve part (*a*).

(*b*) $\mathbf{v}_{cm} = \mathbf{v}_i$ from conservation of $\mathbf{p}$ — $\mathbf{v}_{cm} = 6\ \text{m/s}\ \mathbf{i}$

(*a*) Use $\mathbf{p}_i = \mathbf{p}_f = (3\ \text{kg}) \times \mathbf{v}_{cm}$ — $18\ \text{m/s}\ \mathbf{i} = 1 \times 4\ \text{m/s}\ \mathbf{j} + 2 \times \mathbf{v}_2$; $\mathbf{v}_2 = 9\ \text{m/s}\ \mathbf{i} - 2\ \text{m/s}\ \mathbf{j}$

---

76 •• The beryllium isotope ^{4}Be is unstable and decays into two α particles (helium nuclei of mass $m = 6.68 \times 10^{-27}$ kg) with the release of 1.5×10^{-14} J of energy. Determine the velocities of the two α particles that arise from the decay of a ^{4}Be nucleus at rest.

From momentum conservation it follows that the velocities of the two α particles are equal in magnitude and oppositely directed.

Use energy conservation; solve for v_α — $2(\frac{1}{2}m_\alpha v_\alpha^2) = 1.5 \times 10^{-14}$ J; $v_\alpha = 1.5 \times 10^6$ m/s

---

77* •• The light isotope of lithium, ^{5}Li, is unstable and breaks up spontaneously into a proton (hydrogen nucleus) and an α particle (helium nucleus). In this process, a total energy of 3.15×10^{-13} J is released, appearing as the kinetic energy of the two reaction products. Determine the velocities of the proton and α particle that arise from the decay of a ^{5}Li nucleus at rest. (*Note*: The masses of the proton and alpha particle are $m_p = 1.67 \times 10^{-27}$ kg and $m_\alpha = 4m_p = 6.68 \times 10^{-27}$ kg.)

1. Use $p_i = p_f = 0$ — $4m_pv_\alpha = m_pv_p$; $v_\alpha = v_p/4$
2. Use energy conservation — $\frac{1}{2}m_pv_p^2 + \frac{1}{2}m_\alpha v_\alpha^2 = (5/8)m_pv_p^2 = 3.15 \times 10^{-13}$ J
3. Solve for v_p and v_α — $v_p = 1.74 \times 10^7$ m/s; $v_\alpha = 4.34 \times 10^6$ m/s

---

78 •• Jay and Dave decide that the best way to protest the opening of a new incinerator is to launch a stink bomb into the middle of the ceremony. They calculate that a 6-kg projectile launched with an initial speed of 40 m/s at an angle of 30° will do the trick. The bomb will explode on impact, no one will get hurt, but everyone will stink. Perfect. However, at the top of its flight, the bomb explodes into two fragments, each having a horizontal trajectory. To top it off—this really isn't their day—the 2-kg fragment lands right at the feet of Dave and Jay. (*a*) Where does the 4-kg fragment land? (*b*) Find the energy of the explosion by comparing the kinetic energy of the projectiles just before and just after the explosion.

(*a*) 1. Find h at explosion	$Mgh = \frac{1}{2}Mv_{yi}^2$; $h = (40 \sin 30°)^2/2g$ m $= 20.39$ m
2. Find v_x at explosion	$v_x = 40 \cos 30°$ m/s $= 34.64$ m/s
3. To land at origin, v_x of 2 kg $= -34.64$ m/s	$p_i = p_f$; $6v_x = 4v_{4x} - 2v_x$; $v_{4x} = 8v_x/4 = 69.3$ m/s
4. Find time of flight	$t = 2v_{yi}/g = 40/9.81$ s $= 4.08$ s
5. Find distance traveled by 4 kg	$d = (2.04 \times 34.64 + 2.04 \times 69.3)$ m $= 212$ m
(*b*) 1. Find K_i and K_f at explosion site	$K_i = 3 \times 34.64^2$ J $= 3600$ J; $K_f = (34.64^2 + 2 \times 69.3^2)$ J $= 10800$ J
2. Energy of explosion $= K_f - K_i$	Energy of explosion $= 7200$ J

79 •• A projectile of mass $m = 3$ kg is fired with initial speed of 120 m/s at an angle of 30° with the horizontal. At the top of its trajectory, the projectile explodes into two fragments of masses 1 kg and 2 kg. The 2-kg fragment lands on the ground directly below the point of explosion 3.6 s after the explosion. (*a*) Determine the velocity of the 1-kg fragment immediately after the explosion. (*b*) Find the distance between the point of firing and the point at which the 1-kg fragment strikes the ground. (*c*) Determine the energy released in the explosion.

(*a*) 1. Find $\boldsymbol{v}$ and $\boldsymbol{r}$ at point of explosion	$\boldsymbol{v} = v_{x0}\,\boldsymbol{i}$; $v_{x0} = (120 \cos 30°)$ m/s $= 104$ m/s; $y = v_{y0}^2/2g = 60^2/19.62$ m $= 183.5$ m
Find t_1, time until explosion and x	$t_1 = v_{y0}/g = 6.12$ s; $x = v_{x0}t_1 = 636.5$ m
2. Find $\boldsymbol{v}_2$ of 2 kg after explosion	$v_{x2} = 0$; $183.5 - v_{y2} \times 3.6 - \frac{1}{2} \times 9.81 \times 3.6^2 = 0$; $v_{y2} = -33.3$ m/s
3. Use $\boldsymbol{p}_i = \boldsymbol{p}_f$ to find v_{x1} and v_{y1}	$v_{x1} = 3 \times 104$ m/s $= 312$ m/s; $v_{y1} = 2 \times 33.3$ m/s $= 66.6$ m/s
(*b*) 1. Find t_2, time for 1 kg to reach ground	$\frac{1}{2}gt_2^2 - 66.6t_2 - 183.5 = 0$; $t_2 = 15.94$ s
2. Find distance traveled by 1 kg mass	$d = (636.5 + 312 \times 15.94)$ m $= 5610$ m
(*c*) 1. Find K_i and K_f at explosion;	$K_i = 1.5 \times 104^2$ J $= 16.2$ kJ; $K_f = 33.3^2 + \frac{1}{2}(312^2 + 66.6^2)$ J $= 52$ kJ
$E_{expl.} = \Delta K = K_f - K_i$	$\Delta K = E_{expl.} = 35.8$ kJ

80 ••• The boron isotope ^9B is unstable and disintegrates into a proton and two α particles. The total energy released as kinetic energy of the decay products is 4.4×10^{-14} J. In one such event, with the ^9B nucleus at rest prior to decay, the velocity of the proton is measured to be 6.0×10^6 m/s. If the two α particles have equal energies, find the magnitude and the direction of their velocities with respect to that of the proton.

1. Show a sketch of the velocities. We assume that the proton's velocity is in the negative x direction. Note that $m_\alpha = 4m_p = 6.68 \times 10^{-27}$ kg (see Problem 8-77)

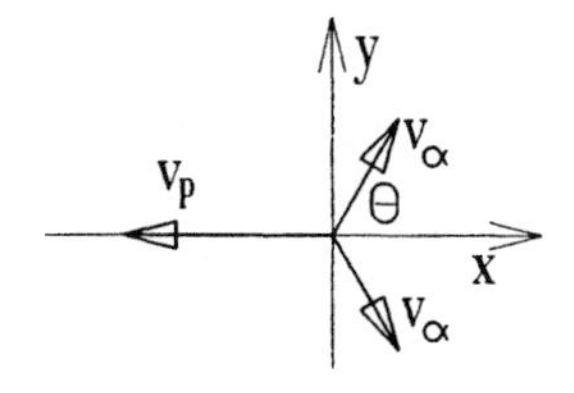

2. Use conservation of energy to find v_α	$\frac{1}{2}m_pv_p^2 + 2(\frac{1}{2}m_\alpha v_\alpha^2) = 4.4 \times 10^{-14}$ J; $m_\alpha v_\alpha^2 = 1.4 \times 10^{-14}$ J; $v_\alpha = 1.45 \times 10^6$ m/s
3. $\Sigma \boldsymbol{p}_i = 0$	$2(4m_p)(1.45 \times 10^6 \cos \theta) = m_p \times 6 \times 10^6$; $\theta = \pm 58.9°$
4. Give the angle with respect to $\boldsymbol{v}_p$	$\theta' = \pm(180° - 58.9°) = \pm 121.1°$

81* • The coefficient of restitution for steel on steel is measured by dropping a steel ball onto a steel plate that is rigidly attached to the earth. If the ball is dropped from a height of 3 m and rebounds to a height of 2.5 m, what is the coefficient of restitution?

Find the ratio v_{rec}/v_{app} and use Equ. 8-31 $\quad e = \dfrac{v_{rec}}{v_{app}} = \sqrt{\dfrac{2gh_{rec}}{2gh_{app}}} = \sqrt{\dfrac{2.5}{3.0}} = 0.913$

82 • According to the official rules of racquetball, a ball acceptable for tournament play must bounce to a height of between 173 and 183 cm when dropped from a height of 254 cm at room temperature. What is the acceptable range of values for the coefficient of restitution for the racquetball–floor system?

See Problem 8-81 $\quad e_{min} = \sqrt{173/254} = 0.825;\ e_{max} = \sqrt{183/254} = 0.849$

83 • A ball bounces to 80% of its original height. (*a*) What fraction of its mechanical energy is lost each time it bounces? (*b*) What is the coefficient of restitution of the ball–floor system?

(*a*) Since $K \propto h$, 20% of energy is lost with each bounce.

(*b*) Since $K \propto v^2$, $e = (0.8)^{1/2} = 0.894$.

84 •• A 2-kg object moving at 6 m/s collides with a 4-kg object that is initially at rest. After the collision, the 2-kg object moves backward at 1 m/s. (*a*) Find the velocity of the 4-kg object after the collision. (*b*) Find the energy lost in the collision. (*c*) What is the coefficient of restitution for this collision?

(*a*) Use $p_i = p_f$ $\quad (2 \times 6)$ kg·m/s $= [-(2 \times 1) + 4 \times v]$ kg·m/s; $v = 3.5$ m/s

(*b*) $E_{loss} = K_i - K_f$ $\quad E_{loss} = 36\ \text{J} - (1 + 2 \times 3.5^2)\ \text{J} = 10.5\ \text{J}$

(*c*) Use Equ. 8-31 $\quad e = (3.5 + 1)/6 = 0.75$

85* •• A 2-kg block moving to the right with speed 5 m/s collides with a 3-kg block that is moving in the same direction at 2 m/s, as in Figure 8-55. After the collision, the 3-kg block moves at 4.2 m/s. Find (*a*) the velocity of the 2-kg block after the collision, and (*b*) the coefficient of restitution for the collision.

(*a*) Use $p_i = p_f$ $\quad (2 \times 5 + 3 \times 2)$ kg·m/s $= (2v_{1f} + 3 \times 4.2)$ kg·m/s;

$v_{1f} = 1.7$ m/s

(*b*) Use Equ. 8-31 $\quad e = (4.2 - 1.7)/(5 - 2) = 0.833$

86 • In a pool game, the cue ball, which has an initial speed of 5 m/s, make an elastic collision with the eight ball, which is initially at rest. After the collision, the eight ball moves at an angle of 30° with the original direction of the cue ball. (*a*) Find the direction of motion of the cue ball after the collision. (*b*) Find the speed of each ball. Assume that the balls have equal mass.

(*a*) Elastic collision, use energy conservation; $m_1 = m_2$ $\quad v_{ci}^2 = v_{cf}^2 + v_8^2$; Note that v_{ci}, v_{cf}, and v_8 form right triangle

$\theta_{cf} + \theta_8 = 90°$ $\quad \theta_{cf} = 60°$

(*b*) 1. Use $\mathbf{p}_i = \mathbf{p}_f$ $\quad 5\ \text{m/s} = v_{cf}\cos 60° + v_8 \cos 30°$; $v_{cf}\sin 60° = v_8 \sin 30°$

2. Solve for v_{cf} and v_8 $\quad v_{cf} = 2.5$ m/s; $v_8 = 4.33$ m/s

87 •• An object of mass $M_1 = m$ collides with velocity $v_0\,\boldsymbol{i}$ into an object of mass $M_2 = 2m$ with velocity $\frac{1}{2}v_0\boldsymbol{j}$. Following the collision, the mass m_2 has a velocity $v_0/4\ \boldsymbol{i}$. (*a*) Determine the velocity of the mass m_1 after the collision. (*b*) Was this an elastic collision? If not, express the energy change in terms of m and v_0.

(*a*) 1. Use $\boldsymbol{p}_\mathrm{i} = \boldsymbol{p}_\mathrm{f}$ — $\boldsymbol{p}_\mathrm{i} = mv_0\,\boldsymbol{i} + mv_0\boldsymbol{j} = \boldsymbol{p}_\mathrm{f} = \frac{1}{2}mv_0\,\boldsymbol{i} + m\boldsymbol{v}_{1\mathrm{f}}$

2. Find $\boldsymbol{v}_{1\mathrm{f}}$ — $\boldsymbol{v}_{1\mathrm{f}} = v_0(\frac{1}{2}\boldsymbol{i} + \boldsymbol{j})$; $v_{1\mathrm{f}}^2 = (5/4)v_0^2$

(*b*) Find $\Delta E = K_\mathrm{i} - K_\mathrm{f}$ — $\Delta E = \frac{1}{2}mv_0^2 + mv_0^2 - mv_0^2/16 - (5/8)mv_0^2 = mv_0^2/16$

88 •• A puck of mass 0.5 kg approaches a second, similar puck that is stationary on frictionless ice. The initial speed of the moving puck is 2 m/s. After the collision, one puck leaves with a speed v_1 at 30° to the original line of motion; the second puck leaves with speed v_2 at 60°, as in Figure 8-56. (*a*) Calculate v_1 and v_2. (*b*) Was the collision elastic?

(*b*) Yes; this is identical to Problem 8-86. Since the angle between pucks is 90°, the collision is elastic.

(*a*) Follow procedure of Problem 8-86(*b*) with $v_{\mathrm{ci}} = 2$ m/s. $v_1 = 1$ m/s, $v_2 = 1.73$ m/s.

89* •• Figure 8-57 shows the result of a collision between two objects of unequal mass. (*a*) Find the speed v_2 of the larger mass after the collision and the angle θ_2. (*b*) Show that the collision is elastic.

(*a*) 1. Use $\boldsymbol{p}_\mathrm{i} = \boldsymbol{p}_\mathrm{f}$; — $3mv_0 = \sqrt{5}mv_0\cos\theta_1 + 2mv_2\cos\theta_2$;

Note: $\sqrt{5}\sin\theta_1 = 2$, $\sqrt{5}\cos\theta_1 = 1$ — $\sqrt{5}mv_0\sin\theta_1 = 2mv_2\sin\theta_2$

2. Solve for θ_2 — $2\cot\theta_2 = 3 - 2$; $\cot\theta_2 = 1$, $\theta_2 = 45°$

3. Solve for v_2 — $v_2 = \sqrt{2}\,v_0$

(*b*) Find K_i and K_f — $K_\mathrm{i} = 4.5mv_0^2$; $K_\mathrm{f} = 2.5mv_0^2 + 2mv_0^2 = 4.5mv_0^2$; Q.E.D.

90 •• A ball moving at 10 m/s makes an off-center elastic collision with another ball of equal mass that is initially at rest. The incoming ball is deflected at an angle of 30° from its original direction of motion. Find the velocity of each ball after the collision.

This is identical to Problem 8-88 except that the initial speed of the object is 10 m/s. It follows that the velocities are 5 m/s at 60° and 8.66 m/s at 30°.

91 ••• A particle has an initial speed v_0. It collides with a second particle that is at rest and is deflected through an angle ϕ. Its speed after the collision is v. The second particle recoils. Its velocity makes an angle θ with the initial direction of the first particle. (*a*) Show that $\tan\theta = (v\sin\phi)/(v_0 - v\cos\phi)$. (*b*) Do you have to assume that the collision is either elastic or inelastic to get the result in part (*a*)?

(*a*) Let m_1 and m_2 be the incoming and struck particles; assume $\boldsymbol{v}_0$ is in the x direction. From $\boldsymbol{p}_\mathrm{i} = \boldsymbol{p}_\mathrm{f}$, one obtains $m_1v_0 = m_1v\cos\phi + m_2v_2\cos\theta$ (1) and $m_1v\sin\phi = m_2v_2\sin\theta$ (2). $m_2v_2\cos\theta = m_1v_0 - m_1v\cos\phi$ (1*a*). Divide (2) by (1*a*) to obtain: $\tan\theta = (v\sin\phi)/(v_0 - v\cos\phi)$.

(*b*) You did not use energy conservation. Therefore, the result is valid for elastic and inelastic collisions.

92 • Describe a perfectly inelastic collision as viewed in the center-of-mass reference frame.

The two objects approach with equal but opposite momenta and remain at rest after the collision.

93* •• A particle with momentum p_1 in one dimension makes an elastic collision with a second particle of momentum $p_2 = -p_1$ in the center-of-mass reference frame. After the collision its momentum is p_1'. Write the total initial and final energies in terms of p_1 and p_1', and show that $p_1' = \pm p_1$. If $p_1' = -p_1$, the particle is merely turned around by the collision and leaves with the speed it had initially. What is the significance of the plus sign in your solution?

$K_{rel} = p_1^2/2m_1 + p_1^2/2m_2 = p_1^2(m_1 + m_2)/2m_1m_2$; $K_{cm} = (2p_1)^2/2(m_1 + m_2) = 2p_1^2/(m_1 + m_2)$; $K = K_{rel} + K_{cm}$, i.e., $K = (p_1^2/2)[(m_1^2 + 6m_1m_2 + m_2^2)/(m_1^2m_2 + m_1m_2^2)]$. In an elastic collision, $K_i = K_f$. Consequently, $(p_1')^2 = (p_1)^2$ and $p_1' = \pm p_1$. If $p_1' = p_1$, the particles do not collide.

94 •• A 3-kg block is traveling to the right at 5 m/s, and a 1-kg block is traveling to the left at 3 m/s. (*a*) Find the velocity v_{cm} of the center of mass. (*b*) Subtract v_{cm} from the velocity of each block to find the velocity of each block in the center-of-mass reference frame. (*c*) After they make an elastic collision, the velocity of each block is reversed in this frame. Find the velocity of each block after an elastic collision. (*d*) Transform back into the original frame by adding v_{cm} to the velocity of each block. (*e*) Check your result by finding the initial and final kinetic energies of the blocks in the original frame.

(*a*) Use Equ. 8-13 — $v_{cm} = [(15 - 3)/4]$ m/s = 3 m/s

(*b*) Follow the indicated procedure — $u_3 = 2$ m/s, $u_1 = -6$ m/s

(*c*) Follow procedure — $u_3' = -2$ m/s, $u_1' = 6$ m/s

(*d*) Transform back to lab system — $v_3' = 1$ m/s, $v_1' = 9$ m/s

(*e*) Evaluate K_i and K_f — $K_i = ½(3 \times 25 + 1 \times 9) = 42$ J; $K_f = ½(3 \times 1 + 1 \times 81) = 42$ J

95 •• Repeat Problem 94 with a second block having a mass of 5 kg and moving to the right at 3 m/s.

Following same procedures one obtains:

(*a*) $v_{cm} = 3.75$ m/s. (*b*) $u_3 = 1.25$ m/s, $u_5 = -0.75$ m/s. (*c*) $u_3' = -1.25$ m/s, $u_5' = 0.75$ m/s.

(*d*) $v_3' = 2.5$ m/s, $v_5' = 4.5$ m/s. (*e*) $K_i = ½(3 \times 5 + 5 \times 9)$ J = 60 J; $K_f = ½(3 \times 6.25 + 5 \times 20.25)$ J = 60 J.

96 •• A rocket burns fuel at a rate of 200 kg/s and exhausts the gas at a relative speed of 6 km/s. Find the thrust of the rocket.

Use Equ. 8-39 — $F_{th} = 6 \times 10^3 \times 200 = 1.2$ MN

97* •• The payload of a rocket is 5% of its total mass, the rest being fuel. If the rocket starts from rest and moves with no external forces acting on it, what is its final velocity if the exhaust velocity of its gas is 5 km/s?

1. No external forces are acting — Equ. 8-42 reduces to $v_f = -u_{ex} \ln(m_f/m_0)$

2. Evaluate v_f for $m_f/m_0 = 1/20$ — $v_f = [5 \ln(20)]$ km/s = 15 km/s

98 •• A rocket moves in free space with no external forces acting on it. It starts from rest and has an exhaust speed of 3 km/s. Find the final velocity if the payload is (*a*) 20%, (*b*) 10%, (*c*) 1%.

From Problem 8-97, $v_f = u_{ex} \ln(m_0/m_f)$.

(*a*) $m_0/m_f = 5$; evaluate v_f — $v_f = 3 \ln(5)$ km/s = 4.83 km/s

(*b*) $m_0/m_f = 10$ — $v_f = 3 \ln(10)$ km/s = 6.91 km/s

(*c*) $m_0/m_f = 100$ — $v_f = 3 \ln(100)$ km/s = 13.8 km/s

99 •• A rocket has an initial mass of 30,000 kg, of which 20% is the payload. It burns fuel at a rate of 200 kg/s and exhausts its gas at a relative speed of 1.8 km/s. Find (*a*) the thrust of the rocket, (*b*) the time until burnout, and (*c*) its final speed assuming it moves upward near the surface of the earth where the gravitational field *g* is constant.

(*a*) Use Equ. 8-39 $F_{th} = 1.8 \times 10^3 \times 200$ N = 360 kN

(*b*) mass of fuel = 0.8 × 30,000 kg = 24,000 kg $t = 24{,}000/200$ s = 120 s

(*c*) Use Equ. 8-42 $v_f = [1.8 \times 10^3 \ln(5) - 9.81 \times 120]$ m/s = 1.72 km/s

---

100 • Why can friction and the force of gravity usually be neglected in collision problems?

The event takes place in such a short time interval that the effects of gravity or friction are negligible.

101* • The condition necessary for the conservation of momentum of a given system is that (*a*) energy is conserved. (*b*) one object is at rest. (*c*) no external force acts. (*d*) internal forces equal external forces. (*e*) the net external force is zero.

(*e*)

---

102 • As a pendulum bob swings back and forth, is the momentum of the bob conserved? Explain why or why not.

No; the net external force is the sum of the force of gravity and the tension in the string; these do not add to zero.

---

103 • A model-train car of mass 250 g traveling with a speed of 0.50 m/s links up with another car of mass 400 g that is initially at rest. What is the speed of the cars immediately after they have linked together? Find the initial and final kinetic energies.

1. Use $p_i = p_f$ $v_f = [(250 \times 0.5)/650]$ m/s = 0.192 m/s

2. Find K_i and K_f $K_i = ½(0.25 \times 0.5^2)$ J = 31.25 mJ; $K_f = ½(0.65 \times 0.192^2)$ = 12 mJ

---

104 • (*a*) Find the total kinetic energy of the two model-train cars of Problem 103 before they couple. (*b*) Find the initial velocities of the two cars relative to the center of mass of the system, and use them to calculate the initial kinetic energy of the system relative to the center of mass. (*c*) Find the kinetic energy of the center of mass. (*d*) Compare your answers for (*b*) and (*c*) with that for (*a*).

(*a*) See Problem 8-103 $K_i = 31.25$ mJ

(*b*) 1. See Problem 8-103; Use Equ. 8-34 $v_{cm} = 0.192$ m/s; $u_1 = 0.308$ m/s, $u_2 = -0.192$ m/s

2. Find $K_{i,rel}$ $K_{i,rel} = ½(0.25 \times 0.308^2 + 0.4 \times 0.192^2)$ J = 19.23 mJ

(*c*) Find $K_{cm} = ½Mv_{cm}^2$ $K_{cm} = ½(0.65 \times 0.192^2)$ J = 12 mJ

(*d*) $K_i = K_{i,rel} + K_{cm}$

105* • A 4-kg fish is swimming at 1.5 m/s to the right. He swallows a 1.2-kg fish swimming toward him at 3 m/s. Neglecting water resistance, what is the velocity of the larger fish immediately after his lunch?

Use Equ. 8-13 $v = v_{cm} = (4 \times 1.5 - 1.2 \times 3)/5.2$ m/s = 0.462 m/s

---

106 • A 3-kg block moves at 6 m/s to the right while a 6-kg block moves at 3 m/s to the right. Find (*a*) the total kinetic energy of the two-block system, (*b*) the velocity of the center of mass, (*c*) the center-of-mass kinetic energy, and (*d*) the kinetic energy relative to the center of mass.

(*a*) $K_t = K_1 + K_2$ $K_t = ½(3 \times 6^2 + 6 \times 3^2)$ J = 81 J

(*b*) Use Equ. 8-13 $v_{cm} = (3 \times 6 + 6 \times 3)/9$ m/s = 4 m/s (*c*) $K_{cm} = \frac{1}{2}Mv_{cm}^2$ $K_{cm} = \frac{1}{2}(9 \times 4^2)$ J = 72 J

(*d*) $K_{rel} = K_t - K_{cm}$ (see Problem 8-104) $K_{rel} = 9$ J

---

107 • A 1500-kg car traveling north at 70 km/h collides at an intersection with a 2000-kg car traveling west at 55 km/h. The two cars stick together. (*a*) What is the total momentum of the system before the collision? (*b*) Find the magnitude and direction of the velocity of the wreckage just after the collision.

(*a*) $\mathbf{p} = \mathbf{p}_1 + \mathbf{p}_2$ $\mathbf{p} = 1.05 \times 10^5 \mathbf{j}$ kg·km/h − $1.1 \times 10^5 \mathbf{i}$ kg·km/h

(*b*) $\mathbf{v}_f = \mathbf{v}_{cm} = \mathbf{p}/M$ $v_f = (1.05^2 + 1.1^2)^{½} \times 10^5/3500$ km/h = 43.4 km/h

$\theta = \tan^{-1}(1.1/1.05) = 46.3°$ west of north

---

108 • The great white shark can have a mass as great as 3000 kg. Suppose such a shark is cruising the ocean when it spots a meal below it: a 200.0-kg fish swimming horizontally at 8.00 m/s. The shark rushes vertically downward at 3.00 m/s and swallows the prey at once. At what angle to the vertical θ will the shark be moving immediately after the snack? What is the final speed of the shark? (Neglect any drag effects of the water.)

1. $\theta = \tan^{-1}(p_x/p_y)$ $\theta = \tan^{-1}[(200 \times 8)/(3000 \times 3)] = 10.1°$
2. $v = v_{cm} = (1/M)(p_x^2 + p_y^2)^{½}$ $v = (1/3200)(9000^2 + 1600^2)^{½}$ m/s = 2.86 m/s

---

109* • Repeat Problem 106 for a 3-kg block moving at 6 m/s to the right and a 6-kg block moving at 3 m/s to the left Follow the procedures outlined in Problem 8-106. One obtains the following results.

(*a*) $K_t = 81$ J (*b*) $v_{cm} = 0$ (*c*) $K_{cm} = 0$ (*d*) $K_{rel} = 81$ J

---

110 • Repeat Problem 106 for a 3-kg block moving at 10 m/s to the right and a 6-kg block moving at 1 m/s to the right.

Follow the procedures outlined in Problem 8-106. One obtains the following results.

(*a*) $K_t = 153$ J. (*b*) $v_{cm} = 4$ m/s. (*c*) $K_{cm} = 72$ J. (*d*) $K_{rel} = 81$ J.

---

111 •• A 60-kg woman stands on the back of a 6-m-long, 120-kg raft that is floating at rest in still water with no friction. The raft is 0.5 m from a fixed pier, as in Figure 8-58. (*a*) The woman walks to the front of the raft and stops. How far is the raft from the pier now? (*b*) While the woman walks, she maintains a constant speed of 3 m/s relative to the raft. Find the total kinetic energy of the system (woman plus raft), and compare with the kinetic energy if the woman walked at 3 m/s on a raft *tied to the pier*. (*c*) Where does this energy come from, and where does it go when the woman stops at the front of the raft? (*d*) On land, the woman can put a lead shot 6 m. She stands at the back of the raft, aims forward, and puts the shot so that just after it leaves her hand, it has the same velocity *relative to her* as it does when she throws it from the ground. Where does the shot land?

We shall use the following convention: Take the origin at initial position of right hand end of raft; measure positive displacement to the left; let Δx be the displacement of raft.

(*a*) $F_{ext} = 0$; consquently x_{cm} does not change $(120 \times 3 + 60 \times 6)$ kg·m

$= [120 \times (3 + \Delta x) + 60 \times \Delta x]$ kg·m

Solve for Δx; distance to dock, $d = (\Delta x + 0.5)$ m $\Delta x = 2.0$ m; $d = 2.5$ m

(*b*) Find u_w, u_r, and v_{cm}; note that the time elapsed is 2 s $u_w = -3$ m/s; $u_r = \Delta x/\Delta t = 1$ m/s

Determine K_t and K on land $K_t = \frac{1}{2}(60 \times 3^2 + 120)$ J = 330 J; K (on land) = 270 J (*c*) 60 J derives from the chemical energy of the woman.

(*d*) The shot will land in the water; the raft shifts to the left.

---

112 •• A 1-kg steel ball and a 2-m cord of negligible mass make up a simple pendulum that can pivot without friction about the point *O*, as in Figure 8-59. This pendulum is released from rest in a horizontal position and when the ball is at its lowest point it strikes a 1-kg block sitting at rest on a rough shelf. Assume that the collision is perfectly elastic and take the coefficient of friction between the block and shelf to be 0.1. (*a*) What is the velocity of the block just after impact? (*b*) How far does the block move before coming to rest?

(*a*) Find v_{ball} at impact — $v_{\text{ball}} = (2gh)^{1/2} = (4 \times 9.81)^{1/2}$ m/s = 6.26 m/s

Since $m_{\text{ball}} = m_{\text{block}}$, $v_{\text{block}} = v_{\text{ball}}$ after collision — $v_{\text{block}} = 6.26$ m/s

(*b*) $a_{\text{block}} = \mu_k g = 0.981$ m/s²; $v^2 = 2as$ — $s = 6.26^2/1.962$ m = 20 m

---

113* •• In World War I, the most awesome weapons of war were huge cannons mounted on railcars. Figure 8-60 shows such a cannon, mounted so that it will project a shall at an angle of 30°. With the car initially at rest, the cannon fires a 200-kg projectile at 125 m/s. Now consider a system composed of a cannon, shell, and railcar, all rolling on the track without frictional losses. (*a*) Will the total vector momentum of that system be the same (i.e., "conserved") before and after the shell is fired? Explain your answer in a few words. (*b*) If the mass of the railcar plus cannon is 5000 kg, what will be the recoil velocity of the car along the track after the firing? (*c*) The shell is observed to rise to a maximum height of 180 m as it moves through its trajectory. At this point, its speed is 80 m/s. On the basis of this information, calculate the amount of thermal energy produced by air friction on the shell on its way from firing to this maximum height.

(*a*) Momentum of system is not conserved; there is an external force, the vertical reaction force of rails.

(*b*) Use $p_{xi} = p_{xf}$ — $200 \times 125 \cos 30° = 5000 v_{\text{rec}}$; $v_{\text{rec}} = 4.33$ m/s

(*c*) 1. Find *h* without air friction; $h = v_y^2/2g$ — $h = 199$ m

2. $W_f = mg\Delta h + \frac{1}{2}m(v_{x0}^2 - v_x^2)$ — $W_f = 200 \times 9.81 \times 19 + 100[(125 \cos 30°)^2 - 80^2)]$ J = 569 kJ

---

114 •• A 15-g bullet traveling at 500 m/s strikes an 0.8-kg block of wood that is balanced on a table edge 0.8 m above ground (Figure 8-61). If the bullet buries itself in the block, find the distance *D* at which the block hits the floor.

1. Find v_{cm} — $v_{\text{cm}} = (500 \times 0.015/0.815)$ m/s = 9.2 m/s

2. Find *t*, time to drop 0.8 m; $D = v_{\text{cm}} t$ — $t = (1.6/9.81)^{1/2}$ s = 0.404 s; $D = 0.404 \times 9.2$ m = 3.72 m

---

115 •• In hand-pumped railcar races, a speed of 32 km/h has been achieved by teams of four. A car of mass 350 kg is moving at that speed toward a river when Carlos, the chief pumper, notices that the bridge ahead is out. All four people (of mass 75 kg each) jump simultaneously backward off the car with a velocity that has a horizontal component of 4 m/s *relative to the car after jumping*. The car proceeds off the bank and falls in the water a distance 25.0 m off the bank. (*a*) Estimate the time of the fall of the railcar. (*b*) What happens to the team of pumpers?

(*a*) 1. Use momentum conservation — $u_{\text{car}} \times 350 = -(4 \times 75)u_p$; $u_p = -(350/300)u_{\text{car}}$

2. Given that $u_{car} - u_p = 4$ m/s; solve for u_{car} $u_{car} = [4/(1 + 350/300)]$ m/s = 1.85 m/s

3. Use Equ. 8-34 to obtain v_{car} $v_{car} = (1.85 + 8.89)$ m/s = 10.74 m/s

4. $t = d/v_{car}$ $t = (25/10.74)$ s = 2.3 s

(*b*) Pumpers hit the ground at 6.74 m/s = 24 km/h They may get bruised a bit

---

116 •• A constant force $\boldsymbol{F} = 12\text{ N }\boldsymbol{i}$ is applied to the 8-kg mass of Problem 5 at $t = 0$. (*a*) What is the velocity of the center of mass of the three-particle system at $t = 5$ s? (*b*) What is the location of the center of mass at $t = 5$ s?

(*a*) $a_{cm} = F/M$; $v_{cm}(t) = a_{cm}t$ $v_{cm}(5) = 5 \times 12/11\text{ m/s }\boldsymbol{i} = 5.45\text{ m/s }\boldsymbol{i}$

(*b*) $x_{cm}(t) = x_{cm}(0) + \frac{1}{2}v_{cm}(t) \times t$ $x_{cm}(5) = [(37/11) + \frac{1}{2}(60/11) \times 5]\text{ m} = 17\text{ m}$

117* •• Two particles of mass m and $4m$ are moving in a vacuum at right angles as in Figure 8-62. A force $\boldsymbol{F}$ acts on both particles for a time T. As a result, the velocity of the particle m is $4v$ in its original direction. Find the new velocity $\boldsymbol{v}'$ of the particle of mass $4m$.

1. Determine $\boldsymbol{F}T = \Delta\boldsymbol{p}$ $\boldsymbol{F}T = 3mv\,\boldsymbol{i}$

2. Find $\boldsymbol{p}_{4m}' = \boldsymbol{p}_{4m}(0) + \Delta\boldsymbol{p}$, and $\boldsymbol{v}'$ $\boldsymbol{p}_{4m}' = 3mv\,\boldsymbol{i} - 4mv\,\boldsymbol{j}$; $\boldsymbol{v}' = 0.75v\,\boldsymbol{i} - v\,\boldsymbol{j}$

---

118 •• An open railroad car of mass 20,000 kg is rolling without friction at 5 m/s along a level track when it starts to rain. After the car has collected 2000 kg of water, the rain stops. (*a*) What is the car's velocity? (*b*) As the car is rolling along, the water begins leaking out of a hole in the bottom at a rate of 5 kg/s. What is the velocity after half the water has leaked out? (*c*) What is the velocity after all the water has leaked out?

(*a*) Assume rain falls vertically; use $p_i = p_f$ $v_f = (20{,}000/22{,}000) \times 5$ m/s = 4.55 m/s

(*b*), (*c*) Drops fall with v of car, so v_{car} unchanged $v = 4.55$ m/s

---

119 •• In the "slingshot effect," the transfer of energy in an elastic collision is used to boost the energy of a space probe so that it can escape from the solar system. Figure 8-63 shows a space probe moving at 10.4 km/s (relative to the sun) toward Saturn, which is moving at 9.6 km/s (relative to the sun) toward the probe. Because of the gravitational attraction between Saturn and the probe, the probe swings around Saturn and heads back in the opposite direction with speed v_f. (*a*) Assuming this collision to be a one-dimensional elastic collision with the mass of Saturn much greater than that of the probe, find v_f. (*b*) By what factor is the kinetic energy of the probe increased? Where does this energy come from?

(*a*) 1. For an elastic collision, $u_{app} = -u_{rec}$ $u_{app} = (10.4 + 9.6)$ km/s = 20 km/s

2. Use Equ. 8-34; $v_{cm} = -9.6$ km/s $v_{rec} = 29.6$ km/s

(*b*) Find $K_f/K_i = v_f^2/v_i^2$ $K_f/K_i = (29.6/10.4)^2 = 8.1$; energy comes from the slowing of Saturn (by an immeasurably small amount)

---

120 •• You (mass 80 kg) and your friend (mass unknown) are in a rowboat (mass 60 kg) on a calm lake. You are at the center of the boat rowing and she is at the back, 2 m from the center. You get tired and stop rowing. She offers to row and after the boat comes to rest, you change places. You notice that after changing places the boat has moved 20 cm relative to a fixed log. What is your friend's mass?

Take the coordinate origin to be at the end of the boat prior to shifting places..

1. Express x_{cm} prior to shift $x_{cm} = [(60 + 80) \times 2/(140 + m)]$ m

2. Write expression for x_{cm}', x_{cm} after shift	$x_{cm}' = [(60 + m) \times (2 \pm 0.2) \pm 80 \times 0.2]/(140 + m)$ m
3. But $F_{ext} = 0$, so $x_{cm}' = x_{cm}$; solve for m	$m = 60$ kg or 104 kg

---

121* •• A small car of mass 800 kg is parked behind a small truck of mass 1600 kg on a level road (Figure 8-64). The brakes of both the car and the truck are off so that they are free to roll with negligible friction. A man sitting on the tailgate of the truck shoves the car away by exerting a constant force on the car with his feet. The car accelerates at 1.2 m/s^2. (*a*) What is the acceleration of the truck? (*b*) What is the magnitude of the force exerted on either the truck or the car?

(*a*) $F_{ext} = 0$, $\therefore a_{cm} = 0$	800×1.2 kg·m/s^2 $= 1600a_t$; $a_t = 0.6$ m/s^s
(*b*) $F = ma$	$F = 960$ N

---

122 •• A 13-kg block is at rest on a level floor. A 400-g glob of putty is thrown at the block such that it travels horizontally, hits the block, and sticks to it. The block and putty slide 15 cm along the floor. If the coefficient of sliding friction is 0.4, what is the initial speed of the putty?

1. Find K_i of block + glob; use energy conservation	$K_i = f_k s = \mu_k Mgs = 0.4 \times 13.4 \times 9.81 \times 0.15$ J $= 7.89$ J
2. Find Mv_{cm} after collision; then use $p_i = p_f$	$Mv_{cm} = \sqrt{2MK_i} = m_{gl}v_{gl}$ $v_{gl} = \sqrt{2\times7.89\times13.4}/0.4$ m/s $= 36.4$ m/s

---

123 •• A careless driver rear-ends a car that is halted at a stop sign. Just before impact, the driver slams on his brakes, locking the wheels. The driver of the struck car also has his foot solidly on the brake pedal, locking his brakes. The mass of the struck car is 900 kg, and that of the initially moving vehicle is 1200 kg. On collision, the bumpers of the two cars mesh. Police determine from the skid marks that after the collision the two cars moved 0.76 m together. Tests revealed that the coefficient of sliding friction between the tires and pavement was 0.92. The driver of the moving car claims that he was traveling at less than 15 km/h as he approached the intersection. Is he telling the truth?

1. Use $v = \sqrt{2as}$ and $a = \mu_s g$ to find $v = v_{cm}$	$v_{cm} = (2 \times 0.92 \times 9.81 \times 0.76)^{1/2}$ m/s $= 3.7$ m/s
2. Use $p_i = p_f$ to find v_{ci}, initial speed of moving car	$v_{ci} = (2100 \times 3.7/1200)$ m/s $= 6.475$ m/s $= 23.3$ km/h; No

---

124 •• A pendulum consists of a 0.4-kg bob attached to a string of length 1.6 m. A block of mass M rests on a horizontal frictionless surface (Figure 8-65). The pendulum is released from rest at an angle of 53° with the vertical and the bob collides elastically with the block. Following the collision, the maximum angle of the pendulum with the vertical is 5.73°. Determine the mass M.

1. Use energy consevation to find K_M	$K_M = 0.4 \times 9.81 \times 1.6[\cos 5.73° - \cos 53°]$ J $= 2.47$ J
2. Find v_{mi} and v_{mf} at collision	$v_{mi} = [2 \times 9.81 \times 1.6(1 - \cos 53°)]^{1/2}$ m/s $= 3.544$ m/s $v_{mf} = [2 \times 9.81 \times 1.6(1 - \cos 5.73°)]^{1/2}$ m/s $= 0.396$ m/s
3. Use $p_i = p_f$ to find p_M after collision	$3.544 \times 0.4 = p_M \pm 0.396 \times 0.4$; $p_M = 1.576$ or 1.26 kg·m/s
4. Use $K_M = p_M^2/2M$ to determine M	$M = 0.32$ kg or $M = 0.50$ kg

125* •• Initially, mass $m = 1.0$ kg and mass M are both at rest on a frictionless inclined plane (Figure 8-66). Mass M rests against a spring that has a spring constant of 11,000 N/m. The distance along the plane between m and M is 4.0 m. Mass m is released, makes an elastic collision with mass M, and rebounds a distance of 2.56 m back up the inclined plane. Mass M comes to rest momentarily 4.0 cm from its initial position. Find the mass M.

1. Use conservation of energy — $mg\Delta h = \frac{1}{2}kx^2 - Mgx \sin 30°$; $\Delta h = 1.44 \sin 30° = 0.72$ m
2. Find M: $x = 0.04$ m, $m = 1$ kg, $k = 11\times10^4$ N/m — $M = kx/g - 2m\Delta h/x$; $M = 8.85$ kg

---

126 •• A circular plate of radius r has a circular hole cut out of it having radius $r/2$ (Figure 8-67). Find the center of mass of the plate. *Hint:* The hole can be represented by two disks superimposed, one of mass m and the other of mass $-m$.

By symmetry, $x_{cm} = 0$. Let σ be the mass per unit area of the disk. Then the mass of the complete disk is $\pi r^2\sigma = M$, and the mass of the material removed is $\pi r^2\sigma/4 = M/4$. Thus, $y_{cm} = (-M/4)(-r/2)/(3M/4) = r/6$.

---

127 •• Using the hint from Problem 126, find the center of mass of a solid sphere of radius r that has a spherical cavity of radius $r/2$, as in Figure 8-68.

Since $V \propto r^3$, the mass of the material removed is $M/8$. Thus, $x_{cm} = y_{cm} = 0$, and $z_{cm} = (-M/8)(-r/2)/(7M/8) = r/14$.

---

128 •• A neutron of mass m makes an elastic head-on collision with a stationary nucleus of mass M. (*a*) Show that the energy of the nucleus after the collision is $K_{nucleus} = [4mM/(m + M)^2]K_n$, where K_n is the initial energy of the neutron. (*b*) Show that the fraction of energy lost by the neutron in this collision is

$$\frac{-\Delta K_n}{K_n} = \frac{4mM}{(m+M)^2} = \frac{4(m/M)}{(1+m/M)^2}.$$

(*a*) 1. Write the expression for energy conservation — $p_{ni}^2/2m = p_M^2/2M + p_{nf}^2/2m$ (1)
2. Use momentum conservation — $p_{ni} = p_{nf} + p_M$; $p_{nf} = p_{ni} - p_M$ (2)
3. Eliminate p_{nf} in (1) using (2); simplify — $p_M/2M + p_M/2m - p_{ni}/m = 0$ (3)
4. Use (3) to write $p_{ni}^2/2m$ in terms of p_M — $p_{ni}^2/2m = K_n = p_M^2(M + m)^2/8M^2m$ (4)
5. Express $K_M = p_M^2/2M$ in terms of K_n — $K_M = p_M^2/2M = K_n[4Mm/(M + m)^2]$; Q.E.D. (5)

(*b*) From energy conservation, $\Delta K_n = -K_M$ — From (5): $\Delta K_n/K_n = 4Mm/(M + m)^2$

129* •• The mass of a carbon nucleus is approximately 12 times the mass of a neutron. (*a*) Use the results of Problem 128 to show that after N head-on collisions of a neutron with carbon nuclei at rest, the energy of the neutron is approximately $0.716^N E_0$, where E_0 is its original energy. Neutrons emitted in the fission of a uranium nucleus have an energy of about 2 MeV. For such a neutron to cause the fission of another uranium nucleus in a reactor, its energy must be reduced to about 0.02 eV. (*b*) How many head-on collisions are needed to reduce the energy of a neutron from 2 MeV to 0.02 eV, assuming elastic head-on collisions with stationary carbon nuclei?

(*a*) 1. Write $K_{nf}/K_{ni} = (K_{ni} - \Delta K_n)/K_{ni}$ (see 8-128*b*) — $K_{nf}/K_{ni} = (M - m)^2/(M + m)^2$ = fractional loss per collision
2. In this case $K_{nf}/K_{ni} = 0.716$ — After N collisions, $K_{nf} = K_0 \times 0.716^N$

(*b*) In this case $(0.716)^N = 10^{-8}$; solve for N — $-8 = N \log(0.716)$; $N = 55$

130 •• On average, a neutron loses 63% of its energy in an elastic collision with a hydrogen atom and 11% of its energy in an elastic collision with a carbon atom. The numbers are lower than the ones we have been using in earlier problems because most collisions are not head-on. Calculate the number of collisions, on average, needed to reduce the energy of a neutron from 2 MeV to 0.02 eV (a desirable outcome for reasons explained in Problem 129) if the neutron collides with (*a*) hydrogen atoms and (*b*) carbon atoms.

(*a*) In this case, $K_{nf}/K_{ni} = 0.37$ per collisions. (see Problem 8-129*b*) $-8 = N\log(0.37)$; $N = 19$

(*b*) In this case, $K_{nf}/K_{ni} = 0.89$ per collision. $-8 = N\log(0.89)$; $N = 158$

---

131 •• A rope of length L and mass M lies coiled on a table. Starting at $t = 0$, one end of the rope is lifted from the table with a force F such that it moves with a constant velocity v. (*a*) Find the height of the center of mass of the rope as a function of time. (*b*) Differentiate your result in (*a*) twice to find the acceleration of the center of mass. (*c*) Assuming that the force exerted by the table equals the weight of the rope still there, find the force F you exert on the top of the rope.

(*a*) Let $\lambda = L/M$ be the mass per unit length. Then $y_{cm} = \frac{1}{2}y(\lambda y)/\lambda L = y^2/2L$; but $y = vt$, so $y_{cm} = v^2t^2/2L$.

(*b*) $a_{cm} = d^2y_{cm}/dt^2 = v^2/L$.

(*c*) $F = ma_{cm} + mg$, where $m = \lambda y = \lambda vt$. Thus, $F = (M/L)(vt)(v^2/L) + (M/L)vtg = Mv^3t/L^2 + Mvtg/L$.

---

132 •• A tennis ball of mass m_t is held a small distance above a basketball of mass m_b. Both are dropped from a height h above the floor. (Take h to be the distance to the center of the basketball.) The basketball collides elastically with the floor. Find the speed v_t of the tennis ball after it then collides elastically with the basketball. Calculate the height reached by the tennis ball if $m_b = 0.480$ kg, $m_t = 0.060$ kg, and $h = 2$ m. (*Caution:* If you try this experimentally, get out of the way of the tennis ball!)

The basketball (subscript *b*) will collide with the floor and rebound first. As it moves up it collides with the tennis ball (subscript *t*) which is moving down. We shall first find the velocities of the two balls prior to their collision, and then use the procedure for elastic collisions to determine the velocity of the tennis ball after the collision.

1. Find v_{bi}, v_{ti}, and v_{cm}. Take up as positive. $v_{bi} = \sqrt{2gh}$, $v_{ti} = -v_{bi}$; $v_{cm} = (m_b - m_t)v_{bi}/(m_b + m_t)$
2. Find the relative velocity u_{ti} $u_{ti} = v_{ti}[1 + (m_b - m_t)/(m_b + m_t)] = -u_{tf}$
3. Find $v_{tf} = u_{tf} + v_{cm}$ $v_{tf} = \sqrt{2gh}[1 + 2(m_b - m_t)/(m_b + m_t)]$
4. $H = v_{tf}^2/2g$; use the numerical values given $H = 2[1 + 2(0.48-0.06)/0.54]^2$ m $= 13.1$ m

---

133* •• Repeat Problem 24 if the cup has a mass m_c and the ball collides with it inelastically.

(*a*) The same as before, i.e., $F = kd + m_pg = mg + m_pg$.

(*b*) 1. Find v_{bi}; use $p_i = p_f$ to find v_{cm} of ball + cup $v_{bi} = \sqrt{2gh}$; $v_{cm} = \sqrt{2gh}[m_b/(m_c + m_b)]$

2. Apply energy conservation $kx^2 = (m_c+m_b)[m_b/(m_c+m_b)]^2(2gh) = 2m_b^2gh/(m_b+m_c)$

3. Solve for the compression x; then multiply by k to find the force the spring exerts on the platform. $x = m_b\sqrt{\dfrac{2gh}{k(m_c + m_b)}}$; $kx = m_bg\sqrt{\dfrac{2kh}{g(m_c + m_b)}}$

4. $F = m_pg + kx$ $F = g\left(m_p + m_b\sqrt{\dfrac{2kh}{g(m_c + m_b)}}\right)$

(*c*) Since the collision is inelastic, the ball never returns to its original position

134 •• Two astronauts at rest face each other in space. One, with mass m_1, throws a ball of mass m_b to the other, whose mass is m_2. She catches the ball and throws it back to the first astronaut. If they each throw the ball with a speed of v relative to themselves, how fast are they moving after each has made one throw and one catch?

Note that each collision is perfectly inelastic. Also, $v_{cm} = 0$. Let v_b be velocity of ball.

1. Find v_1 and v_b after the first throw; use $p_i = p_f$ — $v_1 = -m_b v/(m_1 + m_b)$; $v_b = m_1 v/(m_1 + m_b)$
2. Find v_2 and v_b after m_2 catches the ball — $v_2(m_2 + m_b) = m_b v_b$; $v_2 = v m_b m_1/[(m_1 + m_b)(m_2 + m_b)]$
3. Find v_2 after m_2 throws the ball; let this be v_{2f} — $v_{2f} = v[m_b/(m_2 + m_b)][1 + m_1/(m_1 + m_b)]$
4. Find v_{bf}, velocity of ball after m_2 throws it back — $v_{bf} = -v\{1 - [m_b/(m_2 + m_b)][1 + m_1/(m_1 + m_b)]$
5. Apply momentum conservation to find v_{1f} — $v_{1f}(m_1 + m_b) = v_{bf} m_b + v_1 m_1$
6. Simplify the expression for v_{1f} — $v_{1f} = -m_2 m_b v(2m_1 + m_b)/[(m_1 + m_b)^2(m_2 + m_b)]$

135 •• The ratio of the mass of the earth to the mass of the moon is $M_e/m_m = 81.3$. The radius of the earth is about 6370 km, and the distance from the earth to the moon is about 384,000 km. (*a*) Locate the center of mass of the earth–moon system relative to the surface of the earth. (*b*) What external forces act on the earth–moon system? (*c*) In what direction is the acceleration of the center of mass of this system? (*d*) Assume that the center of mass of this system moves in a circular orbit around the sun. How far must the center of the earth move in the radial direction (toward or away from the sun) during the 14 days between the time the moon is farthest from the sun (full moon) and the time it is closest to the sun (new moon)?

(*a*) Use Equ. 8-4; take origin at center of the earth — $r_{cm} = m_m r_{em}/(M_e + m_m) = r_{em}/82.3 = 4670$ km

(*b*) Gravitational force of sun (and other planets)

(*c*) Acceleration is toward the sun

(*d*) Note that CM is at a fixed distance from sun — $d = 2 \times 4670 \text{ km} = 9340$ km

136 •• You wish to enlarge a skating surface so you stand on the ice at one end and aim a hose horizontally to spray water on the schoolyard pavement. Water leaves the hose at 2.4 kg/s with a speed 30 m/s. If your mass is 75 kg, what is your recoil acceleration? (Neglect friction and the mass of the hose.)

1. Differentiate $p_1 + p_2 = 0$ — $(dm_1/dt)v_1 + m_1 a_1 = -(dm_2/dt)v_2 - m_2 a_2$
2. Here $dm_1/dt = 2.4$ kg/s, $a_1 = 0$; $dm_2/dt = 0$ — $a_2 = -(2.4 \times 30/75)\text{ m/s}^2 = -0.96\text{ m/s}^2$

137* •• A neutron at rest decays into a proton plus an electron. The conservation of momentum implies that the electron and proton should have equal and opposite momentum. However, experimentally they do not. This apparent nonconservation of momentum led Wolfgang Pauli to suggest in 1931 that there was a third, unseen particle emitted in the decay. This particle is called a neutrino, and it was finally observed directly in 1957. Suppose that the electron has momentum $p = 4.65 \times 10^{-22}$ kg.m/s along the negative x direction and the proton ($m = 1.67 \times 10^{-27}$ kg) moves with speed 2.93×10^5 m/s at an angle 17.9° above the x axis. Find the momentum of the neutrino. (The kinetic energy of the electron is comparable to its rest energy, so its energy and momentum are related relativistically rather than classically. However, the rest energy of the proton is large compared with its kinetic energy so the classical relation $E = ½mv^2 = p^2/2m$ is valid.)

1. Momentum conservation: $\mathbf{p}_e + \mathbf{p}_p + \mathbf{p}_\nu = 0$ — $p_p = 1.67 \times 10^{-27} \times 2.93 \times 10^5$ kg·m/s $= 4.89 \times 10^{-22}$ kg·m/s

2. Since $p_{px} + p_e = 0, \boldsymbol{p}_v = -p_{py}\boldsymbol{j}$

$p_{px} = 4.89 \times 10^{-22}\cos 17.9° = 4.65 \times 10^{-22}$ kg·m/s $= -p_e$

$\boldsymbol{p}_v = -4.89 \times 10^{-22}\sin 17.9°\,\boldsymbol{j}$ kg·m/s

$= -1.5 \times 10^{-22}\,\boldsymbol{j}$ kg·m/s

---

138 ••• A stream of glass beads, each with a mass of 0.5 g, comes out of a horizontal tube at a rate of 100 per second (Figure 8-69). The beads fall a distance of 0.5 m to a balance pan and bounce back to their original height. How much mass must be placed in the other pan of the balance to keep the pointer at zero?

1. Find v_y of the bead as it hits the pan — $v_y = (2gh)^{½} = 3.13$ m/s
2. Find Δp_y per bead — $\Delta p_y = 2mv_y$
3. $F = N\Delta p/\Delta t = Mg$; solve for M — $M = (100 \times 6.26/9.81)$ N $= 63.8$ g

---

139 ••• A dumbbell consisting of two balls of mass m connected by a massless rod of length L rests on a frictionless floor against a frictionless wall until it begins to slide down the wall as in Figure 8-70. Find the speed v of the bottom ball at the moment when it equals the speed of the top one.

By symmetry, the speeds will be equal when the angle with the vertical is 45°.

1. Use energy conservation; $E_i = E_f$ — $E_i = mgL/2 = E_f = mgL/2\sqrt{2} + ½(2m)v^2$
2. Solve for v — $v = [Lg(\sqrt{2} - 1)/2\sqrt{2}]^{½} = 1.2\sqrt{L}$

---

140 ••• A chain of length L and mass m is held vertically so that the bottom link just touches the floor. It is then dropped. (*a*) Show that the acceleration of the top end of the chain is g. (*b*) If the chain is moving downward with speed v at time t, and speed $v + \Delta v$ at time $t + \Delta t$, find an expression for the change in momentum of the chain during the interval Δt. (*c*) Find the force exerted on the chain by the floor.

(*a*) Since the chain is in free fall until it hits the floor, the acceleration of each part (other than that which rests already on the floor) is g.

(*b*) Let $\mu = m/L$. Then a length dy just above the bottom has a downward speed of gt and momentum given by $dp = \mu gt\,dy = \mu gtv_y\,dt = \mu g^2t^2\,dt$. Thus, $\Delta p = \mu g^2t^2\Delta t$

(*c*) The force exerted on the floor is $dp/dt + m'g$, where $m' = ½\mu g^2t^2$ is the mass resting on the floor. $F = (3/2)\mu g^2t^2$.

CHAPTER 9

Rotation

1* • Two points are on a disk turning at constant angular velocity, one point on the rim and the other halfway between the rim and the axis. Which point moves the greater distance in a given time? Which turns through the greater angle? Which has the greater speed? The greater angular velocity? The greater tangential acceleration? The greater angular acceleration? The greater centripetal acceleration?

1. The point on the rim moves the greater distance. 2. Both turn through the same angle. 3. The point on the rim has the greater speed 4. Both have the same angular velocity. 5. Both have zero tangential acceleration. 6. Both have zero angular acceleration. 7. The point on the rim has the greater centripetal acceleration.

---

2 • True or false: (*a*) Angular velocity and linear velocity have the same dimensions. (*b*) All parts of a rotating wheel must have the same angular velocity. (*c*) All parts of a rotating wheel must have the same angular acceleration.

(*a*) False (*b*) True (*c*) True

---

3 •• Starting from rest, a disk takes 10 revolutions to reach an angular velocity ω. At constant angular acceleration, how many additional revolutions are required to reach an angular velocity of 2ω? (*a*) 10 rev (*b*) 20 rev (*c*) 30 rev (*d*) 40 rev (*e*) 50 rev.

From Equ. 9-9; $\omega^2 \propto \theta$ $\theta_2 = 4\theta_1$; $\Delta\theta = 3\theta_1 = 30$ rev; (*c*)

---

4 • A particle moves in a circle of radius 90 m with a constant speed of 25 m/s. (*a*) What is its angular velocity in radians per second about the center of the circle? (*b*) How many revolutions does it make in 30 s?

(*a*) $\omega = v/r$ $\omega = (25/90)$ rad/s $= 0.278$ rad/s

(*b*) $\theta = \omega t$ $\theta = 8.33$ rad $= 1.33$ rev.

5* • A wheel starts from rest with constant angular acceleration of 2.6 rad/s². After 6 s, (*a*) What is its angular velocity? (*b*) Through what angle has the wheel turned? (*c*) How many revolutions has it made? (*d*) What is the speed and acceleration of a point 0.3 m from the axis of rotation?

(*a*) $\omega = \alpha t$ $\omega = (2.6 \times 6)$ rad/s $= 15.6$ rad/s

(*b*), (*c*) $\theta = \frac{1}{2}\alpha t^2$ $\theta = 46.8$ rad $= 7.45$ rev

(*d*) $v = \omega r$, $a_c = r\omega^2$, $a_t = r\alpha$; $a = (a_t^2 + a_c^2)^{1/2}$ $v = (15.6 \times 0.3)$ m/s $= 4.68$ m/s;

$a = [(0.3 \times 15.6^2)^2 + (0.3 \times 2.6)^2]^{1/2}$ m/s² $= 73$ m/s²

6 • When a turntable rotating at $33\frac{1}{3}$ rev/min is shut off, it comes to rest in 26 s. Assuming constant angular acceleration, find (*a*) the angular acceleration, (*b*) the average angular velocity of the turntable, and (*c*) the number of revolutions it makes before stopping.

(*a*) $\alpha = \omega/t$ — $\alpha = (33.3 \times 2\pi/60 \times 26)$ rad/s^2 = 0.134 rad/s^2

(*b*) $\omega_{av} = ½\omega_0$ — $\omega_{av} = ½(33.3 \times 2\pi/60)$ rad/s = 1.75 rad/s

(*c*) $\theta = \omega_{av}t$ — $\theta = (1.75 \times 26)$ rad = 45.4 rad = 7.22 rev

7 • A disk of radius 12 cm, initially at rest, begins rotating about its axis with a constant angular acceleration of 8 rad/s^2. At t = 5 s, what are (*a*) the angular velocity of the disk, and (*b*) the tangential acceleration a_t and the centripetal acceleration a_c of a point on the edge of the disk?

(*a*) $\omega = \alpha t$ — $\omega = (8 \times 5)$ rad/s = 40 rad/s

(*b*) $a_t = r\alpha$; $a_c = r\omega^2$ — $a_t = (0.12 \times 8)$ m/s^2 = 0.96 m/s^2;

$a_c = (0.12 \times 40^2)$ m/s^2 = 192 m/s^2

8 • Radio announcers who still play vinyl records have to be careful when cuing up live recordings. While studio albums have blank spaces between the songs, live albums have audiences cheering. If the volume levels are left up when the turntable is turned on, it sounds as though the audience has suddenly burst through the wall. If a turntable begins at rest and rotates through 10° in 0.5 s, how long must the announcer wait before the record reaches the required angular speed of 33.3 rev/min? Assume constant angular acceleration.

1. Determine α; $\theta = ½\alpha t^2$ — $(10 \times 360/2\pi) = ½\alpha \times 0.5^2$ rad; α = 1.4 rad/s^2

2. Find $T = \omega/\alpha$ — $T = (33.3 \times 2\pi/60 \times 1.4)$ s = 2.5 s

9* • A Ferris wheel of radius 12 m rotates once in 27 s. (*a*) What is its angular velocity in radians per second? (*b*) What is the linear speed of a passenger? What is the centripetal acceleration of a passenger?

(*a*) $\omega = 2\pi/27$ rad/s = 0.233 rad/s.

(*b*) $v = r\omega = 12 \times 0.233$ /s = 2.8 m/s. $a_c = r\omega^2 = 12 \times 0.233^2$ m/s^2 = 0.65 m/s^2.

10 • A cyclist accelerates from rest. After 8 s, the wheels have made 3 rev. (*a*) What is the angular acceleration of the wheels? (*b*) What is the angular velocity of the wheels after 8 s?

(*a*) $\theta = ½\alpha t^2$; $\alpha = 2\theta/t^2$ — $\alpha = (2 \times 3 \times 2\pi/8^2)$ rad/s^2 = 0.59 rad/s^2

(*b*) $\omega = \alpha t$ — $\omega = (0.59 \times 8)$ rad/s = 4.72 rad/s

11 • What is the angular velocity of the earth in rad/s as it rotates about its axis?

$\omega = 2\pi$ rad/day = $(2\pi/24 \times 60 \times 60)$ rad/s = 7.27×10^{-5} rad/s.

12 • A wheel rotates through 5.0 radians in 2.8 seconds as it is brought to rest with constant angular acceleration. The initial angular velocity of the wheel before braking began was (*a*) 0.6 rad/s. (*b*) 0.9 rad/s. (*c*) 1.8 rad/s. (*d*) 3.6 rad/s. (*e*) 7.2 rad/s.

$\omega_{av} = ½\omega_0 = \theta/t$; $\omega_0 = 2\theta/t$ = 3.57 rad/s; (*d*)

13* • A circular space station of radius 5.10 km is a long way from any star. Its rotational speed is controllable to some degree, and so the apparent gravity changes according to the tastes of those who make the decisions. Dave the Earthling puts in a request for artificial gravity of 9.8 m/s^2 at the circumference. His secret agenda is to give the Earthlings a home-gravity advantage in the upcoming interstellar basketball tournament. Dave's request would require an angular speed of (*a*) 4.4×10^{-2} rad/s. (*b*) 7.0×10^{-3} rad/s. (*c*) 0.28 rad/s. (*d*) −0.22 rad/s.

(*e*) 1300 rad/s.

(*a*) Use $a_c = r\omega^2$ and solve for ω.

---

14 • A bicycle has wheels of 1.2 m diameter. The bicyclist accelerates from rest with constant acceleration to 24 km/h in 14.0 s. What is the angular acceleration of the wheels?

$a = a_t = r\alpha$; $\alpha = a/r$ $\alpha = (24/3.6 \times 14)/0.6$ rad/s^2 = 0.794 rad/s^2

---

15 •• The tape in a standard VHS videotape cassette has a length $L = 246$ m; the tape plays for 2.0 h (Figure 9-36). As the tape starts, the full reel has an outer radius of about $R = 45$ mm, and an inner radius of about $r = 12$ mm. At some point during the play, both reels have the same angular speed. Calculate this angular speed in rad/s and rev/min.

1. At the instant both reels have the same area, $2(R_f^2 - r^2) = R^2 - r^2$

   Solve for R_f $R_f = 32.9$ mm = 3.29 cm

1. Determine the linear speed v $v = 246/2$ m/h = 123 m/h = 3.42 cm/s

2. Find $\omega = v/r$ $\omega = 3.42/3.29$ rad/s = 1.04 rad/s = 9.93 rev/min

---

16 • The dimension of torque is the same as that of (*a*) impulse. (*b*) energy. (*c*) momentum. (*d*) none of the above.

(*b*)

17* • The moment of inertia of an object of mass M (*a*) is an intrinsic property of the object. (*b*) depends on the choice of axis of rotation, (*c*) Is proportional to M regardless of the choice of axis. (*d*) both (*b*) and (*c*) are correct.

(*d*)

---

18 • Can an object continue to rotate in the absence of torque?

Yes

---

19 • Does an applied net torque always increase the angular speed of an object?

No; it may cause a rotating object to come to rest.

---

20 • True or false: (*a*) If the angular velocity of an object is zero at some instant, the net torque on the object must be zero at that instant. (*b*) The moment of inertia of an object depends on the location of the axis of rotation. (*c*) The moment of inertia of an object depends on the angular velocity of the object.

(*a*) False (*b*) True (*c*) False

21* • A disk is free to rotate about an axis. A force applied a distance d from the axis causes an angular acceleration α. What angular acceleration is produced if the same force is applied a distance $2d$ from the axis? (*a*) α (*b*) 2α (*c*) $\alpha/2$ (*d*) 4α (*e*) $\alpha/4$

(*b*) $\alpha \propto \tau = F\ell$.

---

22 • A disk-shaped grindstone of mass 1.7 kg and radius 8 cm is spinning at 730 rev/min. After the power is shut off, a woman continues to sharpen her ax by holding it against the grindstone for 9 s until the grindstone stops rotating. (*a*) What is the angular acceleration of the grindstone? (*b*) What is the torque exerted by the ax on the grindstone? (Assume constant angular acceleration and a lack of other frictional torques.)

(*a*) $\alpha = \omega/t$ $\alpha = (730 \times 2\pi/60 \times 9)$ rad/s^2 = 8.49 rad/s^2

(*b*) $\tau = I\alpha$, $I = \frac{1}{2}MR^2$ $\tau = \frac{1}{2}(1.7 \times 0.08^2) \times 8.49$ N·m = 0.046 N·m

23 • A 2.5-kg cylinder of radius 11 cm is initially at rest. A rope of negligible mass is wrapped around it and pulled with a force of 17 N. Find (*a*) the torque exerted by the rope, (*b*) the angular acceleration of the cylinder, and (*c*) the angular velocity of the cylinder at $t = 5$ s.

(*a*) $\tau = F\ell$ — $\tau = 17 \times 0.11$ N·m $= 1.87$ N·m

(*b*) $\alpha = \tau/I$; $I = \frac{1}{2}MR^2$ — $\alpha = 1.87/(\frac{1}{2} \times 2.5 \times 0.11^2)$ rad/s² $= 124$ rad/s²

(*c*) $\omega = \alpha t$ — $\omega = 124 \times 5$ rad/s $= 620$ rad/s

24 •• A wheel mounted on an axis that is not frictionless is initially at rest. A constant external torque of 50 N·m is applied to the wheel for 20 s, giving the wheel an angular velocity of 600 rev/min. The external torque is then removed, and the wheel comes to rest 120 s later. Find (*a*) the moment of inertia of the wheel, and (*b*) the frictional torque, which is assumed to be constant.

(*a*) $\alpha = \omega/t = \tau/I$; $I = \tau t/\omega$ — $I = (50 \times 20)/(600 \times 2\pi/60)$ kg·m² $= 15.9$ kg·m²

(*b*) $\tau_{fr} = \tau/6$ — $\tau_{fr} = 2.65$ N·m

25* •• A pendulum consisting of a string of length L attached to a bob of mass m swings in a vertical plane. When the string is at an angle θ to the vertical, (*a*) what is the tangential component of acceleration of the bob? (*b*) What is the torque exerted about the pivot point? (*c*) Show that $\tau = I\alpha$ with $a_t = L\alpha$ gives the same tangential acceleration as found in part (*a*).

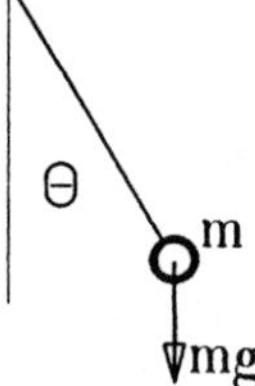

(*a*) The pendulum and the forces acting on it are shown. The tangential force is $mg \sin \theta$. Therefore, the tangential acceleration is $a_t = g \sin \theta$.

(*b*) The tension causes no torque. The torque due to the weight about the pivot is $mgL \sin \theta$.

(*c*) Here $I = mL^2$; so $\alpha = mgL \sin \theta/mL^2 = g \sin \theta/L$, and $a_t = g \sin \theta$.

26 ••• A uniform rod of mass M and length L is pivoted at one end and hangs as in Figure 9-37 so that it is free to rotate without friction about its pivot. It is struck by a horizontal force F_0 for a short time Δt at a distance x below the pivot as shown. (*a*) Show that the speed of the center of mass of the rod just after being struck is given by $v_0 = 3F_0x\Delta t/2ML$. (*b*) Find the force delivered by the pivot, and show that this force is zero if $x = 2L/3$. (*Note:* The point $x = 2L/3$ is called the *center of percussion* of the rod.)

(*a*) The torque due to F_0 is $F_0x = I\alpha = (ML^2/3)\alpha$, thus, $\alpha = 3F_0x/ML^2$, and $\omega = \alpha\Delta t = 3F_0x\Delta t/ML^2$. The center of mass is a distance $L/2$ from the pivot, so $v_{cm} = \omega L/2 = 3F_0x\Delta t/2ML$.

(*b*) Let P_p be the impulse exerted by the pivot on the rod. Then $P_p + F_0\Delta t = Mv_{cm}$ and $P_p = Mv_{cm} - F_0\Delta t$. Using the result from part (*a*) one finds that $P_p = F_0\Delta t(3x/2L - 1)$ and $F_p = F_0(3x/2L - 1)$. If $x = 2L/3$, $F_p = 0$.

27 ••• A uniform horizontal disk of mass M and radius R is rotating about its vertical axis with an angular velocity ω. When it is placed on a horizontal surface, the coefficient of kinetic friction between the disk and surface is μ_k. (*a*) Find the torque $d\tau$ exerted by the force of friction on a circular element of radius r and width dr. (*b*) Find the total torque exerted by friction on the disk. (*c*) Find the time required to bring the disk to a halt.

(*a*) The force of friction, df_k, is the product of μ_k and $g\, dm$, where $dm = 2\pi r\sigma\, dr$, and σ is the mass per unit area. Here $\sigma = M/\pi R^2$. The torque exerted by the friction force df_k is $r df_k$. Combining these quantities we find that: $d\tau = 2\pi\mu_k\sigma g r^2\, dr = 2(M/R^2)\mu_k g r^2\, dr$.

(*b*) To obtain the total torque we have to integrate $d\tau$: $\tau = 2(M/R^2)\mu_k g \int_0^R r^2 dr = (2/3)MR\mu_k g$.

(*c*) $t = \omega/\alpha$, and $\alpha = \tau/I$, where $I = \frac{1}{2}MR^2$. So $t = 3R\omega/4\mu_k g$.

28 • The moment of inertia of an object about an axis that does not pass through its center of mass is ____ the moment of inertia about a parallel axis through its center of mass. (*a*) always less than (*b*) sometimes less than (*c*) sometimes equal to (*d*) always greater than

(*d*)

29* • A tennis ball has a mass of 57 g and a diameter of 7 cm. Find the moment of inertia about its diameter. Assume that the ball is a thin spherical shell.

$I = (2/3)MR^2$ (see Table 9-1) $I = (2/3) \times 0.057 \times 0.035^2$ kg·m² $= 4.66 \times 10^{-5}$ kg·m²

30 • Four particles at the corners of a square with side length $L = 2$ m are connected by massless rods (Figure 9-38). The masses of the particles are $m_1 = m_4 = 3$ kg and $m_2 = m_3 = 4$ kg. Find the moment of inertia of the system about the z axis.

Use Equ. 9-17 $I = [2 \times 3 \times 2^2 + 4 \times (2\sqrt{2})^2]$ kg·m² $= 56$ kg·m²

31 • Use the parallel-axis theorem and your results for Problem 30 to find the moment of inertia of the four-particle system in Figure 9-38 about an axis that is perpendicular to the plane of the masses and passes through the center of mass of the system. Check your result by direct computation.

1. Distance to center of mass = $\sqrt{2}$ m; $M = 14$ kg; by parallel axis theorem $I_{cm} = (56 - 2 \times 14)$ kg·m² = 28 kg·m².
2. By direct computation: $I_{cm} = (4+4+3+3) \times (\sqrt{2})^2$ kg·m² $= 2 \times 14$ kg·m² = 28 kg·m².

32 • For the four-particle system of Figure 9-38, (*a*) find the moment of inertia I_x about the x axis, which passes through m_3 and m_4, and (*b*) Find I_y about the y axis, which passes through m_1 and m_4.

(*a*) $I_x = (3 \times 2^2 + 4 \times 2^2)$ kg·m² = 28 kg·m². (*b*) By symmetry, $I_y = I_x = 28$ kg·m².

33* • Use the parallel-axis theorem to find the moment of inertia of a solid sphere of mass M and radius R about an axis that is tangent to the sphere (Figure 9-39).

$I_{cm} = (2/5)MR^2$ (see Table 9-1); use Equ. 9-21 $I = (2/5)MR^2 + MR^2 = (7/5)MR^2$

34 •• A 1.0-m-diameter wagon wheel consists of a thin rim having a mass of 8 kg and six spokes each having a mass of 1.2 kg. Determine the moment of inertia of the wagon wheel for rotation about its axis.

Use Table 9-1 for I_{rim} and I_{spoke} and add $I = [(8 \times 0.5^2) + (6 \times 1.2 \times 0.5^2/3)]$ kg·m² = 2.6 kg·m²

35 •• Two point masses m_1 and m_2 are separated by a massless rod of length L. (*a*) Write an expression for the moment of inertia about an axis perpendicular to the rod and passing through it at a distance x from mass m_1. (*b*) Calculate dI/dx and show that I is at a minimum when the axis passes through the center of mass of the system.

(*a*) $I = m_1x^2 + m_2(L - x)^2$.

(*b*) $dI/dx = 2m_1x + 2m_2(L - x)(-1) = 2(m_1x + m_2x - m_2L)$; $dI/dx = 0$ when $x = m_2L/(m_1+m_2)$. This is, by definition, the distance of the center of mass from m_1.

36 •• A uniform rectangular plate has mass m and sides of lengths a and b. (*a*) Show by integration that the moment of inertia of the plate about an axis that is perpendicular to the plate and passes through one corner is $m(a^2 + b^2)/3$. (*b*) What is the moment of inertia about an axis that is perpendicular to the plate and passes through its center of mass?

(*a*) The element of mass is $\sigma\,dxdy$, where $\sigma = m/ab$. The distance of the element dm from the corner, which we designate as our origin, is given by $r^2 = x^2 + y^2$.

$$I = \sigma\int_0^a\int_0^b (x^2 + y^2)\,dxdy = \frac{1}{3}\sigma(a^3b + ab^3) = \frac{1}{3}m(a^2 + b^2).$$

(*b*) The distance from the origin to the center of mass is $d = [(\frac{1}{2}a)^2 + (\frac{1}{2}b)^2]^{1/2}$. Using Equ. 9-21 one obtains: $I_{cm} = (1/3)m(a^2 + b^2) - (1/4)m(a^2 + b^2) = (1/12)m(a^2 + b^2)$.

37* •• Tracey and Corey are doing intensive research on theoretical baton-twirling. Each is using "The Beast" as a model baton: two uniform spheres, each of mass 500 g and radius 5 cm, mounted at the ends of a 30-cm uniform rod of mass 60 g (Figure 9-40). Tracey and Corey want to calculate the moment of inertia of The Beast about an axis perpendicular to the rod and passing through its center. Corey uses the approximation that the two spheres can be treated as point particles that are 20 cm from the axis of rotation, and that the mass of the rod is negligible. Tracey, however, makes her calculations without approximations. (*a*) Compare the two results. (*b*) If the spheres retained the same mass but were hollow, would the rotational inertia increase or decrease? Justify your choice with a sentence or two. It is not necessary to calculate the new value of *I*.

(*a*) 1. Use point mass approximation for I_{app} — $I_{app} = (2 \times 0.5 \times 0.2^2)\ \text{kg·m}^2 = 0.04\ \text{kg·m}^2$

2. Use Table 9-1 and Equ. 9-21 to find *I* — $I = [2(2/5)(0.5\times0.05^2) + I_{app} + (1/12)(0.06\times0.3^2)]\ \text{kg·m}^2 = 0.04145\ \text{kg·m}^2$; $I_{app}/I = 0.965$

(*b*) The rotational inertia would increase because I_{cm} of a hollow sphere > I_{cm} of a solid sphere.

---

38 •• The methane molecule (CH_4) has four hydrogen atoms located at the vertices of a regular tetrahedron of side length 1.4 nm, with the carbon atom at the center of the tetrahedron (Figure 9-41). Find the moment of inertia of this molecule for rotation about an axis that passes through the carbon atom and one of the hydrogen atoms.

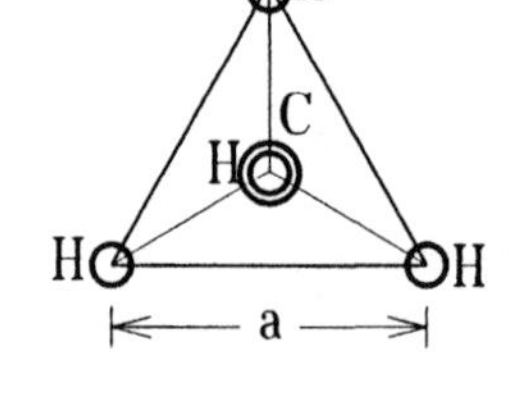

1. The axis of rotation passes through the center of the base of the tetrahedron. The carbon atom and the hydrogen atom at the apex of the tetrahedron do not contribute to *I* because the distance of their nuclei from the axis of rotation is zero.

2. From the geometry, the distance of the three H nuclei from the rotation axis is $a/\sqrt{3}$, where *a* is the side length of the tetrahedron.

3. Apply Equ. 9-17 with $m = 1.67 \times 10^{-27}$ kg — $I = 3m(a/\sqrt{3})^2 = ma^2 = 3.27 \times 10^{-45}\ \text{kg·m}^2$

---

39 ••• A hollow cylinder has mass *m*, an outside radius R_2, and an inside radius R_1. Show that its moment of inertia about its symmetry axis is given by $I = \frac{1}{2}m(R_2^2 + R_1^2)$.

Let the element of mass be $dm = \rho\, dV = 2\pi\rho h r\, dr$, where *h* is the height of the cylinder. The mass *m* of the hollow cylinder is $m = \pi\rho h(R_2^2 - R_1^2)$, so $\rho = m/[\pi h(R_2^2 - R_1^2)]$. The element $dI = r^2\, dm = 2\pi\rho h r^3\, dr$. Integrate *dI* from R_1 to R_2 and obtain $I = \frac{1}{2}\pi\rho h(R_2^4 - R_1^4) = \frac{1}{2}\pi\rho h(R_2^2 + R_1^2)(R_2^2 - R_1^2) = \frac{1}{2}m(R_2^2 + R_1^2)$.

---

40 ••• Show that the moment of inertia of a spherical shell of radius *R* and mass *m* is $2mR^2/3$. This can be done by direct integration or, more easily, by finding the increase in the moment of inertia of a solid sphere when its radius changes. To do this, first show that the moment of inertia of a solid sphere of density ρ is $I = (8/15)\pi\rho R^5$. Then compute the change *dI* in *I* for a change *dR*, and use the fact that the mass of this shell is $dm = 4\pi R^2 \rho\, dR$.

From Table 9-1, $I = (2/5)mR^2$, and $m = (4/3)\pi\rho R^3$. So $I = (8/15)\pi\rho R^5$. Then, $dI = (8/3)\pi\rho R^4\, dR$. We can express this in terms of the mass increase $dm = 4\pi\rho R^2\, dR$: $dI = (2/3)R^2\, dm$. Therefore, the moment of inertia of the spherical shell of mass *m* is $(2/3)mR^2$.

41* ••• The density of the earth is not quite uniform. It varies with the distance r from the center of the earth as $\rho = C(1.22 - r/R)$, where R is the radius of the earth and C is a constant. (*a*) Find C in terms of the total mass M and the radius R. (*b*) Find the moment of inertia of the earth. (See Problem 40.)

(*a*) $M = \int_0^R dm = \int_0^R 4\pi\rho r^2 dr = 4\pi C\int_0^R 1.22r^2 dr - \frac{4\pi C}{R}\int_0^R r^3 dr = \frac{4\pi}{3}1.22CR^3 - \pi CR^3.\ C = 0.508M/R^3.$

(*b*) $I = \int dI = \frac{8\pi}{3}\int_0^R \rho r^4 dr = \frac{8\pi\times 0.508M}{3R^3}\left[\int_0^R 1.22r^4 dr - \frac{1}{R}\int_0^R r^5 dr\right] = \frac{4.26M}{R^3}\left[\frac{1.22}{5}R^5 - \frac{1}{6}R^5\right] = 0.329MR^2.$

42 ••• Use integration to determine the moment of inertia of a right circular homogeneous cone of height H, base radius R, and mass density ρ about its symmetry axis.

We take our origin at the apex of the cone, with the z axis along the cone's symmetry axis. Then the radius, a distance z from the apex is $r = zR/H$. Consider a disk at z of thickness dz. Its mass is $\pi\rho r^2\,dz$ and its mass is

$$M = \pi\rho\int_0^H r^2 dz = \pi\rho\int_0^H \frac{R^2}{H^2}z^2 dz = \frac{\pi\rho R^2 H}{3}.$$

Likewise,

$$I = ½\pi\rho\int_0^H r^4 dz = \frac{\pi\rho R^4}{2H^4}\int_0^H z^4 dz = \frac{\pi\rho R^4 H}{10} = \frac{3}{10}MR^2.$$

43 ••• Use integration to determine the moment of inertia of a hollow, thin-walled, right circular cone of mass M, height H, and base radius R about its symmetry axis.

Use the same coordinates as in Problem 9-42. The element of length along the cone is $[(H^2+R^2)^{½}/H]\,dz$, so

$$M = \frac{2\pi\sigma R\sqrt{H^2+R^2}}{H^2}\int_0^H z\,dz = \pi\sigma R\sqrt{H^2+R^2};$$

likewise,

$$I = \frac{2\pi\sigma R^3\sqrt{H^2+R^2}}{H^4}\int_0^H z^3 dz = ½\pi\sigma R^2\sqrt{H^2+R^2}.\ \text{Thus } I = ½MR^2.$$

44 ••• Use integration to determine the moment of inertia of a thin uniform disk of mass M and radius R for rotation about a diameter. Check your answer by referring to Table 9-1.

The element of mass, dm is $2\sigma\sqrt{R^2-z^2}\,dz$ (See the Figure)

The moment of inertia about the diameter is then

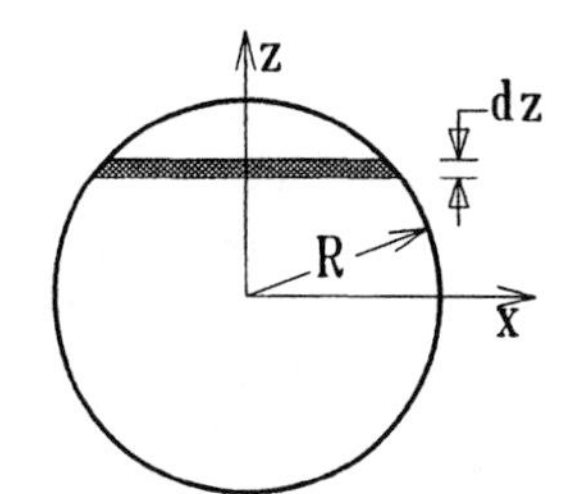

$$I = 2\sigma\int_{-R}^{R} z^2\sqrt{R^2-z^2}\,dz = \frac{\sigma\pi R^4}{4} = \frac{1}{4}MR^2$$

in agreement with the expression given in Table 9-1 for a cylinder of length $L = 0$.

45* ••• Use integration to determine the moment of inertia of a thin circular hoop of radius R and mass M for rotation about a diameter. Check your answer by referring to Table 9-1.

Here, $dm = \lambda R\, d\theta$, and $dI = z^2\, dm$, where $z = R \sin\theta$. Thus, $I = \lambda R^3 \int_{-\pi}^{\pi} \sin^2\theta\, d\theta = \lambda\pi R^3 = ½MR^2$, in agreement with Table 9-1 for a hollow cylinder of length $L = 0$.

---

46 ••• A roadside ice-cream stand uses rotating cones to catch the eyes of travelers. Each cone rotates about an axis perpendicular to its axis of symmetry and passing through its apex. The sizes of the cones vary, and the owner wonders if it would be more energy-efficient to use several smaller cones or a few big ones. To answer this, he must calculate the moment of inertia of a homogeneous right circular cone of height H, base radius R, and mass density ρ. What is the result?

The element of mass is, as in Problem 9-42, $dm = \pi\rho r^2\, dz$. Each elemental disk rotates about an axis that is parallel to its diameter but removed from it by a distance z. We can now use the result of Problem 9-44 and the parallel axis theorem to obtain the expression for the element dI; as before, $r = Rz/H$.

$$dI = \pi\rho\left[\frac{1}{4}\left(\frac{R^2z^2}{H^2}\right)^2 + \frac{R^2z^4}{H^2}\right]dz.$$

Integrate from $z = 0$ to $z = H$ and use the result $M = \pi\rho R^2H/3$: $I = 3M(H^2/5 + R^2/20)$.

---

47 • A constant torque acts on a merry-go-round. The power input of the torque is (*a*) constant. (*b*) proportional to the angular speed of the merry-go-round. (*c*) zero. (*d*) none of the above.

(*b*)

---

48 • The particles in Figure 9-42 are connected by a very light rod whose moment of inertia can be neglected. They rotate about the y axis with angular velocity $\omega = 2$ rad/s. (*a*) Find the speed of each particle, and use it to calculate the kinetic energy of this system directly from $\Sigma ½ m_i v_i^2$. (*b*) Find the moment of inertia about the y axis, and calculate the kinetic energy from $K = ½I\omega^2$.

(*a*) 1. Use $v = r\omega$	$v_3 = (0.2 \times 2)$ m/s $= 0.4$ m/s; $v_1 = (0.4 \times 2)$ m/s $= 0.8$ m/s
2. Find K	$K = (2 \times ½ \times 3 \times 0.4^2 + 2 \times ½ \times 1 \times 0.8^2)$ J $= 1.12$ J
(*b*) 1. Find I using Equ. 9-2	$I = (2 \times 3 \times 0.2^2 + 2 \times 1 \times 0.4^2)$ kg·m^2 $= 0.56$ kg·m^2
2. Find $K = ½I\omega^2$	$K = ½ \times 0.56 \times 2^2$ J $= 1.12$ J

49* • Four 2-kg particles are located at the corners of a rectangle of sides 3 m and 2 m as shown in Figure 9-43. (*a*) Find the moment of inertia of this system about the z axis. (*b*) The system is set rotating about this axis with a kinetic energy of 124 J. Find the number of revolutions the system makes per minute.

(*a*) Use Equ. 9-2	$I = 2[2^2 + 3^2 + (2^2 + 3^2)]$ kg·m^2 $= 52$ kg·m^2
(*b*) Find $\omega = (2K/I)^{½}$	$\omega = (2 \times 124/52)^{½}$ rad/s $= 2.18$ rad/s $= 20.9$ rev/min

---

50 • A solid ball of mass 1.4 kg and diameter 15 cm is rotating about its diameter at 70 rev/min. (*a*) What is its kinetic energy? (*b*) If an additional 2 J of energy are supplied to the rotational energy, what is the new angular speed of the ball?

(*a*) $I = (2/5)MR^2$; $K = ½I\omega^2 = MR^2\omega^2/5$	$K = (1.4 \times 0.075^2 \times 7.33^2/5)$ J $= 0.0846$ J
(*b*) $K = 2.0846$; $\omega \propto K^{½}$	$\omega = [70 \times (2.0846/0.0846)^{½}]$ rev/min $= 347$ rev/min

---

51 • An engine develops 400 N·m of torque at 3700 rev/min. Find the power developed by the engine.

Use Equ. 9-27	$P = (400 \times 3700 \times 2\pi/60)$ W $= 155$ kW

52 •• Two point masses m_1 and m_2 are connected by a massless rod of length L to form a dumbbell that rotates about its center of mass with angular velocity ω. Show that the ratio of kinetic energies of the masses is $K_1/K_2 = m_2/m_1$.

Let r_1 and r_2 be the distances of m_1 and m_2 from the center of mass. Then, by definition, $r_1m_1 = r_2m_2$. Since $K \propto mr^2\omega^2$, $K_1/K_2 = m_1r_1^2/m_2r_2^2 = m_2/m_1$.

53* •• Calculate the kinetic energy of rotation of the earth, and compare it with the kinetic energy of motion of the earth's center of mass about the sun. Assume the earth to be a homogeneous sphere of mass 6.0×10^{24} kg and radius 6.4×10^6 m. The radius of the earth's orbit is 1.5×10^{11} m.

1. Find K_{rot}; use result of Problem 9-11 and Table 9-1 — $K_{rot} = (\frac{1}{2} \times 0.4 \times 6 \times 10^{24} \times 6.4^2 \times 10^{12} \times 7.27^2 \times 10^{-10})$ J $= 2.6 \times 10^{29}$ J
2. Find K_{orb}; $I = M_ER_{orb}^2$; $\omega_{orb} = 2\pi/3.156 \times 10^7$ rad/s — $K_{orb} = (\frac{1}{2} \times 6 \times 10^{24} \times 1.5^2 \times 10^{22} \times 2^2 \times 10^{-14})$ J $= 2.7 \times 10^{33}$ J

$K_{orb} \approx 10^4K_{rot}$

54 •• A 2000-kg block is lifted at a constant speed of 8 cm/s by a steel cable that passes over a massless pulley to a motor-driven winch (Figure 9-44). The radius of the winch drum is 30 cm. (*a*) What force must be exerted by the cable? (*b*) What torque does the cable exert on the winch drum? (*c*) What is the angular velocity of the winch drum? (*d*) What power must be developed by the motor to drive the winch drum?

(*a*) $T = mg$ — $T = (2000 \times 9.81)$ N $= 19.62$ kN

(*b*) $\tau = Tr$ — $\tau = (19.62 \times 0.3)$ kN·m $= 5.89$ kN·m

(*c*) $\omega = v/r$ — $\omega = (0.08/0.3)$ rad/s $= 0.267$ rad/s

(*d*) $P = Fv = Tv$ — $P = (19620 \times 0.08)$ W $= 1.57$ kW

55 •• A uniform disk of mass M and radius R is pivoted such that it can rotate freely about a horizontal axis through its center and perpendicular to the plane of the disk. A small particle of mass m is attached to the rim of the disk at the top, directly above the pivot. The system is given a gentle start, and the disk begins to rotate. (*a*) What is the angular velocity of the disk when the particle is at its lowest point? (*b*) At this point, what force must be exerted on the particle by the disk to keep it on the disk?

(*a*) Use energy conservation for ω; $I = \frac{1}{2}MR^2 + mR^2$ — $2mgR = \frac{1}{2}[\frac{1}{2}MR^2+mR^2]\omega^2$; $\omega = \sqrt{\dfrac{8mg/R}{2m+M}}$

(*b*) $F = mg + mR\omega^2$ — $F = mg\left(1 + \dfrac{8m}{2m+M}\right)$

56 •• A ring 1.5 m in diameter is pivoted at one point on its circumference so that it is free to rotate about a horizontal axis. Initially, the line joining the support and center is horizontal. (*a*) If released from rest, what is its maximum angular velocity? (*b*) What must its initial angular velocity be if it is to just make a complete revolution?

(*a*) Apply energy conservation — $mgR = \frac{1}{2}I\omega^2$; $I = 2mR^2$;

Solve for ω; $R = 0.75$ m — $\omega = \sqrt{g/R} = 3.62$ rad/s

(*b*) Now CM must rise a height R — $\frac{1}{2}I\omega_i^2 = mgR$; $\omega_i = 3.62$ rad/s

57* •• You set out to design a car that uses the energy stored in a flywheel consisting of a uniform 100-kg cylinder of radius R. The flywheel must deliver an average of 2 MJ of mechanical energy per kilometer, with a maximum angular velocity of 400 rev/s. Find the least value of R such that the car can travel 300 km without the flywheel having to be recharged.

1. Find total energy — $K = (2 \times 10^6 \times 300)\text{ J} = 6 \times 10^8\text{ J} = ½ \times 50 \times R^2 \times \omega^2$
2. Solve for R with $\omega = 800\pi$ rad/s — $R = \sqrt{24\times10^6/(800\pi)^2}\text{ m} = 1.95\text{ m}$

58 •• A ladder that is 8.6 m long and has mass 60 kg is placed in a nearly vertical position against the wall of a building. You stand on a rung with your center of mass at the top of the ladder. Assume that your mass is 80 kg. As you lean back slightly, the ladder begins to rotate about its base away from the wall. Is it better to quickly step off the ladder and drop to the ground or to hold onto the ladder and step off just before the top end hits the ground?

We shall solve this problem for the general case of a ladder of length L, mass M, and person of mass m. If the person falls off the ladder at the top, the speed with which he strikes the ground is given by $v_f^2 = 2gL$. Now consider what happens if the person holds on and rotates with the ladder. We shall use conservation of energy. This gives $(m + M/2)gL = ½(m + M/3)L^2\omega^2 = ½(m + M/3)v_r^2$. We find that the ratio $v_r^2/v_f^2 = (m+M/2)/(m+M/3)$. Evidently, unless M, the mass of the ladder, is zero, $v_r > v_f$. It is therefore better to let go and fall to the ground.

59 ••• Consider the situation in Problem 58 with a ladder of length L and mass M. Find the ratio of your speed as you hit the ground if you hang on to the ladder to your speed if you immediately step off as a function of the mass ratio M/m, where m is your mass.

See Problem 9-58. We obtain $\dfrac{v_r}{v_f} = \sqrt{\dfrac{1 + M/2m}{1 + M/3m}}$, where v_r is the speed for hanging on, v_f for stepping off the ladder.

60 •• A 4-kg block resting on a frictionless horizontal ledge is attached to a string that passes over a pulley and is attached to a hanging 2-kg block (Figure 9-45). The pulley is a uniform disk of radius 8 cm and mass 0.6 kg. (*a*) Find the speed of the 2-kg block after it falls from rest a distance of 2.5 m. (*b*) What is the angular velocity of the pulley at this time?

(*a*) Use energy conservation — $mgh = ½(M+m)v^2 + ½I\omega^2 = ½(M+m+I/R^2)v^2$

Solve for and evaluate v; $m = 2$ kg, $M = 4$ kg, $h = 2.5$ m, and $I/R^2 = ½M_p = 0.3$ kg — $v = \sqrt{\dfrac{2mgh}{M + m + ½M_p}} = 3.95\text{ m/s}$

(*b*) $\omega = v/R$, where $R = 0.08$ m — $\omega = (3.95/0.08)\text{ rad/s} = 49.3\text{ rad/s}$

61* •• For the system in Problem 60, find the linear acceleration of each block and the tension in the string.

1. Write the equations of motion for the three objects — $4a = T_1$; $2a = 2g - T_2$; $0.08(T_2 - T_1) = ½ \times 0.6 \times 0.08^2\alpha$
2. Use $\alpha = a/r$ and solve for a — $T_2 - T_1 = 2g - 6a = 0.3a$; $a = 2g/6.3 = 3.11\text{ m/s}^2$
3. Find T_1 (acting on 4 kg) and T_2 (acting on 2 kg). — $T_1 = 12.44$ N; $T_2 = T_1 + 0.3a = 13.37$ N

62 •• Work Problem 60 for the case in which the coefficient of friction between the ledge and the 4-kg block is 0.25.

(*a*) Use energy conservation; see Problem 9-60 — $mgh = ½(M+m+I/R^2)v^2 + \mu_k Mgh$

Solve for and evaluate v for $m = 2$ kg, $M = 4$ kg, $h = 2.5$ m, $M_p = 0.6$ kg, $\mu_k = 0.25$ $\quad$ $v = \sqrt{\dfrac{2h(mg - \mu_k Mg)}{M + m + \frac{1}{2}M_p}} = 2.79$ m/s

(*b*) $\omega = v/R$, where $R = 0.08$ m $\quad$ $\omega = 34.6$ rad/s

---

63 •• Work Problem 61 for the case in which the coefficient of friction between the ledge and the 4-kg block is 0.25.

1. Now $4a = T_1 - 0.25 \times 4g$; also see Prob. 9-61 $\quad$ $4a = T_1 - g$; $2a = 2g - T_2$; $T_2 - T_1 = 0.3a$
2. Solve for a $\quad$ $a = g/6.3 = 1.56$ m/s^2
3. Find T_1 and T_2 $\quad$ $T_1 = 16.0$ N; $T_2 = 16.5$ N

---

64 •• In 1993, a giant yo-yo of mass 400 kg and measuring about 1.5 m in radius was dropped from a crane 57 m high. Assuming the axle of the yo-yo had a radius of $r = 0.1$ m, find the velocity of the descent v at the end of the fall.

1. Write the equations of motion $\quad$ $ma = mg - T$; $a = r\alpha = r\tau/I = r^2T/\frac{1}{2}mR^2$; $T = mR^2a/2r^2$
2. Solve for and evaluate a; $m = 400$ kg, $R = 1.5$ m $\quad$ $a = g/(1 + R^2/2r^2) = 0.0872$ m/s^2
3. Use $v = (2as)^{1/2}$ $\quad$ $v = (2 \times 0.0872 \times 57)^{1/2}$ m/s $= 3.15$ m/s

---

65* •• A 1200-kg car is being unloaded by a winch. At the moment shown in Figure 9-46, the gearbox shaft of the winch breaks, and the car falls from rest. During the car's fall, there is no slipping between the (massless) rope, the pulley, and the winch drum. The moment of inertia of the winch drum is 320 kg·m^2 and that of the pulley is 4 kg·m^2. The radius of the winch drum is 0.80 m and that of the pulley is 0.30 m. Find the speed of the car as it hits the water.

1. Use energy conservation and $\omega = v/r$ $\quad$ $mgh = \frac{1}{2}mv^2 + \frac{1}{2}I_w\omega_w^2 + \frac{1}{2}I_p\omega_p^2 = \frac{1}{2}v^2(m + I_w/r_w^2 + I_p/r_p^2)$
2. Solve for and evaluate v $\quad$ $v = [2mgh/(m + I_w/r_w^2 + I_p/r_p^2)]^{1/2} = 8.2$ m/s

---

66 •• The system in Figure 9-47 is released from rest. The 30-kg block is 2 m above the ledge. The pulley is a uniform disk with a radius of 10 cm and mass of 5 kg. Find (*a*) the speed of the 30-kg block just before it hits the ledge, (*b*) the angular speed of the pulley at that time, (*c*) the tensions in the strings, and (*d*) the time it takes for the 30-kg block to reach the ledge. Assume that the string does not slip on the pulley.

(*a*) 1. m_1=20 kg, m_2=30 kg; use energy conservation $\quad$ $m_2gh = m_1gh + \frac{1}{2}(m_1v^2 + m_2v^2 + I\omega^2)$

2. $I = \frac{1}{2}mr^2$; $\omega^2 = v^2/r^2$; so $I\omega^2 = \frac{1}{2}mv^2$; $m = 5$ kg $\quad$ $v = [2gh(m_2 - m_1)/(m_1 + m_2 + \frac{1}{2}m)]^{1/2} = 2.73$ m/s

(*b*) Use $\omega = v/r$ $\quad$ $\omega = (2.73/0.1)$ rad/s $= 27.3$ rad/s

(*c*) 1. Find acceleration; $a = v^2/2h$ $\quad$ $a = 1.87$ m/s^2

2. $T_1 = m_1(g + a)$; $T_2 = m_2(g - a)$ $\quad$ $T_1 = 234$ N; $T_2 = 238$ N

(*d*) Use $t = h/v_{av} = 2h/v$ $\quad$ $t = (4/2.73)$ s $= 1.47$ s

---

67 •• A uniform sphere of mass M and radius R is free to rotate about a horizontal axis through its center. A string is wrapped around the sphere and is attached to an object of mass m as shown in Figure 9-48. Find (*a*) the acceleration of the object, and (*b*) the tension in the string.

(*a*) The equations of motion for the two objects are $mg - T = ma$ and $I\alpha = \tau$. Now $\tau = RT$, $I = (2/5)MR^2$, and $\alpha = a/R$. Thus, $T = (2/5)Ma$ and $a = g/[1 + (2M/5m)]$.

(*b*) As obtained in (*a*), $T = (2/5)Ma = 2mMg/(5m+2M)$.

68 •• An Atwood's machine has two objects of mass $m_1 = 500$ g and $m_2 = 510$ g, connected by a string of negligible mass that passes over a frictionless pulley (Figure 9-49). The pulley is a uniform disk with a mass of 50 g and a radius of 4 cm. The string does not slip on the pulley. (*a*) Find the acceleration of the objects. (*b*) What is the tension in the string supporting m_1? In the string supporting m_2? By how much do they differ? (*c*) What would your answers have been if you had neglected the mass of the pulley?

Note that this problem is identical to Problem 9-66. We use the result for v^2 and $a = v^2/2h$.

(*a*) $a = (m_2 - m_1)g/(m_1 + m_2 + ½m)$; use the given values — $a = (10/1035)g = 9.478$ cm/s^2

(*b*) $T_1 = m_1(a + g)$; $T_2 = m_2(g - a)$ — $T_1 = 4.9524$ N; $T_2 = 4.9548$ N; $\Delta T = 0.0024$ N

(*c*) If $m = 0$; $a = (m_2 - m_1)g/(m_1 + m_2)$; $T_1 = T_2$ — $a = 9.713$ cm/s^2; $T = 4.9536$ N; $\Delta T = 0$

69* •• Two objects are attached to ropes that are attached to wheels on a common axle as shown in Figure 9-50. The total moment of inertia of the two wheels is 40 kg·m^2. The radii of the wheels are $R_1 = 1.2$ m and $R_2 = 0.4$ m. (*a*) If $m_1 = 24$ kg, find m_2 such that there is no angular acceleration of the wheels. (*b*) If 12 kg is gently added to the top of m_1, find the angular acceleration of the wheels and the tensions in the ropes.

(*a*) Find τ_{net} and set equal to 0 — $\tau = m_1gR_1 - m_2gR_2 = 0$; $m_2 = m_1R_1/R_2 = 72$ kg

(*b*) 1. Write the equations of motion — $T_1 = m_1(g - R_1\alpha)$; $T_2 = m_2(g + R_2\alpha)$; $\alpha = (T_1R_1 - T_2R_2)/I$

2. Solve for and find α with $m_1 = 36$ kg, $m_2 = 72$ kg — $\alpha = (m_1R_1 - m_2R_2)g/(m_1R_1^2 + m_2R_2^2 + I) = 1.37$ rad/s^2

3. Substitute $\alpha = 1.37$ rad/s^2 to find T_1 and T_2 — $T_1 = 294$ N; $T_2 = 745$ N

70 •• A uniform cylinder of mass M and radius R has a string wrapped around it. The string is held fixed, and the cylinder falls vertically as shown in Figure 9-51. (*a*) Show that the acceleration of the cylinder is downward with a magnitude $a = 2g/3$. (*b*) Find the tension in the string.

(*a*) The equation of motion is $\tau = I\alpha = RT = ½MR^2a/R$; $T = ½Ma$. But $Mg - T = Ma$. Thus, $a = (2/3)g$.

(*b*) $T = ½Ma = Mg/3$.

Note that we could have obtained the result also from Problem 9-64, setting $r = R$.

71 •• The cylinder in Figure 9-51 is held by a hand that is accelerated upward so that the center of mass of the cylinder does not move. Find (*a*) the tension in the string, (*b*) the angular acceleration of the cylinder, and (*c*) the acceleration of the hand.

(*a*) Since $a = 0$, $T = Mg$. (*b*) Use $\alpha = RT/I = RMg/½MR^2 = 2g/R$. (*c*) $a = R\alpha = 2g$.

72 •• A 0.1-kg yo-yo consists of two solid disks of radius 10 cm joined together by a massless rod of radius 1 cm and a string wrapped around the rod. One end of the string is held fixed and is under constant tension T as the yo-yo is released. Find the acceleration of the yo-yo and the tension T.

See Problem 9-64 — $a = g(1 + R^2/2r^2) = 0.192$ m/s^2; $T = m(g - a) = 0.902$ N

73* •• A uniform cylinder of mass m_1 and radius R is pivoted on frictionless bearings. A massless string wrapped around the cylinder connects to a mass m_2, which is on a frictionless incline of angle θ as shown in Figure 9-52. The system is released from rest with m_2 a height h above the bottom of the incline. (*a*) What is the acceleration of m_2? (*b*) What is the tension in the string? (*c*) What is the total energy of the system when m_2 is at height h?

(*d*) What is the total energy when m_2 is at the bottom of the incline and has a speed v? (*e*) What is the speed v? (*f*) Evaluate your answers for the extreme cases of $\theta = 0°$, $\theta = 90°$, and $m_1 = 0$.

(*a*) 1. Write the equations of motion — $m_2a = m_2g \sin\theta - T$; $\tau = RT = \frac{1}{2}m_1R^2\alpha$, $T = \frac{1}{2}m_1a$

2. Solve for a — $a = (g \sin\theta)/(1 + m_1/2m_2)$

(*b*) Solve for T — $T = (\frac{1}{2}m_1g \sin\theta)/(1 + m_1/2m_2)$

(*c*) Take $U = 0$ at $h = 0$ — $E = K + U = m_2gh$

(*d*) This is a conservative system — $E = m_2gh$

(*e*) $U = 0$; $E = K = \frac{1}{2}m_2v^2 + \frac{1}{2}I\omega^2$; $\omega = v/R$ — $m_2gh = \frac{1}{2}(m_2 + \frac{1}{2}m_1)v^2$; $v = \sqrt{(2gh)/(1 + m_1/2m_2)}$

(*f*) 1. For $\theta = 0$ — $a = T = 0$

2. For $\theta = 90°$ — $a = g/(1 + m_1/2m_2)$; $T = \frac{1}{2}m_1a$; $v = \sqrt{(2gh)/(1 + m_1/2m_2)}$

3. For $m_1 = 0$ — $a = g \sin\theta$, $T = 0$, $v = \sqrt{2gh}$

---

74 •• A device for measuring the moment of inertia of an object is shown in Figure 9-53. A circular platform has a concentric drum of radius 10 cm about which a string is wound. The string passes over a frictionless pulley to a weight of mass M. The weight is released from rest, and the time for it to drop a distance D is measured. The system is then rewound, the object placed on the platform, and the system again released from rest. The time required for the weight to drop the same distance D then provides the data needed to calculate I. With $M = 2.5$ kg, and $D = 1.8$ m, the time is 4.2 s. (*a*) Find the combined moment of inertia of the platform, drum, shaft, and pulley. (*b*) With the object placed on the platform, the time is 6.8 s for $D = 1.8$ m. Find I of that object about the axis of the platform.

Let r be the radius of the concentric drum (10 cm) and let I_0 be the moment of inertia of the drum plus platform.

(*a*) 1. Write the equations of motion, empty platform — $Ma = Mg - T$; $rT = I_0\alpha = I_0a/r$; $T = I_0a/r^2$

2. Solve for I_0 — $I_0 = Mr^2(g - a)/a$

3. Use $a = 2D/t^2$ and evaluate I_0 — $I_0 = 1.177\ \text{kg}\cdot\text{m}^2$

(*b*) Now $I_{tot} = I_0 + I$; $I_{tot} = Mr^2(g - a)/a$; $a = 2D/t^2$ — $I_{tot} = 3.125\ \text{kg}\cdot\text{m}^2$; $I = 1.948\ \text{kg}\cdot\text{m}^2$

---

75 • True or false: When an object rolls without slipping, friction does no work on the object.

True

---

76 • A wheel of radius R is rolling without slipping. The velocity of the point on the rim that is in contact with the surface, relative to the surface, is (*a*) equal to $R\omega$ in the direction of motion of the center of mass. (*b*) equal to $R\omega$ opposite the direction of motion of the center of mass. (*c*) zero. (*d*) equal to the velocity of the center of mass and in the same direction. (*d*) equal to the velocity of the center of mass but in the opposite direction.

(*c*)

77* •• A solid cylinder and a solid sphere have equal masses. Both roll without slipping on a horizontal surface. If their kinetic energies are the same, then (*a*) the translational speed of the cylinder is greater than that of the sphere. (*b*) the translational speed of the cylinder is less than that of the sphere. (*c*) the translational speeds of the two objects are the same. (*d*) (*a*), (*b*), or (*c*) could be correct depending on the radii of the objects.

$K_c = (3/4)mv_c^2$; $K_s = (7/10)mv_s^2$. If $K_c = K_s$, then $v_c < v_s$. (*b*)

---

78 •• Starting from rest at the same time, a coin and a ring roll down an incline without slipping. Which of the following is true? (*a*) The ring reaches the bottom first. (*b*) The coin reaches the bottom first. (*c*) The coin and ring arrive at the bottom simultaneously. (*d*) The race to the bottom depends on their relative masses. (*e*) The race to the bottom depends on their relative diameters.

$K_r = K_c$; $m_r v_r^2 = m_r gh$; $v_r^2 = gh$. For the coin, $v_c^2 = (4/3)gh$. $v_c > v_r$. (*b*)

---

79 •• For a hoop of mass M and radius R that is rolling without slipping, which is larger, its translational kinetic energy or its rotational kinetic energy? (*a*) Translational kinetic energy is larger. (*b*) Rotational kinetic energy is larger. (*c*) Both are the same size. (*d*) The answer depends on the radius. (*e*) The answer depends on the mass.

(*c*)

---

80 •• For a disk of mass M and radius R that is rolling without slipping, which is larger, its translational kinetic energy or its rotational kinetic energy? (*a*) Translational kinetic energy is larger. (*b*) Rotational kinetic energy is larger. (*c*) Both are the same size. (*d*) The answer depends on the radius. (*e*) The answer depends on the mass.

(*a*)

81* •• A ball rolls without slipping along a horizontal plane. Show that the frictional force acting on the ball must be zero. *Hint:* Consider a possible direction for the action of the frictional force and what effects such a force would have on the velocity of the center of mass and on the angular velocity.

Let us assume that $f \neq 0$ and acts along the direction of motion. Now consider the acceleration of the center of mass and the angular acceleration about the point of contact with the plane. Since $F_{net} \neq 0$, $a_{cm} \neq 0$. However, $\tau = 0$ since $\ell = 0$, so $\alpha = 0$. But $\alpha = 0$ is not consistent with $a_{cm} \neq 0$. Consequently, $f = 0$.

---

82 • A homogeneous solid cylinder rolls without slipping on a horizontal surface. The total kinetic energy is K. The kinetic energy due to rotation about its center of mass is (*a*) $\frac{1}{2}K$. (*b*) $\frac{1}{3}K$. (*c*) $\frac{4}{7}K$. (*d*) none of the above.

(*b*)

---

83 • A homogeneous cylinder of radius 18 cm and mass 60 kg is rolling without slipping along a horizontal floor at 5 m/s. How much work is needed to stop the cylinder?

$W = K = (3/4)mv^2$ $\quad$ $W = (0.75 \times 60 \times 5^2)$ J $= 1125$ J

---

84 • Find the percentages of the total kinetic energy associated with rotation and translation, respectively, for an object that is rolling without slipping if the object is (*a*) a uniform sphere, (*b*) a uniform cylinder, or (*c*) a hoop.

(*a*) For a sphere, $K_{tot} = 0.7mv^2$; $K_{trans} = 0.5mv^2$ $\quad$ $K_{trans} = 71.4\%\ K_{tot}$; $K_{rot} = 28.6\%\ K_{tot}$

(*b*) For a cylinder, $K_{tot} = 0.75mv^2$; $K_{trans} = 0.5mv^2$ $\quad$ $K_{trans} = 66.7\%\ K_{tot}$; $K_{rot} = 33.3\%\ K_{tot}$

(*c*) For a hoop, $K_{tot} = mv^2$ $\quad$ $K_{trans} = 50\%\ K_{tot}$; $K_{rot} = 50\%\ K_{tot}$

85* • A hoop of radius 0.40 m and mass 0.6 kg is rolling without slipping at a speed of 15 m/s toward an incline of slope 30°. How far up the incline will the hoop roll, assuming that it rolls without slipping?

1. Find the energy at the bottom of the slope $\quad$ $K = mv^2$
2. Use energy conservation; $mgL \sin 30° = K$ $\quad$ $L = 2v^2/g = 45.9$ m

---

86 • A ball rolls without slipping down an incline of angle θ. The coefficient of static friction is μ_s. Find (*a*) the acceleration of the ball, (*b*) the force of friction, and (*c*) the maximum angle of the incline for which the ball will roll without slipping.

We assume that the ball is a solid sphere.

The free-body diagram is shown. Note that both the force mg and the normal reaction force F_n act through the center of mass, so their torque about the center of mass is zero.

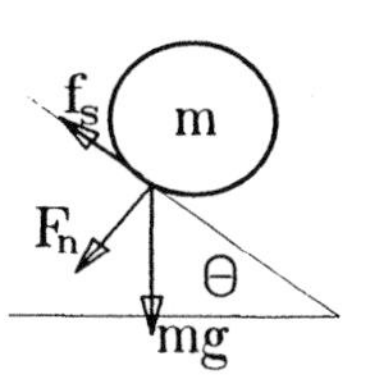

(*a*) 1. Write the equations of motion; use $\alpha = a/r$ — $ma = mg\sin\theta - f_s$; $\tau = f_s r = I\alpha = (2/5)mr^2$; $f_s = (2/5)ma$

2. Solve for a — $a = (5/7)g\sin\theta$

(*b*) Find f_s using the results in part (*a*) — $f_s = (2/7)mg\sin\theta$

(*c*) Use $f_{s,max} = \mu_s F_n = \mu_s mg\cos\theta$ — $\mu_s\cos\theta = (2/7)\sin\theta$, $\theta_{max} = \tan^{-1}(7\mu_s/2)$

---

87 •• An empty can of total mass $3M$ is rolling without slipping. If its mass is distributed as in Figure 9-54, what is the value of the ratio of kinetic energy of translation to the kinetic energy of rotation about its center of mass?

1. Find the total moment of inertia — $I = 2(\frac{1}{2}MR^2) + MR^2 = 2MR^2$
2. $K_{trans} = \frac{1}{2}(3Mv^2)$; $K_{rot} = \frac{1}{2}(2MR^2)v^2/R^2 = Mv^2$ — $K_{trans}/K_{rot} = 3/2$

---

88 •• A bicycle of mass 14 kg has 1.2-m diameter wheels, each of mass 3 kg. The mass of the rider is 38 kg. Estimate the fraction of the total kinetic energy of bicycle and rider associated with rotation of the wheels.

Assume the wheels are hoops, i.e., neglect the mass of the spokes. Then the total kinetic energy is $K = \frac{1}{2}Mv^2 + 2(\frac{1}{2}I_w\omega^2) = \frac{1}{2}Mv^2 + m_w v^2 = [\frac{1}{2}(52) + 3]v^2 = 29v^2$. $K_{rot} = 3v^2$. $K_{rot}/K = 3/29 \approx 10\%$.

89* •• A hollow sphere and uniform sphere of the same mass m and radius R roll down an inclined plane from the same height H without slipping (Figure 9-55). Each is moving horizontally as it leaves the ramp. When the spheres hit the ground, the range of the hollow sphere is L. Find the range L' of the uniform sphere.

1. Find v of each object as it leaves ramp. Use energy conservation. — $mgH = \frac{1}{2}mv_h^2 + \frac{1}{2}(2/3)mv_h^2$; $v_h^2 = 6gH/5$; $mgH = \frac{1}{2}mv_u^2 + \frac{1}{2}(2/5)mv_u^2$; $v_u^2 = 10gH/7$
2. Since distance $\propto v$, $L'/L = v_u/v_h$ — $L' = L(25/21)^{1/2} = 1.09L$

---

90 •• A hollow cylinder and a uniform cylinder are rolling horizontally without slipping. The speed of the hollow cylinder is v. The cylinders encounter an inclined plane that they climb without slipping. If the maximum height they reach is the same, find the initial speed v' of the uniform cylinder.

Since they climb the same height, $K_h = \frac{1}{2}m_h v^2 + \frac{1}{2}I_h\omega_h^2 = m_h v_h^2 = m_h gh = K_u = \frac{1}{2}m_u(v')^2 + \frac{1}{2}I_u\omega_u^2 = (3/4)m_u(v')^2 = m_u gh$. Consequently, $v' = \sqrt{4/3}\,v$.

---

91 •• A hollow, thin-walled cylinder and a solid sphere start from rest and roll without slipping down an inclined plane of length 3 m. The cylinder arrives at the bottom of the plane 2.4 s after the sphere. Determine the angle between the inclined plane and the horizontal.

1. Find a_c and a_s; see Problem 9-86. — $a_s = (5/7)g\sin\theta$, similarly, one obtains $a_c = \frac{1}{2}g\sin\theta$
2. Use $s = \frac{1}{2}at^2$ — $a_s t_s^2 = a_c t_c^2$; $t_c^2 = (t_s + 2.4)^2 = t_s^2 + 4.8t_s + 5.76$
3. Write the quadratic equation for t_s — $t_s^2 + 4.8t_s + 5.76 = (10/7)t_s^2$
4. Solve for t_s — $t_s = 12.3$ s
5. Use steps 1 and 2 and solve for θ — $\sin\theta = 42/(5 \times 9.81 \times 12.3^2) = 0.00567$; $\theta = 0.325°$

92 •• A uniform solid sphere of radius r starts from rest at a height h and rolls without slipping along the loop-the-loop track of radius R as shown in Figure 9-56. (*a*) What is the smallest value of h for which the sphere will not leave the track at the top of the loop? (*b*) What would h have to be if, instead of rolling, the ball slides without friction?

We shall assume that h is the initial height of the center of the sphere of radius r. To just remain in contact with the track, the centripetal acceleration of the sphere's center of mass must equal mg.

(*a*) 1. Note radius of loop for center of mass = $R - r$ — $mv^2/(R - r) = mg$ (1)

2. Use energy conservation — $mg(h - 2R + r) = \frac{1}{2}mv^2 + \frac{1}{2}(2mv^2/5)$ (2)

3. Use Equ. (1) for mv^2 and solve for h — $h = 2.7R - 1.7r$

(*b*) Now $\frac{1}{2}(2mv^2/5)$ term in (2) is absent. — $h = 2.5R - 1.5r$

93* ••• A wheel has a thin 3.0-kg rim and four spokes each of mass 1.2 kg. Find the kinetic energy of the wheel when it rolls at 6 m/s on a horizontal surface.

1. Find I of the wheel — $I = M_{rim}R^2 + 4[(1/3)M_{spoke}R^2]$

2. Write $K = K_{trans} + K_{rot}$; use $v = R\omega$ — $K = \frac{1}{2}(7.8 + 3 + 1.6) \times 6^2$ J = 223 J

94 ••• Two uniform 20-kg disks of radius 30 cm are connected by a short rod of radius 2 cm and mass 1 kg. When the rod is placed on a plane inclined at 30°, such that the disks hang over the sides, the assembly rolls without slipping. Find (*a*) the linear acceleration of the system, and (*b*) the angular acceleration of the system. (*c*) Find the kinetic energy of translation of the system after it has rolled 2 m down the incline starting from rest. (*d*) Find the kinetic energy of rotation of the system at the same point.

(*a*) 1. As in Problem 9-86, $\tau = fr$. Write the equations of motion. — $Mg \sin\theta - f = Ma$, where $M = 41$ kg; $fr = I\alpha$, where $\alpha = a/r$

2. Write a, eliminating f — $a = (Mg \sin\theta)/(M + I/r^2)$

3. Determine I — $I = (2 \times \frac{1}{2} \times 20 \times 0.3^2 + \frac{1}{2} \times 1 \times 0.02^2)$ kg·m^2 = 1.80 kg·m^2

4. Evaluate a — $a = (41 \times 9.81 \times 0.5)/(41 + 1.80/0.02^2)$ m/s^2 = 0.0443 m/s^2

(*b*) $\alpha = a/r$ — $\alpha = (0.0443/0.02)$ rad/s^2 = 2.21 rad/s^2

(*c*) Use $v^2 = 2as$ and $K_{trans} = \frac{1}{2}Mv^2$ — $K_{trans} = (\frac{1}{2} \times 41 \times 2 \times 0.0443 \times 2)$ J = 3.63 J

(*d*) $K_{rot} = Mgh - K_{trans}$; $h = 2 \sin 30°$ m = 1 m — $K_{rot} = [(41 \times 9.81 \times 1) - 3.63]$ J = 399 J

95 ••• A wheel of radius R rolls without slipping at a speed V. The coordinates of the center of the wheel are X, Y. (*a*) Show that the x and y coordinates of point P in Figure 9-57 are $X + r_0 \cos\theta$ and $R + r_0 \sin\theta$, respectively,. (*b*) Show that the total velocity $\boldsymbol{v}$ of point P has the components $v_x = V + (r_0V \sin\theta)/R$ and $v_y = -(r_0V \cos\theta)/R$. (*c*) Show that at the instant that $X = 0$, $\boldsymbol{v}$ and $\boldsymbol{r}$ are perpendicular to each other by calculating $\boldsymbol{v}\cdot\boldsymbol{r}$. (*d*) Show that $v = r\omega$, where $\omega = V/R$ is the angular velocity of the wheel. These results demonstrate that, in the case of rolling without slipping, the motion is the same as if the rolling object were instantaneously rotating about the point of contact with an angular speed $\omega = V/R$.

(*a*) From the figure it is evident that $x = r_0 \cos\theta$ and $y = r_0 \sin\theta$ relative to the center of the wheel. Therefore, if the coordinates of the center are X and R, those of point P are as stated.

(*b*) $v_{Px} = d(X + r_0 \cos\theta)/dt = dX/dt - r_0 \sin\theta\, d\theta/dt$. Note that $dX/dt = V$ and $d\theta/dt = -\omega = -V/R$; therefore, $v_{Px} = V + (r_0V \sin\theta)/R$, $v_{Py} = d(R + r_0 \sin\theta)/dt = r_0 \cos\theta\, d\theta/dt$ $(dR/dt = 0)$. Again, $d\theta/dt = -\omega$, so $v_{Py} = -(r_0V\cos\theta)/R$.

(*c*) $\boldsymbol{v}\cdot\boldsymbol{r} = v_{Px}r_x + v_{Py}r_y = (V + r_0V\sin\theta/R)(r_0\cos\theta) - (r_0V\cos\theta/R)(R + r_0\sin\theta) = 0$.

(*d*) $v^2 = v_x^2 + v_y^2 = V^2[1 + (2r_0/R)\sin\theta + r_0^2/R^2]$; $r^2 = r_x^2 + r_y^2 = R^2[1 + (2r_0/R)\sin\theta + r_0^2/R^2]$; so $v/r = V/R = \omega$.

96 ••• A uniform cylinder of mass M and radius R is at rest on a block of mass m, which in turn rests on a horizontal, frictionless table (Figure 9-58). If a horizontal force $\boldsymbol{F}$ is applied to the block, it accelerates and the cylinder rolls without slipping. Find the acceleration of the block.

We begin by drawing the two free-body diagrams.

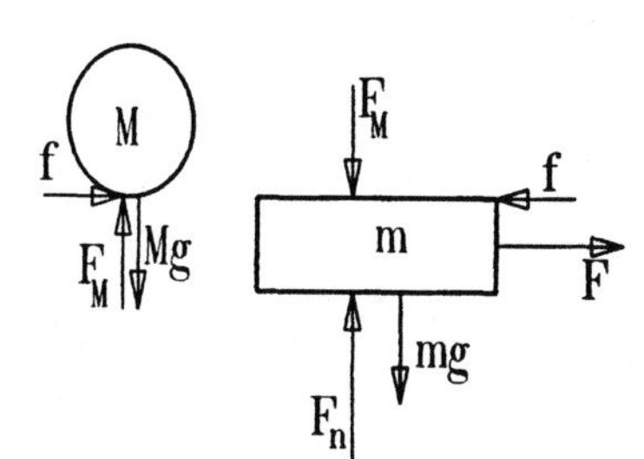

For the block,

$F - f = ma_B$ (1)

For the cylinder,

$f = Ma_C$ (2)

Also, $fR = ½MR^2\alpha$ and $f = ½MR\alpha$. But $a_C = a_B - R\alpha$ or $R\alpha = a_B - a_C$.

Using Equs. (1) and (2) we now obtain $2f/M = a_B - f/M$ and

$3f/M = 3a_C = a_B$ (3)

Equs. (1) and (3) yield $F - Ma_B/3 = ma_B$ and solving for a_B we obtain $a_B = 3F/(M + 3m)$ and $a_C = F/(M + 3m)$.

97* ••• (*a*) Find the angular acceleration of the cylinder in Problem 96. Is the cylinder rotating clockwise or counterclockwise? (*b*) What is the cylinder's linear acceleration relative to the table? Let the direction of $\boldsymbol{F}$ be the positive direction. (*c*) What is the linear acceleration of the cylinder relative to the block?

(*a*) From Problem 9-96, $\alpha = (a_B - a_C)/R = 2F/[R(M + 3m)]$. From the free body diagram of the preceding problem it is evident that the torque and, therefore, α is in the counterclockwise direction.

(*b*) The linear acceleration of the cylinder relative to the table is $a_C = F/(M + 3m)$. (see Problem 96)

(*c*) The acceleration of the cylinder relative to the block is $a_C - a_B = -2F/(M + 3m)$.

98 ••• If the force in Problem 96 acts over a distance d, find (*a*) the kinetic energy of the block, and (*b*) the kinetic energy of the cylinder. (*c*) Show that the total kinetic energy is equal to the work done on the system.

(*a*) $K_m = ½mv_m^2 = ma_md = Fdm/(m + ½M)$.

(*b*) $K_{cyl} = K_{trans} + K_{rot} = ½Mv_M^2 + ½I\omega^2 = ½FdM/(m + ½M) + (1/4)FdM/(m + ½M)$.

(*c*) The total $K = Fd$ which is the work done by the force F.

99 ••• A marble of radius 1 cm rolls from rest without slipping from the top of a large sphere of radius 80 cm, which is held fixed (Figure 9-59). Find the angle from the top of the sphere to the point where the marble breaks contact with the sphere.

Use energy conservation to find $v^2(\theta)$. $\Delta U = -mg(R + r)(1 - \cos\theta) = ½mv^2 + ½I\omega^2 = ½mv^2(1 + 2/5) = 7mv^2/10$. $v^2 = 10g(R + r)(1 - \cos\theta)/7$. The marble will separate from the sphere when $mg\cos\theta = mv^2/(R+r)$. The condition is $\cos\theta = 10/17$; $\theta = 54°$. (Note that θ does not depend on the radii of the sphere and marble.)

100 • True or false: When a sphere rolls and slips on a rough surface, mechanical energy is dissipated.

True

101*• A cue ball is hit very near the top so that it starts to move with topspin. As it slides, the force of friction (*a*) increases v_{cm}. (*b*) decreases v_{cm}. (*c*) has no effect on v_{cm}.

(*a*)

102 • A bowling ball of mass M and radius R is thrown such that at the instant it touches the floor it is moving horizontally with a speed v_0 and is not rotating. It slides for a time t_1 a distance s_1 before it begins to roll without slipping. (*a*) If μ_k is the coefficient of sliding friction between the ball and the floor, find s_1, t_1, and the final speed v_1 of the ball. (*b*) Find the ratio of the final mechanical energy to the initial mechanical energy of the ball. (*c*) Evaluate these quantities for $v_0 = 8$ m/s and $\mu_k = 0.06$.

Part (*a*) of this problem is identical to Example 9-16. From Example 9-16, we have:

(*a*) $s_1 = (12/49)v_0^2/\mu_k g$; $t_1 = 2v_0/7\mu_k g$; $v_1 = (5/2)\mu_k g t_1 = 5v_0/7$.

(*b*) $K_f = ½Mv_1^2 + ½[(2/5)Mv_1^2] = (7/10)Mv_1^2 = (5/14)Mv_0^2$; $K_i = ½Mv_0^2$. $K_f/K_i = 5/7$.

(*c*) Inserting the appropriate numerical values: $s_1 = 26.6$ m; $t_1 = 3.88$ s; $v_1 = 5.71$ m/s.

103 •• A cue ball of radius r is initially at rest on a horizontal pool table (Figure 9-60). It is struck by a horizontal cue stick that delivers a force of magnitude P_0 for a very short time Δt. The stick strikes the ball at a point h above the ball's point of contact with the table. Show that the ball's initial angular velocity ω_0 is related to the initial linear velocity of its center of mass v_0 by $\omega_0 = 5v_0(h - r)/2r^2$.

The translational impulse $P_t = P_0\Delta t = mv_0$. The rotational impulse about the center of mass is $P_\tau = P_t(h - r) = I\omega_0$. With $I = (2/5)mr^2$ one then obtains $\omega_0 = 5v_0(h-r)/2r^2$.

104 •• A uniform spherical ball is set rotating about a horizontal axis with an angular speed ω_0 and is placed on the floor. If the coefficient of sliding friction between the ball and the floor is μ_k, find the speed of the center of mass of the ball when it begins to roll without slipping.

1. f_k gives the ball a forward acceleration a	$a = \mu_k g$; $v = at = \mu_k gt$
2. The torque $\tau = f_k r$ results in a reduction of ω	$\omega = \omega_0 - \alpha t$; $\alpha = \tau/I = \mu_k mrg/[(2/5)mr^2] = (5/2)\mu_k g/r$
3. The ball rolls without slipping when $\omega r = v$	$\omega_0 r - (5/2)\mu_k gt = \mu_k gt$; $t = 2r\omega_0/7\mu_k g$
4. Find v at $t = 2r\omega_0/7\mu_k g$	$v = 2r\omega_0/7$

105* •• A uniform solid ball resting on a horizontal surface has a mass of 20 g and a radius of 5 cm. A sharp force is applied to the ball in a horizontal direction 9 cm above the horizontal surface. The force increases linearly from 0 to a peak value of 40,000 N in 10^{-4} s and then decreases linearly to 0 in 10^{-4} s. (*a*) What is the velocity of the ball after impact? (*b*) What is the angular velocity of the ball after impact? (*c*) What is the velocity of the ball when it begins to roll without sliding? (*d*) For how long does the ball slide on the surface? Assume that $\mu_k = 0.5$.

(*a*) Find the translational impulse; then use $P_t = mv$	$F_{av} = 20{,}000$ N, $\Delta t = 2 \times 10^{-4}$ s; $v_0 = (4/0.02)$ m/s $= 200$ m/s
(*b*) Proceed as in Problem 9-103	$\omega_0 = 5 \times 200 \times (.09 - .05)/(2 \times .05^2)$ rad/s $= 8000$ rad/s
(*c*), (*d*) Note that $\omega_0 r = 400$ m/s $> v_0$; proceed as in Problem 9-104	$\omega = \omega_0 - (5/2)\mu_k gt/r$; $v = v_0 + \mu_k gt$; set $\omega r = v$; find t $t = 2(\omega_0 r - v_0)/7\mu_k g = 11.6$ s; $v = 257$ m/s

106 •• A 0.3-kg billiard ball of radius 3 cm is given a sharp blow by a cue stick. The applied force is horizontal and passes through the center of the ball. The initial velocity of the ball is 4 m/s. The coefficient of kinetic friction is

0.6. (*a*) For how many seconds does the ball slide before it begins to roll without slipping? (*b*) How far does it slide? (*c*) What is its velocity once it begins rolling without slipping?

Since the impulse passes through the CM, $\omega_0 = 0$. We use the results of Problem 9-102.

(*a*) $t = 2v_0/7\mu_k g = 0.194$ s. (*b*) $s = 12v_0^2/49\mu_k g = 0.666$ m. (*c*) $v = 5v_0/7 = 2.86$ m/s.

---

107 •• A billiard ball initially at rest is given a sharp blow by a cue stick. The force is horizontal and is applied at a distance $2R/3$ below the centerline, as shown in Figure 9-61. The initial speed of the ball is v_0, and the coefficient of kinetic friction is μ_k. (*a*) What is the initial angular speed ω_0? (*b*) What is the speed of the ball once it begins to roll without slipping? (*c*) What is the initial kinetic energy of the ball? (*d*) What is the frictional work done as it slides on the table?

(*a*) Use rotation impulse, $P_\tau = mv_0 r$; $r = 2R/3$ — $P_\tau = I\omega_0$; $\omega_0 = (2mv_0R/3)/[(2/5)mR^2] = 5v_0/3R$

(*b*) Since F is below the center line, the spin is backward, i.e., the ball will slow down. Proceed as in Problem 9-105, with $\omega_0 = -5v_0/R$. — $\omega = \omega_0 + (5/2)\mu_k gt/R$; $v = v_0 - \mu_k gt$; set $\omega R = v$

$v_0 - \mu_k gt = -(5/3)v_0 + (5/2)\mu_k gt$; $t = (16/21)v_0/\mu_k g$

$v = (5/21)v_0 = 0.238v_0$

(*c*) $K_i = \frac{1}{2}mv_0^2 + \frac{1}{2}I\omega_0^2$ — $K_i = \frac{1}{2}mv_0^2 + \frac{1}{2}(50/45)mv_0^2 = (19/18)mv_0^2 = 1.056mv_0^2$

(*d*) Find K_f; then $W_{fr} = K_i - K_f$ — $K_f = (7/10)mv^2 = (0.7 \times 0.238^2)mv_0^2 = 0.0397mv_0^2$

$W_{fr} = 1.016mv_0^2$

---

108 •• A bowling ball of radius R is given an initial velocity v_0 down the lane and a forward spin $\omega_0 = 3v_0/R$. The coefficient of kinetic friction is μ_k. (*a*) What is the speed of the ball when it begins to roll without slipping? (*b*) For how long does the ball slide before it begins to roll without slipping? (*c*) What distance does the ball slide down the lane before it begins rolling without slipping?

(*a*) Apply conditions for rolling; see Problem 9-108 — $v = v_0 + \mu_k gt$; $\omega R = 3v_0 - (5/2)\mu_k gt = v$

(*a*) and (*b*) Find t and v — $t = 2v_0/3.5\mu_k g$; $v = 1.57v_0$

(*c*) $s = v_{av}t = \frac{1}{2}(v + v_0)t$ — $s = 0.735v_0^2/\mu_k g$

109* •• A solid cylinder of mass M resting on its side on a horizontal surface is given a sharp blow by a cue stick. The applied force is horizontal and passes through the center of the cylinder so that the cylinder begins translating with initial velocity v_0. The coefficient of sliding friction between the cylinder and surface is μ_k. (*a*) What is the transla-tional velocity of the cylinder when it is rolling without slipping? (*b*) How far does the cylinder travel before it rolls without slipping? (*c*) What fraction of its initial mechanical energy is dissipated in friction?

This Problem is identical to Example 9-16 except that now $I = \frac{1}{2}MR^2$. Follow the same procedure.

(*a*) Set $\omega R = v$; $v = v_0 - \mu_k gt$; $\omega R = 2\mu_k gt$ — $t = v_0/3\mu_k g$; $v = (2/3)v_0$

(*b*) $s = v_{av}t$ — $s = 5v_0^2/18\mu_k g$

(*c*) $W_{fr}/K_i = (K_i - K_f)/K_i$ — $K_i = \frac{1}{2}mv_0^2$; $K_f = (3/4)mv^2 = (1/3)mv_0^2$; $W_{fr}/K_i = 1/3$

---

110 • The torque exerted on an orbiting communications satellite by the gravitational pull of the earth is (*a*) directed toward the earth. (*b*) directed parallel to the earth's axis and toward the north pole. (*c*) directed parallel to the earth's axis and toward the south pole. (*d*) directed toward the satellite. (*e*) zero.

(*e*)

---

111 • The moon rotates as it revolves around the earth so that we always see the same side. Use this fact to find the angular velocity (in rad/s) of the moon about its axis. (The period of revolution of the moon about the earth is 27.3 days.)

$\omega = 1/27.3$ rev/day $= 2\pi/(27.3 \times 24 \times 60 \times 60)$ rad/s $= 2.7 \times 10^{-6}$ rad/s.

---

112 • Find the moment of inertia of a hoop about an axis perpendicular to the plane of the hoop and through its edge.

Use the parallel axis theorem $\qquad I = MR^2 + MR^2 = 2MR^2$

113* •• The radius of a park merry-go-round is 2.2 m. To start it rotating, you wrap a rope around it and pull with a force of 260 N for 12 s. During this time, the merry-go-round makes one complete rotation. (*a*) Find the angular acceleration of the merry-go-round. (*b*) What torque is exerted by the rope on the merry-go-round? (*c*) What is the moment of inertia of the merry-go-round?

(*a*) $\alpha = 2\theta/t^2$ $\qquad \alpha = 4\pi/12^2$ rad/s^2 $= 0.0873$ rad/s^2

(*b*) $\tau = Fr$ $\qquad \tau = (260 \times 2.2)$ N·m $= 572$ N·m

(*c*) $I = \tau/\alpha$ $\qquad I = (572/0.0873)$ kg·m^2 $= 6552$ kg·m^2

---

114 •• A uniform disk of radius 0.12 m and mass 5 kg is pivoted such that it rotates freely about its central axis (Figure 9-62). A string wrapped around the disk is pulled with a force of 20 N. (*a*) What is the torque exerted on the disk? (*b*) What is the angular acceleration of the disk? (*c*) If the disk starts from rest, what is its angular velocity after 5 s? (*d*) What is its kinetic energy after 5 s? (*e*) What is the total angle θ that the disk turns through in 5 s? (*f*) Show that the work done by the torque $\tau\Delta\theta$ equals the kinetic energy.

(*a*) $\tau = FR$ $\qquad \tau = (20 \times 0.12)$ N·m $= 2.4$ N·m

(*b*) $\alpha = \tau/I$; $I = \frac{1}{2}MR^2$ $\qquad \alpha = 2\tau/MR^2 = 66.7$ rad/s^2

(*c*) $\omega = \alpha t$ $\qquad \omega = 333$ rad/s

(*d*) $K = \frac{1}{2}I\omega^2$ $\qquad K = (\frac{1}{2} \times 0.036 \times 333^2)$ J $= 2000$ J

(*e*) $\theta = \frac{1}{2}\alpha t^2$ $\qquad \theta = (\frac{1}{2} \times 66.7 \times 5^2)$ rad $= 833$ rad

(*f*) Express K in terms of τ and θ $\qquad K = \frac{1}{2}(\tau/\alpha)(\alpha t)^2 = \frac{1}{2}\alpha\tau t^2 = \tau\theta$, Q.E.D.

---

115 •• A 0.25-kg rod of length 80 cm is suspended by a frictionless pivot at one end. It is held horizontal and released. Immediately after it is released, what is (*a*) the acceleration of the center of the rod, and (*b*) the initial acceleration of a point on the end of the rod? (*c*) Find the linear velocity of the center of mass of the rod when it is vertical.

(*a*) 1. Find τ and I about the pivot $\qquad \tau = (0.25 \times 9.81 \times 0.4)$ N·m $= 0.981$ N·m

$I = (0.25 \times 0.8^2/3)$ kg·m^2 $= 0.0533$ kg·m^2

2. Find α and $a = \alpha\ell = \alpha L/2$ $\qquad \alpha = \tau/I = 18.4$ rad/s^2; $a_{cm} = (18.4 \times 0.4)$ m/s^2 $= 7.36$ m/s^2

(*b*) $a_{end} = L\alpha$ $\qquad a_{end} = (18.4 \times 0.8)$ m/s^2 $= 14.7$ m/s^2

(*c*) 1. Use energy conservation; $\frac{1}{2}I\omega^2 = mg\Delta h$ $\qquad \omega = (2 \times 0.25 \times 9.81 \times 0.4/0.0533)^{1/2}$ rad/s $= 6.07$ rad/s

2. $v = R\omega = \frac{1}{2}L\omega$ $\qquad v = (0.4 \times 6.07)$ m/s $= 2.43$ m/s

---

116 •• A uniform rod of length $3L$ is pivoted as shown in Figure 9-63 and held in a horizontal position. What is the initial angular acceleration α of the rod upon release?

1. The CM is $0.5L$ from the support; find τ and I — $\tau = 0.5mgL$; $I = m(3L)^2/12 + m(0.5L)^2 = mL^2$

2. $\alpha = \tau/I$ — $\alpha = 0.5mgL/mL^2 = 0.5g/L$

117* •• A uniform rod of length L and mass m is pivoted at the middle as shown in Figure 9-64. It has a load of mass $2m$ attached to one of the ends. If the system is released from a horizontal position, what is the maximum velocity of the load?

1. Find I — $I = mL^2/12 + 2mL^2/4 = 7mL^2/12$

2. $½I\omega^2 = 2mgL/2$; $v = \omega L/2$; solve for v — $v = (2mgL/I)^{½}(L/2) = (6gL/7)^{½}$

---

118 •• A marble of mass M and radius R rolls without slipping down the track on the left from a height h_1 as shown in Figure 9-65. The marble then goes up the *frictionless* track on the right to a height h_2. Find h_2.

1. Find K at the bottom; find v^2 at the bottom — $K = Mgh_1 = ½Mv^2 + (1/5)Mv^2$; $v^2 = 10gh_1/7$

2. There is no friction; so $v^2 = 2gh_2$ — $h_2 = v^2/2g = 5h_1/7$

---

119 •• A uniform disk with a mass of 120 kg and a radius of 1.4 m rotates initially with an angular speed of 1100 rev/min. (*a*) A constant tangential force is applied at a radial distance of 0.6 m. What work must this force do to stop the wheel? (*b*) If the wheel is brought to rest in 2.5 min, what torque does the force produce? What is the mag-nitude of the force? (*c*) How many revolutions does the wheel make in these 2.5 min?

(*a*) Find K_i; $W = K_i$ — $W = ½I\omega^2 = [½(½ \times 120 \times 1.4^2)(1100 \times 2\pi/60)^2]$ J $= 780$ kJ

(*b*) Use $P_{av} = \tau\omega_{av} = W/t$; — $\tau = [780 \times 10^3/2.5 \times 60 \times ½ \times (1100 \times 2\pi/60)]$ N·m $= 90.4$ N·m

$F = \tau/R$ — $F = (90.4/0.6)$ N $= 150.7$ N

(*c*) $\theta = \omega_{av}t$ — $\theta = [2.5 \times 60 \times ½(1100/60)]$ rev $= 1375$ rev

---

120 •• A park merry-go-round consists of a 240-kg circular wooden platform 4.00 m in diameter. Four children running alongside push tangentially along the platform's circumference until, starting from rest, the merry-go-round reaches a steady speed of one complete revolution every 2.8 s. (*a*) If each child exerts a force of 26 N, how far does each child run? (*b*) What is the angular acceleration of the merry-go-round? (*c*) How much work does each child do? (*e*) What is the kinetic energy of the merry-go-round?

(*a*) Use energy conservation; $K_f = 4Fs = ½I\omega^2$ — $s = I\omega^2/8F = [½ \times 240 \times 4 \times (2\pi/2.8)^2/8 \times 26]$ m $= 11.6$ m

(*b*) $\alpha = \tau/I$; $\tau = 4FR$ — $\alpha = (4 \times 26 \times 2/480)$ rad/s^2 $= 0.433$ rad/s^2

(*c*) W per child $= Fs$ — $W = (26 \times 11.6)$ J $= 302$ J

(*d*) $K = 4Fs$ — $K = 1208$ J

121* •• A hoop of mass 1.5 kg and radius 65 cm has a string wrapped around its circumference and lies flat on a horizontal frictionless table. The string is pulled with a force of 5 N. (*a*) How far does the center of the hoop travel in 3 s? (*b*) What is the angular velocity of the hoop about its center of mass after 3 s?

(*a*) $F_{net} = F = ma_{cm}$; $s = ½a_{cm}t^2 = Ft^2/2m$ — $s = (5 \times 3^2/2 \times 1.5)$ m $= 15$ m

(*b*) $\alpha = \tau/I$; $\omega = \alpha t = FRt/mR^2 = Ft/mR$ — $\omega = (5 \times 3/1.5 \times 0.65)$ rad/s $= 15.4$ rad/s

122 •• A vertical grinding wheel is a uniform disk of mass 60 kg and radius 45 cm. It has a handle of radius 65 cm of negligible mass. A 25-kg load is attached to the handle when it is in the horizontal position. Neglecting friction, find (*a*) the initial angular acceleration of the wheel, and (*b*) the maximum angular velocity of the wheel.

(*a*) Find I and τ, $I = \frac{1}{2}MR^2 + mr^2$; $\tau = mgr$; $\alpha = \tau/I$. Here $M = 60$ kg, $m = 25$ kg, $R = 0.45$ m, $r = 0.65$ m.
$I = (\frac{1}{2} \times 60 \times 0.45^2 + 25 \times 0.65^2)\ \text{kg·m}^2 = 16.64\ \text{kg·m}^2$
$\tau = (25 \times 9.81 \times 0.65)\ \text{N·m} = 159.4\ \text{N·m}$; $\alpha = 9.58\ \text{rad/s}^2$

(*b*) Use energy conservation; $mgr = \frac{1}{2}I\omega^2$
$\omega = [2 \times 25 \times 9.81 \times 0.65/16.64]^{1/2}\ \text{rad/s} = 4.38\ \text{rad/s}$

---

123 •• In this problem, you are to derive the perpendicular-axis theorem for planar objects, which relates the moments of inertia about two perpendicular axes in the plane of Figure 9-66 to the moment of inertia about a third axis that is perpendicular to the plane of figure. Consider the mass element dm for the figure shown in the xy plane. (*a*) Write an expression for the moment of inertia of the figure about the z axis in terms of dm and r. (*b*) Relate the distance r of dm to the distances x and y, and show that $I_z = I_y + I_x$. (*c*) Apply your result to find the moment of inertia of a uniform disk of radius R about a diameter of the disk.

(*a*), (*b*) $I_z = \int r^2 dm = \int (x^2 + y^2) dm = \int x^2 dm + \int y^2 dm = I_y + I_x$.

(*c*) Let the z axis be the axis of rotation of the disk. By symmetry, $I_x = I_y$. So $I_x = \frac{1}{2}I_z = (1/4)MR^2$. (see Table 9-1)

---

124 •• A uniform disk of radius r and mass M is pivoted about a horizontal axis parallel to its symmetry axis and passing through its edge such that it can swing freely in a vertical plane (Figure 9-67). It is released from rest with its center of mass at the same height as the pivot. (*a*) What is the angular velocity of the disk when its center of mass is directly below the pivot? (*b*) What force is exerted by the pivot at this time?

(*a*) Use energy conservation; $\frac{1}{2}I\omega^2 = Mgh = Mgr$
$I = \frac{1}{2}Mr^2 + Mr^2 = 3Mr^2/2$; $\omega = \sqrt{4g/3r}$ rad/s

(*b*) $F = Mg + Mr\omega^2$
$F = Mg + 4Mg/3 = 7Mg/3$

125* •• A spool of mass M rests on an inclined plane at a distance D from the bottom. The ends of the spool have radius R, the center has radius r, and the moment of inertia of the spool about its axis is I. A long string of negligible mass is wound many times around the center of the spool. The other end of the string is fastened to a hook at the top of the inclined plane such that the string always pulls parallel to the slope as shown in Figure 9-68. (*a*) Suppose that initially the slope is so icy that there is *no* friction. How does the spool move as it slips down the slope? Use energy considerations to determine the speed of the center of mass of the spool when it reaches the bottom of the slope. Give your answer in terms of M, I, r, R, g, D, and θ. (*b*) Now suppose that the ice is gone and that when the spool is set up in the same way, there is enough friction to keep it from slipping on the slope. What is the direction and magnitude of the friction force in this case?

(*a*) The spool will move down the plane at constant acceleration, spinning in a counterclockwise direction as string unwinds. From energy conservation, $MgD \sin\theta = \frac{1}{2}Mv^2 + \frac{1}{2}I\omega^2$; $v = r\omega$.

$$v = \sqrt{\frac{2MgD\sin\theta}{M + I/r^2}}.$$

(*b*) 1. The direction of the friction force is up along the plane

2. Since $a_{cm} = 0$ and $\alpha = 0$, $F_{net} = 0$ and $\tau = 0$ — $Mg \sin\theta = T + f_s$; $Tr = f_s R$

3. Solve for f_s — $f_s = (Mg \sin\theta)/(1 + R/r)$

---

126 •• Ian has suggested another improvement for the game of hockey. Instead of the usual two-minute penalty, he would like to see an offender placed in a barrel at mid-ice and then spun in a circle by the other team. When the offender is silly with dizziness, he is put back into the game. Assume that a penalized player in a barrel approximates a uniform, 100-kg cylinder of radius 0.60 m, and that the ice is smooth (Figure 9-69). Ropes are wound around the barrel, so that pulling them causes rotation. If two players simultaneously pull the ropes with forces of 40 N and 60 N for 6 s, describe the motion of the barrel. Give its acceleration, velocity, and the position of its center of mass as functions of time.

The barrel will translate to the right and rotate as indicated in the figure. We first consider $t \le 6$ s.

1. $a_{cm} = F_{net}/m$; $v_{cm} = a_{cm}t$; $x_{cm} = \frac{1}{2}a_{cm}t^2$ — $a_{cm} = 0.2$ m/s^2; $v_{cm} = 0.2t$ m/s; $x_{cm} = 0.1t^2$ m

2. $\alpha = \tau/I$; $\omega = \alpha t$; $\theta = \frac{1}{2}\alpha t^2$ — $\tau = 100 \times 0.6$ N·m $= 60$ N·m;
$I = 50 \times 0.6^2$ kg·m^2 $= 18$ kg·m^2;
$\alpha = 3.33$ rad/s^2; $\omega = 3.33t$ rad/s; $\theta = 1.67t^2$ rad

For $t > 6$ s: $a_{cm} = \alpha = 0$; $v_{cm} = 1.2$ m/s; $x_{cm} = [3.6 + 1.2(t - 6)]$ m; $\omega = 20$ rad/s; $\theta = [60 + 20(t - 6)]$ rad.

---

127 •• A solid metal rod 1.5 m long is free to rotate without friction about a fixed, horizontal axis perpendicular to the rod and passing through one end. The other end is held in a horizontal position. Small coins of mass m are placed on the rod 25 cm, 50 cm, 75 cm, 1 m, 1.25 m, and 1.5 m from the bearing. If the free end is now released, calculate the initial force exerted on each coin by the rod. Assume that the mass of the coins may be neglected in comparison to the mass of the rod.

1. Determine α; $\alpha = \tau/I$; $I = ML^2/3$ — $\alpha = (MgL/2)/(ML^2/3) = 3g/2L = |g|$ rad/s^2

2. Determine $a(x)$, where x = distance from pivot — $a(x) = gx$

3. $ma = mg - F$; $F = m(g - a)$ — $F(0.25) = 0.75mg$; $F(0.50) = 0.5mg$; $F(0.75) = 0.25mg$;
$F(1.0) = F(1.25) = F(1.5) = 0$

---

128 •• A thin rod of length L and mass M is supported in a horizontal position by two strings, one attached to each end as shown in Figure 9-70. If one string is cut, the rod begins to rotate about the point where it connects to the other string (point A in the figure). (*a*) Find the initial acceleration of the center of mass of the rod. (*b*) Show that the initial tension in the string is $mg/4$ and that the initial angular acceleration of the rod about an axis through the point A is $3g/2L$. (*c*) At what distance from point A is the initial linear acceleration equal to g?

It is tempting to assume that the tension in the string above A is the same as before the other string is cut, namely $Mg/2$. However, the tension can change instantaneously. What cannot change instantaneously due to its inertia is the position of the rod. Thus point A is momentarily fixed.

(*a*), (*b*), and (*c*) From Problem 9-127 we have $\alpha = 3g/2L$. The center of mass of the rod is at a distance $L/2$ from A; consequently, $a_{cm} = \alpha L/2 = 3g/4$. Now $Ma_{cm} = Mg - T$, and solving for T one obtains $T = Mg/4$. To find the distance from A where $a = g$, set $\alpha x = g$ and solve for x: $x = g/\alpha = 2L/3$.

129* •• Figure 9-71 shows a hollow cylinder of length 1.8 m, mass 0.8 kg, and radius 0.2 m. The cylinder is free to rotate about a vertical axis that passes through its center and is perpendicular to the cylinder's axis. Inside the cylinder are two masses of 0.2 kg each, attached to springs of spring constant k and unstretched lengths 0.4 m. The inside walls of the cylinder are frictionless. (*a*) Determine the value of the spring constant if the masses are located 0.8 m from the center of the cylinder when the cylinder rotates at 24 rad/s. (*b*) How much work was needed to bring the system from $\omega = 0$ to $\omega = 24$ rad/s?

Let $m = 0.2$ kg mass, $M = 0.8$ kg mass of cylinder, $L = 1.8$ m, and x = distance of m from center = $x_0 + \Delta x$.

(*a*) We have $k\Delta x = m(x_0 + \Delta x)\omega^2$; solve for k — $k = (0.2 \times 0.8 \times 24^2/0.4)$ N/m = 230.4 N/m

(*b*) $K = K_{rot} + ½k\Delta x^2$; determine I of system when $x = 0.8$ m — $I_M = ½Mr^2 + ML^2/12 = 0.232$ kg·m²; $I_{2m} = 2(mr^2/4 + mx^2) = 0.13$ kg·m²; $I = 0.362$ kg·m²

Evaluate $K = ½I\omega^2 + ½k\Delta x^2 = W$ — $W = (½ \times 0.362 \times 24^2 + ½ \times 230.4 \times 0.4^2)$ J = 122.7 J

---

130 •• Suppose that for the system described in Problem 129, the spring constants are each $k = 60$ N/m. The system starts from rest and slowly accelerates until the masses are 0.8 m from the center of the cylinder. How much work was done in the process?

1. Proceed as in Problem 9-129*a* and find ω — $\omega = [(60 \times 0.4)/(0.2 \times 0.8)]^{½}$ rad/s = 12.25 rad/s
2. Determine W as in Problem 9-129*b* — $W = (½ \times 0.362 \times 12.25^2 + ½ \times 60 \times 0.4^2)$ J = 32 J

---

131 •• A string is wrapped around a uniform cylinder of radius R and mass M that rests on a horizontal frictionless surface. The string is pulled horizontally from the top with force F. (*a*) Show that the angular acceleration of the cylinder is twice that needed for rolling without slipping, so that the bottom point on the cylinder slides backward against the table. (*b*) Find the magnitude and direction of the frictional force between the table and cylinder needed for the cylinder to roll without slipping. What is the acceleration of the cylinder in this case?

(*a*) The only force is F; therefore, $a_{cm} = F/M$. The torque about the center of mass is $\tau = FR$ and $I = ½MR^2$. Thus $\alpha = \tau/I = 2F/MR$. If the cylinder rolls without slipping, $a_{cm} = \alpha R$. Here, $\alpha = 2a_{cm}/R$.

(*b*) Take the point of contact with the floor as the "pivot" point. The torque about that point is $\tau = 2FR$ and the moment of inertia about that point is $I = ½MR^2 + MR^2 = 3MR^2/2$. Thus, $\alpha = \tau/I = 4F/3MR$, and the linear acceleration of the center of the cylinder is $\alpha R = a_{cm} = 4F/3M$. But $Ma_{cm} = F + f$, where f is the frictional force. We find that the frictional force is $f = F/3$, and is in the same direction as F.

---

132 •• Figure 9-72 shows a solid cylinder of mass M and radius R to which a hollow cylinder of radius r is attached. A string is wound about the hollow cylinder. The solid cylinder rests on a horizontal surface. The coefficient of static friction between the cylinder and surface is μ_s. If a light tension is applied to the string in the vertical direction, the cylinder will roll to the left; if the tension is applied with the string horizontally, the cylinder rolls to the right. Find the angle of the string with the horizontal that will allow the cylinder to remain stationary when a small tension is applied to the string.

1. First, we note that if the tension is small, then there can be no slipping, and the system must roll.
2. Now consider the point of contact of the cylinder with the surface as the "pivot" point. If τ about that point is zero, the system will not roll. This will occur if the line of action of the tension passes through the pivot point. We see from the figure that the angle θ is given by $\theta = \cos^{-1}(r/R)$.

133* ••• A heavy, uniform cylinder has a mass m and a radius R (Figure 9-73). It is accelerated by a force T, which is applied through a rope wound around a light drum of radius r that is attached to the cylinder. The coefficient of static friction is sufficient for the cylinder to roll without slipping. (*a*) Find the frictional force. (*b*) Find the acceleration a of the center of the cylinder. (*c*) Is it possible to choose r so that a is greater than T/m? How? (*d*) What is the direction of the frictional force in the circumstances of part (*c*)?

(*a*) 1. Write the equations for translation and rotation $T + f = ma$ (1)

$Tr - fR = I\alpha = \frac{1}{2}mRa$ (2)

2. Solve (2) for f $f = Tr/R - \frac{1}{2}ma$ (3)

3. Use (3) in (1) to find a $a = (2T/3m)(1 + r/R)$ (4)

4. Use (4) in (3) to find f in terms of T, r, and R $f = (T/3)(2r/R - 1)$ (5)

(*b*) See Equ. (4) above Note: for $r = R$, results agree with Problem 9- 131*b*

(*c*) Find r so that $a > T/m$ From Equ. (4) above, $a > T/m$ if $r > \frac{1}{2}R$

(*d*) If $r > \frac{1}{2}R$ then $f > 0$, i.e., in the direction of T

---

134 ••• A uniform stick of length L and mass M is hinged at one end. It is released from rest at an angle θ_0 with the vertical. Show that when the angle with the vertical is θ, the hinge exerts a force F_r along the stick and a force F_t perpendicular to the stick given by $F_r = \frac{1}{2}Mg(5 \cos \theta - 3 \cos \theta_0)$ and $F_t = (Mg/4) \sin \theta$.

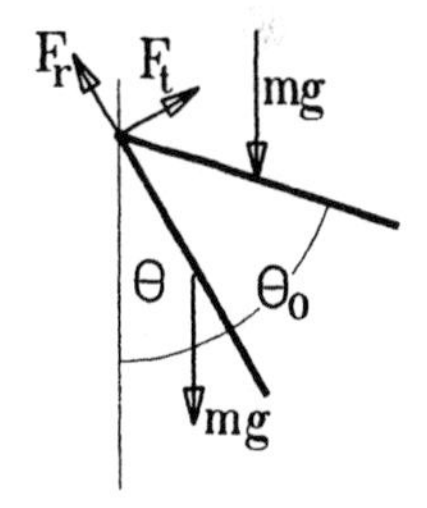

The system is shown in the drawing in two positions, with angles θ_0 and θ with the vertical. We also show all the forces that act on the stick. These forces result in a rotation of the stick—and its center of mass—about the pivot, and a tangential acceleration of the center of mass given by $a_t = \frac{1}{2}L\alpha$. As the stick's angle changes from θ_0 to θ, its potential energy decreases by Mgh, where h is the distance the center of mass falls. Using energy conservation and $I = ML^2/3$, we obtain $\frac{1}{2}(ML^2/3)\omega^2 = (MgL/2)(\cos \theta - \cos \theta_0)$. Thus we have $\omega^2 = (3g/L)(\cos \theta - \cos \theta_0)$. The centripetal force that must act radially on the center of mass is $\frac{1}{2}ML\omega^2$. This is part of the radial component of the force at the pivot. In addition to the centripetal force, gravity also acts on the center of mass. The radial component of **Mg** is $Mg \cos \theta$. Hence the total radial force at the hinge is $F_r = \frac{1}{2}ML(3g/L)(\cos \theta - \cos \theta_0) + Mg \cos \theta = \frac{1}{2}Mg(5 \cos \theta - 3 \cos \theta_0)$. The mass M times the tangential acceleration of the center of mass must equal the sum of the tangential component of **Mg** and the tangential component of the force at the pivot. The tangential acceleration of the center of mass is $a_t = \frac{1}{2}L\alpha$, where $\alpha = \tau/I = (\frac{1}{2}MgL \sin \theta)/(ML^2/3) = (3g \sin \theta)/2L$. Thus, $a_t = (3/4)g \sin \theta = g \sin \theta + F_t/M$, which gives $F_t = -(1/4)Mg \sin \theta$. Here the minus sign indicates that the force F_t is directed opposite to the tangential component of **Mg**.

CHAPTER 10

Conservation of Angular Momentum

1* • True or false: (*a*) If two vectors are parallel, their cross product must be zero. (*b*) When a disk rotates about its symmetry axis, $\boldsymbol{\omega}$ is along the axis. (*c*) The torque exerted by a force is always perpendicular to the force.

(*a*) True (*b*) True (*c*) True

2 • Two vectors **A** and **B** have equal magnitude. Their cross product has the greatest magnitude if **A** and **B** are (*a*) parallel. (*b*) equal. (*c*) perpendicular. (*d*) antiparallel. (*e*) at an angle of 45° to each other.

(*c*)

3 • A force of magnitude *F* is applied horizontally in the negative *x* direction to the rim of a disk of radius *R* as shown in Figure 10-29. Write **F** and **r** in terms of the unit vectors **i**, **j**, and **k**, and compute the torque produced by the force about the origin at the center of the disk.

$\boldsymbol{F} = -F\,\boldsymbol{i}$; $\boldsymbol{r} = R\boldsymbol{j}$; $\boldsymbol{\tau} = \boldsymbol{r} \times \boldsymbol{F} = FR\,\boldsymbol{j} \times -\boldsymbol{i} = FR\,\boldsymbol{i} \times \boldsymbol{j} = FR\,\boldsymbol{k}$.

4 • Compute the torque about the origin for the force $\boldsymbol{F} = -mg\,\boldsymbol{j}$ acting on a particle at $\boldsymbol{r} = x\,\boldsymbol{i} + y\boldsymbol{j}$, and show that this torque is independent of the *y* coordinate.

Use Equs. 10-1 and 10-7 $\quad \boldsymbol{\tau} = -mgx\,\boldsymbol{i} \times \boldsymbol{j} - mgy\,\boldsymbol{j} \times \boldsymbol{j} = -mgx\,\boldsymbol{k}$

5* • Find $\boldsymbol{A} \times \boldsymbol{B}$ for (*a*) $\boldsymbol{A} = 4\,\boldsymbol{i}$ and $\boldsymbol{B} = 6\,\boldsymbol{i} + 6\boldsymbol{j}$, (*b*) $\boldsymbol{A} = 4\,\boldsymbol{i}$ and $\boldsymbol{B} = 6\,\boldsymbol{i} + 6\,\boldsymbol{k}$, and (*c*) $\boldsymbol{A} = 2\,\boldsymbol{i} + 3\boldsymbol{j}$ and $\boldsymbol{B} = -3\,\boldsymbol{i} + 2\boldsymbol{j}$.

Use Equ. 10-7; Note that $\boldsymbol{i} \times \boldsymbol{i} = \boldsymbol{j} \times \boldsymbol{j} = \boldsymbol{k} \times \boldsymbol{k} = 0$

(*a*) $\boldsymbol{A} \times \boldsymbol{B} = 24\,\boldsymbol{i} \times \boldsymbol{j} = 24\,\boldsymbol{k}$. (*b*) $\boldsymbol{A} \times \boldsymbol{B} = 24\,\boldsymbol{i} \times \boldsymbol{k} = -24\boldsymbol{j}$. (*c*) $\boldsymbol{A} \times \boldsymbol{B} = 4\,\boldsymbol{i} \times \boldsymbol{j} - 9\boldsymbol{j} \times \boldsymbol{i} = 13\,\boldsymbol{k}$.

6 • Under what conditions is the magnitude of $\boldsymbol{A} \times \boldsymbol{B}$ equal to $\boldsymbol{A}\cdot\boldsymbol{B}$?

$|\boldsymbol{A} \times \boldsymbol{B}| = |AB \sin\theta| = |\boldsymbol{A}\cdot\boldsymbol{B}| = |AB \cos\theta|$ if $|\sin\theta| = |\cos\theta|$ or $\tan\theta = \pm 1$; $\theta = \pm 45°$ or $\theta = \pm 135°$.

7 •• A particle moves in a circle of radius *r* with an angular velocity ω. (*a*) Show that its velocity is $\boldsymbol{v} = \boldsymbol{\omega} \times \boldsymbol{r}$. (*b*) Show that its centripetal acceleration is $\boldsymbol{a}_c = \boldsymbol{\omega} \times \boldsymbol{v} = \boldsymbol{\omega} \times (\boldsymbol{\omega} \times \boldsymbol{r})$.

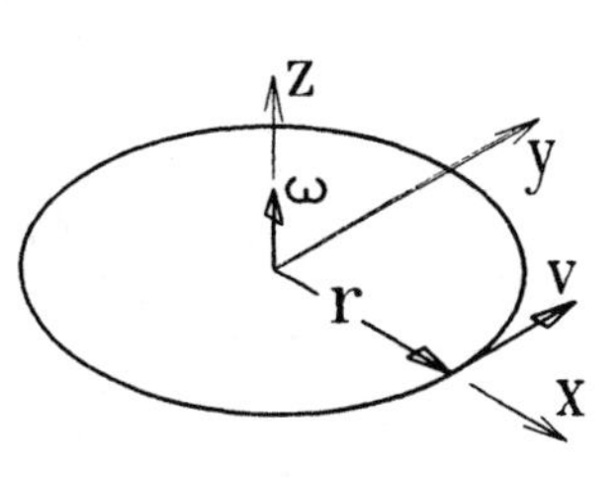

(*a*) Let **r** be in the *xy* plane. Then if $\boldsymbol{\omega}$ points in the positive *z* direction, i.e., $\boldsymbol{\omega} = \omega\,\boldsymbol{k}$, the particle's velocity is in the **j** direction when $\boldsymbol{r} = r\,\boldsymbol{i}$ (see Figure) and has the magnitude $r\omega$. Thus, $\boldsymbol{v} = \boldsymbol{\omega} \times \boldsymbol{r} = r\omega\boldsymbol{j}$.

(*b*) $\boldsymbol{a} = d\boldsymbol{v}/dt = (d\boldsymbol{\omega}/dt) \times \boldsymbol{r} + \boldsymbol{\omega} \times (d\boldsymbol{r}/dt) = (d\boldsymbol{\omega}/dt) \times \boldsymbol{r} + \boldsymbol{\omega} \times \boldsymbol{v} = \boldsymbol{a}_t + \boldsymbol{\omega} \times (\boldsymbol{\omega} \times \boldsymbol{r}) = \boldsymbol{a}_t + \boldsymbol{a}_c$, where $\boldsymbol{a}_t$ and $\boldsymbol{a}_c$ are the tangential and centripetal accelerations, respectively.

---

8 •• If $\boldsymbol{A} = 4\,\boldsymbol{i}$, $B_z = 0$, $|\boldsymbol{B}| = 5$, and $\boldsymbol{A} \times \boldsymbol{B} = 12\,\boldsymbol{k}$, determine $\boldsymbol{B}$.

$\boldsymbol{B} = B_x\boldsymbol{i} + B_y\boldsymbol{j}$ $(B_z = 0)$; write $\boldsymbol{A} \times \boldsymbol{B}$ — $12\,\boldsymbol{k} = 4B_y\,\boldsymbol{i} \times \boldsymbol{j} = 4B_y\,\boldsymbol{k}$; $B_y = 3$

$B_x^2 + B_y^2 = B^2$; solve for B_x — $B_x = 4$; $\boldsymbol{B} = 4\,\boldsymbol{i} + 3\,\boldsymbol{j}$

9* •• If $\boldsymbol{A} = 3\,\boldsymbol{j}$, $\boldsymbol{A} \times \boldsymbol{B} = 9\,\boldsymbol{i}$, and $\boldsymbol{A}\cdot\boldsymbol{B} = 12$, find $\boldsymbol{B}$.

Let $\boldsymbol{B} = B_x\boldsymbol{i} + B_y\boldsymbol{j} + B_z\boldsymbol{k}$; write $\boldsymbol{A}\cdot\boldsymbol{B}$ and find B_y — $\boldsymbol{A}\cdot\boldsymbol{B} = 3B_y = 12$; $B_y = 4$

Write $\boldsymbol{A} \times \boldsymbol{B}$ and determine B_x and B_y — $9\,\boldsymbol{i} = 3B_x\boldsymbol{j} \times \boldsymbol{i} + 3B_z\boldsymbol{j} \times \boldsymbol{k} = -3B_x\,\boldsymbol{k} + 3B_z\,\boldsymbol{i}$; $B_x = 0$, $B_z = 3$

$\boldsymbol{B} = 4\,\boldsymbol{j} + 3\,\boldsymbol{k}$

---

10 • What is the angle between a particle's linear momentum $\boldsymbol{p}$ and its angular momentum $\boldsymbol{L}$?

From Equ. 10-8 it follows that $\boldsymbol{L}$ and $\boldsymbol{p}$ are mutually perpendicular; i.e., the angle is 90°.

---

11 • A particle of mass *m* is moving with speed *v* along a line that passes through point *P*. What is the angular momentum of the particle about point *P*? (*a*) *mv* (*b*) zero (*c*) It changes sign as the particle passes through point *P*. (*d*) It depends on the distance of point *P* from the origin of coordinates.

(*b*)

---

12 •• A particle travels in a circular path. (*a*) If its linear momentum $\boldsymbol{p}$ is doubled, how is its angular momentum affected? (*b*) If the radius of the circle is doubled but the speed is unchanged, how is the angular momentum of the particle affected?

(*a*) $\boldsymbol{L}$ is doubled. (*b*) $\boldsymbol{L}$ is doubled.

13* •• A particle moves along a straight line at constant speed. How does its angular momentum about any point vary over time?

$\boldsymbol{L} = m\boldsymbol{r} \times \boldsymbol{p}$ is constant.

---

14 • A particle moving at constant velocity has zero angular momentum about a particular point. Show that the particle either has passed through that point or will pass through it.

$\boldsymbol{L} = 0$, therefore $\boldsymbol{r} \times \boldsymbol{v} = 0$. Since neither *r* nor *v* is zero, $\sin\theta = 0$, where θ is the angle between $\boldsymbol{r}$ and $\boldsymbol{v}$. $\theta = 0°$ or 180°.

---

15 • A 2-kg particle moves at a constant speed of 3.5 m/s around a circle of radius 4 m. (*a*) What is its angular momentum about the center of the circle? (*b*) What is its moment of inertia about an axis through the center of the circle and perpendicular to the plane of the motion? (*c*) What is the angular speed of the particle?

(*a*) $\boldsymbol{L} = \boldsymbol{r} \times \boldsymbol{p}$; $L = rmv$ — $L = (4 \times 2 \times 3.5)\ \text{kg·m}^2/\text{s} = 28\ \text{kg·m}^2/\text{s}$

(*b*) $I = mr^2$ — $I = (2 \times 4^2)\ \text{kg·m}^2 = 32\ \text{kg·m}^2$

(*c*) $\omega = L/I$ — $\omega = 0.875$ rad/s

---

16 • A 2-kg particle moves at constant speed of 4.5 m/s along a straight line. (*a*) What is the magnitude of its angular momentum about a point 6 m from the line? (*b*) Describe qualitatively how its angular speed about that point varies with time.

(*a*) $L = rmv\sin\theta = (6 \times 2 \times 4.5)\ \text{kg·m}^2/\text{s} = 54\ \text{kg·m}^2/\text{s}$.

(*b*) ω increases as the particle approaches the point, then decreases as it recedes.

17* •• A particle is traveling with a constant velocity v along a line that is a distance b from the origin O (Figure 10-30). Let dA be the area swept out by the position vector from O to the particle in time dt. Show that dA/dt is constant in time and equal to $\frac{1}{2}L/m$, where L is the angular momentum of the particle about the origin.

The area at $t = t_1$ is $A_1 = \frac{1}{2}br_1 \cos \theta_1 = \frac{1}{2}bx_1$, where θ_1 is the angle between r_1 and v and x_1 is the component of r_1 in the direction of v. At $t = t_1 + dt$, $A = A_1 + dA = \frac{1}{2}b(x + dx) = \frac{1}{2}b(x + v\,dt)$. Thus, $dA/dt = \frac{1}{2}bv$ = constant. Note that $r \sin \theta = b$; consequently, $\frac{1}{2}bv = \frac{1}{2}L/m$.

18 •• A 15-g coin of diameter 1.5 cm is spinning at 10 rev/s about a vertical diameter at a fixed point on a tabletop. (*a*) What is the angular momentum of the coin about its center of mass? (*b*) What is its angular momentum about a point on the table 10 cm from the coin? If the coin spins about a vertical diameter at 10 rev/s while its center of mass travels in a straight line across the tabletop at 5 cm/s, (*c*) what is the angular momentum of the coin about a point on the line of motion? (*d*) What is the angular momentum of the coin about a point 10 cm from the line of motion? (There are two answers to this question. Explain why and give both.)

(*a*) $L = L_{spin} = I_{cm}\omega_{spin}$; $I = mr^2/4$ (Problem 9-44) — $L = (0.015 \times 0.0075^2/4)(20\pi)$ kg·m²/s $= 1.33 \times 10^{-5}$ kg·m²/s

(*b*) $L = L_{orbit} + L_{spin}$; $L_{orbit} = 0$ — $L = 1.33 \times 10^{-5}$ kg·m²/s

(*c*) $L_{orbit} = 0$ (Problem 10-14) — $L = 1.33 \times 10^{-5}$ kg·m²/s

(*d*) $L_{orbit} = \pm mvr = \pm 7.5 \times 10^{-5}$ kg·m²/s — $L = 8.83 \times 10^{-5}$ kg·m²/s; $L = -6.17 \times 10^{-5}$ kg·m²/s

19 •• Two particles of masses m_1 and m_2 are located at r_1 and r_2 relative to some origin O as in Figure 10-31. They exert equal and opposite forces on each other. Calculate the resultant torque exerted by these internal forces about the origin O and show that it is zero if the forces F_1 and F_2 lie along the line joining the particles.

$\tau = \Sigma\tau_i = r_1 \times F_1 + r_2 \times F_2 = (r_1 - r_2) \times F_1$ since $F_2 = -F_1$. But $r_1 - r_2$ points along $-F_1$ so $(r_1 - r_2) \times F_1 = 0$.

20 • True or false: (*a*) The rate of change of a system's angular momentum is always parallel to the net external torque. (*b*) If the net torque on a body is zero, the angular momentum must be zero.

(*a*) True (see Equ. 10-13) (*b*) False

21* • A 1.8-kg particle moves in a circle of radius 3.4 m. The magnitude of its angular momentum relative to the center of the circle depends on time according to $L = (4\ \text{N·m})t$. (*a*) Find the magnitude of the torque acting on the particle. (*b*) Find the angular speed of the particle as a function of time.

(*a*) $\tau = dL/dt$ — $\tau = 4$ N·m

(*b*) $\omega = \alpha t$; $\alpha = \tau/I$; $I = mr^2$; $\omega = \tau t/mr^2$ — $\omega = (4/1.8 \times 3.4^2)t$ rad/s $= 0.192t$ rad/s

22 •• A uniform cylinder of mass 90 kg and radius 0.4 m is mounted so that it turns without friction on its fixed symmetry axis. It is rotated by a drive belt that wraps around its perimeter and exerts a constant torque. At time $t = 0$, its angular velocity is zero. At time $t = 25$ s, its angular velocity is 500 rev/min. (*a*) What is its angular momentum at $t = 25$ s? (*b*) At what rate is the angular momentum increasing? (*c*) What is the torque acting on the cylinder? (*d*) What is the magnitude of the force acting on the rim of the cylinder?

(*a*) $L = I\omega$; $I = \frac{1}{2}mr^2$ — $L = (\frac{1}{2} \times 90 \times 0.4^2 \times 500 \times 2\pi/60)$ kg·m²/s $= 377$ kg·m²/s

(*b*) $dL/dt = I\,d\omega/dt = I\alpha$ = constant — $dL/dt = (377/25)$ kg·m²/s² $= 15.1$ kg·m²/s²

(*c*) $\tau = dL/dt$ — $\tau = 15.1$ N·m

(*d*) $\tau = Fr$ — $F = (15.1/0.4)$ N $= 3.77$ N

23 •• In Figure 10-32, the incline is frictionless and the string passes through the center of mass of each block. The pulley has a moment of inertia I and a radius r. (*a*) Find the net torque acting on the system (the two masses, string, and pulley) about the center of the pulley. (*b*) Write an expression for the total angular momentum of the system about the center of the pulley when the masses are moving with a speed v. (*c*) Find the acceleration of the masses from your results for parts (*a*) and (*b*) by setting the net torque equal to the rate of change of the angular momentum of the system.

(*a*) The only external forces acting on the system are m_1g and $m_2g \sin\theta$. (The normal reaction force of the plane on m_2 is balanced by the component of m_2g normal to the plane.) Thus $\tau_{net} = rg(m_2 \sin\theta - m_1)$, where we have taken clockwise to be positive to be consistent with a positive upward velocity of m_1 as indicated in the figure.

(*b*) $L = I\omega + m_1vr + m_2vr = vr(I/r^2 + m_1 + m_2)$.

(*c*) $\tau = dL/dt = ar(I/r^2 + m_1 + m_2) = rg(m_2 \sin\theta - m_1)$; $a = (m_2 \sin\theta - m_1)g/(I/r^2 + m_1 + m_2)$.

24 •• From her elevated DJ booth at a dance club, Caroline is lowering a 2-kg speaker using a 0.6-kg disk of radius 8 cm as a pulley (Figure 10-33). The speaker wire runs straight up from the speaker, over the pulley, and then horizontally across the table. She attaches the wire to the 4-kg amplifier on her tabletop, and then turns to get the other speaker. The table, however, is nearly frictionless, and the whole system begins to move when she lets go. (*a*) What is the net torque about the center of the pulley? (*b*) What is the total angular momentum of the system 3.5 s after release? (*c*) What is the angular momentum of the pulley at this time? (*d*) What is the ratio of the angular momentum of each piece of equipment to the angular momentum of the pulley?

(*a*) Proceed as in Problem 23	$\tau = (2 \times 9.81 \times 0.08)$ N·m = 1.57 N·m
(*b*) $L = \int \tau\, dt$	$L = (1.57 \times 3.5)$ kg·m²/s = 5.5 kg·m²/s
(*c*) From Problem 23, find a ($\theta = 0$) and v; then find I_p and ω.	$a = [(2 \times 9.81)/(\frac{1}{2} \times 0.6 + 6)]$ m/s² = 3.11 m/s²; $v = 10.9$ m/s; $I = (0.3 \times 0.08^2)$ kg·m² = 0.00192 kg·m²; $\omega = v/r = 136$ rad/s; $L_p = I_p\omega = 0.262$ kg·m²/s
(*d*) Find $L(m_1)$ and $L(m_2)$; m_1 = speaker, m_2 = amplifier	$L(m_1) = (2 \times 10.9 \times 0.08)$ kg·m²/s = 1.74 kg·m²/s; $L(m_2) = 3.48$ kg·m²/s
Determine the ratios	$L(m_1)/L_p = 6.64$; $L(m_2)/L_p = 13.3$

25* •• Work Problem 24 for the case in which the coefficient of friction between the table and the 4-kg amplifier is 0.25.

(*a*) In this case, $\tau = m_1gr - \mu_k m_2 gr = (1.57 - 0.25 \times 4 \times 9.81 \times 0.08) = 0.785$ N·m. Now proceed as in Problem 24.

(*b*) $L = 2.75$ kg·m²/s. (*c*) $L_p = 0.131$ kg·m²/s. (*d*) The ratios are the same as in Problem 24.

26 •• Figure 10-34 shows the rear view of a spaceship that is rotating about its longitudinal axis at 6 rev/min. The occupants wish to stop this rotation. They have small jets mounted tangentially, at a distance $R = 3$ m from the axis, as indicated, and can eject 10 g/s of gas from each jet with a nozzle velocity of 800 m/s. For how long must they turn on these jets to stop the rotation? The rotational inertia of the ship around its axis (assumed to be constant) is 4000 kg·m².

$t = \Delta L/\tau = \Delta\omega I/\tau$, $\tau = 2 \times FR$; $F = (10^{-2} \times 800)$ N $\quad t = [(12\pi/60) \times 4000/(2 \times 10^{-2} \times 800 \times 3)]$ s = 52.4 s

27 • True or false: If the net torque on a rotating system is zero, the angular velocity of the system cannot change.

False; the moment of inertia could change.

28 • Folk wisdom says that a cat always lands on its feet. If a cat starts falling with its feet up, how can it land on its feet without violating the law of conservation of angular momentum?

If the cat stretches its feet sideways, air friction provides torque to turn the animal about.

29* • If the angular momentum of a system is constant, which of the following statements must be true? (*a*) No torque acts on any part of the system. (*b*) A constant torque acts on each part of the system. (*c*) Zero net torque acts on each part of the system. (*d*) A constant external torque acts on the system. (*e*) Zero net torque acts on the system.

(*e*)

30 • Two identical cylindrical disks have a common axis. Initially, one of the disks is spinning. When the two disks are brought into contact they stick together. Which of the following statements is true? (*a*) The total kinetic energy and the total angular momentum are unchanged from their initial values. (*b*) Both the total kinetic energy and the total angular momentum are reduced to half of their original values. (*c*) The total angular momentum is unchanged, but the total kinetic energy is reduced to half its original value. (*d*) The total angular momentum is reduced to half of its original value, but the total kinetic energy is unchanged. (*e*) The total angular momentum is unchanged, and the total kinetic energy is reduced to one-quarter of its original value.

(*c*)

31 •• In Example 10-4, does the force exerted by the merry-go-round on the child do work?

No; in each "inelastic collision" the force of static friction does not act through any distance.

32 •• Is it easier to crawl radially outward or radially inward on a rotating merry-go-round? Why?

It is easier to crawl out; in the rotating reference frame the effective force on the person is outward.

33* •• A block sliding on a frictionless table is attached to a string that passes through a hole in the table. Initially, the block is sliding with speed v_0 in a circle of radius r_0. A student under the table pulls slowly on the string. What happens as the block spirals inward? Give supporting arguments for your choice. (*a*) Its energy and angular momentum are conserved. (*b*) Its angular momentum is conserved, and its energy increases. (*c*) Its angular momentum is conserved, and its energy decreases. (*d*) Its energy is conserved, and its angular momentum increases. (*e*) Its energy is conserved, and its angular momentum decreases.

(*b*) $\tau = 0$, so L is conserved. The student does work, $Fs \neq 0$, so the energy of the block must increase.

34 • A planet moves in an elliptical orbit about the sun with the sun at one focus of the ellipse as in Figure 10-35. (*a*) What is the torque produced by the gravitational force of attraction of the sun for the planet? (*b*) At position A, the planet is a distance r_1 from the sun and is moving with a speed v_1 perpendicular to the line from the sun to the planet. At position B, it is at distance r_2 and is moving with speed v_2, again perpendicular to the line from the sun to the planet. What is the ratio of v_1 to v_2 in terms of r_1 and r_2?

(*a*) $\boldsymbol{r} \times \boldsymbol{F} = 0$ since $\boldsymbol{F}$ acts along the direction of $\boldsymbol{r}$ $\quad \tau = 0$

(*b*) $\tau = 0 = dL/dt$; L is constant $\quad$ At A and B, $\boldsymbol{r} \times \boldsymbol{v} = rv$; so $v_1/v_2 = r_2/r_1$

35 • Under gravitational collapse (all forces on various pieces are inward toward the center), the radius of a spinning spherical star of uniform density shrinks by a factor of 2, with the resulting increased density remaining uniform throughout as the star shrinks. What will be the ratio of the final angular speed ω_2 to the initial angular speed ω_1? (*a*) 2 (*b*) 0.5 (*c*) 4 (*d*) 0.25 (*e*) 1.0

(*c*) Since L is constant, $I_1\omega_1 = I_2\omega_2$; $I \propto R^2$, so $\omega_2/\omega_1 = R_1^2/R_2^2 = 4$.

---

36 •• A man stands at the center of a platform that rotates without friction with an angular speed of 1.5 rev/s. His arms are outstretched, and he holds a heavy weight in each hand. The moment of inertia of the man, the extended weights, and the platform is 6 kg·m^2. When the man pulls the weights inward toward his body, the moment of inertia decreases to 1.8 kg·m^2. (*a*) What is the resulting angular speed of the platform? (*b*) What is the change in kinetic energy of the system? (*c*) Where did this increase in energy come from?

(*a*) $dL/dt = 0$, so $\omega_f = \omega_i I_i/I_f$ — $\omega_{fin} = (1.5 \times 6/1.8)$ rev/s $= 5$ rev/s

(*b*) $K_i = \frac{1}{2}I_i\omega_i^2$; $K_f = K_i(I_f\omega_f^2/I_i\omega_i^2)$ $= K_i\omega_f/\omega_i$ — $K_f = 3.33K_i$; $\Delta K = 2.33K_i = 621.8$ J

(*c*) The energy comes from the internal energy of the man

37* •• A small blob of putty of mass m falls from the ceiling and lands on the outer rim of a turntable of radius R and moment of inertia I_0 that is rotating freely with angular speed ω_i about its vertical fixed symmetry axis. (*a*) What is the postcollision angular speed of the turntable plus putty? (*b*) After several turns, the blob flies off the edge of the turntable. What is the angular speed of the turntable after the blob flies off?

(*a*) $\tau_{ext} = 0$; $I_0\omega_i = (I_0 + mR^2)\omega_f$ — $\omega_f = \omega_i/(1 + mR^2/I_0)$

(*b*) When m flies off, its angular momentum does not change — $\omega' = \omega_f$

---

38 •• Two disks of identical mass but different radii (r and $2r$) are spinning on frictionless bearings at the same angular speed ω_0 but in opposite directions (Figure 10-36). The two disks are brought slowly together. The resulting frictional force between the surfaces eventually brings them to a common angular velocity. What is the magnitude of that final angular velocity in terms of ω_0?

$\tau = 0$; $L_i = L_f$; so $\omega_f = \omega_0(I_1 - I_2)/(I_1 + I_2)$; $I \propto R^2$ — $I_1 = 4I_2$; $\omega_f = (3/5)\omega_0 = 0.3\omega_0$

---

39 •• A block of mass m sliding on a frictionless table is attached to a string that passes through a hole in the table. Initially, the block is sliding with speed v_0 in a circle of radius r_0. Find (*a*) the angular momentum of the block, (*b*) the kinetic energy of the block, and (*c*) the tension in the string. A student under the table now pulls slowly on the string. How much work is required to reduce the radius of the circle from r_0 to $r_0/2$?

(*a*) $L_0 = r_0mv_0$. (*b*) $K_0 = \frac{1}{2}mv_0^2$. (*c*) $T = mv_0^2/r_0$.

(*d*) Find I_f and $K_f = L_f^2/2I_f = L_0^2/2I_f$ — $I_f = mr_0^2/4 = I_0/4$; so $K_f = 4K_0$

$W = K_f - K_0$ — $W = 3K_0 = (3/2)mv_0^2$

---

40 •• At the beginning of each term, a physics professor named Dr. Zeus shows the class his expectations of them through a demonstration that he calls "Lesson #1." He stands at the center of a turntable that can rotate without friction. He then takes a 2-kg globe of the earth and swings it around his head at the end of a 0.8-m chain. The world revolves around him every 3 s, and the professor and the platform have a moment of inertia of 0.5 kg·m^2.

(a) What is the angular speed of the professor? (b) What is the total kinetic energy of the globe, professor, and platform?

We shall consider the 2-kg globe to be a point mass.

(a) $\tau = 0$; $L_i = 0 = L_f$. Write an expression for L_f. $L_f = (2 \times 0.8^2) \times (2\pi/3) + 0.5\omega = 0$

Solve for ω. $\omega = -5.36$ rad/s, direction opposite to that of globe

(b) $K = ½M_{gl}v_{gl}^2 + L_p^2/2I_p$ $K = [½ \times 2 \times (0.8 \times 2\pi/3)^2 + (0.5 \times 5.36)^2/2 \times 0.5]$ J $= 10$ J

41* •• The sun's radius is 6.96×10^8 m, and it rotates with a period of 25.3 d. Estimate the new period of rotation of the sun if it collapses with no loss of mass to become a neutron star of radius 5 km.

$\omega_2 = \omega_1(R_1^2/R_2^2)$ (see Problem 35); $T_2 = T_1(R_2/R_1)^2$; $T_2 = [25.3(5/6.96 \times 10^5)^2]$ days $= 1.31 \times 10^{-9}$ days $= 0.11$ ms.

Note: This assumes that the mass distribution in the sun and neutron star are the same. However, the sun's mass is concentrated near its center, whereas the density of the neutron star is nearly constant. The correct period will be substantially greater than 0.11 ms.

---

42 •• Arriving at the baggage claim area in a small airport, Alan (mass m) discovers a large turntable (radius R and moment of inertia I) that is spinning out of control. Not wanting to pass up an opportunity for magnificence, Alan leaps onto the edge of the turntable, which continues to spin freely with an angular speed of 7.5 rad/s. He struggles on his hands and knees to the center, and then rises up into a pose that resembles a hood ornament and spins like a figure skater in finale. Security is notified, but passengers applaud. Assume that $mR^2 = 2.8I$, and that Alan has a moment of inertia of $I/10$ in his final pose. What is his final angular speed if friction is neglected?

$\tau = 0$; $L_i = L_f$. $L_i = (1 + 2.8)I \times 7.5 = (1 + 0.1)I\omega_f$. $\omega_f = (3.8 \times 7.5/1.1)$ rad/s $= 25.9$ rad/s.

---

43 •• A 0.2-kg point mass moving on a frictionless horizontal surface is attached to a rubber band whose other end is fixed at point P. The rubber band exerts a force $F = bx$ toward P, where x is the length of the rubber band and b is an unknown coefficient. The mass moves along the dotted line in Figure 10-37. When it passes point A, its velocity is 4 m/s directed as shown. The distance AP is 0.6 m and BP is 1.0 m (a) Find the velocity of the mass at points B and C. (b) Find b.

(a) $\tau = 0$; $L_A = L_B = L_C$; $L = mr^2\omega = mvr$ $v_B = v_A r_A/r_B = 2.4$ m/s; $v_C = v_A = 4$ m/s

(b) $E_i = E_f$; $E = ½mv^2 + ½bx^2$ $½ \times 0.2 \times 4^2 + ½b \times 0.6^2 = ½ \times 0.2 \times 2.4^2 + ½b \times 1^2$

Solve for b $b = 3.2$ N/m

---

44 • A 2-g particle moves at a constant speed of 3 mm/s around a circle of radius 4 mm. (a) Find the magnitude of the angular momentum of the particle. (b) If $L = \sqrt{\ell(\ell+1)}\ \hbar$, find the value of $\ell(\ell+1)$ and the approximate value of ℓ. (c) Explain why the quantization of angular momentum is not noticed in macroscopic physics.

(a) $L = mvr$ $L = (2 \times 10^{-3} \times 3 \times 10^{-3} \times 4 \times 10^{-3})$ kg·m²/s $= 2.4 \times 10^{-8}$ kg·m²/s

(b) $\ell(\ell+1) = L^2/\hbar^2$ $\ell(\ell+1) = (2.4 \times 10^{-8}/1.05 \times 10^{-34})^2 = 5.22 \times 10^{52}$; $\ell = 2.29 \times 10^{26}$

(c) One can't tell between $\ell = 2 \times 10^{26}$ and $\ell = 2 \times 10^{26} + 1$

45* • The z component of the spin of an electron is $\frac{1}{2}\hbar$, but the magnitude of the spin vector is $\sqrt{0.75}\,\hbar$. What is the angle between the electron's spin angular momentum vector and the z axis?

$\cos\theta = 0.5/(0.75)^{1/2}$; $\theta = 54.7°$.

46 •• Show that the energy difference between one rotational state and the next higher state is proportional $\ell + 1$ (see Equation 10-20*a*).

$\Delta E = (\ell + 1)(\ell + 2)E_{r0} - \ell(\ell + 1)E_{r0} = 2(\ell + 1)E_{r0}$.

47 •• In the HBr molecule, the mass of the bromine nucleus is 80 times that of the hydrogen nucleus (a single proton); consequently, in calculating the rotational motion of the molecule, one may, to a good approximation, assume that the Br nucleus remains stationary as the H atom (mass 1.67×10^{-27} kg) revolves around it. The separation between the H atom and bromine nucleus is 0.144 nm. Calculate (*a*) the moment of inertia of the HBr molecule about the bromine nucleus, and (*b*) the rotational energies for $\ell = 1$, $\ell = 2$, and $\ell = 3$.

(*a*) $I = m_p r^2$ — $I = (1.67 \times 10^{-27} \times 1.44^2 \times 10^{-20})\ \text{kg·m}^2 = 3.46 \times 10^{-47}\ \text{kg·m}^2$

(*b*) Use Equ. 10-20*b* to find E_{r0} — $E_{r0} = 1.61 \times 10^{-22}\ \text{J} = 1.0\ \text{meV}$

Use Equ. 10-20*a* to find E_ℓ for $\ell = 1, 2, 3$ — $E_1 = 2$ meV, $E_2 = 6$ meV, $E_3 = 12$ meV

48 •• The equilibrium separation between the nuclei of the nitrogen molecule is 0.11 nm. The mass of each nitrogen nucleus is 14 u, where u = 1.66×10^{-27} kg. We wish to calculate the energies of the three lowest angular momentum states of the nitrogen molecule. (*a*) Approximate the nitrogen molecule as a rigid dumbbell of two equal point masses, and calculate the moment of inertia about its center of mass. (*b*) Find the rotational energy levels using the relation $E_\ell = \ell(\ell + 1)\hbar^2/2I$.

(*a*) $I = 2m_N r^2$, where $r = 0.11/2$ nm — $I = [28 \times 1.66 \times 10^{-27} \times (0.55)^2 \times 10^{-20}]\ \text{kg·m}^2 = 1.41 \times 10^{-46}\ \text{kg·m}^2$

(*b*) Find $E_{r0} = \hbar^2/2I$ and use Equ. 10-20*a* — $E_{r0} = 3.96 \times 10^{-23}\ \text{J} = 0.25\ \text{meV}$; $E_\ell = 0.25\ell(\ell + 1)$ meV

49* •• A 16.0-kg, 2.4-m-long rod is supported on a knife edge at its midpoint. A 3.2-kg ball of clay is dropped from rest from a height of 1.2 m and makes a perfectly inelastic collision with the rod 0.9 m from the point of support (Figure 10-38). Find the angular momentum of the rod-and-clay system immediately after the inelastic collision.

$L_i = L_f$; find L just prior to collision — $L = mvr = [3.2 \times (2 \times 9.81 \times 1.2)^{1/2} \times 0.9]\ \text{J·s} = 14\ \text{J·s}$

50 •• Figure 10-39 shows a thin bar of length L and mass M, and a small blob of putty of mass m. The system is supported on a frictionless horizontal surface. The putty moves to the right with velocity v, strikes the bar at a distance d from the center of the bar, and sticks to the bar at the point of contact. Obtain expressions for the velocity of the system's center of mass and for the angular velocity of the system about its center of mass.

1. There are no external forces or torques acting on the system. Therefore $v_{cm,i} = v_{cm,f}$ and $L_i = L_f$.
2. $y_{cm} = md/(M+m)$ below the center of the bar.
3. $v_{cm} = mv/(M+m)$.
4. L (about CM) $= mv(d - y_{cm}) = mMvd/(M+m)$.
5. Determine I_{cm}, using the parallel axis theorem: $I_{cm} = ML^2/12 + My_{cm}^2 + m(d-y_{cm})^2 = \dfrac{ML^2}{12} + \dfrac{(M^2+m^2)d^2}{M+m}$.

6. Use $L = I\omega$: $\omega = \frac{L}{I} = \frac{mMvd}{(ML^2/12)(M+m) + (M^2+m^2)d^2}$.

---

51 •• In Problem 50, replace the blob of putty with a small hard sphere of negligible size that collides elastically with the bar. Find d such that the sphere is at rest after the collision.

1. As in Problem 50, $\tau_{ext} = F_{ext} = 0$. 2. Let v' and V be the final velocities of m and M, respectively. Use conservation of linear momentum and angular momentum: $mv = mv' + MV'$. (1)

2. $mvd = mv'd + ML^2\omega/12$. (2)

3. Set $v' = 0$; $V' = mv/M$. (3)

4. Use energy conservation: $mv^2 = M(V')^2 + ML^2\omega^2/12$. (4)

5. Use (2) and (3) in (4) and simplify: $1 = m/M + (12m/M)(d^2/L^2)$.

6. Solve for d: $d = L\sqrt{\frac{M-m}{12m}}$.

---

52 •• Figure 10-40 shows a uniform rod of length L and mass M pivoted at the top. The rod, which is initially at rest, is struck by a particle of mass m at a point $d = 0.8L$ below the pivot. Assume that the collision is perfectly inelastic. What must be the speed v of the particle so that the maximum angle between the rod and the vertical is 90°?

Use conservation of L about pivot to find ω immediately after collision, then use energy conservation to determine v of mass m before collision for an arbitrary angle θ. Then set $\theta = 90°$.

1. Conservation of angular momentum: $0.8Lmv = I\omega = (ML^2/3 + 0.64L^2m)\omega$. $\omega = (0.8Lmv)/(ML^2/3 + 0.64mL^2)$.

2. Conservation of energy: $[MgL/2 + mg(0.8L)](1 - \cos\theta) = \frac{1}{2}I\omega^2 = \frac{0.32(Lmv)^2}{ML^2/3 + 0.64mL^2}$.

3. Solve for v: $v = \sqrt{\frac{(0.5M + 0.8m)(ML^2/3 + 0.64mL^2)g(1-\cos\theta)}{0.32Lm^2}}$.

4. For $\theta = 90°$, $1 - \cos\theta = 1$.

53* •• If, for the system of Problem 52, $L = 1.2$ m, $M = 0.8$ kg, and $m = 0.3$ kg, and the maximum angle between the rod and the vertical is 60°, find the speed of the particle before impact.

Substitute numerical values in result for v of Problem 52, using $(1 - \cos 60°) = 0.5$. $v = 7.75$ m/s.

---

54 •• A projectile of mass m_p is traveling at a constant velocity v_0 toward a stationary disk of mass M and radius R that is free to rotate about a pivot through its axis O (Figure 10-41). Before impact, the projectile is traveling along a line displaced a distance b below the axis. The projectile strikes the disk and sticks to point B. Treat the projectile as a point mass. (*a*) Before impact, what is the total angular momentum L_0 of the projectile and disk about the O axis? (*b*) What is the angular speed ω of the disk and projectile system just after the impact? (*c*) What is the kinetic energy of the disk and projectile system after impact? (*d*) How much mechanical energy is lost in this collision?

(*a*) $L_0 = m_pv_0b$.

(*b*) L about the pivot is conserved; $I = \frac{1}{2}MR^2 + m_pb^2$; $\omega = L/I = 2m_pv_0b/(MR^2 + m_pb^2)$.

(*c*) $K_f = L^2/2I = (m_pv_0b)^2/(MR^2 + 2m_pb^2)$.

(*d*) Energy loss is $K_i - K_f$. $K_i = \frac{1}{2}m_pv_0^2$; $\Delta E = K_i - K_f = \frac{\frac{1}{2}m_pv_0^2}{1 - 2m_pb^2/MR^2}$.

55 •• A uniform rod of length L_1 and mass $M = 0.75$ kg is supported by a hinge at one end and is free to rotate in the vertical plane (Figure 10-42). The rod is released from rest in the position shown. A particle of mass $m = 0.5$ kg is supported by a thin string of length L_2 from the hinge. The particle sticks to the rod on contact. What should be the ratio L_2/L_1 so that $\theta_{max} = 60°$ after the collision?

1. Find ω of rod at impact using energy conservation: $MgL_1/2 = ½(ML_1^2/3)\omega^2$; $\omega = (3g/L_1)^{½}$.
2. Use conservation of L to find ω', where ω' is angular speed after impact: $\omega' = \omega(ML_1^2/3)/(ML_1^2/3 + mL_2^2)$.
3. Use energy conservation: $(3g/L_1)(ML_1^2/3)^2/(ML_1^2/3 + mL_2^2) = (ML_1/2 + mL_2)g$.
4. Simplify above, using $\alpha = m/M$ and $\beta = L_2/L_1$: $3\alpha^2\beta^3 + \alpha\beta^2 + \alpha\beta = 2$; $(4/3)\beta^3 + (2/3)\beta^2 + (2/3)\beta - 2 = 0$.
5. Solve the cubic equation for β: $\beta = 0.88$.

56 •• Returning to Figure 10-42, this time set $L_1 = 1.2$ m, $M = 2.0$ kg, and $L_2 = 0.8$ m. After the inelastic collision, $\theta_{max} = 37°$. Find m. How much energy is dissipated in this inelastic collision?

Follow steps 1 and 2 of Problem 55 to obtain an expression for ω' using appropriate numerical values. This gives $\omega' = 4.75/(0.96 + 0.64m)$ rad/s. Now use energy conservation, keeping in mind that $(1 - \cos 37°) = 0.2$; Thus, $½ \times 4.75^2/(0.96 + 0.64m)$ J $= (2 \times 0.6 + 0.8m) \times 0.2 \times 9.81$ J. Simplify to $m^2 + 3.01m - 9.02 = 0$ and solve for m: $m = 1.85$ kg. $\Delta E = U_i - U_f$; $U_i = MgL_1/2 = 11.8$ J; $U_f = 0.2(MgL_1/2 + mgL_2) = 5.26$ J; $\Delta E = 6.54$ J.

57* •• Suppose that in Figure 10-42, $m = 0.4$ kg, $M = 0.75$ kg , $L_1 = 1.2$ m, and $L_2 = 0.8$ m. What minimum initial angular velocity must be imparted to the rod so that the system will revolve completely about the hinge following the inelastic collision? How much energy is then dissipated in the inelastic collision?

Let ω_i and ω_f be the angular velocities of the rod immediately before and immediately after the inelastic collision with the mass m. Let ω_0 be the initial angular velocity of the rod. We proceed as follows: 1. We apply energy conservation to determine ω_f. 2. Next we apply conservation of angular momentum to determine ω_i. 3. Next, we again apply energy conservation to determine ω_0. 4. Finally, we find the energies of the system immediately before and immediately after the collision and, thereby, the energy loss.

1. Set K immediately after collision equal to potential energy after 180° rotation.
   $½(ML_1^2/3 + mL_2^2)\omega_f^2 = MgL_1 + 2mgL_2$; evaluate ω_f: $\omega_f = 7.0$ rad/s.
2. $(ML_1^2/3)\omega_i = (ML_1^2/3 + mL_2^2)\omega_f$; solve for and evaluate ω_i; $\omega_i = [(0.96 + 0.256)/0.96]\omega_f = 8.87$ rad/s.
3. $½(ML_1^2/3)\omega_i^2 = ½(ML_1^2/3)\omega_0^2 + MgL_1/2$; $\omega_0^2 = \omega_i^2 - 3g/L_1$; $\omega_0 = 7.355$ rad/s.
4. $E_i = ½(ML_1^2/3)\omega_i^2 = 40.95$ J; $E_f = MgL_1 + 2mgL_2 = 29.82$ J. $\Delta E = 11.13$ J.

58 ••• Repeat Problem 56 if the collision between the rod and particle is elastic.

Let v be the speed of m immediately after collision and let ω_i and ω_f be the angular speed of the rod immediately before and immediately after the collision.

1. Use energy conservation at collision	$MgL_1/2 - 0.2MgL_1/2 = ½mv^2$; $mv^2 = 0.8MgL_1 = 18.86$ J
2. Use conservation of energy before the collision	$½(ML_1^2/3)\omega_i^2 = MgL_1/2$; $\omega_i = (3g/L_1)^{½} = 4.95$ rad/s
3. Use conservation of energy of rod after collision	$½(ML_1^2/3)\omega_f^2 = 0.2MgL_1/2$; $\omega_f = (0.6g/L_1)^{½} = 2.215$ rad/s
4. Use conservation of angular momentum	$ML_1^2\omega_i/3 = ML_1^2\omega_f/3 + mvL_2$; $mv = 3.82$ kg·m/s
5. From the results of 1. and 4. find v and m	$v = (18.86/3.82)$ m/s $= 5.75$ m/s; $m = 0.665$ kg
6. The collision is elastic; $\Delta E = 0$	

---

59 • True or false: (*a*) Nutation and precession are the same phenomenon. (*b*) The direction of precession is the direction of the net torque. (*c*) When the gyroscope is not spinning, $\tau = dL/dt$ does not hold.

(*a*) False (*b*) True (*c*) False

---

60 •• The angular momentum vector for a spinning wheel lies along its axle and is pointed east. To make this vector point south, it is necessary to exert a force on the east end of the axle in which direction? (*a*) Up (*b*) Down (*c*) North (*d*) South (*e*) East

(*b*)

61* •• A man is walking north carrying a suitcase that contains a spinning gyroscope mounted on an axle attached to the front and back of the case. The angular velocity of the gyroscope points north. The man now begins to turn to walk east. As a result, the front end of the suitcase will (*a*) resist his attempt to turn and will try to remain pointed north. (*b*) fight his attempt to turn and will pull to the west. (*c*) rise upward. (*d*) dip downward. (*e*) cause no effect whatsoever.

(*d*)

---

62 •• The angular momentum of the propeller of a small airplane points forward. (*a*) As the plane takes off, the nose lifts up and the airplane tends to veer to one side. To which side does it veer and why? (*b*) If the plane is flying horizontally and suddenly turns to the right, does the nose of the plane tend to move up or down? Why?

(*a*) It veers to the right; this is the direction of the torque associated with the lifting of the nose.

(*b*) In turning to the right, the torque points down. The nose will tend to move down.

---

63 •• A car is powered by the energy stored in a single flywheel with an angular momentum **L**. Discuss the problems that would arise for various orientations of **L** and various maneuvers of the car. For example, what would happen if **L** points vertically upward and the car travels over a hilltop or through a valley? What would happen if **L** points forward or to one side and the car attempts to turn to the left or right? In each case that you examine, consider the direction of the torque exerted on the car by the road.

If **L** points up and the car travels over a hill or through a valley, the force on the wheels on one side (or the other) will increase and car will tend to tip. If **L** points forward and car turns left or right, the front (or rear) of the car will tend to lift. These problems can be averted by having two identical flywheels that rotate on the same shaft in opposite directions.

---

64 •• A bicycle wheel of radius 28 cm is mounted at the middle of an axle 50 cm long. The tire and rim weigh 30 N. The wheel is spun at 12 rev/s, and the axle is then placed in a horizontal position with one end resting on a pivot. (*a*) What is the angular momentum due to the spinning of the wheel? (Treat the wheel as a hoop.) (*b*) What is the angular velocity of precession? (*c*) How long does it take for the axle to swing through 360° around the pivot? (*d*) What is the angular momentum associated with the motion of the center of mass, that is, due to the precession? In what direction is this angular momentum?

(*a*) $L = I\omega$; $I = MR^2$ — $L = (30/9.81) \times 0.28^2 \times (24\pi)$ J·s = 18.1 J·s

(*b*) $\omega_p = MgD/L$ — $\omega_p = (30 \times 0.25/18.1)$ rad/s = 0.414 rad/s

(*c*) $T = 2\pi/\omega_p$ — $T = 2\pi/0.414$ s = 15.2 s

(*d*) $L_p = MD^2\omega_p$ — $L_p = 0.079$ J·s; up or down, depending on the direction of L

65* •• A uniform disk of mass 2.5 kg and radius 6.4 cm is mounted in the center of a 10-cm axle and spun at 700 rev/min. The axle is then placed in a horizontal position with one end resting on a pivot. The other end is given an initial horizontal velocity such that the precession is smooth with no nutation. (*a*) What is the angular velocity of precession? (*b*) What is the speed of the center of mass during the precession? (*c*) What are the magnitude and direction of the acceleration of the center of mass? (*d*) What are the vertical and horizontal components of the force exerted by the pivot?

(*a*) $\omega_p = MgD/I_s\omega_s$; $I_s = \frac{1}{2}MR^2$; $\omega_p = 2gD/R^2\omega_s$ $\omega_p = [2 \times 9.81 \times 0.05/(0.064^2 \times 700 \times 2\pi/60)]$ $= 3.27$ rad/s

(*b*) $v_{cm} = \omega_p D$ $v_{cm} = 3.27 \times 0.05$ m/s $= 0.163$ m/s

(*c*) $a_{cm} = D\omega_p^2$ $a_{cm} = 0.535$ m/s^2

(*d*) $F_v = Mg$; $F_h = Ma_{cm}$ $F_v = 24.5$ N; $F_h = 1.34$ N

---

66 • An object of mass M is rotating about a fixed axis with angular momentum L. Its moment of inertia about this axis is I. What is its kinetic energy? (*a*) $IL^2/2$ (*b*) $L^2/2I$ (*c*) $ML^2/2$ (*d*) $IL^2/2M$

(*b*)

---

67 • Explain why a helicopter with just one main rotor has a second smaller rotor mounted on a horizontal axis at the rear as in Figure 10-43. Describe the resultant motion of the helicopter is this rear rotor fails during flight.

To prevent the body of the helicopter from rotating. If the rear rotor fails, the body of the helicopter will tend to rotate on the main axis.

---

68 •• A woman sits on a spinning piano stool with her arms folded. When she extends her arms out to the side, her kinetic energy (*a*) increases. (*b*) decreases. (*c*) remains the same.

(*b*) L constant but I increases.

69* •• In tetherball, a ball is attached to a string that is attached to a pole. When the ball is hit, the string wraps around the pole and the ball spirals inward. Neglecting air resistance, what happens as the ball swings around the pole? Give supporting arguments for your choice. (*a*) The mechanical energy and angular momentum of the ball are conserved. (*b*) The angular momentum of the ball is conserved, but the mechanical energy of the ball increases. (*c*) The angular momentum of the ball is conserved, and the mechanical energy of the ball decreases. (*d*) The mechanical energy of the ball is conserved and the angular momentum of the ball increases. (*e*) The mechanical energy of the ball is conserved and the angular momentum of the ball decreases.

(*e*) Consider the situation shown in the adjoining figure. The ball rotates counterclockwise. The torque about the center of the pole is clockwise and of magnitude RT, where R is the pole's radius and T is the tension. So L must decrease.

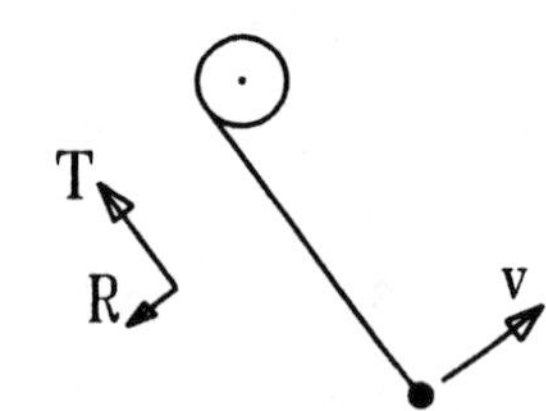

---

70 •• A uniform rod of mass M and length L lies on a horizontal frictionless table. A piece of putty of mass $m = M/4$ moves along a line perpendicular to the rod, strikes the rod near its end, and sticks to the rod. Describe qualitatively the subsequent motion of the rod and putty.

The center of mass of the rod-and-putty system moves in a straight line, and the system rotates about the center of mass.

---

71 • A particle of mass 3 kg moves with velocity $\boldsymbol{v} = 3$ m/s $\boldsymbol{i}$ along the line $z = 0$, $y = 5.3$ m. (*a*) Find the angular momentum $\boldsymbol{L}$ relative to the origin when the particle is at $x = 12$ m, $y = 5.3$ m. (*b*) A force $\boldsymbol{F} = -3$ N $\boldsymbol{i}$ is applied to the particle. Find the torque relative to the origin due to this force.

(*a*) $\boldsymbol{L} = \boldsymbol{r} \times \boldsymbol{p}$ — $\boldsymbol{L} = (5.3 \times 3 \times 3)\boldsymbol{j} \times \boldsymbol{i} = -47.7\,\boldsymbol{k}$

(*b*) $\boldsymbol{\tau} = \boldsymbol{r} \times \boldsymbol{F}$ — $\boldsymbol{\tau} = (5.3 \times 3)\boldsymbol{j} \times -\boldsymbol{i} = 15.9\,\boldsymbol{k}$ N·m

---

72 • The position vector of a particle of mass 3 kg is given by $\boldsymbol{r} = 4\,\boldsymbol{i} + 3t^2\,\boldsymbol{j}$, where $\boldsymbol{r}$ is in meters and t is in seconds. Determine the angular momentum and torque acting on the particle about the origin.

$\boldsymbol{L} = m\boldsymbol{r} \times \boldsymbol{v} = m\boldsymbol{r} \times d\boldsymbol{r}/dt$; find $d\boldsymbol{r}/dt$ and $\boldsymbol{L}$ — $d\boldsymbol{r}/dt = 6t\,\boldsymbol{j}$; $\boldsymbol{L} = (3 \times 4 \times 6)t\,\boldsymbol{i} \times \boldsymbol{j}$ J·s $= 72t\,\boldsymbol{k}$ J·s

$\boldsymbol{\tau} = d\boldsymbol{L}/dt$ — $\boldsymbol{\tau} = 72\,\boldsymbol{k}$ N·m

---

73* •• An ice skater starts her pirouette with arms outstretched, rotating at 1.5 rev/s. Estimate her rotational speed (in revolutions per second) when she brings her arms flat against her body.

Assume I of body, minus arms, $= \frac{1}{2} \times 50 \times 0.2^2 = 1.0$ kg·m². The mass of each arm = 4 kg, and length = 1.0 m. Then I_{tot}(arms out) $= 1.0 + 2 \times 4/3 = 3.7$ kg·m² and I_{tot}(arms in) $= 1.0 + 2 \times 4 \times 0.2^2 = 1.32$ kg·m². With L constant, one then finds that $\omega = 1.5(3.7/1.32) = 4.2$ rev/s.

---

74 •• Two ice skaters hold hands and rotate, making one revolution in 2.5 s. Their masses are 55 kg and 85 kg, and they are separated by 1.7 m. Find (*a*) the angular momentum of the system about their center of mass, and (*b*) the total kinetic energy of the system.

(*a*) 1. Locate the center of mass from 85 kg mass — $x_{cm} = 1.7 \times 55/140$ m $= 0.668$ m

2. Determine I and L — $I = (55 \times 1.032^2 + 85 \times 0.668^2)$ kg·m² $= 96.5$ kgm²

$L = (96.5 \times 2\pi/2.5)$ J·s $= 242.5$ J·s

(*b*) $K = L^2/2I$ — $K = (242.5^2/193)$ J $= 305$ J

---

75 •• A 2-kg ball attached to a string of length 1.5 m moves in a horizontal circle as a conical pendulum (Figure 10-44). The string makes an angle $\theta = 30°$ with the vertical. (*a*) Show that the angular momentum of the ball about the point of support P has a horizontal component toward the center of the circle as well as a vertical component, and find these components. (*b*) Find the magnitude of dL/dt, and show that it equals the magnitude of the torque exerted by gravity about the point of support.

Take the coordinate origin at point P. Then $\boldsymbol{r} = (1.5 \sin 30°)(\cos \omega t\,\boldsymbol{i} + \sin \omega t\,\boldsymbol{j}) - 1.5 \cos 30°\,\boldsymbol{k}$; $\boldsymbol{\omega} = \omega\boldsymbol{k}$.

(*a*) 1. Use Newton's laws to determine v; — $T \cos \theta = mg$; $T \sin \theta = mv^2/(r \sin\theta)$;

$v = (rg \sin \theta \tan \theta)^{1/2}$

$v = (1.5 \times 9.81/2 \times \sqrt{3})^{1/2}$ m/s $= 2.06$ m/s

$\boldsymbol{v} = d\boldsymbol{r}/dt$ — $\boldsymbol{v} = -2.06 \sin \omega t\,\boldsymbol{i} + 2.06 \cos \omega t\,\boldsymbol{j}$

2. Find $\boldsymbol{L} = \boldsymbol{r} \times \boldsymbol{p} = m\boldsymbol{r} \times \boldsymbol{v}$ — $\boldsymbol{L} = 6.18[\sin 30°(\cos\omega t\,\boldsymbol{i} + \sin\omega t\,\boldsymbol{j}) - \cos 30°\,\boldsymbol{k}] \times (-\sin\omega t\,\boldsymbol{i} + \cos\omega t\,\boldsymbol{j})$

$= [(3.09\,\boldsymbol{k} + 5.35(\sin\omega t\,\boldsymbol{j} + \cos\omega t\,\boldsymbol{i})]$ J·s (*b*) 1. Find dL/dt; only the horizontal component of L depends on t — $dL/dt = 5.35\omega(\cos\omega t\, \mathbf{j} - \sin\omega t\, \mathbf{i})$ J; $\omega = v/(r \sin 30°) = 2.75$ rad/s; $|dL/dt| = 14.7$ J

2. Find $\tau = mgr \sin 30°$ — $\tau = 2 \times 9.81 \times 0.75 = 14.7$ N·m

76 •• A mass m on a horizontal, frictionless surface is attached to a string that wraps around a vertical cylindrical post so that when it is set into motion it follows a path that spirals inward. (*a*) Is the angular momentum pf the mass conserved? (*b*) Is the energy of the mass conserved? (*c*) If the speed of the mass is v_0 when the length of the string is r, what is its speed when the unwrapped length has shortened to $r/2$?

See Problem 69. (*a*) No (*b*) Yes (*c*) Since K is conserved, v is constant at v_0.

77* •• Figure 10-45 shows a hollow cylindrical tube of mass M, length L, and moment of inertia $ML^2/10$. Inside the cylinder are two masses m, separated a distance ℓ and tied to a central post by a thin string. The system can rotate about a vertical axis through the center of the cylinder. With the system rotating at ω, the strings holding the masses suddenly break. When the masses reach the end of the cylinder, they stick. Obtain expressions for the final angular velocity and the initial and final energies of the system. Assume that the inside walls of the cylinder are frictionless.

1. $\tau = 0$; $L_f = L_i$; $\omega_f = \omega(I_i/I_f)$. Obtain expressions for I_i and I_f and ω_f.

$$I_i = (ML^2/10 + 2m\ell^2/4);\ I_f = (ML^2/10 + 2mL^2/4) \quad (1)$$

$$\omega_f = [(M + 5m\ell^2/L^2)/(M + 5m)]\omega \quad (2)$$

2. $K_i = ½I_i\omega^2$; $K_f = ½I_f\omega_f^2$

$$K_i = (ML^2 + 5m\ell^2)\omega^2/20 \quad (3)$$

$$K_f = [(ML^2 + 5m\ell^2)^2/(ML^2 + 5mL^2)]\omega^2/20 \quad (4)$$

78 •• Repeat Problem 77, this time adding friction between the masses and walls of the cylinder. However, the coefficient of friction is not enough to prevent the masses from reaching the ends of the cylinder. Can the final energy of the system be determined without knowing the coefficient of kinetic friction?

Again, there is no net external torque. Since the masses, m, come to the same final position, the initial and final configurations are the same as in Problem 77. Therefore, the answers are the same as for Problem 77.

79 •• Suppose that in Figure 10-45, $\ell = 0.6$ m, $L = 2.0$ m, $M = 0.8$ kg, and $m = 0.4$ kg. The system rotates at ω such that the tension in the string is 108 N just before it breaks. Determine the initial and final angular velocities and initial and final energies of the system. Assume that the inside walls of the cylinder are frictionless.

Here we need only to determine ω. We can then use the expressions derived in Problem 77.

$m(\ell/2)\omega^2 = T$; $\omega = \sqrt{2T/m\ell} = 30$ rad/s.

Substitute numerical values into Equ. (2) of Problem 77: $\omega_f = 10.5$ rad/s.

Substitute numerical values into Equ. (3) and Equ. (4) of Problem 77: $K_i = 176.4$ J; $K_f = 61.74$ J.

80 •• For Problem 77, determine the radial velocity of each mass just before it reaches the end of the cylinder.

Until the inelastic collision of the masses m at the ends of the cylinder, both angular momentum and energy are conserved. To determine v_r we set $K' = K_i$, where K' is the kinetic energy of the system just before the masses m reach the end of the cylinder.

1. Write expressions for K' and K_i; solve for v_r — $K' = ½I_f\omega_f^2 + ½(2mv_r^2) = ½I_i\omega^2$; $v_r = \sqrt{\dfrac{I_i\omega^2 - I_f\omega_f^2}{2m}}$

2. Use the results of Problem 77 and simplify — $v_r = (\ell\omega/2L)\sqrt{L^2 - \ell^2}$

81* •• Given the numerical values of Problem 79, suppose the coefficient of friction between the masses and the walls of the cylinder is such that the masses cease sliding 0.2 m from the ends of the cylinder. Determine the initial and final angular velocities of the system and the energy dissipated in friction.

As before, angular momentum is conserved. The masses m now come to rest at $r = 0.8$ m. $\omega_i = 30$ rad/s as before.

1. Use $L_i = L_f$; Find I_i and I_f	$I_i = 0.392$ kg·m^2; $I_f = ML^2/10 + 2m \times 0.8^2 = 0.832$ kg·m^2
2. $\omega_f = \omega_i(I_i/I_f)$	$\omega_f = 14.13$ rad/s
3. $\Delta E = K_i - K_f$; $K_i = 176.4$ J (see Problem 79)	$K_f = ½ \times 0.832 \times 14.13^2$ J $= 83.1$ J; $\Delta E = 93.3$ J

82 •• Kepler's second law states: *The radius vector from the sun to a planet sweeps out equal areas in equal times.* Show that this law follows directly from the law of conservation of angular momentum and the fact that the force of gravitational attraction between a planet and the sun acts along the line joining the two celestial objects.

The drawing shows an elliptical orbit. The triangular element of the area is $dA = ½r(r\,d\theta) = ½r^2\,d\theta$. So $dA/dt = ½r^2\omega$. Since the force acts along r, $\tau = 0$ and $Mr^2\omega = L$ is constant. Consequently dA/dt is constant.

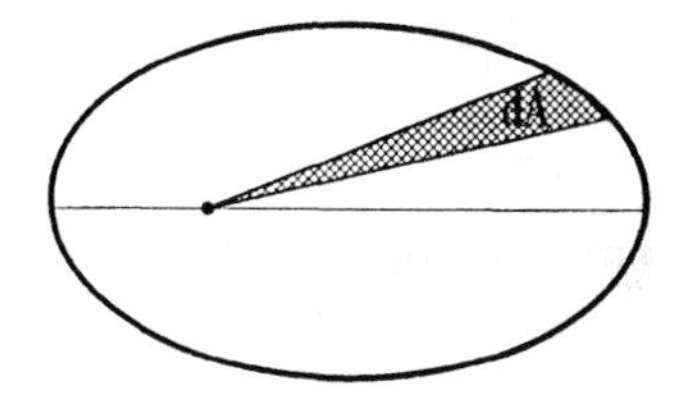

83 •• Figure 10-46 shows a hollow cylinder of length 1.8 m, mass 0.8 kg, and radius 0.2 m that is free to rotate about a vertical axis through its center and perpendicular to the cylinder's axis. Inside the cylinder are two thin disks of 0.2 kg each, attached to springs of spring constant k and unstretched lengths 0.4 m. The system is brought to a rotational speed of 8 rad/s with the springs clamped so they do not stretch. The springs are then suddenly unclamped. When the disks have stopped their radial motion due to friction between the disks and the wall, they come to rest 0.6 m from the central axis. What is the angular velocity of the cylinder when the disks have stopped their radial motion? How much energy was dissipated in friction between the disks and cylinder wall?

1. Find initial and final I of system in kg·m^2; let x be the radial distance of each disk	$I = I_{cyl} + 2I_d$; $I_{cyl} = 0.8(1.8^2/12 + 0.2^2/2) = 0.232$; $I_d = 0.2(0.2^2/4 + x^2)$; $I_i = 0.3$; $I_f = 0.38$
2. $\omega_f = \omega_i(I_i/I_f)$	$\omega_f = 8 \times 0.3/0.38$ rad/s $= 6.316$ rad/s
3. Find the spring constant k;	$0.2k = 0.2 \times 0.6 \times 6.316^2$; $k = 23.9$ N/m
4. Find initial and final energies of system	$E_i = ½I_i\omega_i^2 = 9.6$ J; $E_f = ½I_f\omega_f^2 + ½k\Delta x^2 = 8.06$ J
5. $W_{fr} = E_i - E_f$	$W_{fr} = 1.54$ J

84 •• (*a*) Assuming the earth to be a homogeneous sphere of radius r and mass m, show that the period T of the earth's rotation about its axis is related to its radius by $T = (4\pi m/5L)r^2$, where L is the angular momentum of the earth due to its rotation. (*b*) Suppose that the radius r changes by a very small amount Δr due to some internal effect such as thermal expansion. Show that the fractional change in the period ΔT is given approximately by $\Delta T/T = 2\Delta r/r$. *Hint:* Use the differentials dr and dT to approximate the changes in these quantities. (*c*) By how many kilometers would the earth need to expand for the period to change by $(1/4)d/y$ so that leap years would no longer be necessary?

(*a*) $T = 2\pi/\omega$; $\omega = L/I$; $I = 2mr^2/5$	$T = 2\pi I/L = (4\pi m/5L)r^2$
(*b*) Find dT/dr; then dT/T	$dT/dr = 2(4\pi m/5L)r$; $dT/T = 2dr/r$ or $\Delta T/T = 2\Delta r/r$
(*c*) $\Delta T/T = 1/(4 \times 365)$	$\Delta r = r\Delta T/2T = ½(6.37 \times 10^3)(1/1460)$ km = 2.18 km

85* •• The polar ice caps contain about 2.3×10^{19} kg of ice. This mass contributes negligibly to the moment of inertia of the earth because it is located at the poles, close to the axis of rotation. Estimate the change in the length of the day that would be expected if the polar ice caps were to melt and the water were distributed uniformly over the surface of the earth. (The moment of inertia of a spherical shell of mass m and radius r is $2mr^2/3$.)

1. $T = 2\pi I/L$ (see Problem 84); L is constant	$dT/T = dI/I$ or $\Delta T = T\Delta I/I$
2. $I = 2M_ER_E^2/5$; $M_E = 6 \times 10^{24}$ kg; $\Delta I = 2mR_E^2/3$	$\Delta T = (1\ d)[(5/3)(2.3 \times 10^{19}/6 \times 20^{24}) = 6.4 \times 10^{-6}\ d$ = 0.55 s

86 ••• Figure 10-47 shows a hollow cylinder of mass $M = 1.2$ kg and length $L = 1.6$ m that is free to rotate about a vertical axis through its center. Inside the cylinder are two disks, each of mass 0.4 kg that are tied to a central post by a thin string and separated by a distance $\ell = 0.8$ m. The string breaks if the tension exceeds 100 N. Staring from rest, a torque is applied to the system until the string breaks. Assuming the disks are point masses and the radius of the cylinder is negligible, find the amount of work done up to that instant. Suppose that at that instant, the applied torque is removed, and that the walls of the cylinder are frictionless. Obtain an expression for the angular velocity of the system as a function of x for $x < L/2$, where x is the distance between each mass and the central post.

1. Find $I_i = I_{cyl} + 2I_m$ in kg·m^2	$I_{cyl} = 1.2 \times 1.6^2/12 = 0.256$; $2I_m = 2 \times 0.4 \times x^2 = 0.8x^2$; $I_i = I(x = 0.4) = 0.384$; $I(x) = 0.256 + 0.8x^2$
2. Find ω_i when string breaks; $mr\omega_i^2 = T_c$	$\omega_i = (100/0.4 \times 0.4)^{½} = 25$ rad/s
3. Work done = K; $K = ½I_i\omega_i^2$	$W = ½(0.384 \times 25^2) = 120$ J
4. Find L; $L = I_i\omega_i = I(x)\omega(x)$; solve for $\omega(x)$	$L = 9.6$ J·s; $\omega(x) = [9.6/(0.256 + 0.8x^2)]$ rad/s

87 ••• For the system of Problem 86, find the angular velocity of the system just before and just after the point masses pass the ends of the cylinder.

Use the result of Problem 86: $x = 0.8$ m; $\omega = [9.6/(0.256 + 0.8^3)]$ rad/s = 12.5 rad/s. This is the angular velocity in both instances because the masses leave the cylinder with a tangential velocity of $½L\omega$ so the angular momentum of the system remains constant.

88 ••• Repeat Problem 86 with the radius of the hollow cylinder as 0.4 m and the masses treated as thin disks rather than point masses.

1. To find I_{cyl} and I_m we must refer to Table 9.1 and also use the parallel axis theorem.	$I_{cyl} = M(L^2/12 + R^2/2) = 0.352$; $2I_m = 2m(R^2/4 + x^2)$ $I_i = I_{cyl} + 2I_m(x = 0.4) = 0.512$
2. ω_i is the same as in Problem 86	$\omega_i = 25$ rad/s
3. $W = ½I_i\omega_i^2$	$W = ½ \times 0.512 \times 625$ J = 160 J
4. Find L; $L = I_i\omega_i = I(x)\omega(x)$; solve for $\omega(x)$	$\omega(x) = [12.8/(0.384 + 0.8x^2)]$ rad/s

89* ••• Figure 10-48 shows a pulley in the shape of a uniform disk with a heavy rope hanging over it. The circumference of the pulley is 1.2 m and its mass is 2.2 kg. The rope is 8.0 m long and its mass is 4.8 kg. At the instant shown in the figure, the system is at rest and the difference in height of the two ends of the rope is 0.6 m. (*a*) What is the angular velocity of the pulley when the difference in height between the two ends of the rope is 7.2 m? (*b*) Obtain an expression for the angular momentum of the system as a function of time while neither end of the rope is above the center of the pulley. There is no slippage between rope and pulley.

(*a*) Take $y = 0$ at center of pulley; write U_i and U_f for the free part of the rope. Let $\lambda = M_r/L = 0.6$ kg/m.	$U_i = -\frac{1}{2}L_{1i}(L_{1i}\lambda g) - \frac{1}{2}L_{2i}(L_{2i}\lambda g) = -\frac{1}{2}(L_{1i}^2 + L_{2i}^2)\lambda g$ $U_f = -\frac{1}{2}(L_{1f}^2 + L_{2f}^2)\lambda g$; $L_{1i}+L_{2i} = 7.4$ m; $L_{2i}-L_{2i} = 0.6$ m;
Let L_1 and L_2 be the lengths of the hanging parts.	$L_{1i} = 3.4$ m, $L_{2i} = 4.0$ m; also $L_{1f} = 0.1$ m, $L_{2f} = 7.3$ m
Use conservation of energy: $K + \Delta U = 0$.	$K = \frac{1}{2}I_p\omega^2 + \frac{1}{2}Mv^2$; $v = R\omega$; $R = (1.2/2\pi)$ m $= 0.6/\pi$ m
Write K in terms of ω.	$K = \frac{1}{2} \times \frac{1}{2} \times 2.2(0.6/\pi)^2\omega^2 + \frac{1}{2} \times 8(0.6/\pi)^2\omega^2$ $= 0.166\omega^2$
Find ΔU	$\Delta U = U_f - U_i = -\frac{1}{2}(0.1^2 - 3.4^2 + 7.3^2 - 4.0^2) \times 0.6g$ $= -75.75$ J
Solve for ω	$\omega = (75.75/0.166)^{1/2}$ rad/s $= 21.36$ rad/s
(*b*) Now write $U(\theta)$ and ΔU, where θ is the angle through which the pulley has turned. This will reduce L_1 by $R\theta$ and increase L_2 by $R\theta$.	$U(\theta) = -\frac{1}{2}[(L_{1i} - R\theta)^2 + (L_{2i} + R\theta)^2]\lambda g$ $\Delta U = -U(\theta) + \frac{1}{2}(L_{1i}^2 + L_{2i}^2)\lambda g = R^2\theta^2\lambda g$
Use energy conservation and result for K	$0.166\omega^2 = [(0.6/\pi)^2 \times 0.6 \times 9.81]\theta^2 = 0.215\theta^2$; $\omega = 1.14\theta$
Recall that $\omega = d\theta/dt$	$d\theta/dt = 1.14\theta$; $d\theta/\theta = 1.14\,dt$
Integrate	$\ln(\theta) = 1.14t$; $\theta(t) = e^{1.14t} - 1$
Find $\omega(t)$	$\omega(t) = d\theta/dt = 1.14e^{1.14t}$
$L = L_p + L_r = (I_p + M_rR^2)\omega$; note that the angular momentum of each portion of the rope is the same	$L = [\frac{1}{2} \times 2.2(0.6/\pi)^2 + 8.0(0.6/\pi)^2] \times 1.14e^{1.14t}$ J·s $L = 0.378e^{1.14t}$ J·s

CHAPTER 11

Gravity

1* • True or false: (*a*) Kepler's law of equal areas implies that gravity varies inversely with the square of the distance. (*b*) The planet closest to the sun, on the average, has the shortest period of revolution about the sun.

(*a*) False (*b*) True

2 • If the mass of a satellite is doubled, the radius of its orbit can remain constant if the speed of the satellite (*a*) increases by a factor of 8. (*b*) increases by a factor of 2. (*c*) does not change. (*d*) is reduced by a factor of 8. (*e*) is reduced by a factor of 2.

(*c*)

3 • One night, Lucy picked up a strange message on her ham radio. "Help! We ran away from Earth to live in peace and serenity, and we got disoriented. All we know is that we are orbiting the sun with a period of 5 years. Where are we?" Lucy did some calculations and told the travelers their mean distance from the sun. What is it?

Use Equ. 11-2; $R = R_{ES}(5/1)^{2/3}$; $R_{ES} = 1.5 \times 10^{11}$ m $\quad R = (1.5 \times 10^{11})(5)^{2/3}$ m $= 4.39 \times 10^{11}$ m

4 • Halley's comet has a period of about 76 y. What is its mean distance from the sun?

$R_{mean} = (1\text{ AU})(76)^{2/3}$ (see Problem 3) $\quad R_{mean} = 1.5 \times 10^{11} \times 76^{2/3}$ m $= 26.9 \times 10^{11}$ m

5* • A comet has a period estimated to be about 4210 y. What is its mean distance from the sun? (4210 y was the estimated period of the comet Hale–Bopp, which was seen in the Northern Hemisphere in early 1997. Gravitational interactions with the major planets that occurred during this apparition of the comet greatly changed its period, which is now expected to be about 2380 y.)

$R_{mean} = (1\text{ AU})(4210)^{2/3}$ (see Problem 4) $\quad R_{mean} = 1.5 \times 10^{11} \times 4210^{2/3}$ m $= 3.91 \times 10^{13}$ m

6 • The radius of the earth's orbit is 1.496×10^{11} m and that of Uranus is 2.87×10^{12} m. What is the period of Uranus?

Use Equ. 11-2; $T_U = (1\text{ y})(R_U/1\text{ AU})^{3/2}$ $\quad T_U = (2.87 \times 10^{12}/1.5 \times 10^{11})^{3/2}$ y $= 83.7$ y

7 • The asteroid Hektor, discovered in 1907, is in a nearly circular orbit of radius 5.16 AU about the sun. Determine the period of this asteroid.

$T_H = (1\text{ y})(5.16/1)^{3/2}$ (see Problem 6) $\quad T_H = 11.7$ y

8 •• The asteroid Icarus, discovered in 1949, was so named because its highly eccentric elliptical orbit brings it close to the sun at perihelion. The eccentricity e of an ellipse is defined by the relation $d_p = a(1 - e)$, where d_p is the perihelion distance and a is the semimajor axis. Icarus has an eccentricity of 0.83. The period of Icarus is 1.1 years. (*a*) Determine the semimajor axis of the orbit of Icarus. (*b*) Find the perihelion and aphelion distances of the orbit of Icarus.

(*a*) Use Kepler's third law; $a = (1\text{ AU})(T_I)^{2/3}$ — $a = 1.5 \times 10^{11} \times 1.1^{2/3}\text{ m} = 1.6 \times 10^{11}\text{ m}$

(*b*) $d_p = a(1 - e)$; $d_a + d_p = 2a$; $d_a = 2a - d_p$ — $d_p = (1.6 \times 10^{11} \times 0.17)\text{ m} = 2.72 \times 10^{10}\text{ m}$; $d_a = 2.93 \times 10^{11}\text{ m}$

9* • Why don't you feel the gravitational attraction of a large building when you walk near it?

The mass of the building is insignificant compared to the mass of the earth.

10 • Astronauts orbiting in a satellite 300 km above the surface of the earth feel weightless. Why? Is the force of gravity exerted by the earth on them negligible at this height?

They feel weightless because the force of gravity provides just their centripetal acceleration. The force of gravity is not negligible. (see Example 11-2)

11 •• The distance from the center of the earth to a point where the acceleration due to gravity is $g/4$ is (*a*) R_E. (*b*) $4R_E$. (*c*) $R_E/2$. (*d*) $2R_E$. (*e*) none of the above.

(*d*) $g \propto 1/R^2$.

12 •• At the surface of the moon, the acceleration due to the gravity of the moon is a. At a distance from the center of the moon equal to four times the radius of the moon, the acceleration due to the gravity of the moon is (*a*) $16a$. (*b*) $a/4$. (*c*) $a/3$. (*d*) $a/16$. (*e*) none of the above.

(*d*) $a \propto 1/R^2$.

13* • One of Jupiter's moons, Io, has a mean orbital radius of 4.22×10^8 m and a period of 1.53×10^5 s. (*a*) Find the mean orbital radius of another of Jupiter's moons, Callisto, whose period is 1.44×10^6 s. (*b*) Use the known value of G to compute the mass of Jupiter.

(*a*) Use Equ. 11-2; $R_C = R_I(T_C/T_I)^{2/3}$ — $R_C = (4.22 \times 10^8)(14.4/1.53)^{2/3}\text{ m} = 18.8 \times 10^8\text{ m}$

(*b*) Use Equ. 11-15; $M_J = 4\pi^2 R_I^3/GT_I^2$ — $M_J = 4\pi^2(4.22 \times 10^8)^3/[6.67 \times 10^{-11}(1.53 \times 10^5)^2]$ $= 1.9 \times 10^{27}\text{ kg}$

14 • The mass of Saturn is 5.69×10^{26} kg. (*a*) Find the period of its moon Mimas, whose mean orbital radius is 1.86×10^8 m. (*b*) Find the mean orbital radius of its moon Titan, whose period is 1.38×10^6 s.

(*a*) Use Equ. 11-15; $T = (4\pi^2 R^3/GM_S)^{1/2}$ — $T_M = [4\pi^2(1.86 \times 10^8)^3/(6.67 \times 10^{-11} \times 5.68 \times 10^{26})]^{1/2}\text{ s}$ $= 8.18 \times 10^4\text{ s}$

(*b*) Use Equ. 11-2 — $R_T = 1.86 \times 10^8(138/8.18)^{2/3}\text{ m} = 12.2 \times 10^8\text{ m}$

15 • Calculate the mass of the earth from the period of the moon $T = 27.3$ d, its mean orbital radius $r_m = 3.84 \times 10^8$ m, and the known value of G.

Convert T to s: $T = 2.36 \times 10^6$ s; Use Equ. 11-15 $M_E = 4\pi^2(3.84 \times 10^8)^3/G(2.36 \times 10^6)^2$ kg $= 6.02 \times 10^{24}$ kg

Note: This neglects the mass of the moon; consequently, the mass calculated here is slightly too great.

---

16 • Use the period of the earth (1 y), its mean orbital radius (1.496×10^{11} m), and the value of G to calculate the mass of the sun.

Proceed as in Problem 15. 1 y $= 3.156 \times 10^7$ s; substituting in the numerical values yields $M_S = 1.99 \times 10^{30}$ kg.

17* • An object is dropped from a height of 6.37×10^6 m above the surface of the earth. What is its initial acceleration?

Since $R = 2R_E$, $a = g_E/4 = (9.81/4)$ m/s^2 = 2.45 m/s^2.

---

18 • Suppose you leave the solar system and arrive at a planet that has the same mass per unit volume as the earth but has 10 times the earth's radius. What would you weigh on this planet compared with what you weigh on earth?

1. Find M_P, mass of the planet, in units of M_E For same ρ, $M \propto R^3$; $M = 10^3 M_E$
2. Find g_P in units of g_E Since $g \propto M/R^2$, $g_P = g_E \times 10^3/10^2 = 10g_E$
3. Weight on planet = 10 × weight on earth

---

19 • Suppose that the earth retained its present mass but was somehow compressed to half its present radius. What would be the value of g, the acceleration due to gravity, at the surface of this new, compact planet?

Since g is proportional to $1/R^2$, g on planet would be 4×9.81 m/s^2 = 39.24 m/s^2.

---

20 • A planet moves around a massive sun with constant angular momentum. When the planet is at perihelion, it has a speed of 5×10^4 m/s and is 1.0×10^{15} m from the sun. The orbital radius increases to 2.2×10^{15} m at aphelion. What is the planet's speed at aphelion?

$L = \text{constant} = mv_pr_p = mv_ar_a$; $v_a = v_pr_p/r_a$ $v_a = (5 \times 10^4 \times 10^{15}/2.2 \times 10^{15})$ m/s $= 2.27 \times 10^4$ m/s

---

21* • A comet orbits the sun with constant angular momentum. It has a maximum radius of 150 AU, and at aphelion its speed is 7×10^3 m/s. The comet's closest approach to the sun is 0.4 AU. What is its speed at perihelion?

$v_p = v_ar_a/r_p$ (see Problem 20) $v_p = 7 \times 10^3 \times 150/0.4$ m/s = 2625 km/s

---

22 •• The speed of an asteroid is 20 km/s at perihelion and 14 km/s at aphelion. Determine the ratio of the aphelion to perihelion distance.

$r_a/r_p = v_p/v_a$ (see Problem 20) $r_a/r_p = 20/14 = 1.43$

---

23 •• A satellite with a mass of 300 kg moves in a circular orbit 5×10^7 m above the earth's surface. (*a*) What is the gravitational force on the satellite? (*b*) What is the speed of the satellite? (*c*) What is the period of the satellite?

(*a*) $F_g = Mg(R) = Mg \times [R_E/(R_E + h)]^2$; $R_E = 6.37 \times 10^6$ m $F_g = 300 \times 9.81 \times (6.37/56.37)^2$ N = 37.58 N

(*b*) $Mv^2/R = F_g$; $v = (F_gR/M)^{1/2}$ $v = (37.58 \times 56.37 \times 10^6/300)^{1/2}$ m/s = 2.657 km/s

(*c*) $T = 2\pi R/v$ $T = 2\pi \times 56.37 \times 10^6/2657$ s $= 1.333 \times 10^5$ s = 37 h

24 •• At the airport, a physics student weighs 800 N. The student boards a jet plane that rises to an altitude of 9500 m. What is the student's loss in weight?

1. Find w at 9.5 km; $w(h) = w_E \times [R_E/(R_E + h)]^2$ $\quad w(h) = 800(6370/6379.5)^2$ N = 797.6 N

2. $\Delta w = w_E - w(h)$ $\quad \Delta w = 2.4$ N

25* •• Suppose that Kepler had found that the period of a planet's circular orbit is proportional to the square of the orbit radius. What conclusion would Newton have drawn concerning the dependence of the gravitational attraction on distance between two masses?

Take $F = CR^n$, where C is a constant. Then, for a stable circular orbit, $v^2/R = F = CR^n$. The period of the orbit is given by $T = 2\pi R/v$, and so $T = 2\pi R/C^{½}R^{(n+1)/2}$. Therefore, if $T \propto R^2$, $1 - (n + 1)/2 = 2$, $n = -3$, and $F \propto 1/R^3$.

26 •• A superconducting gravity meter can measure changes in gravity of the order $\Delta g/g = 10^{-11}$. (*a*) Estimate the maximum range at which an 80-kg person can be detected by this gravity meter. Assume that the gravity meter is stationary, and that the person's mass can be considered to be concentrated at his or her center of gravity. (*b*) What vertical change in the position of the gravity meter in the earth's gravitational field is detectable?

(*a*) Find the gravitational field, $g(r)$ due to mass m $\quad g(r) = Gm/r^2$; given: $g(r) = 10^{-11}g_E = 10^{-11}(GM_E/R_E^2)$

Solve for r; $r = R_E(10^{11}m/M_E)^{½}$ $\quad r = 6.37 \times 10^6(8 \times 10^{12}/6 \times 20^{24})^{½}$ m = 7.36 m

(*b*) Write $g(r) = C/r^2$ and differentiate $\quad dg/dr = -2C/r^3$; $dg/g = -2\,dr/r = 10^{-11}$

Find $dr = \Delta r$ for $r = R_E$ $\quad \Delta r = ½ \times 10^{-11} \times 6.37 \times 10^6$ m $\approx 3.2 \times 10^{-5}$ m = 0.032 mm

27 •• During a solar eclipse, when the moon is between the earth and the sun, the gravitational pull of the moon and the sun on a student are in the same direction. (*a*) If the pull of the earth on the student is 800 N, what is the force of the moon on the student? (*b*) What is the force of the sun on the student? (*c*) What percentage correction due to the sun and moon when they are directly overhead should be applied to the reading of a very accurate scale to obtain the student's weight?

(*a*) F due to moon is $F_M = GmM_M/R_{EM}^2$. Write F_M in terms of $F_E = GmM_E/R_E^2 = 800$ N and evaluate F_M. $\quad F_M = 800(M_M/M_E)(R_E/R_{EM})^2$ N; $F_M = 800(7.35/598)(6.37/384.4)^2$ N = 0.0027 N

(*b*) Likewise, $F_S = 800(M_S/M_E)(R_E/R_{ES})^2$ $\quad F_S = 800(1.99 \times 10^6/5.98)(6.37/1.496 \times 10^5)^2 = 0.483$ N

(*c*) Percentage correction = $(-0.485/800) \times 100$ $\quad$ correction = -0.061%

28 •• Suppose that the attractive interaction between a star of mass M and a planet of mass $m << M$ were of the form $F = KMm/r$, where K is the gravitational constant. What would be the relation between the radius of the planet's circular orbit and its period?

$KMm/r = mv^2/r$; so $v^2 = KM$, independent of r $\quad T = 2\pi r/v = 2\pi r/(KM)^{½}$, i.e., proportional to r

29* •• The mass of the earth is 5.98×10^{24} kg and its radius is 6370 km. The radius of the moon is 1738 km. The acceleration of gravity at the surface of the moon is 1.62 m/s^2. What is the ratio of the average density of the moon to that of the earth?

1. Write g_E and g_M in terms of ρ_E and ρ_M $\quad g_E = G(4\pi\rho_E R_E^3/3)/R_E^2$; $g_M = G(4\pi\rho_M R_M^3/3)/R_M^2$

2. Find g_M/g_E and solve for and evaluate ρ_M/ρ_E — $g_M/g_E = \rho_M R_M/\rho_E R_E$; $\rho_M/\rho_E = (1.62/9.81)(6.37/1.738) = 0.605$

---

30 ••• A plumb bob near a large mountain is slightly deflected from the vertical by the gravitational attraction of the mountain. Estimate the order of magnitude of the angle of deflection using any assumptions you like.

First, we note that the two force fields are approximately at right angles. Their magnitudes are GM_E/R_E^2 and Gm_m/D_m^2, where M_m is the mass of the mountain and D_m the distance to its center. Thus the angle of deflection is given by $\theta = \tan^{-1}(M_m R_E^2/M_E D^2)$. Assume that the density of the mountain is the average density of the earth. Let the "mountain" be a cylinder 10 km in diameter and 4 km high. Its mass is $\rho_E \pi r^2 h$. We consider the plumb bob to be adjacent to the mountain cylinder so that $D = r = 5$ km. Now $M_m/M_E = \pi r^2 h/(4\pi R_E^3/3) \approx 3 \times 10^{-10}$ and $R_E^2/D^2 \approx 1.6 \times 10^6$. The angle of deflection is then approximately $\theta = \tan^{-1}(5 \times 10^{-4}) = 5 \times 10^{-4}$ rad $\approx 0.03°$.

---

31 • Why is G so difficult to measure?

Measurement of G is difficult because masses accessible in the laboratory are very small compared to the mass of the earth.

---

32 • The masses in a Cavendish apparatus are $m_1 = 10$ kg and $m_2 = 10$ g, the separation of their centers is 6 cm, and the rod separating the two small masses is 20 cm long. (*a*) What is the force of attraction between the large and small masses? (*b*) What torque must be exerted by the suspension to balance these forces?

(*a*) Use Equs. 11-3 and 11-4 — $F = (6.67 \times 10^{-11} \times 10 \times 10^{-2}/36 \times 10^{-4})$ N $= 1.85 \times 10^{-9}$ N

(*b*) $\tau = 2Fr$, $r = 0.1$ m — $\tau = 3.7 \times 10^{-10}$ N·m

---

33* • The masses in a Cavendish apparatus are $m_1 = 12$ kg and $m_2 = 15$ g, and the separation of their centers is 7 cm. (*a*) What is the force of attraction between these two masses? (*b*) If the rod separating the two small masses is 18 cm long, what torque must be exerted by the suspension to balance the torque exerted by gravity?

(*a*) See Problem 32 — $F = (6.67 \times 10^{-11} \times 12 \times 1.5 \times 10^{-2}/49 \times 10^{-4})$ N $= 2.45 \times 10^{-9}$ N

(*b*) $\tau = 2Fr$, $r = 0.09$ m — $\tau = 4.41 \times 10^{-10}$ N·m

---

34 •• How would everyday life change if gravitational and inertial mass were not identical?

The force required to accelerate an object (car, person, etc.) would not be proportional to its weight. Also, the period of a pendulum clock would depend on the mass of the bob.

---

35 •• If gravitational and inertial mass were not identical, what would change for (*a*) an offensive lineman on a football team? (*b*) a car? (*c*) a paperweight?

(*a*) His effectiveness would depend on his mass rather than his weight. (*b*) The power requirement would not be determined by the car's weight, but by its mass. (*c*) There would be no significant effect.

---

36 • A standard object defined as having a mass of exactly 1 kg is given an acceleration of 2.6587 m/s^2 when a certain force is applied to it. A second object of unknown mass acquires an acceleration of 1.1705 m/s^2 when the same force is applied to it. (*a*) What is the mass of the second object? (*b*) Is the mass that you determined in part (*a*) gravitational or inertial mass?

(*a*) From $F = ma$, $m_2 = (2.6587/1.1705)$ kg $= 2.2714$ kg. (*b*) It is the inertial mass of m_2.

37* • The weight of a standard object defined as having a mass of exactly 1 kg is measured to be 9.81 N. In the same laboratory, a second object weighs 56.6 N. (*a*) What is the mass of the second object? (*b*) Is the mass you determined in part (*a*) gravitational or inertial mass?

(*a*) As in Problem 36, $m_2 = (56.6/9.81)$ kg = 5.77 kg. (*b*) This is the gravitational mass of m_2—determined by the effect on m_2 of the earth's gravitational field.

38 • (*a*) Taking the potential energy to be zero at infinite separation, find the potential energy of a 100-kg object at the surface of the earth. (Use 6.37×10^6 m for the earth's radius.) (*b*) Find the potential energy of the same object at a height above the earth's surface equal to the earth's radius. (*c*) Find the escape speed for a body projected from this height.

(*a*) Use Equ. 11-18; $U(R_E) = -GM_Em/R_E = -gmR_E$ $\quad U(R_E) = -(9.81 \times 100 \times 6.37 \times 10^6)$ J $= -6.25 \times 10^9$ J

(*b*) G, M_E, and m are unchanged, $U(2R_E) = U(R_E)/2$ $\quad U(2R_E) = -3.12 \times 10^9$ J

(*c*) From (*b*), $K_{esc}(2R_E) = \frac{1}{2}K_{esc}(R_E)$; $K \propto v^2$ $\quad v_{esc}(2R_E) = v_{esc}(R_E)/\sqrt{2} = (11.2/\sqrt{2})$ km/s $= 7.92$ km/s

39 • A point mass m_0 is initially at the surface of a large sphere of mass M and radius R. How much work is needed to remove it to a very large distance away from the large sphere?

$W = \Delta U = U_f - U_i$. $U_f = 0$, $U_i = -GMm_0/R$. So $W = GMm_0/R$.

40 • Suppose that in space there is a duplicate earth, except that it has no atmosphere, is not rotating, and is not in motion around any sun. What initial velocity must a spacecraft on its surface have to travel vertically upward a distance above the surface of the planet equal to one earth radius?

From the results of Problem 38, we know that the change in gravitational potential is exactly half the gravitational potential difference between $R = R_E$ and $R = \infty$. Consequently, $v = v_{esc}/\sqrt{2} = 7.92$ km/s.

41* •• An object is dropped from rest from a height of 4×10^6 m above the surface of the earth. If there is no air resistance, what is its speed when it strikes the earth?

1. Use $\frac{1}{2}mv^2 = -\Delta U = U(4 \times 10^6 + R_E) - U(R_E)$ $\quad \frac{1}{2}v^2 = (3.99\times10^{14})[(1/6.37\times10^6) - (1/10.37\times10^6)]$ J/kg

$v = 6.95$ km/s

2. Solve for v

42 •• An object is projected upward from the surface of the earth with an initial speed of 4 km/s. Find the maximum height it reaches.

1. $U_f = K_i + U_i$; divide both sides by m $\quad -GM_E/(R_E + h) = \frac{1}{2}v_i^2 - GM_E/R_E$

2. Solve for and evaluate h $\quad h = (Rv_i^2)/[(2GM_E/R_E) - v_i^2] = 9.3 \times 10^5$ m

43 •• A spherical shell has a radius R and a mass M. (*a*) Write expressions for the force exerted by the shell on a point mass m_0 when m_0 is outside the shell and when it is inside the shell. (*b*) What is the potential-energy function $U(r)$ for this system when the mass m_0 is at a distance r ($r \geq R$) if $U = 0$ at $r = \infty$? Evaluate this function at $r = R$. (*c*) Using the general relation for $dU = -\mathbf{F}\cdot d\mathbf{r} = -F_r dr$, show that U is constant everywhere inside the shell. (*d*) Using the fact that U is continuous everywhere, including at $r = R$, find the value of the constant U inside the shell. (*e*) Sketch $U(r)$ versus r for all possible values of r.

(*a*) $\mathbf{F} = m_0\mathbf{g}$; use Equ. 11-24 $\quad$ Outside: $F = GMm_0/r^2$, radially in; inside: $\mathbf{F} = 0$

(*b*) $U(r) = -\int_{\infty}^{r} F_r dr$ for $r > R$

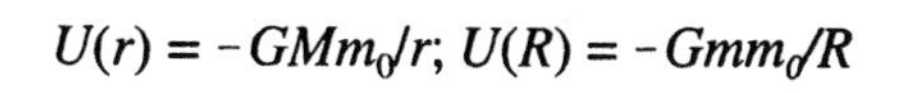

$U(r) = -GMm_0/r$; $U(R) = -Gmm_0/R$

(*c*), (*d*) $F = 0$ for $r < R$, $dU/dr = 0$, U = constant.
Since U is continuous, then for $r < R$, $U(r) = U(R) = -Gmm_0/R$.

(*e*) A sketch of $|U(r)|$ is shown

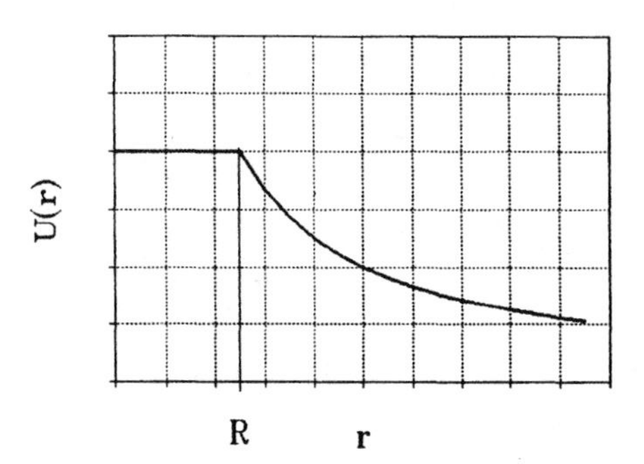

---

44 ••• Our galaxy can be considered to be a large disk of radius R and mass M of approximately uniform mass density. (*a*) Consider a ring element of radius r and thickness dr of such a disk. Find the gravitational potential energy of a 1-kg mass on the axis of this element a distance x from its center. (*b*) Integrate your result for part (*a*) to find the total gravitational potential energy of a 1-kg mass at a distance x due to the disk. (*c*) From $F_x = -dU/dx$ and your result for part (*b*), find the gravitational field g_x on the axis of the disk.

Let σ = mass/unit area = $M/\pi R^2$. Let $d = \sqrt{x^2+r^2}$ be the distance from a point at a radius r from the center of the disk to the point a distance x along its axis.

(*a*) Write an expression for dU $\qquad dU = -\dfrac{2\pi r\sigma G\,dr}{\sqrt{x^2+r^2}}$

(*b*) $U = -\displaystyle\int_0^R \frac{2\pi r\sigma G\,dr}{\sqrt{x^2+r^2}} = -2\pi\sigma G(\sqrt{x^2+R^2} - x)$ $\qquad U(x) = -\dfrac{2GM}{R^2}(\sqrt{x^2+R^2} - x)$

(*c*) $F_x = -dU(x)/dx$ $\qquad F_x = -\dfrac{2GM}{R^2}\left(1 - \dfrac{x}{\sqrt{x^2+R^2}}\right) = g(x)$

45* ••• The assumption of uniform mass density in Problem 44 is rather unrealistic. For most galaxies, the mass density increases greatly toward the center of the galaxy. Repeat Problem 44 using a surface mass density of the form $\sigma(r) = C/r$, where $\sigma(r)$ is the mass per unit area of the disk at a distance r from the center. First determine the constant C in terms of R and M; then proceed as in Problem 44.

(*a*) $M = \displaystyle\int_0^R 2\pi r\sigma\,dr = 2\pi C\int_0^R dr = 2\pi CR$ $\qquad C = M/2\pi R$

(*b*) $dU = -\dfrac{GM\,dr}{R\sqrt{x^2+r^2}}$ $\qquad U(x) = -\dfrac{GM}{R}\displaystyle\int_0^R \frac{dr}{\sqrt{x^2+r^2}} = -\frac{GM}{R}\ln\left(\frac{R+\sqrt{x^2+R^2}}{x}\right)$

(*c*) $F_x = -dU/dx = g(x)$ $\qquad F_x = \dfrac{GM}{x\sqrt{x^2+R^2}} = g(x)$

---

46 • What is the effect of air resistance on the escape speed near the earth's surface?

It causes an increase in the escape speed.

47 • Would it be possible in principle for the earth to escape from the solar system?

Yes, if it interacted with a huge comet.

48 • If the mass of a planet is doubled with no increase in its size, the escape speed for that planet will be (*a*) increased by a factor of 1.4. (*b*) increased by a factor of 2. (*c*) unchanged. (*d*) reduced by a factor of 1.4. (*e*) reduced by a factor of 2.

(*a*) See Equ. 11-19.

49* • The planet Saturn has a mass 95.2 times that of the earth and a radius 9.47 times that of the earth. Find the escape speed for objects near the surface of Saturn.

$v_e(\text{Sat}) = (2GM_S/R_S)^{½} = v_e(\text{earth})[(M_S/M_E)(R_E/R_S)]^{½}$ $v_e(\text{Sat}) = 11.2(95.2/9.47)^{½}$ km/s = 35.5 km/s

50 • Find the escape speed for a rocket leaving the moon. The acceleration of gravity on the moon is 0.166 times that on earth, and the moon's radius is $0.273R_E$.

$v_e(\text{moon}) = (2g_mR_m)^{½} = v_e(\text{earth})[g_mR_m/g_ER_E]^{½}$ $v_e(\text{moon}) = 11.2(0.166 \times 0.273)^{½}$ km/s = 2.38 km/s

51 •• A particle is projected from the surface of the earth with a speed equal to twice the escape speed. When it is very far from the earth, what is its speed?

Since its speed is twice the escape speed, its initial kinetic energy is four times the energy needed to escape. Therefore, when it has escaped, it has an energy that is three times the escape energy and its speed is $v_e\sqrt{3}$ = 19.4 km/s.

52 •• What initial speed should a particle be given if it is to have a final speed when it is very far from the earth equal to its escape speed?

Using the same reasoning as in Problem 51, one finds that $v_i = v_e\sqrt{2}$ = 15.8 km/s.

53* •• A space probe launched from the earth with an initial speed v_i is to have a speed of 60 km/s when it is very far from the earth. What is v_i?

$½mv_i^2 = ½mv_f^2 + ½mv_e^2$; $v_i = (v_f^2 + v_e^2)^{½}$ $v_i = (60^2 + 11.2^2)^{½}$ km/s = 61.04 km/s

54 •• (*a*) Calculate the energy in joules necessary to launch a 1-kg mass from the earth at escape speed. (*b*) Convert this energy to kilowatt-hours. (*c*) If energy can be obtained at 10 cents per kilowatt-hour, what is the minimum cost of giving an 80-kg astronaut enough energy to escape the earth's gravitational field?

(*a*) $E = ½mv_e^2$ $E = ½(11.2 \times 10^3)^2$ J = 62.72 MJ

(*b*) 1 kW·h = 3.6 MJ $E = (62.72/3.6)$ kW·h = 17.42 kW·h

(*c*) Cost = \$ $(M \times 0.1 \times E)$ Cost = \$ 139.36

55 •• Show that the escape speed from a planet is related to the speed of a circular orbit just above the surface of the planet by $v_e = \sqrt{2}v_c$, where v_c is the speed of the object in the circular orbit.

From Example 11-6, the kinetic energy of a mass m in a circular orbit is $K = ½|U|$. Note that this result is true for any circular orbit of a mass m about a massive center. But $|U|$ is the escape energy of $½mv_e^2$. Consequently, $K = ½mv_c^2 = ½(½mv_e^2)$ and $v_c = v_e/\sqrt{2}$.

56 •• Find the speed of the earth v_c as it orbits the sun, assuming a circular orbit. Use this and the result of Problem 55 to calculate the speed v_{eS} needed by the earth to escape from the sun.

1. $v_c = 2\pi R_{ES}/T$; $R_{ES} = 1.5 \times 10^{11}$ m, $T = 3.156 \times 10^7$ s	$v_c = (2\pi \times 1.5 \times 10^{11}/3.156 \times 10^7)$ m/s = 29.9 km/s
2. $v_e = v_c\sqrt{2}$	$v_e = 42.2$ km/s

57* •• If an object has just enough energy to escape from the earth, it will not escape from the solar system because of the attraction of the sun. Use Equation 11-19 with M_S replacing M_E and the distance to the sun r_S replacing R_E to calculate the speed v_{eS} needed to escape from the sun's gravitational field for an object at the surface of the earth. Neglect the attraction of the earth. Compare your answer with that in Problem 56. Show that if v_e is the speed needed to escape from the earth, neglecting the sun, then the speed of an object at the earth's surface needed to escape from the solar system is given by $v_{e,solar}^2 = v_e^2 + v_{eS}^2$, and calculate $v_{e,solar}$.

From Equ. 11-19, $v_{eS} = (2GM_S/r_S)^{1/2} = [2 \times (6.67 \times 10^{-11}) \times (2 \times 10^{30})/1.5 \times 10^{11}]^{1/2} = 42.2$ km/s; this is the same speed calculated in Problem 56, as it should be. The energy needed to escape the solar system, starting from the surface of the earth, is the sum of the energy needed to escape the earth's gravity plus that required to escape from the sun, starting at the earth's orbit radius. These energies are proportional to the squares of the corresponding escape velocities. Therefore, $v_{e,solar}^2 = v_e^2 + v_{eS}^2$; $v_{e,solar} = (11.2^2 + 42.2^2)^{1/2}$ km/s = 43.7 km/s.

58 •• Why is it reasonable to neglect the other planets in calculating the speed needed to escape from the solar system? Would you expect the actual value of this speed to be greater or less than that calculated in Problem 57?

The masses of the planets are only about 0.1% of the sun's mass. The actual value of $v_{e,solar}$ is slightly greater because the total mass of the solar system is slightly larger than that of the sun.

59 •• An object is projected vertically from the surface of the earth. Show that the maximum height reached by the object is $H = R_E H'/(R_E - H')$, where H' is the height that it would reach if the gravitational field were constant.

$H' = v^2/2g$. To determine H use energy conservation: $\frac{1}{2}mv^2 = -\Delta U = GMm[1/R_E - 1/(R_E+H)]$ or $v^2 = 2gR_E^2[1/R_E - 1/(R_E+H)]$. So $H' = R_E H/(R_E + H)$ and, solving for H, $H = R_E H'/(R_E - H')$.

60 •• An object (say, a newly discovered comet) enters the solar system and makes a pass around the sun. How can we tell if the object will return many years later, or if it will never return?

If the path is an ellipse, it will return; if its path is hyperbolic or parabolic, it will not return.

61* •• A spacecraft of 100 kg mass is in a circular orbit about the earth at a height $h = 2R_E$. (*a*) What is the period of the spacecraft's orbit about the earth? (*b*) What is the spacecraft's kinetic energy? (*c*) Express the angular momentum L of the spacecraft about the earth in terms of its kinetic energy K and find its numerical value.

We will use the result of Problem 55, i.e., $v_c = v_e/\sqrt{2}$, where v_c is the speed of a circular orbit just above the surface.

(*a*) 1. $K(3R_E) = K(R_E)/3$; $v = v_c/\sqrt{3} = v_e/\sqrt{6}$	$v = 11.2/\sqrt{6}$ km/s = 4.57 km/s
2. $T = 2\pi(3R_E)/v$	$T = 2\pi \times 3 \times 6370/(4.57 \times 3600)$ h = 7.3 h
(*b*) $K = \frac{1}{2}mv^2$	$K = 50 \times (4.57 \times 10^3)^2 = 1.04 \times 10^9$ J
(*c*) $K = L^2/2I$; $L = (2KI)^{1/2}$; $I = m(3R_E)^2$	$L = 3R_E(2Km)^{1/2} = 3R_E mv = 8.73 \times 10^{10}$ J·s

62 •• Many satellites orbit the earth about 1000 km above the earth's surface. Geosynchronous satellites orbit at a distance of 4.22×10^7 m from the center of the earth. How much more energy is required to launch a 500-kg satellite into a geosynchronous orbit than into an orbit 1000 km above the surface of the earth?

Let E_O be the total energy of a satellite in orbit. Then, from Problem 55, $E_O = K_O + U_O = \frac{1}{2}U_O$.

$\Delta E = E_{geo} - E_{1000} = \frac{1}{2}(U_{geo} - U_{1000})$ $\Delta E = -\frac{1}{2}GM_Em(1/4.22 \times 10^7 - 1/7.73 \times 10^6)$
$= 1.05 \times 10^{10}$ J

---

63 •• It is theoretically possible to place a satellite at a position between the earth and the sun on the line joining them, where the gravitational forces of the sun and the earth on the satellite combine in such a way that the satellite will execute a circular orbit around the sun that is synchronous with the earth's orbit around the sun. (In other words, the satellite and the earth have the same orbital period about the sun, even though they are at different distances from the sun. The satellite always remains on the line joining the earth and the sun.) Write an expression that relates the appropriate circular orbital speed v of a satellite in such a situation to its distance r from the sun. Your expression may also contain quantities shown in Figure 11-20 plus the gravitational constant G.

Write $F_{net} = F_{centrip} = mv^2/r = GmM_S/r^2 - GmM_E/(D - r)^2$; $v = \sqrt{GM_S/r - GM_E r/(D-r)^2}$.

---

64 • A 3-kg mass experiences a gravitational force of 12 N $\boldsymbol{i}$ at some point P. What is the gravitational field at that point?

Use Equ. 11-21 $\boldsymbol{g} = (12/3)\,\boldsymbol{i}$ N/kg $= 4\,\boldsymbol{i}$ N/kg

---

65* • The gravitational field at some point is given by $\boldsymbol{g} = 2.5 \times 10^{-6}$ N/kg$\boldsymbol{j}$. What is the gravitational force on a mass of 4 g at that point?

Use Equ. 11-21 $\boldsymbol{F} = m\boldsymbol{g} = (4 \times 10^{-3} \times 2.5 \times 10^{-6})\boldsymbol{j}$ N $= 10^{-8}\boldsymbol{j}$ N

---

66 •• A point mass m is on the x axis at $x = L$ and a second equal point mass m is on the y axis at $y = L$. (*a*) Find the gravitational field at the origin. (*b*) What is the magnitude of this field?

(*a*), (*b*) Use Equ. 11-24 and vector addition $\boldsymbol{g} = (Gm/L^2)(\boldsymbol{i} + \boldsymbol{j})$; $g = \sqrt{2}\,Gm/L^2$

---

67 •• Five equal masses M are equally spaced on the arc of a semicircle of radius R as in Figure 11-21. A mass m is located at the center of curvature of the arc. (*a*) If M is 3 kg, m is 2 kg, and R is 10 cm, what is the force on m due to the five masses? (*b*) If m is removed, what is the gravitational field at the center of curvature of the arc?

(*a*) By symmetry, $g_x = 0$; write expression for F_y $F_y = (GMm/R^2)(2 \sin 45° + 1)$

Substitute numerical values $\boldsymbol{F} = 9.66 \times 10^{-8}$ N$\boldsymbol{j}$

(*b*) $\boldsymbol{g} = \boldsymbol{F}/m$ $\boldsymbol{g} = 4.83 \times 10^{-8}$ N/kg$\boldsymbol{j}$

---

68 •• A point mass $m_1 = 2$ kg is at the origin and a second point mass $m_2 = 4$ kg is on the x axis at $x = 6$ m. Find the gravitational field at (*a*) $x = 2$ m, and (*b*) $x = 12$ m. (*c*) Find the point on the x axis for which $g = 0$.

The configuration is shown below.

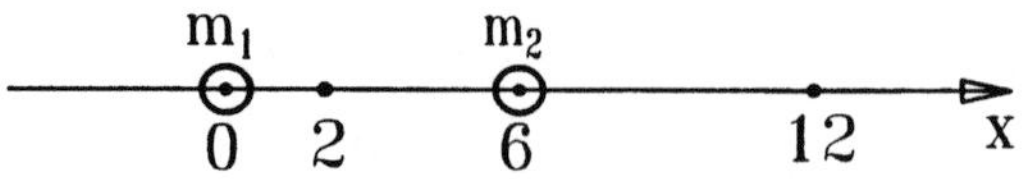

m1 m2
0 2 6 12 x

(*a*) $\boldsymbol{g} = \boldsymbol{g}_1 + \boldsymbol{g}_2$; $\boldsymbol{g}_1 = -Gm_1/4\,\boldsymbol{i}$; $\boldsymbol{g}_2 = Gm_2/16\,\boldsymbol{i}$ $\boldsymbol{g} = G(1/4 - 1/2)\,\boldsymbol{i} = -1.67 \times 10^{-11}\,\boldsymbol{i}$ N/kg

(*b*) $\boldsymbol{g} = (-Gm_1/144 - Gm_2/36)\,\boldsymbol{i}$ $\boldsymbol{g} = -G(1/72 + 1/9)\,\boldsymbol{i} = -G/8\,\boldsymbol{i} = -8.34 \times 10^{-12}\,\boldsymbol{i}$ N/kg (*c*) $g = 0$ when $2/x^2 = 4/(6-x)^2$; solve for x — $x^2 + 72x - 36 = 0$; $x = 2.484$ m, -8.48 m. From the diagram it is clear that only at $x = 2.484$ m is $\mathbf{g} = 0$.

69* •• (*a*) Show that the gravitational field of a ring of uniform mass is zero at the center of the ring. (*b*) Figure 11-22 shows a point *P* in the plane of the ring but not at its center. Consider two elements of the ring of length s_1 and s_2 at distances of r_1 and r_2, respectively. 1. What is the ratio of the masses of these elements? 2. Which produces the greater gravitational field at point *P*? 3. What is the direction of the field at point *P* due to these elements? (*c*) What is the direction of the gravitational field at point *P* due to the entire ring? (*d*) Suppose that the gravitational field varied as $1/r$ rather than $1/r^2$. What would be the net gravitational field at point *P* due to the two elements? (*e*) How would your answers to parts (*b*) and (*c*) differ if point *P* were inside a spherical shell of uniform mass rather than inside a plane circular ring?

Let λ = mass per unit length of the ring.

(*a*) $\mathbf{g}$ of opposite elements of mass $R\lambda\, d\theta$ cancel — By symmetry $\mathbf{g} = 0$ at center

(*b*) 1. $m_1 = r_1\lambda\, d\theta$; $m_2 = r_2\lambda\, d\theta$ — $m_1/m_2 = r_1/r_2$

2. $g = Gm/r^2 = Gr\lambda\, d\theta/r^2 = G\lambda\, d\theta/r$ — $r_1 < r_2$, therefore $g_1 > g_2$

3. By symmetry, $\mathbf{g}$ points along *OP* — $\mathbf{g}$ points toward m_1, i.e., in direction of *OP*

(*c*) Take x along *OP* — By symmetry, $g_y = 0$; $\mathbf{g}$ points in the direction OP

(*d*) For $g \propto 1/r$, $g_1 = g_2 \propto \lambda\, d\theta$ — $\mathbf{g}_1 = -\mathbf{g}_2$; $\mathbf{g} = 0$

(*e*) Now m_1 and $m_2 \propto r^2$, so $g_1 = g_2$ — $\mathbf{g}_1 = -\mathbf{g}_2$; $\mathbf{g} = 0$; note: $\mathbf{g} = 0$ everywhere inside the shell

---

70 •• Show that the maximum value of $|g_x|$ for the field of Example 11-7 occurs at the points $x = \pm a/\sqrt{2}$.

From Example 11-7, $g_x = -2GMx/(x^2 + a^2)^{3/2}$. To find maximum, differentiate with respect to x and equate to 0. $dg_x/dx = -2GM[(x^2 + a^2)^{-3/2} - 3x^2(x^2 + a^2)^{-5/2}] = 0$. Solve for x: $x = \pm a/\sqrt{2}$.

---

71 •• A nonuniform stick of length *L* lies on the *x* axis with one end at the origin. Its mass density λ (mass per unit length) varies as $\lambda = Cx$, where *C* is a constant. (Thus, an element of the stick has mass $dm = \lambda\, dx$.) (*a*) What is the total mass of the stick? (*b*) Find the gravitational field due to the stick at a point $x_0 > L$.

(*a*) $$M = \int_0^L \lambda\, dx = C\int_0^L x\, dx = \frac{1}{2}CL^2.$$

(*b*) $$d\mathbf{g} = -G\, dm/(x_0 - x)^2\, \mathbf{i};\quad \mathbf{g} = -GC\int_0^L \frac{x\, dx}{(x_0 - x)^2} = \frac{2GM}{L^2}\left[\ln\left(\frac{x_0}{x_0 - L}\right) - \left(\frac{L}{x_0 - L}\right)\right]\mathbf{i}.$$

---

72 ••• A uniform rod of mass *M* and length *L* lies along the *x* axis with its center at the origin. Consider an element of length *dx* at a distance *x* from the origin. (*a*) Show that this element produces a gravitational field at a point x_0 on the *x* axis ($x_0 > ½L$) given by

$$dg_x = -\frac{GM}{L(x_0 - x)^2}dx$$

(*b*) Integrate this result over the length of the rod to find the total gravitational field at the point x_0 due to the rod. (*c*) What is the force on an object of mass m_0 at x_0? (*d*) Show that for $x_0 >> L$, the field is approximately equal to that of a point mass *M*.

(*a*), (*b*) See Example 11-8.

(*c*) $\mathbf{F} = m_0\mathbf{g} = -GMm_0/(x_0^2 - L^2/4)\,\mathbf{i}$ N.

(*d*) For $x_0 >> L$, $g_x = -GM/x_0^2$, which is the result for a point mass M at the origin.

73* •• Explain why the gravitational field increases with r rather than decreasing as $1/r^2$ as one moves out from the center inside a solid sphere of uniform mass.

g is proportional to the mass within the sphere and inversely proportional to the radius, i.e., proportional to $r^3/r^2 = r$.

---

74 • A spherical shell has a radius of 2 m and a mass of 300 kg. What is the gravitational field at the following distances from the center of the shell: (*a*) 0.5 m; (*b*) 1.9 m; (*c*) 2.5 m?

(*a*) $g = 0$ [see Problem 69(*e*)]. (*b*) $g = 0$. (*c*) $g = GM/r^2 = 3.2 \times 10^{-9}$ N/kg.

---

75 • A spherical shell has a radius of 2 m and a mass of 300 kg, and its center is located at the origin of a coordinate system. Another spherical shell with a radius of 1 m and mass 150 kg is inside the larger shell with its center at 0.6 m on the x axis. What is the gravitational force of attraction between the two shells?

The gravitational attraction is zero. The gravitational field inside the 2 m shell due to that shell is zero; therefore, it exerts no force on the 1 m shell, and, by Newton's third law, that shell exerts no force on the larger shell.

---

76 • Two spheres, S_1 and S_2, have equal radii R and equal masses M. The density of sphere S_1 is constant, whereas that of sphere S_2 depends on the radial distance according to $\rho(r) = C/r$. If the acceleration of gravity at the surface of sphere S_1 is g_1, what is the acceleration of gravity at the surface of sphere S_2?

The accelerations of gravity are the same for both spheres; they depend only on M and R.

77* •• Two homogeneous spheres, S_1 and S_2, have equal masses but different radii, R_1 and R_2. If the acceleration of gravity on the surface of sphere S_1 is g_1, what is the acceleration of gravity on the surface of sphere S_2?

$g \propto M/r^2$ $\qquad g_2 = g_1(R_1^2/R_2^2)$

---

78 •• Two concentric uniform spherical shells have masses M_1 and M_2 and radii a and $2a$ as in Figure 11-23. What is the magnitude of the gravitational force on a point mass m located (*a*) a distance $3a$ from the center of the shells? (*b*) a distance $1.9a$ from the center of the shells? (*c*) a distance $0.9a$ from the center of the shells?

(*a*) At $r = 3a$, both masses contribute to g $\qquad F = Gm(M_1 + M_2)/9a^2$

(*b*) At $r = 1.9a$, g due to $M_2 = 0$ $\qquad F = GmM_1/3.61a^2$

(*c*) At $r = 0.9a$, $g = 0$ $\qquad F = 0$

---

79 •• The inner spherical shell in Problem 78 is shifted such that its center is now at $x = 0.8a$. The points $3a$, $1.9a$, and $0.9a$ lie along the same radial line from the center of the larger spherical shell. (*a*) What is the force on m at $x = 3a$? (*b*) What is the force on m at $x = 1.9a$? (*c*) What is the force on m at $x = 0.9a$?

The configuration is shown on the right. The centers of the spheres are indicated by the center-lines. The locations of the mass m for parts (a), (b), and (c) are shown by small dots along the x axis.

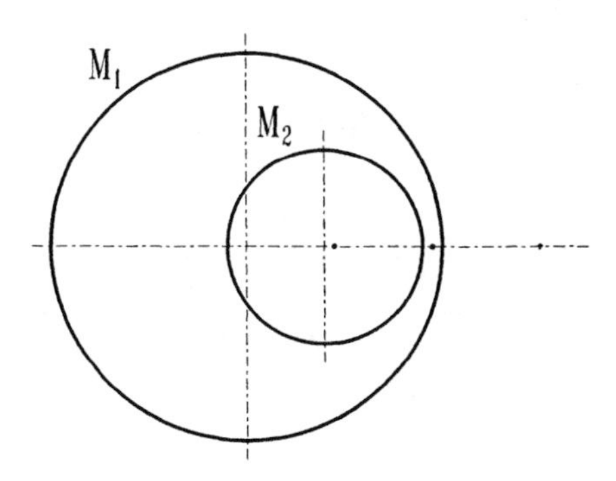

(a) At $x = 3a$, $g_1 = GM_1/(2.2a)^2$ and $g_2 = GM_2/(3a)^2$ $\quad$ $F = (Gm/a^2)(M_1/4.84 + M_2/9)$

(b) At $x = 1.9a$, $g_1 = GM_1/(1.1a)^2$ and $g_2 = 0$ $\quad$ $F = GmM_1/1.21a^2$

(c) At $x = 0.9a$, $g_1 = g_2 = 0$ $\quad$ $F = 0$

---

80 •• Suppose the earth were a sphere of uniform mass. If there were a deep elevator shaft going 15,000 m into the earth, what would be the loss in weight at the bottom of this deep shaft for a student who weighs 800 N at the surface of the earth?

1. Write an expression for M inside $R = R_E - 15$ km $\quad$ $M = M_E[(R_E - 15)/R_E]^3$, where R_E is in km
2. Write an expression for g at $R = R_E - 15$ km $\quad$ $g = GM/R^2 = GM_E(R_E - 15)^3/[(R_E - 15)^2 R_E^3]$ $= g_E(R_E - 15)/R_E$
3. $w(R) = mg(R) = w(R_E)[(R_E - 15)/R_E]$ $\quad$ $\Delta w = (800\text{ N})(15/R_E) = (800 \times 15/6370)\text{ N} = 1.88\text{ N}$

81* •• A sphere of radius R has its center at the origin. It has a uniform mass density ρ_0, except that there is a spherical cavity in it of radius $r = \frac{1}{2}R$ centered at $x = \frac{1}{2}R$ as in Figure 11-24. Find the gravitational field at points on the x axis for $|x| > R$. (*Hint:* The cavity may be thought of as a sphere of mass $m = (4/3)\pi r^3\rho_0$ plus a sphere of mass $-m$.)

Write $g(x)$ using the hint. That is, find the sum of g of the solid sphere plus the field of a sphere of radius $\frac{1}{2}R$ of negative mass centered at $x = \frac{1}{2}R$.

$$g(x) = G\left(\frac{4\pi\rho_0}{3}\right)\left(\frac{R^3}{x^2} - \frac{R^3/8}{(x - \frac{1}{2}R)^2}\right)$$

---

82 ••• For the sphere with the cavity in Problem 81, show that the gravitational field inside the cavity is uniform, and find its magnitude and direction.

1. Find the x and y components of $\mathbf{g}_1$, where $\mathbf{g}_1$ is the field due to a solid sphere of radius R and density ρ_0.
2. Find the x and y components of $\mathbf{g}_2$, where $\mathbf{g}_2$ is a sphere of radius $\frac{1}{2}R$ and negative density ρ_0 centered at $\frac{1}{2}R$.
3. Add the x and y components to obtain the components of the field $\mathbf{g} = \mathbf{g}_1 + \mathbf{g}_2$.

1. $g_1 = 4\pi\rho_0 Gr^3/3r^2 = 4\pi\rho_0 Gr/3$ $\quad$ $g_{1x} = -g_1 \cos\theta = -g_1(x/r) = -4\pi\rho_0 Gx/3$; $g_{1y} = -4\pi\rho_0 Gy/3$ (the negative sign because the field points inward)
2. $g_2 = 4\pi\rho_0 Gr_2^3/3r_2^2 = 4\pi\rho_0 Gr_2/3$, where $r_2 = [(x - \frac{1}{2}R)^2 + y^2]^{1/2}$ $\quad$ $g_{2x} = g_2[(x - \frac{1}{2}R)/r_2] = 4\pi\rho_0 G(x - \frac{1}{2}R)/3$; $g_{2y} = g_2(y/r_2) = 4\pi\rho_0 Gy/3$
3. $g_x = g_{1x} + g_{2x}$; $g_y = g_{1y} + g_{2y}$ $\quad$ $g_x = -2\pi\rho_0 GR/3$; $g_y = 0$. $g = g_x$, a constant

83 ••• A straight, smooth tunnel is dug through a spherical planet whose mass density ρ_0 is constant. The tunnel passes through the center of the planet and is perpendicular to the planet's axis of rotation, which is fixed in space. The planet rotates with an angular velocity ω such that objects in the tunnel have no acceleration relative to the tunnel. Find ω.

$F_g = 4\pi\rho_0 Gmr/3$ (see Problem 82); set $F_r = mr\omega^2$ $\quad$ $\omega^2 = 4\pi\rho_0 G/3$; $\omega = (4\pi\rho_0 G/3)^{1/2}$

84 ••• The density of a sphere is given by $\rho(r) = C/r$. The sphere has a radius of 5 m and a mass of 1011 kg. (*a*) Determine the constant *C*. (*b*) Obtain expressions for the gravitational field for (1) $r > 5$ m, and (2) $r < 5$ m.

(*a*) $\int_0^5 4\pi\rho r^2\,dr = 4\pi C\int_0^5 r\,dr = M$ $\quad$ $2\pi C \times 25 = 1011$ kg; $C = 6.436$ kg/m^2

(*b*) 1. For $r > 5$ m, $g = GM/r^2$ $\quad$ $g = 6.67 \times 10^{-11} \times 1011/r^2 = 6.74 \times 10^{-8}/r^2$ N/kg

2. For $r < 5$ m, $g = G\dfrac{\int_0^r 4\pi Cr\,dr}{r^2}$ $\quad$ $g = 2\pi GC = 2.7 \times 10^{-9}$ N/kg

(Note that g is continuous at $r = 5$ m)

85* ••• A hole is drilled into the sphere of Problem 84 toward the center of the sphere to a depth of 2 km below the sphere's surface. A small mass is dropped from the surface into the hole. Determine the speed of the small mass as it strikes the bottom of the hole.

1. Write the work per kg done by g between $r = 5$ m and $r = 3$ m; note that g points inward $\quad$ $E = \int_5^3 g\,dr = 2\pi GC(5-3) = 5.4 \times 10^{-9}$ J $= \frac{1}{2}v^2$

2. Evaluate v $\quad$ $v = 0.104$ mm/s

86 ••• The solid surface of the earth has a density of about 3000 kg/m^3. A spherical deposit of heavy metals with a density of 8000 kg/m^3 and radius of 1000 m is centered 2000 m below the surface. Find $\Delta g/g$ at the surface directly above this deposit, where Δg is the increase in the gravitational field due to the deposit.

1. Determine g' at $r = 2000$ m due to a sphere of radius 1000 m with density $\Delta\rho = 5000$ kg/m^3 $\quad$ $M = 4\pi\Delta\rho R^3/3 = 2.09 \times 10^{13}$ kg; $g' = GM/r^2$; $g' = \Delta g = 3.49 \times 10^{-4}$ N/kg

2. Evaluate $\Delta g/g$ were $g = 9.81$ N/kg $\quad$ $\Delta g/g = 3.56 \times 10^{-5}$

87 ••• Two identical spherical hollows are made in a lead sphere of radius *R*. The hollows have a radius *R*/2. They touch the outside surface of the sphere and its center as in Figure 11-25. The mass of the lead sphere before hollowing was *M*. (*a*) Find the force of attraction of a small sphere of mass *m* to the lead sphere at the position shown in the figure. (*b*) What is the attractive force if *m* is located right at the surface of the lead sphere?

1. First find the force F_S due to the solid lead sphere. 2. Find the force F_C due to each sphere of negative mass. 3. Add the forces.

(*a*) 1. Write the force due to the solid sphere $\quad$ $F_S = Gmm/d^2$; F_S acts in the negative x direction

2. Write F_C $\quad$ $F_C = G(M/8)m/(d^2 + R^2/4)$

3. Find the x and y component of the two forces F_C $\quad$ $2F_{Cx} = 2F_C\cos\theta = 2F_C d/(d^2 + R^2/4)^{1/2}$ in the positive x direction; y components add to 0 by symmetry

4. Find $\boldsymbol{F} = F_x \boldsymbol{i} = (-F_S + 2F_{Cx})\,\boldsymbol{i}$ $\boldsymbol{F} = -(GMm/d^2)[1 - (d^3/4)/(d^2 + R^2/4)^{3/2}]\,\boldsymbol{i}$

(*b*) Set $d = R$ $\boldsymbol{F} = -0.821(GMm/R^2)\,\boldsymbol{i}$

---

88 • If K is the kinetic energy of the moon in its orbit around the earth, and U is the potential energy of the earth–moon system, what is the relationship between K and U?

$K = -\frac{1}{2}U$. (see Example 11-6)

89* •• A woman whose weight on earth is 500 N is lifted to a height two earth radii above the surface of the earth. Her weight will (*a*) decrease to one-half of the original amount. (*b*) decrease to one-quarter of the original amount. (*c*) decrease to one-third of the original amount. (*d*) decrease to one-ninth of the original amount.

(*d*) g depends on $1/r^2$.

---

90 • The mean distance of Pluto from the sun is 39.5 AU. Find the period of Pluto.

Proceed as in Problem 7 $T_P = (39.5)^{3/2}$ y = 248 y

---

91 • The semimajor axis of Ganymede, a moon of Jupiter discovered by Galileo, is 1.07×10^6 km, and its period is 7.155 days. Determine the mass of Jupiter.

Use the same procedure as for Problem 15 $M_J = 4\pi^2(1.07 \times 10^9)^3/G(6.18 \times 10^5)^2 = 1.9 \times 10^{27}$ kg

---

92 • Calculate the mass of the earth using the known values of G, g, and R_E.

Use Equ. 11-27; $M_E = gR_E^2/G$ $M_E = [9.81(6.37 \times 10^6)^2/6.67 \times 10^{-11}]$ kg

$= 5.97 \times 10^{24}$ kg

93* • Uranus has a moon, Umbriel, whose mean orbital radius is 2.67×10^8 m and whose period is 3.58×10^5 s. (*a*) Find the period of another of Uranus's moons, Oberon, whose mean orbital radius is 5.86×10^8 m. (*b*) Use the known value of G to find the mass of Uranus.

(*a*) Use Equ. 11-2; $T_O = T_U(R_O/R_U)^{3/2}$ $T_O = (3.58 \times 10^5\text{ s})(5.86/2.67)^{3/2} = 1.16 \times 10^6$ s

(*b*) $M = 4\pi^2R^3/GT^2$ $M_U = 4\pi^2(2.67 \times 10^8)^3/G(3.58 \times 10^5)^2$ kg

$= 8.79 \times 10^{25}$ kg

---

94 •• Joe and Sally learn that there is a point between the earth and the moon where the gravitational effects of the two bodies balance each other. Being of a New Age bent, they decide to try to conceive a child free from the bondage of gravity, so they book an earth-to-moon trip. How far from the center of the earth should they try to conceive Zerog, the first zero-gravity baby?

1. If $M_E/r^2 = M_m/(R_{EM} - r)^2$, $g = 0$; solve for r $r = \beta R_{EM}/(1 + \beta)$, where $\beta = (M_E/M_m)^{1/2}$

2. Evaluate r for $M_E/M_m = 81.36$, $R_{EM} = 3.844 \times 10^8$ m $r = 3.46 \times 10^8$ m

---

95 •• The force exerted by the earth on a particle of mass m a distance r from the center of the earth has the magnitude $GM_Em/r^2 = mgR_E^2/r^2$. (*a*) Calculate the work you must do against gravity to move the particle from a distance r_1 to r_2. (*b*) Show that when $r_1 = R_E$ and $r_2 = R_E + h$, the result can be written $W = mgR_E^2[(1/R_E) - 1/(R_E + h)]$. (*c*) Show that when $h \ll R_E$, the work is given approximately by $W = mgh$.

(*a*) $W = -\int_{r_1}^{r_2} F_g\, dr = GM_E m \int_{r_1}^{r_2} \frac{dr}{r} = -GM_E m\left(\frac{1}{r_2} - \frac{1}{r_1}\right) = GM_E m\left(\frac{1}{r_1} - \frac{1}{r_2}\right)$.

(*b*) In the above expression, replace $GM_E m$ by mgR_E^2, r_1 by R_E, and r_2 by $R_E + h$ to obtain the result given.

(*c*) $[(1/R_E) - 1/(R_E + h)] = h/[R_E(R_E + h)]$; if $h << R_E$, the denominator $\approx R_E^2$ and $W = mgh$.

---

96 •• Suppose that the gravitational force of attraction depended not on $1/r^2$ but was proportional to the distance between the two masses, like the force of a spring. In a planetary system like the solar system, what would then be the relation between the period of a planet and its orbit radius, assuming all orbits were circular?

If $F = Cr$, where C is a constant, then $r\omega^2 \propto r$. Thus ω = constant, and T is constant, independent of r.

97* •• A uniform sphere of radius 100 m and density 2000 kg/m³ is in free space far from other massive objects. (*a*) Find the gravitational field outside of the sphere as a function of r. (*b*) Find the gravitational field inside the sphere as a function of r.

(*a*) $g = -GM/r^2$; $M = 4\pi\rho R^3/3$ — $M = 8.38 \times 10^9$ kg; $g = -0.559/r^2$

(*b*) Use Equ. 11-27 — $g = -5.59 \times 10^{-7} r$

---

98 •• Two spherical planets have identical mass densities. Planet P_1 has a radius R_1, and planet P_2 has a radius R_2. If the acceleration of gravity at the surface of planet P_1 is g_1, what is the acceleration of gravity at the surface of planet P_2?

$g \propto M/R^2 \propto R^3/R^2 = R$ — $g_2 = g_1 R_2/R_1$

---

99 •• Jupiter has a mass 320 times that of Earth and a volume 1320 times that of Earth. A "day" on Jupiter is 9 h 50 min long. Find the height h above Jupiter at which a satellite must be revolving to have a period equal to 9 h 50 min.

1. Use Equ. 11-15 with M_J replacing M_S — $R = (T^2GM_J/4\pi^2)^{1/3}$; $M_J = 1.91 \times 10^{27}$ kg
2. Determine $R_J = R_E(1320)^{1/3}$ — $R_J = 6.37 \times 10^6(1320)^{1/3}$ m $= 6.99 \times 10^7$ m
3. Find R for $T = 35400$ s; $h = R - R_J$ — $R = 1.59 \times 10^8$ m; $h = R - R_J = 8.94 \times 10^7$ m

---

100 •• The average density of the moon is $\rho = 3340$ kg/m³. Find the minimum possible period T of a spacecraft orbiting the moon.

The minimum period is when the orbit radius equals the object's radius, i.e., orbit just above the surface of the moon.

1. Write the condition for a stable orbit, $R_o = R_m = R$ — $mR\omega^2 = mMG/R^2 = 4\pi\rho R^3 mG/3R^2 = 4\pi\rho mGR/3$
2. Solve for $T^2 = 4\pi^2/\omega^2$ and T — $T^2 = 3\pi/\rho G$; $T = (3\pi/\rho G)^{1/2}$
3. Evaluate $T = T_{min}$ — $T_{min} = (3\pi/3340 \times 6.67 \times 10^{-11})^{1/2}$ s $= 6500$ s $= 1$h 48 min

---

101* •• A satellite is circling around the moon (radius 1700 km) close to the surface at a speed v. A projectile is launched from the moon vertically up at the same initial speed v. How high will it rise?

1. From Problem 55, $v^2 = v_e^2/2$; $\frac{1}{2}mv_e^2 = GmM_m/R_m$ — $\frac{1}{2}v^2 = \frac{1}{2}GM_m/R_m = GM_m[1/R_m - 1/(R_m + h)]$
2. Solve for h — $h = R_m = 1700$ m

---

102 •• Two space colonies of equal mass orbit a star (Figure 11-26). The Yangs in m_1 move in a circular orbit of radius 10^{11} m with a period of 2 y. The Yins in m_2 move in an elliptical orbit with a closest distance $r_1 = 10^{11}$ m and a farthest distance $r_2 = 1.8 \times 10^{11}$ m. (*a*) Using the fact that the mean radius of an elliptical orbit is the length of the semimajor axis, find the length of the Yin year. (*b*) What is the mass of the star? (*c*) Which colony moves faster at point P in Figure 11-26? (*d*) Which colony has the greater total energy? (*e*) How does the speed of the Yins at point P compare with their speed at point A?

(*a*) 1. Find R_2, the semi-major axis of Yins	$R_2 = \frac{1}{2} \times 2.8 \times 10^{11}$ m $= 1.4 \times 10^{11}$ m
2. Use Equ. 11-2; $T_{Yin} = T_2 = T_{Yang}(R_2/r_1)^{3/2}$	$T_2 = 2(1.4)^{3/2}$ y $= 3.31$ y
(*b*) From Equ. 11-15, $M = 4\pi^2 r^3/GT^2$	$M = 4\pi^2 \times 10^{33}/[6.67 \times 10^{-11}(6.31 \times 10^7)^2]$ kg $= 1.49 \times 10^{29}$ kg
(*c*) Since $R_2 > r_1$, $E_2 > E_1$; $U_{2P} = U_1$; $E = K + U$	$K_2 > K_1$; $v_2 > v_1$
(*d*) See above	$E_2 > E_1$
(*e*) L is conserved; $r_2 v_A = r_1 v_P$	$v_P = 1.8 v_A$

103 •• In a binary star system, two stars orbit about their common center of mass. If the stars have masses m_1 and m_2 and are separated by a distance r, show that the period of rotation is related to r by $T^2 = 4\pi r^3/[G(m_1 + m_2)]$.

Take the coordinate origin at the center of mass. Then $r_1 m_1 = r_2 m_2$ and $r = r_1 + r_2$. The force holding m_2 in orbit is $Gm_1m_2/(r_1 + r_2)^2 = m_2 r_2 \omega^2$. $\omega^2 = Gm_1/r_2(r_1 + r_2)^2$. Now $r_2 = rm_1/(m_1 + m_2)$, so $\omega^2 = 4\pi^2/T^2 = G(m_1 + m_2)/r^3$ and $T^2 = 4\pi^2 r^3/G(m_1 + m_2)$.

104 •• Two particles of mass m_1 and m_2 are released from rest with infinite separation. Find their speeds v_1 and v_2 when their separation distance is r.

1. Use energy conservation	$Gm_1m_2/r = \frac{1}{2}m_1v_1^2 + \frac{1}{2}m_2v_2^2$	(1)
2. Use conservation of linear momentum	$m_1v_1 = -m_2v_2$; $v_1 = -v_2(m_2/m_1)$	(2)
3. Use (2) in (1)	$v_2^2(m_2 + m_2^2/m_1) = 2Gm_1m_2/r$	(3)
4. Solve for v_2	$v_2 = [2Gm_1^2/r(m_1 + m_2)]^{1/2}$	(4)
5. To find v_1 permute subscripts in (4)	$v_1 = [2Gm_2^2/r(m_1 + m_2)]^{1/2}$	

105* •• A hole is drilled from the surface of the earth to its center as in Figure 11-27. Ignore the earth's rotation and air resistance. (*a*) How much work is required to lift a particle of mass m from the center of the earth to the earth's surface? (*b*) If the particle is dropped from rest at the surface of the earth, what is its speed when it reaches the center of the earth? (*c*) What is the escape speed for a particle projected from the center of the earth? Express your answers in terms of m, g, and R_E.

(*a*) From Equ. 11-27, $F = GmM_Er/R_E^3 = gmr/R_E$	$W = \int_0^{R_E} F\,dr = \frac{gm}{R_E}\int_0^{R_E} r\,dr = \frac{gmR_E}{2}$
(*b*) Use energy conservation; $\frac{1}{2}mv^2 = W$	$v = \sqrt{gR_E}$
(*c*) $E_{esc} = W + \frac{1}{2}mv_e^2 = \frac{1}{2}mv_{esc}^2$	$v_{esc}^2 = gR_E + 2gR_E$; $v_{esc} = (3gR_E)^{1/2} = 13.7$ km/s

106 •• A thick spherical shell of mass M and uniform density has an inner radius R_1 and an outer radius R_2. Find the gravitational field g_r as a function of r for all possible values of r. Sketch a graph of g_r versus r.

1. For $r < R_1$, $g = 0$ (see Equ. 11-24*b*). 2. For $r > R_2$, $g(r)$ is that of a mass M centered at the origin, i.e., $g(r) = GM/r^2$. 3. For $R_1 < r < R_2$, $g(r)$ is determined by the mass within the shell of radius r. The mass density is $\rho = 3M/[4\pi(R_2^3 - R_1^3)]$, and the mass within a radius r is given by $4\pi\rho(r^3 - R_1^3)/3 = M(r^3 - R_1^3)/(R_2^3 - R_1^3)$. So in this region $g(r) = GM(r^3 - R_1^3)/r^2(R_2^3 - R_1^3)$.

A graph of $g(r)$ is shown alongside.
Here we have set $R_1 = 2$, $R_2 = 3$, and $GM = 1$.

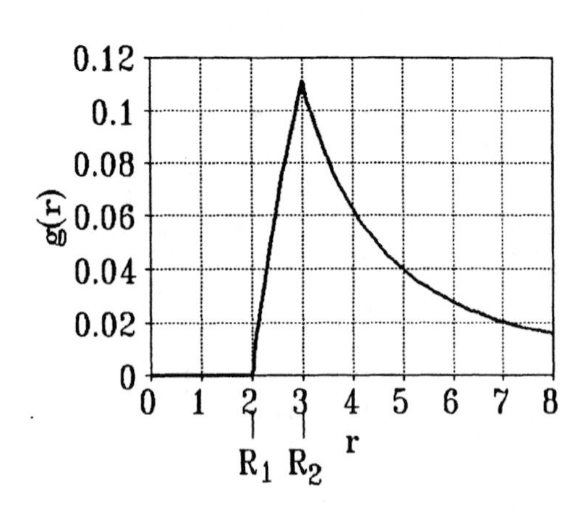

107 •• (*a*) Sketch a plot of the gravitational field g_x versus x due to a uniform ring of mass M and radius R whose axis is the x axis. (*b*) At what points is the magnitude of g_x maximum?

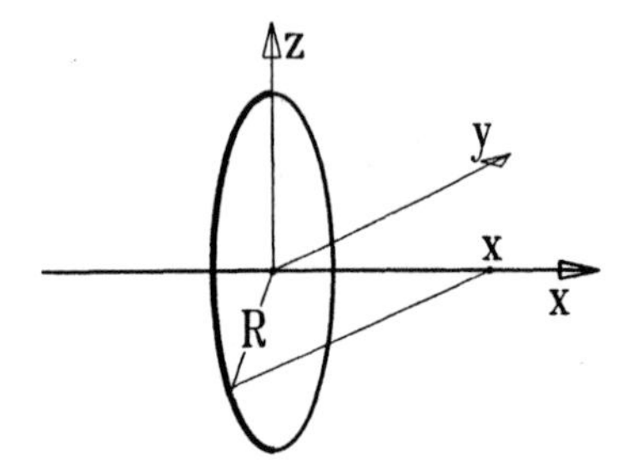

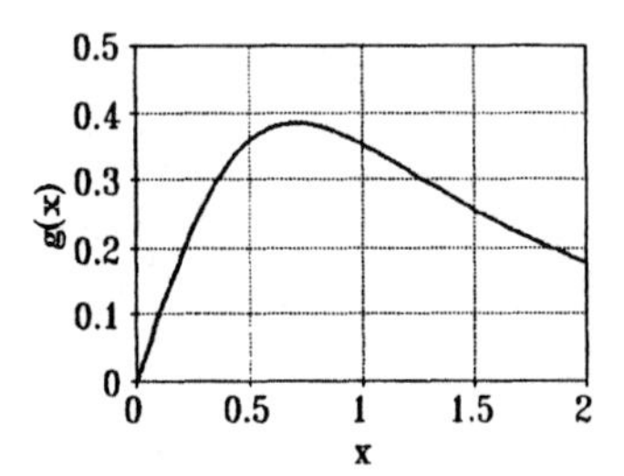

(*a*) The geometry of the problem is shown at the left above. Consider an element of the ring of length dL. The element of field at a point x is $dg = G\lambda\, dL/(x^2 + R^2)$. By symmetry, the y and z components of g vanish. The x component of $dg = dg \cos\theta = G\lambda\, dL[x/(x^2 + R^2)^{3/2}]$. Since all parts of the ring contribute equally and $2\pi\lambda R = M$, the field $g(x) = Gmx/(x^2 + R^2)^{3/2}$. A plot of $g(x)$ is shown to the right above. The curve is normalized for $R = 1$ and $GM = 1$.

(*b*) Differentiate $g(x)$ and set $dg/dx = 0$; $(x^2 + R^2)^{3/2} - 3x^2(x^2 + R^2)^{½} = 0$; $x = \pm R/\sqrt{2}$. This agrees with the sketch shown.

108 ••• In this problem, you are to find the gravitational potential energy of the stick in Example 11-8 and a point mass m_0 that is on the x axis at x_0. (*a*) Show that the potential energy of an element of the stick dm and m_0 is given by

$$dU = -\frac{Gm_0 dm}{x_0 - x} = \frac{GMm_0}{L(x_0 - x)}dx$$

where $U = 0$ at $x_0 = \infty$. (*b*) Integrate your result for part (*a*) over the length of the rod to find the total potential energy for the system. Write your result as a general function $U(x)$ by setting x_0 equal to a general point x. (*c*) Compute the force on m_0 at a general point x from $F_x = -dU/dx$ and compare your result with m_0g, where g is the field at x_0 calculated in Example 11-8.

(*a*) Let $U = 0$ at $x = \infty$. Let $\lambda = M/L$. Then the potential energy of the masses m_0 at x_0 and $dm = \lambda\, dx$ at x is given by $dU = -Gm_0 dm/r$ where $r = x_0 - x$. Thus $dU = -Gmm_0\, dx/[L(x_0 - x)]$.

(*b*) $U = -\dfrac{GMm_0}{L}\displaystyle\int_{-L/2}^{L/2}\dfrac{dx}{x_0 - x} = \dfrac{GMm_0}{L}[\ln(x_0 - L/2) - \ln(x_0 + L/2)] = -\dfrac{GMm_0}{L}\ln\left(\dfrac{x_0 + L/2}{x_0 - L/2}\right)$.

(*c*) Since x_0 is a general point along the *x* axis, $F(x_0) = -dU/dx_0 = (Gmm_0/L)[1/(x_0 + L/2) - 1/(x_0 - L/2)]$; this simplifies to $F(x_0) = -Gmm_0/(x^2 - L^2/4)$ in agreement with the result of Example 11-8.

109* ••• A uniform sphere of mass *M* is located near a thin, uniform rod of mass *m* and length *L* as in Figure 11-28. Find the gravitational force of attraction exerted by the sphere on the rod. (see Problem 72)

We shall determine the force exerted by the rod on the sphere and then use Newton's third law. The sphere is equivalent to a point mass *m* located at the sphere's center.

1. Use the result of Problem 72; $x_0 = a + L/2$ — $F = GmM/[(a + L/2)^2 - L^2/4] = GmM/[a(a + L)]$
2. By Newton's third law, this is the force on the rod

---

110 ••• A uniform rod of mass $M = 20$ kg and length $L = 5$ m is bent into a semicircle. What is the gravitational force exerted by the rod on a point mass $m = 0.1$ kg located at the center of the circular arc?

The semicircular rod is shown in the figure. We shall use an element of length $R\,d\theta = (L/\pi)\,d\theta$ of mass $(M/\pi)\,d\theta$. By symmetry, $F_y = 0$. So we first find dF_x and then integrate over θ from $-\pi/2$ to $\pi/2$ to determine F_x.

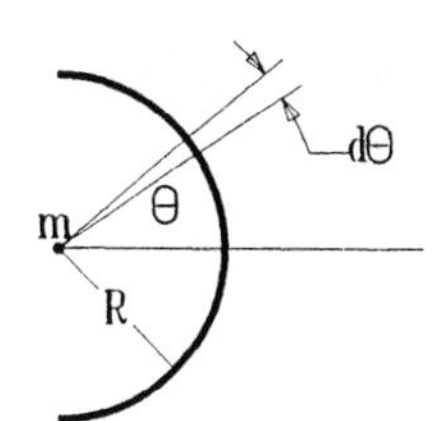

1. Obtain an expression for dF_x — $dF_x = [GMm/\pi(L/\pi)^2]d\theta\cos\theta$
2. Integrate from $\theta = -\pi/2$ to $\pi/2$ — $F_x = 2\pi GMm/L^2$
3. Evaluate F_x — $F_x = 3.35 \times 10^{-11}$ N

---

111 ••• Both the sun and the moon exert gravitational forces on the oceans of the earth, causing tides. (*a*) Show that the ratio of the force exerted by the sun to that exerted by the moon is $M_S r_m^2/M_m r_S^2$, where M_S and M_m are the masses of the sun and moon and r_S and r_m are the distances from the earth to the sun and to the moon. Evaluate this ratio. (*b*) Even though the sun exerts a much greater force on the oceans than the moon does, the moon has a greater effect on the tides because it is the difference in the force from one side of the earth to the other that is important. Differentiate the expression $F = Gm_1m_2/r^2$ to calculate the change in *F* due to a small change in *r*. Show that $dF/F = (-2\,dr)/r$. (*c*) During one full day, the rotation of the earth can cause the distance from the sun or moon to an ocean to change by at most the diameter of the earth. Show that for a small change in distance, the change in the force exerted by the sun is related to the change in the force exerted by the moon by $\Delta F_S/\Delta F_m \approx (M_S r_m^3)/(M_m r_S^3)$ and calculate this ratio.

(*a*) Force on a mass *m* is $F = GMm/r^2$ — $F_S/F_m = M_S r_m^2/M_m r_S^2 = 179$ (see Problem 27)

(*b*) Find dF/dr and dF/F — $dF/dr = -2Gm_1m_2/r^3 = -2F/r$; $dF/F = -2\,dr/r$

(*c*) $\Delta F = -2(F/r)\Delta r$ — $\Delta F_S/\Delta F_m = (F_S/F_m)(r_m/r_S) = (M_S r_m^3)/(M_m r_S^3)$

Insert numerical values — $$\frac{\Delta F_S}{\Delta F_m} = \frac{(1.99\times10^{30})(3.844\times10^8)^3}{(7.35\times10^{22})(1.496\times10^{11})^3} = 0.46$$

CHAPTER 12

Static Equilibrium and Elasticity

1* • True or false: (*a*) $\Sigma F = 0$ is sufficient for static equilibrium to exist. (*b*) $\Sigma F = 0$ is necessary for static equilibrium to exist. (*c*) In static equilibrium, the net torque about any point is zero. (*d*) An object is in equilibrium only when there are no forces acting on it.

(*a*) False (*b*) True (*c*) True (*d*) False

---

2 • A seesaw consists of a 4-m board pivoted at the center. A 28-kg child sits on one end of the board. Where should a 40-kg child sit to balance the seesaw?

Apply $\Sigma\tau = 0$ about the pivot — $2 \times 28 = 40d$; $d = 1.4$ m from pivot

---

3 • In Figure 12-23, Misako is about to do a push-up. Her center of gravity lies directly above point *P* on the floor, which is 0.9 m from her feet and 0.6 m from her hands. If her mass is 54 kg, what is the force exerted by the floor on her hands?

Apply $\Sigma\tau = 0$ about her feet asa pivot — $0.9 \times 54 \times 9.81 = F \times 1.5$ N·m; $F = 318$ N

---

4 • Juan and Bettina are carrying a 60-kg block on a 4-m board as shown in Figure 12-24. The mass of the board is 10 kg. Since Juan spends most of his time reading cookbooks, whereas Bettina regularly does push-ups, they place the block 2.5 m from Juan and 1.5 m from Bettina. Find the force in newtons exerted by each to carry the block.

Apply $\Sigma\tau = 0$ about the right end of the board — $(60 \times 1.5 + 10 \times 2)g = 4 \times F_J$; $F_J = 270$ N

Apply $\Sigma F = 0$ — $70g = 270\text{ N} + F_B$; $F_B = 417$ N

5* • Misako wishes to measure the strength of her biceps muscle by exerting a force on a test strap as shown in Figure 12-25. The strap is 28 cm from the pivot point at the elbow, and her biceps muscle is attached at a point 5 cm from the pivot point. If the scale reads 18 N when she exerts her maximum force, what force is exerted by the biceps muscle?

Apply $\Sigma\tau = 0$ about the pivot — $28 \times 18 = 5 \times F$; $F = 101$ N

---

6 • A crutch is pressed against the sidewalk with a force $\mathbf{F}_c$ along its own direction as in Figure 12-26. This force is balanced by the normal force $\mathbf{F}_n$ and a frictional force f_s. (*a*) Show that when the force of friction is at its maximum value, the coefficient of friction is related to the angle θ by $\mu_s = \tan\theta$. (*b*) Explain how this result applies to the forces on your foot when you are not using a crutch. (*c*) Why is it advantageous to take short steps when walking on ice?

(*a*) $f_{s,max} = \mu_s F_n$; $F_n = F_c \cos\theta$. For equilibrium, $f_s = F_c \sin\theta$. If $f_s = f_{s,max} = F_c \sin\theta = \mu_s F_c \cos\theta$, $\mu_s = \tan\theta$.

(*b*) Taking long strides requires a large coefficient of static friction because θ is then large.

(*c*) If μ_s is small, i.e., there is ice on the surface, θ must be small to avoid slipping.

---

7 • True or false: The center of gravity is always at the geometric center of a body.

False; the location depends on the mass distribution.

---

8 • Must there be any material at the center of gravity of an object?

No

9* • If the acceleration of gravity is not constant over an object, is it the center of mass or the center of gravity that is the pivot point when the object is balanced?

The center of gravity is then the pivot point for balance.

---

10 • Two spheres of radius R rest on a horizontal table with their centers a distance $4R$ apart. One sphere has twice the weight of the other sphere. Where is the center of gravity of this system?

Take the origin at the center of sphere of mass M $\qquad Md_{cm} = 2M(4R - d_{cm})$; $d_{cm} = 8R/3$

---

11 • An automobile has 58% of its weight on the front wheels. The front and back wheels are separated by 2 m. Where is the center of gravity located with respect to the front wheels?

Proceed as in Problem 10 $\qquad 0.58Md_{cm} = 0.42M(2 - d_{cm})$; $d_{cm} = 0.84$ m

---

12 • Each of the objects shown in Figure 12-27 is suspended from the ceiling by a thread attached to the point marked + on the object. Describe the orientation of each suspended object with a diagram.

The figures are shown on the right. The center of mass for each is indicated by a small +. At static equilibrium, the center of gravity is directly below the point of support.

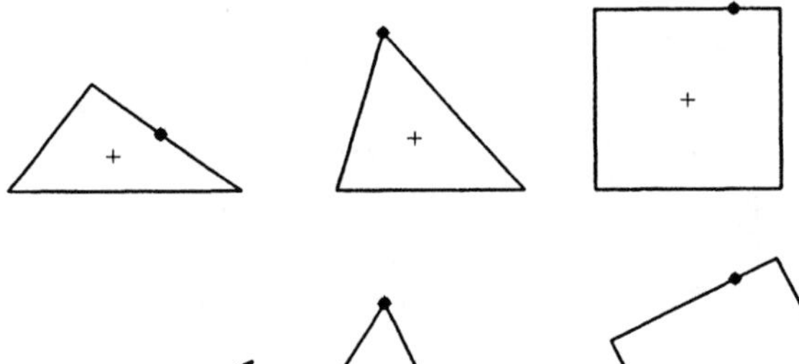

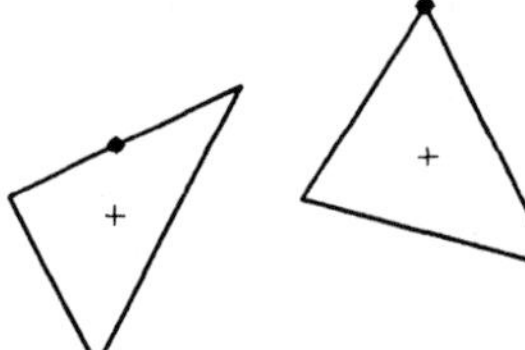

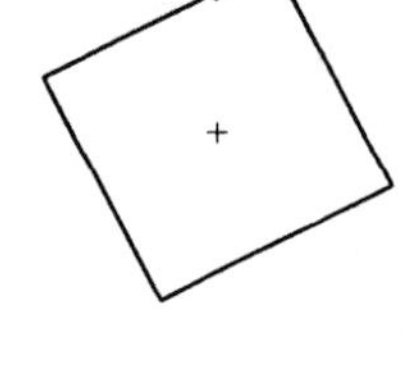

13* •• A square plate is produced by welding together four smaller square plates, each of side a as shown in Figure 12-28. Plate 1 weighs 40 N; plate 2, 60 N; plate 3, 30 N; and plate 4, 50 N. Find the center of gravity (x_{cg}, y_{cg}).

1. Use Equ. 12-3 to find x_{cg} $\qquad (1/2)a \times 100 + (3/2)a \times 80 = 180 \times x_{cg}$; $x_{cg} = 0.944a$
2. Similarly, find y_{cg} $\qquad (1/2)a \times 90 + (3/2)a \times 90 = 180 \times y_{cg}$; $y_{cg} = a$

14 •• A uniform rectangular plate has a circular section of radius R cut out as shown in Figure 12-29. Find the center of gravity of the system. *Hint:* Do not integrate. Use superposition of a rectangular plate minus a circular plate.

Take the coordinate origin at the lower left corner of the plate. Let a and b be the length and width of the plate. By symmetry, $y_{cg} = b/2$. Next, determine the missing mass. Let σ be the mass per unit area. Then the missing mass, m_h, is given by $m_h = \pi R^2\sigma$. The center of mass of the plate before the hole was cut was at $x = a/2$. Now apply Equ. 12-3.

$\frac{1}{2}a(ab\sigma) - (a - R)\pi R^2\sigma = (ab\sigma - \pi R^2\sigma)x_{cg}$; $x_{cg} = (\frac{1}{2}a^2b - \pi aR^2 + \pi R^3)/(ab - \pi R^2)$.

15 • When the tree in front of his house was cut down to widen the road, Jay did not want it to go without ceremony, so he hauled out his electric guitar and amplifier. All that remained was a uniform 10-m log of mass 100 kg resting on two supports, waiting to be cut up and taken away the next day. One support was 2 m from the left end, and the other was 4 m from the right end. Find the forces exerted on the log by the supports as Jay played his ear-splitting "Requiem for a Fallen Tree."

1. Take moments about the left hand support	3×981 N·m $= 4F_R$; $F_R = 736$ N
2. Use $\Sigma F = 0$ to find F_L	$F_L = (981 - 736)$ N $= 245$ N

16 • Bubba uses a crowbar that is 1 m long to lift a heavy crate off the ground. The crowbar rests on a rigid fulcrum 10 cm from one end as shown in Figure 12-30. (*a*) If Bubba exerts a downward force of 600 N on one end of the crowbar, what is the upward force exerted on the crate by the other end? (*b*) The ratio of the forces at the ends of the crowbar is called the mechanical advantage of the crowbar. What is the mechanical advantage here?

(*a*) Apply $\Sigma\tau = 0$ at the fulcrum	0.90×600 N·m $= 0.10F$ N·m; $F = 5400$ N
(*b*) $F/600 = D_2/D_1$ = mechanical advantage	Mechanical advantage = 9

17* • Figure 12-31 shows a 25-foot sloop. The mast is a uniform pole of 120 kg and is supported on the deck and held fore and aft by wires as shown. The tension in the forestay (wire leading to the bow) is 1000 N. Determine the tension in the backstay and the force that the deck exerts on the mast. Is there a tendency for the mast to slide forward or aft? If so, where should a block be placed to prevent the mast from moving?

1. Find θ_F, the angle of the forestay with vertical	$\theta_F = \tan^{-1}(2.74/4.88) = 29.3°$
2. Apply $\Sigma\tau = 0$ with pivot at bottom of the mast	$2.74 \times 1000 \times \cos 29.3° = 4.88T_B \cos 45°$; $T_B = 692$ N
3. Apply $\Sigma F = 0$ to mast; F_D = force exerted by the deck on the mast	$692 \times \sin 45° - 1000 \times \sin 29.3° = F_{Dx}$; $F_{Dx} = 0$ $F_{Dy} = 692 \times \cos 45° + 1000 \times \cos 29.3° + 120g = 2539$ N

18 • The sloop in Figure 12-32 is rigged slightly differently from the one in Problem 17. The mass of the mast is 150 kg and the tension in the forestay is again 1000 N. Find the tension in the backstay and the force that the deck exerts on the mast. Is there a tendency for the mast to slide forward or aft? If so, where should a block be placed on the deck to prevent the mast from moving?

1. Find the angles θ_B and θ_F of the backstay and the forestay	$\theta_B = \tan^{-1}(4.57/6.1) = 36.84°$; $\theta_F = \tan(0.5) = 26.57°$
2. Apply $\Sigma\tau = 0$ with pivot at bottom of mast	$2.44 \times 1000 \times \cos 26.57° = 4.57T_B \cos 36.84°$; $T_B = 596.7$ N

3. Apply $\Sigma F = 0$ to the mast; F_{Dx}, F_{Dy} are horizontal and vertical components of the force of the deck on the mast — $596.7 \sin 36.84° - 1000 \sin 26.57° + F_{Dx} = 0$; $F_{Dx} = 89.5$ N; $F_{Dy} = 596.7 \cos 36.84° + 1000 \cos 26.57° + 150g = 2843$ N
4. F_{Dx} points toward the stern — Tendency to slide forward; place block in front of mast

---

19 •• A 10-m beam of mass 300 kg extends over a ledge as in Figure 12-33. The beam is not attached, but simply rests on the surface. A 60-kg student intends to position the beam so that he can walk to the end of it. How far from the edge of the ledge can the beam extend?

Apply $\Sigma\tau = 0$ with the pivot at the ledge — $60x = 300(5 - x)$; $x = 4.17$ m

---

20 •• A gravity board for locating the center of gravity of a person consists of a horizontal board supported by a fulcrum at one end and by a scale at the other end. A physics student lies horizontally on the board with the top of his head above the fulcrum point as shown in Figure 12-34. The scale is 2 m from the fulcrum. The student has a mass of 70 kg, and when he is on the gravity board, the scale advances 250 N. Where is the center of gravity of the student?

Apply $\Sigma\tau = 0$ about the fulcrum — $70gx_{cg} = 2 \times 250$ N·m; $x_{cg} = 72.8$ cm from the fulcrum

---

21* •• A 3-m board of mass 5 kg is hinged at one end. A force $\boldsymbol{F}$ is applied vertically at the other end to lift a 60-kg block, which rests on the board 80 cm from the hinge, as shown in Figure 12-35. (*a*) Find the magnitude of the force needed to hold the board stationary at $\theta = 30°$. (*b*) Find the force exerted by the hinge at this angle. (*c*) Find the magnitude of the force $\boldsymbol{F}$ and the force exerted by the hinge if $\boldsymbol{F}$ is exerted perpendicular to the board when $\theta = 30°$.

(*a*) Apply $\Sigma\tau = 0$ about the hinge; cos 30° factors cancel — $3F = (0.8 \times 60 + 1.5 \times 5)9.81$ N·m; $F = 181.5$ N

(*b*) Use $\Sigma F = 0$ — $F_H + 181.5 - 65 \times 9.81 = 0$; $F_H = 456$ N

(*c*) Apply $\Sigma\tau = 0$ — $3F = (48 + 7.5) \times 9.81 \times \cos 30°$; $F = 157.2$ N

Use $\Sigma F_x = 0$, $\Sigma F_y = 0$ — $F_{Hy} = 65 \times 9.81 - 157.2 \cos 30° = 501.5$ N; $F_{Hx} = 157.2 \sin 30° = 78.6$ N; $\boldsymbol{F}_H = (78.6\,\boldsymbol{i} + 501.5\,\boldsymbol{j})$ N

---

22 •• A cylinder of weight W is supported by a frictionless trough formed by a plane inclined at 30° to the horizontal on the left and one inclined at 60° on the right as shown in Figure 12-36. Find the force exerted by each plane on the cylinder.

The planes are frictionless; therefore, the force exerted by each plane must be perpendicular to that plane. Let $\boldsymbol{F}_1$ be the force exerted by the 30° plane, and let $\boldsymbol{F}_2$ be the force exerted by the 60° plane.

1. Use $\Sigma F_x = 0$ — $F_1 \sin 30° = F_2 \sin 60°$; $F_1 = F_2 \times \sqrt{3}$
2. Use $\Sigma F_y = 0$ — $F_1 \cos 30° + F_2 \cos 60° = W = 2F_2$; $F_2 = ½W$, $F_1 = (\sqrt{3}/2)W$

---

23 •• An 80-N weight is supported by a cable attached to a strut hinged at point A as in Figure 12-37. The strut is supported by a second cable under tension T_2. The mass of the strut is negligible. (*a*) What are the three forces acting on the strut? (*b*) Show that the vertical component of the tension T_2 must equal 80 N. (*c*) Find the force exerted on the strut by the hinge.

(*a*) The forces acting on the strut are the tensions T_1 and T_2 and F_H, the force exerted on the strut by the hinge.

(*b*) Take moments about the hinge. $\Sigma\tau = 0$ then requires that $T_{2v} = T_1$.

(*c*) Since $T_{2v} = T_1 = 80$ N, $T_{2h} = 80 \tan 60° = 138.6$ N. Using $\Sigma F = 0$, we find $F_H = T_{2h} = 138.6$ N, to the right.

---

24 •• A horizontal board 8.0 m long is used by pirates to make their victims walk the plank. A pirate of mass 105 kg stands on the shipboard end of the plank to prevent it from tipping. Find the maximum distance the plank can overhang for a 63-kg victim to be able to walk to the end if (*a*) the mass of the plank is negligible, and (*b*) the mass of the plank is 25 kg.

(*a*) Apply $\Sigma\tau = 0$; let x be the overhang $\quad 63x = 105(8.0 - x);\ x = 5.0$ m

(*b*) Apply $\Sigma\tau = 0$ $\quad 63x + 25(x - 4) = 105(8 - x);\ x = 4.87$ m

---

25* •• As a farewell prank on their alma mater, Sharika and Chico decide to liberate thousands of marbles in the hallway during final exams. They place a 2-m × 1-m × 1-m box on a hinged board, as in Figure 12-38, and fill it with marbles. When the building is perfectly silent, they slowly lift one end of the plank, increasing θ, the angle of the incline. If the coefficient of static friction is large enough to prevent the box from slipping, at what angle will the box tip? (Assume that the marbles stay in the box until it tips over.)

The box will tip when its center of mass is no longer above the base of the box. So $\theta = \tan^{-1}(0.5) = 26.6°$.

---

26 •• A uniform 18-kg door that is 2.0 m high by 0.8 m wide is hung from two hinges that are 20 cm from the top and 20 cm from the bottom. If each hinge supports half the weight of the door, find the magnitude and direction of the horizontal components of the forces exerted by the two hinges on the door.

The drawing shows the door and its two supports. The center of gravity of the door is 0.8 m above (and below) the hinge, and 0.4 m from the hinges horizontally. Denote the horizontal and vertical components of the hinge force by F_{Hh} and F_{hv}.

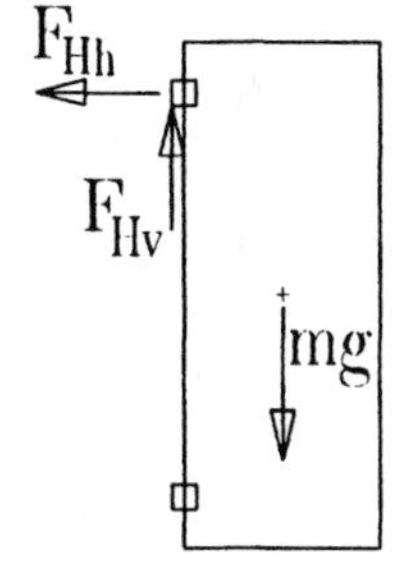

Take moments about the lower hinge, and apply $\Sigma\tau = 0$ $\quad 18 \times 9.81 \times 0.4$ N·m $= F_{Hh} \times 1.6$ m; $F_{Hh} = 44.15$ N

The upper hinge pulls on the door; the lower pushes

---

27 •• Find the force exerted by the corner on the wheel in Example 12-4, just as the wheel lifts off the surface.

Applying $\Sigma F = 0$ and using the results of Example 12-4, one obtains $\boldsymbol{F} = Mg(2Rh - h^2)^{1/2}/(h - R)\,\boldsymbol{i} + Mg\,\boldsymbol{j}$.

---

28 •• Lou is promoting the grand opening of Roswell's, a new nightclub with an alien theme. One end of a uniform 100-kg beam, 10 m long, is hinged to a wall, and the other end sticks out horizontally over the dance floor. A cable connects to the beam 6 m from the wall, as in Figure 12-39. Lou sits at the controls of a mock UFO, which hangs from the free end of the beam. From there, he sends down abduction beams, hypnotic light effects, and spaceship noises to the patrons below. If the combined weight of Lou and his UFO is 400 kg, (*a*) what is the tension in the cable? (*b*) What is the horizontal force on the hinge? What is the vertical force of the beam on the hinge?

(*a*) $\Sigma\tau = 0$ about the hinge $\quad (400 \times 10 + 100 \times 5)g = 0.8T \times 6;\ T = 9200$ N (*b*) Horizontal force on the hinge = $0.6T$ — $F_h = 5520$ N, acting to the left

(*c*) Vertical force on the hinge = $0.8T - 500g$ — $F_v = 2455$ N, acting upward

---

29* •• The diving board shown in Figure 12-40 has a mass of 30 kg. Find the force on the supports when a 70-kg diver stands at the end of the diving board. Give the direction of each support force as a tension or a compression.

1. Use $\Sigma\tau = 0$ about the end support as a pivot to find the force on the middle support — $(4.2 \times 70 + 2.1 \times 30)g = 1.2F$; $F = 2920$ N, compression
2. $\Sigma\tau = 0$ about the right support to find F of the end support — $1.2F_{end} = (0.9 \times 30 + 3 \times 70)g$; $F_{end} = 1940$ N, tension

---

30 •• Find the force exerted on the strut by the hinge at *A* for the arrangement in Figure 12-41 if (*a*) the strut is weightless, and (*b*) the strut weighs 20 N.

Let T be the tension in the line attached to the wall and L be the length of the strut.

(*a*) Take moments about the hinge — $60L \sin 45° = TL$; $T = 42.43$ N

Apply $\Sigma \boldsymbol{F} = 0$ to the strut; here $\boldsymbol{F}_H$ is the force the hinge exerts on the strut — $F_{Hx} - T \sin 45° = 0$; $F_{Hy} + T \cos 45° - 60 = 0$; $F_{Hx} = 30$ N, $F_{Hy} = 30$ N; $\boldsymbol{F}_H = (30\,\boldsymbol{i} + 30\,\boldsymbol{j})$ N

(*b*) Proceed as in part (*a*) — $(60L + 20L/2)\sin 45° = TL$; $T = 49.5$ N; $F_{Hx} = T \sin 45° = 35$ N; $F_{Hy} + T \cos 45° - 80 = 0$; $F_{Hy} = 45$ N; $\boldsymbol{F}_H = (35\,\boldsymbol{i} + 45\,\boldsymbol{j})$ N

---

31 •• Julie has been hired to help paint the trim of a building, but she is not convinced of the safety of the apparatus. A 5.0-m plank is suspended horizontally from the top of the building by ropes attached at each end. She knows from previous experience, however, that the ropes being used will break if the tension exceeds 1 kN. Her 80-kg boss dismisses Julie's worries and begins painting while standing 1 m from the end of the plank. If Julie's mass is 60 kg and the plank has a mass of 20 kg, then over what range of positions can Julie stand if a colorful plummet is to be avoided?

See the diagram. Note that if the 60 kg mass is at the far left end of the plank, T_1 and T_2 are less than 1 kN. Let x be the distance of the 60 kg mass from T_1.

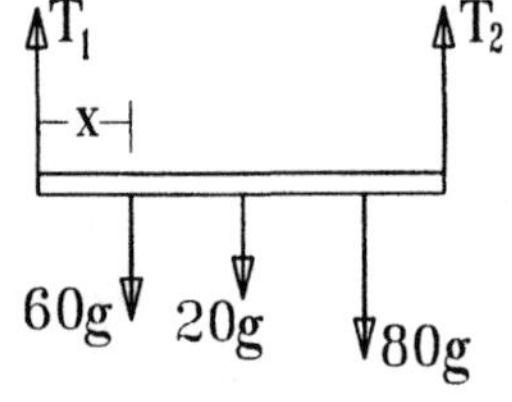

1. Take moments about the left end of the plank — $(60x + 20 \times 2.5 + 80 \times 4)g = 5T_2$
2. Set $T_2 = 1$ kN and solve for x — $x = 2.33$ m; safe for $0 < x < 2.33$ m

---

32 •• The cable in Figure 12-39 must remain attached to the wall 8 m above the hinge, but its length can vary so that it can be connected to the beam at various distances x from the wall. How far from the wall should it be attached so that the force on the hinge has no vertical component?

It should be attached to the end of the beam. (see Problem 23)

33* •• A cylinder of mass M and radius R rolls against a step of height h as shown in Figure 12-42. When a horizontal force $\boldsymbol{F}$ is applied to the top of the cylinder, the cylinder remains at rest. (*a*) What is the normal force exerted by the floor on the cylinder? (*b*) What is the horizontal force exerted by the corner of the step on the cylinder? (*c*) What is the vertical component of the force exerted by the corner on the cylinder?

(*a*) Take moments about the step's corner; see Example 12-4 for the moment arm of Mg and F_n — $F(2R - h) = (Mg - F_n)(2Rh - h^2)^{1/2}$; solve for F_n

$F_n = Mg - F[(2R - h)/h]^{1/2}$

(*b*) Use $\Sigma F_x = 0$ — $F_{x,corn} = -F$

(*c*) Use $\Sigma F_y = 0$ — $F_n - Mg + F_{y,corn} = 0$; $F_{y,corn} = F[(2R - h)/h]^{1/2}$

---

34 •• For the cylinder in Problem 33, find the minimum horizontal force $\boldsymbol{F}$ that will roll the cylinder over the step if the cylinder does not slide on the corner.

To roll over the step, the cylinder must lift off the floor, i.e., $F_n = 0$. From 32(*a*), $F = Mg\sqrt{\dfrac{h}{2R - h}}$.

---

35 •• A strong man holds one end of a 3-m rod of mass 5 kg at rest in a horizontal position. (*a*) What total force does the man exert on the rod? (*b*) What total torque does the man exert on the rod? (*c*) If you approximate the effort of the man with two forces that act in opposite directions and are separated by the width of the man's hand, which is taken to be 10 cm, what are the magnitudes and directions of the two forces?

(*a*) Use $\Sigma \boldsymbol{F} = 0$ — $F_{man} = mg = 49.1$ N

(*b*) $\Sigma\tau = 0$ — $\tau_{man} = mgL/2 = 73.6$ N·m

(*c*) Take the moment about the point of application of the force acting down. Set $\tau_{man} = F_u \times d$; F_u is the force of the couple acting up; F_d is the force of the couple acting down $= F_u - mg$ — $F_u = (73.6/0.1)$ N $= 736$ N, acting up; $F_d = 687$ N, acting down

---

36 •• A large gate weighing 200 N is supported by hinges at the top and bottom and is further supported by a wire as shown in Figure 12-44. (*a*) What must the tension in the wire be for the force on the upper hinge to have no horizontal component? (*b*) What is the horizontal force on the lower hinge? (*c*) What are the vertical forces on the hinges?

(*a*) Set $\Sigma\tau = 0$; pivot is at the lower hinge — $mg \times 1.5\text{ m} = [(T\cos 45°) + (T\sin 45°)] \times 1.5\text{ m}$

Solve for T — $T = 141$ N

(*b*) Use $\Sigma F_x = 0$; let 1 denote the lower hinge — $F_{x,1} = T\cos 45° = 100$ N

(*c*) Use $\Sigma F_y = 0$ — $F_{y,1} + F_{y,2} - mg + T\sin 45° = 0$; $F_{y,1} + F_{y,2} = 100$ N

$F_{y,1}$ and $F_{y,2}$ cannot be determined independently

37* ••• A uniform log with a mass of 100 kg, a length of 4 m, and a radius of 12 cm is held in an inclined position, as shown in Figure 12-45. The coefficient of static friction between the log and the horizontal surface is 0.6. The log is on the verge of slipping to the right. Find the tension in the support wire and the angle the wire makes with the vertical wall.

We shall use the following nomenclature: T = the tension in the wire; F_n = the normal force of the surface; $f_s = \mu_s F_n$ = the force of static friction. We shall also use $\Sigma\tau = 0$ and use the point where the wire is attached to the log as the pivot. Taking this as the origin, the center of mass of the log is at the coordinates

$(-2\cos 20° + 0.12\sin 20°, -2\sin 20° - 0.12\cos 20°) = (-1.838, -0.797)$.

The point of contact with the floor is at $(-3.676, -1.594)$.

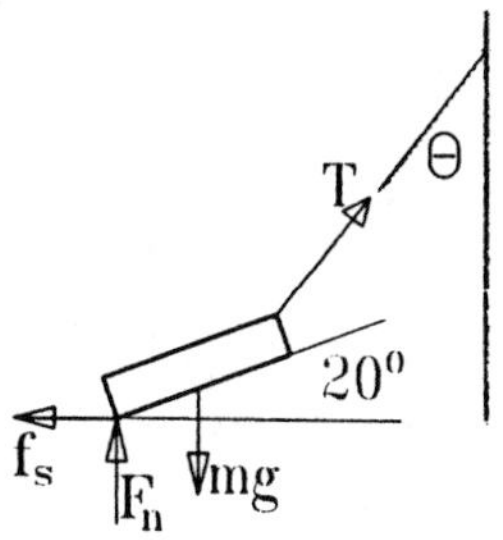

1. Use $\Sigma \boldsymbol{F} = 0$; see the free body diagram	$T\sin\theta - \mu_s F_n = 0$; $T\cos\theta + F_n - mg = 0$
2. Apply $\Sigma\tau = 0$ about the origin	$1.838mg - 3.676F_n - 1.142\mu_s F_n = 0$
3. Solve for F_n; $m = 100$ kg, $\mu_s = 0.6$	$F_n = 389$ N
4. Insert 225 N = F_n into the force equations	$T\sin\theta = 233$ N; $T\cos\theta = 592$ N; $\theta = \tan^{-1}(0.394) = 21.5°$
5. Evaluate T	$T = (233/\sin 21.5°)$ N = 636 N

38 ••• A tall, uniform, rectangular block sits on an inclined plane as shown in Figure 12-46. A cord is attached to the top of the block to prevent it from falling down the incline. What is the maximum angle θ for which the block will not slide on the incline? Let b/a be 4 and $\mu_s = 0.8$.

Consider what happens just as θ increases beyond θ_{max}. Since the top of the block is fixed by the cord, the block will in fact rotate with only the lower right edge of the block remaining in contact with the plane. It follows that just prior to this slipping, F_n and $f_s = \mu_s F_n$ act at the lower right edge of the block.

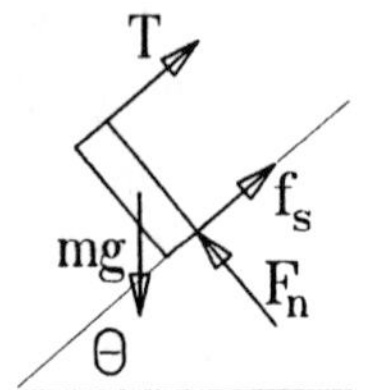

1. Apply $\Sigma \boldsymbol{F} = 0$; see the free-body diagram	$T + \mu_s F_n - mg\sin\theta = 0$; $F_n - mg\cos\theta = 0$
2. Use $\Sigma\tau = 0$ about the lower right edge	$½a(mg\cos\theta) + ½b(mg\sin\theta) - bT = 0$
3. Use $b = 4a$, $\mu_s = 0.8$, and the force equations	$mg(\cos\theta + 4\sin\theta) = 8T$; $F_n + 4(T + 0.8F_n) = 8T$; $T = 1.05F_n$
4. Insert $T = 1.05F_n$ into force equations to find θ	$1.85F_n = mg\sin\theta$; $F_n = mg\cos\theta$; $\theta = \tan^{-1}1.85 = 61.6°$

39 ••• A thin rail of length 10 m and mass 20 kg is supported at a 30° incline. One support is 2 m and the other is 6 m from the lower end of the rail. Friction prevents the rail from sliding off the supports. Find the force (magnitude and direction) exerted on the rail by each support.

The rail and the forces acting on it are shown in the free-body diagram. Supports are indicated by "1" and "2" as shown. Since the x components of the forces at the supports are friction forces, they are proportional to the normal, i.e., y, components of the forces at the supports.

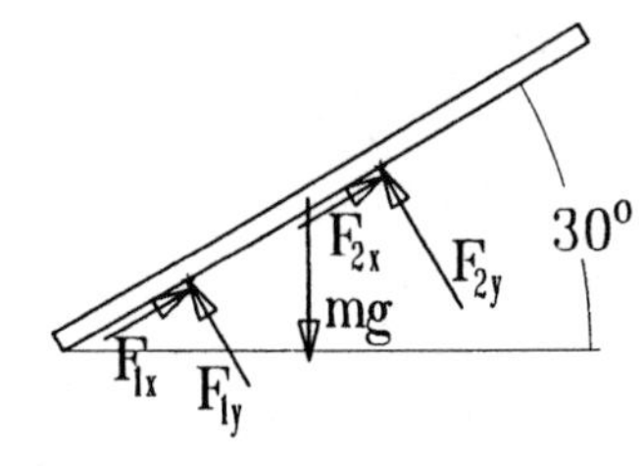

Take x up along the rail, y up and normal to the rail.

1. Apply $\Sigma\tau = 0$ about support 1	$3mg\cos 30° = 4F_{2y}$; $F_{2y} = 127.4$ N
2. Apply $\Sigma\tau = 0$ about support 2	$mg\cos 30° = 4F_{1y}$; $F_{1y} = 42.5$ N; note $F_{2y} = 3F_{1y}$
3. Use $\Sigma F_x = 0$	$F_{1x} + F_{2x} = mg\sin 30° = 98.1$ N
4. Use $F_{1x}/F_{2x} = F_{1y}/F_{2y} = 1/3$; $F_{2x} = 3F_{1x}$	$F_{1x} = 24.5$ N; $F_{2x} = 73.6$ N
5. Find θ_1 and θ_2 (see diagram)	$\theta_1 = \theta_2 = \tan^{-1}(F_y/F_x) = 60°$; forces act vertically up
6. Find the magnitudes of F_1 and F_2	$F_1 = mg/4$; $F_2 = 3mg/4$; $F_1 = 49.1$ N; $F_2 = 147.2$ N

40 • Two 80-N forces are applied to opposite corners of a rectangular plate as shown in Figure 12-47. Find the torque produced by this couple.

Take the torque about the top left corner. $\tau = b(80\cos 30°) - a(80\sin 30°) = 69.3b - 40a$.

41* • A uniform cube of side a and mass M rests on a horizontal surface. A horizontal force $\boldsymbol{F}$ is applied to the top of the cube as in Figure 12-48. This force is not sufficient to move or tip the cube. (a) Show that the force of static friction exerted by the surface and the applied force constitute a couple, and find the torque exerted by the couple. (b) This couple is balanced by the couple consisting of the normal force exerted by the surface and the weight of the cube. Use this fact to find the effective point of application of the normal force when $F = Mg/3$. (c) What is the greatest magnitude of F for which the cube will not tip?

(a) The cube is stationary. Therefore $\boldsymbol{f}_s = -\boldsymbol{F}$, and the torque of that couple is Fa.

(b) Let x = the distance from the point of application of F_n to the center of the cube. Now, $F_n = Mg$, so $Mgx = Fa$; $x = Fa/Mg$. If $F = Mg/3$, then $x = a/3$.

(c) Note that $x_{max} = a/2$. The cube will tip if $F > Mg/2$.

---

42 •• Resolve each force in Problem 40 into its horizontal and vertical components, producing two couples. The algebraic sum of the two component couples equals the resultant couple. Use this result to find the perpendicular distance between the lines of action of the two forces.

The vertical components are $F \cos 30° = F\sqrt{3}/2$; the horizontal components are $F \sin 30° = F/2$. The net torque is, as in Problem 40, $\frac{1}{2}F(\sqrt{3}b - a)$. This is also equal to FD, where D is the moment arm of the couple, i.e., the distance between the two forces. So $D = \frac{1}{2}(\sqrt{3}b - a)$.

---

43 • Is it possible to climb a ladder placed against a wall where the ground is frictionless but the wall is not? Explain.

No. Since the floor can exert no horizontal force, neither can the wall. Consequently, the friction force between the wall and the ladder is zero regardless of the coefficient of friction between the wall and the ladder.

---

44 •• Romeo takes a uniform 10-m ladder and leans it against the smooth wall of the Capulet residence. The ladder's mass is 22.0 kg, and the bottom rests on the ground 2.8 m from the wall. When Romeo, whose mass is 70 kg, gets 90% of the way to the top, the ladder begins to slip. What is the coefficient of static friction between the ground and the ladder?

The ladder and the forces acting on it at the critical moment of slipping are shown in the diagram. The force of static friction is then $f_s = \mu_s F_n$, as indicated. The angle θ is given by $\theta = \cos^{-1} 0.28 = 73.74°$.

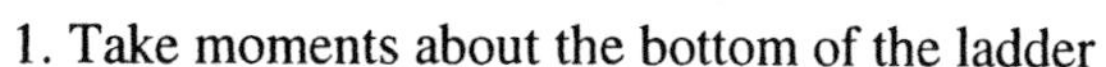

1. Take moments about the bottom of the ladder	$22g \times 1.4 + 70g \times 0.9 \times 2.8 = F_w \times 10 \sin 73.74°$
2. Find F_w; $F_w + f_s = 0$; find f_s	$F_n = 211.7 \text{ N} = f_s$
3. $\Sigma F_y = 0$	$F_n = 92g = 902.5 \text{ N}$
4. Find μ_s	$\mu_s = f_s/F_n = 211.7/902.5 = 0.235$

45* •• A massless ladder of length L leans against a smooth wall making an angle θ with the horizontal floor. The coefficient of friction between the ladder and the floor is μ_s. A man of mass M climbs the ladder. What height h can he reach before the ladder slips?

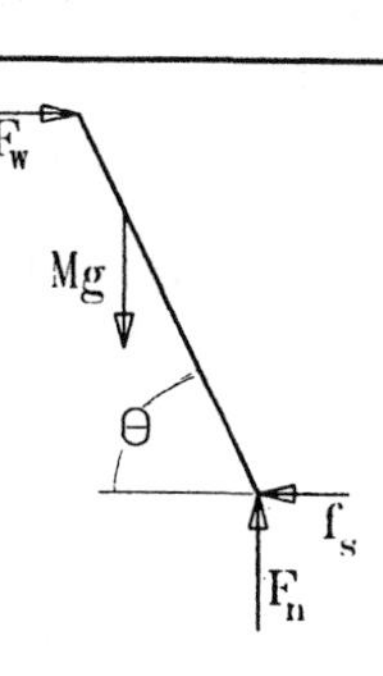

The ladder and the forces acting on it are shown in the diagram. Since the wall is smooth, the force the wall exerts on the ladder must be horizontal.

1. Use $\Sigma F = 0$	$F_n = Mg$; $f_s = F_w$

2. Apply $\Sigma\tau = 0$ about the bottom of the ladder — $F_wL \sin\theta = Mgx \cos\theta$
3. Solve for x; use $f_s = \mu_s F_n = \mu_s Mg = F_w$ — $x = \mu_s L \tan\theta$; $h = x \sin\theta = \mu_s L \tan\theta \sin\theta$

---

46 •• A uniform ladder of length L and mass m leans against a frictionless vertical wall with its lower end on the ground. It makes an angle of 60° with the horizontal ground. The coefficient of static friction between the ladder and ground is 0.45. If your mass is four times that of the ladder, how far up the ladder can you climb before it begins to slip?

The ladder and the forces acting on it are shown in the drawing. We proceed as in the preceding problems.

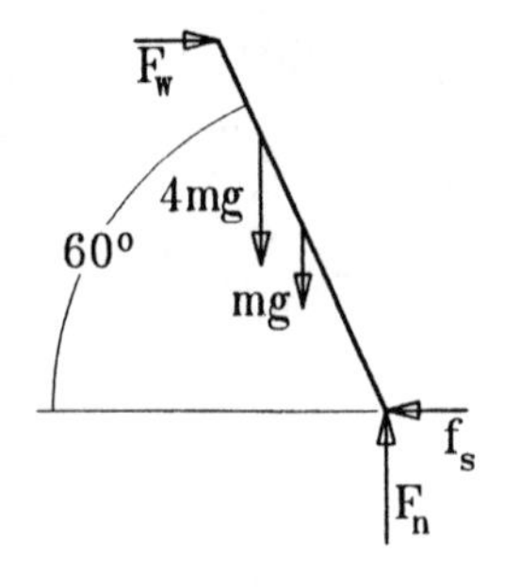

1. Use $\Sigma F = 0$ and solve for F_w — $\mu_s F_n = F_w$; $F_n = 5mg$; $F_w = 2.25mg$
2. Apply $\Sigma\tau = 0$ about the bottom of the ladder — $F_wL \sin 60° = (mgL/2 + 4mgx)\cos 60° = 2.25mgL \sin 60°$
3. Solve for x — $x = 0.85L$; you can go 85% of the way to the top

---

47 •• A ladder of mass m and length L leans against a frictionless, vertical wall making an angle θ with the horizontal. The center of mass is at a height h from the floor. A force F pulls horizontally against the ladder at the midpoint. Find the minimum coefficient of static friction μ_s for which the top end of the ladder will separate from the wall while the lower end does not slip.

Find F required to pull the ladder away from the wall. Apply $\Sigma\tau = 0$ about the bottom of the ladder. The moment due to mg is $mgh/\tan\theta$, that due to F is $\frac{1}{2}LF \sin\theta$. This yields $F = 2mgh/(L \sin\theta \tan\theta)$. Now $F_n = mg$ and $f_s = \mu_s mg = F$. One obtains $\mu_s = 2h/(L \sin\theta \tan\theta)$.

---

48 •• A 900-N boy sits on top of a ladder of negligible weight that rests on a frictionless floor as in Figure 12-49. There is a cross brace halfway up the ladder. The angle at the apex is $\theta = 30°$. (*a*) What is the force exerted by the floor on each leg of the ladder? (*b*) Find the tension in the cross brace. (*c*) If the cross brace is moved down toward the bottom of the ladder (maintaining the same angle θ), will its tension be greater or less?

(*a*) The force exerted by the frictionless floor must be vertical. By symmetry, each leg carries half the total weight. So the force on each leg is 450 N.

(*b*) Consider one of the ladder's legs and take moments about the apex. Then $F_nD/2 = Tx$, where D is the separation between the legs at the bottom and x is the distance of the cross brace from the apex. Clearly, if x is increased, i.e., the brace moved lower, T will decrease.

49* •• A ladder rests against a frictionless vertical wall. The coefficient of static friction between ladder and the floor is 0.3. What is the smallest angle at which the ladder will remain stationary?

Using the notation of Problem 45, we have $F_n = mg$ and $f_s = \mu_s mg = F_w$. Now take moments about the bottom of the ladder. This gives $\frac{1}{2}Lmg \cos\theta = Lf_w \sin\theta = Lmg\mu_s \sin\theta$. Solve for θ; $\theta = \tan^{-1}(1/2\mu_s) = 59°$.

---

50 ••• Having failed in his first attempt, Romeo acquires a new ladder to try once again to get to Juliet's window. This one has a length L and a weight of 200 N. He tries placing it on the other side of the window, where the coefficients of static friction are 0.4 between the ladder and the wall, and 0.7 between the ladder and the ground. Because of bruises suffered in his last fall, Romeo wears heavy padding, which gives him a total mass of 80 kg.

Sure enough, when he is 4/5 of the way up the ladder, it begins to slip. What was the angle between the ladder and the ground when Romeo was making his ascent?

The ladder and the forces acting on it are shown in the drawing.

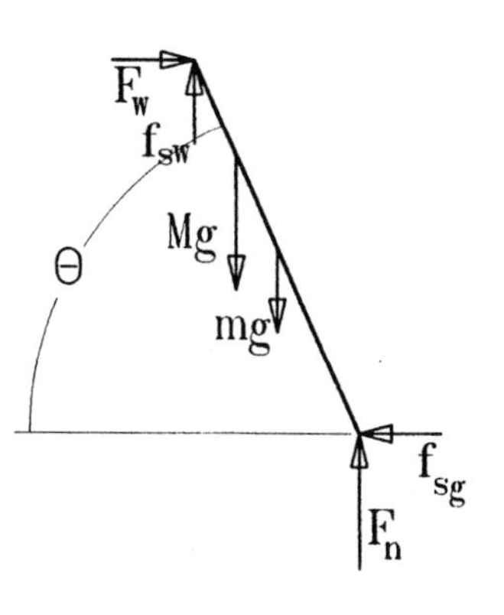

1. Use $\Sigma F = 0$ $\quad (M + m)g = F_n + f_{sw} = F_n + \mu_w F_w$; $F_w = f_{sg} = \mu_g F_n$
2. Solve for and evaluate F_n $\quad F_n = [(M + m)g]/(1 + \mu_w\mu_g)$; $F_n = 769$ N
3. Apply $\Sigma\tau = 0$ to the top of the ladder $\quad (MgL/5 + mgL/2)\cos\theta = F_nL\cos\theta - F_n\mu_gL\sin\theta$
4. Use numerical values to get an equation for θ $\quad 512\cos\theta = 538.3\sin\theta$; $\theta = \tan^{-1}(0.951) = 43.6°$

---

51 ••• A ladder leans against a large smooth sphere of radius R that is fixed in place on a horizontal surface. The ladder makes an angle of 60° with the horizontal surface and has a length $5R/2$. (*a*) What is the force that the sphere exerts on the ladder? (*b*) What is the frictional force that prevents the ladder from slipping? (*c*) What is the normal force that the horizontal surface exerts on the ladder?

The drawing shows the ladder and sphere. The distance between the point where the ladder touches the ground and touches the sphere is $D = R/\tan 30° = \sqrt{3}R$.

The sphere is smooth; hence the force F_s must be radial.

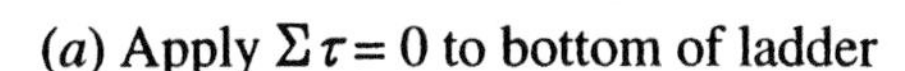

(*a*) Apply $\Sigma\tau = 0$ to bottom of ladder $\quad (5R/4)(mg)\cos 60° = \sqrt{3}Rf_s$; $F_s = 0.361mg$

(*b*) Use $\Sigma F_x = 0$ to find f_s $\quad f_s = F_s\cos 30° = 0.313mg$

(*c*) Use $\Sigma F_y = 0$ to find F_n $\quad F_n = mg - F_s\sin 30° = 0.820mg$

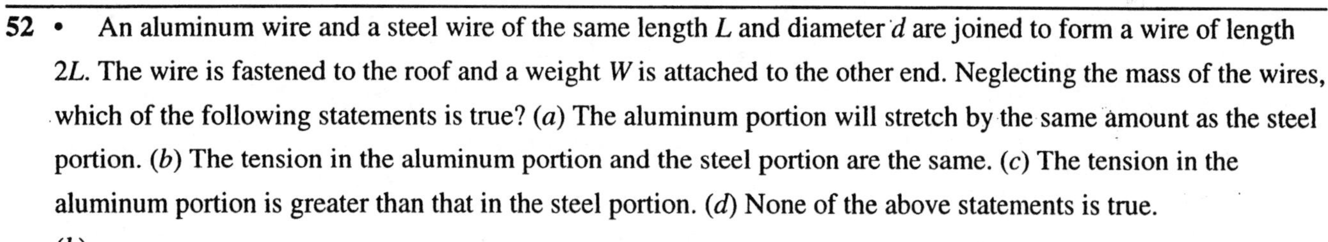

52 • An aluminum wire and a steel wire of the same length L and diameter d are joined to form a wire of length $2L$. The wire is fastened to the roof and a weight W is attached to the other end. Neglecting the mass of the wires, which of the following statements is true? (*a*) The aluminum portion will stretch by the same amount as the steel portion. (*b*) The tension in the aluminum portion and the steel portion are the same. (*c*) The tension in the aluminum portion is greater than that in the steel portion. (*d*) None of the above statements is true.

(*b*)

53* • A 50-kg ball is suspended from a steel wire of length 5 m and radius 2 mm. By how much does the wire stretch?

1. Find the stress $\quad F = mg = 490.5$ N; $A = \pi r^2 = 1.26 \times 10^{-5}$ m²; $F/A = 3.9 \times 10^7$ N/m²
2. From Equ. 12-7 and Table 12-1 find ΔL $\quad \Delta L = (5 \times 3.9 \times 10^7/2 \times 10^{11})$ m $= 0.976$ mm

---

54 • Copper has a breaking stress of about 3×10^8 N/m². (*a*) What is the maximum load that can be hung from a copper wire of diameter 0.42 mm? (*b*) If half this maximum load is hung from the copper wire, by what percentage of its length will it stretch?

(*a*) Use Equ. 12-6 $\quad F_{max} = [3 \times 10^8 \times \pi \times (2.1 \times 0^{-4})^2]$ N $= 41.6$ N; $M_{max} = 4.24$ kg (*b*) Use Equ. 12-7 to find $\Delta L/L$ $\Delta L/L = 1.5 \times 10^8/1.1 \times 10^{11} = 1.36 \times 10^{-3} = 0.136\%$

---

55 • A 4-kg mass is supported by a steel wire of diameter 0.6 mm and length 1.2 m. How much will this wire stretch under this load?

1. Find the stress $F/A = [4 \times 9.81/\pi \times (3 \times 10^{-4})^2]$ N/m^2 = 1.39×10^8 N/m^2
2. Use Equ. 12-7 and Table 12-1 to find ΔL $\Delta L = (1.2 \times 1.39 \times 10^8/2 \times 10^{11})$ m = 0.833 mm

---

56 • As a runner's foot touches the ground, the shearing force acting on an 8-mm-thick sole is as shown in Figure 12-50. If the force of 25 N is distributed over an area of 15 cm^2, find the angle of shear θ shown, given that the shear modulus of the sole is 1.9×10^5 N/m^2.

Use Equ. 12-10 to find tan θ $\tan\theta = (25/15 \times 10^{-4})/1.9 \times 10^5 = 0.0877$; $\theta = 5.01°$

57* •• A steel wire of length 1.5 m and diameter 1 mm is joined to an aluminum wire of identical dimensions to make a composite wire of length 3.0 m. What is the length of the composite wire if it is used to support a mass of 5 kg?

1. Find the stress in each wire $F/A = (5 \times 9.81 \times 4/\pi \times 10^{-6})$ N/m^2 = 6.245×10^7 N/m^2
2. Find $\Delta L = \Delta L_S + \Delta L_A$ $\Delta L_S = 1.5 \times 6.245 \times 10^7/2 \times 10^{11}$ m = 0.468 mm

$\Delta L_A = 0.468(200/70)$ mm = 1.338 mm; ΔL = 1.81 mm

$L = 3.0$ m + ΔL = 3.0018 m

---

58 •• A force F is applied to a long wire of length L and cross-sectional area A. Show that if the wire is considered to be a spring, the force constant k is given by $k = AY/L$ and the energy stored in the wire is $U = \frac{1}{2}F\Delta L$, where Y is Young's modulus and ΔL is the amount the wire has stretched.

For a spring, $\Delta L = F/k$ or $k = F/\Delta L$. From Equ. 12-7, $F/\Delta L = AY/L$, which is k. The energy stored in a spring is $U = \frac{1}{2}kx^2$, or in this case, $U = \frac{1}{2}AY(\Delta L)^2/L = \frac{1}{2}F\Delta L$.

---

59 •• The steel E string of a violin is under a tension of 53 N. The diameter of the string is 0.20 mm, and its length under tension is 35.0 cm. Find (*a*) the unstretched length of this string, and (*b*) the work needed to stretch the string. (see Problem 58)

(*a*) $L' = L + \Delta L$; $\Delta L = LF/AY$; solve for L $L = L'/(1 + F/AY)$

$= [0.35/(1 + 53/\pi \times 10^{-8} \times 2 \times 10^{11})]$ m

$L = 0.347$ m [ΔL = 2.93 mm]

(*b*) $W = \frac{1}{2}F\Delta L$ $W = (53 \times 2.93 \times 10^{-3}/2)$ J = 0.0776 J

---

60 •• When a rubber strip with a cross section of 3 mm × 1.5 mm is suspended vertically and various masses are attached to it, a student obtains the following data for length versus load:

Load, g	0	100	200	300	400	500
Length, cm	5.0	5.6	6.2	6.9	7.8	10.0

(*a*) Find Young's modulus for the rubber strip for small loads. (*b*) Find the energy stored in the strip when the load is 150 g. (see Problem 58)

(*a*) Find $\Delta L/F$ for small loads — For loads < 200 g, $\Delta L/F = 1.2 \times 10^{-2}/0.2 \times 9.81$ m/N

$\Delta L/F = 6.12 \times 10^{-3}$ m/N

$Y = FL/\Delta LA$ — $Y = (5 \times 10^{-2}/4.5 \times 10^{-6} \times 6.12 \times 10^{-3})$ N/m^2

$= 1.82 \times 10^6$ N/m^2

(*b*) $U = \frac{1}{2}F\Delta L$ — $U = 0.15 \times 9.81 \times 9 \times 10^{-3}/2 = 6.62$ mJ

61* •• A building is to be demolished by a 400-kg steel ball swinging on the end of a 30-m steel wire of diameter 5 cm hanging from a tall crane. As the ball is swung through an arc from side to side, the wire makes an angle of 50° with the vertical at the top of the swing. Find the amount by which the wire is stretched at the bottom of the swing.

1. Find the tension at bottom of the swing — $F = mg + mv^2/R$; $v^2/R = 2g(1 - \cos\theta) = 0.714g$

   $F = 1.714mg = 6727$ N

2. $\Delta L = FL/AY$; $A = \pi \times 25 \times 10^{-4}/4$ m^2 $= 1.96 \times 10^{-3}$ m^2 — $\Delta L = (6727 \times 30/1.96 \times 10^{-3} \times 2 \times 10^{11})$ m $= 0.515$ mm

---

62 •• A large mirror is hung from a nail as shown in Figure 12-51. The supporting steel wire has a diameter of 0.2 mm and an unstretched length of 1.7 m. The distance between the points of support at the top of the mirror's frame is 1.5 m. The mass of the mirror is 2.4 kg. What is the distance between the nail and the top of the frame when the mirror is hung?

1. Find the stress in the wire. The right triangle has sides 0.75 m, 0.4 m, and hypothenuse 0.85 m — $T(0.4/0.85) = mg/2$; $T = 25$ N; $A = \pi \times 10^{-8}$ m^2

   $T/A = 7.96 \times 10^8$ N/m^2

2. Find $\Delta L/L$ of the hypothenuse — $\Delta L/L = 7.96 \times 10^8/2 \times 10^{11} = 3.98 \times 10^{-3}$
3. Using $a^2 + h^2 = L^2$ and differentiating — $h\Delta h = L\Delta L$; $\Delta h = 0.85^2 \times 3.98 \times 10^{-3}/0.4 = 7.2$ mm
4. $h = 0.4$ m $+ \Delta h$ — $h = 40.72$ cm

---

63 •• Two masses, M_1 and M_2, are supported by wires of equal length when unstretched. The wire supporting M_1 is an aluminum wire 0.7 mm in diameter, and the one supporting M_2 is a steel wire 0.5 mm in diameter. What is the ratio M_1/M_2 if the two wires stretch by the same amount?

Note that $\Delta L_1 = M_1gL_1/A_1Y_1$ and $\Delta L_2 = M_2gL_2/A_2Y_2$. Since $L_1 = L_2$ and $\Delta L_1 = \Delta L_2$, $(M_1/d_1^2Y_1) = (M_2/d_2^2Y_2)$. With the data given, $M_1/M_2 = (49 \times 70)/(25 \times 200) = 0.686$.

---

64 •• A mass of 0.5 kg is attached to an aluminum wire having a diameter of 1.6 mm and an unstretched length of 0.7 m. The other end of the wire is fixed to a post. The mass rotates about the post in a horizontal plane at a rotational speed such that the angle between the wire and the horizontal is 5.0°. Find the tension in the wire and its length.

1. Find the tension; see Example 5-10 — $T = mg/\sin 5° = 56.3$ N
2. Find ΔL and $L = L_0 + \Delta L$ — $\Delta L = (56.3 \times 0.7/\pi \times 64 \times 10^{-8} \times 7 \times 10^{10})$ m

   $= 0.28$ mm; $L = 70.03$ cm

65* ••• It is apparent from Table 12-2 that the tensile strength of most materials is two to three orders of magnitude smaller than Young's modulus. Consequently, these materials, e.g., aluminum, will break before the strain exceeds even 1%. For nylon, however, the tensile strength and Young's modulus are approximately equal. If a nylon line of unstretched length L_0 and cross section A_0 is subjected to a tension T, the cross section may be substantially less than A_0 before the line breaks. Under these conditions, the tensile stress T/A may be significantly greater than T/A_0. Derive an expression that relates the area A to the tension T, A_0, and Young's modulus Y.

Assume constant volume of the line. Then LA = constant, and taking differentials $L\Delta A + A\Delta L = 0$ or $\Delta L/L = -\Delta A/A$. But $\Delta L/L = T/AY$, so $\Delta A = -T/Y$. Thus $A = A_0 + \Delta A = A_0 - T/Y$.

---

66 • If the net torque about some point is zero, must it be zero about any other point? Explain.

Yes; if it were otherwise, angular momentum conservation would depend on the choice of coordinates.

---

67 • The horizontal bar in Figure 12-52 will remain horizontal if (*a*) $L_1 = L_2$ and $R_1 = R_2$. (*b*) $L_1 = L_2$ and $M_1 = M_2$. (*c*) $R_1 = R_2$ and $M_1 = M_2$. (*d*) $L_1M_1 = L_2M_2$. (*e*) $R_1L_1 = R_2L_2$.

(*c*)

---

68 • Which of the following could not have units of N/m^2? (*a*) Young's modulus (*b*) Shear modulus (*c*) Stress (*d*) Strain

(*d*) Strain is dimensionless.

69* •• Sit in a chair with your back straight. Now try to stand up without leaning forward. Explain why you cannot do it.

The body's center of gravity must be above the feet.

---

70 • A 90-N board 12 m long rests on two supports, each 1 m from the end of the board. A 360-N block is placed on the board 3 m from one end as shown in Figure 12-53. Find the force exerted by each support on the board.

$\Sigma\tau = 0$ about the right support; obtain F_L, the force on the left support.	$2 \times 360 + 5 \times 90 = F_L \times 10$; $F_L = 117$ N
Use $\Sigma F_y = 0$ to find F_R.	$F_R = (450 - 117)$ N $= 333$ N

---

71 • The height of the center of gravity of a man standing erect is determined by weighing the man as he lies on a board of negligible weight supported by two scales as shown in Figure 12-54. If the man's height is 188 cm and the left scale reads 445 N while the right scale reads 400 N, where is his center of gravity relative to his feet?

Take moments about feet and use definition of CG	$d \times 845 = 188 \times 445$; $d = 99$ cm

---

72 • Figure 12-55 shows a mobile consisting of four weights hanging on three rods of negligible mass. Find the value of each of the unknown weights if the mobile is to balance. *Hint:* Find the weight w_1 first.

Apply the balance condition $\Sigma\tau = 0$ successively, starting with the lowest part of the mobile.

1. $4w_1 = 6$; $w_1 = 1.5$ N. 2. $2w_2 = 4 \times 3.5$; $w_2 = 7$ N. 3. $6w_3 = 2 \times 10.5$; $w_3 = 3.5$ N.

73* • A block and tackle is used to support a mass of 120 kg as shown in Figure 12-56. (*a*) What is the tension in the rope? (*b*) What is the mechanical advantage of this device?

(*a*) With this arrangement, the mass is supported by three ropes. Thus $T = 120g/3 = 392$ N.

(*b*) The mechanical advantage is 3.

---

74 •• A plate in the shape of an equilateral triangle of mass M is suspended from one corner and a mass m is suspended from another corner. What should be the ratio m/M so that the base of the triangle makes an angle of 6.0° with the horizontal?

The figures below show the equilateral triangle without the mass m, and then the same triangle with the mass m and rotated through a small angle θ. We take the side length of the triangle to be $2a$. Then the center of mass of the triangle is at a distance of $2a/\sqrt{3}$ from each vertex. As the triangle rotates, its CM shifts by $(2a/\sqrt{3})\theta$, for $\theta << 1$. Also the vertex to which m is attached moves toward the plumb line by the amount $\sqrt{3}a\theta$ (see drawing).

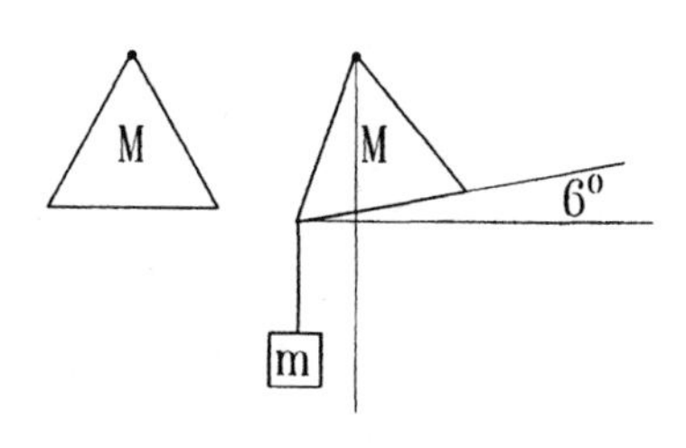

Take moments about the point of suspension $M(2a/\sqrt{3})\theta = ma(1 - \sqrt{3}\,\theta)$; $\theta = 6° = 0.105$ rad

Solve for and evaluate m/M $m/M = (2\theta/\sqrt{3})/(1 - \sqrt{3}\,\theta)$; $m/M = 0.148$

75 •• A standard six-sided pencil is placed on a paper pad (Figure 12-57). Find the minimum coefficient of static friction μ_s such that it rolls down rather than slides if the pad is inclined.

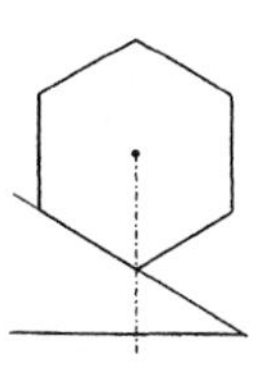

If the hexagon is to roll rather than slide, the incline's angle must be sufficient that the center of mass falls just beyond the support base. From the geometry of the hexagon, one can see that the critical angle is 30°. It follows (see Chapter 5) that $\mu_s \geq \tan 30° = 0.577$.

76 •• Having lost his job at the post office, Barry decides to explore the possibility that he might be a brilliant sculptor. Not one to start at the bottom, he borrows the money for a marble slab 3 m × 1 m × 1 m. After loading the marble onto the back of his truck, he drives off with the slab resting on its square end. But on the way home, a confused squirrel runs into his path, and Barry slams on the brakes. What deceleration will cause the uniform slab to tip over?

The block and forces on it are shown. At the critical condition for tipping, F_n acts at the edge of the block, as indicated in the drawing.

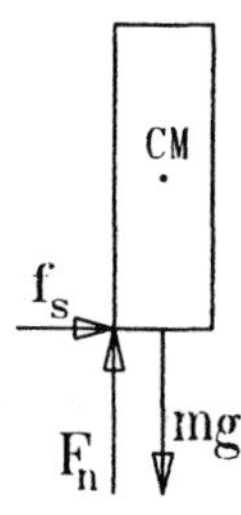

Take moments about the block's center of mass $0.5F_n = 1.5f_s$; $F_n = mg$ and $f_s = ma$

Solve for a $a = g/3 = 3.27\ \mathrm{m/s^2}$

77* •• A uniform box of mass 8 kg that is twice as tall as it is wide rests on the floor of a truck. What is the maximum coefficient of static friction between the box and floor so that the box will slide toward the rear of the truck rather than tip when the truck accelerates on a level road?

Proceed as in the previous problem. The box will tip if $\mu_s > 0.5$, so it must have $\mu_s < 0.5$.

78 •• Barry's art exhibit contains many tiny marble sculptures placed around a central piece called "Politics." The central piece consists of three identical bars, each of length L and mass m, joined as in Figure 12-58. Two bars form a fixed T, and the third bar is suspended on a hinge. Asked to explain the name, Barry said, "It swings from left to right with hinges flapping, but no matter where you start it from, you end up in the same place." When the system is in equilibrium, what is the value of θ?

The figure shows the system. We use the angle α rather than θ for convenience. The locations of the centers of mass of the three parts of the figure are indicated, and we can now determine the moment arms and torques about the point of support. These are also shown in the drawing.

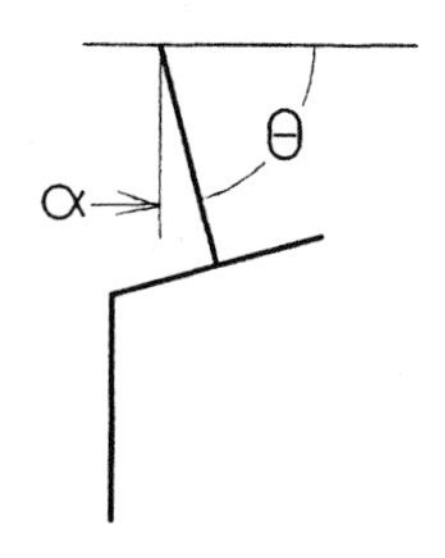

Take moments about the pivot point $(mgL/2)\sin\alpha + mgL\sin\alpha = (mgL/2)\cos\alpha - mgL\sin\alpha$

This yields the following $5\sin\alpha = \cos\alpha$; $\alpha = \tan^{-1} 0.2 = 11.3°$; $\theta = 90° - \alpha = 78.7°$

79 •• In the 1996 Olympics, the Russian super-heavyweight weightlifter Andrei Chemerkin broke the world record with a lift of mass 260 kg. Suppose his grip was slightly asymmetrical as shown in Figure 12-59. Find the maximum mass of the barbell Chemerkin could have handled with a symmetrical grip, assuming that his arms are equally strong.

1. From the figure it is evident that $F > F'$; use $\Sigma\tau = 0$ about the right hand of the weightlifter — $0.6F + 0.45 \times 130g = 1.15 \times 130g$; $F = 152g$
2. Find the total mass if each arm supports 152 kg — $M_{tot} = 304$ kg

80 •• A balance scale has unequal arms. A 1.5-kg block appears to have a mass of 1.95 kg on the left pan of the scale (Figure 12-60). Find its apparent mass if the block is placed on the right pan.

1. Find L_1/L_2 — $1.5L_1 = 1.95L_2$; $L_1/L_2 = 1.3$
2. Find M with 1.5 kg at L_2 — $1.5L_2 = ML_1 = 1.3ML_2$; $M = 1.15$ kg

81* •• A cube of mass M leans against a frictionless wall making an angle θ with the floor as shown in Figure 12-61. Find the minimum coefficient of static friction μ_s between the cube and the floor that allows the cube to stay at rest.

The figure alongside shows the location of the cube's center of mass and the forces acting on the cube. The moment arm of the couple formed by the normal force, F_n, and Mg is $d = (a/\sqrt{2})\sin(45° - \theta)$. The opposing couple is formed by the friction force f_s and the force exerted by the wall.

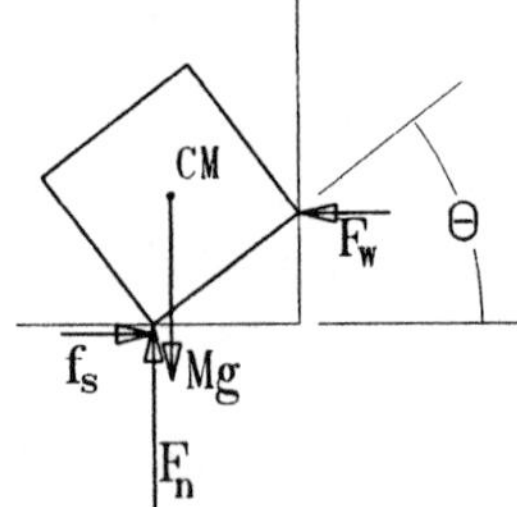

1. Set $\Sigma\tau = 0$ — $(Mga/\sqrt{2})\sin(45° - \theta) = f_s a\sin\theta = \mu_s Mga\sin\theta$
2. Solve for μ_s; $\sin(\alpha + \beta) = \sin\alpha\cos\beta + \sin\beta\cos\alpha$ — $\mu_s = ½(\cot\theta - 1)$

82 •• Figure 12-62 shows a steel meter stick hinged to a vertical wall and supported by a thin wire. The wire and meter stick make angles of 45° with the vertical. The mass of the meter stick is 5.0 kg. When a mass $M = 10.0$ kg is suspended from the midpoint of the meter stick, the tension T in the supporting wire is 52 N. If the wire will break should the tension exceed 75 N, what is the maximum distance along the meter stick at which the 10.0-kg mass can be suspended?

1. Use $\Sigma\tau = 0$ about the hinge — $75 \times 1 = 5g \times 0.5 \times \cos 45° + 10g \times d \times \cos 45°$
2. Solve for d — $d = 0.83$ m

83 •• Figure 12-63 shows a 20-kg ladder leaning against a frictionless wall and resting on a frictionless horizontal surface. To keep the ladder from slipping, the bottom of the ladder is tied to the wall with a thin wire; the tension in the wire is 29.4 N. The wire will break if the tension exceeds 200 N. (*a*) If an 80-kg person climbs halfway up the ladder, what force will be exerted by the ladder against the wall? (*b*) How far up can an 80-kg person climb this ladder?

(*a*) Use $\Sigma\tau = 0$ about the bottom of the ladder	$(20 + 80) \times 9.81 \times 0.75 = 5 \times F_w$; $F_w = 147$ N
(*b*) Note that the tension in the wire = F_w; let f be the fraction of the total length climbed.	$20 \times 9.81 \times 0.75 + 80 \times 9.81 \times 1.5f = 5 \times 200$ $f = 0.724$; can climb to height of 3.62 m

---

84 •• Suppose that the bar hanging from the end of the T in Problem 78 is of a different length $\ell \neq L$, although its mass per unit length is the same as that of the bars of the T. Find the ratio L/ℓ such that $\theta = 75°$.

The only change here from Problem 78 is that the mass of the free-hanging bar is not m but $M = (\ell/L)m$. Set $\theta = 75°$, $\alpha = 15°$ and write the torque equation. $m \sin 15° + 2m \sin 15° = M \cos 15° - 2M \sin 15°$. This gives $M/m = 1.73$ or $\ell = 1.73L$.

85* •• A uniform cube can be moved along a horizontal plane either by pushing the cube so that it slips or by turning it over ("rolling"). What coefficient of kinetic friction μ_k between the cube and the floor makes both ways equal in terms of the work needed?

To "roll" the cube one must raise its center of mass from $y = a/2$ to $y = \sqrt{2}a/2$, where a is the cube length. Thus the work done is $W = ½mga(\sqrt{2} - 1) = 0.207mga$. Since no work is done as the cube flops down, this is the work done to move the cube a distance a. Now set $0.207mga = Fa = \mu_k mga$ = work done against friction in moving a distance a. Thus $\mu_k = 0.207$.

---

86 •• A tall, uniform, rectangular block sits on an inclined plane as shown in Figure 12-64. If $\mu_s = 0.4$, does the block slide or fall over as the angle θ is slowly increased?

1. The condition for sliding (see Chapter 5) is $\mu_s \leq \tan\theta$, so $\theta \geq \tan^{-1}0.4 = 21.8°$.

2. The condition for tipping is that the plumb line from the center of mass pass outside of the base. In this case, that requirement gives $\tan\theta = 1/3$ or $\theta = 18.4°$. So as θ is increased, the block will tip before it slides.

---

87 •• A 360-kg mass is supported on a wire attached to a 15-m-long steel bar that is pivoted at a vertical wall and supported by a cable as shown in Figure 12-65. The mass of the bar is 85 kg. (*a*) With the cable attached to the bar 5.0 m from the lower end as shown, find the tension in the cable and the force exerted by the wall on the steel bar. (*b*) Repeat if a somewhat longer cable is attached to the steel bar 5.0 m from its upper end, maintaining the same angle between the bar and the wall.

(*a*) The cable is normal to the bar; use $\Sigma\tau = 0$ about the hinge	$85g \times 7.5 \times \cos 30° + 360g \times 15 \times \cos 30° = 5T$; $T = 10260$ N
Use $\Sigma F = 0$	$F_y = (445 \times 9.81 - 10260 \sin 60°)$ N $= -4520$ N $F_x = (10260 \cos 60°)$ N $= 5130$ N
(*b*) Use $\Sigma\tau = 0$; angle between cable and bar = 60°	$51300 = (10 \sin 60°)T'$; $T' = 5924$ N
Use $\Sigma F = 0$	$F_y = (4365 - 5924 \sin 30°)$ N $= 1403$ N $F_x = (5924 \cos 30°)$ N $= 5130$ N

---

88 •• Repeat Problem 77 if the truck accelerates up a hill that makes an angle of 9.0° with the horizontal.

Following the same procedure as in Problem 77 gives $\mu_s \leq 0.5 - \tan 9° = 0.342$.

89* •• A thin rod 60 cm long is balanced 20 cm from one end when a mass of $2m + 2$ g is at the end nearest the pivot and a mass of m at the opposite end (Figure 12-66*a*). Balance is again achieved if the mass $2m + 2$ g is replaced by the mass m and no mass is placed at the other end (Figure 12-66*b*). Determine the mass of the rod.

1. Take moments about pivot for initial condition	$20(m + 2) = 40m + 10M$

2. Take moments about pivot for second condition — $20m = 10M$; $M = 2m$

3. Solve for m and M using first equation — $m = 1$ g; $M = 2$ g

---

90 •• The planet Mars has two satellites, Phobos and Deimos, in nearly circular orbits. The orbit radii of Phobos and Deimos are 9.38×10^3 km and 23.46×10^3 km, respectively. The mass of Mars is 6.42×10^{23} kg, that of Phobos is 9.63×10^{15} kg, and that of Deimos is 1.93×10^{15} kg. Find the center of gravity and the center of mass of the two-satellite system using the center of Mars as the origin when (*a*) the satellites are in opposition (i.e., on exactly opposite sides of Mars), and (*b*) the satellites are in conjunction (i.e., in line on the same side of Mars).

Note: The two satellites (taken as point masses) are in different gravitational fields. Therefore, the center of gravity of the two-satellite sysem is not the same as the center of mass. We shall consider case (*b*) first.

(*b*) 1. The gravitational field on the two satelites is in the same direction. Write Equ. 12-3. — $(m_P + m_D)g(x_{CG})x_{CG} = m_Pg(x_P)x_P + m_Dg(x_D)x_D$, where $g(x) = GM_M/x$

2. Solve for x_{CG} — $x_{CG} = x_Px_D(m_P + m_D)/(m_Px_D + m_Dx_P)$

3. Substitute numerical values — $x_{CG} = 10.43 \times 10^3$ km

(*a*) Now g at Phobos and Deimos are in opposite directions — $x_{CG} = x_Px_D(m_P - m_D)/(m_Px_D - m_Dx_P)$ and substituting, $x_{CG} = 8.15 \times 10^3$ km, in the direction of Phobos

---

91 •• When a picture is hung on a smooth vertical wall using a wire and a nail, as in Figure 12-51, the picture almost always tips slightly forward, i.e., the plane of the picture makes a small angle with the vertical. (*a*) Explain why pictures supported in this manner generally do not hang flush against the wall. (*b*) A framed picture 1.5 m wide and 1.2 m high and having a mass of 8.0 kg is hung as in Figure 12-51 using a wire of 1.7 m length. The ends of the wire are fastened to the sides of the frame at the rear and 0.4 m below the top. When the picture is hung, the angle between the plane of the frame and the wall is 5.0°. Determine the force that the wall exerts on the bottom of the frame.

(*a*) Since the center of gravity of the picture is in front of the wall, the torque due to mg about the nail must be balanced by an opposing torque due to the force of the wall on the picture, acting horizontally. So that $\Sigma F_x = 0$, the tension in the wire must have a horizontal component, and the picture must therefore tilt forward.

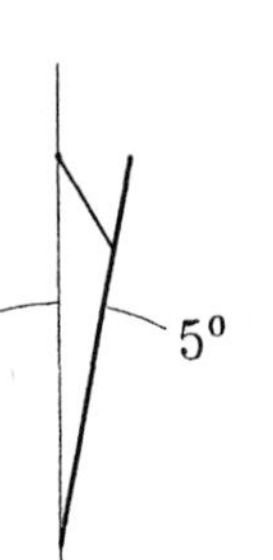

(*b*) Note that 0.75, 0.4, and 0.85 form a Pythagorean triad. Thus, the nail will be at the same level as the top of the frame. The drawing now shows the end view of the system. Since the frame's width is not specified, we assume it to be negligible.

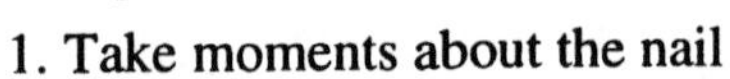

1. Take moments about the nail — $8g \times 0.6 \sin 5° = F_w \times 1.2 \cos 5°$

2. Solve for F_w — $F_w = (8 \times 9.81 \times 0.6 \tan 5°)/1.2$ N $= 3.43$ N

---

92 •• Repeat Problem 76 if Barry is driving (*a*) uphill on a road inclined at 10° with the horizontal or (*b*) downhill on a road inclined at 10° with the horizontal.

The figures on the next page show the block and the forces acting on it when traveling (*a*) uphill and decelerating and (*b*) downhill and decelerating. At the critical acceleration the normal force acts at the edge of the block as indicated. The force mg acts through the center of mass, CM.

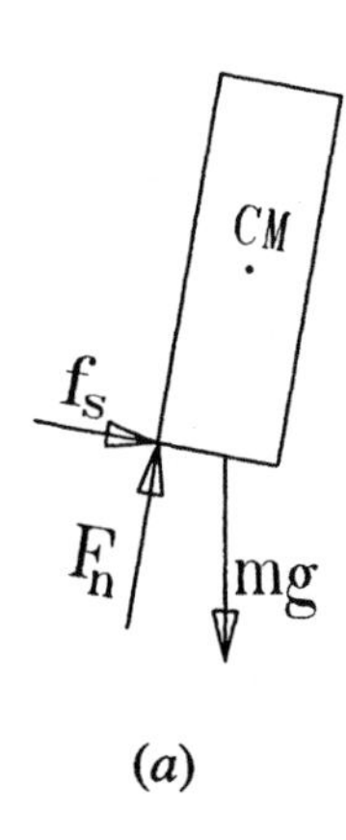

(*a*)

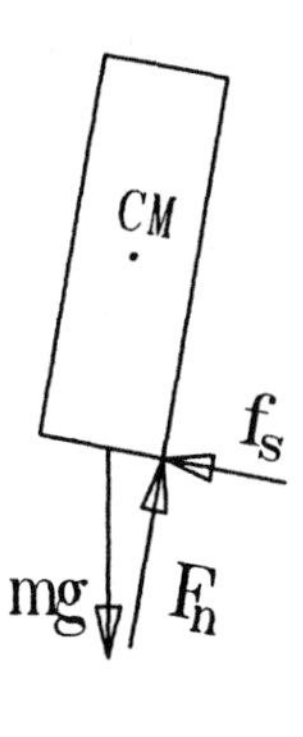

(*b*)

(*a*) 1. Take moments about the block's CM	$0.5(mg\cos 10°) - 1.5f_s = 0; f_s = (mg\cos 10°)/3$
2. $F_{net} = ma$	$f_s + mg\sin 10° = ma$
3. Solve for a	$a = g[(\cos 10°)/3 + \sin 10°] = 4.92\text{ m/s}^2$
(*b*) 1. Take moments about the block's CM	$f_s = (mg\cos 10°)/3$
2. $F_{net} = ma$	$f_s - mg\sin 10° = ma$
3. Solve for a	$a = g[(\cos 10°)/3 - \sin 10°] = 1.52\text{ m/s}^2$

93* •• If a train travels around a bend in the railbed too fast, the freight cars will tip over. Assume that the cargo portions of the freight cars are regular parallelepipeds of uniform density and 1.5×10^4 kg mass, 10 m long, 3.0 m high, and 2.20 m wide, and that their base is 0.65 m above the rails. The axles are 7.6 m apart, each 1.2 m from the ends of the boxcar. The separation between the rails is 1.55 m. Find the maximum safe speed of the train if the radius of curvature of the bend is (*a*) 150 m, and (*b*) 240 m.

The box car and rail are shown in the drawing. At the critical speed, the normal force is entirely on the outside rail. As indicated, the center of gravity is 0.775 m from that rail and 2.15 m above it. To find the speed at which this situation prevails, we take moments about the center of gravity.

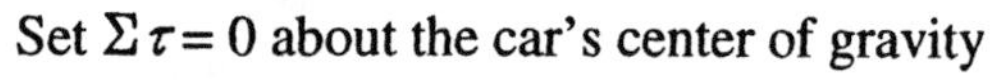

Set $\Sigma\tau = 0$ about the car's center of gravity	$0.775mg = 2.15mv^2/R$
(*a*) Solve for v with $R = 150$ m	$v = (0.775 \times 150 \times 9.81/2.15)^{½}$ m/s
	$= 23$ m/s $= 83$ km/h
(*b*) Find v for $R = 240$ m$0.775mg = 2.15mv^2/R$	$v = 29.1$ m/s $= 105$ km/h

---

94 •• For balance, a tightrope walker uses a thin rod 8 m long and bowed in a circular arc shape. At each end of the rod is a lead mass of 20 kg. The tightrope walker, whose mass is 58 kg and whose center of gravity is 0.90 m above the rope, holds the rod tightly at its center 0.65 m above the rope. What should the radius of curvature of the arc of the rod be so that he will be in stable equilibrium as he slowly makes his way across the rope? Neglect the mass of the rod.

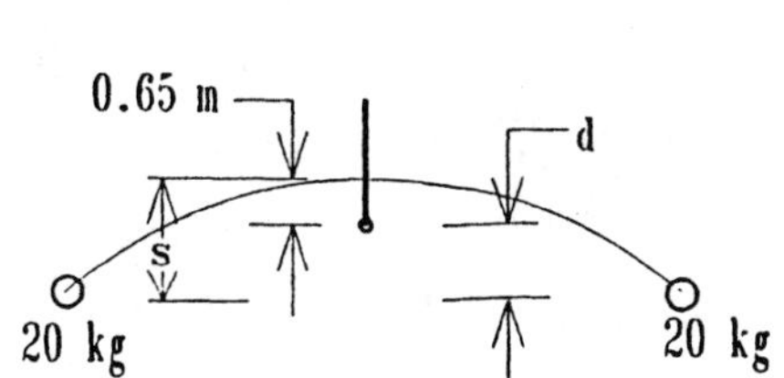

1. For stable equilibrium, the center of mass of the system must be below the foot of the tightrope walker. The system is shown in the drawing. We take the coordinate origin at the rope and will determine the distance d such that $y_{cm} = 0$. We

then determine the angle θ (not shown on the diagram) subtended by one half of the long rod.

2. Find distance d — $0 = 58 \times 0.9 - 40d;\ d = 1.305$ m
3. Length of $s = 0.65$ m $+ d$ — $s = 1.955$ m
4. Find expression for θ — $R(1 - \cos\theta) = 1.955$ m; $R\theta = 4$ m; $\dfrac{1 - \cos\theta}{\theta} = 0.489$
5. Find θ by trial and error — $\theta = 1.08$ rad
6. Find the radius of curvature R — $R = (4/1.08)$ m $= 3.7$ m

---

95 •• A large crate weighing 4500 N rests on four 12-cm-high blocks on a horizontal surface (Figure 12-67). The crate is 2 m long, 1.2 m high, and 1.2 m deep. You are asked to lift one end of the crate using a long steel bar. The coefficient of static friction between the blocks and the supporting surface is 0.4. Estimate the length of the steel bar you will need to lift the end of the crate.

1. Note that when the crowbar lifts the crate, only half the weight of the crate is supported by the bar. Also, that force has a horizontal component that must be balanced by the frictional force on the other end of the crate's support. Assume that the maximum force you can apply is 500 N (about 110 lb). Let ℓ be the distance between the points of contact of the steel bar with the floor and the crate, and let L be the total length of the bar. We denote the force that the steel bar exerts on the crate by F.
2. Find F_x and F_y; $F_y = ½(4500$ N$)$ — $F_y = 2250$ N $= F\cos\theta$; $F_x = 2250\mu_s = 900$ N $= F\sin\theta$
3. Find θ and F — $\theta = \tan^{-1}0.4 = 21.8°$; $F = 2423$ N
4. Determine the mechanical advantage, ME — ME $= 2423/500 = 4.85$
5. Determine the lengths ℓ and L of the crowbar — $\ell = (12/\tan\theta)$ cm $= 30$ cm; $L = \text{ME} \times \ell \approx 1.5$ m

---

96 ••• Six identical bricks are stacked one on top of the other lengthwise and slightly offset to produce a stepped tower with the maximum offset that will still allow the tower to stand. (*a*) Starting from the top, give the maximum possible offset for each successive brick. (*b*) What is the total protrusion or offset of the six bricks?

(*a*) Let each brick have a length L. The maximum offset of the top brick is $L/2$ so that its CM is just at the point of support. The distance of the CM of the top and next lower brick from the center of the lower brick is $x_{cm} = (mL/2)/2m = L/4$. It follows that the maximum overhang of the two-brick combination is $L/2 - L/4 = L/4$. Next, use the same procedure for the top three bricks, locating the CM of the three-brick combination relative to the center of the third brick. That distance is $L/3$ so the overhang of the three-brick combination is $L/2 - L/3 = L/6$. The next overhang is $L/8$, and the following one is $L/10$.

(*b*) The total offset is $½L(1 + 1/2 + 1/3 + 1/4 + 1/5) = 1.14L$. The total offset is greater than the length of one brick, so the left edge of the top brick is to the right of the right edge of the bottom brick.

97* ••• A uniform sphere of radius R and mass M is held at rest on an inclined plane of angle θ by a horizontal string, as shown in Figure 12-68. Let $R = 20$ cm, $M = 3$ kg, and $\theta = 30°$. (*a*) Find the tension in the string. (*b*) What is the normal force exerted on the sphere by the inclined plane? (*c*) What is the frictional force acting on the sphere?

There are four forces acting on the sphere: its weight, mg; the normal force of the plane, F_n; the frictional force, f, acting parallel to the plane; and the tension in the string, T.

(*a*) 1. Take moments about the center of the sphere — $TR = fR;\ T = f$

2. Set $\Sigma F_x = 0$, where x is along the plane	$T \cos \theta + f = mg \sin \theta$; $T = mg \sin \theta/(1 + \cos \theta)$
3. Evaluate T	$T = 3 \times 9.81 \times 0.5/(1 + 0.866) = 7.89$ N
(*b*) Set $\Sigma F_y = 0$	$F_n = mg \cos \theta + T \sin \theta = 29.4$ N
(*c*) $f = T$	$f = 7.89$ N

98 ••• The legs of a tripod make equal angles of 90° with each other at the apex, where they join together. A 100-kg block hangs from the apex. What are the compressional forces in the three legs?

The three legs of the tripod form three sides of a cube. The length of the diagonal is $\sqrt{3}L$, where L is the lengh of each leg. The downward force of mg is equally distributed over the three legs. Consequently, the compressive force in each leg is $(\sqrt{3}/3)mg = 566$ N.

99 ••• Figure 12-69 shows a 20-cm-long uniform beam resting on a cylinder of 4 cm radius. The mass of the beam is 5.0 kg, and that of the cylinder is 8.0 kg. The coefficient of friction between beam and cylinder is zero. (*a*) Find the forces that act on the beam and on the cylinder. (*b*) What must the minimum coefficients of static friction be between beam and floor and between the cylinder and floor to prevent slipping?

The forces that act on the beam are its weight, mg; the force of the cylinder, F_c, acting along the radius of the cylinder; the normal force of the ground, F_n; and the friction force $f_s = \mu_s F_n$. The forces acting on the cylinder are its weight, Mg; the force of the beam on the cylinder, $F_{cb} = F_c$ in magnitude, acting inward radially; the normal force of the ground on the cylinder, F_{nc}; and the force of friction, $f_{sc} = \mu_{sc} F_{nc}$.

(*a*) Take moments about the end of right beam to find F_c	$mg \times 10 \times \cos 30° = 15F_c$; $F_c = 28.3$ N
Use $\Sigma F_y = 0$ to find F_n	$F_c \cos 30° + F_n = mg$; $F_n = 24.5$ N
Use $\Sigma F_x = 0$ to find f_s	$f_s = F_c \sin 30° = 14.15$ N, to the left
$F_{cb} = F_c$, acting radially and downward	$F_{cb} = 28.3$ N in magnitude
Use $\Sigma F_y = 0$ to find F_{nc}	$F_{nc} = Mg + F_{cb} \cos 30° = 103$ N
Use $\Sigma F_x = 0$ to find f_{sc}	$f_{sc} = F_{cb} \sin 30° = 14.15$ N, to the right
(*b*) μ_s(beam–floor) $= f_s/F_n$	μ_s(beam–floor) $= 0.577 = \tan 30°$
μ_s(cylinder–floor) $= f_{sc}/F_{nc}$	μ_s(cylinder–floor) $= 0.137$

100 ••• Two solid smooth spheres of radius r are placed inside a cylinder of radius R as in Figure 12-70. The mass of each sphere is m. Find the force exerted by the bottom of the cylinder, the force exerted by the wall of the cylinder, and the force exerted by one sphere on the other.

The geometry of the system is shown in the drawing We denote the angle between the vertical center line and the line joining the two centers by θ. The distance x in the drawing is then $x = R - r$, and $\sin \theta = (R - r)/r$. Also, $\tan \theta = (R - r)/[R(2r - R)]^{½}$. The force exerted by the bottom of the cylinder is just $2mg$. Let F be the force that the top sphere exerts on the lower sphere. Since the cylinder wall is smooth, $F \cos \theta = mg$, and $F = mg/\cos \theta$. Its x component is $F_x = F \sin \theta = mg \tan \theta$. This is the force that the wall of the cylinder exerts. Thus $F_w = mg(R - r)/[R(2r - R)]^{½}$. Note that as r approaches $R/2$, $F_w \rightarrow \infty$.

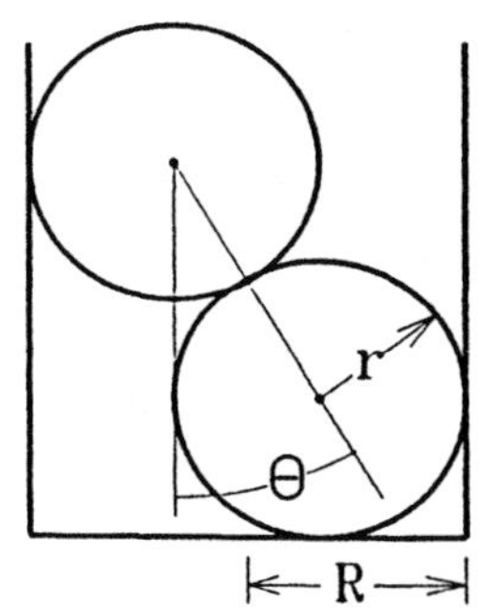

101*••• A solid cube of side length a balanced atop a cylinder of diameter d is in unstable equilibrium if $d << a$ and is in stable equilibrium if $d >> a$ (Figure 12-71). Determine the minimum value of d/a for which the cube is in stable equilibrium.

Consider a small rotational displacement, $\delta\theta$, of the cube of the figure below from equilibrium. This shifts the point of contact between cube and cylinder by $R\delta\theta$, where $R = d/2$. As a result of that motion, the cube itself is rotated through the same angle $\delta\theta$, and so its center is shifted in the same direction by the amount $(a/2)\delta\theta$, neglecting higher order terms in $\delta\theta$. If the displacement of the cube's center of mass is less than that of the point of contact, the torque about the point of contact is a restoring torque, and the cube will return to its equilibrium position. If, on the other hand, $(a/2)\delta\theta > (d/2)\delta\theta$, then the torque about the point of contact due to mg is in the direction of $\delta\theta$, and will cause the displacement from equilibrium to increase. We see that the minimum value of d/a for stable equilibrium is $d/a = 1$.

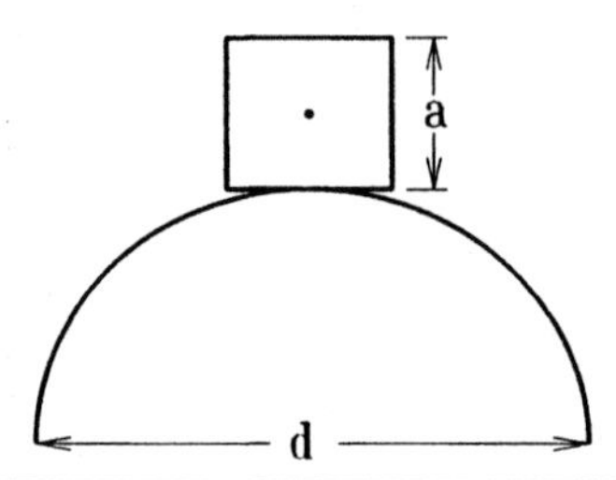

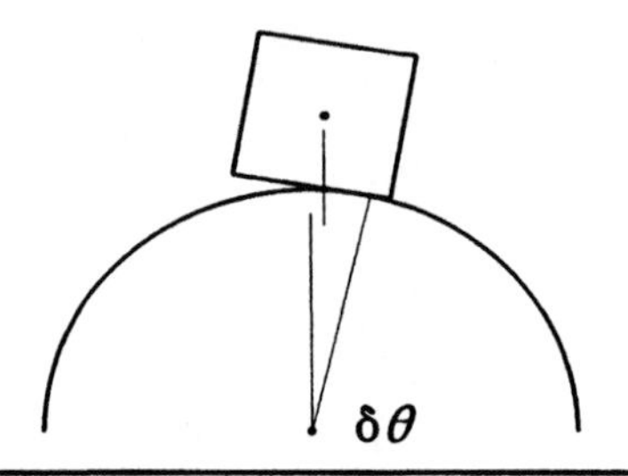

CHAPTER 13

Fluids

1* • A copper cylinder is 6 cm long and has a radius of 2 cm. Find its mass.

Find the volume, then $m = \rho V$ $V = \pi r^2 h = 75.4 \times 10^{-6}$ m^3; $m = 8.93 \times 10^3 V = 0.673$ kg

2 • Find the mass of a lead sphere of radius 2 cm.

Find the volume, then $m = \rho V$ $V = 4\pi r^3/3 = 3.35 \times 10^{-5}$ m^3; $m = 11.3 \times 10^3 V = 0.379$ kg

3 • Find the mass of air in a room 4 m × 5 m × 4 m.

$m = \rho V$ $m = (1.293 \times 80)$ kg $= 103$ kg

4 • A solid oak door is 200 cm high, 75 cm wide, and 4 cm thick. How much does it weigh?

$w = mg = \rho Vg$; oak is dense wood, take $\rho = 0.9 \times 10^3$ $w = (0.9 \times 10^3 \times 2 \times 0.75 \times 0.04 \times 9.81)$ N $= 530$ N

5* •• A 60-mL flask is filled with mercury at 0°C (Figure 13-22). When the temperature rises to 80°C, 1.47 g of mercury spills out of the flask. Assuming that the volume of the flask is constant, find the density of mercury at 80°C if its density at 0°C is 13,645 kg/m^3.

1. Write ρ' in terms of ρ_0, V, and Δm $m = \rho_0 V$; $m' = m - \Delta m = \rho' V$; $\rho' = \rho_0 - \Delta m/V$
2. Evaluate ρ' $\rho' = 13{,}621$ kg/m^3

6 • If the gauge pressure is doubled, the absolute pressure will be (*a*) halved. (*b*) doubled. (*c*) unchanged. (*d*) squared. (*e*) Not enough information is given to determine the effect.

(*e*)

7 • Barometer readings are commonly given in inches of mercury. Find the pressure in inches of mercury equal to 101 kPa.

See Equ. 13-9 101 kPa = 29.8 inHg

8 • The pressure on the surface of a lake is atmospheric pressure $P_{at} = 101$ kPa. (*a*) At what depth is the pressure twice atmospheric pressure? (*b*) If the pressure at the top of a deep pool of mercury is P_{at}, at what depth is the pressure $2P_{at}$?

(*a*) $P = P_{at} + \rho gh = 2P_{at}$; $h = P_{at}/\rho g$ $h = [1.01 \times 10^5/(1 \times 10^3 \times 9.81)]$ m = 10.3 m

(*b*) $h_{Hg} = P_{at}/\rho_{Hg}g$ $h_{Hg} = 10.3/13.6$ m = 75.7 cm

---

9* • Find (*a*) the absolute pressure and (*b*) the gauge pressure at the bottom of a swimming pool of depth 5.0 m.

(*a*) $P = P_{at} + \rho gh$ $P = (1.01 \times 10^5 + 9.81 \times 10^3 \times 5) = 1.5 \times 10^5$ N/m² = 1.5 atm

(*b*) $P_{gauge} = P - P_{at}$ $P_{gauge} = 0.5$ atm

---

10 • When a woman in high heels takes a step, she momentarily places her entire weight on one heel of her shoe, which has a radius of 0.4 cm. If her mass is 56 kg, what is the pressure exerted on the floor by her heel?

$P = mg/A = mg/\pi r^2$ $P = [56 \times 9.81/(\pi \times 16 \times 10^{-6})]$ N/m² $= 1.09 \times 10^7$ N/m² ≈ 100 atm

---

11 • A hydraulic lift is used to raise an automobile of mass 1500 kg. The radius of the shaft of the lift is 8 cm and that of the piston is 1 cm. How much force must be applied to the piston to raise the automobile?

Use Equ. 13-3 $F = [1500 \times 9.81 \times (1/8)^2]$ N = 230 N

---

12 • Blood flows into the aorta through a circular opening of radius 0.9 cm. If the blood pressure is 120 torr, how much force must be exerted by the heart?

$F = PA$; 120 torr = 120 × 133.3 N/m² $F = 120 \times 133.3 \times \pi \times (9 \times 10^{-3})^2$ N = 4.07 N

13* • What pressure is required to reduce the volume of 1 kg of water from 1.00 L to 0.99 L?

Use Equ. 13-6 $\Delta P = 2.0 \times 10^9 \times 10^{-2} = 2 \times 10^7$ Pa ≈ 200 atm

---

14 • A 1500-kg car rests on four tires, each of which is inflated to a gauge pressure of 200 kPa. What is the area of contact of each tire with the road, if the four tires support the weight equally?

$F = P_{gauge}A$; $A = F/P_{gauge}$ $A = 1500 \times 9.81/(4 \times 2 \times 10^5) = 1.84 \times 10^{-2}$ m² = 184 cm²

---

15 •• In the seventeenth century, Pascal performed the experiment shown in Figure 13-23. A wine barrel filled with water was coupled to a long tube. Water was added to the tube until the barrel burst. (*a*) If the radius of the lid was 20 cm and the height of the water in the tube was 12 m, calculate the force exerted on the lid. (*b*) If the tube had an inner radius of 3 mm, what mass of water in the tube caused the pressure that burst the barrel?

(*a*) $F = PA$; $P = \rho gh$; $F = \rho_w gh\pi R^2$ $F = (1 \times 10^3 \times 9.81 \times 12 \times \pi \times 0.2^2)$ N = 14,800 N

(*b*) $m = \rho h\pi r^2$ $m = 1 \times 10^3 \times 12 \times \pi \times 9 \times 10^{-6}$ kg = 0.34 kg = 340 g

---

16 •• Blood plasma flows from a bag through a tube into a patient's vein, where the blood pressure is 12 mmHg. The specific gravity of blood plasma at 37°C is 1.03. What is the minimum elevation the bag must have so that the pressure of the plasma as it flows into the vein is at least 12 mmHg?

$P = \rho gh$; $h = P/g\rho$; $P = 12 \times 133.3$ Pa $h = [12 \times 133.3/(9.81 \times 1.03 \times 10^3)]$ m = 0.158 m = 15.8 cm

17* •• Many people have imagined that if they were to float the top of a flexible snorkel tube out of the water, they would be able to breathe through it while walking underwater (Figure 13-24). However, they generally do not reckon with just how much water pressure opposes the expansion of the chest and the inflation of the lungs. Suppose you can just breathe while lying on the floor with a 400-N weight on your chest. How far below the surface of the water could your chest be for you still to be able to breathe, assuming your chest has a frontal area of 0.09 m^2?

$P = \rho_w gh = F/A$; $h = F/\rho_w gA$ $h = (400/1 \times 10^3 \times 9.81 \times 0.09)$ m $= 0.453$ m $= 45.3$ cm

18 •• When the ground becomes saturated with water during floods, pressure develops similar to the pressure of water contained in a vessel of the same volume. This pressure forces water through the joints in concrete-block cellar walls. If this happens quickly enough to fill up the cellar with water, there may be no further damage. Otherwise, the upward pressure on the cellar floor may float the house like a ship. What upward force would be exerted on a 10-m × 10-m basement floor if the floor were 2 m below the surface of the water?

$F = PA = \rho ghA$ $F = (1 \times 10^3 \times 9.81 \times 2 \times 100)$ N $= 1.96$ MN

19 •• In Example 13-3, a force of 147 N is applied to a small piston to lift a car that weighs 14,700 N. Demonstrate that this does not violate the law of conservation of mechanical energy by showing that, when the car is lifted some distance h, the work done by the force acting on the small piston equals the work done by the large piston on the car.

$W = Fh$; $F_1 = F_2(A_1/A_2)$; $h_1A_1 = h_2A_2$; $h_1 = h_2(A_2/A_1)$ $W_1 = F_1h_1 = F_2(A_1/A_2)h_2(A_2/A_1) = F_2h_2 = W_2$

20 •• A hollow cube with edge a is half-filled with water of density ρ. Find the force exerted on a side of the cube by the water.

$F = \int_0^{a/2} \rho gha\, dh = ½\rho ga(a^2/4) = \rho ga^3/8$. (see Example 13-2)

21* ••• The volume of a cone of height h and base radius r is $V = \pi r^2h/3$. A conical vessel of height 25 cm resting on its base of radius 15 cm is filled with water. (*a*) Find the volume and weight of the water in the vessel. (*b*) Find the force exerted by the water on the base of the vessel. Explain how this force can be greater than the weight of the water.

(*a*) $w = \rho gV$; V is given $V = 5.89 \times 10^{-3}$ m^3; $w = 5.89 \times 9.81 = 57.8$ N

(*b*) $F = PA = \rho ghA$ $F = \rho gh\pi r^2 = 3\rho gV = 3 \times 57.8$ N $= 173$ N

This occurs in the same way that the force on Pascal's barrel >> the weight of water in the tube. The downward force on the base is also the result of the downward component of the force exerted by the slanting walls of the cone on the water.

22 • Does Archimedes' principle hold in a satellite orbiting the earth in a circular orbit? Explain.

No. In an environment where $g_{eff} = 0$, there is no buoyant force; there is no "up" or "down."

23 •• A rock of mass M with a density twice that of water is sitting on the bottom of an aquarium tank filled with water. The normal force exerted on the rock by the bottom of the tank is (*a*) $2Mg$. (*b*) Mg. (*c*) $Mg/2$. (*d*) zero. (*e*) impossible to determine from the given information.

(*c*)

---

24 •• A rock is thrown into a swimming pool filled with water of uniform temperature. Which of the following statements is true? (*a*) The buoyant force on the rock is zero as it sinks. (*b*) The buoyant force on the rock increases as it sinks. (*c*) The buoyant force on the rock decreases as it sinks. (*d*) The buoyant force on the rock is constant as it sinks. (*e*) The buoyant force on the rock as it sinks is nonzero at first but becomes zero once the terminal velocity is reached.

(*b*) The density of the water increases with depth; therefore, $F_B = \rho g V$ increases with depth.

25* •• A fishbowl rests on a scale. The fish suddenly swims upward to get food. What happens to the scale reading?

Nothing. The fish is in neutral buoyancy, so the upward acceleration of the fish is balanced by the downward acceleration of the displaced water.

---

26 •• Two objects are balanced as in Figure 13-25. The objects have identical volumes but different masses. Will the equilibrium be disturbed if the entire system is completely immersed in water? Explain.

Yes. When submerged, the downward force on each side is reduced by the same amount, not in proportion to the masses. That is, if $m_1L_1 = m_2L_2$ and $L_1 \neq L_2$, then $(m_1 - c)L_1 \neq (m_2 - c)L_2$.

---

27 •• A 200-g block of lead and a 200-g block of copper rest at the bottom of an aquarium filled with water. Which of the following is true? (*a*) The buoyant force is greater on the lead than on the copper. (*b*) The buoyant force is greater on the copper than on the lead. (*c*) The buoyant force is the same on both blocks. (*d*) More information is needed to choose from the above.

(*b*) The volume of copper > volume of lead.

---

28 •• A 20-cm^3 block of lead and a 20-cm^3 block of copper rest at the bottom of an aquarium filled with water. Which of the following is true? (*a*) The buoyant force is greater on the lead than on the copper. (*b*) The buoyant force is greater on the copper than on the lead. (*c*) The buoyant force is the same on both blocks. (*d*) More information is needed to choose from the above.

(*c*)

29* • A 500-g piece of copper (specific gravity 9.0) is suspended from a spring scale and is submerged in water (Figure 13-26). What force does the spring scale read?

$w = \rho_{Cu}Vg$; $w' = \rho_{Cu}Vg - \rho_W Vg = (\rho_{Cu} - \rho_W)w/\rho_{Cu}$ $w' = 0.5 \times 9.81(7.93/8.93)$ N $= 4.36$ N

---

30 • When a 60-N stone is attached to a spring scale and is submerged in water, the spring scale reads 40 N. What is the specific gravity of the stone?

sp. gravity $= \rho_{stone}/\rho_W = w/(w - w')$ (see Problem 29) sp. gravity $= 60/20 = 3.0$

---

31 • A block of an unknown material weighs 5 N in air and 4.55 N when submerged in water. (*a*) What is the density of the material? (*b*) Of what material is the block made?

$\rho = \rho_W w/(w - w')$ (see Problem 30) $\rho = 1 \times 10^3(5/0.45)$ kg/m^3 $= 11.1 \times 10^3$ kg/m^3 $\approx \rho_{Pb}$

---

32 • A solid piece of metal weighs 90 N in air and 56.6 N when submerged in water. Determine the specific gravity of this metal.

sp. gravity $= w(w - w')$ (see Problem 30) sp. gravity $= 90/33.4 = 2.69$

33* •• An object floats on water with 80% of its volume below the surface. The same object when placed in another liquid floats on that liquid with 72% of its volume below the surface. Determine the density of the object and the specific gravity of the liquid.

$\rho = \rho_W(V'/V) = \rho_L(V''/V)$ (see Example 13-7) $\rho = 0.8\rho_W = 800\ \text{kg/m}^3$; $\rho_L/\rho_W = 0.8/0.72 = 1.11 = \text{sp. gravity}$

34 •• A 5-kg iron block is attached to a spring scale and is submerged in a fluid of unknown density. The spring scale reads 6.16 N. What is the density of the fluid?

$(\rho_{obj} - \rho_{liqu})/\rho_{obj} = w'/w$ (see Problem 29) $\rho_{liqu} = \rho_{Fe}(1 - w'/w)$; $\rho_{liqu} = 6.96 \times 10^3\ \text{kg/m}^3$

35 •• A large piece of cork weighs 0.285 N in air. When held submerged underwater by a spring scale as shown in Figure 13-27, the spring scale reads 0.855 N. Find the density of the cork.

1. Write an expression for F of spring scale $F = F_B - w = \rho_W gV - \rho_C gV$
2. Solve for ρ_C using $gV = F_B/\rho_W$ $\rho_C = \rho_W(1 - F/F_B) = \rho_W w/(F + w) = 250\ \text{kg/m}^3$

36 •• As you step onto the *Icarus* spacecraft, you are supposed to give a very accurate report of your weight. As you approach the front of the line, you realize that you've forgotten to subtract the buoyant force exerted on you by the earth's atmosphere. Estimate the correction that you'll have to make to the spring-scale reading of your weight.

$w = \rho gV$; since a body floats in water, take $\rho = 0.95\rho_W$. $\Delta w = F_B = \rho_{air}gV = (\rho_{air}/\rho)w = (1.293/950)w$. So the percent correction is $100\Delta w/w = 0.136\%$.

37* •• A helium balloon lifts a basket and cargo of total weight 2000 N under standard conditions, in which the density of air is 1.29 kg/m³ and the density of helium is 0.178 kg/m³. What is the minimum volume of the balloon?

1. Set F_B = total weight $= \rho_{He}gV + 2000\ \text{N}$ $1.29gV = 0.178gV + 2000\ \text{N}$
2. Solve for and evaluate V $V = [2000/9.81(1.29 - 0.178)]\ \text{m}^3 = 183\ \text{m}^3$

38 •• Zoe is packing her belongings and moving in with Margaret. Her books are in boxes, which she plans to float down the river to Margaret's shack on a square raft that is 3 m on each side and 11 cm thick. It is made of wood having a specific gravity of 0.6. If each box has a mass of 20 kg, how many boxes can be placed on the raft if the books are to remain dry? Assume that the water remains calm.

Let A be the area of the raft and d its thickness. Let n be the largest integer number of 20 kg boxes.

$F_B = \rho_W Adg = 0.6\rho_W Adg + 20ng$ $n = 0.4\rho_W Ad/20 = 0.4 \times 10^3 \times 9 \times 0.11/20 = 19.8$; $n = 19$

39 •• An object has neutral buoyancy when its density equals that of the liquid in which it is submerged, which means that it neither floats nor sinks. If the average density of an 85-kg diver is 0.96 kg/L, what mass of lead should be added to give him neutral buoyancy?

Let V = volume of diver, ρ_D the density of the diver, V_{Pb} volume of added lead, and $m_{Pb} = \rho_{Pb}V_{Pb}$ the mass of lead.

1. Write the condition for neutral buoyancy $(V + V_{Pb})\rho_W g = (m_D + m_{Pb})g$
2. Rewrite in terms of masses and densities $(m_D/\rho_D + m_{Pb}/\rho_{Pb})\rho_W = (m_D + m_{Pb})$
3. Solve for and evaluate m_{Pb} $m_{Pb} = m_D[(\rho_W - \rho_D)\rho_{Pb}/\rho_D(\rho_{Pb} - \rho_W)] = 3.89\ \text{kg}$

---

40 •• A beaker of mass 1 kg containing 2 kg of water rests on a scale. A 2-kg block of aluminum (specific gravity 2.70) suspended from a spring scale is submerged in the water as in Figure 13-28. Find the readings of both scales.

As before, we let w' be the reading of the spring scale.

1. $w' = w(\rho_{Al} - \rho_W)/\rho_{Al}$ (see Problem 34) — $w' = (2 \times 9.81 \times 1.7/2.7)$ N = 12.35 N
2. The bottom scale reads $M_{tot}g - w'$ — Bottom scale reading = $(5 \times 9.81 - 12.35)$ N = 36.7 N

41* ••• A ship sails from seawater (specific gravity 1.025) into fresh water and therefore sinks slightly. When its load of 600,000 kg is removed, it returns to its original level. Assuming that the sides of the ship are vertical at the water line, find the mass of the ship before it was unloaded.

Let V = displacement of ship in the two cases, m be the mass of ship without load, Δm be the load.

1. Write condition for floating in both cases — $(m + \Delta m) = \rho_{SW}V$; $m = \rho_W V$
2. Solve for and evaluate $m + \Delta m$ — $m + \Delta m = \Delta m\rho_{SW}/(\rho_{SW} - \rho_W) = 2.06 \times 10^7$ kg

---

42 ••• The hydrometer shown in Figure 13-29 is a device for measuring the density of liquids. The bulb contains lead shot, and the density can be read directly from the liquid level on the stem after the hydrometer has been calibrated. The volume of the bulb is 20 mL, the stem is 15 cm long and has a diameter of 5.00 mm, and the mass of the glass is 6.0 g. (*a*) What mass of lead shot must be added so that the least density of liquid that can be measured is 0.9 kg/L? (*b*) What is the maximum density of liquid that can be measured?

(*a*) 1. For ρ_{min}, we have neutral buoyancy of the device — $\rho_{min}V = m_{tot}$; $0.9(20 + 15\pi \times 0.25/4)$ g = 6 g + m_{Pb}

2. Find m_{Pb} — $m_{Pb} = 14.65$ g

(*b*) Now only the bulb is submerged — $\rho_{max} \times 20 = m_{tot} = 20.65$; $\rho_{max} = 1.03$ kg/L

For Problems 43 - 56 neglect the viscosity of the fluids.

---

43 • In a department store, a beach ball is supported by the airstream from a hose connected to the exhaust of a vacuum cleaner. Does the air blow under or over the ball to support it? Why?

It blows over the ball, reducing the pressure above the ball to below atomspheric pressure.

---

44 • A horizontal pipe narrows from a diameter of 10 cm to 5 cm. For a fluid flowing from the larger diameter to the smaller, (*a*) the velocity and pressure both increase. (*b*) the velocity increases and the pressure decreases. (*c*) the velocity decreases and the pressure increases. (*d*) the velocity and pressure both decrease. (*e*) either the velocity or pressure changes but not both.

(*b*)

45* •• When water emerges from a faucet, the stream narrows as the water falls. Explain why.

Pressure within the stream diminishes as the velocity of the stream increases.

---

46 • Water flows at 0.65 m/s through a hose with a diameter of 3 cm. The diameter of the nozzle is 0.30 cm. (*a*) At what speed does the water pass through the nozzle? (*b*) If the pump at one end of the hose and the nozzle at the other end are at the same height, and if the pressure at the nozzle is atmospheric pressure, what is the pressure at the pump?

(*a*) Av = constant; $A \propto d^2$; $v_N = v_H(d_H/d_N)^2$ $v_N = 100 \times 0.65$ m/s = 65 m/s

(*b*) Find P_P using Equ. 13-18 $P_P = P_{atm} + \frac{1}{2}\rho_W(65^2 - 0.65^2) = 22.1$ kPa = 21.8 atm

---

47 • Water is flowing at 3 m/s in a horizontal pipe under a pressure of 200 kPa. The pipe narrows to half its original diameter. (*a*) What is the speed of flow in the narrow section? (*b*) What is the pressure in the narrow section? (*c*) How do the volume flow rates in the two sections compare?

(*a*) $A_1v_1 = A_2v_2$; $v_2 = 4v_1$ $v_2 = 12$ m/s

(*b*) Use Equ. 13-18; $P_2 = P_1 + \frac{1}{2}\rho_W(v_1^2 - v_2^2)$ $P_2 = [2 \times 10^5 + 500(9 - 144)] = 132.5$ kPa

(*c*) $I_{V1} = I_{V2}$ The volume flows are equal

---

48 • The pressure in a section of horizontal pipe with a diameter of 2 cm is 142 kPa. Water flows through the pipe at 2.80 L/s. If the pressure at a certain point is to be reduced to 101 kPa by constricting a section of the pipe, what should the diameter of the constricted section be?

1. Find v_1 $v_1 = I_V/A = (2.8 \times 10^{-3} \times 4/\pi \times 4 \times 10^{-4}) = 8.91$ m/s

2. Use Equ. 13-18 to find v_2 $v_2 = [v_1^2 + 2(P_1 - P_2)/\rho_W]^{1/2} = 12.7$ m/s

3. $A_2 = A_1(v_1/v_2)$; $d_2 = d_1(v_1/v_2)^{1/2}$ $d_2 = 2(8.91/12.7)^{1/2}$ cm = 1.68 cm

---

49* •• Blood flows in an aorta of radius 9 mm at 30 cm/s. (*a*) Calculate the volume flow rate in liters per minute. (*b*) Although the cross-sectional area of a capillary is much smaller than that of the aorta, there are many capillaries, so their total cross-sectional area is much larger. If all the blood from the aorta flows into the capillaries and the speed of flow through the capillaries is 1.0 mm/s, calculate the total cross-sectional area of the capillaries.

(*a*) $I_V = Av$ $I_V = (\pi \times 81 \times 10^{-6} \times 0.3)$ m³/s $= 7.63 \times 10^{-5}$ m³/s = 4.58 L/min

(*b*) $A_{cap}v_{cap} = I_V$; $A_{cap} = I_V/v_{cap}$ $A_{cap} = 7.63 \times 10^{-5}/10^{-3}$ m² $= 7.63 \times 10^{-2}$ m² = 763 cm²

---

50 •• Dorothy is up on her 15-m × 15-m roof enjoying the view of Kansas. Suddenly, a strong wind blows down her ladder, leaving her stranded. She knows that a high wind reduces the air pressure on the roof, and that there is a danger that the atmospheric pressure inside the house will blow the roof off. As the wind reaches a speed of 30 m/s, she calls to her Auntie Em for help. Calculate the force on the roof.

1. Determine the pressure difference $\Delta P = \frac{1}{2}\rho_{air}v^2 = 582$ Pa

2. $F = A\Delta P$ $F = (582 \times 225)$ N = 131 kN

---

51 •• A large tank of water is tapped a distance h below the water surface by a small pipe as in Figure 13-30. Find the distance x reached by the water flowing out the pipe.

1. Find v as water leaves hole (see Example 13-8) $v = \sqrt{2gh}$

2. Find time t required to fall $y = H - h$ $t = \sqrt{2(H-h)/g}$

3. $x = vt$ $x = 2\sqrt{h(H-h)}$

---

52 •• The \$8-billion, 800-mile long Alaskan Pipeline has a capacity of 240,000 m³ of oil per day. It has a standard radius of 60 cm. Find the pressure P' at a point where the pipe has half the standard radius. Take the standard pressure to be $P = 180$ kPa and the density of oil to be 800 kg/m³.

1. Determine $v_S = I_V/A_S$ — $v_S = (2.4 \times 10^5/8.64 \times 10^4)/\pi \times 0.6^2$ m/s $= 2.456$ m/s

2. $P' = P_S + \frac{1}{2}\rho[v_S{}^2 - (v')^2]$; $v' = 4v_S$ — $P' = [1.8 \times 10^5 - 400 \times 2.456^2 \times 15]$ Pa $= 144$ kPa

---

53* •• Water flows through a Venturi meter like that in Example 13-9 with a pipe diameter of 9.5 cm and a constriction diameter of 5.6 cm. The U-tube manometer is partially filled with mercury. Find the flow rate of the water in the pipe of 9.5-cm diameter if the difference in the mercury level in the U-tube is 2.40 cm.

$v_1 = \sqrt{\dfrac{2\rho_L gh}{\rho_F(r-1)}}$; $r = \dfrac{R_1^2}{R_2^2}$ (see Example 13-9) — $v_1 = \sqrt{\dfrac{2 \times 13.6 \times 10^3 \times 9.81 \times 2.4 \times 10^{-2}}{10^3(2.88 - 1)}}$ m/s

$v_1 = 1.85$ m/s

$I_v = \pi r^2 v_1$ — $I_v = (\pi \times 0.095^2 \times 1.85/4)$ m^3/s $= 13.1$ L/s

---

54 •• A firefighter holds a hose with a bend in it as in Figure 13-31. Water is expelled from the hose in a stream of radius 1.5 cm at a speed of 30 m/s. (*a*) What mass of water emerges from the hose in 1 s? (*b*) What is the horizontal momentum of this water? (*c*) Before reaching the bend, the water has momentum upward, whereas afterward, its momentum is horizontal. Draw a vector diagram of the initial and final momentum vectors, and find the change in the momentum of the water at the bend in 1 s. From this, find the force exerted on the water by the hose.

(*a*) $m = \rho A v$ — $m = 10^3 \times \pi \times 2.25 \times 10^{-4} \times 30$ kg/s $= 21.2$ kg/s

(*b*) $p = mv$ — $p = 636$ kg·m/s

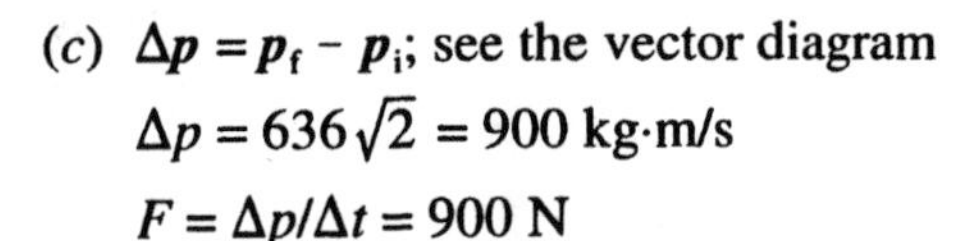

(*c*) $\Delta\mathbf{p} = \mathbf{p}_f - \mathbf{p}_i$; see the vector diagram

$\Delta p = 636\sqrt{2} = 900$ kg·m/s

$F = \Delta p/\Delta t = 900$ N

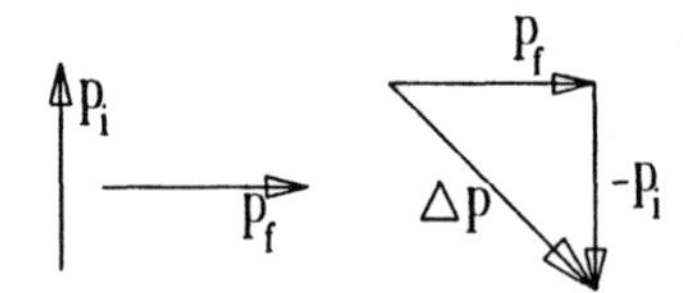

---

55 •• A fountain designed to stray a column of water 12 m into the air has a 1-cm-diameter nozzle at ground level. The water pump is 3 m below the ground. The pipe to the nozzle has a diameter of 2 cm. Find the necessary pump pressure.

1. Use Equ. 13-17*a*; $v_N = 4v_P$ — $P_P - 3\rho g + \frac{1}{2}\rho v_N{}^2/16 = P_{atm} + \frac{1}{2}\rho v_N{}^2$

2. $v_N{}^2 = 2gh$; solve for P_P with $h = 12$ m — $P_P = 241$ kPa

---

56 ••• In Figure 13-30, (*a*) find the distance x at which the water strikes the ground as a function of h and H. (*b*) Show that there are two values of h that are equidistant from the point $h = \frac{1}{2}H$ that give the same distance x. (*c*) Show that x is a maximum when $h = \frac{1}{2}H$. What is the value of this maximum distance x?

(*a*) See Problem 51.

(*b*) From Problem 51, $x^2 = 4hH - 4h^2$. Solving the quadratic equation for h, $h = \frac{1}{2}H \pm \frac{1}{2}\sqrt{H^2 - x^2}$; Q.E.D.

(*c*) Set $dx/dh = 0$. $dx/dh = (H - 2h)/\sqrt{h(H-h)} = 0$ if $h = H/2$. For $h = H/2$, $x = H$.

57* • A horizontal tube with an inside diameter of 1.2 mm and a length of 25 cm has water flowing through it at 0.30 mL/s. Find the pressure difference required to drive this flow if the viscosity of water is 1.00 mPa·s.

Use Equ. 13-23 — $\Delta P = (8 \times 10^{-3} \times 0.25 \times 0.3 \times 10^{-6}/\pi \times 0.6^4 \times 10^{-12})$ Pa

$= 1.47$ kPa

58 • Find the diameter of a tube that would give double the flow rate for the pressure difference in Problem 57.
Must double r^4 or d^4 to double I_V while keeping ΔP constant. Thus, $d' = 2^{1/4} \times 1.2$ mm = 1.43 mm.

59 • Blood takes about 1.0 s to pass through a 1-mm-long capillary of the human circulatory system. If the diameter of the capillary is 7 μm and the pressure drop is 2.60 kPa, find the viscosity of blood.

From Equ. 13-23, $\eta = \pi r^4 \Delta P/8LI_V = r^2\Delta P/8Lv$ $\quad \eta = 3.5^2 \times 10^{-12} \times 2.6 \times 10^3/8 \times 10^{-3} \times 10^{-3}$ Pa·s
$= 3.98$ mPa·s

60 • True or false: The buoyant force on a submerged object depends on the shape of the object.
False

61* • A glass of water has ice cubes floating in it. What happens to the water level when the ice melts?
The water level remains constant.

62 • Why is it easier to float in salt water than in fresh water?
The density of salt water is greater than that of fresh water.

63 • Smoke usually rises from a smokestack, but it may sink on a very humid day. What can be concluded about the relative densities of humid air and dry air?
The density of humid air is less than that of dry air.

64 •• A certain object has a density just slightly less than that of water so that it floats almost completely submerged. However, the object is more compressible than water. What happens if the floating object is given a slight push to submerge it?
It will be compressed, its density will increase, and so it will sink to the bottom.

65* •• In Example 13-9, the fluid is accelerated to a greater speed as it enters the narrow part of the pipe. Identify the forces that act on the fluid to produce this acceleration?
The force acting on the fluid is the difference in pressure between the wide and narrow parts times the area of the narrow part.

66 •• A glass of water is accelerating to the right along a horizontal surface. What is the origin of the force that produces the acceleration on a small element of water in the middle of the glass? Explain with a picture.
The drawing shows the beaker and a strip within the water. As is readily established by a simple demonstration, the surface of the water is not level while the beaker is accelerated, showing that there is a pressure gradient. That pressure gradient results in a force on the small element shown in the figure.

67 •• A 0.5-kg mass of lead is submerged in a container filled to the brim with water and a block of wood floats on top. The lead mass is slowly lifted from the container by a thin wire, and as it emerges into air the level of the water in the container drops a bit. The lead mass is now placed on the block of wood, which remains afloat. As the lead is placed on the block of wood, (*a*) some water spills over the edge of the container. (*b*) the water level rises exactly to the brim as before. (*c*) the water level rises but does not reach the brim of the container. (*d*) there is not enough information provided to decide between the three options.
(*a*) Water spills over because the amount displaced is now greater than the volume of the lead block.

68 •• You are sitting in a boat floating on a very small pond. You take the anchor out of the boat and drop it into the water. What happens to the water level in the pond?

The water level in the pond drops slightly.

69* • The top of a card table is 80 cm × 80 cm. What is the force exerted on it by the atmosphere? Why doesn't the table collapse?

The net force is zero. Neglecting the thickness of the table, the atmospheric pressure is the same above and below the surface of the table.

70 • A 4.0-g Ping-Pong ball is attached by a thread to the bottom of a container. When the container is filled with water so that the ball is totally submerged, the tension in the thread is 2.8×10^{-2} N. Determine the diameter of the ball.

$T = F_B - mg$; $F_B = \rho_W g(\pi d^3/6)$; solve for r — $r = [6(T + mg)/\pi \times 10^3 g]^{1/3} = 0.0505$ m $= 5.05$ cm

71 • Seawater has a bulk modulus of 2.3×10^9 N/m^2. Find the density of seawater at a depth where the pressure is 800 atm if the density at the surface is 1025 kg/m^3.

1. $m = \rho V =$ constant; $d\rho/\rho = -\Delta V/V$ — $\Delta\rho = \rho_0\Delta P/B$; $\rho = \rho_0(1 + \Delta P/B)$
2. Evaluate ρ using the data given — $\rho = [1025(1 + 800 \times 1.01 \times 10^5/2.3 \times 10^9)]$ $= 1061$ kg/m^3

72 • Your car misses a turn and sinks into a small lake to a depth of 8 m. Your quick mind tells you that the chances of driving out are slim, so you'd better swim for it. The car door, however, is not budging, even though it seems undamaged. (*a*) If the outside area of the car door is 0.9 m^2, what is the force exerted on the door by the water? (*b*) What is the force exerted on the inside of the door by the air, assuming atmospheric pressure inside? (*c*) What will you have to do to get the door open and save your skin?

(*a*) $F = A\Delta P$; $\Delta P = \rho_W gh$ — $F = [0.9 \times 1000 \times 9.81 \times 8]$ N $= 70.6$ kN

(*b*) $F' = AP_{atm}$ — $F' = (0.9 \times 1.01 \times 10^5)$ N $= 90.9$ kN

(*c*) The net force acting on the door is 70.6 kN — Roll down the window; let in water to equalize the pressure

73* • A solid cubical block of side length 0.6 m is suspended from a spring balance. When the block is in water, the spring balance reads 80% of the reading when the block is in air. Determine the density of the block.

Use Equ. 13-12; $\rho = \rho_W(w_0/\Delta w)$ — $\rho = 1000(1/0.2) = 5000$ kg/m^3

74 • When submerged in water, a block of copper has an apparent weight of 56 N. What fraction of this copper block will be submerged if floated on a pool of mercury?

$w = \rho_{Cu} gV = \rho_{Hg} gV'$, where V' is submerged volume — $V'/V = \rho_{Cu}/\rho_{Hg} = 0.657$; 65.7% is submerged

75 • A 4.5-kg block of material floats on ethanol with 10% of its volume above the liquid surface. What fraction of this block will be submerged if floated on water?

$\rho V = 0.9\rho_{eth} V = f\rho_W V$, where f is the fraction underwater — $f = 0.9(\rho_{eth}/\rho_W) = 0.7245$; 72.45% of the block is submerged

---

76 • What is the buoyant force on your body when floating (*a*) in a freshwater lake (specific gravity = 1.0) and (*b*) in the ocean (specific gravity = 1.03)?

F_B = body weight (150 lb) in both (*a*) and (*b*).

77* • Suppose that when you are floating in fresh water, 96% of your body is submerged. What is the volume of water your body displaces when it is fully submerged?

The mass of your body divided by the density of water; e.g., (60 kg)/(1000 kg/m^3) = 0.06 m^3.

---

78 •• A block of wood of 1.5-kg mass floats on water with 68% of its volume submerged. A lead block is placed on the wood and the wood is then fully submerged. Find the mass of the lead block.

1. Find the volume of wood; $\rho_{wood} = 680$ kg/m^3 $\qquad V = 1.5/680$ m^3 $= 2.206 \times 10^{-3}$ m^3
2. Set $F_B = \rho_W Vg = m_{tot}g = (1.5 \text{ kg} + m_{Pb})g$ $\qquad m_{Pb} = (2.206 - 1.5)$ kg $= 0.706$ kg

---

79 •• A Styrofoam cube, 25 cm on a side, is weighed on a simple beam balance. The balance is in equilibrium when a 20-g mass of brass is placed on the opposite pan of the balance. Find the true mass of the Styrofoam cube.

We shall neglect the buoyancy of the brass.

$20 \times 10^{-3}g = mg - F_B$; $F_B = \rho_{air}gV$; solve for m $\qquad m = 20 \times 10^{-3} + 1.293 \times 0.25^3 = 0.0402$ kg $= 40.2$ g

---

80 •• A spherical shell of copper with an outer diameter of 12 cm floats on water with half its volume above the water surface. Determine the inner diameter of the shell.

$\frac{1}{2}(\pi d_{out}^3/6)\rho_W g = mg$; $m = \pi/6(d_{out}^3 - d_{in}^3)\rho_{Cu}$ $\qquad 8.93d_{out}^3 - 0.5d_{out}^3 = 8.93d_{in}^3$; $d_{in} = d_{out}(8.43/8.93)^{1/3}$ $= 11.8$ cm

---

81* •• A beaker filled with water is balanced on the left cup of a scale. A cube 4 cm on a side is attached to a string and lowered into the water so that it is completely submerged. The cube is not touching the bottom of the beaker. A weight m is added to the system to retain equilibrium. What is the weight m and on which cup of the balance is it added?

The additional weight on the beaker side equals the weight of the displaced water, i.e., 64 g. This is the mass that must be placed on the other cup to maintain balance.

---

82 •• Crude oil has a viscosity of about 0.8 Pa·s at normal temperature. A 50-km pipeline is to be constructed from an oil field to a tanker terminal. The pipeline is to deliver oil at the terminal at a rate of 500 L/s and the flow through the pipeline is to be laminar to minimize the pressure needed to push the fluid through the pipeline. Estimate the diameter of the pipeline that should be used.

1. Since the density of crude oil is not given, we assume $\rho_{oil} = 700$ kg/m^3.
2. Use Equ. 13-24; take $N_R = 1000$. Note that $v = I_V/\pi r^2$, so $N_R = 2\rho I_V/\pi r\eta$. Solve for r; $r = 28$ cm.
3. Now determine $\Delta P = 8\eta L I_V/\pi r^4 = 8 \times 0.8 \times 5 \times 10^4 \times 0.5/\pi \times 0.28^4 = 3.3 \times 10^7$ Pa ≈ 328 atm. This is much too large a pressure to maintain in the pipe. So, a more reasonable radius might be 50 cm, giving $\Delta P \approx 30$ atm.

---

83 •• Water flows through the pipe in Figure 13-32 and exits to the atmosphere at C. The diameter of the pipe is 2.0 cm at A, 1.0 cm at B, and 0.8 cm at C. The gauge pressure in the pipe at A is 1.22 atm and the flow rate is 0.8 L/s. The vertical pipes are open to the air. Find the level of the liquid–air interfaces in the two vertical pipes.

We measure the height of the liquid–air interfaces relative to the centerline of the pipe.

1. Determine the flow velocity v_A — $v_A = I_V/A_A = 8.0 \times 10^{-4}/\pi \times 10^{-4}$ m/s $= 8/\pi$ m/s
2. Apply Equ. 13-7 — $P_A - P_C = 1.22 \times 1.01 \times 10^5 = 10^3 \times 9.81h_A$; $h_A = 12.6$ m
3. Use Equ. 13-18 to find P_B; $v_B = 4v_A$ — $P_B = P_A - ½(15 \times 10^3 \times 64/\pi^2) = 1.53 \times 10^5$ Pa $= 1.51$ atm
4. Apply Equ. 13-7 to find h_B — $0.51 \times 1.01 \times 10^5 = 10^3 \times 9.81h_B$; $h_B = 5.3$ m

---

84 •• Repeat Problem 83 with the flow rate reduced to 0.6 L/s and the size of the opening at C reduced so that the pressure in the pipe at A remains unchanged .

1. Determine v_A and v_B (see Problem 83) — $v_A = 6/\pi$ m/s; $v_B = 4v_A$
2. Use Equ. 13-18 to find P_B — $P_B = (2.24 \times 10^5 - 500 \times 15 \times 36/\pi^2)$ Pa $= 1.97 \times 10^5$ Pa
3. Use Equ. 13-7 to find h_B — $0.97 \times 10^5 = 10^3 \times 9.81h_B$; $h_B = 9.89$ m
4. $P_A - P_C = 1.22$ atm, as in Problem 83 — $h_A = 12.6$ m

85* •• Figure 13-33 is a sketch of an *aspirator,* a simple device that can be used to achieve a partial vacuum in a reservoir connected to the vertical tube at B. An aspirator attached to the end of a garden hose may be used to deliver soap or fertilizer from the reservoir. Suppose that the diameter at A is 2.0 cm and at C, where the water exits to the atmosphere, it is 1.0 cm. If the flow rate is 0.5 L/s and the gauge pressure at A is 0.187 atm, what diameter of the constriction at B will achieve a pressure of 0.1 atm in the container?

Since it is not given, we shall neglect the difference in height between the centers of the pipes at A and B.

1. Find v_A — $v_A = I_V/A_A = 5 \times 10^{-4}/\pi \times 10^{-4}$ m/s $= 5/\pi$ m/s
2. Use Equ. 13-18 to find v_B^2 — $v_B^2 = (1.087 \times 1.01 \times 10^5 + 500 \times 25/\pi^2)/500$ m²/s² $= 222$ m²/s²
3. $v_B = 5 \times 10^{-4}/\pi r_B^2 = 14.9$ m/s; solve for r_B — $r_B = (5 \times 10^{-4}/14.9\pi)^{½}$ m $= 3.27$ mm; $d = 6.54$ mm

---

86 •• A cylindrical buoy at the entrance of a harbor has a diameter of 0.9 m and a height of 2.6 m. The mass of the buoy is 600 kg. It is attached to the bottom of the sea with a nylon cable of negligible mass. The specific gravity of the seawater is 1.025. (*a*) How much of the buoy is visible when the cable is slack? (*b*) If a tidal wave completely submerges the buoy, what is the tension in the taut cable? (*c*) If the cable breaks, what is the initial upward acceleration of the buoy?

(*a*) Find ρ_{av} of the buoy — $\rho_{av} = m/V = (600/2.6 \times \pi \times 0.45^2)$ kg/m³ $= 363$ kg/m³

Let V' be the submerged volume; see Example 13-7 — $V'/V = \rho_{av}/\rho_{sw} = 363/1025 = 0.354$

Visible height is $2.6(1 - 0.354)$ m — Visible height is 1.68 m or 64.6% of the volume

(*b*) $T = F_B - mg$ — $T = (2.6 \times \pi \times 0.45^2 \times 1025 \times 9.81 - 600 \times 9.81)$ N $= 10.75$ kN

(*c*) $a = T/m$ — $a = 10750/600$ m/s² $= 17.9$ m/s²

---

87 •• Two communicating vessels contain a liquid of density ρ_0 (Figure 13-34). The cross-sectional areas of the vessels are A and $3A$. Find the change in elevation of the liquid level if an object of mass m and density $\rho' = 0.8\rho_0$ is put into one of the vessels.

Since the object floats, the volume of the displaced liquid is $m/\rho_0 = 4A\Delta h$; $\Delta h = m/4\rho_0 A$.

88 •• If an oil-filled manometer (ρ = 900 kg/m^3) can be read to ±0.05 mm, what is the smallest pressure change that can be detected?

$\Delta P = \rho g \Delta h = 5 \times 10^{-5} \times 900 \times 9.81$ Pa = 0.44 Pa $\approx 3.3 \times 10^{-3}$ mmHg; $\Delta P \approx 3\mu$mHg.

89* •• A rectangular dam 30 m wide supports a body of water to a height of 25 m. (*a*) Neglecting atmospheric pressure, find the total force due to water pressure acting on a thin strip of the dam of height dy located at a depth y. (*b*) Integrate your result in part (*a*) to find the total horizontal force exerted by the water on the dam. (*c*) Why is it reasonable to neglect atmospheric pressure?

This problem is identical to Example 13-2.

(*a*) $dF = \rho g y L\, dy$.

(*b*) $F = \frac{1}{2}\rho g L h^2 = 9.20 \times 10^7$ N.

(*c*) Atmospheric pressure is exerted on each side of the dam and it can, therefore, be neglected.

90 •• A U-tube is filled with water until the liquid level is 28 cm above the bottom of the tube. An oil of specific gravity 0.78 is now poured into one arm of the U-tube until the level of the water in the other arm of the tube is 34 cm above the bottom of the tube. Find the level of the oil–water and oil–air interfaces in the other arm of the tube.

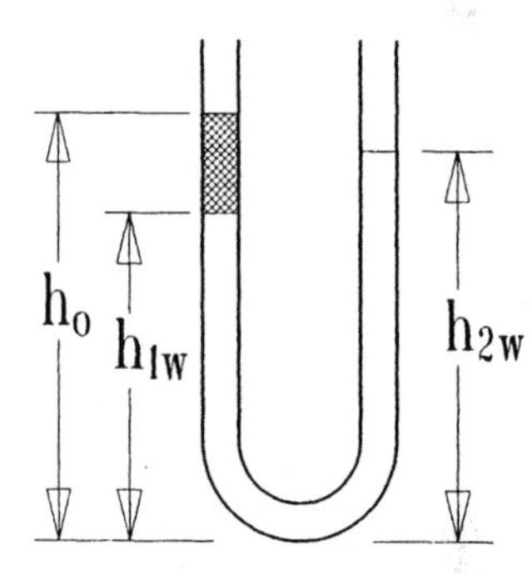

The pressure at the bottom of the U-tube due to one side is the same as that due to the other.

Let h be the height of the oil above the water. Since the amount of water is constant, $h_{1W} + h_{2W}$ = 56 cm. So, after oil is added, h_{1W} = 56 − 34 = 22 cm. This is the height of the oil–water interface.

The height of the oil–air interface is determined from

$34\rho_W g = 22\rho_W g + 0.78\rho_W g h_{oil}$; h_{oil} = 15.4 cm. So the oil–air interface is at (22 + 15.4) cm = 37.4 cm.

91 •• A U-tube contains liquid of unknown specific gravity. An oil of density 800 kg/m^3 is poured into one arm of the tube until the oil column is 12 cm high. The oil–air interface is then 5.0 cm above the liquid level in the other arm of the U-tube. Find the specific gravity of the liquid.

We designate the specific gravity by σ. Using the reasoning outlined in the previous problem we have $\sigma_L g h = \sigma_L g(h - 7) + 12 \times 0.8g$; thus, $\sigma_L = 12 \times 0.8/7 = 1.37$.

92 •• A lead block is suspended from the underside of a 0.5-kg block of wood of specific gravity of 0.7. If the upper surface of the wood is just level with the water, what is the mass of the lead block?

1. Write the condition for neutral buoyancy	$V_{Wood}\rho_W g = 0.5g + m_{Pb}g - V_{Pb}\rho_W g$	(1)
2. $V_{Wood} = 0.5/\rho_{Wood}$; $V_{Pb} = m_{Pb}/\rho_{Pb}$; rewrite (1)	$0.5\rho_W/\rho_{Wood} + m_{Pb}\rho_W/\rho_{Pb} = 0.5 + m_{Pb}$	(2)
3. Evaluate m_{Pb} using $\rho_{Wood}/\rho_W = 0.7$; $\rho_{Pb}/\rho_W = 11.3$	m_{Pb} = 0.235 kg	

93* •• A helium balloon can just lift a load of 750 N. The skin of the balloon has a mass of 1.5 kg. (*a*) What is the volume of the balloon? (*b*) If the volume of the balloon is twice that found in part (*a*), what is the initial acceleration of the balloon when it carries a load of 900 N?

(*a*) Write the condition for neutral buoyancy	$V\rho_{air}g = 750 + 1.5g + V\rho_{He}g$
Solve for V with $\rho_{air} = 1.293$, $\rho_{He} = 0.1786$	V = 70 m^3

(*b*) $F_{net} = F_B - mg$; $mg = 900 + 1.5g$ $F_{net} = [140(1.293 - 0.1786)g - 915]$ N $= 616$ N

$a = F_{net}/m$ $a = [616/(900/g + 1.5)]$ m/s^2 $= 6.61$ m/s^2

---

94 •• A hollow sphere with an inner radius R and outer radius $2R$ is made of material of density ρ_0 and is floating in a liquid of density $2\rho_0$. The interior is now filled with material of density ρ' so that the sphere just floats completely submerged. Find ρ'.

1. Apply the condition for neutral buoyancy $(4\pi/3)(2R)^3(2\rho_0) = (4\pi/3)[(2R)^3 - R^3]\rho_0 + (4\pi/3)R^3\rho'$
2. Solve for ρ' $\rho' = 9\rho_0$

---

95 •• A balloon is filled with helium at atmospheric pressure. The skin of the balloon has a mass of 2.8 kg and the volume of the balloon is 16 m^3. What is the greatest weight that this balloon can lift?

1. Write the condition for neutral buoyancy $16(1.293 - 0.1786)g = 2.8g + w$
2. Solve for the weight w $w = 147$ N; $m \approx 15$ kg

---

96 •• As mentioned in the discussion of *the law of atmospheres*, the fractional decrease in atmospheric pressure is proportional to the change in altitude. Expressed in mathematical terms we have $dP/P = -C\,dh$, where C is a constant. (*a*) Show that $P(h) = P_0e^{-Ch}$ is a solution of the differential equation. (*b*) Show that if $\Delta h \ll h_0$, then $P(h + \Delta h) \approx P(h)(1 - \Delta h/h_0)$, where $h_0 = 1/C$. (*c*) Given that the pressure at $h = 5.5$ km is half that at sea level, find the constant C.

(*a*) For $P(h) = P_0e^{-Ch}$, $dP/dh = -CP_0e^{-Ch} = -CP$; thus, $dP/P = -C\,dh$.

(*b*) For $Ch \ll 1$, $e^{-Ch} \approx 1 - Ch = 1 - h/h_0$. Therefore, $P(h + \Delta h) \approx P(h)(1 - \Delta h/h_0)$.

(*c*) Take the logarithm of the equation $P_0/P(5.5\text{ km}) = 2 = e^{5.5C}$; $5.5C = \ln 2$; $C = 0.126$ km^{-1}; $h_0 = 7.93$ km.

97* •• A submarine has a total mass of 2.4×10^6 kg, including crew and equipment. The vessel consists of two parts, the pressure hull, which has a volume of 2×10^3 m^3, and the diving tanks, which have a volume of 4×10^2 m^3. When the sub cruises on the surface, the diving tanks are filled with air; when cruising below the surface, seawater is admitted into the tanks. (*a*) What fraction of the submarine's volume is above the water surface when the tanks are filled with air? (*b*) How much water must be admitted into the tanks to give the submarine neutral buoyancy? Neglect the mass of air in the tanks and use 1.025 as the specific gravity of seawater.

(*a*) Apply the expression of Example 13-7 $V'/V = \rho'/\rho_{SW}$; $V'/V = 1/1.025 = 0.9756$

The fraction above the surface $= 1 - V'/V$ fraction above the surface $= 2.44\%$

(*b*) Apply the condition for neutral buoyancy $2.4 \times 10^3 \times 1.025 \times 10^3 = 2.4 \times 10^6 + 1.025 \times 10^3V_{SW}$

Solve for V_{SW} $V_{SW} = 58.5$ m^3

---

98 •• A marine salvage crew raises a crate that measures 1.4 m × 0.75 m × 0.5 m. The average density of the empty crate is the same as seawater, 1.025×10^3 kg/m^3, and its mass when empty is 32 kg. The crate contains gold bullion that fills 36% of its volume; the remaining volume is filled with seawater. (*a*) What is the tension in the cable that raises the crate and bullion while the crate is below the surface of the sea? (*b*) What is the tension in the cable while the crate is lifted to the deck of the ship if (1) none of the seawater leaks out of the crate, and (2) the crate is lifted so slowly that all of the seawater leaks out of the crate?

(*a*) $T = m_{Au}g - F_B(Au)$; the crate has neutral buoyancy; $T = (V_{Au}\rho_{Au} - V_{au}\rho_{SW})g$

$= 0.189 \times 9.81(19.3 - 1.025) \times 10^3$ N = 33.9 kN

$V_{au} = 0.36 \times 1.4 \times 0.75 \times 0.5 = 0.189$ m^3

(*b*) 1. $T = V_{Au}\rho_{Au}g + V_{SW}\rho_{SW}g + 32g$ $T = (0.189 \times 19.3 + 0.336 \times 1.025 + 32) \times 9.81 \times 10^3$

$= 39.5$ kN

2. $T = V_{Au}\rho_{Au}g + 32g$ $T = 36.1$ kN

---

99 ••• When the hydrometer of Problem 42 is placed in a liquid whose specific gravity is greater than some minimum value, the device floats with part of the glass tube above the liquid level. Consider a hydrometer that has a spherical bulb 2.4 cm in diameter. The glass tube attached to the bulb is 20 cm long and has a diameter of 7.5 mm. The mass of the hydrometer before lead pellets are dropped into the bulb and the tube is sealed is 7.28 g. (*a*) What mass of lead should be placed in the bulb so that the hydrometer just floats in a liquid of specific gravity 0.78? (*b*) If the hydrometer is now placed in water, what is the length of the tube that shows above the surface of the water? (*c*) The hydrometer is placed in a liquid of unknown specific gravity; the length of the tube above the surface of the liquid is 12.2 cm. Determine the specific gravity of the liquid.

(*a*) 1. Find the volumes of the bulb and tube $V_{bulb} = \pi d^3/6 = 7.238$ cm^3; $V_{tube} = \pi d^2L/4 = 8.836$ cm^3

2. Write the condition for neutral buoyancy $1.6074 \times 10^{-5} \times 780 = 7.28 \times 10^{-3} + m_{Pb}$; $m_{Pb} = 5.26$ g

(*b*) Find the submerged volume in water for $m = 12.54$ g $1.254 \times 10^{-2} = V \times 10^3$; $V = 1.254 \times 10^{-5}$ m^3 = 12.54 cm^3

Find V_{tube} and the length h' that is submerged $V_{tube,sub} = 5.3$ cm^3 $= \pi(0.75)^2h'/4$; $h' = 12$ cm

The length of the tube above the water is 20 cm $- h'$ $h = 8$ cm above water

(*c*) Find the displaced volume $V = [7.238 + \pi(0.75)^2 \times 7.8/4]$ cm^3 = 10.68 cm^3

$F_B = mg = V\rho_L g$ $1.068 \times 10^{-5}\rho_L = 1.254 \times 10^{-2}$ kg;

$\rho_L = 1.174 \times 10^3$ kg/m^3

The specific gravity = $\rho_L/10^3$ specific gravity = 1.174

---

100 ••• A large beer keg of height H and cross-sectional area A_1 is filled with beer. The top is open to atmospheric pressure. At the bottom is a spigot opening of area A_2, which is much smaller than A_1. (*a*) Show that when the height of the beer is h, the speed of the beer leaving the spigot is approximately $\sqrt{2gh}$. (*b*) Show that for the approximation $A_2 << A_1$, the rate of change of the height h of the beer is given by $dh/dt = -(A_2/A_1)\sqrt{2gh}$. (*c*) Find h as a function of time if $h = H$ at $t = 0$. (*d*) Find the total time needed to drain the keg if $H = 2$ m, $A_1 = 0.8$ m^2, and $A_2 = 10^{-4}A_1$.

(*a*) v_2 = velocity of flow at A_2 (see Problem 41) $v_2 = \sqrt{2gh}$

(*b*) Note that $A_1v_1 = A_2v_2$, and $v_1 = -dh/dt$ $dh/dt = -(A_2/A_1)\sqrt{2gh}$

(*c*) Write the differenial equation for h

$$-\frac{(A_1/A_2)}{\sqrt{2g}}\frac{dh}{\sqrt{h}} = dt$$

Integrate the differential equation

$$-\frac{(A_1/A_2)}{\sqrt{2g}}\int_H^h \frac{dh}{\sqrt{h}} = \int_0^t dt;\quad \frac{2(A_1/A_2)}{\sqrt{2g}}(\sqrt{H} - \sqrt{h}) = t$$

Solve for h $$h = \left(\sqrt{H} - \frac{A_2}{2A_1}\sqrt{2g}t\right)^2$$

(d) Find the time t' to drain the keg; $h = 0$ $t' = (A_1/A_2)\sqrt{2H/g}$

Evaluate t' for the parameters given $t' = 6.39 \times 10^3\text{ s} \approx 1\text{h } 46\text{ min}$

CHAPTER 14

Oscillations

1* • Deezo the Clown slept in again. As he roller-skates out the door at breakneck speed on his way to a lunchtime birthday party, his superelastic suspenders catch on a fence post, and he flies back and forth, oscillating with an amplitude A. What distance does he move in one period? What is his displacement over one period?

In one period, he moves a distance $4A$. Since he returns to his initial position, his displacement is zero.

---

2 • A neighbor takes a picture of the oscillating Deezo (from Problem 1) at a moment when his speed is zero. What is his displacement from the fence post at that time?

His displacement is then a maximum.

---

3 • What is the magnitude of the acceleration of an oscillator of amplitude A and frequency f when its speed is maximum? When its displacement is maximum?

When $v = v_{max}$, $a = 0$; when $x = x_{max}$, $a = \omega^2 A = 4\pi^2 f^2 A$.

---

4 • Can the acceleration and the displacement of a simple harmonic oscillator ever be in the same direction? The acceleration and the velocity? The velocity and the displacement? Explain.

Acceleration and displacement are always oppositely directed; $F = -kx$. v and a can be in the same direction, as can v and x; see Equs. 14-4, 14-5, and 14-6.

5* • True or false: (*a*) In simple harmonic motion, the period is proportional to the square of the amplitude. (*b*) In simple harmonic motion, the frequency does not depend on the amplitude. (*c*) If the acceleration of a particle is proportional to the displacement and oppositely directed, the motion is simple harmonic.

(*a*) False (*b*) True (*c*) True

---

6 • The position of a particle is given by $x = (7\text{ cm}) \times \cos 6\pi t$, where t is in seconds. What is (*a*) the frequency, (*b*) the period, and (*c*) the amplitude of the particle's motion? (*d*) What is the first time after $t = 0$ that the particle is at its equilibrium position? In what direction is it moving at that time?

(*a*) Compare the expression with Equ. 14-4 $\quad f = \omega/2\pi = 3$ Hz

(*b*) $T = 1/f$ $\quad T = 0.33$ s

(*c*) See Equ. 14-4 $\quad A = 7$ cm

(*d*) $x = 0$ when $\cos \omega t = 0$ $\quad \omega t = \pi/2$; $t = \pi/12\pi = 1/12$ s; v is then negative

7 • (*a*) What is the maximum speed of the particle in Problem 6? (*b*) What is its maximum acceleration?

(*a*) $v_{max} = A\omega = 42\pi$ cm/s = 1.32 m/s. (*b*) $a_{max} = A\omega^2 = 252\pi^2$ cm/s^2 = 24.9 m/s^2.

8 • What is the phase constant δ in Equation 14-4 if the position of the oscillating particle at time $t = 0$ is (*a*) 0, (*b*) $-A$, (*c*) A, (*d*) $A/2$?

Compare to Equ. 14-4: (*a*) cos $\delta = 0$; $\delta = \pi/2, 3\pi/2$. (*b*) cos $\delta = -1$; $\delta = \pi$. (*c*) cos $\delta = 1$; $\delta = 0$. (*d*) cos $\delta = ½$; $\delta = \pi/3$.

9* • A particle of mass m begins at rest from $x = +25$ cm and oscillates about its equilibrium position at $x = 0$ with a period of 1.5 s. Write equations for (*a*) the position x versus the time t, (*b*) the velocity v versus t, and (*c*) the acceleration a versus t.

(*a*) $x = A\cos[(2\pi/T)t + \delta]$ — $x = 25\cos(4.19t)$ cm

(*b*) $v = -A\omega\sin(\omega t)$ — $v = -105\sin(4.19T)$ cm/s

(*c*) $a = -\omega^2 x$ — $a = -439\cos(4.19t)$ cm/s^2

10 • Find (*a*) the maximum speed, and (*b*) the maximum acceleration of the particle in Problem 6. (*c*) What is the first time that the particle is at $x = 0$ and moving to the right?

(*a*), (*b*) See Problem 7. (*c*) $x = 0$ when $\omega t = \pi/2$ and $3\pi/2$; for $\omega t = \pi/2$, $v = -A\omega$, i.e., moving to the left; for $\omega t = 3\pi/2$, $v = A\omega$, i.e., moving to the right. So $\omega t = 3\pi/2$, and $t = 3\pi/12\pi = 0.25$ s.

11 •• Work Problem 9 with the particle initially at $x = 25$ cm and moving with velocity $v_0 = +50$ cm/s.

(*a*) Find δ using Equs. 14-4 and 14-5 — $2 = -\omega\cot\delta = -4.19\cot\delta$; $\delta = -64.5° = -1.125$ rad

Determine A from Equ. 14-4 — $A = [25/\cos(-64.5°)]$ cm = 58 cm

Write the equation for $x(t)$ — $x = [58\cos(4.19t - 1.125)]$ cm

(*b*) Write the expression for v — $v = [-243\sin(4.19t - 1.125)]$ cm/s

(*c*) Write the expression for a — $a = [-10.18\cos(4.19t - 1.125)]$ m/s^2

12 •• The period of an oscillating particle is 8 s, and its amplitude is 12 cm. At $t = 0$, it is at its equilibrium position. Find the distance traveled during the interval (*a*) $t = 0$ to $t = 2$ s, (*b*) $t = 2$ s to $t = 4$ s, (*c*) $t = 0$ to $t = 1$ s, and (*d*) $t = 1$ s to $t = 2$ s.

(*a*) $\Delta t = T/4$; find Δx — $\Delta x = [12(\sin\pi/2 - \sin 0)]$ cm = 12 cm

(*b*) At $t = 4$ s, the particle is again at $x = 0$ — distance traveled = 12 cm

(*c*) $\Delta t = T/8$ — $\Delta x = [12(\sin\pi/4 - \sin 0)]$ cm = 8.49 cm

(*d*) $\Delta x = x(2) - x(1)$ — $\Delta x = 3.51$ cm

13* •• The period of an oscillating particle is 8 s. At $t = 0$, the particle is at rest at $x = A = 10$ cm. (*a*) Sketch x as a function of t. (*b*) Find the distance traveled in the first second, the next second, the third second, and the fourth second after $t = 0$.

(*a*) A sketch of $x = 10\cos(\pi t/4)$ cm is shown

(*b*) In each case, $\Delta x = 10[\cos(\pi t_f/4) - \cos(\pi t_i/4)]$ cm

For $t_f = 1$ s, $t_i = 0$, $\Delta x = 2.93$ cm

For $t_f = 2$ s, $t_i = 1$ s, $\Delta x = 7.07$ cm

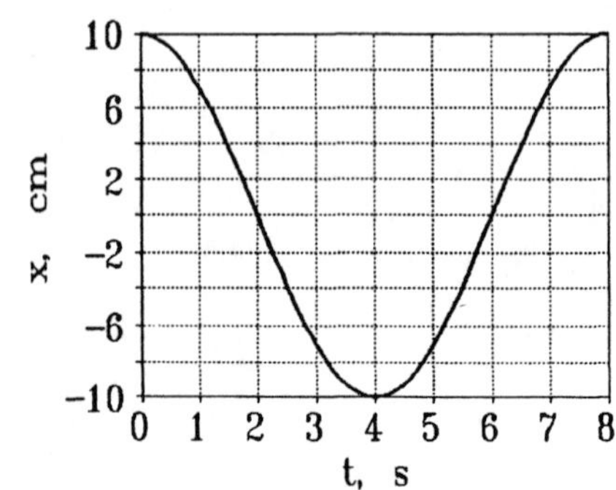

For $t_f = 3$ s, $t_i = 2$ s, $\Delta x = 7.07$ cm

For $t_f = 4$ s, $t_i = 3$ s, $\Delta x = 2.93$ cm

---

14 •• Military specifications often call for electronic devices to be able to withstand accelerations of $10g = 98.1$ m/s^2. To make sure that their products meet this specification, manufacturers test them using a shaking table that can vibrate a device at various specified frequencies and amplitudes. If a device is given a vibration of amplitude 1.5 cm, what should its frequency be in order to test for compliance with the $10g$ military specification?

$a_{max} = A\omega^2 = 98.1$ m/s^2 $\omega = (98.1/0.015)^{½}$ rad/s $= 80.9$ rad/s; $f = 12.9$ Hz

---

15 •• The position of a particle is given by $x = 2.5 \cos \pi t$, where x is in meters and t is in seconds. (*a*) Find the maximum speed and maximum acceleration of the particle. (*b*) Find the speed and acceleration of the particle when $x = 1.5$ m.

(*a*) $v_{max} = A\omega$; $a_{max} = A\omega^2$; $\omega = \pi$ rad/s $v_{max} = 2.5\pi$ m/s $= 7.85$ m/s; $a_{max} = 24.7$ m/s^2

(*b*) 1. Find t when $x = 1.5$ m $\pi t = \cos^{-1}(0.6) = 0.927$ rad

2. $v = v_{max} \sin(0.927)$; $a = -a_{max} \cos(0.927)$ $v = 6.28$ m/s; $a = -14.8$ m/s^2

---

16 •• The bow of a destroyer undergoes a simple harmonic vertical pitching motion with a period of 8.0 s and an amplitude of 2.0 m. (*a*) What is the maximum vertical velocity of the destroyer's bow? (*b*) What is its maximum acceleration? (*c*) An 80-kg sailor is standing on a scale in the bunkroom in the bow. What are the maximum and minimum readings on the scale in newtons?

(*a*) $v_{max} = A\omega = 2\pi A/T$ $v_{max} = (4\pi/8)$ m/s $= 1.57$ m/s

(*b*) $a_{max} = A\omega^2 = A(2\pi/T)^2$ $a_{max} = 2(\pi/4)^2$ m/s^2 $= 1.23$ m/s^2

(*c*) $g_{eff} = g \pm a_{max}$; $w = mg_{eff}$ $w_{min} = 80 \times 8.58$ N $= 686$ N; $w_{max} = 883$ N

17* • A particle moves in a circle of radius 40 cm with a constant speed of 80 cm/s. Find (*a*) the frequency of the motion, and (*b*) the period of the motion. (*c*) Write an equation for the x component of the position of the particle as a function of time t, assuming that the particle is on the positive x axis at time $t = 0$.

(*b*) $T = 2\pi r/v$ $T = \pi$ s $= 3.14$ s

(*a*) $f = 1/T$ $f = 1/\pi$ Hz $= 0.318$ Hz

(*c*) $x = 40 \cos(2\pi ft + \delta)$ cm $x = 40 \cos(2t + \delta)$ cm, where $\delta < \pi/2$

---

18 • A particle moves in a circle of radius 15 cm, making 1 revolution every 3 s. (*a*) What is the speed of the particle? (*b*) What is its angular velocity ω? (*c*) Write an equation for the x component of the position of the particle as a function of time t, assuming that the particle is on the positive x axis at time $t = 0$.

(*a*) $v = 2\pi r/T$ $v = 30\pi/3$ cm/s $= 31.4$ cm/s

(*b*) $\omega = 2\pi f = 2\pi/T$ $\omega = 2\pi/3$ rad/s

(*c*) $x = r \cos(\omega t + \delta)$ $x = 15 \cos(2\pi t/3 + \delta)$ cm, $\delta < \pi/2$

---

19 • If the amplitude of a simple harmonic oscillator is tripled, by what factor is the energy changed?

$E \propto A^2$; energy is increased by a factor of 9.

---

20 •• An object attached to a spring has simple harmonic motion with an amplitude of 4.0 cm. When the object is 2.0 cm from the equilibrium position, what fraction of its total energy is potential energy? (*a*) one-quarter (*b*) one-third (*c*) one-half (*d*) two-thirds (*e*) three-quarters

(*a*) $U \propto x^2$.

21* • A 2.4-kg object is attached to a horizontal spring of force constant $k = 4.5$ kN/m. The spring is stretched 10 cm from equilibrium and released. Find its total energy.

$U = ½kx^2$ $\quad$ $U = 4.5 \times 10^3 \times 0.1^2/2$ J $= 22.5$ J

---

22 • Find the total energy of a 3-kg object oscillating on a horizontal spring with an amplitude of 10 cm and a frequency of 2.4 Hz.

$E = ½mv_{max}^2 = ½mA^2\omega^2 = ½mA^2 \times 4\pi^2f^2 = 2mA^2\pi^2f^2$ $\quad$ $E = 2 \times 3 \times 0.1^2 \times \pi^2 \times 2.4^2$ J $= 3.41$ J

---

23 • A 1.5-kg object oscillates with simple harmonic motion on a spring of force constant $k = 500$ N/m. Its maximum speed is 70 cm/s. (*a*) What is the total energy? (*b*) What is the amplitude of the oscillation?

(*a*) $E = ½mv_{max}^2 = 1.5 \times 0.7^2/2$ J $= 0.368$ J. (*b*) $E = 0.386$ J $= ½kA^2$; $A = 0.0383$ m $= 3.83$ cm.

---

24 • A 3-kg object oscillating on a spring of force constant 2 kN/m has a total energy of 0.9 J. (*a*) What is the amplitude of the motion? (*b*) What is the maximum speed?

(*a*) $U_{max} = ½kA^2$ $\quad$ $A = (2 \times 0.9/2 \times 10^3)^{½}$ m $= 3$ cm

(*b*) $E_{max} = ½mv_{max}^2 = U_{max}$ $\quad$ $v_{max} = (2 \times 0.9/3)^{½}$ m/s $= 0.775$ m/s

---

25* • An object oscillates on a spring with an amplitude of 4.5 cm. Its total energy is 1.4 J. What is the force constant of the spring?

$E = ½kA^2$ $\quad$ $k = 2E/A^2 = 2.8/0.045^2$ N/m $= 1383$ N/m

---

26 •• A 3-kg object oscillates on a spring with an amplitude of 8 cm. Its maximum acceleration is 3.50 m/s². Find the total energy.

$\omega^2 = k/m = a_{max}/A$; $E = ½mv_{max}^2 = ½mA^2\omega^2$ $\quad$ $E = ½mAa_{max} = 3 \times 0.08 \times 3.5/2$ J $= 0.42$ J

---

27 • True or false: (*a*) For a given object on a given spring, the period is the same if the spring is vertical or horizontal. (*b*) For a given object oscillating with amplitude *A* on a given spring, the maximum speed is the same if the spring is vertical or horizontal.

(*a*) True (*b*) True

---

28 • Herb plans to ring in the new year by playing trombone while oscillating up and down on a spring that hangs down from a building at Times Square in New York City. He intends to oscillate with a period of one second to synchronize with the crowd as it counts down to midnight. If he uses a spring with a spring constant of 3000 N/m, Herb must be sure that the total of his vibrating mass adds up to (*a*) 3000 kg. (*b*) $\sqrt{3000}$ kg. (*c*) $4\pi^2(3000)$ kg. (*d*) $3000/4\pi^2$ kg. (*e*) none of the above.

(*d*)

29* • A 2.4-kg object is attached to a horizontal spring of force constant $k = 4.5$ kN/m. The spring is stretched 10 cm from equilibrium and released. Find (*a*) the frequency of the motion, (*b*) the period, (*c*) the amplitude, (*d*) the maximum speed, and (*e*) the maximum acceleration. (*f*) When does the object first reach its equilibrium position? What is its acceleration at this time?

(*a*) $f = (1/2\pi)(k/m)^{1/2}$ — $f = (1/2\pi)(4.5 \times 10^3/2.4)^{1/2}$ Hz $= 6.89$ Hz

(*b*) $T = 1/f$ — $T = 0.145$ s

(*c*) A is given — $A = 0.1$ m

(*d*) $v_{max} = A\omega = 2\pi Af$ — $v_{max} = 4.33$ m/s

(*e*) $a_{max} = A\omega^2 = \omega v_{max}$ — $a_{max} = 187$ m/s^2

(*f*) It reaches equilibrium after 1/4 period — $t = T/4 = 36.25$ ms; at equilibrium, $a = 0$

---

30 • Answer the questions in Problem 29 for a 5-kg object attached to a spring of force constant $k = 700$ N/m when the spring is initially stretched 8 cm from equilibrium.

Follow the same procedures as in the preceding problem using $m = 5$ kg, $k = 700$ N/m, $A = 0.08$ m.

(*a*) $f = 1.88$ Hz. (*b*) $T = 0.531$ s. (*c*) $A = 0.08$ m. (*d*) $v_{max} = 0.947$ m/s. (*e*) $a_{max} = 11.2$ m/s. (*f*) $t = 0.133$ s; $a = 0$.

---

31 • A 3-kg object attached to a horizontal spring oscillates with an amplitude $A = 10$ cm and a frequency $f = 2.4$ Hz. (*a*) What is the force constant of the spring? (*b*) What is the period of the motion? (*c*) What is the maximum speed of the object? (*d*) What is the maximum acceleration of the object?

(*a*) $k/m = \omega^2 = 4\pi^2 f^2$; $k = 4\pi^2 f^2 m$ — $k = 4\pi^2 \times 2.4^2 \times 3$ N/m $= 682$ N/m

(*b*) $T = 1/f$ — $T = 1/2.4$ s $= 0.417$ s

(*c*) $v_{max} = A\omega = 2\pi fA$ — $v_{max} = 2\pi \times 2.4 \times 0.1$ m/s $= 1.508$ m/s

(*d*) $a_{max} = 4\pi^2 f^2 A$ — $a_{max} = 4\pi^2 \times 2.4^2 \times 0.1$ m/s^2 $= 22.7$ m/s^2

---

32 • An 85-kg person steps into a car of mass 2400 kg, causing it to sink 2.35 cm on its springs. Assuming no damping, with what frequency will the car and passenger vibrate on the springs?

1. Determine k — $k = 85 \times 9.81/0.0235 = 35.5$ kN/m

2. Find f of the system — $f = (1/2\pi)(k/m)^{1/2} = (1/2\pi)(35500/2485)^{1/2}$ Hz $= 0.601$ Hz

---

33* • A 4.5-kg object oscillates on a horizontal spring with an amplitude of 3.8 cm. Its maximum acceleration is 26 m/s^2. Find (*a*) the force constant k, (*b*) the frequency, and (*c*) the period of the motion.

(*a*) $k = ma_{max}/A$ (see Problem 26) — $k = 4.5 \times 26/0.038$ N/m $= 3079$ N/m

(*b*) $\omega^2 = a_{max}/A$; $f = (a_{max}/A)^{1/2}/2\pi$ — $f = (26/0.038)^{1/2}/2\pi$ Hz $= 4.16$ Hz

(*c*) $T = 1/f$ — $T = 0.24$ s

---

34 • An object oscillates with an amplitude of 5.8 cm on a horizontal spring of force constant 1.8 kN/m. Its maximum speed is 2.20 m/s. Find (*a*) the mass of the object, (*b*) the frequency of the motion, and (*c*) the period of the motion.

(*b*) $v_{max} = A\omega$; $f = \omega/2\pi = v_{max}/2\pi A$ — $f = 2.2/(2\pi \times 0.058)$ Hz $= 6.04$ Hz

(*a*) $k/m = \omega^2$; $m = k/\omega^2 = kA^2/v_{max}^2$ — $m = 1800 \times 0.058^2/2.2^2$ kg $= 1.25$ kg

(*c*) $T = 1/f$ — $T = 0.166$ s

---

35 •• A 0.4-kg block attached to a spring of force constant 12 N/m oscillates with an amplitude of 8 cm. Find (*a*) the maximum speed of the block, (*b*) the speed and acceleration of the block when it is at $x = 4$ cm from the equilibrium position, and (*c*) the time it takes the block to move from $x = 0$ to $x = 4$ cm.

(*a*) $\omega = (k/m)^{1/2} = (30)^{1/2}$ rad/s; $v_{max} = A\omega$ — $v_{max} = 0.08(30)^{1/2}$ m/s $= 0.438$ m/s (*b*) Let $x = A\cos\omega t$; then $x = A/2$ for $\omega t = \pi/3$ — $v = v_{max}\sin(\pi/3) = 0.379$ m/s; $a = 0.5v_{max}\omega = 1.2$ m/s^2

(*c*) Δt = time to go from $\omega t = \pi/3$ to $\omega t = \pi/2$ — $\Delta t = \pi/6\omega = 0.0956$ s

---

36 •• An object of mass *m* is supported by a vertical spring of force constant 1800 N/m. When pulled down 2.5 cm from equilibrium and released from rest, the object oscillates at 5.5 Hz. (*a*) Find *m*. (*b*) Find the amount the spring is stretched from its natural length when the object is in equilibrium. (*c*) Write expressions for the displacement *x*, the velocity *v*, and the acceleration *a* as functions of time *t*.

(*a*) $4\pi^2f^2 = k/m$; $m = k/4\pi^2f^2$ — $m = 1.51$ kg

(*b*) $x_0 = F/k$; $F = mg$ — $x_0 = 1.51 \times 9.81/1800$ m $= 8.23$ mm

(*c*) $x = -A\cos(2\pi ft)$, taking up as positive — $x = -2.5\cos(34.6t)$ cm; $v = 86.4\sin(34.6t)$ cm/s; $a = 29.85\cos(34.6t)$ m/s^2

37* •• An object of unknown mass is hung on the end of an unstretched spring and is released from rest. If the object falls 3.42 cm before first coming to rest, find the period of the motion.

1. Take $E_i = 0$; $E_f = U_g + U_{spring}$ — $-mg \times 0.0342 + ½k \times 0.0342^2 = 0$
2. Solve for $k/m = \omega^2 = 4\pi^2/T^2$ — $\omega^2 = 573.7$ rad^2/s^2; $T = 0.262$ s

---

38 •• A spring of force constant $k = 250$ N/m is suspended from a rigid support. An object of mass 1 kg is attached to the unstretched spring and the object is released from rest. (*a*) How far below the starting point is the equilibrium position for the object? (*b*) How far down does the object move before it starts up again? (*c*) What is the period of oscillation? (*d*) What is the speed of the object when it first reaches its equilibrium position? (*e*) When does it first reach its equilibrium position?

(*a*) At equilibrium, $ky_0 = mg$ — $y_0 = mg/k = 9.81/250$ m $= 3.92$ cm

(*b*) Let $E_i = 0$; $E_f = U_g + U_{spring}$ — $-mgy_f + ½ky_f^2 = 0$; $y_f = 2mg/k = 7.84$ cm

(*c*) $T = 2\pi(m/k)^{½}$ — $T = 0.397$ s

(*d*) At equilibrium, $v = v_{max} = A\omega$ — $v_{max} = A(k/m)^{½} = 3.92(250)^{½}$ cm/s $= 62$ cm/s

(*e*) It reaches equilibrium at $t = T/4$ — $t = 99.3$ ms

---

39 •• The St. Louis Arch has a height of 192 m. Suppose a stunt woman of mass 60 kg jumps off the top of the arch with an elastic band attached to her feet. She reaches the ground at zero speed. Find her kinetic energy *K* after 2.00 s of the flight. (Assume that the elastic band obeys Hooke's law, and neglect its length when relaxed.)

1. Find *k* of band using energy conservation — $-mgh + ½kh^2 = 0$; $k = 2mg/h = 6.13$ N/m
2. Find ω of motion — $\omega = (k/m)^{½} = 0.32$ rad/s
3. Write $v(t)$; $A = ½(192\text{ m}) = 96$ m — $v(t) = -A\omega\sin(\omega t) = 30.7\sin(0.32t)$ m/s
4. Evaluate $½mv^2$ at $t = 2$ s — $K = 30[30.7\sin(0.64)]^2 = 10.1$ kJ

---

40 •• A 0.12-kg block is suspended from a spring. When a small stone of mass 30 g is placed on the block, the spring stretches an additional 5 cm. With the stone on the block, the spring oscillates with an amplitude of 12 cm. (*a*) What is the frequency of the motion? (*b*) How long does the block take to travel from its lowest point to its highest point? (*c*) What is the net force of the stone when it is at a point of maximum upward displacement?

(*a*) 1. Determine *k* — $kx = mg$; $k = 0.03 \times 9.81/0.05$ N/m $= 5.89$ N/m

2. $f = (1/2\pi)(k/m)^{½}$ — $f = (1/2\pi)(5.89/0.15)^{½}$ Hz $= 0.997$ Hz

(*b*) Time of travel is $\frac{1}{2}T = 1/2f$ — $t = 0.502$ s

(*c*) At this point, the block is momentarily at rest — $F = mg = 0.294$ N

41* •• In Problem 40, find the maximum amplitude of oscillation such that the stone will remain on the block.

To remain on the block, the block's maximum downward acceleration must not exceed g.

1. Find a_{max} for the amplitude A — $a_{max} = kA/m = 47.9A$
2. Set $a_{max} = g$ to determine A_{max} — $A_{max} = 9.81/47.9$ m $= 20.5$ cm

---

42 •• An object of mass 2.0 kg is attached to the top of a vertical spring that is anchored to the floor. The uncompressed length of the spring is 8.0 cm, and the equilibrium position of the object on the spring is 5.0 cm from the floor. When the object is resting at its equilibrium position, it is given a downward impulse with a hammer such that its initial speed is 0.3 m/s. (*a*) To what maximum height above the floor does the object eventually rise? (*b*) How long does it take for the object to reach its maximum height the first time? (*c*) Does the spring ever become uncompressed? What minimum initial velocity must be given to the object for the spring to be uncompressed at some time?

(*a*) 1. Determine k — $k = mg/y = 2 \times 9.81/0.03$ N/m $= 654$ N/m

2. $A = \sqrt{mv_{max}^2/k}$; $h = A + 5.0$ cm — $A = 1.66$ cm; $h = 6.66$ cm

height above floor $= 5.0$ cm $+ h = 5.46$ cm

(*b*) 1. Find $T = 2\pi(m/k)^{1/2}$ — $T = 2\pi(2/654)^{1/2}$ s $= 0.347$ s

2. Time of motion = 3/4 period — $t = 0.261$ s

(*c*) 1. The spring is never uncompressed

2. To be uncompressed, $h \geq 3$ cm — $v_i = (2gh)^{1/2} = 0.767$ m/s

---

43 •• Lou has devised a new kiddie ride and is testing it for safety. A child is placed on a large block that is attached to a horizontal spring. When pulled back and released, the child and block oscillate with a period of 2 s. (*a*) If the coefficient of static friction between the child and the block is 0.25, will an amplitude of 1 m cause the child to slip? (*b*) What is the maximum amplitude that will avoid slipping?

(*b*) 1. Condition for slipping: $mg\mu_s < ma_{max}$ — $a_{max} = g\mu_s$

2. Express a_{max} in terms of A_{max} and T — $a_{max} = 4\pi^2 A_{max}/T^2$; $A_{max} = g\mu_s T^2/4\pi^2 = 24.8$ cm

(*a*) Note that $A = 1$ m $> A_{max}$ — The child will slip on the block

---

44 •• A 2.5-kg object hanging from a vertical spring of force constant 600 N/m oscillates with an amplitude of 3 cm. When the object is at its maximum downward displacement, find (*a*) the total energy of the system, (*b*) the gravitational potential energy, and (*c*) the potential energy in the spring. (*d*) What is the maximum kinetic energy of the object? Choose $U = 0$ when the object is at equilibrium.

We can set $U_{grav} = 0$ by selecting our coordinate origin to be at y_0, where y_0 is the equilibrium position of the object. Since $F_{net} = 0$ at equilibrium, the extension of the spring is then $y_0 = mg/k$, and the potential energy stored in the spring is $U_{spring} = \frac{1}{2}ky_0^2$. A further extension of the spring by an amount y increases U_{spring} to $\frac{1}{2}k(y + y_0)^2 = \frac{1}{2}ky^2 + kyy_0 + \frac{1}{2}ky_0^2 = \frac{1}{2}ky^2 + mgy + \frac{1}{2}ky_0^2$. Consequently, if we set $U = U_{grav} + U_{spring} = 0$, a further extension of the spring by y increases U_{spring} by $\frac{1}{2}ky^2 + mgy$ while decreasing U_{grav} by mgy. Therefore, if $U = 0$ at the equilibrium position, the change in U is given by $\frac{1}{2}k(y')^2$, where $y' = y - y_0$.

(*a*) $E = \frac{1}{2}kA^2$ — $E = 300 \times 0.03^2$ J $= 0.27$ J

(*b*) $U_g = -mgA$ — $U_g = -0.736$ J

(*c*) $U_{spring} = \frac{1}{2}kA^2 + mgA$ — $U_{spring} = 0.27 + 0.736$ J $= 1.006$ J

(*d*) $K_{max} = \frac{1}{2}kA^2$ — $K_{max} = 0.27$ J

45* •• A 1.5-kg object that stretches a spring 2.8 cm from its natural length when hanging at rest oscillates with an amplitude of 2.2 cm. (*a*) Find the total energy of the system. (*b*) Find the gravitational potential energy at maximum downward displacement. (*c*) Find the potential energy in the spring at maximum downward displacement. (*d*) What is the maximum kinetic energy of the object? (Choose $U = 0$ when the object is in equilibrium.)

Find k and then proceed as in the preceding problem. $k = mg/y_0 = 1.5 \times 9.81/0.028$ N/m $= 526$ N/m.

(*a*) $E = 263 \times 0.022^2$ J $= 0.127$ J. (*b*) $U_g = -1.5 \times 9.81 \times 0.022$ J $= -0.324$ J.

(*c*) $U_{spring} = (0.127 + 0.324)$ J $= 0.451$ J. (*d*) $K_{max} = \frac{1}{2}kA^2 = 263 \times 0.022^2$ J $= 0.127$ J.

---

46 •• A 1.2-kg object hanging from a spring of force constant 300 N/m oscillates with a maximum speed of 30 cm/s. (*a*) What is its maximum displacement? When the object is at its maximum displacement, find (*b*) the total energy of the system, (*c*) the gravitational potential energy, and (*d*) the potential energy in the spring. (Choose $U = 0$ when the object is in equilibrium.)

(*a*) Find A using $v_{max} = \omega A$, $\omega = (k/m)^{1/2}$ — $\omega = (300/1.2)^{1/2}$ rad/s $= 15.8$ rad/s; $A = 0.3/15.8$ m $= 1.9$ cm

(*b*) $E = \frac{1}{2}kA^2$ — $E = 150 \times 0.019^2$ J $= 0.0542$ J

(*c*) $U_g = \pm mgA$ — $U_g = \pm 0.225$ J

(*d*) $U_{spring} = \frac{1}{2}kA^2 + mgA$ — $U_{spring} = 0.279$ J

---

47 • True or false: The motion of a simple pendulum is simple harmonic for any initial angular displacement.

False

---

48 • True or false: The motion of a simple pendulum is periodic for any initial angular displacement.

True

49* •• The length of the string or wire supporting a pendulum increases slightly when its temperature is raised. How would this affect a clock operated by a simple pendulum?

The clock would run slow.

---

50 • Find the length of a simple pendulum if the period is 5 s at a point where $g = 9.81$ m/s^2.

Use Equ. 14-27; $T^2 = 4\pi^2 L/g$ — $L = gT^2/4\pi^2 = 9.81 \times 25/4\pi^2$ m $= 6.21$ m

---

51 • What would be the period of the pendulum in Problem 50 if the pendulum were on the moon, where the acceleration due to gravity is one-sixth that on earth?

Since $T \propto 1/\sqrt{g}g$, $T_{moon} = 5\sqrt{6} = 12.2$ s.

---

52 • If the period of a pendulum 70 cm long is 1.68 s, what is the value of g at the location of the pendulum?

$g = 4\pi^2 L/T^2$ — $g = 4\pi^2 \times 0.7/1.68^2$ m/s^2 $= 9.79$ m/s^2

53* • A pendulum set up in the stairwell of a 10-story building consists of a heavy weight suspended on a 34.0-m wire. If $g = 9.81$ m/s^2, what is the period of oscillation?

Use Equ. 14-27 $T = 2\pi(34/9.81)^{1/2}$ s = 11.7 s

---

54 •• Show that the total energy of a simple pendulum undergoing oscillations of small amplitude ϕ_0 is approximately $E \approx \frac{1}{2}mgL\phi_0^2$. (*Hint*: Use the approximation $\cos\phi \approx 1 - \phi^2/2$ for small ϕ.)

Find U_g at maximum displacement $U_g = mgL(1 - \cos\phi_0) \approx \frac{1}{2}mgL\phi_0^2$

---

55 •• A simple pendulum of length L is attached to a cart that slides without friction down a plane inclined at angle θ with the horizontal as shown (Figure 14-28). Find the period of oscillation of the pendulum on the sliding cart.

1. Find the effective acceleration $g_{eff} = g - g\sin\theta = g(1 - \sin\theta)$
2. Apply Equ. 14-27 $T = 2\pi[L/g(1 - \sin\theta)]^{1/2}$

---

56 •• A simple pendulum of length L is released from rest from an angle ϕ_0. (*a*) Assuming that the pendulum undergoes simple harmonic motion, find its speed as it passes through $\phi = 0$. (*b*) Using the conservation of energy, find this speed exactly. (*c*) Show that your results for (*a*) and (*b*) are the same when ϕ_0 is small. (*d*) Find the difference in your results for $\phi_0 = 0.20$ rad and $L = 1$ m.

(*a*) $\phi = \phi_0\cos\omega t$; $v = L(d\phi/dt)$; $\omega = (g/L)^{1/2}$ $v = -L\phi_0\omega\sin\omega t$; $v_{max} = L\phi_0\omega = \phi_0(gL)^{1/2}$

(*b*) $\frac{1}{2}mv_{max}^2 = mgL(1 - \cos\phi_0)$ $v_{max} = [2gL(1 - \cos\phi_0)]^{1/2}$

(*c*) For $\phi_0 << 1$, $1 - \cos\phi_0 \approx \phi_0^2/2$ $v_{max} \approx [2gL(\phi_0^2/2)]^{1/2} = \phi_0(gL)^{1/2}$

(*d*) Evaluate expressions (*b*) and (*a*) $v_{max,b} = 0.6254$ m/s; $v_{max,a} = 0.6264$ m/s; $\Delta v = 1$ mm/s

---

57* • A thin disk of mass 5 kg and radius 20 cm is suspended by a horizontal axis perpendicular to the disk through its rim. The disk is displaced slightly from equilibrium and released. Find the period of the subsequent simple harmonic motion.

1. Find I through the pivot; use parallel axis theorem $I = \frac{1}{2}MR^2 + MR^2 = 3MR^2/2 = 0.3$ kg·m^2
2. Apply Equ. 14-31 $T = 2\pi(I/MgD)^{1/2} = 2\pi(0.3/5 \times 9.81 \times 0.2)^{1/2}$ s = 1.1 s

---

58 • A circular hoop of radius 50 cm is hung on a narrow horizontal rod and allowed to swing in the plane of the hoop. What is the period of its oscillation, assuming that the amplitude is small?

1. Find I through the pivot $I = MR^2 + MR^2 = 2MR^2$
2. Apply Equ. 14-31 $T = 2\pi(2MR^2/MgR)^{1/2} = 2\pi(2R/g) = 2.01$ s

---

59 • A 3-kg plane figure is suspended at a point 10 cm from its center of mass. When it is oscillating with small amplitude, the period of oscillation is 2.6 s. Find the moment of inertia I about an axis perpendicular to the plane of the figure through the pivot point.

1. Write the expression for I using Equ. 14-31 $I = MgDT^2/4\pi^2 = 3 \times 9.81 \times 0.1 \times 2.6^2/4\pi^2$ $= 0.504$ kg·m^2
2. Use the parallel axis theorem to find I_{cm} $I_{cm} = (0.504 - 0.03)$ kg·m^2 = 0.474 kg·m^2

---

60 •• Figure 14-29 shows a dumbbell with two equal masses (to be considered as point masses) attached to a very thin (massless) rod of length L. (*a*) Show that the period of this pendulum is a minimum when the pivot point P is at one of the masses. (*b*) Find the period of this physical pendulum if the distance between P and the upper mass is $L/4$.

(a) $I_{cm} = 2m(L/2)^2 = mL^2/2$. Let x be distance of pivot from center of rod. Then $I = ½mL^2 + 2mx^2$. The period is

$$T = 2\pi\sqrt{\frac{L^2/4 + x^2}{gx}}.$$

Set $dT/dx = 0$ to find the condition for minimum T.

$$\frac{d}{dx}\sqrt{\frac{L^2/4 + x^2}{x}} = \frac{2x^2 - (L^2/4 + x^2)}{x^2\sqrt{\frac{L^2/4 + x^2}{x}}}.$$

This is zero if $x^2 = L^2/4$ or $x = L/2$.

(b) If $x = L/4$, using the above expression for T, one obtains $T = \pi\sqrt{5L/g} = 3.17$ s for $L = 2.0$ m.

61* •• Suppose the rod in Problem 60 has a mass of $2m$ (Figure 14-30). Determine the distance between the upper mass and the pivot point P such that the period of this physical pendulum is a minimum.

Folllow the same procedure as in Problem 60(a). Here $I_{cm} = mL^2/2 + 2mL^2/12 = 2mL^2/3$ and $I = 2mL^2/3 + 4mx^2$, where x is again the distance of the pivot from the center of the rod. The period is then $T = C[(2L^2/3 + 4x^2)/x]^{½}$, where C is a constant. Setting $dT/dx = 0$ gives $4x^2 = 2L^2/3$ and $x = L/\sqrt{6}$. The distance to the pivot from the nearer mass is then $d = L/2 - L/\sqrt{6} = 0.0918L$.

---

62 •• You are given a meter stick and asked to drill a hole in it so that when pivoted about the hole the period of the pendulum will be a minimum. Where should you drill the hole?

See Problem 60. In this case, $I_{cm} = mL^2/12$ and $I = mL^2/12 + mx^2$. Now $T = C[(L^2/12 + x^2)/x]^{½}$, setting $dT/dx = 0$, one obtains $x = L/2\sqrt{3} = 0.289L = 28.9$ cm. So, the hole should be drilled at 21.1 cm.

---

63 •• An irregularly shaped plane object of mass 3.2 kg is suspended by a thin rod of adjustable length and is free to swing in the plane of the object (Figure 14-31). When the length of the supporting rod is 1.0 m, the period of this pendulum for small oscillations is 2.6 s. When the rod is shortened to 0.8 m, the period decreases to 2.5 s. What will be the period of this physical pendulum if the length of the rod is 0.5 m?

Let d be the distance between lower end of rod and center of mass of object.

1. Write initial condition for $T_i^2 = 6.76$ s^2	$[I_{cm} + 3.2(1.0 + d)^2]/(1.0 + d) = 5.375$ kg·m	(1)
2. Write condition for length of 0.8 m	$[I_{cm} + 3.2(0.8 + d)^2]/(0.8 + d) = 4.97$ kg·m	(2)
3. Solve for d and I_{cm}	$d = 0.283$ m; $I_{cm} = 1.63$ kg·m^2	
4. Find D and I for $\ell = 0.5$ m	$D = 0.783$ m; $I = (1.63 + 3.2 \times 0.783^2)$ kg·m^2 $= 3.59$ kg·m^2	
5. $T = 2\pi(I/MgD)^{½}$	$T = 2.40$ s	

---

64 •• When a short person and a tall person walk together at the same speed, the short person will take more steps. Consider the leg to be a physical pendulum that swings about the hip joint. Estimate the natural frequency of this pendulum for a person of your height, and compare the result with the rate at which you take steps when walking in a leisurely manner.

The length of the leg is approximately half the height, say 1 m. The moment of inertia about the joint is then approximately $M/3$, where M is the mass of the leg. As a physical pendulum, $T = 2\pi(M/3/Mg \times 0.5)^{½} = 1.6$ s. Each step corresponds to half a period, or about 0.8 s. This agrees fairly well with the walking rate.

65* •• Figure 14-32 shows a uniform disk of radius $R = 0.8$ m and a 6-kg mass with a small hole a distance d from the disk's center that can serve as a pivot point. (*a*) What should be the distance d so that the period of this physical pendulum is 2.5 s? (*b*) What should be the distance d so that this physical pendulum will have the shortest possible period? What is this period?

(*a*) 1. Use Equ. 14-31; $T^2(mg/4\pi^2)d = I_{cm} + md^2$ — $9.32d = \frac{1}{2} \times 6 \times 0.64 + 6d^2 = 1.92 + 6d^2$

2. Solve the quadratic equation for d — $d = 1.31$ m, $d = 0.244$ m; $d = 24.4$ cm

(*b*) Set $dT^2/dd = 0$ and solve for d — $d^2 - \frac{1}{2}R^2 = 0$; $d = R/\sqrt{2} = 56.6$ cm

$$T = 2\pi\sqrt{\frac{I}{mgR/\sqrt{2}}};\qquad T = 2\pi\sqrt{\frac{R\sqrt{2}}{g}} = 2.1\ \text{s}$$

$$I = \tfrac{1}{2}mR^2 + m(\tfrac{1}{2}R^2) = mR^2$$

66 ••• A plane object has moment of inertia I about its center of mass. When pivoted at point P_1, as shown in Figure 14-33, it oscillates about the pivot with a period T. There is a second point P_2 on the opposite side of the center of mass about which the object can be pivoted so that the period of oscillation is also T. Show that $h_1 + h_2 = gT^2/4\pi^2$.

From Equ. 14-31 we have $T^2(mg/4\pi^2) = I_{cm}/h_1 + mh_1 = I_{cm}/h_2 + mh_2$. Solve for I_{cm}: $I_{cm} = mh_1h_2$. Substitute this result into the first equation to obtain $gT^2/4\pi^2 = h_1 + h_2$.

67 ••• A physical pendulum consists of a spherical bob of radius r and mass m suspended from a string (Figure 14-34). The distance from the center of the sphere to the point of support is L. When r is much less than L, such a pendulum is often treated as a simple pendulum of length L. (*a*) Show that the period for small oscillations is given by

$$T = T_0\sqrt{1 + \frac{2r^2}{5L^2}},$$

where $T_0 = 2\pi\sqrt{L/g}$ is the period of a simple pendulum of length L. (*b*) Show that when r is much smaller than L, the period is approximately $T \approx T_0(1 + r^2/5L^2)$. (*c*) If $L = 1$ m and $r = 2$ cm, find the error when the approximation $T = T_0$ is used for this pendulum. How large must the radius of the bob be for the error to be 1%?

(*a*) $T = 2\pi[(2mr^2/5 + mL^2)/mgL]^{1/2} = 2\pi(L/g)^{1/2}[1 + 2r^2/5L^2]^{1/2} = T_0[1 + 2r^2/5L^2]^{1/2}$.

(*b*) Use binomial expansion — $T = T_0[1 + \frac{1}{2}(2r^2/5L^2) + \cdots] \approx T_0(1 + r^2/5L^2)$

(*c*) 1. $\Delta T/T \approx \Delta T/T_0 = r^2/5L^2$ — $\Delta T/T_0 = 4 \times 10^{-4}/5 = 8 \times 10^{-5} = 0.008\%$

2. Set $r^2/5L^2 = 0.01$; $L = 1.0$ m; find r — $r = 0.05^{1/2}$ m $= 22.4$ cm

68 ••• Figure 14-35 shows the pendulum of a clock. The uniform rod of length $L = 2.0$ m has a mass $m = 0.8$ kg. Attached to the rod is a disk of mass $M = 1.2$ kg and radius 0.15 m. The clock is constructed to keep perfect time if the period of the pendulum is exactly 3.50 s. (*a*) What should be the distance d so that the period of this pendulum is 3.50 s? (*b*) Suppose that the pendulum clock loses 5.0 min per day. How far and in what direction should the disk be moved to ensure that the clock will keep perfect time?

(*a*) 1. Find I as a function of d — $I = 0.8 \times 2^2/3 + \frac{1}{2} \times 1.2 \times 0.15^2 + 1.2d^2$
$= (1.08 + 1.2d^2)$ kg·m^2

2. Locate the center of mass from pivot — $1.0 \times 0.8 + 1.2d = 2.0x_{cm}$; $x_{cm} = (0.4 + 0.6d)$ m
3. Write expression for $T^2g/4\pi^2$ — $T^2g/4\pi^2 = 3.04 = (1.08 + 1.2d^2)/(0.4 + 0.6d)$
4. Solve quadratic equation for d — $d = 1.59$ m

(*b*) $\Delta T/T = -0.0035$; Find dT/dd at $d = 1.59$ m — $dT = 1.145\,dd$; $\Delta T/T = 1.145\Delta d/d = 0.72\Delta d$

Evaluate Δd — $\Delta d = -2.52$ mm; move the disk up by 2.52 mm

---

69* •• Two clocks have simple pendulums of identical lengths L. The pendulum of clock A swings through an arc of 10°; that of clock B swings through an arc of 5°. When the two clocks are compared one will find that (*a*) A runs slow compared to B. (*b*) A runs fast compared to B. (*c*) both clocks keep the same time. (*d*) the answer depends on the length L.

(*a*) The period of A is longer. (see Equ. 14-28)

---

70 •• A simple-pendulum clock keeps accurate time when its length is L. If the length is increased a small amount, how will the accuracy of the clock be affected? (*a*) The clock will lose time. (*b*) The clock will gain time. (*c*) The clock will continue to keep accurate time. (*d*) The answer cannot be determined without knowing the original length of the pendulum. (*e*) The answer cannot be determined without knowing the percent increase in the length of the pendulum.

(*a*) It will run slow.

---

71 •• A pendulum clock loses 48 s per day when the amplitude of the pendulum is 8.4°. What should be the amplitude of the pendulum so that the clock keeps perfect time?

1. Apply Equ. 14-28 — $[\sin^2(8.4/2) - \sin^2(\phi/2)]/4 = 48/86400$
2. Solve for ϕ — $\phi = 6.43°$

---

72 •• A pendulum clock that has run down to a very small amplitude gains 5 min each day. What angular amplitude should the pendulum have to keep the correct time?

From Equ. 14-28, $\sin^2(\phi/2) = 4 \times 5/1440$ — Solving for ϕ: $\phi = 13.5°$

73* • True or false: The energy of a damped, undriven oscillator decreases exponentially with time.

True

---

74 • Show that the dampening constant, b, has units of kg/s.

$b = m/\tau$; dimensionally, $b = [M]/[T] = $ kg/s.

---

75 • An oscillator has a Q factor of 200. By what percentage does its energy decrease during one period?

From Equ. 14-41, $\Delta E/E = 2\pi/Q$ — $\Delta E/E = 2\pi/200 = 3.14\%$

---

76 • A 2-kg object oscillates with an initial amplitude of 3 cm on a spring of force constant $k = 400$ N/m. Find (*a*) the period, and (*b*) the total initial energy. (*c*) If the energy decreases by 1% per period, find the damping constant b and the Q factor.

(*a*) $T = 2\pi(m/k)^{1/2}$ (Q is large, so $\omega = \omega_0$) — $T = 2\pi(1/200)^{1/2}$ s $= 0.444$ s

(*b*) $E = \frac{1}{2}kA^2$ — $E = 200 \times 0.03^2$ J $= 0.18$ J

(*c*) $Q = 2\pi/(\Delta E/E)$ — $Q = 2\pi/0.01 = 628$

$b = \omega_0 m/Q = 2\pi m/QT$ — $b = 2\pi \times 2/628 \times 0.444$ kg/s $= 0.0451$ kg/s

77* •• Show that the ratio of the amplitudes for two successive oscillations is constant for a damped oscillator.

From Equ. 14-36, $A(t)/A(t+T) = e^{-T/2\tau}$.

78 •• An oscillator has a period of 3 s. Its amplitude decreases by 5% during each cycle. (*a*) By how much does its energy decrease during each cycle? (*b*) What is the time constant τ? (*c*) What is the Q factor?

(*a*) Since $E \propto A^2$; $dE/E = 2\ dA/A$ — $\Delta E/E = 10\%$

(*b*) Use Equ. 14-40 — $0.1 = (3\text{ s})/\tau$, $\tau = 30$ s

(*c*) $Q = 2\pi\tau/T$ — $Q = 62.8$

79 •• An oscillator has a Q factor of 20. (*a*) By what fraction does the energy decrease during each cycle? (*b*) Use Equation 14-35 to find the percentage difference between ω' and ω_0. (*Hint*: Use the approximation $(1+x)^{½} \approx 1 + ½x$ for small x.)

(*a*) $\Delta E/E = 2\pi/Q$ — $\Delta E/E = 0.314$

(*b*) $\omega' = \omega_0(1 - 1/4Q^2)^{½}$ (Equ. 14-42) — With the approximation, $\omega_0 - \omega' = 1/8Q^2$ $= 3.13 \times 10^{-2}$ %

80 •• For a child on a swing, the amplitude drops by a factor of $1/e$ in about eight periods if no energy is fed in. Estimate the Q factor for this system.

Since $A = A_0e^{-t/2\tau}$, the problem statement means that $1/e = e^{-8T/2\tau}$ or $4T = \tau$. $Q = \omega_0\tau = 2\pi\tau/T = 8\pi$.

81* •• A damped mass–spring system oscillates at 200 Hz. The time constant of the system is 2.0 s. At $t = 0$, the amplitude of oscillation is 6.0 cm and the energy of the oscillating system is then 60 J. (*a*) What are the amplitudes of oscillation at $t = 2.0$ s and $t = 4.0$ s? (*b*) How much energy is dissipated in the first 2-s interval and in the second 2-s interval?

(*a*) $A(t) = A_0e^{-t/2\tau}$ — $A(2) = 6e^{-0.5}$ cm $= 3.64$ cm; $A(4) = 6e^{-1}$ cm $= 2.21$ cm

(*b*) $E(t) = E_0e^{-t/\tau}$; $\Delta E = E_0(1 - e^{-t/\tau})$ — $\Delta E_{0\text{-}2} = 60 \times 0.632\text{ J} = 37.9$ J; $\Delta E_{2\text{-}4} = 37.9 \times 0.632\text{ J} = 24$ J

82 •• It has been stated that the vibrating earth has a resonance period of 54 min and a Q factor of about 400 and that after a large earthquake, the earth "rings" (continues to vibrate) for about 2 months. (*a*) Find the percentage of the energy of vibration lost to damping forces during each cycle. (*b*) Show that after n periods, the energy is $E_n = (0.984)^nE_0$, where E_0 is the original energy. (*c*) If the original energy of vibration of an earthquake is E_0, what is the energy after 2 days?

(*a*) $\Delta E/E = 2\pi/Q$ — $\Delta E/E = 2\pi/400 = 1.57\%$

(*b*) Each cycle reduces E by factor $(1 - 0.0157)$ — $E_n = E_0(1 - 0.0157)^n = E_0(0.9843)^n$

(*c*) 2 d = 2880 min = $53.3T$ — $E(2\text{ d}) = E_0(0.9843)^{53.3} = 0.43E_0$

83 ••• A 3-kg sphere dropped through air has a terminal speed of 25 m/s. (Assume that the drag force is $-bv$.) Now suppose the sphere is attached to a spring of force constant $k = 400$ N/m, and that it oscillates with an initial amplitude of 20 cm. (*a*) What is the time constant τ? (*b*) When will the amplitude be 10 cm? (*c*) How much energy will have been lost when the amplitude is 10 cm?

(*a*) Determine b from $v_t = mg/b$; $\tau = m/b = v_t/g$ — $\tau = 25/9.81\text{ s} = 2.55$ s (*b*) $A(t) = A_0e^{-t/2\tau}$ $2 = e^{t/5.1}$; $t = 5.1 \ln(2) = 3.54$ s

(*c*) $E_0 = \frac{1}{2}kA_0^2$; since $E \propto A^2$, $E(3.54\text{ s}) = E_0/4$ Energy loss = $(3/4)E_0 = 3 \times 8/4$ J = 6 J

---

84 • True or false: (*a*) Resonance occurs when the driving frequency equals the natural frequency. (*b*) If the Q value is high, the resonance is sharp.

(*a*) True (*b*) True

85* • Give some examples of common systems that can be considered to be driven oscillators.

The pendulum of a clock, a violin string when bowed, and the membrane of a loudspeaker can be considered driven oscillators.

---

86 • A crystal wineglass shattered by an intense sound is an example of (*a*) resonance. (*b*) critical damping. (*c*) an exponential decrease in energy. (*d*) overdamping.

(*a*)

---

87 • Find the resonance frequency for each of the three systems shown in Figure 14-36.

(*a*) $\omega_0 = (k/m)^{1/2} = 6.32$ rad/s; $f_0 = 1.01$ Hz. (*b*) Similarly, $f_0 = 2.01$ Hz. (*c*) $f_0 = (1/2\pi)(g/L)^{1/2} = 0.352$ Hz.

---

88 • A damped oscillator loses 2% of its energy during each cycle. (*a*) What is its Q factor? (*b*) If its resonance frequency is 300 Hz, what is the width of the resonance curve $\Delta\omega$ when the oscillator is driven?

(*a*) Use Equ. 14-41 $Q = 2\pi/0.02 = 314$

(*b*) $\Delta\omega = \omega_0/Q = 2\pi f_0/Q$ $\Delta\omega = 2\pi \times 300/100\pi$ rad/s = 6 rad/s

---

89* •• A 2-kg object oscillates on a spring of force constant $k = 400$ N/m. The damping constant has a value of $b = 2.00$ kg/s. The system is driven by a sinusoidal force of maximum value 10 N and angular frequency $\omega = 10$ rad/s. (*a*) What is the amplitude of the oscillations? (*b*) If the driving frequency is varied, at what frequency will resonance occur? (*c*) What is the amplitude of oscillation at resonance? (*d*) What is the width of the resonance curve $\Delta\omega$?

(*a*) 1. Determine ω_0 $\omega_0 = (k/m)^{1/2} = 14.14$ rad/s

2. Find A; use Equ. 14-49 $A = 10/[4(200 - 100)^2 + 4 \times 100]^{1/2}$ m = 0.05 m = 5.0 cm

(*b*) Resonance is at $\omega = \omega_0$ $\omega_{res} = 14.14$ rad/s

(*c*) Use Equ. 14-49 to find A_{res} $A_{res} = 10/(4 \times 200)^{1/2} = 35.4$ cm

(*d*) From Equ. 14-39 and 14-45, $\Delta\omega = b/m$ $\Delta\omega = 1$ rad/s

---

90 •• A damped oscillator loses 3.5% of its energy during each cycle. (*a*) How many cycles elapse before half of its original energy is dissipated? (*b*) What is its Q factor? (*c*) If the natural frequency is 100 Hz, what is the width of the resonance curve when the oscillator is driven?

(*a*) $E_n = E_0(1 - \Delta E/E)^n$ (see Problem 82); find n $0.5 = (0.965)^n$; $n = 19.5$, or 20 complete cycles

(*b*) $Q = 2\pi(E/\Delta E)$ $Q = 2\pi/0.035 = 180$

(*c*) $\Delta\omega = \omega_0/Q$ $\Delta\omega = 2\pi \times 100 \times 0.035/2\pi = 3.5$ rad/s

---

91 •• Tarzan is depressed again. He ties a vine to his ankle and swings upside-down with a period of 3 s as he contemplates his troubles. Cheetah the chimpanzee pushes him so that the amplitude remains constant. Tarzan's mass is 90 kg and his speed at the bottom of the swing is 2.0 m/s. (*a*) What is Tarzan's total energy? (*b*) If Q = 20, how much energy is dissipated during each oscillation? (*c*) What is Cheetah's power input? (*Note*: Pushing a swing is usually not done sinusoidally. However, to maintain a steady amplitude, the energy lost per cycle due to damping must be replaced by an external energy source.)

(*a*) $E = \frac{1}{2}mv_{max}^2$ — $E = 45 \times 4$ J = 180 J

(*b*) $\Delta E = E_0(2\pi/Q)$ — $\Delta E = 180 \times 0.314 = 56.5$ J

(*c*) $P = \Delta E/\Delta t$ — $P = 56.5/3$ W = 18.8 W

---

92 •• Peter lays his jack-in-the-box on its side with the lid open, so that Jack, a painted 0.4-kg clown, sticks out horizontally at the end of a spring. Peter then takes a 0.6-kg wad of putty, places it in his favorite slingshot, and fires it at the top of Jack's head. The putty sticks to the clown's head, and the clown and putty oscillate with an amplitude of 16 cm and a frequency of 0.38 Hz. Assuming that the box remains immobile, determine (*a*) the putty's speed before the collision, and (*b*) the spring constant.

(*a*) 1. Find velocity v_f of the total mass after collision — $v_f = A\omega = 0.16 \times 2\pi \times 0.38$ m/s = 0.382 m/s

2. Apply momentum conservation — $1.0 \times 0.382 = 0.6 \times v_i$; $v_i = 0.637$ m/s

(*b*) $(k/m) = \omega^2$ — $k = 1.0 \times (2\pi \times 0.38)^2 = 5.7$ N/m

---

93* ••• Figure 14-37 shows a vibrating mass–spring system supported on a frictionless surface and a second equal mass that is moving toward the vibrating mass with velocity v. The motion of the vibrating mass is given by $x(t) = (0.1\text{ m})\cos(40\text{ s}^{-1}\,t)$, where x is the displacement of the mass from its equilibrium position. The two masses collide elastically just as the vibrating mass passes through its equilibrium position traveling to the right. (*a*) What should be the velocity v of the second mass so that the mass–spring system is at rest following the elastic collision? (*b*) What is the velocity of the second mass after the elastic collision?

(*a*) 1. Apply conservation of energy and momentum — $Mv_{1i}^2 + Mv_{2i}^2 = Mv_{2f}^2$; $Mv_{1i} + Mv_{2i} = Mv_{2f}$

2. Since masses cancel we have — $(v_{2f} + v_{2i})(v_{2f} - v_{2i}) = v_{1i}^2$; $v_{2f} - v_{2i} = v_{1i}$

3. Solve for v_{1i} — $v_{1i} = 0$; the mass is at rest initially

(*b*) $v_{2f} = v_{1i}$; $v_{1i} = A\omega = 4$ m/s — $v_{2f} = 4$ m/s

---

94 ••• Following the elastic collision in Problem 93, the energy of the recoiling mass is 8.0 J. Find the masses m and the spring constant k.

1. $\frac{1}{2}Mv_{1i}^2 = E$ — $M = (8/8)$ kg = 1 kg

2. $k/m = \omega^2$ — $k = 1 \times 40^2$ N/m = 1600 N/m

---

95 ••• An object of mass 2 kg resting on a frictionless horizontal surface is attached to a spring of force constant 600 N/m. A second object of mass 1 kg slides along the surface toward the first object at 6 m/s. (*a*) Find the amplitude of oscillation if the objects make a perfectly inelastic collision and remain together on the spring. What is the period of oscillation? (*b*) Find the amplitude and period of oscillation if the collision is elastic. (*c*) For each type of collision, write an expression for the position x as a function of time t for the object attached to the spring, assuming that the collision occurs at time $t = 0$.

(*a*) 1. Use momentum conservation to find v_{max} — $v_{max} = 1 \times 6/3$ m/s = 2 m/s = $A\omega$

2. Determine $\omega = (k/M)^{1/2}$ — $\omega = (600/3)^{1/2}$ rad/s = 14.14 rad/s; $T = 2\pi/\omega = 0.444$ s

(*c*) Find A and write $x(t)$ — $A = 14.1$ cm; $x(t) = 14.1\sin(14.1t)$ cm

(*b*) 1. Let $m_1 = 2$ kg stationary mass; use Equ. 8-30	$v_{2f} = 4$ m/s $= A\omega$
2. Determine $\omega = (k/m_1)^{1/2}$ and T	$\omega = 300^{1/2}$ rad/s $= 17.32$ rad/s; $T = 2\pi/\omega = 0.363$ s
(*c*) Find A and write $x(t)$	$A = 4/17.32$ m $= 23.1$ cm; $x(t) = 23.1 \sin(17.3t)$ cm

---

96 • The effect of the mass of a spring on the motion of an object attached to it is usually neglected. Describe qualitatively its effect when it is not neglected.

The frequency of vibration will be reduced.

97* •• A lamp hanging from the ceiling of the club car in a train oscillates with period T_0 when the train is at rest. The period will be (match left and right columns)

1. greater than T_0 when	B. the train rounds a curve of radius R with speed v.
2. less than T_0 when	D. the train goes over a hill of radius of curvature R with constant speed.
3. equal to T_0 when	A. the train moves horizontally with constant velocity.
	C. the train climbs a hill of inclination θ at constant speed.

---

98 •• Two mass–spring systems oscillate at frequencies f_A and f_B. If $f_A = 2f_B$ and the spring constants of the two springs are equal, it follows that the masses are related by (*a*) $M_A = 4M_B$. (*b*) $M_A = M_B/\sqrt{2}$. (*c*) $M_A = M_B/2$. (*d*) $M_A = M_B/4$.

(*d*) $4 = M_B/M_A$.

---

99 •• Two mass–spring systems A and B oscillate so that their energies are equal. If $M_A = 2M_B$, then which formula below relates the amplitudes of oscillation? (*a*) $A_A = A_B/4$. (*b*) $A_A = A_B/\sqrt{2}$. (*c*) $A_A = A_B$. (*d*) Not enough information is given to determine the ratio of the amplitudes.

(*d*) It is necessary to know ω or k.

---

100 •• Two mass–spring systems A and B oscillate so that their energies are equal. If $k_A = 2k_B$, then which formula below relates the amplitudes of oscillation? (*a*) $A_A = A_B/4$. (*b*) $A_A = A_B/\sqrt{2}$. (*c*) $A_A = A_B$. (*d*) Not enough information is given to determine the ratio of the amplitudes.

(*b*) $2k_BA_A^2 = k_BA_B^2$.

101* •• Pendulum A has a bob of of mass M_A and a length L_A; pendulum B has a bob of mass M_B and a length L_B. If the period of A is twice that of B, then (*a*) $L_A = 2L_B$ and $M_A = 2M_B$. (*b*) $L_A = 4L_B$ and $M_A = M_B$. (*c*) $L_A = 4L_B$ whatever the ratio M_A/M_B. (*d*) $L_A = \sqrt{2}L_B$ whatever the ratio M_A/M_B.

(*c*) T is independent of M; $L \propto T^2$.

---

102 • A particle has a displacement $x = 0.4 \cos(3t + \pi/4)$, where x is in meters and t is in seconds. (*a*) Find the frequency f and period T of the motion. (*b*) Where is the particle at $t = 0$? (*c*) Where is the particle at $t = 0.5$ s?

(*a*) Compare with Equ. 14-4	$f = 3/2\pi$ Hz $= 0.477$ Hz; $T = 1/f = 2.09$ s
(*b*) At $t = 0$, $x = 0.4 \cos(\pi/4)$ m	$x(0) = 0.283$ m
(*c*) At $t = 0.5$ s, $x = 0.4 \cos(1.5 + \pi/4)$	$x(1.5) = 0.4 \cos(2.285) = -0.262$ m

103 • (a) Find an expression for the velocity of the particle whose position is given in Problem 102. (b) What is the velocity at time $t = 0$? (c) What is the maximum velocity? (d) At what time after $t = 0$ does this maximum velocity first occur?

(a) $v = -A\omega \sin(\omega t + \delta)$ — $v = -1.2 \sin(3t + \pi/4)$ m/s

(b) $v(0) = -A\omega \sin(\pi/4)$ — $v = -0.849$ m/s

(c) $v_{max} = A\omega$ — $v_{max} = 1.2$ m/s

(d) Find t for $3t + \pi/4 = 3\pi/2$ — $t = 5\pi/12$ s $= 1.31$ s

---

104 • An object on a horizontal spring oscillates with a period of 4.5 s. If the object is suspended from the spring vertically, by how much is the spring stretched from its natural length when the object is in equilibrium?

$m/k = (T/2\pi)^2$; $kx = mg$; $x = mg/k = (T/2\pi)^2 g$ — $x = 9.81(4.5/2\pi)^2$ m $= 5.03$ m

---

105* •• A small particle of mass m slides without friction in a spherical bowl of radius r. (a) Show that the motion of the particle is the same as if it were attached to a string of length r. (b) Figure 14-38 shows a particle of mass m_1 that is displaced a small distance s_1 from the bottom of the bowl, where s_1 is much smaller than r. A second particle of mass m_2 is dislaced in the opposite direction a distance $s_2 = 3s_1$, where s_2 is also much smaller than r. If the particles are released at the same time, where do they meet? Explain.

(a) Since there is no friction, the only forces acting on the particle are mg and the normal force acting radially inward; the normal force is identical to the tension in a string of length r that keeps the particle moving in a circular path.

(b) The particles meet at the bottom. Since s_1 and s_2 are both much smaller than r, the particles behave like the bobs of simple pendulums of equal length and, therefore, have the same periods.

---

106 •• As your jet plane speeds down the runway on take-off, you measure its acceleration by suspending your yo-yo as a simple pendulum and noting that when the bob (mass 40 g) is at rest relative to you, the string (length 70 cm) makes an angle of 22° with the vertical. Find the period T for small oscillations of this pendulum.

1. Find g_{eff} — $g_{eff} = g/\cos 22° = 10.6$ m/s^2

2. Find $T = 2\pi\sqrt{L/g_{eff}}$ — $T = 2\pi\sqrt{0.70/10.6} = 1.62$ m/s^2

---

107 •• Two identical blocks placed one on top of the other rest on a frictionless horizontal air track. The lower block is attached to a spring of spring constant $k = 600$ N/m. When displaced slightly from its equilibrium position, the system oscillates with a frequency of 1.8 Hz. When the amplitude of oscillation exceeds 5 cm, the upper block starts to slide relative to the lower one. (a) What are the masses of the two blocks? (b) What is the coefficient of static friction between the two blocks?

(a) $\omega^2 = k/M = k/2m = 4\pi^2 f^2$ — $m = k/8\pi^2 f^2 = 2.35$ kg; this is the mass of each block

(b) $\mu_s = a_{max}/g$; $a_{max} = A\omega^2 = 4\pi^2 Af^2$ — $\mu_s = 4\pi^2 Af^2/g = 4\pi^2 \times 0.05 \times 1.8^2/9.81 = 0.652$

---

108 •• Two atoms are bound together in a molecule. The potential energy U resulting from their interaction is shown in Figure 14-39. The variable r is the distance between the atom centers, and E_0 is the lowest (ground-state) energy. (a) As a result of a collision, the molecule acquires a kinetic energy of vibration whose maximum value is 1.0 eV. With this kinetic energy, over what range of separation distance will the molecule vibrate? (b) Give an approximate value for the force $F(r)$ between the two atoms at $r = 0.4$ nm. Express your answer in units appropriate to those used on the graph. (*c*) Calculate the force in (*b*) in newtons. Is this force atttractive or repulsive?

(*a*) Adding 1.0 eV to E_0, the range of r is between 0.28 nm and 0.35 nm.

(*b*), (*c*) $F = -dU/dr$; the slope at $r = 0.4$ nm $\approx$ (3 eV/0.1 nm) = 30 eV/nm = 4.8×10^{-9} J/m. $F \approx -4.8 \times 10^{-9}$ N. This force is attractive, pointing toward the origin.

109* •• A wooden cube with edge a and mass m floats in water with one of its faces parallel to the water surface. The density of the water is ρ. Find the period of oscillation in the vertical direction if it is pushed down slightly.

1. Find the change in the buoyant force — $dF_B = -\rho Vg = -a^2\rho gy$
2. Write the equation of motion — $m(d^2y/dt^2) = -a^2\rho gy$; $d^2y/dt^2 = -(a^2\rho g/m)y$
3. Compare with Equs. 14-2 and 14-7 — $\omega = a\sqrt{\rho g/m}$; $T = 2\pi/\omega = (2\pi/a)\sqrt{m/\rho g}$

---

110 •• A spider of mass 0.36 g sits in the middle of its horizontal web, which sags 3.00 mm under its weight. Estimate the frequency of vertical vibration for this system.

$kx = mg$; $k/m = g/x$. Neglecting the mass of the web, $f = \omega/2\pi = (1/2\pi)(9.81/3 \times 10^{-3})^{1/2}$ Hz = 9.1 Hz.

---

111 •• A clock with a pendulum keeps perfect time on the earth's surface. In which case will the error be greater: if the clock is placed in a mine of depth h or if the clock is elevated to a height h? Assume that $h << R_E$.

If placed in a mine, $g' = (GM_E/R_E^3)(R_E - h) = g(1 - h/R_E)$. If elevated, $g' = GM_E/(R_E + h)^2 = g/(1 + h/R_E)^2$. Use the binomial expansion: $g' \approx g(1 - 2h/R_E)$. Evidently, the error is greater if the clock is elevated.

---

112 •• Figure 14-40 shows a pendulum of length L with a bob of mass M. The bob is attached to a spring of spring constant k as shown. When the bob is directly below the pendulum support, the spring is at its equilibrium length. (*a*) Derive an expression for the period of this oscillating system for small amplitude vibrations. (*b*) Suppose that $M = 1$ kg and L is such that in the absence of the spring the period is 2.0 s. What is the spring constant k if the period of the oscillating system is 1.0 s?

(*a*) For small displacements, $ML(d^2\phi/dt^2) = -Mg\phi - kL\phi = -(Mg + kL)\phi = -\omega^2\phi$. Thus, $\omega = \sqrt{(g/L) + (k/M)}$ and $T = 2\pi/\sqrt{(g/L) + (k/M)}$.

(*b*) To reduce T by a factor of 2 increases ω by a factor of 2, increasing ω^2 by a factor of 4. Hence $k/M = 3g/L$. But $g/L = 4\pi^2/T^2 = \pi^2$. So $k = 3\pi^2 M = 3\pi^2$ N/m = 29.6 N/m.

113* •• An object of mass m_1 sliding on a frictionless horizontal surface is attached to a spring of force constant k and oscillates with an amplitude A. When the spring is at its greatest extension and the mass is instantaneously at rest, a second object of mass m_2 is placed on top of it. (*a*) What is the smallest value for the coefficient of static friction μ_s such that the second object does not slip on the first? (*b*) Explain how the total energy E, the amplitude A, the angular frequency ω, and the period T of the system are changed by placing m_2 on m_1, assuming that the friction is great enough so that there is no slippage.

(*a*) $\mu_{s,\min} = a_{\max}/g = \omega^2 A/g\mu_s = Ak/g(m_1 + m_2)$.

(*b*) A is unchanged; E is unchanged since $E = \frac{1}{2}kA^2$; ω is reduced by increasing the total mass; T is increased.

---

114 •• The acceleration due to gravity g varies with geographical location because of the earth's rotation and because the earth is not exactly spherical. This was first discovered in the seventeenth century, when it was noted that a pendulum clock carefully adjusted to keep correct time in Paris lost about 90 s per day near the equator. (*a*) Show that a small change in the acceleration of gravity Δg produces a small change in the period ΔT of a pendulum given by $\Delta T/T \approx -½(\Delta g/g)$. (Use differentials to approximate ΔT and Δg.) (*b*) How great a change in g is needed to account for a change in the period of 90 s per day?

(*a*) $T = 2\pi(L/g)^{½}$; so $dT/dg = -\pi L^{½}g^{-3/2} = -½T/g$. So $dT/T = -½\, dg/g$ and $\Delta T/T \approx -\Delta g/2g$.

(*b*) 90 s/d = 1.5/1440 = $1.04 \times 10^{-3} = \Delta T/T$; so $\Delta g = -2 \times 9.81 \times 1.04 \times 10^{-3}$ m/s² = 0.0204 m/s² = 2.04 cm/s².

---

115 •• Figure 14-41 shows two equal masses of 0.6 kg glued to each other and connected to a spring of spring constant k = 240 N/m. The masses, which rest on a frictionless horizontal surface, are displaced 0.6 m from their equilibrium position and released. Before being released, a few drops of solvent are deposited on the glue. (*a*) Find the frequency of vibration and total energy of the vibrating system before the glue has dissolved. (*b*) Find the frequency, amplitude of vibration, and energy of the vibrating system if the glue dissolves when the spring is (1) at maximum compression and (2) at maximum extension.

(*a*) 1. $f = (1/2\pi)\sqrt{k/2m}$, where m = 0.6 kg	$f = (1/2\pi)(240/1.2)^{½}$ Hz = 2.25 Hz
2. $E = ½kA^2$	$E = 240 \times 0.36/2$ J = 43.2 J
(*b*) 1. Now $f = f_1 = (1/2\pi)\sqrt{k/m}$	$f_1 = (1/2\pi) \times 20 = 3.18$ Hz
2. The second block detaches at the equilibrium point	$v_{max} = A\omega = 0.6 \times 2\pi \times 2.25$ m/s = 8.48 m/s
3. New amplitude, $A_1 = v_{max}/\omega_1$	$A_1 = 8.48/20$ m = 42.4 cm
4. $E_1 = ½kA_1^{\,2}$	$E_1 = 120 \times 0.424^2$ J = 21.6 J
(2) 1. Again f is increased as in case (1)	$f_2 = 3.18$ Hz
2. Now the second block is at rest, E and A are unchanged	$A_2 = 0.6$ m; $E_2 = 43.2$ J

---

116 •• Show that for the situations in Figures 14-42*a* and *b*, the object oscillates with a frequncy $f = [1/(2\pi)]\sqrt{k_{eff}/m}$, where k_{eff} is given by (*a*) $k_{eff} = k_1 + k_2$ and (*b*) $1/k_{eff} = 1/k_1 + 1/k_2$. (*Hint*: Find the net force F on the object for a small displacement x and write $F = -k_{eff}x$. Note that in (*b*) the springs stretch by different amounts, the sum of which is x.)

(*a*) Find the force for the displacement x	$F = -k_1x - k_2x = -(k_1 + k_2)x$; $k_{eff} = k_1 + k_2$
(*b*) 1. The same force acts on each spring	$F = -k_1x_1 = -k_2x_2$; $x_2 = x_1(k_1/k_2)$
2. The total extension = $x_1 + x_2 = x = -F/k_{eff}$	$k_{eff} = -F/(x_1 + x_2) = k_1/(1 + k_1/k_2) = k_1k_2/(k_1 + k_2)$
3. Write $1/k_{eff}$	$1/k_{eff} = 1/k_1 + 1/k_2$

117* •• A small block of mass m_1 rests on a piston that is vibrating vertically with simple harmonic motion given by $y = A \sin \omega t$. (*a*) Show that the block will leave the piston if $\omega^2 A > g$. (*b*) If $\omega^2 A = 3g$ and A = 15 cm, at what time will the block leave the piston?

(*a*) At maximum upward extension, the block is momentarily at rest. Its downward acceleration is g. The downward acceleration of the piston is $\omega^2 A$. Therefore, if $\omega^2 A > g$, the block will separate from the piston.

(*b*) 1. $y = A \sin(\omega t)$; find a and critical ωt	$a = -\omega^2 A \sin(\omega t) = -3g \sin(\omega t) = -g$; $\omega t = 0.34$ rad
2. $\omega = (3g/0.15)^{½}$; find t	$t = 0.34/14 = 0.0243$ s

---

118 •• The plunger of a pinball machine has mass m_p and is attached to a spring of force constant k (Figure 14-43). The spring is compressed a distance x_0 from its equilibrium position $x = 0$ and released. A ball of mass m_b is next to the plunger. (*a*) Where does the ball leave the plunger? (*b*) What is the speed v_s of the ball when it separates?

(*c*) At what distance x_f does the plunger come to rest momentarily? (Assume that the surface is horizontal and frictionless so that the ball slides rather than rolls.)

(*a*) The ball leaves the plunger at $x = 0$. Thereafter, a of plunger is negative.

(*b*) $v_s = A\omega = x_0[k/(m_p + m_b)]^{1/2}$.

(*c*) $\frac{1}{2}kx_f^2 = \frac{1}{2}m_p v_s^2 = \frac{1}{2}m_p x_0^2 k/(m_p + m_b)$; $x_f = x_0[m_p/(m_p + m_b)]^{1/2}$.

---

119 •• A level platform vibrates horizontally with simple harmonic motion with a period of 0.8 s. (*a*) A box on the platform starts to slide when the amplitude of vibration reaches 40 cm; what is the coefficient of static friction between the body and the platform? (*b*) If the coefficient of friction between the box and platform were 0.40, what would be the maximum amplitude of vibration before the box would slip?

(*a*) $\mu_s = a_{max}/g$; $a_{max} = A\omega^2$; $\mu_s = A\omega^2/g$ $\qquad$ $\mu_s = 0.4 \times 4\pi^2/0.64 \times 9.81 = 2.52$

(*b*) $A_{max} = \mu_s g/\omega^2$ $\qquad$ $A_{max} = 0.4 \times 9.81 \times 0.64/4\pi^2$ m $= 6.36$ cm

---

120 ••• The potential energy of a mass m as a function of position is given by $U(x) = U_0(\alpha + 1/\alpha)$, where $\alpha = x/a$ and a is a constant. (*a*) Plot $U(x)$ versus x for $0.1a < x < 3a$. (*b*) Find the value of $x = x_0$ at stable equilibrium. (*c*) Write the potential energy $U(x)$ for $x = x_0 + \varepsilon$, where ε is a small displacement from the equilibrium position x_0. (*d*) Approximate the $1/x$ term using the binomial expansion

$$(1 + r)^n = 1 + nr + \frac{n(n-1)}{2 \times 1}r^2 + \frac{n(n-1)(n-2)}{3 \times 2 \times 1}r^3 + \ldots,$$

with $r = \varepsilon/x_0 << 1$ and discarding all terms of power greater than r^2. (*e*) Compare your result with the potential for a simple harmonic oscillator. Show that the mass will undergo simple harmonic motion for small displacements from equilibrium and determine the frequency of this motion.

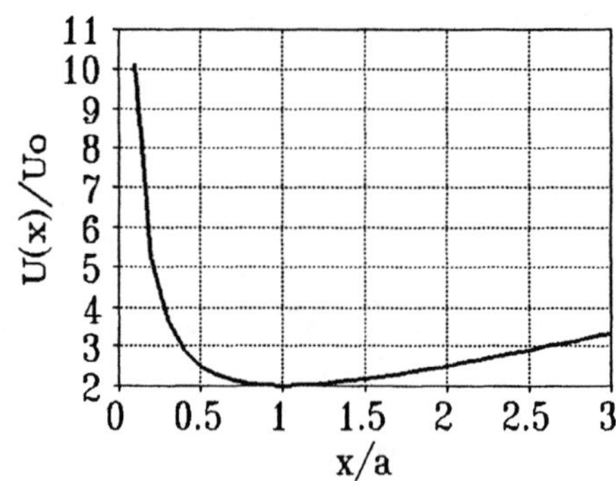

(*a*) A plot of $U(x)$ versus $\alpha = x/a$ is shown.

(*b*) $F = 0 = dU/dx = (dU/d\alpha)(d\alpha/dx) = (U_0/a)(1 - \alpha^2)$; $\alpha_0 = 1$; $x_0 = a$

(*c*) $U(x_0 + \varepsilon) = U_0[1 + \beta + (1 + \beta)^{-1}]$, where $\beta = \varepsilon/a$.

(*d*) $U(x_0 + \varepsilon) \approx U_0(1 + \beta + 1 - \beta + \beta^2) = \text{constant} + U_0\varepsilon^2/a^2$

(*e*) From the plot we see that U is a minimum at $x = x_0$. For SHO, $U = \text{constant} + \frac{1}{2}k\varepsilon^2$, so $k = 2U_0/a^2$. The frequency is $f = (1/2\pi)(k/m)^{1/2} = (1/2\pi a)(2U_0/m)^{1/2}$

121* ••• Repeat Problem 120 with $U(x) = U_0(\alpha^2 + 1/\alpha^2)$.

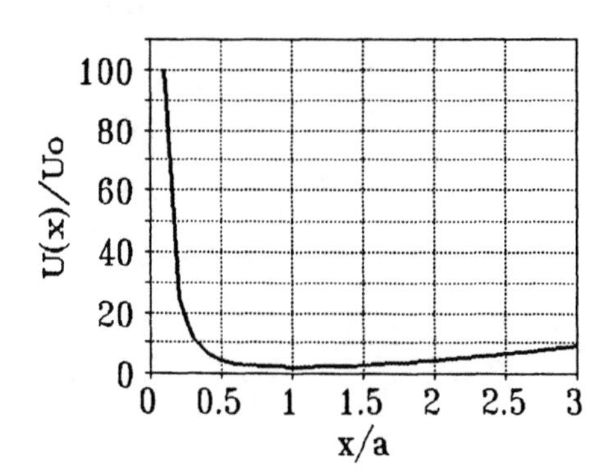

(*a*) A plot of $U(x)$ versus x/a is shown on the next page.

(*b*) $dU/dx = 0 = (2U_0/a)(\alpha - 1/\alpha^3)$; $\alpha_0 = 1$, $x_0 = a$.

(*c*) $U(x_0 + \varepsilon) = U_0[(1 + \beta)^2 + (1 + \beta)^{-2}]$; $\beta = \varepsilon/a$

(*d*) $U(x_0 + \varepsilon) \approx U_0(1 + 2\beta + \beta^2 + 1 - 2\beta + 3\beta^2) = \text{constant} + 4U_0\beta^2$;
$U(x_0 + \varepsilon) = \text{constant} + 4U_0\varepsilon^2/a^2$

(*e*) See Problem 120; $k = 8U_0/a^2$; $f = (1/\pi a)(2U_0/m)^{1/2}$.

122 ••• A solid cylindrical drum of mass 6.0 kg and diameter 0.06 m rolls without slipping on a horizontal surface (Figure 14-44). The axle of the drum is attached to a spring of spring constant $k = 4000$ N/m as shown. (*a*) Determine the frequency of oscillation of this system for small displacements from equilibrium. (*b*) What is the minimum value of the coefficient of static friction such that the drum will not slip when the vibrational energy is 5.0 J?

(*a*), (*b*) We shall first do this problem for the general case and then substitute numerical values.
Find K_{max} and set it equal to $\frac{1}{2}kA^2$. $K = \frac{1}{2}I\omega^2 + \frac{1}{2}Mv^2$, where $\omega = R/v$; $I = \frac{1}{2}MR^2$. $K_{max} = 3Mv_{max}^2/4$. We now replace v_{max}^2 by $A^2\omega^2$ (here ω is the angular frequency of the oscillator). So, $\frac{1}{2}kA^2 = 3MA^2\omega^2/4$ and $\omega = (2k/3M)^{1/2}$. The frequency is $\omega/2\pi$. To avoid slipping, $kA \le \mu_s Mg$. So, critical $\mu_s = kA/Mg$. But $kA = (2Ek)^{1/2}$, so $\mu_s = (2Ek)^{1/2}/Mg$.

(*a*) $f = (1/2\pi)(2k/3M)^{1/2}$ $\qquad$ $f = (1/2\pi)(8000/18)^{1/2}$ Hz $= 3.36$ Hz

(*b*) $\mu_s = (2Ek)^{1/2}/Mg$ $\qquad$ $\mu_s = (40{,}000)^{1/2}/6 \times 9.81 = 3.4$

123 ••• Figure 14-45 shows a solid half-cylinder of mass M and radius R resting on a horizontal surface. If one side of this cylinder is pushed down slightly and then released, the object will oscillate about its equilibrium position. Determine the period of this oscillation.

The system acts like a physical pendulum. We need to determine its moment of inertia about the pivot point and the distance from the pivot point to the center of mass. The CM of a semicircular disk was found to be $4R/3\pi$ above the base of the diameter of the disk (see Problem 8-10). This distance D in Equ. 14-31 is given by $D = R(1 - 4/3\pi)$.

To obtain I about the pivot point we proceed as follows. The moment of inertia about the center of the diameter of a complete disk is $\frac{1}{2}(2M)R^2$, where M is the mass of the half-cylinder. The moment of inertia of the half-cylinder about that point is therefore $\frac{1}{2}[\frac{1}{2}(2M)R^2] = \frac{1}{2}MR^2$. We can now find I_{cm} using the parallel axis theorem. We obtain $I_{cm} = \frac{1}{2}MR^2 - (16/9\pi^2)MR^2$. We apply the parallel axis theorem once more to find I_p, the moment of inertia about the pivot point. $I_p = I_{cm} + MD^2 = MR^2[(3/2) - (8/3\pi)]$. Substituting these results into Equ. 14-31 one now finds

$$T = 2\pi\sqrt{\frac{R}{g}\left(\frac{9\pi/8 - 1}{6\pi/8 - 1}\right)} = 8.59\sqrt{R/g}.$$

124 ••• Repeat Problem 123 replacing the half-cylinder with a half-sphere.

The CM of the hemisphere was found in Problem 8-12 to be $3R/8$ above the center. Thus $D = 5R/8$. The moment of inertia of the hemisphere about the center is obtained using the same argument as in Problem 123, namely $2MR^2/5$. Then $I_{cm} = 2MR^2/5 - 9MR^2/64$, and $I_p = I_{cm} + MD^2 = 9MR^2/20$. We find $T = 2\pi(0.72R/g)^{1/2}$.

125* ••• A straight tunnel is dug through the earth as shown in Figure 14-46. Assume that the walls of the tunnel are frictionless. (*a*) The gravitational force exerted by the earth on a particle of mass m at a distance r from the center of the earth when $r < R_E$ is $F_r = -(GmM_E/R_E^3)r$, where M_E is the mass of the earth and R_E is its radius. Show that the net force on a particle of mass m at a distance x from the middle of the tunnel is given by $F_x = -(GmM_E/R_E^3)x$, and that the motion of the particle is therefore simple harmonic motion. (*b*) Show that the period of the motion is given by $T = 2\pi\sqrt{R_E/g}$ and find its value in minutes. (This is the same period as that of a satellite orbiting near the surface of the earth and is independent of the length of the tunnel.)

(*a*) From Figure 14-46, $F_x = F_r \sin\theta$, $\sin\theta = x/R$	$F_x = -(GmM_E/R_E^3)x$
(*b*) 1. Here $k_{eff} = (GmM_E/R_E^3)$; $T = 2\pi(m/k_{eff})^{1/2}$	$T = 2\pi(R_E^3/GM_E)^{1/2} = 2\pi(R_E/g)$
2. Substitute appropriate numerical values	$T = 2\pi(6.37\times10^6/9.81)^{1/2}$ s $= 5.06\times10^3$ s $= 84.4$ min

---

126 ••• A damped oscillator has a frequency ω' that is 10% less than its undamped frequency. (*a*) By what factor is the amplitude of the oscillator decreased during each oscillation? (*b*) By what factor is its energy reduced during each oscillation?

(*a*) Use Equ. 14-35	$1 - (b/2m\omega_0)^2 = 0.81$; $b/2m\omega_0 = 0.435$
Use Equ. 14-36; $t = T = 2\pi/\omega'$	$A = A_0 \exp(-0.436\omega_0 \times 2\pi/0.9\omega_0) = 0.048A_0$
(*b*) $E \propto A^2$	$E = E_0(0.048^2) = 0.0023E_0$

---

127 ••• Show by direct substitution that Equation 14-48 is a solution of Equation 14-47.

Equation 14-47 reads

$$m\frac{d^2x}{dt^2} + b\frac{dx}{dt} + m\omega_0^2 x = F_0 \cos\omega t$$

and its proposed solution is $x = A\cos(\omega t - \delta)$. We begin by obtaining expressions for the first and second derivatives of $x(t)$.

$$\frac{dx}{dt} = -A\omega\sin(\omega t - \delta);\quad \frac{d^2x}{dt^2} = -A\omega^2\cos(\omega t - \delta).$$

Substitution into the differential equation gives

$$-mA\omega^2\cos(\omega t - \delta) - bA\omega\sin(\omega t - \delta) + mA\omega_0^2\cos(\omega t - \delta) = F_0\cos(\omega t).$$

We now use the trigonometric identities $\cos(\alpha - \beta) = \cos\alpha\cos\beta + \sin\alpha\sin\beta$ and $\sin(\alpha - \beta) = \sin\alpha\cos\beta - \cos\alpha\sin\beta$. The equation now reads

$$-mA(\omega^2 - \omega_0^2)(\cos\omega t\cos\delta + \sin\omega t\sin\delta) - bA\omega(\sin\omega t\cos\delta - \cos\omega t\sin\delta) = F_0\cos\omega t.$$

A bit of algebra now yields

$$mA(\omega_0^2 - \omega^2)(\cos\omega t\cos\delta)(1 + \tan\omega t\tan\delta) - bA\omega(\sin\omega t\cos\delta)(1 - \tan\delta/\tan\omega t) = F_0\cos\omega t$$

Divide both sides by $m(\omega_0^2 - \omega^2)$ and use Equation 14-50 for $\tan\delta$. The above expression now reduces to

$$A(\cos\omega t\cos\delta)(1 + \tan^2\delta) = \frac{F_0\cos\delta}{m(\omega_0^2 - \omega^2)};$$

we now employ the trigonometric identity $1 + \tan^2\delta = 1/\cos^2\delta$. This results in

$$A\cos\omega t = \frac{F_0\cos\delta\cos\omega t}{m(\omega_0^2 - \omega^2)}$$

which is evidently a solution to the differential equation for all values of the variable t, provided

$$A = \frac{F_0\cos\delta}{m(\omega_0^2 - \omega^2)},$$ which reduces to Equ. 14-49.

128 ••• A block of mass m on a horizontal table is attached to a spring of force constant k as shown in Figure 14-47. The coefficient of kinetic friction between the block and the table is μ_k. The spring is stretched a distance A and released. (*a*) Apply Newton's second law to the block to obtain an equation for its acceleration d^2x/dt^2 for the first half-cycle, during which the block is moving to the left. Show that the resulting equation can be written $d^2x'/dt^2 = -\omega^2x'$, where $x = 0$ at the equilibrium position of the spring, and $x' = x - x_0$, with $x_0 = \mu_k mg/k = \mu_k g/\omega^2$. (*b*) Repeat part (*a*) for the second half-cycle as the block moves to the right, and show that $d^2x''/dt^2 = -\omega^2x''$, where $x'' = x + x_0$ and x_0 has the same value. (*c*) Sketch $x(t)$ for the first few cycles for $A = 10x_0$.

(*a*) Let $x = 0$ at equilibrium position. Then $d^2x/dt^2 = -(k/m)x + \mu_k g = -\omega^2x'$, where $x' = x - \mu_k g/\omega^2$; but since $\mu_k g/\omega^2$ is a constant, $d^2x/dt^2 = d^2x'/dt^2$. So $d^2x'/dt^2 = -\omega^2x'$.

(*b*) Now both the spring force and friction are reversed. Thus, $d^2x/dt^2 = -(k/m)x - \mu_k g = -\omega^2(x + x_0)$. Let $x + x_0 = x''$, then $d^2x''/dt^2 = -\omega^2x''$.

(*c*) We have $x' = A' \cos \omega t = x(t) - x_0$. At $t = 0$, $x = 10x_0 = x_0 + A'$; $A' = 9x_0$ and $x(t) = x_0(1 + 9 \cos \omega t)$. Now, do the same for the second half-cycle and obtain $x(t) = -x_0(1 + 7 \cos \omega t')$ where $t' = t - T/2$. Proceeding, one gets the plot shown.

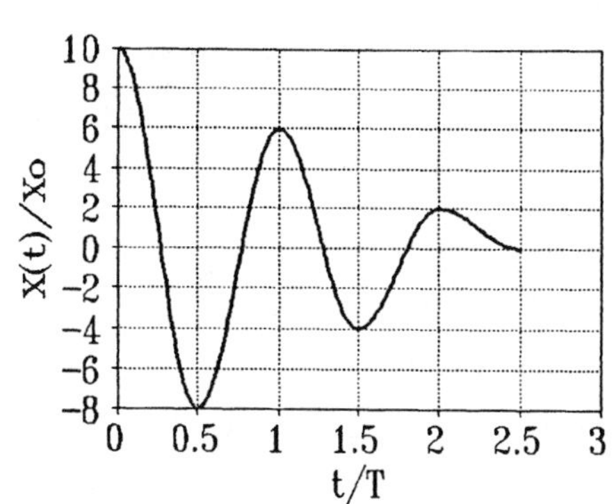

129* ••• In this problem, you will derive the expression for the average power delivered by a driving force to a driven oscillator (Figure 14-25). (*a*) Show that the instantaneous power input of the driving force is given by $P = Fv = -A\omega F_0 \cos \omega t \sin(\omega t - \delta)$. (*b*) Use the trigonometric identity $\sin(\theta_1 - \theta_2) = \sin \theta_1 \cos \theta_2 - \cos \theta_1 \sin \theta_2$ to show that the equation in (*a*) can be written $P = A\omega F_0 \sin \delta \cos^2 \omega t - A\omega F_0 \cos \delta \cos \omega t \sin \omega t$. (*c*) Show that the average value of the second term in your result for (*b*) over one or more periods is zero and that therefore $P_{av} = ½A\omega F_0 \sin \delta$. (*d*) From Equation 14-50 for $\tan \delta$, construct a right triangle in which the side opposite the angle δ is $b\omega$ and the side adjacent is $m(\omega_0^2 - \omega^2)$, and use this triangle to show that

$$\sin \delta = \frac{b\omega}{\sqrt{m^2(\omega_0^2 - \omega^2)^2 + b^2\omega^2}} = \frac{b\omega A}{F_0}.$$

(*e*) Use your result for (*d*) to eliminate ωA so that the average power input can be written

$$P_{av} = \frac{1}{2}\frac{F_0^2}{b}\sin^2 \delta = \frac{1}{2}\left[\frac{b^2\omega^2F_0^2}{m^2(\omega_0^2 - \omega^2)^2 + b^2\omega^2}\right] \quad (14\text{-}51).$$

(*a*) $F = F_0 \cos \omega t$; $x(t) = A \cos(\omega t - \delta)$. So $v(t) = dx/dt = -\omega A \sin(\omega t - \delta)$. $P = Fv = -A\omega F_0 \cos(\omega t) \sin(\omega t - \delta)$.

(*b*) Perform the appropriate substitution which gives the expression quoted in the problem statement.

(*c*) $\int \cos \theta \sin \theta \, d\theta = ½\sin^2 \theta$; for the limits $\theta = 0$ and $\theta = 2\pi$, this gives zero.

$\cos^2 \theta = ½[1 + \cos(2\theta)]$. The average of $\cos(2\theta) = 0$ over a complete cycle, so $\langle\cos^2 \theta\rangle = ½$ and $P_{av} = ½A\omega F_0 \sin \delta$

(d) Note that the hypotenuse of this triangle is the expression in the denominator, which gives the first equation. We can use Equ. 14-49 to reduce this result to the simpler form $\sin\delta = bA\omega/F_0$.

(e) From the above, $A\omega = (F_0 \sin\delta)/b$. Thus $P_{av} = \frac{1}{2}F_0^2 \sin^2\delta/b = \frac{1}{2}\{(bF_0^2\omega^2)/[m^2(\omega_0^2 - \omega^2)^2 + b^2\omega^2]\}$.

130 ••• In this problem, you are to use the result of Problem 129 to derive Equation 14-45, which relates the width of the resonance curve to the Q value when the resonance is sharp. At resonance, the denominator of the fraction in brackets in Equation 14-51 is $b^2\omega_0^2$ and P_{av} has its maximum value. For a sharp resonance, the variation in ω in the numerator in Equation 14-51 can be neglected. Then the power input will be half its maximum value at the values of ω, for which the denominator is $2b^2\omega_0^2$. (a) Show that ω then satisfies $m^2(\omega - \omega_0)^2(\omega + \omega_0)^2 = b^2\omega_0^2$. ($b$) Using the approximation $\omega + \omega_0 \approx 2\omega_0$, show that $\omega - \omega_0 \approx \pm(b/2m)$. ($c$) Express b in terms of Q. (d) Combine the results of (b) and (c) to show that there are two values of ω for which the power input is half that at resonance, and that they are given by $\omega_1 = \omega_0 - \omega_0/2Q$ and $\omega_2 = \omega_0 + \omega_0/2Q$. Therefore, $\omega_2 - \omega_1 = \Delta\omega = \omega_0/Q$, which is equivalent to Equation 14-45.

(a) Set $m^2(\omega_0^2 - \omega^2)^2 + b^2\omega^2 = 2b^2\omega_0^2$; so $m^2(\omega_0^2 - \omega^2)^2 \approx b^2\omega_0^2$, if the resonance is sharp. Note that $\omega_0^2 - \omega^2 = (\omega_0 + \omega)(\omega_0 - \omega)$, so $m^2(\omega_0^2 - \omega^2)^2 = m^2(\omega_0 - \omega)^2(\omega_0 + \omega)^2$.

(b) From (a) and using $\omega_0 + \omega = 2\omega_0$, $m^2(\omega_0^2 - \omega^2)^2 = m^2(\omega_0 - \omega)^2(4\omega_0^2) = b^2\omega_0^2$ which gives $\omega_0 - \omega = \pm b/2m$.

(c) $Q = \omega_0 m/b$; $b = \omega_0 m/Q$.

(d) $\delta\omega = \pm\omega_0/2Q$, where $\delta\omega = |\omega_0 - \omega_{1/2}|$ and $\omega_{1/2}$ is the value of ω at half the maximum. The width of the resonance is twice $\delta\omega$, i.e., $\Delta\omega = \omega_0/Q$.

CHAPTER 15

Wave Motion

1* • A rope hangs vertically from the ceiling. Do waves on the rope move faster, slower, or at the same speed as they move from bottom to top? Explain.

They move faster as they move up because the tension increases due to the weight of the rope below.

---

2 • (*a*) The bulk modulus for water is 2.0×10^9 N/m^2. Use it to find the speed of sound in water. (*b*)The speed of sound in mercury is 1410 m/s. What is the bulk modulus for mercury ($\rho = 13.6 \times 10^3$ kg/m^3)?

(*a*) Use Equ. 15-4 $\quad v = \sqrt{2 \times 10^9/10^3}$ m/s $= 1.41 \times 10^3$ m/s

(*b*) From Equ. 15-4, $B = v^2\rho$ $\quad B = (1410^2 \times 1.36 \times 10^3)$ N/m$^2 = 2.70 \times 10^{10}$ N/m^2

---

3 • Calculate the speed of sound waves in hydrogen gas at $T = 300$ K. (Take $M = 2$ g/mol and $\gamma = 1.4$.)

Use Equ. 15-5; $M = 2 \times 10^{-3}$ kg/mol $\quad v = \sqrt{1.4 \times 8.314 \times 300/2 \times 10^{-3}}$ m/s $= 1320$ m/s

---

4 • A steel wire 7 m long has a mass of 100 g. It is under a tension of 900 N. What is the speed of a transverse wave pulse on this wire?

Use Equ. 15-3 $\quad v = \sqrt{\dfrac{900}{0.1/7.0}}$ m/s $= 251$ m/s

---

5* • Transverse waves travel at 150 m/s on a wire of length 80 cm that is under a tension of 550 N. What is the mass of the wire?

From Equ. 15-3, $\mu = F/v^2$; $m = \mu L = FL/v^2$ $\quad m = (550 \times 0.8/150^2)$ kg $= 0.0196$ kg $= 19.6$ g

---

6 • A wave pulse propagates along a wire in the positive x direction at 20 m/s. What will the pulse velocity be if we (*a*) double the length of the wire but keep the tension and mass per unit length constant? (*b*) double the tension while holding the length and mass per unit length constant? (*c*) double the mass per unit length while holding the other variables constant?

See Equ. 15-3. (*a*) 20 m/s. (*b*) $20\sqrt{2}$ m/s $= 28.8$ m/s. (*c*) $20/\sqrt{2}$ m/s $= 14.1$ m/s.

---

7 • A steel piano wire is 0.7 m long and has a mass of 5 g. It is stretched with a tension of 500 N. (*a*) What is the speed of transverse waves on the wire? (*b*) To reduce the wave speed by a factor of 2 without changing the tension, what mass of copper wire would have to be wrapped around the steel wire?

(*a*) Use Equ. 15-3 $v = \sqrt{\dfrac{500}{0.005/0.7}}$ m/s = 265 m/s

(*b*) From Equ. 15-3, $m_f = 4m_i$ $\Delta m = 3m_i = 15$ g

---

8 • The cable of a ski lift runs 400 m up a mountain and has a mass of 80 kg. When the cable is struck with a transverse blow at one end, the return pulse is detected 12 s later. (*a*) What is the speed of the wave? (*b*) What is the tension in the cable?

$v = \Delta x/\Delta t$ $v = 800/12$ m/s = 66.7 m/s

From Equ. 15-3, $F = v^2m/L$ $F = (66.7^2 \times 80/400)$ N = 889 N

9* •• A common method for estimating the distance to a lightning flash is to begin counting when the flash is observed and continue until the thunder clap is heard. The number of seconds counted is then divided by 3 to get the distance in kilometers. (*a*) What is the velocity of sound in kilometers per second? (*b*) How accurate is this procedure? (*c*) Is a correction for the time it takes for the light to reach you important? (The speed of light is 3×10^8 m/s.)

(*a*) $v = 340$ m/s = 0.340 km/s.

(*b*) $s = vt = 0.340t \approx t/3 = 0.333t$; error = 7/340 = 2%.

(*c*) No; e.g., if $t = 3$ s, $s \approx 1$ km, and the time required for light to travel 1 km is only 10^{-5} s.

---

10 •• A method for measuring the speed of sound using an ordinary watch with a second hand is to stand some distance from a large flat wall and clap your hands rhythmically in such a way that the echo from the wall is heard halfway between every two claps. (*a*) Show that the speed of sound is given by $v = 4LN$, where L is the distance to the wall and N is the number of claps per second. (*b*) What is a reasonable value for L for this experiment to be feasible? (If you have access to a flat wall outdoors somewhere, try this method and compare your result with the standard value for the speed of sound.)

(*a*) Time of travel = $\Delta t = 2L/v = \frac{1}{2}(1/N)$. So $v = 4LN$.

(*b*) Assume $N = 1$ clap/s; then $L = 340/4$ m = 85 m. So a distance of about 80 m is appropriate.

---

11 •• A man drops a stone from a high bridge and hears it strike the water below exactly 4 s later. (*a*) Estimate the distance to the water based on the assumption that the travel time for the sound to reach the man is negligible. (*b*) Improve your estimate by using your result from part (*a*) for the distance to the water to estimate the time it takes for sound to travel this distance and then calculate the distance the rock falls in 4 s minus this time. (*c*) Calculate the exact distance and compare your result with your previous estimates.

(*a*) $d = \frac{1}{2}gt^2$ $d = \frac{1}{2} \times 9.81 \times 16$ m = 78.5 m

(*b*) 1. $\Delta t = d/v_s$ $\Delta t = 78.5/340$ s = 0.23 s

2. $d' = \frac{1}{2}g(t - \Delta t)^2$ $d' = \frac{1}{2} \times 9.81 \times 3.77^2$ m = 69.7 m

(*c*) $t = \sqrt{2d/g} + d/v_s$; $2d/g = t^2 - 2dt/v_s + d^2/v_s^2$ For $t = 4$ s, $v_s = 340$ m/s, $d = 70.5$ m

---

12 •• (*a*) Compute the derivative of the speed of a wave on a string with respect to the tension dv/dF, and show that the differentials dv and dF obey $dv/v = \frac{1}{2}dF/F$. (*b*) A wave moves with a speed of 300 m/s on a wire that is under a tension of 500 N. Using dF to approximate a change in tension, determine how much the tension must be changed to increase the speed to 312 m/s.

(*a*) $\dfrac{dv}{dF} = \dfrac{d}{dF}\sqrt{\dfrac{F}{\mu}} = \dfrac{1}{2}\sqrt{\dfrac{1}{F\mu}} = \dfrac{1}{2}\dfrac{v}{F};\ \dfrac{dv}{v} = \dfrac{1}{2}\dfrac{dF}{F}.$

(*b*) $\Delta F = 2F(\Delta v/v) = 2 \times 500 \times 12/300\ \text{N} = 40\ \text{N}.$

13* •• (*a*) Compute the derivative of the velocity of sound with respect to the absolute temperature, and show that the differentials dv and dT obey $dv/v = \frac{1}{2}dT/T$. (*b*) Use this result to compute the percentage change in the velocity of sound when the temperature changes from 0 to 27°C. (*c*) If the speed of sound is 331 m/s at 0°C, what is it (approximately) at 27°C? How does this approximation compare with the result of an exact calculation?

(*a*) Follow the same procedure as in Problem 12. Since $v \propto \sqrt{T}$, $dv/v = ½dT/T$.

(*b*) $\Delta T/T = 27/273$; $\Delta v/v = 4.95\%$.

(*c*) $v_{300} = v_{273}(1.0495) = 347$ m/s. Using the fact that $v \propto \sqrt{T}$ we obtain $v_{300} = 331\sqrt{300/273}$ m/s = 347 m/s.

---

14 ••• In this problem, you will derive a convenient formula for the speed of sound in air at temperature t in Celsius degrees. Begin by writing the temperature as $T = T_0 + \Delta T$, where $T_0 = 273$ K corresponds to 0°C and $\Delta T = t$, the Celsius temperature. The speed of sound is a function of T, $v(T)$. To a first-order approximation, you can write

$$v(T) \approx v(T_0) + (dv/dT)_{T_0}\Delta T$$

where $(dv/dT)_{T_0}$ is the derivative evaluated at $T = T_0$. Compute this derivative, and show that the result leads to

$$v = (331\ \text{m/s})\left(1 + \frac{t}{2T_0}\right) = (331 + 0.606t)\ \text{m/s}$$

Since t differs from T only by an additive constant, $dv/dt = ½(v/T) = ½[v/(t + 273)]$. So $v(t) = v(0°\ \text{C}) + \Delta v$, where $\Delta v = ½(331\ \text{m/s})[t/(t + 273)]$. For $t << 273$, one then obtains $v = (331 + 0.606t)$ m/s.

---

15 ••• While studying physics in her dorm room, a student is listening to a live radio broadcast of a baseball game. She is 1.6 km due south of the baseball field. Over her radio, the student hears a noise generated by the electromagnetic pulse of a lightning bolt. Two seconds later, she hears over the radio the thunder picked up by the microphone at the baseball field. Four seconds after she hears the noise of the electromagnetic pulse over the radio, thunder rattles her windows. Where, relative to the ballpark, did the lightning bolt occur?

The locations of the lighning strike (L), dorm room (R), and baseball park (P) are indicated in the diagram. We can neglect the time required for the electromagnetic pulse to reach the source of the radio transmission, which is the ball park (see Problem 9).

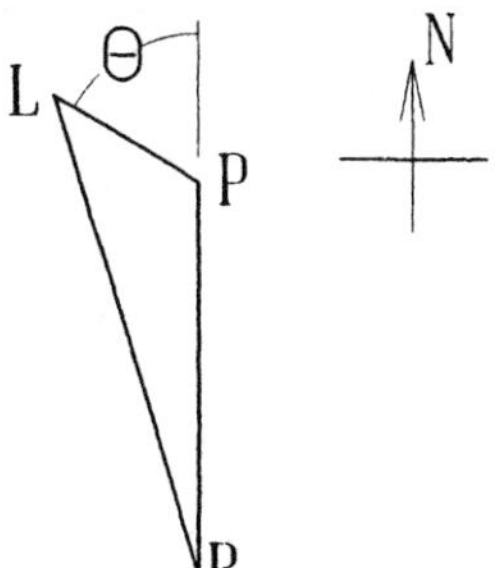

1. From the data $d_{LP} = 2 \times 340\ \text{m} = 680\ \text{m}$
2. We know that $d_{RP} = 1600$ m
3. Also, $d_{LR} = 6 \times 340\ \text{m} = 2040\ \text{m}$

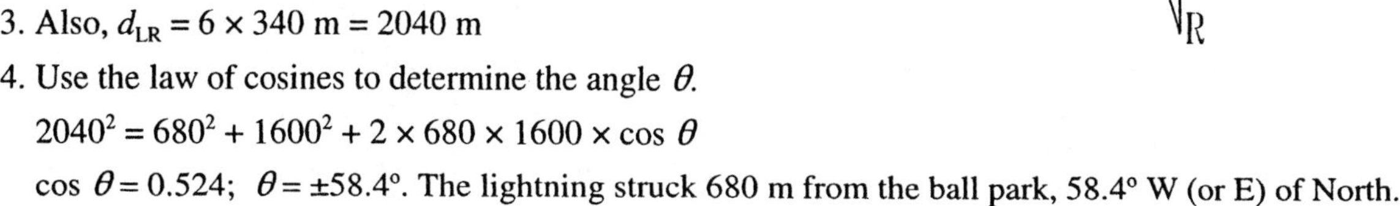

4. Use the law of cosines to determine the angle θ.

$2040^2 = 680^2 + 1600^2 + 2 \times 680 \times 1600 \times \cos\theta$

$\cos\theta = 0.524$; $\theta = \pm 58.4°$. The lightning struck 680 m from the ball park, 58.4° W (or E) of North.

16 ••• A coiled spring, such as a Slinky, is stretched to a length L. It has a force constant k and a mass m. (*a*) Show that the velocity of longitudinal compression waves along the spring is given by $v = L\sqrt{k/m}$ (*b*) Show that this is also the velocity of transverse waves along the spring if the natural length of the spring is much less than L.

(*a*) For longitudinal waves, $v = \sqrt{B/\rho}$; for the slinky, $\rho = m/V$ and $B = -P/(\Delta V/V)$. Let the cross-sectional area of the slinky be A. Then $\rho = m/AL$, $P = -k(\Delta L/A)$, $\Delta V = A\Delta L$. Thus one obtains the result given above.

(*b*) For transverse waves, we use $v = \sqrt{F/\mu}$. Now $\mu = m/L$ and $F = k\Delta L \approx L$ if $\Delta L >> L_0$. Then $v = L\sqrt{k/m}$.

17* • Show explicitly that the following functions satisfy the wave equation: (*a*) $y(x,t) = k(x + vt)^3$; (*b*) $y(x,t) = Ae^{ik(x - vt)}$, where A and k are constants and $i = \sqrt{-1}$; and (*c*) $y(x,t) = \ln k(x - vt)$.

(*a*) $\partial y/\partial x = 3k(x + vt)^2$; $\partial^2 y/\partial x^2 = 6k(x + vt)$; $\partial y/\partial t = 3kv(x + vt)^2$; $\partial^2 y/\partial t^2 = 6kv^2(x + vt)$. $v^2(\partial^2 y/\partial x^2) = \partial^2 y/\partial t^2$.

(*b*) $\partial y/\partial x = ikAe^{ik(x - vt)}$; $\partial^2 y/\partial x^2 = -k^2Ae^{ik(x - vt)}$; $\partial y/\partial t = -ikvAe^{ik(x - vt)}$; $\partial^2 y/\partial t^2 = -k^2v^2Ae^{ik(x - vt)} = v^2(\partial^2 y/\partial x^2)$.

(*c*) $\partial y/\partial x = 1/(x - vt)$; $\partial^2 y/\partial x^2 = -1/(x - vt)^2$; $\partial y/\partial t = -v/(x - vt)$; $\partial^2 y/\partial t^2 = -v^2/(x - vt)^2 = v^2(\partial^2 y/\partial x^2)$.

---

18 • Show that the function $y = A \sin kx \cos \omega t$ satisfies the wave equation.

$\partial^2 y/\partial x^2 = -Ak^2 \sin kx \cos \omega t$; $\partial^2 y/\partial t^2 = -A\omega^2 \sin kx \cos \omega t$. $\partial^2 y/\partial x^2 = (1/v^2)\partial^2 y/\partial t^2$ if $v = \omega/k$.

---

19 ••• Consider the following equation:

$$\frac{\partial^2 y}{\partial x^2} + i\alpha\frac{\partial y}{\partial t} = 0, \quad i = \sqrt{-1}$$

where α is a constant. Show that $y(x, t) = A \sin(kx - \omega t)$ is not a solution of this equation but that the functions $y(x,t) = Ae^{i(kx - \omega t)}$ and $y(x,t) = Ae^{i(kx + \omega t)}$ do satisfy that equation.

1. For $y(x, t) = A \sin(kx - \omega t)$, $\partial^2 y/\partial x^2 = -Ak^2 \sin(kx - \omega t)$ and $i(\partial y/\partial t) = -Ai\omega \cos(kx - \omega t)$. Evidently, $\partial^2 y/\partial x^2 + i\alpha(\partial y/\partial t) \neq 0$.

2. For $y(x, t) = Ae^{i(kx - \omega t)}$, $\partial^2 y/\partial x^2 = -k^2 y$ and $i(\partial y/\partial t) = \omega y$. Likewise, for $y(x,t) = Ae^{i(kx + \omega t)}$. Both functions are solutions of the equation provided $k^2 = \alpha\omega$.

---

20 • A traveling wave passes a point of observation. At this point, the time between successive crests is 0.2 s. Which of the following is true?

(*a*) The wavelength is 5 m.

(*b*) The frequency is 5 Hz.

(*c*) The velocity of propagation is 5 m/s.

(*d*) The wavelength is 0.2 m.

(*e*) There is not enough information to justify any of these statements.

(*b*) is correct; $T = 1/f$.

21* • True or false: The energy in a wave is proportional to the square of the amplitude of the wave.

True; see Equ. 15-24.

---

22 • A rope hangs vertically. You shake the bottom back and forth, creating a sinusoidal wave train. Is the wavelength at the top the same as, less than, or greater than the wavelength at the bottom?

The wavelength is greater at the top; $\lambda = v/f$, f is constant and v is greater (see Problem 1).

---

23 • One end of a string 6 m long is moved up and down with simple harmonic motion at a frequency of 60 Hz. The waves reach the other end of the string in 0.5 s. Find the wavelength of the waves on the string.

1. Find the wave velocity $v = (6/0.5)$ m/s = 12 m/s
2. $\lambda = v/f$ $\lambda = 12/60$ m = 20 cm

---

24 • Equation 15-13 expresses the displacement of a harmonic wave as a function of x and t in terms of the wave parameters k and ω. Write the equivalent expressions that contain the following pairs of parameters instead of k and ω: (*a*) k and v, (*b*) λ and f, (*c*) λ and T, (*d*) λ and v, and (*e*) f and v.

We have the following relations: $k = 2\pi/\lambda$; $\omega = 2\pi f$; $v = f\lambda$ so $\omega = 2\pi v/\lambda = kv$. For a wave traveling to the right, the expressions are then:

(*a*) $y = A \sin k(x - vt)$; (*b*) $y = A \sin 2\pi(x/\lambda - ft)$; (*c*) $y = A \sin 2\pi(x/\lambda - t/T)$; (*d*) $y = A \sin[(2\pi/\lambda)(x - vt)]$; (*e*) $y = A \sin 2\pi f(x/v - t)$. For a wave traveling to the left, change the − to a + sign.

25* • Equation 15-10 applies to all types of periodic waves, including electromagnetic waves such as light waves and microwaves, which travel at 3×10^8 m/s in a vacuum. (*a*) The range of wavelengths of light to which the eye is sensitive is about 4×10^{-7} to 7×10^{-7}m. What are the frequencies that correspond to these wavelengths? (*b*) Find the frequency of a microwave that has a wavelength of 3 cm.

(*a*) $f = v/\lambda$; $v = c = 3 \times 10^8$ m/s $f_{min} = (3 \times 10^8/7 \times 10^{-7})$ Hz $\approx 4.3 \times 10^{14}$ Hz;

$f_{max} = 7.5 \times 10^{14}$ Hz

(*b*) $f = c/\lambda$ $f = 3 \times 10^8/3 \times 10^{-2}$ Hz = 10^{10} Hz

---

26 • A harmonic wave on a string with a mass per unit length of 0.05 kg/m and a tension of 80 N has an amplitude of 5 cm. Each section of the string moves with simple harmonic motion at a frequency of 10 Hz. Find the power propagated along the string.

Find $v = (F/\mu)^{1/2}$ $v = \sqrt{80/0.05}$ m/s = 40 m/s

$P = \frac{1}{2}\mu\omega^2A^2v$; $\omega = 2\pi f$ $P = (0.05 \times 400\pi^2 \times 25 \times 10^{-4} \times 40)/2$ W = 9.87 W

---

27 • A rope 2 m long has a mass of 0.1 kg. The tension is 60 N. A power source at one end sends a harmonic wave with an amplitude of 1 cm down the rope. The wave is extracted at the other end without any reflection. What is the frequency of the power source if the power transmitted is 100 W?

1. Determine the wave velocity $v = \sqrt{F/\mu} = \sqrt{60/(0.1/2)}$ m/s = 34.6 m/s
2. From Equation 15-19, $f = \dfrac{1}{2\pi A}\sqrt{\dfrac{2P}{\mu v}}$ $f = [(1/2\pi \times 0.01)(200/34.6 \times 0.05)^{1/2}]$ Hz = 171 Hz

---

28 •• The wave function for a harmonic wave on a string is $y(x, t) = (0.001 \text{ m}) \sin(62.8 \text{ m}^{-1}x + 314\text{s}^{-1}t)$. (*a*) In what direction does this wave travel, and what is its speed? (*b*) Find the wavelength, frequency, and period of this wave. (*c*) What is the maximum speed of any string segment?

(*a*) The wave travels to the left (−x direction); $v = \omega/k = 314/62.8$ m/s = 5.0 m/s.

(*b*) $\lambda = 2\pi/k = 0.1$ m; $f = \omega/2\pi = 50$ Hz; $T = 1/f = 0.02$ s.

(*c*) $v_{max} = A\omega = 0.314$ m/s.

29* •• A harmonic wave with a frequency of 80 Hz and an amplitude of 0.025 m travels along a string to the right with a speed of 12 m/s. (*a*) Write a suitable wave function for this wave. (*b*) Find the maximum speed of a point on the string. (*c*) Find the maximum acceleration of a point on the string.

(*a*) See Problem 24(*e*). $y(x, t) = 0.025 \sin[(160\pi)(x/12 - t)]$ m = $0.025 \sin(41.9x - 503t)$ m.

(*b*) $v_{max} = A\omega = (0.025 \times 503)$ m/s = 12.6 m/s.

(*c*) $a_{max} = A\omega^2 = \omega v_{max} = 6321$ m/s^2.

---

30 •• Waves of frequency 200 Hz and amplitude 1.2 cm move along a 20-m string that has a mass of 0.06 kg and a tension of 50 N. (*a*) What is the average total energy of the waves on the string? (*b*) Find the power transmitted past a given point on the string.

(*a*) Apply Equ. 15-18 with $\Delta x = L$ — $E = \frac{1}{2}[(0.06/20)(4\pi^2 \times 200^2)(0.012^2)(20)]$ J = 6.82 J

(*b*) 1. Determine $v = \sqrt{F/\mu}$ — $v = \sqrt{50/(0.06/20)}$ m/s = 129 m/s

2. Evaluate P using Equs. 15-19 and 15-18 — $P = Ev/L = (6.82 \times 129/20)$ W = 44 W

---

31 •• In a real string, a wave loses some energy as it travels down the string. Such a situation can be described by a wave function whose amplitude $A(x)$ depends on x:

$$y = A(x)\sin(kx - \omega t) = (A_0e^{-bx})\sin(kx - \omega t)$$

(*a*) What is the original power carried by the wave at the origin? (*b*) What is the power transported by the wave at point x?

(*a*) At $x = 0$, $A = A_0$; $P(0) = \frac{1}{2}\mu\omega^2A_0^2v$.

(*b*) At x, $A^2 = A_0^2e^{-2bx}$; $P(x) = \frac{1}{2}\mu\omega^2A_0^2ve^{-2bx}$.

---

32 •• Power is to be transmitted along a stretched wire by means of transverse harmonic waves. The wave speed is 10 m/s, and the linear mass density of the wire is 0.01 kg/m. The power source oscillates with an amplitude of 0.50 mm. (*a*) What average power is transmitted along the wire if the frequency is 400 Hz? (*b*) The power transmitted can be increased by increasing the tension in the wire, the frequency of the source, or the amplitude of the waves. How would each of these quantities have to be changed to effect an increase in power by a factor of 100 if it is the only quantity changed? (*c*) Which of the quantities would probably be the easiest to change?

(*a*) Apply Equ. 15-19 — $P = \frac{1}{2}(0.01)(2\pi \times 400 \times 0.0005)^2(10)$ W $= 0.079$ W

(*b*) Since P is proportional to f^2, A^2, and v, and v is proportional to $\sqrt{F}$, the frequency or amplitude could be increased by a factor of 10, but the tension would have to be increased by a factor of 10,000. Evidently, it will be simplest to increase the amplitude to 5.0 mm.

33* • A sound wave in air produces a pressure variation given by

$$p(x,t) = 0.75\cos\frac{\pi}{2}(x - 340t)$$

where p is in pascals, x is in meters, and t is in seconds. Find (*a*) the pressure amplitude of the sound wave, (*b*) the wavelength, (*c*) the frequency, and (*d*) the speed?

(*a*) $p_0 = 0.75$ Pa. (*b*) From Problem 24(*d*), we see that $2\pi/\lambda = \pi/2$, $\lambda = 4$ m. (*c*) $f = v/\lambda = 85$ Hz. (*d*) $v = 340$ m/s.

---

34 • (*a*) Middle C on the musical scale has a frequency of 262 Hz. What is the wavelength of this note in air? (*b*) The frequency of the C an octave above middle C is twice that of middle C. What is the wavelength of this note in air?

(*a*) $\lambda = v/f = 340/262$ m = 1.30 m. (*b*) $\lambda' = \lambda/2 = 0.65$ m.

35 • (*a*) What is the displacement amplitude for a sound wave having a frequency of 100 Hz and a pressure amplitude of 10^{-4} atm? (*b*) The displacement amplitude of a sound wave of frequency 300 Hz is 10^{-7} m. What is the pressure amplitude of this wave?

(*a*) From Equ. 15-22, $s_0 = p_0/(\rho\omega v)$ $s_0 = [10.1/(1.29 \times 2\pi \times 100 \times 340)]$ m $= 3.67 \times 10^{-5}$ m

(*b*) $p_0 = s_0\rho\omega v$ $p_0 = 1.29 \times 300 \times 2\pi \times 340 \times 10^{-7}$ Pa $= 8.27 \times 10^{-2}$ Pa

---

36 • (*a*) Find the displacement amplitude of a sound wave of frequency 500 Hz at the pain-threshold pressure amplitude of 29 Pa. (*b*) Find the displacement amplitude of a sound wave with the same pressure amplitude but a frequency of 1 kHz.

(*a*) See Problem 35. $s_0 = 3.67 \times 10^{-5} \times 29/(10.1 \times 5)$ m $= 2.11 \times 10^{-5}$ m.

(*b*) $s_0 = (2.11 \times 10^{-5}/2)$ m $= 1.05 \times 10^{-5}$ m.

37* • A typical loud sound wave with a frequency of 1 kHz has a pressure amplitude of about 10^{-4} atm. (*a*) At $t = 0$, the pressure is a maximum at some point x_1. What is the displacement at that point at $t = 0$? (*b*) What is the maximum value of the displacement at any time and place? (Take the density of air to be 1.29 kg/m³.)

(*a*) When p is a maximum, $s = 0$. (*b*) $s_0 = p_0/\rho\omega v = 3.67 \times 10^{-6}$ m.

---

38 • (*a*) Find the displacement amplitude of a sound wave of frequency 500 Hz at the threshold-of-hearing pressure amplitude of 2.9×10^{-5} Pa. (*b*) Find the displacement amplitude of a wave of the same pressure amplitude but a frequency of 1 kHz.

From Problem 36 it follows that (*a*) $s_0 = 2.11 \times 10^{-11}$ m; (*b*) $s_0 = 1.05 \times 10^{-11}$ m.

---

39 • A piston at one end of a long tube filled with air at room temperature and normal pressure oscillates with a frequency of 500 Hz and an amplitude of 0.1 mm. The area of the piston is 100 cm². (*a*) What is the pressure amplitude of the sound waves generated in the tube? (*b*) What is the intensity of the waves? (*c*) What average power is required to keep the piston oscillating (neglecting friction)?

(*a*) $p_0 = s_0\rho\omega v$ $p_0 = (10^{-4} \times 1.29 \times 2\pi \times 500 \times 340)$ Pa $= 138$ Pa

(*b*) $I = \frac{1}{2}\rho\omega^2 s_0^2 v = \frac{1}{2}p_0^2/\rho v$ $I = \frac{1}{2}(138^2/1.29 \times 340)$ W/m² $= 21.7$ W/m²

(*c*) $P_{av} = IA$; $A = 10^{-2}$ m² $P_{av} = 0.217$ W

---

40 • A spherical source radiates sound uniformly in all directions. At a distance of 10 m, the sound intensity level is 10^{-4} W/m². (*a*) At what distance from the source is the intensity 10^{-6} W/m²? (*b*) What power is radiated by this source?

(*a*) $I \propto 1/r^2$; $r(I = 10^{-6}) = 100$ m. (*b*) $P = IA = 4\pi r^2 I = 0.126$ W.

41* • A loudspeaker at a rock concert generates 10^{-2} W/m² at 20 m at a frequency of 1 kHz. Assume that the speaker spreads its energy uniformly in three dimensions. (*a*) What is the total acoustic power output of the speaker? (*b*) At what distance will the intensity be at the pain threshold of 1 W/m²? (*c*) What is the intensity at 30 m?

(*a*) $P = 4\pi r^2 I = 4\pi \times 400 \times 10^{-2}$ W $= 50.3$ W. (*b*) Since $I \propto 1/r^2$, r at pain threshold is 2.0 m.

(*c*) $I = 10^{-2}(4/9) = 4.44 \times 10^{-3}$ W/m².

---

42 •• When a pin of mass 0.1 g is dropped from a height of 1 m, 0.05% of its energy is converted into a sound pulse with a duration of 0.1 s. (*a*) Estimate the range at which the dropped pin can be heard if the minimum audible intensity is 10^{-11} W/m². (*b*) Your result in (*a*) is much too large in practice because of background noise.

If you assume that the intensity must be at least 10^{-8} W/m^2 for the sound to be heard, estimate the range at which the dropped pin can be heard. (In both parts, assume that the intensity is $P/4\pi r^2$.)

(*a*) Sound energy is $5 \times 10^{-4}(mgh) = 5 \times 10^{-4} \times 1 \times 10^{-4} \times 9.81$ J $= 4.9 \times 10^{-7}$ J; $P_{av} = E/\Delta t = 4.9 \times 10^{-6}$ W $= 4\pi r^2 \times 10^{-11}$ W; so $r \approx 200$ m. (*b*) $r \approx 200/\sqrt{1000} = 6.24$ m.

---

43 • True or false: A 60-dB sound has twice the intensity of a 30-dB sound.

False

---

44 • What is the intensity level in decibels of a sound wave of intensity (*a*) 10^{-10} W/m^2, and (*b*) 10^{-2} W/m^2?

Use Equ. 15-29. (*a*) $\beta = 20$ dB. (*b*) $\beta = 100$ dB.

45* • Find the intensity of a sound wave if (*a*) $\beta = 10$ dB, and (*b*) $\beta = 3$ dB. (*c*) Find the pressure amplitudes of sound waves in air for each of these intensities.

(*a*), (*b*) Use Equ. 15-29 — (*a*) $I = 10^{-11}$ W/m^2; (*b*) $I = 2 \times 10^{-12}$ W/m^2

(*c*) $p_0 = \sqrt{2I\rho v}$ — (*a*) $p_0 = 9.37 \times 10^{-5}$ Pa; (*b*) $p_0 = 4.19 \times 10^{-5}$ Pa

---

46 • The sound level of a dog's bark is 50 dB. The intensity of a rock concert is 10,000 times that of the dog's bark. What is the sound level of the rock concert?

$\beta = \beta_{dog} + 40$ dB $= 90$ dB.

---

47 • Two sounds differ by 30 dB. The intensity of the louder sound is I_L and that of the softer sound is I_S. The value of the ratio I_L/I_S is (*a*) 1000. (*b*) 30. (*c*) 9. (*d*) 100. (*e*) 300.

(*a*) is correct.

---

48 • Show that if the intensity is doubled, the intensity level increases by 3.0 dB.

$\Delta\beta = 10 \log 2 = 3.01 \approx 3.0$.

49* • What fraction of the acoustic power of a noise would have to be eliminated to lower its sound intensity level from 90 to 70 dB?

99% must be eliminated so that the power is reduced by factor of 100.

---

50 •• Normal human speech has a sound intensity level of about 65 dB at 1 m. Estimate the power of human speech.

Determine I at 1 m. $I = 10^{-12} \times 10^{6.5}$ W/m$^2 = 10^{-5.5}$ W/m$^2 = 3.16 \times 10^{-6}$ W/m$^2 = P/4\pi$. $P = 4 \times 10^{-5}$ W.

---

51 •• A spherical source radiates sound uniformly in all directions. At a distance of 10 m, the sound intensity level is 80 dB. (*a*) At what distance from the source is the intensity level 60 dB? (*b*) What power is radiated by this source?

(*a*) $I(r) = I(10)/100$; $I \propto 1/r^2$ — $r = 100$ m

(*b*) $P = IA$ — $P = 10^{-4} \times 4\pi \times 10^2$ W $= 0.126$ W

---

52 •• A spherical source of intensity I_0 radiates sound uniformly in all directions. Its intensity level is β_1 at a distance r_1, and β_2 at a distance r_2. Find β_2/β_1.

$$\frac{\beta_2}{\beta_1} = \frac{\log (I_2/I_0)}{\log (I_1/I_0)} = \frac{\log I_2}{\log I_1} = \frac{\log [(r_1^2/r_2^2)I_1]}{\log I_1} = \frac{\log I_1 + 2 \log (r_1/r_2)}{\log I_1} = \frac{\beta_1 + 20 \log (r_1/r_2)}{\beta_1}.$$

53* •• A loudspeaker at a rock concert generates 10^{-2} W/m^2 at 20 m at a frequency of 1 kHz. Assume that the speaker spreads its energy uniformly in all directions. (*a*) What is the intensity level at 20 m? (*b*) What is the total acoustic power output of the speaker? (*c*) At what distance will the intensity level be at the pain threshold of 120 dB? (*d*) What is the intensity level at 30 m?

(*a*) $\beta = 100$ dB at 20 m. (*b*), (*c*), (*d*) See Problem 41; (*d*) 4.44×10^{-3} W/m^2 = 96.5 dB.

---

54 •• An article on noise pollution claims that sound intensity levels in large cities have been increasing by about 1 dB annually. (*a*) To what percentage increase in intensity does this correspond? Does this increase seem reasonable? (*b*) In about how many years will the intensity of sound double if it increases at 1 dB annually?

(*a*) $I/I_0 = 10^{0.1} = 1.26$; increase in intensity is 26% annually. This is not reasonable; if true, the intensity level would increase by a factor of 10 in ten years.

(*b*) To double, $\Delta\beta = 3$, so the intensity level will double in 3 years.

---

55 •• Three noise sources produce intensity levels of 70, 73, and 80 dB when acting separately. When the sources act together, their intensities add. (*a*) Find the sound intensity level in decibels when the three sources act at the same time. (*b*) Discuss the effectiveness of eliminating the two least intense sources in reducing the intensity level of the noise.

(*a*) $I_1 = 10^{-5}$ W/m^2, $I_2 = 2 \times 10^{-5}$ W/m^2, $I_3 = 10^{-4}$ W/m^2. So $I_{tot} = 1.3 \times 10^{-4}$ W/m^2 ; $\beta_{tot} = 81.14$ dB.

(*b*) Eliminating the two least intense sources does not reduce the intensity level significantly.

---

56 •• The equation $I = P_{av} / 4\pi r^2$ is predicated on the assumption that the transmitting medium does not absorb any energy. It is known that absorption of sound by dry air results in a decrease of intensity of approximately 8 dB/km. The intensity of sound at a distance of 120 m from a jet engine is 130 dB. Find the intensity at 2.4 km from the jet engine (*a*) assuming no absorption of sound by air, and (*b*) assuming a diminution of 8 dB/km. (Assume that the sound radiates uniformly in all directions.)

(*a*) $\Delta\beta = 20 \log (r_2/r_1)$ — $20 \log (20) = 26$; $\beta(2.4 \text{ km}) = (130 - 26)$ dB = 104 dB

(*b*) Subtract 2.28×8 dB from result of (*a*) — $\beta = (104 - 18.2)$ dB = 85.8 dB

57* ••• Everyone at a party is talking equally loudly. If only one person were talking, the sound level would be 72 dB. Find the sound level when all 38 people are talking.

$I_{tot} = 38I_1$; $\beta_{tot} = [10 \log (38) + 72]$ dB = (15.8 + 72) dB = 88.8 dB.

---

58 ••• When a violinist pulls the bow across a string, the force with which the bow is pulled is fairly small, about 0.6 N. Suppose the bow travels across the A string, which vibrates at 440 Hz, at 0.5 m/s. A listener 35 m from the performer hears a sound of 60 dB intensity. With what efficiency is the mechanical energy of bowing converted to sound energy? (Assume that the sound radiates uniformly in all directions.)

1. Find the power delivered by the bow to the string — $P_{in} = Fv = 0.6 \times 0.5 \text{ W} = 0.3 \text{ W}$
2. Find the power of the sound emitted — $P_{out} = IA = (10^{-6} \times 4\pi \times 35^2) \text{ W} = 0.0154 \text{ W}$
3. Efficiency = P_{out}/P_{in} — $\eta = 5.13\%$

---

59 ••• The noise level in an empty examination hall is 40 dB. When 100 students are writing an exam, the sounds of heavy breathing and pens traveling rapidly over paper cause the noise level to rise to 60 dB (not counting the occasional groans). Assuming that each student contributes an equal amount of noise power, find the noise level to the nearest decibel when 50 students have left.

1. Find the noise level increase due to one student $\Delta I_{100} = (10^{-6} - 10^{-8})$ W/m^2 $\approx 10^{-6}$ W/m^2;

$\Delta I_1 = 10^{-8}$ W/m^2

2. Find I_{50} and β_{50} $I_{50} = 5.1 \times 10^{-7}$ W/m^2; $\beta_{50} = 57$ dB

---

60 • If the source and receiver are at rest relative to each other but the wave medium is moving relative to them, will there be any Doppler shift in frequency?

No

61* • The frequency of a car horn is f_0. What frequency is observed if both the car and the observer are at rest, but a wind blows toward the observer?

(*a*) f_0

(*b*) Greater than f_0

(*c*) Less than f_0

(*d*) It could be either greater or less than f_0.

(*e*) It could be f_0 or greater than f_0, depending on how wind speed compares to speed of sound.

(*a*) There is no relative motion of the source and receiver.

---

62 •• Stars often occur in pairs revolving around their common center of mass. If one of the stars is a black hole, it is invisible. Explain how the existence of such a black hole might be inferred from the light observed from the other, visible star.

The light from the companion star will be shifted about its mean frequency periodically due to the relative approach to and recession from the earth of the companion star as it revolves about the black hole.

---

63 • A conveyor belt moves to the right with a speed $v = 300$ m/min. A very fast piemaker puts pies on the belt at a rate of 20 per minute, and they are received at the other end by a pie eater. (*a*) If the piemaker is stationary, find the spacing λ between the pies and the frequency f with which they are received by the stationary pie eater. (*b*) The piemaker now walks with a speed of 30 m/min toward the receiver while continuing to put pies on the belt at 20 per minute. Find the spacing of the pies and the frequency with which they are received by the stationary pie eater. (*c*) Repeat your calculations for a stationary piemaker and a pie eater who moves toward the piemaker at 30 m/min.

(*a*) Since 20 pies per minute are placed on the belt, 20 pies per minute will be received. The time between the placing of the pies is 0.05 min, and during that time the belt moves 300×0.05 m $= 15$ m. Consequently, the spacing between the pies is $\lambda = 15$ m.

(*b*) Relative to the piemaker, the belt moves at 270 m/min. Consequently, $\lambda = 270 \times 0.05$ m $= 13.5$ m. Since the pies are traveling toward the receiver at 300 m/min, the number received per minute, i.e., the frequency, is given by $f = 300/13.5$ min^{-1} $= 22.2$ min^{-1}.

(*c*) If the receiver moves toward the piemaker, the spacing of the pies on the belt is, as in (*a*), 15 m. However, the speed of the belt relative to the receiver is 330 m/min, so the frequency $f = 330/15$ min^{-1} $= 22$ min^{-1}.

---

64 • For the situation described in Problem 63, derive general expressions for the spacing of the pies λ and the frequency f with which they are received by the pie eater in terms of the speed of the belt v, the speed of the sender u_s, the speed of the receiver u_r, and the frequency f_0 with which the piemaker places pies on the belt.

Equations 15-31 and 15-35 apply. Since the source is moving in the direction of the belt and the receiver is moving toward the observer, we must use the negative sign in the numerator of Equ. 15-31 and in the denominator of Equ. 15-35.

Thus $\lambda' = \dfrac{v - u_s}{f_0}$; $f' = f_0\dfrac{v + u_r}{v - u_s}$.

In Problems 65 through 70, a source emits sounds of frequency 200 Hz that travel through still air at 340 m/s.

65* • The sound source described above moves with a speed of 80 m/s relative to still air toward a stationary listener. (*a*) Find the wavelength of the sound between the source and the listener. (*b*) Find the frequency heard by the listener.

(*a*) Apply Equ. 15-31 $\lambda' = (260/200)$ m $= 1.3$ m

(*b*) Apply Equ. 15-35 $f' = 200(340/260)$ Hz $= 262$ Hz

---

66 • Consider the situation in Problem 65 from the reference frame in which the source is at rest. In this frame, the listener moves toward the source with a speed of 80 m/s, and there is a wind blowing at 80 m/s from the listener to the source. (*a*) What is the speed of the sound from the source to the listener in this frame? (*b*) Find the wavelength of the sound between the source and the listener. (*c*) Find the frequency heard by the listener.

(*a*) In the reference frame of the source, $v = (340 - 80)$ m/s $= 260$ m/s.

(*b*) Since f is unchanged, $\lambda = 260/f = 1.3$ m.

(*c*) In the moving reference frame, the observer is approaching the source at 80 m/s. Consequently, applying Equ. 15-35, we find $f_r = [200(1 + 80/260)]$ Hz $= 262$ Hz.

---

67 • The source moves away from the stationary listener at 80 m/s. (*a*) Find the wavelength of the sound waves between the source and the listener. (*b*) Find the frequency heard by the listener.

Proceed as in Problem 65, changing signs as appropriate.

(*a*) $\lambda = 420/200$ m $= 2.1$ m. (*b*) $f = 200(340/420)$ Hz $= 162$ Hz.

---

68 • The listener moves at 80 m/s relative to still air toward the stationary source. (*a*) What is the wavelength of the sound between the source and the listener? (*b*) What is the frequency heard by the listener?

(*a*) λ is unaffected by the motion of the observer $\lambda = 340/200$ m $= 1.7$ m

(*b*) Apply Equ. 15-35 $f' = [200(1 + 80/340)]$ Hz $= 247$ Hz

---

69* • Consider the situation in Problem 68 in a reference frame in which the listener is at rest. (*a*) What is the wind velocity in this frame? (*b*) What is the speed of the sound from the source to the listener in this frame, that is, relative to the listener? (*c*) Find the wavelength of the sound between the source and the listener in this frame. (*d*) Find the frequency heard by the listener.

(*a*) Moving at 80 m/s in still air, the observer experiences a wind of 80 m/s.

(*b*) Using the standard Galilean transformation, $v' = v + u_r = 420$ m/s.

(*c*) The distance between wave crests is unchanged, so $\lambda' = \lambda = 1.7$ m.

(*d*) $f' = v'/\lambda' = 247$ Hz.

---

70 • The listener moves at 80 m/s relative to the still air away from the stationary source. Find the frequency heard by the listener.

Apply Equ. 15-35 $f' = [200(1 - 80/340)]$ Hz = 152 Hz

---

71 • A jet is traveling at Mach 2.5 at an altitude of 5000 m. (*a*) What is the angle that the shock wave makes with the track of the jet? (Assume that the speed of sound at this altitude is still 340 m/s.) (*b*) Where is the jet when a person on the ground hears the shock wave?

(*a*) $u/v = 2.5$; $\theta = \sin^{-1}(v/u)$ $\theta = \sin^{-1}(0.4) = 23.6°$

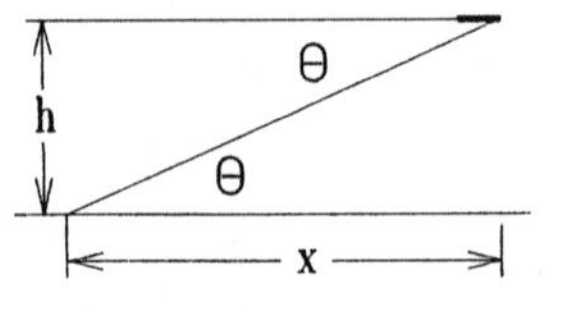

(*b*) From the diagram showing the wave front of the shock wave, it is evident that $x = h/\tan\,\theta$ $x = (5000/\tan\ 23.6°)$ m = 11.5 km

---

72 • If you are running at top speed toward a source of sound at 1000 Hz, estimate the frequency of the sound that you hear. Suppose that you can recognize a change in frequency of 3%. Can you use your sense of pitch to estimate your running speed?

Top speed ≈ 10 m/s. $\Delta f/f_0 = u/v = 10/340$ Hz = 0.0294 or 2.9% . No; it would be impossible to decide between speeds of 8 m/s and 10 m/s.

73* •• A radar device emits microwaves with a frequency of 2.00 GHz. When the waves are reflected from a car moving directly away from the emitter, a frequency difference of 293 Hz is detected. Find the speed of the car.

1. The frequency f received by the car is given by Equ. 15-37*a*.
2. The car now acts as the source, sending signals of frequency f to the stationary radar receiver.
3. Consequently, $f_{rec} = \dfrac{1 + v/c}{1 - v/c} f_0 = (1 + 2v/c)f_0$ since $v << c$.
4. Solve for v; $v = c\Delta f/2f_0$ $v = (3 \times 10^8 \times 293/4 \times 10^9)$ m/s = 22 m/s = 79.2 km/h

---

74 •• A stationary destroyer is equipped with sonar that sends out pulses of sound at 40 MHz. Reflected pulses are received from a submarine directly below with a time delay of 80 ms at a frequency of 39.958 MHz. If the speed of sound in seawater is 1.54 km/s, find (*a*) the depth of the submarine, and (*b*) its vertical speed.

(*a*) $2D = v\Delta t$ $D = (1540 \times 0.080/2)$ m = 61.6 m

(*b*) The submarine acts as a receiver and source (see Problem 73) $u = v\Delta f/2f_0 = (1540 \times 0.042/80)$ m/s = 0.809 m/s
The velocity of the submarine is down

---

75 •• Two airplanes, one flying due east and the other due west, are on a near collision course separated by 15 km when the pilot of one plane, traveling at 900 km/h, observes the other on his Doppler radar. The radar unit emits electromagnetic waves of frequency 3×10^{10} Hz. The radar readout indicates that the other plane's speed is 750 km/h. Determine the frequency of the signal received by the pilot's radar.

1. This is similar to Problem 73; $\Delta f/f_0 = 2v/c$ $v = 1650$ km/h = 458 m/s; $\Delta f = 2f_0 v/c = 9.16$ kHz
2. $f = f_0 + \Delta f$ $f = 30.000916$ GHz

---

76 •• A police radar unit transmits microwaves of frequency 3×10^{10} Hz. The speed of these waves in air is 3.0×10^8 m/s. Suppose a car is receding from the stationary police car at a speed of 140 km/h. What is the frequency difference between the transmitted signal and the signal received from the receding car?

Proceed as in Problem 75 $v = 38.9$ m/s; $\Delta f = 2f_0v/c = 7.78$ kHz

---

77* •• Suppose the police car of Problem 76 is moving in the same direction as the other vehicle at a speed of 60 km/h. What then is the difference in frequency between the emitted and the reflected signals?

Now the relative velocity is 80 km/h $\Delta f = (80 \times 7.78/140)$ kHz $= 4.45$ kHz

---

78 •• At time $t = 0$, a supersonic plane is directly over point P flying due west at an altitude of 12 km and a speed of Mach 1.6. Where is the plane when the sonic boom is heard?

See Problem 71. $x = h/\tan\theta$; $\theta = \sin^{-1}(1/1.6) = 38.7°$; $x = (12/\tan 38.7°)$ km $= 15.0$ km west of P.

---

79 •• A small radio of 0.10 kg mass is attached to one end of an air track by a spring. The radio emits a sound of 800 Hz. A listener at the other end of the air track hears a sound whose frequency varies between 797 and 803 Hz. (*a*) Determine the energy of the vibrating mass–spring system. (*b*) If the spring constant is 200 N/m, what is the amplitude of vibration of the mass and what is the period of the oscillating system?

(*a*) 1. Determine u_{max}; use $\Delta f/f_0 = u/v$ $\Delta f/f_0 = 3/800$; $u_{max} = 3 \times 340/800$ m/s $= 1.275$ m/s

2. $E = \frac{1}{2}mu_{max}^2$ $E = 0.05 \times 1.275^2$ J $= 0.0813$ J

(*b*) 1. $E = \frac{1}{2}kA^2$ $A = \sqrt{2E/k} = \sqrt{2 \times 0.0813/200} = 2.85$ cm

2. $T = 2\pi\sqrt{m/k}$ $T = 2\pi\sqrt{1/2000} = 0.14$ s

---

80 •• A sound source of frequency f_0 moves with speed u_s relative to still air toward a receiver who is moving with speed u_r relative to still air away from the source. (*a*) Write an expression for the received frequency f'. (*b*) Use the result that $(1 - x)^{-1} \approx 1 + x$ to show that if both u_s and u_r are small compared to v, then the received frequency is approximately

$$f' \approx \left(1 + \frac{u_s - u_r}{v}\right)f_0 = \left(1 + \frac{u_{rel}}{v}\right)f_0$$

where u_{rel} is the relative velocity of the source and receiver.

Start with Equ. 15-35. Use the binomial expansion for the denominator: $(1 - u_s/v)^{-1} \approx 1 + u_s/v$. The product $(1 - u_r/v)(1 + u_s/v) \approx 1 + (u_s - u_r)/v$. So Equ. 15-35 reduces to the expression given above.

81* •• Two students with vibrating 440-Hz tuning forks walk away from each other with equal speeds. How fast must they walk so that they each hear a frequency of 438 Hz from the other fork?

See Problem 80. In this case, $f' = (1 - 2u/v)f_0$. $\Delta f = 2f_0u/v$. $u = \Delta fv/2f_0 = (2 \times 340/880)$ m/s $= 0.773$ m/s.

---

82 •• A physics student walks down a long hall carrying a vibrating 512-Hz tuning fork. The end of the hall is closed so that sound reflects from it. The student hears a sound of 516 Hz from the wall. How fast is the student walking?

The student serves as a source moving toward the wall, and a moving receiver for the echo. We can use the result of Problem 80, since $\Delta f/f_0 << 1$. So, $(4/512) = 2u/v$; $u = 2 \times 340/512$ m/s $= 1.33$ m/s.

---

83 •• A small speaker radiating sound at 1000 Hz is tied to one end of an 0.8-m-long rod that is free to rotate about its other end. The rod rotates in the horizontal plane at 4.0 rad/s. Derive an expression for the frequency heard by a stationary observer far from the rotating speaker.

1. Find the velocity of the source — $u_s = r\omega \sin(\omega t) = 3.2 \sin(4t)$ m/s
2. Use Equ. 15-35 — $f' = [1000/(1 - 3.2 \sin(4t)/340)]$ Hz
3. Since 3.2/340 << 1 we can use the binomial approximation — $f' = 1000 + 9.41 \sin(4t)$ Hz

---

84 •• You have won a free trip on the *Queen Elizabeth II* and are in mid-Atlantic steaming due east at 45 km/h as the Concorde passes directly overhead flying due west at Mach 1.6 at an altitude of 12,500 m. Where is the Concorde relative to the *QEII* when you hear the sonic boom?

This problem is identical to Problem 71. The horizontal distance between the Concorde and the point at ground level where the sonic boom is heard is given by $x = h/[\tan \sin^{-1}(v/u)] = 15.6$ km due west.

85* •• A balloon driven by a 36-km/h wind emits a sound of 800 Hz as it approaches a tall building. (*a*) What is the frequency of the sound heard by an observer at the window of this building? (*b*) What is the frequency of the reflected sound heard by a person riding in the balloon?

The simplest way to approach this problem is to transform to a reference frame in which the balloon is at rest. In that reference frame, the speed of sound is $v = 340$ m/s, and $u_r = 36$ km/h $= 10$ m/s.

(*a*) Use Equ. 15-34 — $f' = 800(1 + 10/340)$ Hz $= 823.5$ Hz

(*b*) f' acts as moving source; use Equ. 15-33 — $f' = 823.5/(1 - 10/340)$ Hz $= 848.5$ Hz

---

86 •• A car is approaching a reflecting wall. A stationary observer behind the car hears a sound of frequency 745 Hz from the car horn and a sound of frequency 863 Hz from the wall. (*a*) How fast is the car traveling? (*b*) What is the frequency of the car horn? (*c*) What frequency does the car driver hear reflected from the wall?

(*a*) 1. Let $\alpha = u/v$; express data in terms of f_0 and α — 745 Hz $= f_0/(1 + \alpha)$; 863 Hz $= f_0/(1 - \alpha)$

2. Find α and $u = 340\alpha$ m/s — $(1 + \alpha)/(1 - \alpha) = 1.158$; $\alpha = 0.0734$

$u = 24.95$ m/s $= 89.8$ km/l

(*b*) Evaluate f_0 — $f_0 = 745 \times 1.0734$ Hz $= 800$ Hz

(*c*) Use Equ. 15-34 with $f_0 = 863$ Hz — $f = 863 \times 1.0734$ Hz $= 926$ Hz

---

87 •• The driver of a car traveling at 100 km/h toward a vertical cliff briefly sounds the horn. Exactly one second later she hears the echo and notes that its frequency is 840 Hz. How far from the cliff was the car when the driver sounded the horn and what is the frequency of the horn?

1. d = distance to cliff at $t = 0$; write equation for d — $d + (d - u\Delta t) = v\Delta t = 340$ m
2. Find d for $u = 100$ km/h $= 27.8$ m/s — $d = 183$ m
3. Find f_0 from Equ. 15-35 — $f_0 = 840[(1 - 27.8/340)/(1 + 27.8/340)]$ Hz $= 713$ Hz

---

88 •• You are on a transatlantic flight traveling due west at 800 km/h. A Concorde flying at Mach 1.6 and 3 km to the north of your plane is also on an east-to-west course. What is the distance between the two planes when you hear the sonic boom from the Concorde?

This problem is almost identical to Problem 84. The Concorde is 15.6 km west and 3 km north of your plane. The distance between the planes is $D = \sqrt{15.6^2 + 3^2}$ km $=$ 15.9 km.

89* ••• Astronomers can deduce the existence of a binary star system even if the two stars cannot be visually resolved by noting an alternating Doppler shift of a spectral line. Suppose that an astronomical observation shows that the source of light is eclipsed once every 18 h. The wavelength of the spectral line observed changes from a maximum of 563 nm to a minimum of 539 nm. Assume that the double star system consists of a very massive, dark object and a relatively light star that radiates the observed spectral line. Use the data to determine the separation between the two objects (assume that the light object is in a circular orbit about the massive one) and the mass of the massive object. (Use the approximation $\Delta f/f_0 \approx v/c$.)

1. Determine the maximum and minimum frequencies — $f_{max} = \dfrac{3 \times 10^8}{5.39 \times 10^{-7}}$ Hz $= 5.566 \times 10^{14}$ Hz; $f_{min} = 5.329$ Hz
2. $f_0 = ½(f_{max} + f_{min})$ — $f_0 = 5.4475 \times 10^{14}$ Hz
3. $v = c\Delta f/f_0$ — $v = 6.526 \times 10^6$ m/s
4. Determine R, the radius of the orbit — $R = vT/2\pi = (6.526 \times 10^6 \times 64800/2\pi)$ m $= 6.73 \times 10^{10}$ m
5. From Equ. 11-15, $M = \dfrac{4\pi^2 R^3}{GT^2}$ — $M = \dfrac{4\pi^2(6.73 \times 10^{10})^3}{6.67 \times 10^{-11}(6.48 \times 10^4)^2}$ kg $= 4.3 \times 10^{34}$ kg

---

90 ••• A physics student drops a vibrating 440-Hz tuning fork down the elevator shaft of a tall building. When the student hears a frequency of 400 Hz, how far has the tuning fork fallen?

1. Find the speed of the source from Equ. 15-33 — $u = 340(440/400 - 1)$ m/s $= 34$ m/s
2. Find the location of the source at that time (t_1) — $y = u^2/2g = 58.92$ m; $t_1 = u/g = 3.466$ s
3. Find the time (t_2) for sound to travel 58.92 m — $t_2 = 58.92/340$ s $= 0.173$ s
4. Find the distance the source has fallen in time $t_1 + t_2$ — $d = ½gt^2 = 9.81 \times 3.639^2/2$ m $= 65$ m

---

91 • When a guitar string is plucked, is the wavelength of the wave it produces in air the same as the wavelength of the wave on the string?

No; the frequencies are the same but the speeds of propagation are different.

---

92 • A wave pulse travels along a light string that is attached to a heavier string in which the wave speed is smaller. The reflected pulse is _____, and the transmitted pulse is _____.

(*a*) inverted/ inverted

(*b*) inverted/ not inverted

(*c*) not inverted/ not inverted

(*d*) not inverted/ inverted

(*e*) nonexistent/ not inverted

(*b*) See Section 15-4.

93*• True or false: (*a*) Wave pulses on strings are transverse waves. (*b*) Sound waves in air are transverse waves of compression and rarefaction. (*c*) The speed of sound at 20°C is twice that at 5°C.

(*a*) True (*b*) False (*c*) False

94 • Sound travels at 340 m/s in air and 1500 m/s in water. A sound of 256 Hz is made under water. In the air, the frequency will be

(*a*) the same, but the wavelength will be shorter.

(*b*) higher, but the wavelength will stay the same.

(*c*) lower, but the wavelength will be longer.

(*d*) lower, and the wavelength will be shorter.

(*e*) the same, and the wavelength too will stay the same.

(*a*) from $\lambda = v/f$.

95 •• Figure 15-30 shows a wave pulse at time $t = 0$ moving to the right. At this particular time, which segments of the string are moving up? Which are moving down? Is there any segment of the string at the pulse that is instantaneously at rest? Answer these questions by sketching the pulse at a slightly later time and a slightly earlier time to see how the segments of the string are moving.

The figure shows the pulse at an earlier time ($-\Delta t$) and later time (Δt). One can see that at $t = 0$, the portion of the string between 1 cm and 2 cm is moving down, the portion between 2 cm and 3 cm is moving up, and the string at $x = 2$ cm is instantaneously at rest.

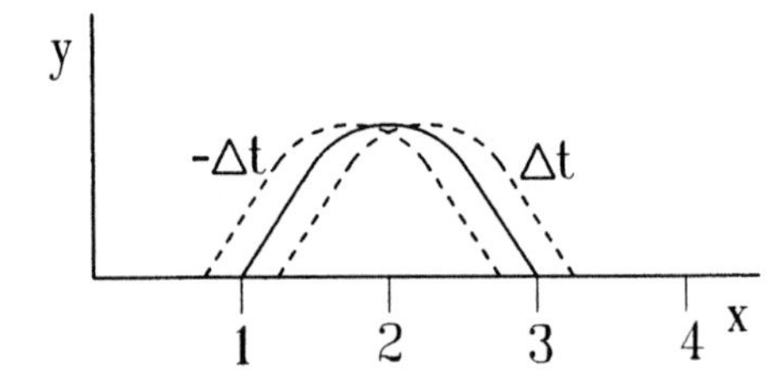

96 •• Make a sketch of the velocity of each string segment versus position for the pulse shown in Figure 15-30.

The velocity of the string at $t = 0$ is shown. Note that the velocity is negative for x between 1 cm and 2 cm and is positive for x between 2 cm and 3 cm. (see Problem 95)

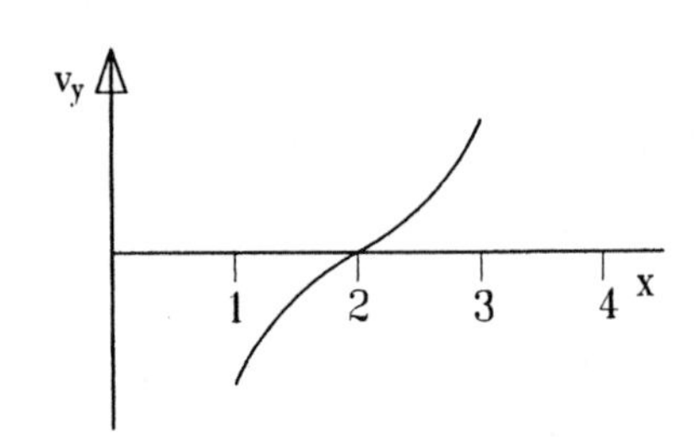

97* •• Consider a long line of cars equally spaced by one car length and moving slowly with the same speed. One car suddenly slows to avoid a dog and then speeds up until it is again one car length behind the car ahead. Discuss how the space between cars propagates back along the line. How is this like a wave pulse? Is there any transport of energy? What does the speed of propagation depend on?

The driver of the car behind slows and then speeds up. This gives rise to a longitudinal wave pulse propagating backwards along the line of cars. There is no transport of energy. The speed of propagation of the pulse depends on the length of a car and on the driver's reaction time.

98 • At time $t = 0$, the shape of a wave pulse on a string is given by the function

$$y(x, 0) = \frac{0.12\ \text{m}^3}{(2.00\ \text{m})^2 + x^2}$$

where x is in meters. (*a*) Sketch $y(x, 0)$ versus x. Give the wave function $y(x, t)$ at a general time t if (*b*) the pulse is moving in the positive x direction with a speed of 10 m/s, and (*c*) the pulse is moving in the negative x direction with a speed of 10 m/s.

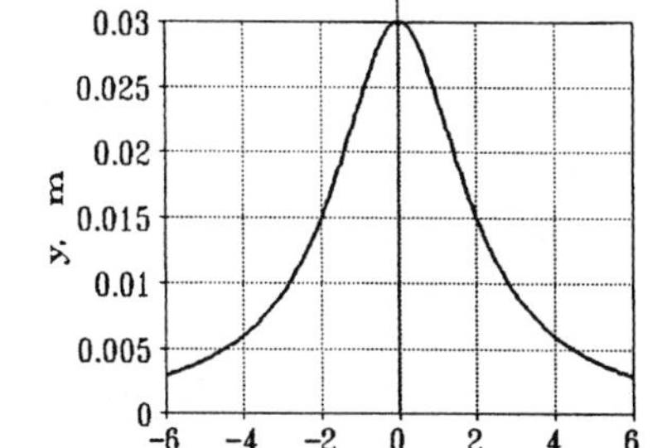

(*a*) The pulse at $t = 0$ is shown on the right

(*b*) The wave function must be of the form $y(x, t) = f(x - vt)$;

$y(x, t) = 0.12/[4 + (x - 10t)^2]$ m

(*c*) In this case, $y(x, t) = 0.12/[4 + (x + 10t)^2]$ m

---

99 • A wave with frequency of 1200 Hz propagates along a wire that is under a tension of 800 N. The wavelength of the wave is 24 cm. What will be the wavelength if the tension is decreased to 600 N and the frequency is kept constant?

$\lambda \propto v \propto \sqrt{T};\ \lambda_2/\lambda_1 = \sqrt{T_2/T_1}$ $\qquad$ $\lambda_2 = 24\sqrt{600/800}$ cm $= 20.8$ cm

---

100 • In a common lecture demonstration of wave pulses, a piece of rubber tubing is tied at one end to a fixed post and is passed over a pulley to a weight hanging at the other end. Suppose that the distance from the fixed support to the pulley is 10 m, the mass of this length of tubing is 0.7 kg, and the suspended weight is 110 N. If the tubing is given a transverse blow at one end, how long will it take the resulting pulse to reach the other end?

$t = L/v = L/\sqrt{F/\mu} = \sqrt{Lm/F}$ $\qquad$ $t = \sqrt{7/110}$ s $= 0.252$ s

---

101*• The following wave functions represent traveling waves:

(*a*) $y_1(x,t) = A \cos k[x + (34 \text{ m/s})t]$,

(*b*) $y_2(x,t) = Ae^{k[x - (20 \text{ m/s})t]}$,

(*c*) $y_3(x,t) = BC + \{k[x - (10 \text{ m/s})t]\}^2$,

where x is in meters, t is in seconds, and A, k, B, and C are constants that have the proper units for y to be in meters. Give the direction of propagation and the speed of the wave for each wave function.

(*a*) Wave propagates to the left ($-x$ direction) at a speed of 34 m/s.

(*b*) Wave propagates to the right at 20 m/s.

(*c*) Wave propagates to the right at 10 m/s.

---

102 • A boat traveling at 10 m/s on a still lake makes a bow wave at an angle of 20° with its direction of motion. What is the speed of the bow wave?

From Equ. 15-38, $v = u \sin\theta$ $\qquad$ $v = 10 \sin 20°$ m/s $= 3.42$ m/s

---

103 • If a wavelength is much larger than the diameter of a loudspeaker, the speaker radiates in all directions, much like a point source. On the other hand, if the wavelength is much smaller than the diameter, the sound travels in an approximately straight line in front of the speaker. Find the frequency of a sound wave that has a wavelength (*a*) 10 times the diameter of a 30-cm speaker, and (*b*) one-tenth the diameter of a 30-cm speaker. (*c*) Repeat this problem for a 6-cm speaker.

In each case use $f = v/\lambda$. (*a*) $f = 340/3$ Hz $= 113$ Hz. (*b*) $f = 11.3$ kHz. (*c*)-(*a*) $f = 567$ Hz. (*c*)-(*b*) $f = 56.7$ kHz.

104 • A whistle of frequency 500 Hz moves in a circle of radius 1 m at 3 rev/s. What are the maximum and minimum frequencies heard by a stationary listener in the plane of the circle and 5 m away from its center?

As shown in the figure, the maximum and minimum frequencies are determined by f_0 and the tangential speed $u_s = 2\pi r/T$. We now apply Equ. 15-33.

$f' = \dfrac{f_0}{1 \pm u_s/v}$; $u_s = 6\pi$ m/s = 18.85 m/s

$f_{max} = 500/(1 - 18.85/340)$ Hz = 529 Hz;

$f_{min} = 500/(1 + 18.85/340)$ Hz = 474 Hz

105*• Ocean waves move toward the beach with a speed of 8.9 m/s and a crest-to-crest separation of 15.0 m. You are in a small boat anchored off shore. (*a*) What is the frequency of the ocean waves? (*b*) You now lift anchor and head out to sea at a speed of 15 m/s. What frequency of the waves do you observe?

Given: $\lambda = 15$ m, $v = 8.9$ m/s.

(*a*) $f_0 = v/\lambda$ — $f_0 = 8.9/15$ Hz = 0.593 Hz

(*b*) $f' = f_0(1 + u_r/v)$ — $f' = 0.593(1 + 15/8.9)$ Hz = 1.59 Hz

---

106 •• Two connected wires with linear mass densities that are related by $\mu_1 = 3\mu_2$ are under the same tension. When the wires oscillate at a frequency of 120 Hz, waves of wavelength 10 cm travel down the first wire with the linear density of μ_1. (*a*) What is the wave speed in the first wire? (*b*) What is the wave speed in the second wire? (*c*) What is the wavelength in the second wire?

(*a*) $v_1 = f\lambda_1$ — $v_1 = 120 \times 0.1$ m/s = 12.0 m/s

(*b*) $v \propto 1/\sqrt{u}$; $\mu_2 = \mu_1/3$ — $v_2 = 12.0\sqrt{3}$ m/s = 20.8 m/s

(*c*) $f_1 = f_2 = f$; $\lambda_2 = \lambda_1(v_2/v_1)$ — $\lambda_2 = 0.1\sqrt{3}$ m = 17.3 cm

---

107 •• A 12.0-m wire of mass 85 g is stretched under a tension of 180 N. A pulse is generated at the left end of the wire, and 25 ms later a second pulse is generated at the right end of the wire. Where do the pulses first meet?

Let *t* be the time of travel of the left hand pulse. Both pulses travel at the same speed.

1. Find the pulse speed — $v = \sqrt{F/\mu} = \sqrt{180 \times 12/0.085}$ m/s = 159 m/s
2. Write the equation for the distance traveled — 12 m = $vt + v(t - 2.5 \times 10^{-2}$ s)
3. Solve for vt = distance from left end — $vt = 8.0$ m

---

108 •• A harmonic wave moves down a string with speed 12.4 m/s. A particle on the string has a maximum displacement of 4.5 cm and a maximum speed of 9.4 m/s. Find (*a*) the wavelength of the wave, and (*b*) the frequency. (*c*) Write an equation for the wave function.

The general expression is $y(x, t) = A \sin(kx - \omega t)$. We are given $A = 0.045$ m, $(dy/dt)_{max} = 9.4$ m/s and $v = 12.4$ m/s.

1. Find ω using $(dy/dt)_{max} = A\omega$ — $\omega = 9.4/0.045$ rad/s = 209 s^{-1}
2. Find k using $v = \omega/k$ — $k = \omega/v = 209/12.4\ m^{-1} = 16.85\ m^{-1}$

(*a*) $\lambda = 2\pi/k$ — $\lambda = 37.3$ cm

(*b*) $f = \omega/2\pi$ — $f = 33.3$ Hz

(*c*) Write $y(x, t)$ using numerical results — $y(x, t) = 0.045 \sin(16.85x - 209t)$

109*•• Find the speed of a car for which the tone of its horn will drop by 10% as it passes you.

Let $\alpha = u_s/v$.

Then we are given $\frac{1+\alpha}{1-\alpha} = 0.9$; solve for α and u_s $\quad \alpha = 0.1/1.9$; $u_s = 34/1.9$ m/s $= 17.9$ m/s $= 64.4$ km/h

---

110 •• A loudspeaker diaphragm 20 cm in diameter is vibrating at 800 Hz with an amplitude of 0.025 mm. Assuming that the air molecules in the vicinity have this same amplitude of vibration, find (*a*) the pressure amplitude immediately in front of the diaphragm, (*b*) the sound intensity, and (*c*) the acoustic power being radiated.

Except for numerical values, this problem is identical to Example 15-6. Following the same procedure, one finds (*a*) $p_0 = 55.1$ N/m^2. (*b*) $I = 3.46$ W/m^2. (*c*) $P = 0.109$ W.

---

111 •• A plane, harmonic, acoustical wave that oscillates in air with an amplitude of 10^{-6} m has an intensity of 10^{-2} W/m^2. What is the frequency of the sound wave?

From Equ. 15-28, $\omega = \sqrt{\frac{2I}{\rho v s_0^2}}$ $\quad \omega = \sqrt{\frac{2 \times 10^{-2}}{1.29 \times 340 \times 10^{-12}}}\ \mathrm{s}^{-1} = 6.75 \times 10^3\ \mathrm{s}^{-1}$

$= 1.07$ kHz

---

112 •• Water flows at 7 m/s in a pipe of radius 5 cm. A plate having an area equal to the cross-sectional area of the pipe is suddenly inserted to stop the flow. Find the force exerted on the plate. Take the speed of sound in water to be 1.4 km/s. (*Hint:* When the plate is inserted, a pressure wave propagates through the water at the speed of sound v_s. The mass of water brought to a stop in time Δt is the water in a length of tube equal to $v_s\Delta t$.)

1. Find Δm of water in length $v\Delta t$ $\quad \Delta m = \rho v A \Delta t$

2. $F = \Delta p/\Delta t = \Delta m v_w/\Delta t = \rho v v_w A$ $\quad F = 1.4 \times 10^3 \times 10^3 \times 7 \times \pi \times 0.05^2$ N $= 77$ kN

113*•• Two wires of different linear mass densities are soldered together end to end and are then stretched under a tension F (the tension is the same in both wires). The wave speed in the second wire is three times that in the first wire. When a harmonic wave traveling in the first wire is reflected at the junction of the wires, the reflected wave has half the amplitude of the incident wave. (*a*) If the amplitude of the incident wave is A, what are the amplitudes of the reflected and transmitted waves?(*b*) Assuming no loss in the wire, what fraction of the incident power is reflected at the junction and what fraction is transmitted? (*c*) Show that the displacement just to the left of the junction equals that just to the right of the junction.

(*a*) 1. From Example 15-9, $A^2/v_1 = A_t^2/v_2 + A_r^2/v_1$ $\quad A^2/v_1 = A_t^2/3v_1 + A_t^2/4v_1$

2. Solve for A_t and A_r $\quad A_t = 3A/2$, $A_r = \frac{1}{2}A$ (given)

(*b*) $P_r = (A_r^2/A^2)P_i$; $P_t = P_i - P_r$ $\quad P_r = P_i/4$; $P_t = 3P_i/4$

(*c*) $A_{\mathrm{left}} = A + A_r$ $\quad A_{\mathrm{left}} = 3A/2 = A_t$

---

114 •• A column of precision marchers keeps in step by listening to the band positioned at the head of the column. The beat of the music is for 100 paces per minute. A television camera shows that only the marchers at the front and the rear of the column are actually in step. The marchers in the middle section are striding forward with the left foot when those at the front and rear are striding forward with the right foot. The marchers are so well trained, however, that they are all certain that they are in proper step with the music. Explain the source of the problem, and calculate the length of the column.

The source of the problem is that it takes a finite time for the sound to travel from the front of the line of marchers to the back. From the data given, the time for the sound to travel the length of the column is 1/100 min = 0.6 s. Therefore, the length of the column is $L = 340 \times 0.6$ m = 204 m.

---

115 •• Hovering over the pit of hell, the devil observes that as a student falls past (with terminal velocity), the frequency of his scream decreases from 842 to 820 Hz. (*a*) Find the speed of descent of the student. (*b*) The student's scream reflects from the bottom of the pit. Find the frequency of the echo as heard by the student. (*c*) Find the frequency of the echo as heard by the devil.

(*a*) 1. 842 Hz $= f_0(1 + u/v)$; 820 Hz $= f_0(1 - u/v)$ — $f_0 = 1662/2$ Hz = 831 Hz

2. Solve for u — $u = 11v/831$ m/s = 4.5 m/s

(*b*) Apply Equ. 15-35 — $f' = [831(1 + 4.5/340)/(1 - 4.5/340)]$ Hz = 853 Hz

(*c*) Here the shift is only due to the moving source — $f' = 842$ Hz

---

116 •• A bat flying toward an obstacle at 12 m/s emits brief, high-frequency sound pulses at a repetition frequency of 80 Hz. What is the time interval between the echo pulses heard by the bat?

1. Use Equ. 15-35 to find the reflected repetition rate — $f' = [80(1 + 12/340)/(1 - 12/340)]\ \mathrm{s^{-1}} = 85.9\ \mathrm{s^{-1}}$

2. Time interval $\Delta t = 1/f'$ — $\Delta t = 11.6$ ms

117*•• A tuning fork attached to a stretched wire generates transverse waves. The vibration of the fork is perpendicular to the wire. Its frequency is 400 Hz, and the amplitude of its oscillation is 0.50 mm. The wire has linear mass density of 0.01 kg/m and is under a tension of 1 kN. Assume that there are no reflected waves. (*a*) Find the period and frequency of waves on the wire. (*b*) What is the speed of the waves? (*c*) What are the wavelength and wave number? (*d*) Write a suitable wave function for the waves on the wire. (*e*) Calculate the maximum speed and acceleration of a point on the wire. (*f*) At what average rate must energy be supplied to the fork to keep it oscillating at a steady amplitude?

(*a*) $f = 400$ Hz (given); $T = 1/f = 2.5$ ms. (*b*) $v = \sqrt{F/\mu} = \sqrt{10^5}$ m/s = 316 m/s. (*c*) $\lambda = f/v = 79$ cm; $k = 2\pi/\lambda = 7.95\ \mathrm{m^{-1}}$. (*d*) $y(x, t) = A \sin(kx - \omega t) = [5 \times 10^{-4} \sin(7.95x - 2.51 \times 10^3 t)]$ m.
(*e*) $v_{max} = A\omega = 1.26$ m/s; $a_{max} = A\omega^2 = 3.16 \times 10^3\ \mathrm{m/s^2}$. (*f*) $P_{av} = \frac{1}{2}\mu\omega^2A^2v = 10^{-2} \times 2\pi^2 \times 16 \times 10^4 \times 25 \times 10^{-8} \times 316$ W = 2.5 W.

---

118 •• A very long wire can be vibrated up and down with a mechanical motor to produce waves traveling down the wire. At the far end of the wire, the traveling waves are absorbed by a clever device that allows no reflected waves to be returned to the motor. The wave speed is observed to be 240 m/s, the maximum transverse displacement of the wire is 1 cm, and the distance between maxima is 3.0 m. (*a*) Write a wave function to represent the wave propagating down this wire. (*b*) What is the frequency of vibration of the motor? (*c*) What is the period of the transverse oscillations of the wire? (*d*) What is the maximum transverse velocity of a small insect clinging to the wire?

(*a*) 1. Determine k and ω; $k = 2\pi/\lambda$; $\omega = kv$ — $k = 2\pi/3\ \mathrm{m^{-1}}$; $\omega = 160\pi$ rad/s

2. $y(x, t) = A \sin(kx - \omega t)$ — $y(x, t) = 0.01 \sin(2\pi x/3 - 160\pi t)$ m

(*b*) $f = \omega/2\pi$ — $f = 80$ Hz

(*c*) $T = 1/f$ $\qquad$ $T = 12.5$ ms

(*d*) $v_{max} = A\omega$ $\qquad$ $v_{max} = 1.6\pi$ m/s $= 5.03$ m/s

---

119 ••• If a loop of chain is spun at high speed, it will roll like a hoop without collapsing. Consider a chain of linear mass density μ that is rolling without slipping at a high speed v_0. (*a*) Show that the tension in the chain is $F = \mu v_0^2$. (*b*) If the chain rolls over a small bump, a transverse wave pulse will be generated in the chain. At what speed will it travel along the chain? (*c*) How far around the loop (in degrees) will a transverse wave pulse travel in the time the hoop rolls through one complete revolution?

Since the chain is rolling at high speed we shall neglect the effect of gravity.

(*a*) The adjacent figure shows a (small) portion of the chain. The angle θ is presumed to be small (although shown large). Let Δm be the mass of that segment of the chain. Then we require that $\Delta m v_0^2/R = F_{net}$. Now $\Delta m = \mu R\theta$ and $F_{net} = 2F\sin(\theta/2)$. For $\theta << 1$, $\sin(\theta/2) = \theta/2$. Thus $\mu v_0^2\theta = F\theta$ and $F = \mu v_0^2$.

(*b*) The wave speed is $(F/\mu)^{1/2} = v_0$, i.e., the same as the speed at which the chain is moving.

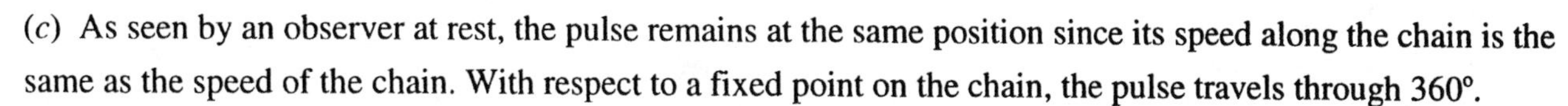

(*c*) As seen by an observer at rest, the pulse remains at the same position since its speed along the chain is the same as the speed of the chain. With respect to a fixed point on the chain, the pulse travels through 360°.

---

120 ••• A long rope with a mass per unit length of 0.1 kg/m is under a constant tension of 10 N. A motor at the point $x = 0$ drives one end of the rope with harmonic motion at 5 oscillations per second and an amplitude of 4 cm. (*a*) What is the wave speed? (*b*) What is the wavelength? (*c*) What is the maximum transverse linear momentum of a 1-mm segment of the rope? (*d*) What is the maximum net force on a 1-mm segment of the rope?

(*a*) $v = \sqrt{F/\mu}$ $\qquad$ $v = 10$ m/s

(*b*) $\lambda = v/f$ $\qquad$ $\lambda = 2$ m

(*c*) $p_{max} = \mu\Delta x A\omega$ $\qquad$ $p_{max} = 2\pi \times 0.1 \times 10^{-3} \times 0.04 \times 5$ kg·m/s $= 1.26 \times 10^{-4}$ kg·m/s

(*d*) $F_{max} = \mu\Delta x A\omega^2 = \omega p_{max}$ $\qquad$ $F_{max} = 10\pi \times 1.26 \times 10^{-4}$ N $= 3.95$ mN

121* ••• A heavy rope 3 m long is attached to the ceiling and is allowed to hang freely. (*a*) Show that the speed of transverse waves on the rope is independent of its mass and length but does depend on the distance y from the bottom according to the formula $v = \sqrt{gy}$. (*b*) If the bottom end of the rope is given a sudden sideways displacement, how long does it take the resulting wave pulse to go to the ceiling, reflect, and return to the bottom of the rope?

(*a*) The speed is given by $v = \sqrt{F/\mu}$. At a distance y from the bottom, $F = \mu gy$. Thus $v = \sqrt{gy}$.

(*b*) $dy/dt = v\ \sqrt{gy}$. The time to travel up and back is two times t, the time to travel from $y = 0$ to $y = 3$ m. We find t by integration.

$$t = \frac{1}{\sqrt{g}}\int_0^3 \frac{dy}{\sqrt{y}} = 2\sqrt{3/9.81}\ \text{s} = 1.106\ \text{s}.$$

Thus the total time is 2.21 s.

122 ••• The linear mass density of a nonuniform wire under constant tension decreases gradually along the wire so that an incident wave is transmitted without reflection. The wire is uniform for $-\infty \le x \le 0$. In this region, a transverse wave has the form $y(x, t) = 0.003 \cos(25x - 50t)$, where y and x are in meters and t is in seconds. From $x = 0$ to $x = 20$ m, the linear mass density decreases gradually from μ_1 to $\mu_1/4$. For $20 \le x \le \infty$, the linear mass density is $\mu = \mu_1/4$. (*a*) Find the wave velocity for large values of x. (*b*) Find the amplitude of the wave for large values of x. (*c*) Give $y(x, t)$ for $20 \le x \le \infty$.

(*a*) 1. Find v_1 from the expression for $y(x, t)$ — $v_1 = \omega/k = 2$ m/s

2. $v \propto 1/\sqrt{\mu}$ — $v_2 = 2v_1 = 4$ m/s

(*b*) No reflection, so use energy conservation;

$P_1 = P_2$. $P \propto \mu A^2 v$, so $A_2 = A_1\sqrt{\mu_1 v_1/\mu_2 v_2}$ — $A_2 = A_1\sqrt{2} = 4.24$ mm

(*c*) Write $y(x, t)$ using above results; $k_2 = k_1/2$. — $y(x,t) = 0.00424 \cos(12.5x - 50t)$

---

123 ••• In this problem you will derive an expression for the potential energy of a segment of a string carrying a traveling wave (Figure 15-31). The potential energy of a segment equals the work done by the tension in stretching the string, which is $\Delta U = F(\Delta\ell - \Delta x)$, where F is the tension, $\Delta\ell$ is the length of the stretched segment, and Δx is its original length. From the figure we see that

$$\Delta\ell \approx \sqrt{(\Delta x)^2 + (\Delta y)^2} = \Delta x[1 + (\Delta y/\Delta x)^2]^{1/2}$$

(*a*) Use the binomial expansion to show that $\Delta\ell - \Delta x \approx \frac{1}{2}(\Delta y/\Delta x)^2\Delta x$, and therefore $\Delta U \approx \frac{1}{2}F(\Delta y/\Delta x)^2\Delta x$.

(*b*) Compute dy/dx from the wave function in Equ. 15-13 and show that $\Delta U \approx \frac{1}{2}Fk^2A^2\cos^2(kx - \omega t)\Delta x$.

(*c*) Use $F = \mu v^2$ and $v = \omega/k$ to show that your result for (*b*) is the same as Equ. 15-16*b*.

(*a*) For $\Delta y/\Delta x \ll 1$, $\Delta\ell = \Delta x[1 + \frac{1}{2}(\Delta y/\Delta x)^2]$; so, $\Delta\ell - \Delta x = \frac{1}{2}(\Delta y/\Delta x)^2\Delta x$ and $\Delta U = \frac{1}{2}F(\Delta y/\Delta x)^2\Delta x$.

(*b*) $(dy/dx)^2 = A^2k^2\cos^2(kx - \omega t)$. So $\Delta U = \frac{1}{2}FA^2k^2\Delta x\cos^2(kx - \omega t)$.

(*c*) Replace F by $\mu v^2 = \mu\omega^2/k^2$. This gives Equ. 15-16*b*.

CHAPTER 16

Superposition and Standing Waves

1* • True or false: (*a*) The waves from two coherent sources that are radiating in phase interfere constructively everywhere in space. (*b*) Two wave sources that are out of phase by 180° are incoherent. (*c*) Interference patterns are observed only for coherent sources.

(*a*) False (*b*) False (*c*) True

2 • Two violinists standing a few feet apart are playing the same notes. Are there places in the room where certain notes are not heard because of destructive interference? Explain.

No; the two sources are incoherent and do not give rise to interference.

3 •• Two rectangular wave pulses are traveling in opposite directions along a string. At $t = 0$, the two pulses are as shown in Figure 16-27. Sketch the wave functions for $t = 1$, 2, and 3 s.

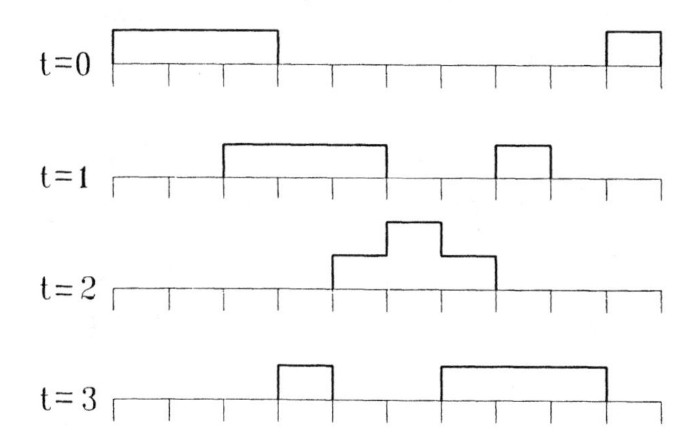

4 •• Repeat Problem 3 for the case in which the pulse on the right is inverted.

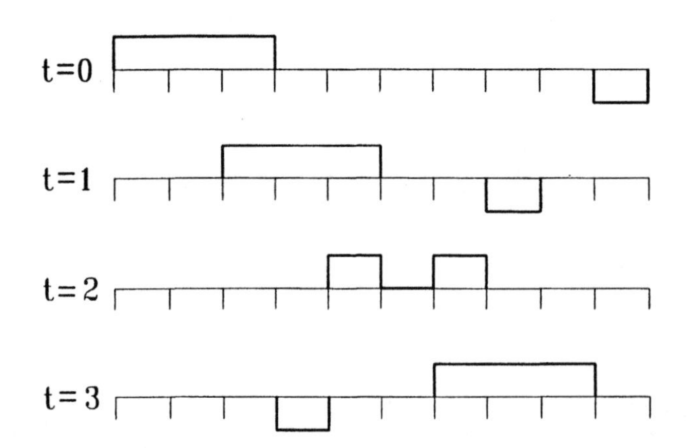

5* • Two waves traveling on a string in the same direction both have a frequency of 100 Hz, a wavelength of 2 cm, and an amplitude of 0.02 m. What is the amplitude of the resultant wave if the original waves differ in phase by (*a*) $\pi/6$, and (*b*) $\pi/3$?

(*a*) $A = 2y_0 \cos \frac{1}{2}\delta$ $\qquad$ $A = 4 \cos \pi/12$ cm = 3.86 cm

(*b*) Repeat for $\delta = \pi/3$ $\qquad$ $A = 4 \cos \pi/6$ cm = 3.46 cm

6 • What is the phase difference between the two waves of Problem 5 if the amplitude of the resultant wave is 0.02 m, the same as the amplitude of each original wave?

Now $\cos \delta/2 = 0.5$ $\qquad$ $\delta = 120° = 2\pi/3$ rad

7 • Two waves having the same frequency, wavelength, and amplitude are traveling in the same direction. If they differ in phase by $\pi/2$ and each has an amplitude of 0.05 m, what is the amplitude of the resultant wave?

$A = 2y_0 \cos \pi/4$ $\qquad$ $A = 7.07$ cm

8 • Two sound sources oscillate in phase with the same amplitude *A*. They are separated in space by $\lambda/3$. What is the amplitude of the resultant wave from the two sources at a point that is on the line that passes through the sources but is not between the sources?

Use Equs. 16-9 and 16-6 $\qquad$ $\delta = 2\pi/3$; $A_{res} = 2A \cos \pi/3 = A$

9* • Two sound sources oscillate in phase with a frequency of 100 Hz. At a point 5.00 m from one source and 5.85 m from the other, the amplitude of the sound from each source separately is *A*. (*a*) What is the phase difference in the sound waves from the two sources at that point? (*b*) What is the amplitude of the resultant wave at that point?

(*a*) Find $\Delta x/\lambda$, then use Equ. 16-9 $\qquad$ $\lambda = 3.4$ m; $\Delta x/\lambda = 0.25$; $\delta = 90°$

(*b*) Apply Equ. 16-6 $\qquad$ $A_{res} = 2A \cos 45° = A\sqrt{2}$

10 • With a compass, draw circular arcs representing wave crests originating from each of two point sources a distance $d = 6$ cm apart for $\lambda = 1$ cm. Connect the intersections corresponding to points of constant path difference, and label the path difference for each line. (see Figure 16-8)

The Figure is shown below. Lines of constructive interference are shown for path differences of 0, λ, 2λ, and 3λ.

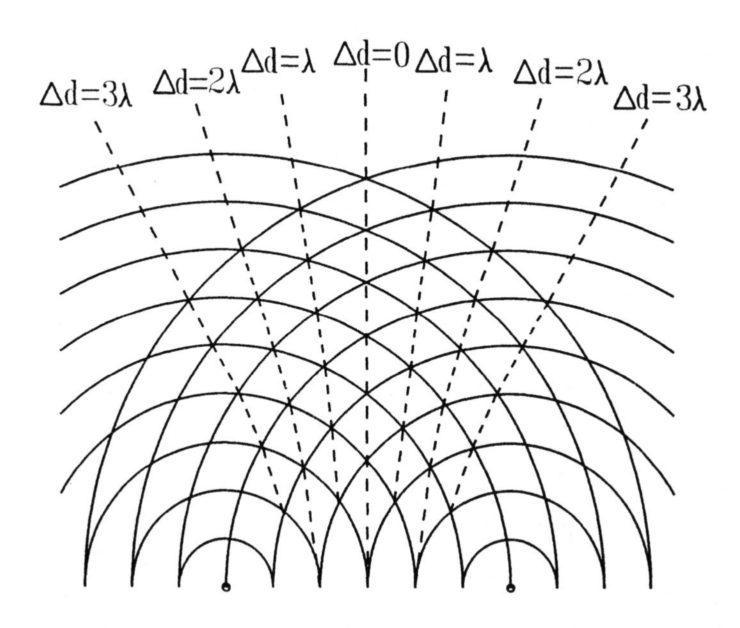

11 • Two loudspeakers are separated by a distance of 6 m. A listener sits directly in front of one speaker at a distance of 8 m so that the two speakers and the listener form a right triangle. (*a*) Find the two lowest frequencies for which the path difference from the speakers to the listener is an odd number of half-wavelengths. (*b*) Why can these frequencies be heard even if the speakers are driven in phase by the same amplifier?

(*a*) 1. Find λ_1 and λ_2; $r_1 = 8$ m, $r_2 = 10$ m — $\Delta s = 2$ m; $\Delta s/\lambda_1 = ½$; $\lambda_1 = 4$ m; $\lambda_2 = \lambda_1/3$

2. Determine f_1 and f_2; $f = v/\lambda$ — $f_1 = 85$ Hz; $f_2 = 255$ Hz

(*b*) The sounds can be heard because of random reflections by the walls of the room.

12 • Two speakers separated by some distance emit sound waves of the same frequency. At some point *P*, the intensity due to each speaker separately is I_0. The path distance from *P* to one of the speakers is $½\lambda$ greater than that from *P* to the other speaker. What is the intensity at *P* if (*a*) the speakers are coherent and in phase, (*b*) the speakers are incoherent, and (*c*) the speakers are coherent but have a phase difference of π rad?

(*a*) Since the path difference is $\lambda/2$, there is complete destructive interference; $I = 0$.

(*b*) The sources are incoherent and the intensities add; $I = 2I_0$.

(*c*) The phase difference is $\pi + 2\pi\Delta x/\lambda = 2\pi$. The waves interfere constructively. $I = 4I_0$.

13* • Answer the questions of Problem 12 for a point *P'* for which the distance to the far speaker is 1λ greater than the distance to the near speaker. Assume that the intensity at point *P'* due to each speaker separately is again I_0.

(*a*) The path difference is λ, so we have constructive interference. $I = 4I_0$.

(*b*) As in Problem 12, $I = 2I_0$.

(*c*) Now the phase difference is π; $I = 0$.

14 • Two speakers separated by some distance emit sound waves of the same frequency, but the speakers are out of phase by 90°. Let r_1 be the distance from some point to speaker 1 and r_2 be the distance from that point to speaker 2. Find the smallest value of $r_2 - r_1$ such that the sound at that point will be (*a*) maximum, and (*b*) minimum. (Express your answers in terms of the wavelength.)

(*a*) A phase difference of 90° corresponds to a path difference of $\lambda/4$; to compensate $r_2 - r_1 = \lambda/4$.

(*b*) In this case, the smallest difference in path is again $\lambda/4$, but now $\lambda/4 = r_1 - r_2$.

---

15 •• Show that if the separation between two sound sources radiating coherently in phase is less than half a wavelength, complete destructive interference will not be observed in any direction.

For destructive interference, $\delta = \pi = 2\pi\Delta x/\lambda$. $\Delta x = d\sin\theta \le d$, where d is the source separation. So if $\Delta x < \lambda/2$, $\delta < \pi$, and there is no complete destructive interference in any direction.

---

16 •• A transverse wave of frequency 40 Hz propagates down a string. Two points 5 cm apart are out of phase by $\pi/6$. (*a*) What is the wavelength of the wave? (*b*) At a given point, what is the phase difference between two displacements for times 5 ms apart? (*c*) What is the wave velocity?

(*a*) $\delta = \pi/6$ corresponds to a distance of $\lambda/12$ $\qquad \lambda = 5 \times 12$ cm $= 60$ cm

(*b*) $T = 1/f = 25$ ms $\qquad$ 5 ms $= T/5$; $\delta = 2\pi/5$ rad

(*c*) $v = f\lambda$ $\qquad v = 40 \times 0.6$ m/s $= 24$ m/s

---

17* •• It is thought that the brain determines the direction toward the source of a sound by sensing the phase difference between the sound waves striking the eardrums. A distant source emits sound of frequency 680 Hz. When you are facing directly toward a sound source there should be no phase difference. Estimate the change in phase difference between the sounds received by the ears as you turn from facing directly toward the source through 90°.

The distance between the ears ≈ 20 cm. $\lambda = 340/680$ m $= 50$ cm. So $\delta \approx 2\pi(20/50) = 0.8\pi$ rad.

---

18 •• Sound source A is located at $x = 0$, $y = 0$, and sound source B is placed at $x = 0$, $y = 2.4$ m. The two sources radiate coherently in phase. An observer at $x = 40$ m, $y = 0$ notes that as she walks along in either the positive or negative y direction away from $y = 0$, the sound intensity diminishes. What is the lowest and the next higher frequency of the sources that can account for that observation?

The path difference between the two sources is $\Delta x = (\sqrt{40^2 + 2.4^2} - 40)$ m $= 0.072$ m. For constructive interference at the point of observation, we must have 0.072 m $= n\lambda$. For $n = 1$, $\lambda = 0.072$ m and $f_1 = 4722$ Hz; for $n = 2$, $f_2 = 2f_1 = 9444$ Hz.

---

19 •• Suppose that the observer in Problem 18 finds herself at a point of minimum intensity at $x = 40$ m, $y = 0$. What is then the lowest and next higher frequency of the sources consistent with this observation?

We now require that the path difference of 0.072 m be $\lambda/2$ or $3\lambda/2$. If $\lambda_1 = 0.144$ m, $f_1 = 2361$ Hz; if $\lambda = \lambda_2 = 3\lambda_1$, $f_2 = 7083$ Hz.

---

20 •• Two point sources that are in phase are separated by a distance d. An interference pattern is detected along a line parallel to the line through the sources and a large distance D from the sources, as shown in Figure 16-28. (*a*) Show that the path difference from the two sources to some point on the line at a small angle θ is given approximately by $\Delta s = d\sin\theta$. (*Hint:* Assume that the lines from the sources to P are approximately parallel.) (*b*) Show that the distance y_m from the central maximum point to the mth interference maximum is given approximately by $y_m = m(D\lambda/d)$.

(*a*) See the diagram. As shown, $\Delta s = d \sin \theta$

(*b*) If $\theta << 1$, then $\Delta x = d \sin \theta \approx d \tan \theta = dy_m/D$. For constructive interference, $\delta = 2\pi m = 2\pi \Delta x/\lambda = 2\pi d y_m/D\lambda$.
Solving for y_m we obtain $y_m = mD\lambda/d$.

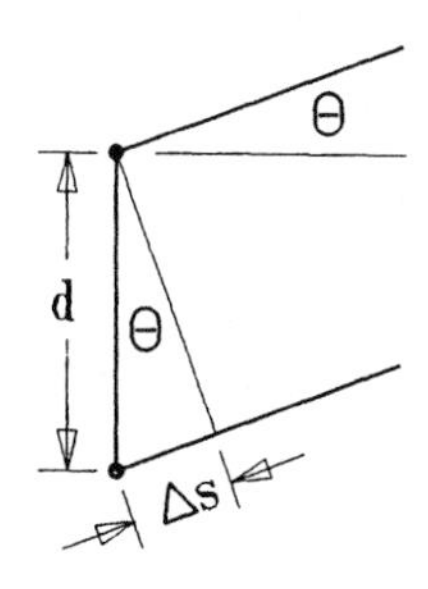

21* •• Two sound sources radiating in phase at a frequency of 480 Hz interfere such that maxima are heard at angles of 0° and 23° from a line perpendicular to that joining the two sources. Find the separation between the two sources and any other angles at which a maximum intensity will be heard. (Use the result of Problem 20.)

Since 23° = 0.401 rad is not a "small" angle, we cannot use the small angle approximation.

$\Delta x = \lambda = 340/480$ m $= 0.708$ m $= d \sin 23°$ — $d = 0.708/\sin 23°$ m $= 1.81$ m

If $d \sin \theta = 2\lambda$, there will be another intensity maximum — $\sin \theta = 1.416/1.81 = 0.782$; $\theta = 51.5°$

---

22 ••• Two loudspeakers are driven in phase by an audio amplifier at a frequency of 600 Hz. The speakers are on the *y* axis, one at $y = +1.00$ m and the other at $y = -1.00$ m. A listener begins at $y = 0$ a very large distance D away and walks along a line parallel to the *y* axis. (See Problem 20.) (*a*) At what angle θ will she first hear a minimum in the sound intensity? (*b*) At what angle will she first hear a maximum (after $\theta = 0$)? (*c*) How many maxima can she possibly hear if she keeps walking in the same direction?

Proceed as in Problem 21. However, for the first intensity minimum we must have $\Delta x = \lambda/2$.

(*a*) 1. Determine λ	$\lambda = 340/600$ m $= 0.567$ m
2. $d \sin \theta = \lambda/2$; here $d = 2.0$ m	$\sin \theta = \lambda/2d = 0.567/4 = 0.142$; $\theta = 8.14°$
(*b*) For maximum, $d \sin \theta = \lambda$	$\sin \theta = 0.567/2$; $\theta = 16.5°$
(*c*) $\sin \theta = n\lambda/d \le 1$; solve for n (integer)	$n = 3$

---

23 ••• Two sound sources driven in phase by the same amplifier are 2 m apart on the *y* axis. At a point a very large distance from the *y* axis, constructive interference is first heard at an angle $\theta_1 = 0.140$ rad with the *x* axis and is next heard at $\theta_2 = 0.283$ rad (see Figure 16-28). (*a*) What is the wavelength of the sound waves from the sources? (*b*) What is the frequency of the sources? (*c*) At what other angles is constructive interference heard? (*d*) What is the smallest angle for which the sound waves cancel?

(*a*) $\sin \theta_1 = \lambda/d$ (see Problem 22)	$\lambda = d \sin \theta_1 = 0.279$ m
(*b*) $f = v/\lambda$	$f = 1219$ Hz
(*c*) $\sin \theta_n = n\lambda/d$	$\theta_3 = 0.432$ rad; $\theta_4 = 0.592$ rad; $\theta_5 = 0.772$ rad; $\theta_6 = 0.992$ rad; $\theta_7 = 1.354$ rad
(*d*) $\sin \theta_{dest} = \lambda/2d$	$\theta_{dest} = 0.0698$ rad

---

24 ••• Two identical sound sources have a frequency of 500 Hz. The coordinates of the sources are (0,1 m) and (0,−1 m) A detector 80 m from the origin is free to revolve in the *xy* plane with a radius of 80 m. The first maximum in intensity is detected when the detector is at (80, 0) m. (*a*) Find the coordinates of the detector for the first five maxima in order of decreasing positive values of x. (*b*) Find the coordinates for the first four minima in order of decreasing positive values of x subject to the condition $\sqrt{x^2 + y^2} = 80\,\text{m}$.

(*a*) Find the angles θ_m from $\sin\theta_m = m\lambda/d$ $\qquad$ $\lambda = 0.68$ m; $\theta_0 = 0$; $\theta_1 = 19.9°$; $\theta_2 = 42.8°$

The configuration of the sources and the detector is shown in the figure to the right. There are ten locations at which a maximum (constructive interference) can be observed. These are at $\theta = 0$, ±19.9°, ±42.8°, ±137.2°, ±160.1°, and 180°. The coordinates corresponding to these angles are given by $x = (80\cos\theta_m)$ m and $y = (80\sin\theta_m)$ m. In order of decreasing x we have: (80, 0), (75.2, ±27.2), (58.7, ±54.4), (−58.7, ±54.4), (−75.2, ±27.2), and (−80, 0). The first five maxima are those for $x > 0$.

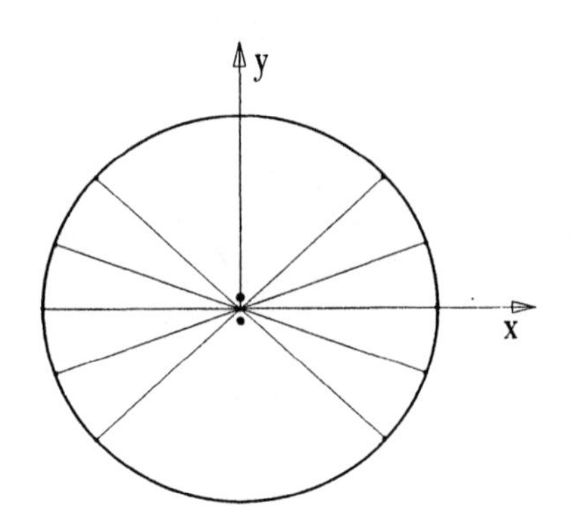

For destructive interference, $\sin\theta_n = (n - ½)\lambda/d$ $\qquad$ $\theta_1 = 9.79°$, $\theta_2 = 30.7°$

Again, positive and negative angles are possible. Thus, the first four minima in diminishing order of x are at coordinates (78.8, ±13.6), (68.8, ±40.8).

25* ••• A radio telescope consists of two antennas separated by a distance of 200 m. Both antennas are tuned to a particular frequency, such as 20 MHz. The signals from each antenna are fed into a common amplifier, but one signal first passes through a phase adjuster that delays its phase by a chosen amount so that the telescope can "look" in different directions. When the phase delay is zero, plane radio waves that are incident vertically on the antennas produce signals that add constructively at the amplifier. What should the phase delay be so that signals coming from an angle $\theta = 10°$ with the vertical (in the plane formed by the vertical and the line joining the antennas) will add constructively at the amplifier?

1. Determine λ and the path difference for $\theta = 10°$ $\qquad$ $\lambda = (3\times10^8/2\times10^7)\text{ m} = 15\text{ m}$
   Note that we can disregard the path of 2λ $\qquad$ $\Delta s = (200\sin 10°)\text{ m} = 34.73\text{ m} = 2.315\lambda = (2 + 0.315)\lambda$
2. Determine the phase difference for $\Delta\lambda = 0.315\lambda$ $\qquad$ $\delta = 0.315\times 2\pi = 1.98\text{ rad} = 113.5°$

---

26 • Beats are produced by the superposition of two harmonic waves only if (*a*) their amplitudes and frequencies are equal. (*b*) their amplitudes are the same but their frequencies differ slightly. (*c*) their frequencies differ slightly even if their amplitudes are not equal. (*d*) their frequencies are equal but their amplitudes differ slightly.

(*c*)

---

27 • True or false: The beat frequency between two sound waves of nearly equal frequencies equals the difference in the frequencies of the individual sound waves.

True

---

28 •• About how accurately do you think you can tune a piano string to a tuning fork?

It is possible to tune to within 0.5 Hz by listening to beats.

29* • Two tuning forks have frequencies of 256 and 260 Hz. If the forks are vibrating at the same time, what is the beat frequency?

$f_{\text{beat}} = \Delta f = 4$ Hz.

30 • When a violin string is played (without fingering) simultaneously with a tuning fork of frequency 440 Hz, beats are heard at the rate of 3 per second. When the tension in the string is increased slightly, the beat frequency decreases. What was the initial frequency of the violin string?

The initial frequency was low since an increase in tension reduces f_{beat}. $f_i = (440 - 3)$ Hz = 437 Hz.

---

31 • When two tuning forks are struck simultaneously, 4 beats per second are heard. The frequency of one fork is 500 Hz. (*a*) What are the possible values for the frequency of the other fork? (*b*) A piece of wax is placed on the 500-Hz fork to lower its frequency slightly. Explain how the measurement of the new beat frequency can be used to determine which of your answers to part (*a*) is the correct frequency of the second fork.

(*a*) $f_2 = f_1 \pm \Delta f = (500 \pm 4)$ Hz; $f_2 = 504$ Hz or 496 Hz.

(*b*) If the beat frequency is increased, then $f_2 = 504$ Hz; if diminished, $f_2 = 496$ Hz.

---

32 • True or false: (*a*) The frequency of the third harmonic is three times that of the first harmonic. (*b*) The frequency of the fifth harmonic is five times that of the fundamental. (*c*) In a pipe that is open at one end and closed at the other, the even harmonics are not excited.

(*a*) True (*b*) True (*c*) True

33* • Standing waves result from the superposition of two waves of (*a*) the same amplitude, frequency, and direction of propagation. (*b*) the same amplitude and frequency and opposite directions of propagation. (*c*) the same amplitude, slightly different frequency, and the same direction of propagation. (*d*) the same amplitude, slightly different frequency, and opposite directions of propagation.

(*b*)

---

34 • An organ pipe open at both ends has a fundamental frequency of 400 Hz. If one end of this pipe is now closed, the fundamental frequency will be (*a*) 200 Hz. (*b*) 400 Hz. (*c*) 546 Hz. (*d*) 800 Hz.

(*a*) Initially, $\lambda_1 = 2L$. If one end is closed, $\lambda_1 = 4L$.

---

35 •• A string fixed at both ends resonates at a fundamental frequency of 180 Hz. Which of the following will reduce the fundamental frequency to 90 Hz? (*a*) Double the tension and double the length. (*b*) Halve the tension and keep the length fixed. (*c*) Keep the tension fixed and double the length. (*d*) Keep the tension fixed and halve the length.

(*c*)

---

36 •• How do the resonance frequencies of an organ pipe change when the air temperature increases?

Since $f = v/\lambda$ and $v \propto \sqrt{T}$, inceasing the temperature will increase the resonance frequency.

37* • A string fixed at both ends is 3 m long. It resonates in its second harmonic at a frequency of 60 Hz. What is the speed of transverse waves on the string?

1. Determine λ_2	$\lambda_2 = L = 3$ m
2. $v = f_2\lambda_2$	$v = 60 \times 3$ m/s = 180 m/s

---

38 • A string 3 m long and fixed at both ends is vibrating in its third harmonic. The maximum displacement of any point on the string is 4 mm. The speed of transverse waves on this string is 50 m/s. (*a*) What are the wavelength and frequency of this wave? (*b*) Write the wave function for this wave.

(*a*) Determine λ_3 and f_3; $\lambda_3 = 2L/3$, $f_3 = v/\lambda_3$	$\lambda_3 = 2$ m; $f_3 = 50/2$ Hz = 25 Hz
(*b*) See Equ. 16-16	$y(x, t) = 4 \sin(x/\pi) \cos(50\pi t)$ mm

39 • Calculate the fundamental frequency for a 10-m organ pipe that is (*a*) open at both ends, and (*b*) closed at one end.

(*a*) and (*b*) Find λ_o and λ_c — $\lambda_o = 2L = 20$ m; $\lambda_c = 4L = 40$ m

$f_o = v/\lambda_o$; $f_c = v/\lambda_c$ — $f_o = 340/20$ Hz $= 17$ Hz; $f_c = 8.5$ Hz

40 • A steel wire having a mass of 5 g and a length of 1.4 m is fixed at both ends and has a tension of 968 N. (*a*) Find the speed of transverse waves on the wire. (*b*) Find the wavelength and frequency of the fundamental. (*c*) Find the frequencies of the second and third harmonics.

(*a*) $v = \sqrt{F/\mu} = \sqrt{FL/m}$ — $v = \sqrt{968 \times 1.4/5 \times 10^{-3}}$ m/s $= 520.6$ m/s

(*b*) $\lambda_1 = 2L$; $f_1 = v/\lambda$ — $\lambda_1 = 2.8$ m; $f_1 = 520.6/2.8$ Hz $= 186$ Hz

(*c*) $f_n = nf_1$ — $f_2 = 372$ Hz; $f_3 = 558$ Hz

41* • A rope 4 m long is fixed at one end; the other end is attached to a light string so that it is free to move. The speed of waves on the rope is 20 m/s. Find the frequency of (*a*) the fundamental, (*b*) the second harmonic, and (*c*) the third harmonic.

(*a*) Find $\lambda_1 = 4L$; $f_1 = v/\lambda_1$ — $\lambda_1 = 16$ m; $f_1 = 20/16$ Hz $= 1.25$ Hz

(*b*) The system does not support a second harmonic — $f_2 = 0$

(*c*) $f_3 = 3f_1$ — $f_3 = 3.75$ Hz

42 • A piano wire without windings has a fundamental frequency of 200 Hz. When it is wound with wire, its linear mass density is doubled. What is its new fundamental frequency, assuming that the tension is unchanged?

Doubling the mass per unit length reduces v by a factor of $1/\sqrt{2}$ since the tension is constant. As the length remains fixed, the new frequency is $f = 200/\sqrt{2}$ Hz $= 141$ Hz.

43 • The normal range of hearing is about 20 to 20,000 Hz. What is the greatest length of an organ pipe that would have its fundamental note in this range if (*a*) it is closed at one end, and (*b*) it is open at both ends?

$\lambda_{max} = 340/20$ m $= 17$ m. (*a*) 17 m $= 4L$, $L = 4.25$ m. (*b*) 17 m $= 2L$, $L = 8.5$ m.

44 •• The length of the B string on a certain guitar is 60 cm. Its fundamental is at 247 Hz. (*a*) What is the speed of transverse waves on the string? (*b*) If the linear mass density of the guitar string is 0.01 g/cm, what should its tension be when it is in tune?

(*a*) $\lambda = 2L$; $v = f\lambda$ — $\lambda = 1.2$ m; $v = 247 \times 1.2$ m/s $= 296.4$ m/s

(*b*) $F = \mu v^2$ — $F = 296.4^2 \times 10^{-3}$ N $= 87.9$ N

45* •• The wave function $y(x, t)$ for a certain standing wave on a string fixed at both ends is given by $y(x,t) = 4.2 \sin 0.20x \cos 300t$, where y and x are in centimeters and t is in seconds. (*a*) What are the wavelength and frequency of this wave? (*b*) What is the speed of transverse waves on this string? (*c*) If the string is vibrating in its fourth harmonic, how long is it?

(*a*) See Equ. 16-16 — $\lambda = 10\pi$ cm $= 31.4$ cm; $f = 300/2\pi$ Hz $= 47.7$ Hz

(*b*) $v = f\lambda$ — $v = 10\pi \times 300/2\pi$ cm/s $= 1500$ cm/s $= 15$ m/s

(*c*) $\lambda_4 = \lambda_1/4 = 2L/4 = L/2$ — $L = 20\pi$ cm $= 62.8$ cm

46 •• The wave function $y(x, t)$ for a certain standing wave on a string fixed at both ends is given by $y(x, t) = (0.05 \text{ m}) \sin 2.5 \text{ m}^{-1} x \cos 500 \text{ s}^{-1} t$. (*a*) What are the speed and amplitude of the two traveling waves that result in this standing wave? (*b*) What is the distance between successive nodes on the string? (*c*) What is the shortest possible length of the string?

(*a*) 1. From Equ. 16-16, find λ and f — $\lambda = 2\pi/2.5$ m = 2.51 m; $f = 500/2\pi$ Hz = 79.6 Hz

2. $A = A_{SW}/2$; $v = f\lambda = \omega/k$ — $A = 0.025$ m; $v = 500/2.5$ m/s = 200 m/s

(*b*) The distance between nodes = $\lambda/2$ — Distance between nodes is $\pi/2.5$ m = 1.26 m

(*c*) Shortest $L = \lambda/2$ — $L_{min} = 1.26$ m

---

47 •• A 2.51-m-long string has the wave function given in Problem 46. (*a*) Sketch the position of the string at the times $t = 0$, $t = T/4$, $t = T/2$, and $t = 3T/4$, where $T = 1/f$ is the period of the vibration. (*b*) Find T in seconds. (*c*) At a time t when the string is horizontal, that is, $y(x) = 0$ for all x, what has become of the energy in the wave?

(*a*) The function $y(x, t)$ for $t = 0$, $T/4$, $T/2$, and $3T/4$ are shown below. Since the functions for $t = T/4$ and $3T/4$ are identical, we show only one graph.

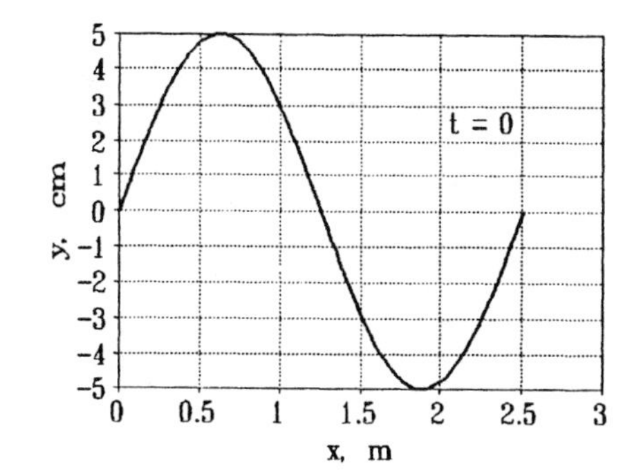

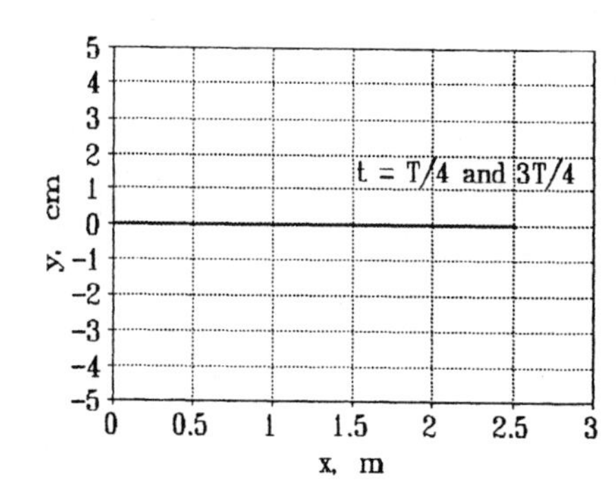

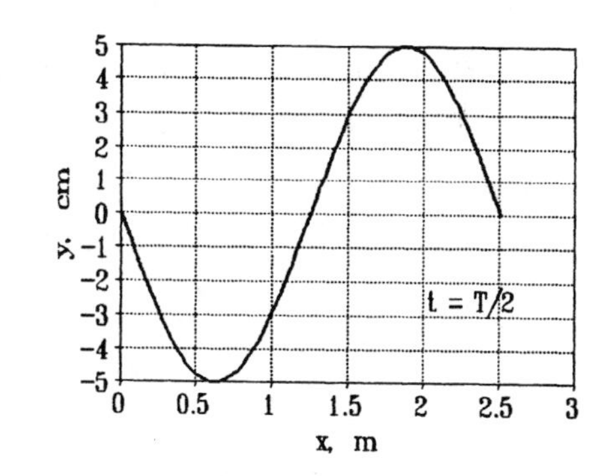

(*b*) $T = 1/f$; $f = 79.6$ Hz (see Problem 46) — $T = 12.6$ ms

(*c*) The energy of the wave is entirely kinetic energy

---

48 •• A string fixed at one end only is vibrating in its fundamental mode. The wave function is $y(x,t) = 0.02 \times \sin 2.36x \cos 377t$, where x and y are in meters and t is in seconds. (*a*) What is the wavelength of the wave? (*b*) What is the length of the string? (*c*) What is the speed of transverse waves on the string?

(*a*) $\lambda = 2\pi/k$; $k = 2.36 \text{ m}^{-1}$ — $\lambda = 2.66$ m

(*b*) In this case, $\lambda = 4L$ — $L = 0.666$ m

(*c*) $v = \omega/k$ — $v = 377/2.36$ m/s = 160 m/s

49* •• Three successive resonance frequencies for a certain string are 75, 125, and 175 Hz. (*a*) Find the ratios of each pair of successive resonance frequencies. (*b*) How can you tell that these frequencies are for a string fixed at one end only rather than for a string fixed at both ends? (*c*) What is the fundamental frequency? (*d*) Which harmonics are these resonance frequencies? (*e*) If the speed of transverse waves on this string is 400 m/s, find the length of the string.

(*a*) The ratios are 3/5 and 5/7. (*b*) There are no even harmonics, so the string must be fixed at one end only.

(*c*) The fundamental frequency is 75/3 Hz = 25 Hz. (*d*) They are the third, fifth, and seventh harmonics. (*e*) For the fundamental frequency, $\lambda_1 = v/f_1$ = 400/25 m = 16 m, and $L = \lambda_1/4$ = 4 m.

---

50 •• The space above the water in a tube like that shown in Example 16-8 is 120 cm long. Near the open end, there is a loudspeaker that is driven by an audio oscillator whose frequency can be varied from 10 to 5000 Hz. (*a*) What is the lowest frequency of the oscillator that will produce resonance within the tube? (*b*) What is the highest frequency that will produce resonance? (*c*) How many different frequencies of the oscillator will produce resonance? (Neglect the end correction.)

(*a*) $\lambda_{max} = 4L$; $f_{min} = v/\lambda_{max} = v/4L$ — f_{min} = 340/4.8 Hz = 70.8 Hz

(*b*) 1. Find the highest harmonic below 5000 Hz — $f_n = (2n + 1)(340/4.8) = 5000$

2. Solve for the lowest integer n — $n = 34$; f_{max} = 69(340/4.8) = 4888 Hz

(*c*) There are 35 resonance frequencies; n = 0 to 34

---

51 •• A 460-Hz tuning fork causes resonance in the tube in Example 16-8 when the top of the tube is 18.3 and 55.8 cm above the water surface. (*a*) Find the speed of sound in air. (*b*) What is the end correction to adjust for the fact that the antinode does not occur exactly at the end of the open tube?

(*a*) The distance between the antinodes is $\lambda/2$ = (55.8 − 18.3) cm. So λ = 75 cm, and $v = f\lambda$ = 345 m/s.

(*b*) Note that 55.8 ≈ 3 × 18.3. So assume 18.3 cm = $\lambda/4$ since the tube is closed at one end. Then λ = 73.2 cm. Assuming that v = 345 m/s is correct, the end correction is (75 − 73.2)/4 cm = 0.45 cm.

---

52 •• At 16°C, the fundamental frequency of an organ pipe is 440.0 Hz. What will be the fundamental frequency of the pipe if the temperature increases to 32°C? Would it be better to construct the pipe with a material that expands substantially as the temperature increases or should the pipe be made of material that maintains the same length at all normal temperatures?

$v \propto \sqrt{T}$ and $f = v/\lambda$; if λ fixed then $f \propto \sqrt{T}$ — $f_{305\text{ K}} = f_{289\text{ K}}\sqrt{305/289}$ = 440 × 1.0273 Hz = 452 Hz

It would be better to have the pipe expand so that v/L is independent of temperature.

53* •• A violin string of length 40 cm and mass 1.2 g has a frequency of 500 Hz when it is vibrating in its fundamental mode. (*a*) What is the wavelength of the standing wave on the string? (*b*) What is the tension in the string? (*c*) Where should you place your finger to increase the frequency to 650 Hz?

(*a*) $\lambda = 2L$ — λ = 80 cm

(*b*) $v = f\lambda = \sqrt{FL/m}$; $F = f^2\lambda^2 m/L$ — F = 480 N

(*c*) $L_{650} = L_{500}(500/650)$ — L_{650} = 30.77 cm; place finger 9.23 cm from scroll bridge

---

54 •• The G string on a violin is 30 cm long. When played without fingering, it vibrates at a frequency of 196 Hz. The next higher notes on the C-major scale are A (220Hz), B (247 Hz), C (262 Hz), and D (294 Hz). How far from the end of the string must a finger be placed to play each of these notes?

Determine the free length of the string from $L(f') = L(f_G)(196/f')$. Position of finger from scroll bridge is then $x(f') = L(f_G) - L(f')$.

1. Determine L(A), L(B), L(C), and L(D) — L(A) = 30 × 196/220 cm = 26.73 cm; L(B) = 23.81 cm; L(C) = 22.44 cm; L(D) = 20.0 cm

2. Determine x_A, x_B, x_C, and x_D — x_A = 3.27 cm; x_B = 6.19 cm; x_C = 7.56 cm; x_D = 10 cm

---

55 •• A string with a mass density of 4×10^{-3} kg/m is under a tension of 360 N and is fixed at both ends. One of its resonance frequencies is 375 Hz. The next higher resonance frequency is 450 Hz. (*a*) What is the fundamental frequency of this string? (*b*) Which harmonics are the ones given? (*c*) What is the length of the string?

(*a*) $nf_1 = 375$ Hz; $(n + 1)f_1 = 450$ Hz; solve for f_1 — $f_1 = 75$ Hz

(*b*) Determine n — $n = 5$, $n + 1 = 6$; fifth and sixth harmonics

(*c*) Find v; $\lambda = 2L = v/f_1$; $L = v/2f_1$ — $v = \sqrt{360/4\times10^{-3}}$ m/s $= 300$ m/s; $L = 2$ m

---

56 •• A string fastened at both ends has successive resonances with wavelengths of 0.54 m for the *n*th harmonic and 0.48 m for the $(n + 1)$th harmonic. (*a*) Which harmonics are these? (*b*) What is the length of the string?

(*a*) $nf_1 = v/0.54$; $(n + 1)f_1 = v/0.48$; solve for n — $(n + 1)/n = 1.125$; $n = 8$, $n + 1 = 9$

(*b*) $L = n\lambda_n/2$ — $L = 4 \times 0.54 = 2.16$ m

57* •• A rubber band with an unstretched length of 0.80 m and a mass of 6×10^{-3} kg stretches to 1.20 m when under a tension of 7.60 N. What is the fundamental frequency of oscillation of this band when stretched between two fixed posts 1.20 m apart?

1. Find v; $v = \sqrt{F/\mu}$; $\mu = m/L$ — $v = \sqrt{7.6\times1.2/6\times10^{-3}}$ m/s $= 39$ m/s

2. $f_1 = v/\lambda = v/2L$ — $f_1 = 39/2.4$ Hz $= 16.2$ Hz

---

58 •• The strings of a violin are tuned to the tones G, D, A, and E, which are separated by a fifth from one another. That is, $f(\text{D}) = 1.5f(\text{G})$, $f(\text{A}) = 1.5f(\text{D}) = 440$ Hz, and $f(\text{E}) = 1.5f(\text{A})$. The distance between the two fixed points, the bridges at the scroll and over the body of the instrument, is 30 cm. The tension on the E string is 90 N. (*a*) What is the mass per meter of the E string? (*b*) To prevent distortion of the instrument over time, it is important that the tension on all strings be the same. Find the masses per meter of the other strings.

(*a*) $v = f_E \times 2L$; $\mu = F/v^2$ — $v = 660 \times 0.6$ m/s $= 396$ m/s; $\mu = 90/396^2$ kg/m $= 0.574$ g/m

(*b*) 1. Determine v_A, v_D, and v_G — $v_A = 440 \times 0.6$ m/s $= 264$ m/s; $v_D = 176$ m/s; $v_G = 117$ m/s

2. Find μ_A, μ_D, and μ_G; $\mu = F/v^2$ — $\mu_A = 1.29$ g/m; $\mu_D = 2.91$ g/m; $\mu_G = 6.54$ g/m

---

59 •• To tune a violin, the violinist first tunes the A string to the correct pitch of 440 Hz and then bows two adjoining strings simultaneously and listens for a beat pattern. While bowing the A and E strings, the violinist hears a beat frequency of 3 Hz and notes that the beat frequency increases as the tension on the E string is increased. (The E string is to be tuned to 660 Hz.) (*a*) Why is a beat produced by these two strings bowed simultaneously? (*b*) What is the frequency of the E string vibration when the beat frequency is 3 Hz? (*c*) If the tension on the E string is 80.0 N when the beat frequency is 3 Hz, what tension corresponds to perfect tuning of that string?

(*a*) The two sounds produce a beat because the third harmonic of 440 Hz equals the second harmonic of 660 Hz, and the original frequency of the E string is slightly greater than 660 Hz. If $f_E = (660 + \Delta f)$ Hz, a beat of $2\Delta f$ will be heard.

(*b*) Since f_{beat} increases with increasing tension, the frequency of the E string is greater than 660 Hz. Thus the frequency of the E string is 661.5 Hz.

(*c*) $f \propto \sqrt{F}$; so $F_{660} = 80(660/661.5)^2$ N = 79.6 N.

---

60 •• (*a*) For the wave function given in Problem 48, find the velocity of a string segment at some point *x* as a function of time. (*b*) Which point has the greatest speed at any time? What is the maximum speed of this point? (*c*) Find the acceleration of a string segment at some point *x* as a function of time. (*d*) Which point has the greatest acceleration? What is the maximum acceleration of this point?

(*a*) $v_y = dy/dt = -(377 \times 0.02 \sin 2.36x \sin 377t)$ m/s $\quad v_y = -7.54 \sin 2.36x \sin 377t$ m/s

(*b*) v_y is maximum when $\sin 2.36x = 1$ $\quad x = \frac{1}{2}\pi/2.36$ m = 0.666 m; $v_{y,max} = 377 \times 0.02$ m/s = 7.54 m/s

(*c*) $a_y = -(377^2 \times 0.02 \sin 2.36x \cos 377t)$ m/s² $\quad a_y = -2843 \sin 2.36x \cos 377t$ m/s²

(*d*) $a_{y,max}$ for $\sin 2.36x = 1$, when $\cos 377t = 1$ $\quad x = 0.666$ m; $a_{y,max} = 2843$ m/s²

61* •• A student carries a small oscillator and speaker as she walks very slowly down a long hall. The speaker emits a sound of frequency 680 Hz which is reflected from the walls at each end of the hall. The student notes that as she walks along, the sound intensity she hears passes through successive maxima and minima. What distance must she walk to pass from one maximum to the next?

The distance between successive maxima corresponds to a path difference of λ for the two reflected waves. As she moves a distance *d*, the path to the nearer wall and back is reduced by 2*d*, that to the farther wall and back is increased by 2*d*. Thus for a distance *d*, the path difference is 4*d*. Therefore, she moves a distance $d = \lambda/4$ between sucessive maxima. Since $\lambda = 340/680$ m = 50 cm, she moves 12.5 cm.

---

62 •• Assume that the rubber band of Problem 57 behaves like an ideal spring. The band is attached to two posts whose separation *D* can be varied. (*a*) Derive an expression for the frequency of the fundamental vibration of this system. (*b*) What should be the separation between the fixed ends of the rubber band so that it will vibrate with a fundamental frequency of 21 Hz?

(*a*) 1. Determine the "sping constant" *k* $\quad k = F/\Delta x = F/(D - 0.8 \text{ m}) = 7.6/0.4$ N/m = 19 N/m

2. Find $v = \sqrt{F/\mu} = \sqrt{k\Delta x D/m}$; $f = v/2D$ $\quad f = \sqrt{19(D - 0.8)/24{\times}10^{-3}D}$ Hz, where *D* is in meters

(*b*) Set $f = 21$ Hz and solve for *D* $\quad D = 1.806$ m

---

63 •• A 2-m string is fixed at one end and is vibrating in its third harmonic with amplitude 3 cm and frequency 100 Hz. (*a*) Write the wave function for this vibration. (*b*) Write an expression for the kinetic energy of a segment of the string of length *dx* at a point *x* at some time *t*. At what time is this kinetic energy maximum? What is the shape of the string at this time? (*c*) Find the maximum kinetic energy of the string by integrating your expression for part (*b*) over the total length of the string.

(*a*) See Equ. 16-61; use $\lambda = 4/3$ m $\quad y(x, t) = 0.03 \sin 3\pi x/4 \cos 200\pi t$ m

(*b*) $dm = \mu\, dx$; $dK = \frac{1}{2}v_y^2\, dm$ $\quad dK = \frac{1}{2}(6\pi \sin 3\pi x/4 \sin 200\pi t)^2 \mu\, dx$

dK is a maximum when $200\pi t = \pi/2, 3\pi/2, \ldots$ $\quad t = 2.5$ ms, 7.5 ms, ...; the string is then a straight line

(*c*) Integrate *dK* from $x = 0$ to $x = L$;

note that $dK_{max} = \frac{1}{2}\mu\omega^2 A^2 \sin^2 kx\, dx$; $\quad K = \frac{1}{2}\mu\omega^2 A^2 \int_0^L \sin^2 kx\, dx$

here $k = 3\pi/L$

$$K = \tfrac{1}{2}\mu\omega^2 A^2 \int_0^L \sin^2 (3\pi x/L)\, dx$$

Insert numerical values for ω and A

$K = m\omega^2A^2/4$, where m is the mass of the string

$K = (200\pi)^2(0.03)^2m/4$ J $= 89m$ J

---

64 • Information for use by computers is transmitted along a cable in the form of short electric pulses at the rate of 10^7 pulses per second. (*a*) What is the maximum duration of each pulse if no two pulses overlap? (*b*) What is the range of frequencies to which the receiving equipment must respond?

(*a*) The maximum duration is $T = 1/f = 0.1\ \mu$s.

(*b*) $\Delta\omega\Delta t \approx 1$. So $\Delta\omega = 10^7$ rad/s and $\Delta f = 10^7/2\pi$ Hz $= 1.6$ MHz.

65* • A tuning fork of frequency f_0 begins vibrating at time $t = 0$ and is stopped after a time interval Δt. The waveform of the sound at some later time is shown as a function of x. Let N be the (approximate) number of cycles in this waveform. (*a*) How are N, f_0, and Δt related? (*b*) If Δx is the length in space of this wave packet, what is the wavelength in terms of Δx and N? (*c*) What is the wave number k in terms of N and Δx? (*d*) The number N is uncertain by about ±1 cycle. Use Figure 16-29 to explain why. (*e*) Show that the uncertainty in the wave number due to the uncertainty in N is $2\pi/\Delta x$.

(*a*) The number of cycles in the interval Δt is N; hence $\Delta t \approx NT = N/f_0$.

(*b*) There are about N complete wavelengths in Δx; hence $\lambda \approx \Delta x/N$.

(*c*) $k = 2\pi/\lambda = 2\pi N/\Delta x$.

(*d*) N is uncertain because the waveform dies out gradually rather than stopping abruptly at some time; hence, where the pulse starts and stops is not well defined.

(*e*) $\Delta k = 2\pi\Delta N/\Delta x = 2\pi/\Delta x$.

---

66 • When two waves moving in opposite directions superimpose as in Figure 16-1, does either impede the progress of the other?

No; they move independently.

---

67 • When a guitar string is plucked, is the wavelength of the wave it produces in air the same as the wavelength of the wave on the string?

No; the velocities of propagation differ.

---

68 • When two waves interfere constructively or destructively, is there any gain or loss in energy? Explain.

No; when averaged over a region in space including one or more wavelengths, the energy is unchanged.

69* • A musical instrument consists of drinking glasses partially filled with water that are struck with a small mallet. Explain how this works.

When the edges of the glass vibrate, sound waves are produced in the air in the glass. The resonance frequency of the air columns depends on the length of the air column, which depends on how much water is in the glass.

---

70 •• During an organ recital, the air compressor that drives the organ pipes suddenly fails. An enterprising physics student in the audience comes to the rescue by connecting a tank of pure nitrogen gas under high pressure to the output of the compressor. What effect, if any, will this change have on the operation of the organ? What if the tank contained helium?

From Equ. 15-5, $v \propto \sqrt{\gamma/M}$. Since $M(N_2) < M(\text{air})$, and $f = v/\lambda$, f will increase for each organ pipe. If helium were used, the effect would be even more pronounced, since $M(\text{He}) \ll M(\text{air})$ and $\gamma_{He} > \gamma_{air}$.

---

71 •• When the tension on a piano wire is increased, which of the following occurs? (*a*) Its wavelength decreases. (*b*) Its wavelength remains the same while its frequency increases. (*c*) Its wavelength and frequency increase. (*d*) None of the above.

(*b*)

---

72 •• The following instructions are given for connecting stereo speakers to an amplifier so that they are in phase: "After both speakers are connected, play a monophonic record or program with the bass control turned up and the treble control turned down. While listening to the speakers, turn the balance control so that first one speaker is heard separately, then the two together, and then the other separately. If the bass is stronger when both speakers play together, they are connected properly. If the bass is weaker when both play together than when each plays separately, reverse the connections on one speaker." Explain why this method works. In particular, explain why a stereo source is not used and why only the bass is compared.

If connected properly, they will oscillate in phase and interfere constructively. If connected incorrectly, they interfere destructively. It would be difficult to detect the interference if the wavelength is short, less than the distance between the ears of the observer. Thus, one should use bass notes of low frequency and long wavelength.

73* •• The constant γ for helium (and all monatomic gases) is 1.67. If a man inhales helium and then commences to speak, he sounds like Alvin of the Chipmunks. Why?

The pitch is determined in large part by the resonant cavity of the mouth. Since $v_{He} > v_{air}$ (see Problem 70), the resonance frequency is higher if helium is the gas in the cavity.

---

74 • Middle C on the equal-temperament scale used by modern instrument makers has a frequency of 261.63 Hz. If a 7-g piano wire that is 80 cm long is to be tuned so that 261.63 is its fundamental frequency, what should the tension in the wire be?

$v = \lambda f = 2Lf$; $F = v^2 m/L = 4Lf^2 m$ $\quad$ $F = 4 \times 0.8 \times 261.63^2 \times 0.007$ N $= 1533$ N

---

75 • The ear canal, which is about 2.5 cm long, roughly approximates a pipe that is open at one end and closed at the other. (*a*) What are the resonance frequencies of the ear canal? (*b*) Describe the possible effect of the resonance modes of the ear canal on the threshold of hearing.

(*a*) $\lambda = 4L/(2n + 1)$; $f = v/\lambda$ $\quad$ $f_1 = 340/0.1$ Hz $= 3400$ Hz; $f_3 = 10200$ Hz, $f_5 = 17000$ Hz

(*b*) Frequencies near 3400 Hz will be most readily perceived

---

76 • A 4-m-long, 160-g rope is fixed at one end and is tied to a light string at the other end. Its tension is 400 N. (*a*) What are the wavelengths of the fundamental and the next two harmonics? (*b*) What are the frequencies of these standing waves?

(*a*) $\lambda_1 = 4L$; only odd harmonics are excited $\quad$ $\lambda_1 = 16$ m; $\lambda_3 = 5.33$ m; $\lambda_5 = 3.2$ m

(*b*) $v = \sqrt{FL/m}$; $f = v/\lambda$ $\quad$ $v = 100$ m/s; $f_1 = 6.25$ Hz; $f_3 = 18.75$ Hz; $f_5 = 31.25$ Hz

77* • The shortest pipes used in organs are about 7.5 cm long. (*a*) What is the fundamental frequency of a pipe this long that is open at both ends? (*b*) For such a pipe, what is the highest harmonic that is within the audible range (see Problem 43)?

(*a*) $\lambda_1 = 2L; f_1 = v/\lambda_1 = v/2L; f_n = nf_1$ — $f_1 = 340/0.15$ Hz $= 2267$ Hz; $f_9 = 20400$ Hz is about at frequency threshold

---

78 •• Two waves from two coherent sources have the same wavelength λ, frequency ω, and amplitude A. What is the path difference if the resultant wave at some point has amplitude A?

From Equ. 16-6, $A = 2A \cos \frac{1}{2}\delta$; $\Delta x = \lambda\delta/2\pi$ — $\delta = 2 \cos^{-1} 0.5 = 2\pi/3$ rad; $\Delta x = \lambda/3$

---

79 •• A 35-m string has a linear mass density of 0.0085 kg/m and is under a tension of 18 N. Find the frequencies of the lowest four harmonics if (*a*) the string is fixed at both ends, and (*b*) the string is fixed at one end and attached to a long, thin, massless thread at the other end.

(*a*) $v = \sqrt{F/\mu}; f_n = nv/2L;\ v = 46.0$ m/s — $f_1 = 0.657$ Hz; $f_2 = 1.31$ Hz; $f_3 = 1.97$ Hz; $f_4 = 2.63$ Hz

(*b*) $f_n = (2n - 1)v/4L$ — $f_1 = 0.329$ Hz; $f_3 = 0.986$ Hz; $f_5 = 1.64$ Hz; $f_7 = 2.30$ Hz

---

80 •• You find an abandoned mine shaft and wish to measure its depth. Using an audio oscillator of variable frequency, you note that you can produce successive resonances at frequencies of 63.58 and 89.25 Hz. What is the depth of the shaft?

1. Shaft is pipe closed at one end; $f_n = (2n - 1)f_1$ — $f_1 = \frac{1}{2}(89.25 - 63.58)$ Hz $= 12.84$ Hz

2. $f_1 = v/4L$; $L = v/4f_1$ — $L = 340/51.34$ m $= 6.62$ m

81* •• A string 5 m long that is fixed at one end only is vibrating in its fifth harmonic with a frequency of 400 Hz. The maximum displacement of any segment of the string is 3 cm. (*a*) What is the wavelength of this wave? (*b*) What is the wave number k? (*c*) What is the angular frequency? (*d*) Write the wave function for this standing wave.

(*a*) $\lambda_1 = 4L$; $\lambda_5 = 4L/5$ — $\lambda_5 = 4$ m

(*b*) $k = 2\pi/\lambda$ — $k = \pi/2\ \text{m}^{-1}$

(*c*) $\omega = 2\pi f$ — $\omega = 800\pi$ rad/s

(*d*) $y(x, t) = A \sin kx \cos \omega t$ — $y(x, t) = 0.03 \sin(\pi x/2) \cos(800\pi t)$

---

82 •• The wave function for a standing wave on a string is described by $y(x,t) = 0.02 \sin 4\pi x \cos 60\pi t$, where y and x are in meters and t is in seconds. Determine the maximum displacement and maximum speed of a point on the string at (*a*) $x = 0.10$ m, (*b*) $x = 0.25$ m, (*c*) $x = 0.30$ m, and (*d*) $x = 0.50$ m.

$y_{max}(x) = 0.02 \sin 4\pi x$. (*a*) $y_{max} = 0.02 \sin (0.4\pi)$ m $= 0.019$ m; (*b*) $y_{max} = 0$; (*c*) $y_{max} = 0.0118$ m; (*d*) $y_{max} = 0$.

$v_{y,max}(x) = 1.2\pi \sin 4\pi x = 60\pi y_{max}(x)$. (*a*) $v_{y,max} = 3.58$ m/s; (*b*) $v_{y,max} = 0$; (*c*) $v_{y,max} = 2.22$ m/s; (*d*) $v_{y,max} = 0$.

---

83 •• A 2.5-m-long wire having a mass of 0.10 kg is fixed at both ends and is under tension of 30 N. When the nth harmonic is excited, there is a node 0.50 m from one end. (*a*) What is n? (*b*) What are the frequencies of the first three allowed modes of vibration?

(*a*) $\lambda_1 = 2L$; $\lambda_n = 2L/n$; distance between nodes $= \lambda/2$ — $\lambda_1 = 5$ m; $\lambda_n = 1$ m; $n = 5$

(*b*) Find $v = \sqrt{FL/m}$; $v = \sqrt{30 \times 2.5/0.1}$ m/s = 27.4 m/s

$f_n = nv/\lambda_1$ $f_1 = 5.48$ Hz; $f_2 = 10.95$ Hz; $f_3 = 16.43$ Hz

---

84 •• In an early method of determining the speed of sound in gases, powder was spread along the bottom of a horizontal, cylindrical glass tube. One end of the tube was closed by a piston that oscillated at a known frequency *f*. The other end was closed by a movable piston whose position was adjusted until resonance occurred. At resonance, the powder collected in equally spaced piles along the bottom of the tube. (*a*) Explain why the powder collects in this way. (*b*) Derive a formula that gives the speed of sound in the gas in terms of *f* and the distance between the piles of powder. (*c*) Give suitable values for the frequency *f* and the distance between the piles of powder. (*d*) Give suitable values for the frequency *f* and the length *L* of the tube for which the speed of sound could be measured in either air or helium.

(*a*) At resonance, standing waves are set up in the tube. At a displacement antinode, the powder is moved about; at a node, the powder is stationary, and so the powder collects at the nodes.

(*b*) Let D = distance between nodes. Then $\lambda = 2D$ and $v = f\lambda = 2fD$.

(*c*) Let $f = 2000$ Hz, $D = 10$ cm; then $v = 400$ m/s.

(*d*) If $L = 1.2$ m, then (neglecting end effects) a resonance will be observed for $\lambda = 10$ cm, corresponding to the third harmonic of the resonance frequency. If $v = 400$ m/s, the frequency should be 4000 Hz. (At the fundamental, the only antinode appears at the open end of the tube and end effects need to be considered; see Problem 51.)

85* •• In a lecture demonstration of standing waves, a string is attached to a tuning fork that vibrates at 60 Hz and sets up transverse waves of that frequency on the string. The other end of the string passes over a pulley, and the tension is varied by attaching weights to that end. The string has approximate nodes at the tuning fork and at the pulley. (*a*) If the string has a linear mass density of 8 g/m and is 2.5 m long (from the tuning fork to the pulley), what must the tension be for the string to vibrate in its fundamental mode? (*b*) Find the tension necessary for the string to vibrate in its second, third, and fourth harmonic.

(*a*) $v^2 = F/\mu = f^2\lambda^2 = 4f^2L^2$; $F = 4f^2L^2\mu$ $F = 4 \times 60^2 \times 2.5^2 \times 0.008$ N = 720 N

(*b*) $f_n = nf_1$ and $F \propto f^2$ $F_2 = 4 \times 720$ N = 2880 N; $F_3 = 6480$ N; $F_4 = 11520$ N

---

86 •• Three successive resonance frequencies in an organ pipe are 1310, 1834, and 2358 Hz. (*a*) Is the pipe closed at one end or open at both ends? (*b*) What is the fundamental frequency? (*c*) What is the length of the pipe?

(*a*) If the pipe is open at both ends then $\Delta f = f_1$, and $f_n = nf_1$, where n is an integer. $\Delta f = 524$ Hz and 1310 Hz = $2.5f_1$. So the pipe is closed at one end.

(*b*) For a pipe closed at one end, $\Delta f = 2f_1$, $f_1 = 262$ Hz.

(*c*) $L = \lambda/4$; $\lambda = 340/262$ m = 1.30 m. $L = 32.4$ cm.

---

87 •• A wire of mass 1 g and length 50 cm is stretched with a tension of 440 N. It is then placed near the open end of the tube in Example 16-8 and stroked with a violin bow so that it oscillates at its fundamental frequency. The water level in the tube is then lowered until a resonance is obtained, which occurs at 18 cm below the top of the tube. Use the data given to determine the speed of sound in air. Why is this method not very accurate?

1. $f_1 = v/2L = \sqrt{F/4Lm}$ $\qquad f_1 = \sqrt{440/2 \times 10^{-3}}$ Hz = 469 Hz

2. $v_s = f\lambda = 4fL$ $\qquad v_s = 338$ m/s

The method is not very accurate because it neglects end effects. (see Problem 51)

88 •• On a windy day, a drain pipe will sometimes resonate. Estimate the resonance frequency of a drain pipe on a single-story house. How much might this frequency change from winter to summer in your region?

Pipe length ≈ 5 m. Pipe open at both ends, so $\lambda_1 \approx 10$ m, and $f \approx v/\lambda = 34$ Hz.

The frequency will be somewhat higher in the summer because $v \propto \sqrt{T}$.

89* •• A 50-cm-long wire fixed at both ends vibrates with a fundamental frequency f_0 when the tension is 50 N. If the tension is increased to 60 N, the fundamental frequency increases by 5 Hz, and a further increase in tension to 70 N results in a fundamental frequency of $(f_0 + 7)$ Hz. Determine the mass of the wire.

1. $f \propto \sqrt{F}$ $\qquad (f_0 + 5)/f_0 = \sqrt{60/50} = 1.0954;\ f_0 = 52.4$ Hz

2. $m = F/4Lf_0^2$ (see Problem 87) $\qquad m = 50/(4 \times 0.5 \times 52.4^2)$ kg = 9.1 g

90 •• A standing wave on a rope is represented by the following wave function:

$$y(x,t) = 0.02 \sin \frac{\pi x}{2} \cos 40\pi t,$$

where x and y are in meters and t is in seconds. (*a*) Write wave functions for two traveling waves that when superimposed will produce the resultant standing-wave pattern. (*b*) What is the distance between the nodes of the standing wave? (*c*) What is the velocity of a segment of the rope at $x = 1$ m? (*d*) What is the acceleration of a segment of the rope at $x = 1$ m?

(*a*) $y_1(x, t) = 0.01 \sin[(\pi x/2) - 40\pi t]$ m; $y_2(x, t) = 0.01 \sin[(\pi x/2) + 40\pi t]$ m.

(*b*) Distance between nodes = $\lambda/2$; $\lambda = 2\pi/k = 4$ m; distance between nodes = 2 m.

(*c*) $x = 1$ m is an antinode. $v_y(1, t) = -0.8\pi \sin(40\pi t)$ m/s = $-2.51 \sin(40\pi t)$ m/s.

(*d*) $a_y(1, t) = -316 \cos(40\pi t)$ m/s^2.

91 •• Two identical speakers emit sound waves of frequency 680 Hz uniformly in all directions. The total audio output of each speaker is 1 mW. A point P is 2.00 m from one speaker and 3.00 m from the other. (*a*) Find the intensities I_1 and I_2 from each speaker separately at point P. (*b*) If the speakers are driven coherently and are in phase, what is the intensity at point P? (*c*) If they are driven coherently but are 180° out of phase, what is the intensity at point P? (*d*) If the speakers are incoherent, what is the intensity at point P?

(*a*) $I = P/4\pi r^2$ $\qquad I_1 = 1/16\pi$ mW/m^2 = 19.9 μW/m^2;

$I_2 = 1/36\pi$ mW/m^2 = 8.84 μW/m^2

(*b*) $\lambda = 0.5$ m; $\Delta x = 2\lambda$; constructive interference $\qquad A_1 = C\sqrt{I_1}$; $A_2 = C\sqrt{I_2}$; $A^2 = C^2(\sqrt{I_1} + \sqrt{I_2})^2$; C is a constant

$I = A^2/C^2 = (\sqrt{I_1} + \sqrt{I_2})^2 = 55.3$ μW/m^2

(*c*) Now we have destructive interference $\qquad I = (\sqrt{I_1} - \sqrt{I_2})^2 = 2.21$ μW/m^2

(*d*) Incoherent sources; intensities add $\qquad I = I_1 + I_2 = 28.7$ μW/m^2

92 •• Three waves with the same frequency, wavelength, and amplitude are traveling in the same direction. The three waves are given by

$$y_1(x,t) = 0.05 \sin(kx - \omega t - \pi/3),\ y_2(x,t) = 0.05 \sin(kx - \omega t),\ y_3(x,t) = 0.05 \sin(kx - \omega t + \pi/3).$$

Find the resultant wave.

In Chapter 14, Section 14.1, it was shown that a harmonic function could be represented by a vector rotating at the angular frequency ω. The simplest way to do this problem is to use that representation. The vector corresponding to y_2 is the one along the horizontal axis. The ones for y_1 and y_3 are at angles of $-60°$ and $60°$, respectively. All vectors are of equal magnitude. It is evident that the "y" component of the sum of the vectors is zero, and that the "x" component has the magnitude $2A$, where A is the magnitude of one vector. Consequently, $y_{res} = 0.1 \sin (kx - \omega t)$.

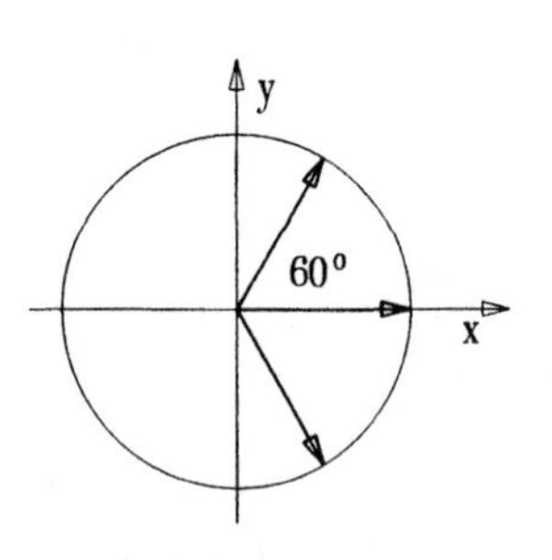

93* •• (*a*) Show that if the temperature changes by a small amount ΔT, the fundamental frequency of an organ pipe changes by approximately Δf, where $\Delta f/f = \frac{1}{2}\Delta T/T$. (*b*) Suppose an organ pipe that is closed at one end has a fundamental frequency of 200 Hz when the temperature is 20°C. What will its fundamental frequency be when the temperature is 30°C? (Ignore any change in the length of the pipe due to thermal expansion.)

(*a*) $f = CT^{1/2}$, where C is a constant; $df/dT = \frac{1}{2}CT^{-1/2} = \frac{1}{2}f/T$; so $df/f = \frac{1}{2}dT/T$ and $\Delta f/f = \frac{1}{2}\Delta T/T$ if $\Delta T << T$.

(*b*) $\Delta f = (200 \times \frac{1}{2} \times 10/293)$ Hz $= 3.41$ Hz; $f_{30} = f_{20} + \Delta f = 203.4$ Hz.

94 •• Two traveling wave pulses on a string are represented by the wave functions

$$y_1(x,t) = \frac{0.02\text{ m}^3}{2\text{ m}^2 + (x - 2t)^2}, \text{ and } y_2(x,t) = \frac{-0.02\text{ m}^3}{2\text{ m}^2 + (x + 2t)^2}$$

where x is in meters and t is in seconds. (*a*) Sketch each wave function separately as a function of x at $t = 0$, and describe the behavior of each as time increases. (*b*) Find the resultant wave function at $t = 0$. (*c*) Find the resultant wave function at $t = 1$ s. (*d*) Sketch the resultant wave function at $t = 1$ s.

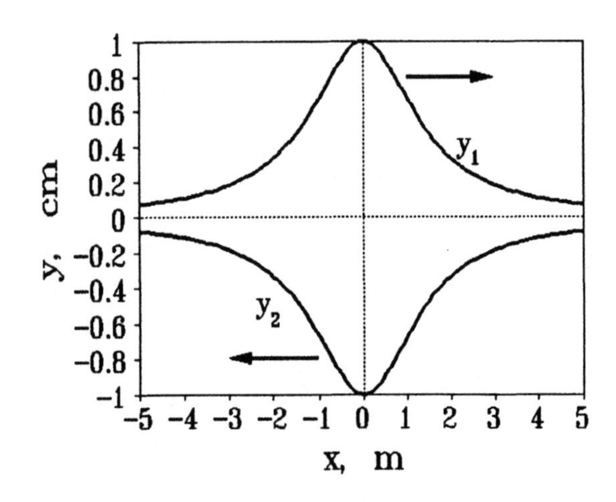

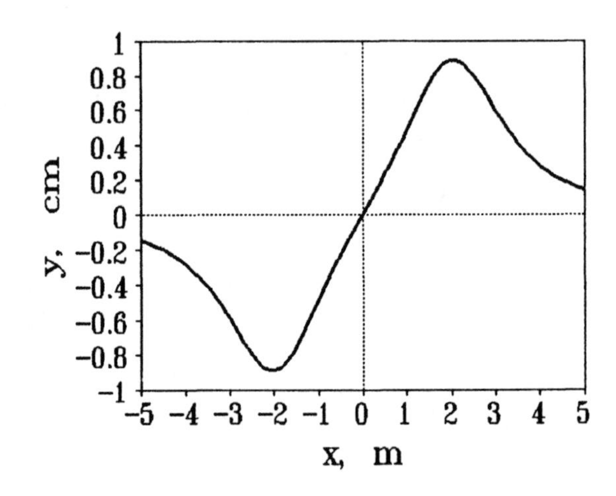

(*a*) The two wave functions at $t = 0$ are shown above.

(*b*) The resultant wave at $t = 0$ is evidently a straight line; $y_1 + y_2 = 0$.

(*c*) At $t = 1$ s, $y_1 + y_2 = [0.16x/(x^4 - 4x^2 + 36)]$ m.

(*d*) The resultant wave at $t = 1$ s is shown above.

95 •• The kinetic energy of a segment Δm of a vibrating string is given by $\Delta K = \frac{1}{2}\Delta m(\partial y/\partial t)^2 = \frac{1}{2}\mu(\partial y/\partial t)^2\Delta x$. (*a*) Find the total kinetic energy of the nth mode of vibration of a string of length L fixed at both ends. (*b*) Give the maximum kinetic energy of the string. (*c*) What is the wave function when the kinetic energy has its maximum value? (*d*) Show that the maximum kinetic energy in the *n*th mode is proportional to $n^2A_n^2$.

(*a*) $(\partial y/\partial t)^2 = \omega_n^2 A_n^2 \sin^2 \omega_n t \sin^2 k_n x$. $dK = \frac{1}{2}\mu\omega_n^2 A_n^2 \sin^2 \omega_n t \sin^2 k_n x$, where $k_n = n\pi x/L$.

$$K = \tfrac{1}{2}\mu\omega_n^2 A_n^2 \sin^2\omega_n t \int_0^L \sin^2 (n\pi x/L)\, dx = (1/4)m\omega_n^2 A_n^2 \sin^2\omega_n t$$

(*b*) $K_{\text{max}} = (1/4)m\omega_n^2 A_n^2$; note that this is exactly the same result as in Problem 63 with ω_n, A_n replacing ω and A.

(*c*) When $K = K_{\text{max}}$, $\omega_n t = \pi/2$, so $\cos \omega_n t = 0$. The wave function is then a straight line, i.e., $y(x, t) = 0$.

(*d*) Since $\omega_n = n\omega_1$, K_{max} is proportional to $n^2A_n^2$.

---

96 •• (*a*) Show that when the tension in a string fixed at both ends is changed by a small amount dF, the frequency of the fundamental is changed by approximately df, where $df/f = \frac{1}{2}dF/F$. Does this result apply to all harmonics? (*b*) Use this result to find the percentage change in the tension needed to increase the frequency of the fundamental of a piano wire from 260 to 262 Hz.

(*a*) Since $f_n = nv/2L$ and $v = CF^{1/2}$, where C is a constant, $df_n/dF = \frac{1}{2}(Cn/2L)F^{-1/2} = \frac{1}{2}f_n/F$; $df_n/f_n = \frac{1}{2}dF/F$. The result, as shown, is valid for all harmonics.

(*b*) $\Delta F/F = 2\Delta f/f = 4/260 = 0.0154 = 1.54\%$.

97* ••• Two sources have a phase difference δ_0 that is proportional to time: $\delta_0 = Ct$, where C is a constant. The amplitude of the wave from each source at some point P is A_0. (*a*) Write the wave functions for each of the two waves at point P, assuming this point to be a distance x_1 from one source and $x_1 + \Delta x$ from the other. (*b*) Find the resultant wave function, and show that its amplitude is $2A_0 \cos \frac{1}{2}(\delta + \delta_0)$, where δ is the phase difference at P due to the path difference. (*c*) Sketch the intensity at point P versus time for a zero path difference. (Let I_0 be the intensity due to each wave separately.) What is the time average of the intensity? (*d*) Make the same sketch for the intensity at a point for which the path difference is $\frac{1}{2}\lambda$.

(*a*) $y_1 = A_0 \cos(kx_1 - \omega t)$; $y_2 = A_0 \cos(kx_1 - \omega t + k\Delta x + \delta_0) = A_0 \cos[kx_1 - \omega t + (\delta + \delta_0)]$, where $\delta = k\Delta x$.

(*b*) Use $\cos \alpha + \cos \beta = 2 \cos \frac{1}{2}(\alpha + \beta) \cos \frac{1}{2}(\alpha - \beta)$; $y_{\text{tot}} = y_1 + y_2 = 2A_0 \cos \frac{1}{2}(\delta + \delta_0) \cos \frac{1}{2}[kx_1 - \omega t + \frac{1}{2}(\delta + \delta_0)]$. The amplitude of the resultant wave is $2A_0 \cos \frac{1}{2}(\delta + \delta_0)$.

(*c*) Note that $I \propto A^2$. With $\delta = 0$ and $\delta_0 = Ct$, $I \propto 4A_0^2 \cos^2 \frac{1}{2}Ct$. The average of $\cos^2 \theta$ over a complete period is ½. So the average intensity is $2I_0$. See the figure to the left, below.

(*d*) If $\Delta x = \frac{1}{2}\lambda$, then at $t = 0$ the two waves interfere destructively. The plot of intensity versus time is shown below in the figure to the right.

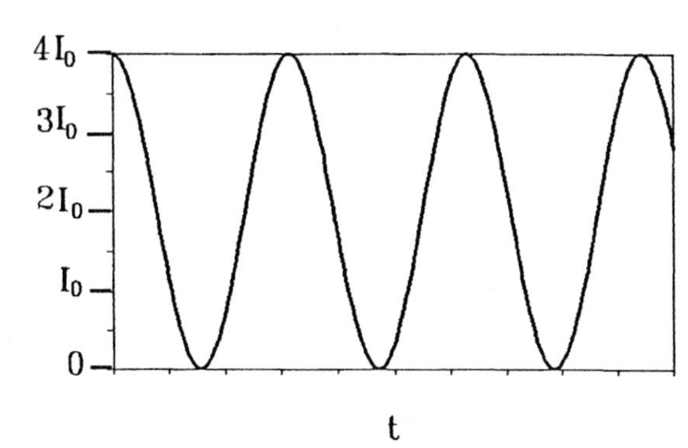

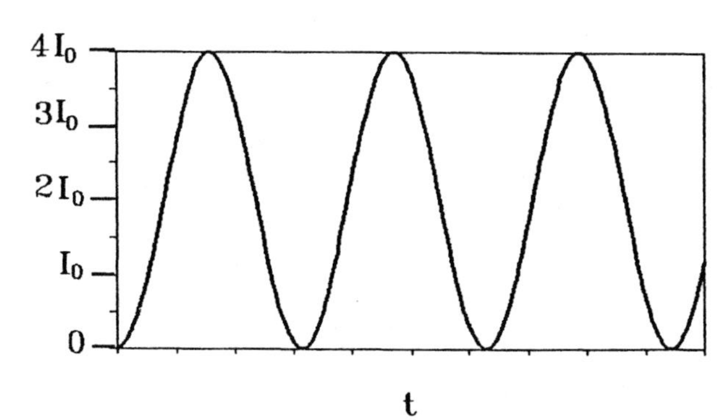

98 ••• The wave functions of two standing waves on a string of length L are $y_1(x,t) = A_1 \cos \omega_1 t \sin k_1 x$ and $y_2(x, t) = A_2 \cos \omega_2 t \sin k_2 x$, where $k_n = n\pi/L$, and $\omega_n = n\omega_1$. The wave function of the resultant wave is $y_r(x, t) = y_1(x, t) + y_2(x, t)$. (*a*) Find the velocity of a segment dx of the string. (*b*) Find the kinetic energy of this segment. (*c*) By integration, find the total kinetic energy of the resultant wave. Notice the disappearance of the cross terms so that the total kinetic energy is proportional to $(n_1A_1)^2 + (n_2A_2)^2$.

(*a*) $v_y = \partial y_r/\partial t = \omega_1 A_1 \sin \omega_1 t \sin k_1 x + \omega_2 A_2 \sin \omega_2 t \sin k_2 x$.

(*b*) $dK = \frac{1}{2}\mu v_y^2 dx = \frac{1}{2}\mu[\omega_1^2 A_1^2 \sin^2 \omega_1 t \sin^2 k_1 x + \omega_2^2 A_2^2 \sin^2 \omega_2 \sin^2 k_2 x +$
$\omega_1 \omega_2 A_1 A_2 \sin \omega_1 t \sin \omega_2 t \sin k_1 x \sin k_2 x]\, dx$.

(*c*) The integral between 0 and L of $\sin^2 k_1 x$ and of $\sin^2 k_2 x$ is ½. The integral of $\sin(n_1 \pi x/L) \sin(n_2 \pi x/L) = 0$ for $n_1 \neq n_2$. Thus, $K = (1/4)m(\omega_1^2 A_1^2 + \omega_2^2 A_2^2) = K_1 + K_2$. (see Problem 95)

99 ••• A 2-m wire fixed at both ends is vibrating in its fundamental mode. The tension in the wire is 40 N and the mass of the wire is 0.1 kg. At the midpoint of the wire, the amplitude is 2 cm. (*a*) Find the maximum kinetic energy of the wire. (*b*) At the instant the transverse displacement is given by (0.02 m) sin ($\pi x/2$), what is the kinetic energy of the wire? (*c*) At what position on the wire does the kinetic energy per unit length have its largest value? (*d*) Where does the potential energy per unit length have its maximum value?

(*a*) Find the frequency f_1 and $\omega_1 = 2\pi f_1$	$f_1 = v/2L = \sqrt{F/4Lm} = 7.07$ Hz; $\omega_1 = 14.1\,\pi$ rad/s
(see Problem 95)	$K_{max} = m\omega^2 A^2/4 = 0.0197$ J
(*b*) When $y(x, t) = y_{max}(x, t)$, $\omega t = 0$; see Equ. 16-61	From Problem 95, $K = 0$
(*c*) See Problem 95	dK is maximum at the midpoint; $x = 1$ m
(*d*) dU_{max} at $\partial y/\partial x = (\partial y/\partial x)_{max}$; (Problem 15-123)	dU_{max} at midpoint; $x = 1$ m

100 ••• A string 3.2 m long and with a linear mass density of 0.008 kg/m is kept under tension so that traveling waves propagate at 48 m/s along the string. The ends of the string are clamped and the string vibrates in its 3rd harmonic with an amplitude of 5.0 cm. How much energy is stored in this vibrating system at that time? If the amplitude of the standing wave diminishes to 3.0 cm in 1.0 s, what is the Q of this vibrating system?

1. $E = K_{max} = m\omega^2 A^2/4$; $\omega^2 = n^2\pi^2 v^2/L^2$;	$E = n^2 A^2 \pi^2 v^2 m/4L^2 = 0.320$ J
2. Use Equ. 14-36 to find τ	$\tau = 1/[2 \ln(5/3)] = 0.979$ s
3. Use Equ. 14-39 to determine Q	$Q = n\pi v \tau/L = 138$

CHAPTER 17

Wave–Particle Duality and Quantum Physics

1* • The quantized character of electromagnetic radiation is revealed by (*a*) the Young double-slit experiment. (*b*) diffraction of light by a small aperture. (*c*) the photoelectric effect. (*d*) the J.J. Thomson cathode-ray experiment.

(*c*)

---

2 •• Two monochromatic light sources, A and B, emit the same number of photons per second. The wavelength of A is $\lambda_A = 400$ nm, and that of B is $\lambda_B = 600$ nm. The power radiated by source B is (*a*) equal to that of source A. (*b*) less than that of source A. (*c*) greater than that of source A. (*d*) cannot be compared to that from source A using the available data.

(*b*)

---

3 • Find the photon energy in joules and in electron volts for an electromagnetic wave of frequency (*a*) 100MHz in the FM radio band, and (*b*) 900 kHz in the AM radio band.

(*a*) $E = hf$; $E = 6.63 \times 10^{-26}$ J $= 4.14 \times 10^{-7}$ eV

$h = 6.626 \times 10^{-34}$ J·s $= 4.136 \times 10^{-15}$ eV·s

(*b*) $E = hf$ $E = 5.96 \times 10^{-28}$ J $= 3.72 \times 10^{-9}$ eV

---

4 • An 80-kW FM transmitter operates at a frequency of 101.1 MHZ. How many photons per second are emitted by the transmitter?

$P = Nhf$; $N = P/hf$ $N = 8 \times 10^4/(6.63 \times 10^{-34} \times 101.1 \times 10^6)$ s^{-1}

$= 1.19 \times 10^{30}$ s^{-1}

5* • What are the frequencies of photons having the following energies? (*a*) 1 eV, (*b*) 1 keV, and (*c*) 1 MeV.

(*a*) $f = 1/4.14 \times 10^{-15}$ Hz $= 2.42 \times 10^{14}$ Hz. (*b*) $f = 2.24 \times 10^{17}$ Hz. (*c*) $f = 2.24 \times 10^{20}$ Hz.

---

6 • Find the photon energy for light of wavelength (*a*) 450 nm, (*b*) 550 nm, and (*c*) 650 nm.

(*a*), (*b*), (*c*) $E = hc/\lambda = (1240$ eV·nm$)/\lambda$ (*a*) $E = 2.76$ eV (*b*) $E = 2.25$ eV (*c*) $E = 1.91$ eV

---

7 • Find the photon energy if the wavelength is (*a*) 0.1 nm (about 1 atomic diameter), and (*b*) 1 fm (1 fm = 10^{-15} m, about 1 nuclear diameter).

(*a*) $E = 1240/0.1$ eV = 12.4 keV. (*b*) $E = 1240/10^{-6}$ eV = 1.24 GeV.

---

8 •• The wavelength of light emitted by a 3-mW He-Ne laser is 632 nm. If the diameter of the laser beam is 1.0 mm, what is the density of photons in the beam?

1. Find the number of photons emitted per second — $N = P/E = (3 \times 10^{-3}/1.6 \times 10^{-19})/(1240/632) = 9.56 \times 10^{15}$
2. Find the volume containing the photons — $V = Ac = (\pi \times 10^{-6}/4) \times 3 \times 10^{8}\ \text{m}^3 = 2.36 \times 10^{2}\ \text{m}^3$
3. Density of photons = N/V — $\rho = 9.56 \times 10^{15}/2.36 \times 10^{2}\ \text{m}^{-3} = 4.05 \times 10^{13}\ \text{m}^{-3}$

9* • True or false: In the photoelectric effect, (*a*) the current is proportional to the intensity of the incident light. (*b*) the work function of a metal depends on the frequency of the incident light. (*c*) the maximum kinetic energy of electrons emitted varies linearly with the frequency of the incident light. (*d*) the energy of a photon is proportional to its frequency.

(*a*) True (*b*) False (*c*) True (*d*) True

---

10 • In the photoelectric effect, the number of electrons emitted per second is (*a*) independent of the light intensity. (*b*) proportional to the light intensity. (*c*) proportional to the work function of the emitting surface. (*d*) proportional to the frequency of the light.

(*b*)

---

11 • The work function of a surface is ϕ. The threshold wavelength for emission of photoelectrons from the surface is (*a*) hc/ϕ. (*b*) ϕ/hf. (*c*) hf/ϕ. (*d*) none of the above.

(*a*)

---

12 •• When light of wavelength λ_1 is incident on a certain photoelectric cathode, no electrons are emitted no matter how intense the incident light is. Yet when light of wavelength $\lambda_2 < \lambda_1$ is incident, electrons are emitted even when the incident light has low intensity. Explain.

hc/λ must be greater than ϕ. Evidently, $hc/\lambda_1 < \phi$, but $hc/\lambda_2 > \phi$.

13* • The work function for tungsten is 4.58 eV. (*a*) Find the threshold frequency and wavelength for the photoelectric effect. (*b*) Find the maximum kinetic energy of the electrons if the wavelength of the incident light is 200 nm, and (*c*) 250 nm.

(*a*) $f_t = \phi/h$; $\lambda = c/f$ — $f_t = 4.58/4.136 \times 10^{-15}$ Hz $= 1.11 \times 10^{15}$ Hz; $\lambda_t = 270$ nm

(*b*), (*c*) $K_m = E - \phi = hc/\lambda - \phi$ — (*b*) $K_m = (1240/200 - 4.58)$ eV $= 1.62$ eV

(*c*) $K_m = 0.38$ eV

---

14 • When light of wavelength 300 nm is incident on potassium, the emitted electrons have maximum kinetic energy of 2.03 eV. (*a*) What is the energy of an incident photon? (*b*) What is the work function for potassium? (*c*) What would be the maximum kinetic energy of the electrons if the incident light had a wavelength of 430 nm? (*d*) What is the threshold wavelength for the photoelectric effect with potassium?

(*a*) $E = hc/\lambda$ — $E = 1240/300$ eV $= 4.13$ eV

(*b*) $\phi = E - K_m$ — $\phi = 2.10$ eV

(*c*) $K_m = E - \phi$ — $K_m = (1240/430 - 2.10)$ eV $= 0.784$ eV

(*d*) $\lambda_t = hc/\phi$ — $\lambda_t = 1240/2.10$ nm $= 590$ nm

15 • The threshold wavelength for the photoelectric effect for silver is 262 nm. (*a*) Find the work function for silver. (*b*) Find the maximum kinetic energy of the electrons if the incident radiation has a wavelength of 175 nm.

(*a*), (*b*) See Problem 14 — (*a*) $\phi = 1240/262$ eV = 4.73 eV

(*b*) $K_m = (1240/175 - 4.73)$ eV = 2.36 eV

16 • The work function for cesium is 1.9 eV. (*a*) Find the threshold frequency and wavelength for the photoelectric effect. Find the maximum kinetic energy of the electrons if the wavelength of the incident light is (*b*) 250 nm, and (*c*) 350 nm.

(*a*) $f_t = \phi/h$; $\lambda_t = hc/\phi$ — $f_t = 1.9/4.136 \times 10^{-15}$ Hz = 4.59×10^{14} Hz; $\lambda_t = 653$ nm

(*b*), (*c*) $K_m = E - \phi$ — (*b*) $K_m = (1240/250 - 1.9)$ eV = 3.06 eV

(*c*) $K_m = 1.64$ eV

17* •• When a surface is illuminated with light of wavelength 512 nm, the maximum kinetic energy of the emitted electrons is 0.54 eV. What is the maximum kinetic energy if the surface is illuminated with light of wavelength 365 nm?

1. Find $\phi = E - K_m$ — $\phi = (1240/512 - 0.54)$ eV = 1.88 eV
2. Find K_m for $\lambda = 365$ nm — $K_m = (1240/365 - 1.88)$ eV = 1.52 eV

18 • Find the shift in wavelength of photons scattered at $\theta = 60°$.

Use Equ. 17-8; $h/m_ec = 2.43$ pm — $\Delta\lambda = 2.43 \times 0.5$ pm = 1.215 pm

19 • When photons are scattered by electrons in carbon, the shift in wavelength is 0.33 pm. Find the scattering angle.

Use Equ. 17-8 — $\theta = \cos^{-1}(1 - 0.33/2.43) = 30.2°$

20 • The wavelength of Compton-scattered photons is measured at $\theta = 90°$. If $\Delta\lambda/\lambda$ is to be 1.5%, what should the wavelength of the incident photons be?

Find $\Delta\lambda$ from Equ. 17-8; $\lambda = \Delta\lambda/0.015$ — $\Delta\lambda = 2.43$ pm; $\lambda = 162$ pm = 0.162 nm

21* • Compton used photons of wavelength 0.0711 nm. (*a*) What is the energy of these photons? (*b*) What is the wavelength of the photon scattered at $\theta = 180°$? (*c*) What is the energy of the photon scattered at this angle?

(*a*) $E = hc/\lambda$ — $E = 1240/0.0711$ eV = 17.44 keV

(*b*) Use Equ. 17-8; $\lambda_f = \lambda_i + \Delta\lambda$ — $\Delta\lambda = 2 \times 2.43$ pm = 0.00486 nm; $\lambda_f = 0.076$ nm

(*c*) $E = hc/\lambda$ — $E = 1240/0.076$ eV = 16.3 keV

22 • For the photons used by Compton, find the momentum of the incident photon and that of the photon scattered at 180°, and use the conservation of momentum to find the momentum of the recoil electron in this case (see Problem 21).

$p_i = h/\lambda_i = 9.32 \times 10^{-24}$ kg·m/s. $p_f = -8.72 \times 10^{-24}$ kg·m/s. $p_e = -(p_f - p_i) = 18.0 \times 10^{-24}$ kg·m/s.

23 •• An X-ray photon of wavelength 6 pm that collides with an electron is scattered by an angle of 90°. (*a*) What is the change in wavelength of the photon? (*b*) What is the kinetic energy of the scattered electron?

(*a*) Use Equ. 17-8 — $\Delta\lambda = 2.43$ pm = 0.00243 nm

(*b*) $\Delta E = hc/\lambda_i - hc/\lambda_f$ — $\Delta E = (1240/0.006 - 1240/0.00843)$ eV = 59.6 keV

---

24 •• How many head-on Compton scattering events are necessary to double the wavelength of a photon having initial wavelength 200 pm?

1. Find $\Delta\lambda$ per collision from Equ.17-8 — $\Delta\lambda$/collision = 4.86 pm
2. Number of collisions = $\Delta\lambda/(\Delta\lambda/\text{collision})$ — $N = 200/4.86 = 42$

25* • True or false: (*a*) The de Broglie wavelength of an electron varies inversely with its momentum. (*b*) Electrons can be diffracted. (*c*) Neutrons can be diffracted. (*d*) An electron microscope is used to look at electrons.

(*a*) True (*b*) True (*c*) True (*d*) False

---

26 • If the de Broglie wavelength of an electron and a proton are equal, then (*a*) the velocity of the proton is greater than that of the electron. (*b*) the velocity of the proton and electron are equal. (*c*) the velocity of the proton is less than that of the electron. (*d*) the energy of the proton is greater than that of the electron. (*e*) both (*a*) and (*d*) are correct.

(*c*) $p_e = p_p$. $m_p > m_e$.

---

27 • A proton and an electron have equal kinetic energies. It follows that the de Broglie wavelength of the proton is (*a*) greater than that of the electron. (*b*) equal to that of the electron. (*c*) less than that of the electron.

(*c*) $p_p^2/2m_p = p_e^2/2m_e$; $p_e < p_p$ so $\lambda_e > \lambda_p$.

---

28 • Use Equation 17-13 to calculate the de Broglie wavelength for an electron of kinetic energy (*a*) 2.5 eV, (*b*) 250 eV, (*c*) 2.5 keV, and (*d*) 25 keV.

(*a*), (*b*), (*c*), (*d*) Use Equ. 17-13 — (*a*) $\lambda = 1.23/2.5^{1/2}$ nm = 0.778 nm (*b*) $\lambda = 0.0778$ nm (*c*) $\lambda = 0.0246$ nm (*d*) $\lambda = 7.78$ pm

---

29* • An electron is moving at $v = 2.5 \times 10^5$ m/s. Find its de Broglie wavelength.

Find $p = mv$; $\lambda = h/p = h/mv$ — $\lambda = 6.626 \times 10^{-34}/9.11 \times 10^{-31} \times 2.5 \times 10^5$ m = 2.91 nm

---

30 • An electron has a wavelength of 200 nm. Find (*a*) its momentum, and (*b*) its kinetic energy.

(*a*) $p = h/\lambda$ — $p = 6.626 \times 10^{-34}/2 \times 10^{-7}$ kg·m/s = 3.31×10^{-27} kg·m/s

(*b*) Use Equ. 17-13; $K = (1.23/\lambda)^2$, λ in nm — $K = 3.78 \times 10^{-5}$ eV

---

31 • Find the energy of an electron in electron volts if its de Broglie wavelength is (*a*) 5 nm, and (*b*) 0.01 nm.

(*a*), (*b*) $K = (1.23/\lambda)^2$, λ in nm — (*a*) $K = 6.05 \times 10^{-2}$ eV (*b*) $K = 15.1$ keV

---

32 • A neutron in a reactor has kinetic energy of about 0.02 eV. Calculate the de Broglie wavelength of this neutron from Equation 17-12, where $mc^2 = 940$ MeV is the rest energy of the neutron.

Use Equ. 17-12;

$$\lambda = \frac{1240}{\sqrt{1880 \times 10^6}\sqrt{K}} = \frac{2.86 \times 10^{-2}}{\sqrt{K}}$$

$\lambda = 0.202$ nm

where λ is in nm and K in eV

33* • Use Equation 17-12 to find the de Broglie wavelength of a proton (rest energy $mc^2 = 938$ MeV) that has a kinetic energy of 2 MeV.

For protons, $\lambda = 2.86 \times 10^{-2}/\sqrt{K}$, λ in nm, K in eV $\lambda = 2.02 \times 10^{-5}$ nm = 20.2 fm

---

34 • A proton is moving at $v = 0.003c$, where c is the speed of light. Find its de Broglie wavelength.

Find p; $\lambda = h/p = h/m_pv = (h/m_ec)(c/v)(m_e/m_p)$ $\lambda = 2.43(1/0.003)(0.511/938)$ pm = 0.441 pm

---

35 • What is the kinetic energy of a proton whose de Broglie wavelength is (*a*) 1 nm, and (*b*) 1 fm?

(*a*), (*b*) $K = (2.86 \times 10^{-2}/\lambda)^2$, K in eV, λ in nm (*a*) $K = 0.818$ meV (*b*) $K = 818$ MeV

---

36 • Find the de Broglie wavelength of a baseball of mass 0.145 kg moving at 30 m/s.

$\lambda = h/mv$ $\lambda = 6.626 \times 10^{-34}/4.35$ m $= 1.52 \times 10^{-34}$ m

---

37* • The energy of the electron beam in Davisson and Germer's experiment was 54 eV. Calculate the wavelength for these electrons.

Use Equ. 17-13 $\lambda = 0.167$ nm

---

38 • The distance between Li^+ and Cl^- ions in a LiCl crystal is 0.257 nm. Find the energy of electrons that have a wavelength equal to this spacing.

$K = (1.23/\lambda)^2$ $K = 22.9$ eV

---

39 • An electron microscope uses electrons of energy 70 keV. Find the wavelength of these electrons.

Use Equ. 17-13 $\lambda = 4.65 \times 10^{-3}$ nm = 4.65 pm

---

40 • What is the de Broglie wavelength of a neutron with speed 10^6 m/s?

$\lambda = h/p = h/mv = 6.626 \times 10^{-34}/(1.67 \times 10^{-27} \times 10^6)$ m $= 3.97 \times 10^{-13}$ m = 0.397 pm.

41* • Suppose you have a spherical object of mass 4 g moving at 100 m/s. What size aperture is necessary for the object to show diffraction? Show that no common objects would be small enough to squeeze through such an aperture.

For diffraction, the diameter of the aperture $d \approx \lambda$. So $d \approx 6.626 \times 10^{-34}/(4 \times 10^{-3} \times 100) = 1.66 \times 10^{-33}$ m. This is many orders of magnitude smaller than even the diameter of a proton.

---

42 • A neutron has a kinetic energy of 10 MeV. What size object is necessary to observe neutron diffraction effects? Is there anything in nature of this size that could serve as a target to demonstrate the wave nature of 10-MeV neutrons?

From Problem 32, $\lambda = 2.86 \times 10^{-2}/\sqrt{K} \approx d$ $d = 9.04 \times 10^{-6}$ nm $\approx$ 10 fm ~ nuclear diameter

---

43 • What is the de Broglie wavelength of an electron of kinetic energy 200 eV? What are some common targets that could demonstrate the wave nature of such an electron?

From Equ. 17-13 one finds that $\lambda = 1.23/14.1$ nm = 0.0872 nm. This is of the order of the size of an atom.

---

44 •• Sketch the wave function $\psi(x)$ and the probability distribution $\psi^2(x)$ for the state $n = 4$ of a particle in a box.

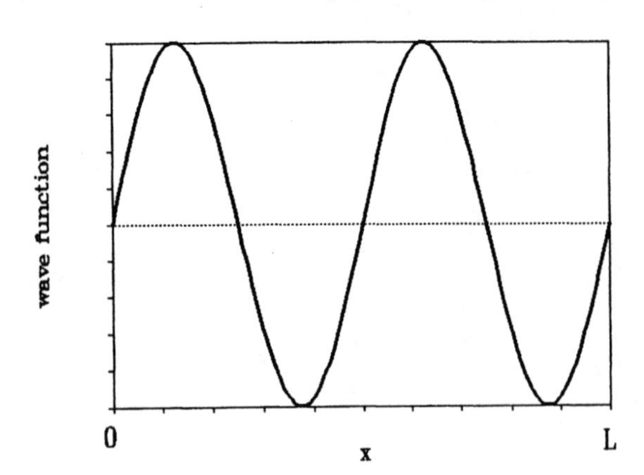

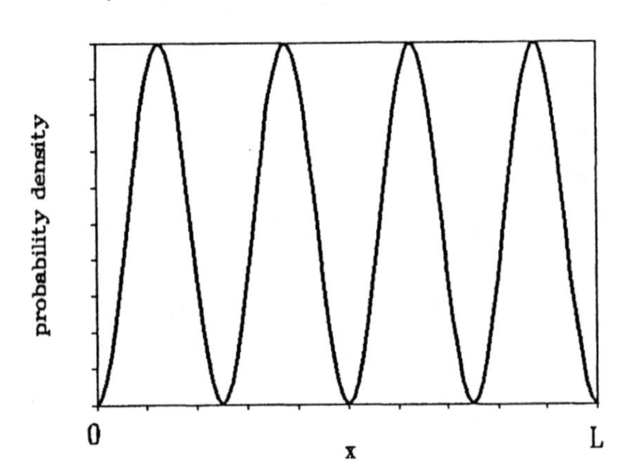

45* •• (*a*) Find the energy of the ground state ($n = 1$) and the first two excited states of a proton in a one-dimensional box of length $L = 10^{-15}$ m = 1 fm. (These are the order of magnitude of nuclear energies.) Make an energy-level diagram for this system and calculate the wavelength of electromagnetic radiation emitted when the proton makes a transition from (*b*) $n = 2$ to $n = 1$, (*c*) $n = 3$ to $n = 2$, and (*d*) $n = 3$ to $n = 1$.

(*a*) $E_1 = h^2/8mL^2 = 3.28 \times 10^{-11}$ J = 205 MeV; the energy level diagram is shown

(*b*) For $n = 2$ to $n = 1$, $\Delta E = 3E_1$ so $\lambda = 1240/615 \times 10^6$ nm = 2.02 fm

(*c*) For $n = 3$ to $n = 2$, $\Delta E = 5E_1$ and $\lambda = 3 \times 2.02/5$ fm = 1.21 fm

(*d*) For $n = 3$ to $n = 1$, $\Delta E = 8E_1$ and $\lambda = 3 \times 2.02/8$ fm = 0.758 fm

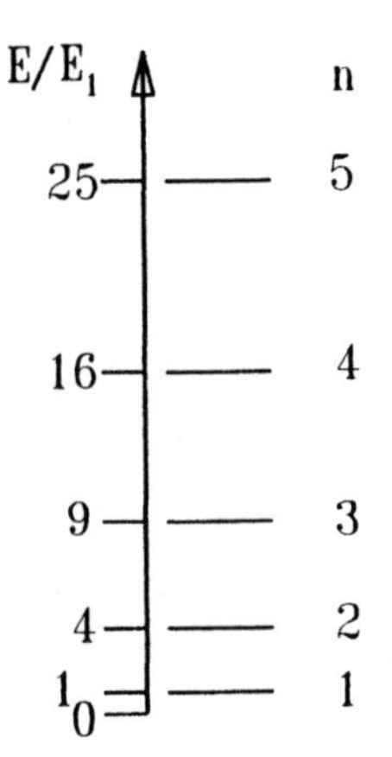

---

46 •• (*a*) Find the energy of the ground state ($n = 1$) and the first two excited states of a proton in a one-dimensional box of length 0.2 nm (about the diameter of a H_2 molecule). Calculate the wavelength of electromagnetic radiation emitted when the proton makes a transition from (*b*) $n = 2$ to $n = 1$, (*c*) $n = 3$ to $n = 2$, and (*d*) $n = 3$ to $n = 1$.

(*a*) $E_1 = \dfrac{(hc)^2}{8(mc^2)L^2}$, L in nm, E in eV $\qquad E_1 = \dfrac{(1240)^2}{8(9.38 \times 10^8)(0.04)} = 5.12$ meV

$E_2 = 4E_1$; $E_3 = 9E_1$ $\qquad E_2 = 20.5$ meV; $E_3 = 46.1$ meV

(*b*), (*c*), (*d*) $\lambda = 1240/\Delta E$ $\qquad \lambda_{2\text{-}1} = 8.07 \times 10^4$ nm = 80.7 μm; $\lambda_{3\text{-}2} = 48.4\ \mu$m; $\lambda_{3\text{-}1} = 30.3\ \mu$m

---

47 •• (*a*) Find the energy of the ground state and the first two excited states of a small particle of mass 1 μg confined to a one-dimensional box of length 1 cm. (*b*) If the particle moves with a speed of 1 mm/s, calculate its kinetic energy and find the approximate value of the quantum number n.

(*a*) $E_1 = \dfrac{(hc)^2}{8(mc^2)L^2}$ $\qquad E_1 = \dfrac{(6.626\times10^{-34}\times3\times10^8)^{-2}}{8(10^{-9}\times9\times10^{16})(10^{-2})^2} = 5.56\times10^{-55}$ J

$E_2 = 4E_1$; $E_3 = 9E_1$ $\qquad E_2 = 2.22\times10^{-54}$ J; $E_3 = 5.03\times10^{-54}$ J

(*b*) $E = \frac{1}{2}mv^2$; $n^2 = \frac{1}{2}mv^2/E_1$ $\qquad E = 5\times10^{-16}$ J; $n = 3\times10^{19}$

---

48 •• A particle is in the ground state of a box of length L. Find the probability of finding the particle in the interval $\Delta x = 0.002L$ at (*a*) $x = L/2$, (*b*) $x = 2L/3$, and (*c*) $x = L$. (Since Δx is very small, you need not do any integration because the wave function is slowly varying.)

$P(x)\Delta x = \psi^2(x)\Delta x$; $\psi(x) = \psi_1(x) = (2/L)^{1/2}\sin(\pi x/L)$ $\qquad P(x) = (2/L)\sin^2(\pi x/L)$

(*a*) Evaluate $P(x)$ at $x = L/2$; $P = P(x)\Delta x$ $\qquad P = (2/L)(0.002L) = 0.004$

(*b*) Repeat as in (*a*) for $x = 2L/3$ $\qquad P = (2/L)(0.75)(0.002L) = 0.003$

(*c*) Repeat as in (*a*) for $x = L$ $\qquad P = 0$

49* •• Do Problem 48 for a particle in the first excited state ($n = 2$).

Repeat procedure of Problem 47 with $P(x) = \psi^2(x) = \psi_2^2(x) = (2/L)\sin^2(2\pi x/L)$. The results are:

(*a*) $P = 0$. (*b*) $P = 0.003$. (*c*) $P = 0$.

---

50 •• Do Problem 48 for a particle in the second excited state ($n = 3$).

Proceed as in Problem 48 with $P(x) = \psi^2(x) = \psi_3^2(x) = (2/L)\sin^2(3\pi x/L)$. The results are:

(*a*) $P = 0.004$. (*b*) $P = 0$. (*c*) $P = 0$.

---

51 •• The classical probability distribution function for a particle in a box of length L is given by $P(x) = 1/L$. Use this to find $\langle x\rangle$ and $\langle x^2\rangle$ for a classical particle in such a box.

$$\langle x\rangle = \int_0^L (x/L)\,dx = L/2;\quad \langle x^2\rangle = \int_0^L (x^2/L)\,dx = L^2/3.$$

---

52 •• (*a*) Find $\langle x\rangle$ for the first excited state ($n = 2$) for a particle in a box of length L, and (*b*) find $\langle x^2\rangle$.

Proceed as in Example 17-8, replacing $\psi_1(x)$ with $\psi_2(x)$.

$$(a)\ \langle x\rangle = \int_0^L \frac{2x}{L}\sin^2\frac{2\pi x}{L}\,dx = \frac{2L}{\pi^2}\int_0^{2\pi}\theta\sin^2\theta\,d\theta = \frac{L}{2}.$$

$$(b)\ \langle x^2\rangle = \int_0^L \frac{2x^2}{L}\sin^2\frac{2\pi x}{L}\,dx = \frac{2L^2}{\pi^3}\int_0^{2\pi}\theta^2\sin^2\theta\,d\theta = \left(\frac{1}{3}-\frac{1}{8\pi^2}\right)L^2 = 0.321L^2.$$ We have used

$$\int\theta\sin^2\theta\,d\theta = \left(\frac{\theta^2}{4}-\frac{\theta\sin2\theta}{4}-\frac{\cos2\theta}{8}\right);\quad \int\theta^2\sin^2\theta\,d\theta = \left[\frac{\theta^3}{6}-\left(\frac{\theta^2}{4}-\frac{1}{8}\right)\sin2\theta-\frac{\theta\cos2\theta}{4}\right].$$

53* •• (*a*) Find $\langle x\rangle$ for the second excited state ($n = 3$) for a particle in a box of length L, and (*b*) find $\langle x^2\rangle$.

We proceed as in the preceding problem. Now the integrals over θ extend from 0 to 3π.

(*a*) $\langle x\rangle = L/2$. (*b*) $\langle x^2\rangle = (1/3 - 1/18\pi^2)L^2 = 0.328L^2$. (Note that $\langle x^2\rangle$ approaches the classical value 1/3 as the quantum number n increases.)

54 •• A particle in a one-dimensional box is in the first excited state ($n = 2$). (*a*) Sketch $\psi^2(x)$ versus x for this state. (*b*) What is the expectation value $\langle x \rangle$ for this state? (*c*) What is the probability of finding the particle in some small region dx centered at $x = \frac{1}{2}L$? (*d*) Are your answers for (*b*) and (*c*) contradictory? If not, explain.

(*a*) The probability density is shown

(*b*) $\langle x \rangle = L/2$ as found in Problem 52

(*c*) Since $P(L/2) = 0$, $P(L/2)\, dx = 0$

(*d*) Parts (*b*) and (*c*) are not contradictory. (*b*) states that the average value of measurements of the position of the particle will yield $L/2$, even though the probability that any one measurement of the position will yield $L/2$ is zero.

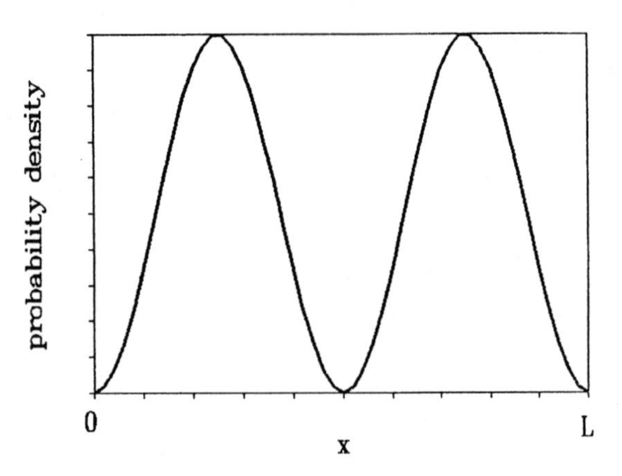

---

55 •• A particle of mass m has a wave function given by $\psi(x) = A\, e^{-|x|/a}$, where A and a are constants. (*a*) Find the normalization constant A. (*b*) Calculate the probability of finding the particle in the region $-a \le x \le a$.

(*a*) We evaluate $\int_{-\infty}^{\infty} \psi^2(x)\, dx = 1 = 2\int_0^{\infty} A^2 e^{-2x/a}\, dx = 2A^2 \int_0^{\infty} e^{-2x/a}\, dx = aA^2;\ A = 1/\sqrt{a}.$

(*b*) $P = \int_{-a}^{a} \psi^2(x)\, dx = 2\int_0^{a} \frac{1}{a} e^{-2x/a}\, dx = 1 - e^{-2} = 0.865.$

---

56 •• A particle in a one-dimensional box of length L is in its ground state. Calculate the probability that the particle will be found in the region (*a*) $0 < x < \frac{1}{2}L$, (*b*) $0 < x < L/3$, and (*c*) $0 < x < 3L/4$.

The probability density is given by $P(x) = (2/L)\sin^2(\pi x/L)$. We must now evaluate the integral of $P(x)$ between the limits specified in (*a*), (*b*), and (*c*). Changing the variable from x to θ as in Example 17-8, we have

(*a*) $P = \frac{2}{L}\frac{L}{\pi}\int_0^{\pi/2} \sin^2\theta\, d\theta = \frac{1}{2}.$

(*b*) $P = \frac{2}{\pi}\int_0^{\pi/3} \sin^2\theta\, d\theta = \frac{1}{3} - \frac{\sqrt{3}}{4\pi} = 0.196.$

(*c*) $P = \frac{2}{\pi}\int_0^{3\pi/4} \sin^2\theta\, d\theta = \frac{3}{4} + \frac{1}{2\pi} = 0.909.$

57* •• Repeat Problem 56 for a particle in the first excited state of the box.

For the first excited state, i.e., for $\psi^2(x) = (2/L)\sin^2(2\pi x/L)$, the integrals over θ go from 0 to π, 0 to $2\pi/3$, and 0 to $3\pi/2$ for parts (*a*), (*b*), and (*c*), respectively. The other change is that the factor (L/π) is replaced by $(L/2\pi)$.

(*a*) $P = \frac{2}{L}\frac{L}{2\pi}\int_0^{\pi} \sin^2\theta\, d\theta = \frac{1}{2}.$

$$(b)\ P = \frac{1}{\pi}\int_0^{2\pi/3} \sin^2\theta\, d\theta = \frac{1}{3} + \frac{\sqrt{3}}{8\pi} = 0.402.$$

$$(c)\ P = \frac{1}{\pi}\int_0^{3\pi/2} \sin^2\theta\, d\theta = \frac{3}{4} = 0.75.$$

---

58 •• (*a*) For the wave functions $\psi_n(x) = \sqrt{2/L}\ \sin\,(n\pi x/L)$, $n = 1, 2, 3, \ldots$ corresponding to a particle in the *n*th state of a one-dimensional box of length *L*, show that $\langle x^2\rangle = (L^2/3) - [L^2/(2n^2\pi^2)]$. (*b*) Compare this result for $n >> 1$ with your answer for the classical distribution of Problem 51.

We proceed as in Problem 52(*b*), now replacing the argument $(n\pi x/L)$ by θ.

$$(a)\ \langle x^2\rangle = \frac{2}{L}\frac{L^3}{(n\pi)^3}\int_0^{n\pi} \theta^2 \sin^2\theta\, d\theta = L^2\left(\frac{1}{3} - \frac{1}{2n^2\pi^2}\right).$$

(*b*) For large values of *n*, the result agrees with the classical value of $L^2/3$ given in Problem 51.

---

59 •• The wave functions for a particle of mass *m* in a one-dimensional box of length *L centered at the origin* (so that the ends are at $x = \pm L/2$) are given by $\psi_n(x) = \sqrt{2/L}\ \cos(n\pi x/L)$, $n = 1, 3, 5, 7, \ldots$ and $\psi_n(x) = \sqrt{2/L}\ \sin(n\pi x/L)$, $n = 2, 4, 6, 8, \ldots$. Calculate $\langle x\rangle$ and $\langle x^2\rangle$ for the ground state.

1. Since $\psi_1^2(x)$ is an even function of *x*, $x\psi_1^2(x)$ is an odd function of *x*. It follows that the integral between $-L/2$ and $L/2$ is zero. Thus $\langle x\rangle = 0$ for all values of *n*.

$$2.\ \langle x^2\rangle = \frac{2}{L}\int_{-L/2}^{L/2} x^2\cos^2\frac{\pi x}{L}\, dx = \frac{2L^2}{\pi^3}\int_{-\pi/2}^{\pi/2} \theta^2(1-\sin^2\theta)\, d\theta = \frac{2L^2}{\pi^3}\left[\frac{\pi^3}{12} - \left(\frac{\pi^3}{24} + \frac{\pi}{4}\right)\right] = L^2\left(\frac{1}{12} - \frac{1}{2\pi^2}\right).$$

Note: The result differs from that of Example 17-8. Since we have shifted the origin by $\Delta x = L/2$, we could have arrived at the above result, without performing the integration, by subtracting $(\Delta x)^2 = L^2/4$ from $\langle x^2\rangle$ as given in Example 17-8.

---

60 •• Calculate $\langle x\rangle$ and $\langle x^2\rangle$ for the first excited state of the box described in Problem 59.

$\langle x\rangle = 0$. (see Problem 59)

$$\langle x^2\rangle = L^2\left(\frac{1}{3} - \frac{1}{8\pi^2}\right) - \frac{L^2}{4} = L^2\left(\frac{1}{12} - \frac{1}{8\pi^2}\right).$$ (see Note of Problem 59)

61* • Can the expectation value of *x* ever equal a value that has zero probability of being measured?

Yes

---

62 • Explain why the maximum kinetic energy of electrons emitted in the photoelectric effect does not depend on the intensity of the incident light, but the total number of electrons emitted does.

In the photoelectric effect, an electron absorbs the energy of a single photon. Therefore, $K_{max} = hf - \phi$, independent of the number of photons incident on the surface. However, the number of photons incident on the surface determines the number of electrons that are emitted.

---

63 •• A six-sided die has the number 1 painted on three sides and the number 2 painted on the other three sides. (*a*) What is the probability of a 1 coming up when the die is thrown? (*b*) What is the expectation value of the number that comes up when the die is thrown?

(*a*) $P(1) = ½$. (*b*) $\langle n\rangle = (3 \times 1 + 3 \times 2)/6 = 1.5$.

---

64 •• True or false: (*a*) It is impossible in principle to know precisely the position of an electron. (*b*) A particle that is confined to some region of space cannot have zero energy. (*c*) All phenomena in nature are adequately described by classical wave theory. (*d*) The expectation value of a quantity is the value that you expect to measure.

(*a*) False (*b*) True (*c*) False (*d*) False; it is the most probable value of the measurement.

65* •• It was once believed that if two identical experiments are done on identical systems under the same conditions, the results must be identical. Explain why this is not true, and how it can be modified so that it is consistent with quantum physics.

According to quantum theory, the average value of many measurements of the same quantity will yield the expectation value of that quantity. However, any single measurement may differ from the expectation value.

---

66 • A light beam of wavelength 400 nm has an intensity of 100 W/m². (*a*) What is the energy of each photon in the beam? (*b*) How much energy strikes an area of 1 cm² perpendicular to the beam in 1 s? (*c*) How many photons strike this area in 1 s?

(*a*) $E_{ph} = hc/\lambda$ $E_{ph} = 1240/400\text{ eV} = 3.1\text{ eV}$

(*b*) $E = IAt$ $E = 100 \times 10^{-4} \times 1 = 0.01\text{ J} = 6.25 \times 10^{16}\text{ eV}$

(*c*) $N = E/E_{ph}$ $N = 6.25 \times 10^{16}/3.1 = 2.02 \times 10^{16}$

---

67 • A mass of 10^{-6} g is moving with a speed of about 10^{-1} cm/s in a box of length 1 cm. Treating this as a one-dimensional particle in a box, calculate the approximate value of the quantum number *n*.

1. Write the energy of the particle $E = ½mv^2$
2. Write the expression for E_n $E_n = n^2h^2/8mL^2$
3. Solve for *n* $n = 2mvL/h = 3.02 \times 10^{19}$

---

68 • (*a*) For the classical particle of Problem 67, find Δx and Δp, assuming that these uncertainties are given by $\Delta x/L = 0.01\%$ and $\Delta p/p = 0.01\%$. (*b*) What is $(\Delta x\Delta p)/\hbar$?

(*a*) $\Delta x = 10^{-4} \times 10^{-2}\text{ m} = 10^{-6}\text{ m}$; $\Delta p = 10^{-4}(mv) = 10^{-4} \times 10^{-9} \times 10^{-3}\text{ kg·m/s} = 10^{-16}\text{ kg·m/s}$.

(*b*) $\Delta x\Delta p/\hbar = 10^{-22}/1.05 \times 10^{-34} = 0.948 \times 10^{12}$.

69* • In 1987, a laser at Los Alamos National Laboratory produced a flash that lasted 1×10^{-12} s and had a power of 5.0×10^{15} W. Estimate the number of emitted photons if their wavelength was 400 nm.

$N = E/E_{ph} = (P\Delta t)/(hc/\lambda)$ $N = (5 \times 10^3 \times 1.6 \times 10^{-19}\text{ eV})/3.1\text{ eV} = 10^{22}$

---

70 • You can't see anything smaller than the wavelength λ used. What is the minimum energy of an electron needed in an electron microscope to see an atom, which has a diameter of about 0.1 nm?

Use Equ. 17-13; $K = 1.23^2/\lambda^2$ eV, λ in nm $K = 151$ eV

---

71 • A common flea that has a mass of 0.008 g can jump vertically as high as 20 cm. Estimate the de Broglie wavelength for the flea immediately after takeoff.

1. $p = m\sqrt{2gh}$ $\quad p = 1.584 \times 10^{-5}$ kg·m/s
2. $\lambda = h/p$ $\quad \lambda = 6.626 \times 10^{-34}/1.584 \times 10^{-5}$ m $= 4.2 \times 10^{-29}$ m

---

72 • The work function for sodium is $\phi = 2.3$ eV. Find the minimum de Broglie wavelength for the electrons emitted by a sodium cathode illuminated by violet light with a wavelength of 420 nm.

1. Use Equ. 17-3 to find K_{max} $\quad K_{max} = (1240/420 - 2.3)$ eV $= 0.652$ eV
2. Use Equ. 17-13 to find λ $\quad \lambda = 1.23/\sqrt{0.652} = 1.52$ nm

---

73* •• Suppose that a 100-W source radiates light of wavelength 600 nm uniformly in all directions and that the eye can detect this light if only 20 photons per second enter a dark-adapted eye having a pupil 7 mm in diameter. How far from the source can the light be detected under these rather extreme conditions?

1. At a distance R from the source, the fraction of the light energy entering the eye is $A_{eye}/4\pi R^2 = r^2/4R^2$.
2. Find the number of photons emitted per second $\quad N = P/E_{ph} = 100/[(1240/600) \times 1.6 \times 10^{-19}]$ $= 3.02 \times 10^{20}$/s
3. Solve for R from $20 = 3.02 \times 10^{20} \times r^2/4R^2$ $\quad R = 6800$ km (neglects absorption by atmosphere)

---

74 •• Data for maximum kinetic energy of the electrons versus wavelength for the photoelectric effect using sodium are

λ, nm	200	300	400	500	600
K_{max}, eV	4.20	2.06	1.05	0.41	0.03

Plot these data so as to obtain a straight line and from your plot find (*a*) the work function, (*b*) the threshold frequency, and (*c*) the ratio h/e.

We plot K_{max} versus frequency, f. The plot is shown below.

(*a*) $\phi = hf_{th}$, where f_{th} is the threshold frequency for emission of electrons. Here $f_{th} = 0.5 \times 10^{15}$ Hz; $\phi = 2.07$ eV.

(*b*) See (*a*). $f_{th} = 0.5$ PHz.

(*c*) h/e is the slope of the straight line divided by e. The slope is $4.2/10^{15}$ eV/Hz; so $h/e = 4.2 \times 10^{-15}$ V/Hz, which gives the value of h as 4.2×10^{-15} eV·s, in fair agreement with the exact value of 4.136×10^{-15} eV·s.

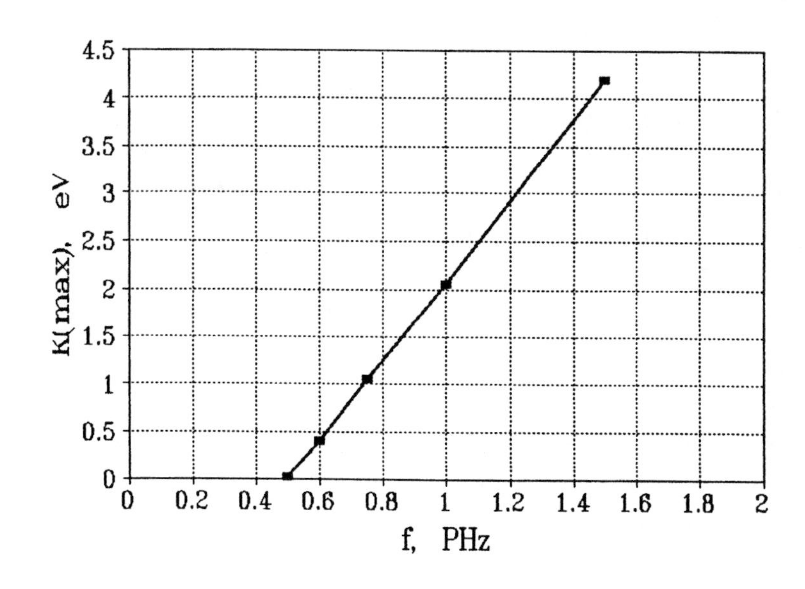

---

75 •• The diameter of the pupil of the eye under room-light conditions is about 5 mm. (It can vary from about 1 to 8 mm.) Find the intensity of light of wavelength 600 nm such that 1 photon per second passes through the pupil.

$I = P/A = E/At = hc/\lambda At$; evaluate I — $I = 1240/(600 \times \pi \times 2.5^2 \times 10^{-6})$ eV/m²·s $= 1.68 \times 10^{-14}$ W/m²

---

76 •• A light bulb radiates 90 W uniformly in all directions. (*a*) Find the intensity at a distance of 1.5 m. (*b*) If the wavelength is 650 nm, find the number of photons per second that strike a surface of area 1 cm² oriented so that the line to the bulb is perpendicular to the surface.

(*a*) $I = P/A = P/4\pi R^2$ — $I = 90/4\pi \times 2.25$ W/m² $= 3.18$ W/m²

(*b*) $N = IA/E_{ph}$ — $N = [3.18 \times 10^{-4}/1.6 \times 10^{-19}(1240/650)] = 1.04 \times 10^{15}$

77* •• When light of wavelength λ_1 is incident on the cathode of a photoelectric tube, the maximum kinetic energy of the emitted electrons is 1.8 eV. If the wavelength is reduced to $\lambda_1/2$, the maximum kinetic energy of the emitted electrons is 5.5 eV. Find the work function ϕ of the cathode material.

1. Use Equ. 17-3 for λ_1 and $\lambda_1/2$ — 1.8 eV $= 1240/\lambda_1 - \phi$; 5.5 eV $= 2480/\lambda_1 - \phi$
2. Solve for ϕ — $\phi = 1.9$ eV

---

78 •• A photon of energy E is scattered at an angle of θ. Show that the energy E' of the scattered photon is given by $E' = \dfrac{E}{(E/m_ec^2)(1 - \cos\theta) + 1}$

From Equ. 17-8, $\lambda' = \dfrac{h}{m_ec}(1 - \cos\theta) + \lambda$, and

$$E' = \frac{hc}{\lambda'} = \frac{hc}{\lambda + \dfrac{h}{m_ec}(1 - \cos\theta)} = \frac{\dfrac{hc}{\lambda}}{1 + \dfrac{hc}{m_ec^2\lambda}(1 - \cos\theta)} = \frac{E}{1 + \dfrac{E}{m_ec^2}(1 - \cos\theta)}.$$

---

79 •• A particle is confined to a one-dimensional box. In making a transition from the state n to the state $n - 1$, radiation of 114.8 nm is emitted; in the transition from the state $n - 1$ to the state $n - 2$, radiation of wavelength 147 nm is emitted. The ground-state energy of the particle is 1.2 eV. Determine n.

1. $\Delta E = n^2E_1 - (n - 1)^2E_1 = (2n - 1)E_1 = hc/\lambda$ — $2n - 1 = 1240/1.2 \times 114.8 = 9$
2. Solve for n — $n = 5$

---

80 •• A particle confined to a one-dimensional box has a ground-state energy of 0.4 eV. When irradiated with light of 206.7 nm it makes a transition to an excited state. When decaying from this excited state to the next lower state it emits radiation of 442.9 nm. What is the quantum number of the state to which the particle has decayed?

1. Find E_f, energy of the final state — $E_f = (0.4 + 1240/206.7 - 1240/442.9)$ eV $= 3.6$ eV
2. $E_f = n^2E_1$; solve for n — $n^2 = 9$; $n = 3$

81* •• When a surface is illuminated with light of wavelength λ the maximum kinetic energy of the emitted electrons is 1.2 eV. If the wavelength $\lambda' = 0.8\lambda$ is used the maximum kinetic energy increases to 1.76 eV, and for wavelength $\lambda'' = 0.6\lambda$ the maximum kinetic energy of the emitted electrons is 2.676 eV. Determine the work function of the surface and the wavelength λ.

1. Use Equ. 17-3 — $1240/\lambda = 1.2 \text{ eV} + \phi$; $1240/0.8\lambda = 1.76 \text{ eV} + \phi$
2. Solve for λ — $(1550 - 1240)/\lambda = 310/\lambda = 0.56 \text{ eV}$; $\lambda = 553.6 \text{ nm}$
3. Evaluate ϕ — $\phi = 1.04 \text{ eV}$

---

82 •• A simple pendulum of length 1 m has a bob of mass 0.3 kg. The energy of this oscillator is quantized to the values $E_n = (n + ½)hf_0$, where n is an integer and f_0 is the frequency of the pendulum. (*a*) Find n if the angular amplitude is 10°. (*b*) Find Δn if the energy changes by 0.01%.

(*a*) Find E of pendulum; $E = mgL(1 - \cos\theta)$ — $E = 0.3 \times 9.81 \times 1(1 - \cos 10^\circ) = 0.0447 \text{ J}$

Set $E = (n + ½)\dfrac{h}{2\pi}\sqrt{\dfrac{g}{L}}$ and solve for n — $n = 1.35 \times 10^{32}$

(*b*) For $\Delta E = 10^{-4}E$, $\Delta n = 10^{-4}n$ — $\Delta n = 1.35 \times 10^{28}$

---

83 •• (*a*) Show that for large n, the fractional difference in energy between state n and state $n + 1$ for a particle in a box is given approximately by $(E_{n+1} - E_n)/E_n \approx 2/n$. (*b*) What is the approximate percentage energy difference between the states $n_1 = 1000$ and $n_2 = 1001$? (*c*) Comment on how this result is related to Bohr's correspondence principle.

(*a*) $$\frac{E_{n+1} - E_n}{E_n} = \frac{(n+1)^2 - n^2}{n^2} = \frac{2n+1}{n^2} \approx \frac{2}{n}.$$

(*b*) Using the above, the percentage difference is 0.2%.

(*c*) Classically, the energy is continuous. For very large values of n (see, e.g., Problem 67) the energy difference between adjacent levels is infinitesimal.

---

84 •• In 1985, a light pulse of 1.8×10^{12} photons was produced in an AT&T laboratory during a time interval of 8×10^{-15} s. The wavelength of the produced light was $\lambda = 2400$ nm. Suppose all of the light was absorbed by the black surface of a screen. Estimate the force exerted by the photons on the screen.

$p = E/c = Nhf/c = Nh/\lambda$; then $F = \Delta p/\Delta t = Nh/\lambda\Delta t$ — $F = (1.8 \times 10^{12} \times 6.626 \times 10^{-34}/2.4 \times 10^{-6} \times 8 \times 10^{-15})\text{ N} = 0.0621 \text{ N}$

85* •• This problem is one of estimating the time lag (expected classically but not observed) in the photoelectric effect. Let the intensity of the incident radiation be 0.01 W/m^2. (*a*) If the area of the atom is 0.01 nm^2, find the energy per second falling on an atom. (*b*) If the work function is 2 eV, how long would it take classically for this much energy to fall on one atom?

(*a*) $P = IA$ — $P = 10^{-2} \times 10^{-20} \text{ J/s} = 6.25 \times 10^{-4} \text{ eV/s}$

(*b*) $t = E/P$ — $t = 2/6.25 \times 10^{-4} \text{ s} = 3200 \text{ s} = 53.3 \text{ min}$

CHAPTER 18

Temperature and the Kinetic Theory of Gases

1* • True or false:

(*a*) Two objects in thermal equilibrium with each other must be in thermal equilibrium with a third object.

(*b*) The Fahrenheit and Celsius temperature scales differ only in the choice of the zero temperature.

(*c*) The kelvin is the same size as the Celsius degree.

(*d*) All thermometers give the same result when measuring the temperature of a particular system.

(*a*) False (*b*) False (*c*) True (*d*) False

---

2 • How can you determine if two bodies are in thermal equilibrium with each other if it is impossible to put them into thermal contact with each other?

Put each in thermal equilibrium with a third body, e.g., a thermometer.

---

3 • Which is greater, an increase in temperature of 1 C° or of 1 F°?

An increase of 1°C is greater.

---

4 • "One day I woke up and it was 20°F in my bedroom," said Mert to his old friend Mort. "That's nothing," replied Mort. "My room was once –5°C." Which room was colder?

Convert 20°F to t_C; Equ. 18-2 $t_C = (5/9)(20 - 32) = -6.67$°C; 20°F is colder

---

5* • A certain ski wax is rated for use between –12 and –7°C. What is this temperature range on the Fahrenheit scale?

Convert -12 and -7°C to t_F using $t_F = (9/5)t_C + 32$ $t_{F1} = 10.4$°F; $t_{F2} = 19.4$°F; between 10.4°F and 19.4°F

---

6 • The melting point of gold (Au) is 1945.4°F. Express this temperature in degrees Celsius.

Use Equ. 18-2 $t_C = (5/9)(1954.4 - 32) = 1068$°C

---

7 • The highest and lowest temperatures ever recorded in the United States are 134°F (in California in 1913) and –80°F (in Alaska in 1971). Express these temperatures using the Celsius scale.

Use Equ. 18-2 $t_{CH} = (5/9)102 = 56.7$°C; $t_{CL} = (5/9)(-112) = -62.2$°C

---

8 • What is the Celsius temperature corresponding to the normal temperature of the human body, 98.6°F?

Use Equ. 18-2 $t_C = (5/9)66.6 = 37.0$°C

9* • The length of the column of mercury in a thermometer is 4.0 cm when the thermometer is immersed in ice water and 24.0 cm when the thermometer is immersed in boiling water. (*a*) What should the length be at room temperature, 22.0°C? (*b*) If the mercury column is 25.4 cm long when the thermometer is immersed in a chemical solution, what is the temperature of the solution?

(*a*), (*b*) $L = [(20/100)t_C + 4]$ cm | (*a*) $L = 8.4$ cm (*b*) $t_C = (5 \times 21.4)$°C = 107°C

---

10 • The temperature of the interior of the sun is about 10^7 K. What is this temperature on (*a*) the Celsius scale, and (*b*) the Fahrenheit scale?

Neglect 32 and 273 compared to 10^7 | (*a*) $t_C \approx t_K = 10^{7}$°C (*b*) $t_F \approx (9/5)t_K = 1.8 \times 10^{7}$°F

---

11 • The boiling point of nitrogen N_2 is 77.35 K. Express this temperature in degrees Fahrenheit.

Convert 77.35 K to °C then to °F | 77.35 K = −198.5°C = (−1.8 × 198.5 + 32)°F = −320.4°F

---

12 • The pressure of a constant-volume gas thermometer is 0.400 atm at the ice point and 0.546 atm at the steam point. (*a*) When the pressure is 0.100 atm, what is the temperature? (*b*) What is the pressure at 444.6°C, the boiling point of sulfur?

(*a*) $P = CT$; find C, then T for $P = 0.100$ atm | $T = (273.15/0.4) \times 0.100$ K = 68.3 K

(*b*) Convert to T, then determine P | $T = 717.75$ K; $P = (0.4/273.15) \times 717.75$ atm = 1.05 atm

---

13* • A constant-volume gas thermometer reads 50 torr at the triple point of water. (*a*) What will the pressure be when the thermometer measures a temperature of 300 K? (*b*) What ideal-gas temperature corresponds to a pressure of 678 torr?

(*a*), (*b*) Use Equ. 18-4 | (*a*) $T = 50(300/273.16)$ torr = 54.9 torr (*b*) $T = 3704$ K

---

14 • A constant-volume gas thermometer has a pressure of 30 torr when it reads a temperature of 373 K. (*a*) What is its triple-point pressure P_3? (*b*) What temperature corresponds to a pressure of 0.175 torr?

(*a*), (*b*) Use Equ. 18-4 | (*a*) $P_3 = 21.97$ torr (*b*) $T = 2.176$ K

---

15 • At what temperature do the Fahrenheit and Celsius temperature scales give the same reading?

Set $t_F = t_C$ in Equ. 18-2 and solve for t_C | $0.8t_C = -32$; $t_C = -40$°C = −40°F

---

16 • Sodium melts at 371 K. What is the melting point of sodium on the Celsius and Fahrenheit temperature scales?

$t_C = T - 273$; then use Equ. 18-2 | $t_C = 98$°C; $t_F = 208.4$°F

17* • The boiling point of oxygen at one atmosphere is 90.2 K. What is the boiling point of oxygen on the Celsius and Fahrenheit scales?

Proceed as in Problem 16 | $t_C = -182.95$°C; $t_F = -297.3$°F

---

18 •• On the Réaumur temperature scale, the melting point of ice is 0°R and the boiling point of water is 80°R. Derive expressions for converting temperatures on the Réaumur scale to the Celsius and Fahrenheit scales.

0°C = 0°R and 100°C = 80°R. Consequently, $t_R = 0.8t_C$ and $t_C = 1.25t_R = (5/4)t_R$. From Equ. 18-2 we have $t_F = 1.8t_C + 32 = 2.25t_R + 32 = (9/4)t_R + 32$.

---

19 ••• A thermistor is a solid-state device whose resistance varies greatly with temperature. Its temperature dependence is given approximately by $R = R_0 e^{B/T}$, where R is in ohms (Ω), T is in kelvins, and R_0 and B are constants that can be determined by measuring R at calibration points such as the ice point and the steam point. (*a*) If R = 7360 Ω at the ice point and 153 Ω at the steam point, find R_0 and B. (*b*) What is the resistance of the thermistor at t = 98.6°F? (*c*) What is the rate of change of the resistance with temperature (dR/dT) at the ice point and the steam point? (*d*) At which temperature is the thermistor most sensitive?

(*a*) Find B from $\dfrac{R_{273}}{R_{373}} = \dfrac{7360}{153} = 48.1$ $\qquad e^{\frac{B}{273} - \frac{B}{373}} = 48.1;\ B = \dfrac{\ln 48.1}{1/273 - 1/373} = 3.94 \times 10^3$ K

Evaluate R_0 from $R_0 = R_T e^{-B/T}$ $\qquad R_0 = 7360 e^{-3940/273}\ \Omega = 3.91 \times 10^{-3}\ \Omega$

(*b*) 98.6°F = 310 K (see Problem 8) $\qquad R = 3.91 \times 10^{-3} \times e^{3940/310} = 1.29$ kΩ

(*c*), (*d*) $\dfrac{dR}{dT} = -\dfrac{R_0 B}{T^2} e^{B/T} = -RB/T^2$ $\qquad (dR/dT)_{ice} = -389$ Ω/K; $(dR/dT)_{steam} = -4.33$ Ω/K

Sensitivity is greater the lower the temperature

---

20 •• Two identical vessels contain different ideal gases at the same pressure and temperature. It follows that

(*a*) the number of gas molecules is the same in both vessels.

(*b*) the total mass of gas is the same in both vessels.

(*c*) the average speed of the gas molecules is the same in both vessels.

(*d*) none of the above is correct.

(*a*)

21* •• Figure 18-15 shows a plot of volume versus temperature for a process that takes an ideal gas from point A to point B. What happens to the pressure of the gas?

The pressure increases.

---

22 •• Figure 18-16 shows a plot of pressure versus temperature for a process that takes an ideal gas from point A to point B. What happens to the volume of the gas?

The volume decreases.

---

23 • A gas is kept at constant pressure. If its temperature is changed from 50 to 100°C, by what factor does the volume change?

At constant pressure, $V_2/V_1 = T_2/T_1$ $\qquad V_2/V_1 = 373/323 = 1.15$

---

24 • A 10-L vessel contains gas at a temperature of 0°C and a pressure of 4 atm. How many moles of gas are in the vessel? How many molecules?

From Equ. 18-12, V/mol at STP = 22.4 L $\qquad n = 10 \times 4/22.4 = 1.79;\ N = nN_A = 1.08 \times 10^{24}$

25* •• A pressure as low as 1×10^{-8} torr can be achieved using an oil diffusion pump. How many molecules are there in 1 cm^3 of a gas at this pressure if its temperature is 300 K?

1. Convert torr to atm and cm³ to L — $P = 10^{-8} \times 1.316 \times 10^{-3}$ atm $= 1.316 \times 10^{-11}$ atm; $V = 1 \times 10^{-3}$ L

2. Use Equs. 18-12 and 18-13 — $N = 1.316 \times 10^{-14} \times 6.022 \times 10^{23}/0.08206 \times 300 = 3.22 \times 10^{8}$

---

26 •• A motorist inflates the tires of her car to a pressure of 180 kPa on a day when the temperature is −8.0°C. When she arrives at her destination, the tire pressure has increased to 245 kPa. What is the temperature of the tires if we assume that (*a*) the tires do not expand, or (*b*) that the tires expand by 7%?

(*a*) From Equ. 18-13, $T_2 = T_1(P_2/P_1)$ — $T_2 = 265 \times 245/180$ K $= 360.7$ K $= 87.7$°C

(*b*) $T_2 = T_1(P_2V_2/P_1V_1)$ — $T_2 = 360.7 \times 1.07 = 386$ K $= 113$°C

---

27 •• A room is 6 m by 5 m by 3 m. (*a*) If the air pressure in the room is 1 atm and the temperature is 300 K, find the number of moles of air in the room. (*b*) If the temperature rises by 5 K and the pressure remains constant, how many moles of air leave the room?

(*a*) V/mol at STP = 22.4 L — $n = 90 \times 10^3 \times 273/22.4 \times 300 = 3.66 \times 10^3$ mol

(*b*) Find $n' = n(305\text{ K})$ and Δn — $n' = n(300/305)$; $\Delta n = n(1 - 0.9836) = 60$ mol

---

28 •• A seafood restaurant hires Lou to run its advertising campaign. Lou figures that snorklers are a great pool of potential customers for seafood, so he prints ads on Mylar balloons that he ties to the coral of an underwater reef. Each balloon has a volume of 4 L and is filled with air at 20°C. At 15 m below the ocean surface, the volume has diminished to 1.60 L. What is the temperature of the water at this depth?

1. Find P_2 = pressure at 15 m below surface — $P_2 = (101 + 1.025 \times 9.81 \times 15)$ kPa $= 251.8$ kPa

2. Use Equ. 18-13 — $T_2 = 293 \times 251.8 \times 1.6/101 \times 4$ K $= 292$ K $= 19$°C

29* •• The boiling point of helium at one atmosphere is 4.2 K. What is the volume occupied by helium gas due to evaporation of 10 g of liquid helium at 1 atm pressure and a temperature of (*a*) 4.2 K, and (*b*) 293 K?

(*a*) 10 g = 2.5 mol; $V = nRT/P$ — $V_{4.2} = 2.5 \times 0.08206 \times 4.2/1$ L $= 0.862$ L

(*b*) $V_2 = V_1(T_2/T_1)$ — $V_{293} = 0.862 \times 293/4.2 = 60.1$ L

---

30 •• A container with a volume of 6.0 L holds 10 g of liquid helium. As the container warms to room temperature, what is the pressure exerted by the gas on its walls?

From Problem 29, $P = 1$ atm for $V = 60.1$ L — $P = 1 \times 60.1/6$ atm $= 10.0$ atm

---

31 •• An automobile tire is filled to a gauge pressure of 200 kPa when its temperature is 20°C. (Gauge pressure is the difference between the actual pressure and atmospheric pressure.) After the car has been driven at high speeds, the tire temperature increases to 50°C. (*a*) Assuming that the volume of the tire does not change, and that air behaves as an ideal gas, find the gauge pressure of the air in the tire. (*b*) Calculate the gauge pressure if the volume of the tire expands by 10%.

(*a*) $P_2 = P_1(T_2/T_1)$ — $P_2 = 301 \times 323/293$ kPa $= 332$ kPa; $P_{2\text{gauge}} = 231$ kPa

(*b*) $P_2' = P_2(V_1/V_2)$ — $P_2' = 332/1.1 = 302$ kPa; $P_{2\text{gauge}} = 201$ kPa

---

32 •• A scuba diver is 40 m below the surface of a lake, where the temperature is 5°C. He releases an air bubble with a volume of 15 cm^3. The bubble rises to the surface, where the temperature is 25°C. What is the volume of the bubble right before it breaks the surface? *Hint:* Remember that the pressure also changes.

1. Determine P_1 — $P_1 = 101 + 1.0 \times 9.81 \times 40$ kPa $= 493$ kPa
2. $V_2 = V_1(T_2/T_1)(P_1/P_2)$; $P_2 = 101$ kPa — $V_2 = 15(298/278)(493/101)$ $cm^3 = 78.5$ cm^3

33* ••• A helium balloon is used to lift a load of 110 N. The weight of the balloon's skin is 50 N, and the volume of the balloon when fully inflated is 32 m^3. The temperature of the air is 0°C and the atmospheric pressure is 1 atm. The balloon is inflated with sufficient helium gas so that the net buoyant force on the balloon and its load is 30 N. Neglect changes of temperature with altitude.

(*a*) How many moles of helium gas are contained in the balloon?

(*b*) At what altitude will the balloon be fully inflated?

(*c*) Does the balloon ever reach the altitude at which it is fully inflated?

(*d*) If the answer to (*c*) is affirmative, what is the maximum altitude attained by the balloon?

(*a*) Find V from $F_B = mg + 30$ N — $\rho_{air}Vg = 190\text{ N} + \rho_{He}Vg$; $V = 17.38$ m^3

$n = PV/RT$ — $n = 1 \times 17.38/0.08206 \times 273 = 776$

(*b*) Find P for $V = 32$ m^3 — $P = 17.32/32$ atm $= 0.543$ atm

Determine h from Problem 13-96 — $0.543 = e^{-h/7.93}$; $h = [7.93 \ln(1/.543)]$ km $= 4.84$ km

(*c*) Determine F_B at 4.84 km; $\rho_{air} = 1.293 \times 0.543$ — $F_B = 1.293 \times 0.543 \times 32 \times 9.81$ N $= 220.4$ N

Find $m_{tot}g$ — $m_{tot}g = (160 + 0.004 \times 776 \times 9.81)$ N $= 190.5$ N

Note that F_B at 4.84 km is greater than $m_{tot}g$ — Yes; the balloon will rise above 4.84 km

(*d*) Find ρ_{air} such that $F_B = 190.5$ N — $\rho_{air} = 190.5/32 \times 9.81$ kg/m^3 $= 0.6068$ kg/m^3

Note that $P = (0.6068/1.293)$ atm $= e^{-h/7.93}$ — $h = [7.93 \ln(1/0.469)]$ km $= 6.0$ km

---

34 • True or false: The absolute temperature of a gas is a measure of the average translational kinetic energy of the gas molecules.

True

---

35 • By what factor must the absolute temperature of a gas be increased to double the rms speed of its molecules?

4; $v_{rms} \propto \sqrt{T}$.

---

36 • How does the average translational kinetic energy of a molecule of a gas change if the pressure is doubled while the volume is kept constant? If the volume is doubled while the pressure is kept constant?

If P is doubled at constant V, T is doubled. Consequently, K_{av} increases by a factor of 2.

If V is doubled at constant P, T is halved. Consequently, K_{av} is reduced by a factor 1/2.

37* • A mole of He molecules is in one container and a mole of CH_4 molecules is in a second container, both at standard conditions. Which molecules have the greater mean free path?

From Equ. 18-25, it follows that the He atoms have the greater mean free path since the diameter of He is smaller than that of CH_4.

---

38 •• A vessel holds an equal number of moles of helium and methane, CH_4. The ratio of the rms speeds of the helium atoms to the CH_4 molecules is ____.

(*a*) 1

(*b*) 2
(*c*) 4
(*d*) 16
(*b*)

---

39 • (*a*) Find v_{rms} for an argon atom if 1 mol of the gas is confined to a 1-L container at a pressure of 10 atm. (For argon, $M = 40 \times 10^{-3}$ kg/mol.) (*b*) Compare this with v_{rms} for a helium atom under the same conditions. (For helium, $M = 4 \times 10^{-3}$ kg/mol.)

(*a*) From Equs. 18-23 and 18-13, $v_{rms} = \sqrt{\dfrac{3PV}{nM}}$ $\quad v_{rms} = \sqrt{\dfrac{3 \times 1.01 \times 10^6 \times 10^{-3}}{1 \times 40 \times 10^{-3}}}$ m/s = 275 m/s

(*b*) $v_{rms}(\mathrm{He}) = v_{rms}(\mathrm{Ar})\sqrt{\dfrac{M_{Ar}}{M_{He}}}$ $\quad v_{rms}(\mathrm{He}) = 870$ m/s

---

40 • Find the total translational kinetic energy of 1 L of oxygen gas held at a temperature of 0°C and a pressure of 1 atm.

Use Equs. 18-22 and 18-13; $K = (3/2)PV$ $\quad K = 1.5 \times 1.01 \times 10^5 \times 10^{-3}$ J = 152 J

41* • Find the rms speed and the average kinetic energy of a hydrogen atom at a temperature of 10^7 K. (At this temperature, which is of the order of the temperature in the interior of a star, the hydrogen is ionized and consists of a single proton.)

Use Equs. 18-22 and 18-23; $M = 10^{-3}$ kg/mol $\quad K = 1.5kT$ J $= 2.07 \times 10^{16}$ J

$$v_{rms} = \sqrt{\frac{3 \times 8.314 \times 10^7}{10^{-3}}} \text{ m/s} = 499 \text{ m/s}$$

---

42 • In one model of a solid, the material is assumed to consist of a regular array of atoms in which each atom has a fixed equilibrium position and is connected by springs to its neighbors. Each atom can vibrate in the *x*, *y*, and *z* directions. The total energy of an atom in this model is

$$E = \frac{1}{2}mv_x^2 + \frac{1}{2}mv_y^2 + \frac{1}{2}mv_z^2 + \frac{1}{2}Kx^2 + \frac{1}{2}Ky^2 + Kz^2$$

What is the average energy of an atom in the solid when the temperature is *T*? What is the total energy of one mole of such a solid?

There are six degrees of freedom. Consequently, $E_{av} = 6(\frac{1}{2}kT) = 3kT$ and E/mol $= 3RT$.

---

43 • Show that the mean free path for a molecule in an ideal gas at temperature *T* and pressure *P* is given by

$$\lambda = \frac{kT}{\sqrt{2}P\pi d^2}$$

We have $\lambda = \dfrac{1}{\sqrt{2}n_v \pi d^2}$ (Equ. 18-25). Here $n_v = N/V = nN_A/V$. From the ideal gas law,

$$V = \frac{nRT}{V} = \frac{nN_AkT}{V}.$$

Thus, $n_v = \frac{P}{kT}$. One obtains the expression given, namely, $\lambda = \frac{kT}{\sqrt{2}P\pi d^2}$.

---

44 •• The escape velocity on Mars is 5.0 km/s, and the surface temperature is typically 0°C. Calculate the rms speeds for (*a*) H_2, (*b*) O_2, and (*c*) CO_2 at this temperature. (*d*) If the rms speed of a gas is greater than about 15% to 20% of the escape velocity of a planet, virtually all of the molecules of that gas will escape the atmosphere of the planet. Based on this criterion, are H_2, O_2, and CO_2 likely to be found in Mars's atmosphere?

(*a*) $v_{rms} = \sqrt{\frac{3RT}{M}}$; M of $H_2 = 2 \times 10^{-3}$ kg/mol $\qquad v_{rms} = \sqrt{\frac{3 \times 8.314 \times 273}{2 \times 10^{-3}}}$ m/s = 1845 m/s

(*b*) M of $O_2 = 32 \times 10^{-3}$ kg/mol $\qquad v_{rms} = 1845/4$ m/s = 461 m/s

(*c*) M of $CO_2 = 44 \times 10^{-3}$ kg/mol $\qquad v_{rms} = 1845/\sqrt{22}$ m/s = 393 m/s

(*d*) $v_{esc}/5 = 1000$ m/s $\qquad$ No H_2, but O_2 and CO_2 should be present

---

45* •• Repeat Problem 44 for Jupiter, whose escape velocity is 60 km/s and whose temperature is typically −150°C.

(*a*), (*b*), (*c*) See Problem 44; use $T = 123$ K $\qquad$ (*a*) $v_{rms} = 1368$ m/s (*b*) $v_{rms} = 342$ m/s (*c*) $v_{rms} = 291$ m/s

(*d*) $v_{esc}/5 = 12000$ m/s $\qquad$ H_2, O_2, and CO_2 should all be found on Jupiter

---

46 •• A pressure as low as $P = 7 \times 10^{-11}$ Pa has been obtained. Suppose a chamber contains helium at this pressure and at room temperature (300 K). Estimate the mean free path λ and the collision time τ for helium in the chamber. Take the diameter of a helium molecule to be 10^{-10} m.

1. Use the result of Problem 43 for λ $\qquad \lambda = \frac{1.38 \times 10^{-23} \times 300}{1.41 \times 7 \times 10^{-11} \times \pi \times 10^{-20}}$ m = 1.33×10^9 m

2. $\tau = \lambda/v_{av} = \lambda\sqrt{\frac{M}{3RT}}$ $\qquad \tau = 1.33 \times 10^9 \sqrt{\frac{4 \times 10^{-3}}{3 \times 8.314 \times 300}}$ s = 9.72×10^5 s

---

47 •• Oxygen (O_2) is confined to a cubic container 15 cm on a side at a temperature of 300 K. Compare the average kinetic energy of a molecule of the gas to the change in its gravitational potential energy if it falls from the top of the container to the bottom.

The average $K = 1.5kT = 1.5 \times 1.38 \times 10^{-23} \times 300$ J = 6.21×10^{-21} J. The change in potential energy is $\Delta U = mgh = 32 \times 10^{-3} \times 9.81 \times 0.15/6.022 \times 10^{23}$ J = 7.82×10^{-26} J. The average kinetic energy of the molecules is the same (within 1 part in 10^5) for all molecules in the container.

---

48 •• The class in Room 101 prepares their traditional greeting for a substitute teacher. Ten toy cars are wound up and released as the teacher arrives. The cars have the following speeds.

Speed, m/s	2	5	6	8
Number of cars	3	3	3	1

Calculate (*a*) the average speed, and (*b*) the rms speed of the cars.

(*a*) $v_{av} = \sum_i n_i v_i / \sum_i n_i$ $\qquad v_{av} = [(6 + 15 + 18 + 8)/10]$ m/s = 4.7 m/s

(*b*) $v_{rms} = \sqrt{\sum_i n_i v_i^2 / \sum_i n_i}$ $\qquad v_{rms} = \sqrt{(12 + 75 + 108 + 64)/10}$ m/s = 5.09 m/s

49* •• Show that $f(v)$ given by Equation 18-37 is maximum when $v = \sqrt{2kT/m}$. *Hint:* Set $df/dv = 0$ and solve for v.

The derivative of f with respect to v is $\frac{df}{dv} = \frac{4}{\sqrt{\pi}}\left(\frac{m}{2kT}\right)^{3/2}\left(2v - \frac{mv^3}{kT}\right)$. Set this equal to zero and solve for v. The result is $v = \sqrt{2kT/m}$.

50 •• Since $f(v)\,dv$ gives the fraction of molecules that have speeds in the range dv, the integral of $f(v)\,dv$ over all the possible ranges of speeds must equal 1. Given the integral

$$\int_0^\infty v^2 e^{-av^2}\,dv = \frac{\sqrt{\pi}}{4}a^{-3/2}$$

show that $\int_0^\infty f(v)\,dv = 1$, where $f(v)$ is given by Equation 18-37.

Follow the procedure indicated. $\int_0^\infty f(v)\,dv = \frac{4}{\sqrt{\pi}}a^{3/2}\int_0^\infty v^2 e^{-av^2}\,dv$, where $a = \left(\frac{m}{2kT}\right)$. The integral has the value $\frac{\sqrt{\pi}}{4}a^{-3/2}$, and it follows that $\int_0^\infty f(v)\,dv = 1$.

51 •• Given the integral

$$\int_0^\infty v^3 e^{-av^2}\,dv = \frac{a^{-2}}{2}$$

calculate the average speed v_{av} of molecules in a gas using the Maxwell-Boltzmann distribution function .

From Problem 50 we know that $f(v)$ is normalized. Therefore, $v_{av} = \int_0^\infty v f(v)\,dv$. We again set $a = (m/2kT)$ and perform the integration using the result given in the problem and obtain;

$$v_{av} = \frac{4}{\sqrt{\pi}}\left(\frac{m}{2kT}\right)^{3/2}\frac{1}{2}\left(\frac{2kT}{m}\right)^2 = \frac{2}{\sqrt{\pi}}\sqrt{\left(\frac{2kT}{m}\right)}.$$

52 ••• In Chapter 11, we found that the escape speed at the surface of a planet of radius R is $v_e = \sqrt{2gR}$, where g is the acceleration due to gravity. (*a*) At what temperature is v_{rms} for O_2 equal to the escape speed for the earth? (*b*) At what temperature is v_{rms} for H_2 equal to the escape speed for the earth? (*c*) Temperatures in the upper atmosphere reach 1000 K. How does this account for the low abundance of hydrogen in the earth's atmosphere? (*d*) Compute the temperatures for which the rms speeds of O_2 and H_2 are equal to the escape velocity at the surface of the moon, where g is about one-sixth of its value on earth and R = 1738 km. How does this account for the absence of an atmosphere on the moon?

(*a*) $v_{rms}^2 = v_{esc}^2 = 2gR_E$; $v_{rms}^2 = 3RT/M$ $\qquad T = (2 \times 32 \times 10^{-3} \times 9.81 \times 6.37 \times 10^6/3 \times 8.314)$ K

$T = 2MgR_E/3R$ $\qquad T = 1.60 \times 10^5$ K

(*b*) For H_2, $M(H_2) = M(O_2)/16$ $\qquad T = 1.0 \times 10^4$ K (*c*) If $v > v_{esc}/5$ or $T/25 \geq T_{atm}$, molecules escape

The more energetic H_2 molecules escape from the upper atmosphere

(*d*) $T = 2Mg_MR_M/3R$

The temperature on the moon with atmosphere might be about 1000 K, so all O_2 and H_2 molecules would escape in the time since the formation of the moon to today.

$T(O_2) = [1.6 \times 10^5(1/6)(1738/6370)]$ K $= 7.28 \times 10^3$ K

$T(H_2) = (7.28 \times 10^3/16)$ K $= 455$ K

53* • True or false: If the pressure of a gas increases, the temperature must increase.

False

---

54 • What is the difference between 1 °C and 1 C°?

1°C is a specific temperature; 1 C° is a temperature interval.

---

55 • Why might the Celsius and Fahrenheit scales be more convenient than the absolute scale for ordinary, nonscientific purposes?

For the Celsius scale, the ice point (0°C) and the boiling point of water at 1 atm (100°C) are more convenient than 273 K and 373 K; temperatures in roughly this range are normally encountered. On the Fahrenheit scale, the temperature of warm-blooded animals is roughly 100°F; this may be a more convenient reference than approximately 300 K. Throughout most of the world, the Celsius scale is the standard for nonscientific purposes.

---

56 • The temperature of the interior of the sun is said to be about 10^7 degrees. Do you think that this is degrees Celsius or kelvins, or does it matter?

It does not matter; $10^7 >> 273$.

57* • If the temperature of an ideal gas is doubled while maintaining constant pressure, the average speed of the molecules

(*a*) remains constant.

(*b*) increases by a factor of 4.

(*c*) increases by a factor of 2.

(*d*) increases by a factor of $\sqrt{2}$.

(*d*)

---

58 • If both temperature and volume of an ideal gas are halved, the pressure

(*a*) diminishes by a factor of 2.

(*b*) remains constant.

(*c*) increases by a factor of 2.

(*d*) diminishes by a factor of $\sqrt{2}$.

(*b*)

---

59 • The average translational kinetic energy of the molecules of an ideal gas depends on

(*a*) the number of moles of the gas and its temperature.

(*b*) the pressure of the gas and its temperature.

(*c*) the pressure of the gas only.

(*d*) the temperature of the gas only.

(*d*)

---

60 • If a vessel contains equal amounts, by weight, of helium and argon, which of the following are true?

(*a*) The pressure exerted by the two gases on the walls of the container is the same.

(*b*) The average speed of a helium atom is the same as that of an argon atom.

(*c*) The number of helium atoms and argon atoms in the vessel are equal.

(*d*) None of the above statements is correct.

(*d*)

61* • Two different gases are at the same temperature. What can you say about the rms speeds of the gas molecules? What can you say about the average kinetic energies of the molecules?

The rms speeds are inversely proportional to the square root of the molecular masses. The average kinetic energies of the molecules are the same.

---

62 •• Explain in terms of molecular motion why the pressure on the walls of a container increases when a gas is heated at constant volume.

The pressure is a measure of the change in momentum per second of a gas molecule on collision with the wall of the container. When the gas is heated, the average velocity and, hence, the average momentum of the molecules increases, and so does the pressure.

---

63 •• Explain in terms of molecular motion why the pressure on the walls of a container increases when the volume of a gas is reduced at constant temperature.

The molecule's time between collisions with the walls of the container is reduced. Consequently, each molecule makes more collisions with the walls per second, and so $\Delta p/\Delta t$ increases.

---

64 •• Oxygen has a molar mass of 32 g/mol, and nitrogen has a molar mass of 28 g/mol. The oxygen and nitrogen molecules in a room have

(*a*) equal average kinetic energies, but the oxygen molecules are faster.

(*b*) equal average kinetic energies, but the oxygen molecules are slower.

(*c*) equal average kinetic energies and speeds.

(*d*) equal average speeds, but the oxygen molecules have a higher average kinetic energy.

(*e*) equal average speeds, but the oxygen molecules have a lower average kinetic energy.

(*f*) None of the above is correct.

(*b*)

65* • At what temperature will the rms speed of an H_2 molecule equal 331 m/s?

From Equ. 18-23, $T = Mv_{rms}^2/3R$ $\quad$ $T = (2 \times 10^{-3} \times 331^2/3 \times 8.314)\ \text{K} = 8.79\ \text{K}$

---

66 • A solid-state temperature transducer is essentially a linear amplifier whose amplification is linearly temperature dependent. If the amplification is 25 times at 20°C and 60 times at 70°C, what is the temperature when the amplification is 45 times?

1. Amplification, $A = a + bt_C$; find a and b $\quad$ $b = 35/50 = 0.7$; $a = 25 - 0.7 \times 20 = 11$

2. Find t_C for $A = 45$ $\quad$ $t_C = [(45 - 11)/0.7]°\text{C} = 48.6°\text{C}$

---

67 •• (*a*) If 1 mol of a gas in a container occupies a volume of 10 L at a pressure of 1 atm, what is the temperature of the gas in kelvins? (*b*) The container is fitted with a piston so that the volume can change. When the gas is heated at constant pressure, it expands to a volume of 20 L. What is the temperature of the gas in kelvins? (*c*) The volume is fixed at 20 L, and the gas is heated at constant volume until its temperature is 350 K. What is the pressure of the gas?

(*a*) At STP, V/mol = 22.4 L — $T = (10/22.4)273\text{ K} = 122\text{ K}$

(*b*) At constant P, $T \propto V$ — $T = 244\text{ K}$

(*c*) At constant V, $P \propto T$ — $P = 1 \times 350/244\text{ atm} = 1.43\text{ atm}$

---

68 •• A cubic metal box with sides of 20 cm contains air at a pressure of 1 atm and a temperature of 300 K. The box is sealed so that the volume is constant, and it is heated to a temperature of 400 K. Find the net force on each wall of the box.

$F = A\Delta P$; $\Delta P = P_{\text{inside}} - P_{\text{outside}}$ — $\Delta P = 101/3\text{ kPa} = 33.7\text{ kPa}$; $F = 33.7 \times 0.04\text{ kN} = 1.35\text{ kN}$

69* •• Water, H_2O, can be converted into H_2 and O_2 gas by electrolysis. How many moles of these gases result from the electrolysis of 2 L of water?

$n(H_2O) \rightarrow n(H_2) + 2n(O_2)$; $M(H_2O) = 18$ — $n(H_2O) = 2000/18 = 111$; $n(H_2) = 111$, $n(O_2) = 55.5$

---

70 •• A massless cylinder 40 cm long rests on a horizontal frictionless table. The cylinder is divided into two equal sections by a membrane. One section contains nitrogen and the other contains oxygen. The pressure of the nitrogen is twice that of the oxygen. How far will the cylinder move if the membrane is removed?

The Figure shows the cylinder before removal of the membrane. The approximate location of the center of mass (CM) is indicated.

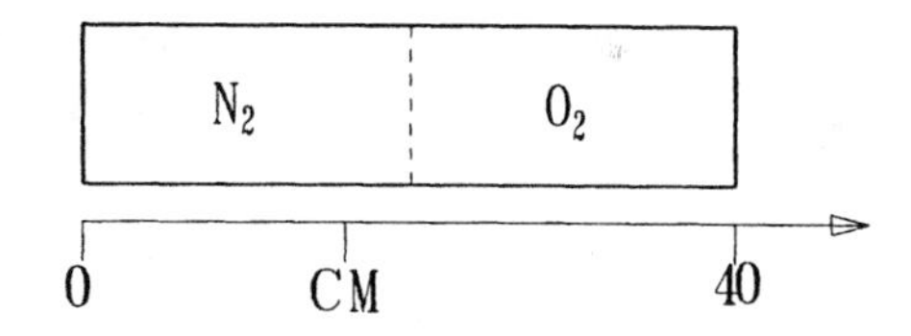

The temperatures are the same on both sides of the membrane. Consequently, $n(N_2) = 2n(O_2)$. The mass of $O_2 = n(O_2)M(O_2)$ and the mass of $N_2 = 2n(O_2)M(N_2)$.

1. Locate the center of mass — $x_{CM} = (2 \times 10 \times 28 + 30 \times 32)/(2 \times 28 + 32)\text{ cm} = 17.3\text{ cm}$

2. After the membrane is removed, the CM is at the center of the cylinder — Cylinder moves $(20 - 17.3)\text{ cm} = 2.7\text{ cm}$ to the left

---

71 •• A cylinder contains a mixture of nitrogen gas (N_2) and hydrogen gas (H_2). At a temperature T_1 the nitrogen is completely dissociated but the hydrogen does not dissociate at all, and the pressure is P_1. If the temperature is doubled to $T_2 = 2T_1$, the pressure is tripled due to complete dissociation of hydrogen. If the mass of hydrogen is m_H, find the mass of nitrogen m_N.

1. Write the ideal gas law for both cases — $P_1V = [2n(N_2) + n(H_2)]RT_1$; $3P_1 = [2n(N_2) + 2n(H_2)]2RT_1$

2. Find $n(H_2)$ in terms of $n(N_2)$ — $n(H_2) = 2n(N_2)$

3. Note that $m_N = 28n(N_2)$ and $m_H = 2n(H_2)$ — $m_N = 7m_H$

72 •• A vertical closed cylinder of cross-sectional area A is divided into two equal parts by a heavy insulating movable piston of mass m_p. The top part contains nitrogen at a temperature T_1 and pressure P_1, and the bottom part is filled with oxygen at a temperature $2T_1$. The cylinder is turned upside-down. To keep the piston in the middle, the oxygen must be cooled to $T_2 = T_1/3$, with the temperature of the nitrogen remaining at T_1. Find the initial pressure of oxygen P_i.

The initial and final configurations of the cylinder and piston of mass m_P are shown. We will set the volume of each half equal to 1 for convenience and let n be the number of moles of O_2.

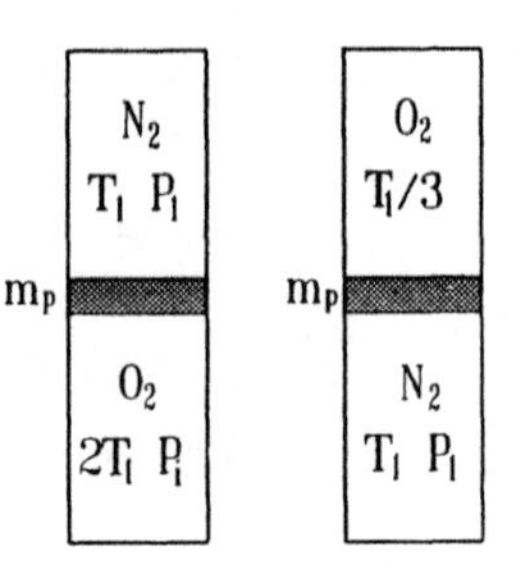

To support the piston at the middle we must have $P_i = P_1 + m_P g/A$, where A is the area of the cylinder. Applying the ideal gas law, $2nRT_1 = P_1 + m_P g/A$.

In the final position of the cylinder, the pressure exerted by the oxygen is $nRT_1/3$.

Since the pressure exerted by N_2 is again P_1 we have $nRT_1/3 + m_P g/A = P_1$.

We can now solve for P_i in terms of P_1 and $m_P g/A$ and find $P_i = (P_1 + 13m_P g/A)/6$.

73* •• Three insulated vessels of equal volumes V are connected by thin tubes that can transfer gas but do not transfer heat. Initially all vessels are filled with the same type of gas at a temperature T_0 and pressure P_0. Then the temperature in the first vessel is doubled and the temperature in the second vessel is tripled. The temperature in the third vessel remains unchanged. Find the final pressure P' in the system in terms of the initial pressure P_0.

Initially, we have $3P_0V = n_0RT_0$. Later, the pressures in the three vessels, each of volume V, are still equal, but the number of moles are not. We can now write

$$P' = P_1 = P_2 = P_3 = \frac{n_1 R 2T_0}{V} = \frac{n_2 R 3T_0}{V} = \frac{n_3 R T_0}{V}.$$ Also,

$$\sum_{i=1}^{3} n_i = n_0 = \frac{3P_0V}{RT_0} = \frac{P'V}{RT_0}\left(\frac{1}{2} + \frac{1}{3} + 1\right) = \frac{P'V}{RT_0}\left(\frac{11}{6}\right),$$ and solving for P', we find $P' = \frac{18}{11}P_0$.

74 •• At the surface of the sun, the temperature is about 6000 K, and all the substances present are gaseous. From data given by the light spectrum of the sun, it is known that most elements are present. (*a*) What is the average kinetic energy of translation of an atom at the surface of the sun? (*b*) What is the range of rms speeds at the surface of the sun if the atoms present range from hydrogen (M = 1 g/mol) to uranium (M = 238 g/mol)?

(*a*) $K_{av} = (3/2)kT$ $\quad$ $K_{av} = 1.5 \times 1.38 \times 10^{-23} \times 6 \times 10^3$ J $= 1.24 \times 10^{-19}$ J $= 0.776$ eV

(*b*) $v_{rms} = \sqrt{3RT/M}; M = 0.001$ to 0.238 kg/mol $\quad$ 1.22×10^4 m/s $\geq v_{rms} \geq 7.92 \times 10^2$ m/s

75 •• A constant-volume gas thermometer with a triple-point pressure $P_3 = 500$ torr is used to measure the boiling point of some substance. When the thermometer is placed in thermal contact with the boiling substance, its pressure is 734 torr. Some of the gas in the thermometer is then allowed to escape so that its triple-point pressure is 200 torr. When it is again placed in thermal contact with the boiling substance, its pressure is 293.4 torr. Again, some of the gas is removed from the thermometer so that its triple-point pressure is 100 torr. When the thermometer is placed in thermal contact with the boiling substance once again, its pressure is 146.65 torr. Find the ideal-gas temperature of the boiling substance.

1. Find the temperature for each measurement	$T_1 = 273.16(734/500)$ K = 401 K; $T_2 = 400.73$ K; $T_3 = 400.59$ K
2. Fit the two pressure ratios to a linear equation Then extrapolate to zero gas pressure	$1.468 = a + 500b$; $1.4665 = a + 100b$; a is ratio for $P = 0$ $a = 1.46613$; so $T = 273.16 \times 1.46613 = 400.49$ K

---

76 ••• A cylinder 2.4 m tall is filled with an ideal gas at standard temperature and pressure (Figure 18-17). The top of the cylinder is then closed with a tight-fitting piston whose mass is 1.4 kg and the piston is allowed to drop until it is in equilibrium. (*a*) Find the height of the piston, assuming that the temperature of the gas does not change as it is compressed. (*b*) Suppose that the piston is pushed down below its equilibrium position by a small amount and then released. Assuming that the temperature of the gas remains constant, find the frequency of vibration of the piston.

(*a*) Let A be the area of the cylinder; find A	At STP, 0.1mol = 0.00224 m^3 = 2.4A m^3; $A = 9.33 \times 10^{-4}$ m^2
$P(\text{inside}) = P_{in} = 1\text{ atm} + mg/A = nRT/hA$	$nRT/hA = (2.4/h)(1\text{ atm}) = 2.424 \times 10^5/h$ Pa; $h = 2.095$ m
(*b*) At equilibrium, F_{net} on m is zero	For $y = 0$, $F_{net} = P_{in}A - mg - 1.01 \times 10^5 A$
Let y be displacement from equilibrium	$P_{in}(y) = P_{in}(0)/(1 + y/h) = P_{in}(0)(1 - y/h)$ if $y << h$
Write F_{net} acting on m and spring constant k	$F_{net}(y) = -y(nRT/h^2)$; $k = nRT/h^2$
Evaluate $f = (1/2\pi)\sqrt{k/m}$	$f = (1/2\pi)\sqrt{\dfrac{0.1 \times 8.314 \times 300}{2.095^2 \times 1.4}}$ Hz = 1.014 Hz

CHAPTER 19

Heat and the First Law of Thermodynamics

1* • Body A has twice the mass and twice the specific heat of body B. If they are supplied with equal amounts of heat, how do the subsequent changes in their temperatures compare?

$M_A = 2M_B$; $c_A = 2c_B$; $C = Mc$; $\Delta T = Q/C$ $\quad\quad$ $C_A = 4C_B$; $\Delta T_A = \Delta T_B/4$

---

2 • The temperature change of two blocks of masses M_A and M_B is the same when they absorb equal amounts of heat. It follows that the specific heats are related by

(*a*) $c_A = (M_A/M_B)c_B$.

(*b*) $c_A = (M_B/M_A)c_B$.

(*c*) $c_A = c_B$.

(*d*) none of the above.

(*b*)

---

3 • The specific heat of aluminum is more than twice that of copper. Identical masses of copper and aluminum, both at 20°C, are dropped into a calorimeter containing water at 40°C. When thermal equilibrium is reached,

(*a*) the aluminum is at a higher temperature than the copper.

(*b*) the aluminum has absorbed less energy than the copper.

(*c*) the aluminum has absorbed more energy than the copper.

(*d*) both (*a*) and (*c*) are correct statements.

(*c*)

---

4 • Sam the shepherd's partner, Bernard, who is a working dog, consumes 2500 kcal of food each day. (*a*) How many joules of energy does Bernard consume each day? (*b*) Sam and Bernard often find themselves sleeping out in the cold night. If the energy consumed by Bernard is dissipated as heat at a steady rate over 24 h, what is his power output in watts as a heater for Sam?

(*a*) $E = Q$ $\quad\quad$ $E = (4.184 \times 2.5 \times 10^6)$ J = 10.46 MJ

(*b*) $P = E/t$ $\quad\quad$ $P = (10.46 \times 10^6/8.64 \times 10^4)$ W = 121 W

5* • A solar home contains 10^5 kg of concrete (specific heat = 1.00 kJ/kg·K). How much heat is given off by the concrete when it cools from 25 to 20°C?

$Q = C\Delta T = mc\Delta T$ $\quad Q = (10^5 \times 10^3 \times 5)$ J = 500 MJ

---

6 • How many calories must be supplied to 60 g of ice at −10°C to melt it and raise the temperature of the water to 40°C?

$Q = C_{ice}\Delta T_{ice} + mL_f + C_{water}\Delta T_{water}$ $\quad Q = 60(10 \times 0.49 + 79.7 + 40)$ cal = 7.48 kcal

---

7 •• How much heat must be removed when 100 g of steam at 150°C is cooled and frozen into 100 g of ice at 0°C. (Take the specific heat of steam to be 2.01 kJ/kg·K.)

$Q = m(c_{steam}\Delta T_{steam} + L_v + c_w\Delta T_w + L_f)$ $\quad Q = 100(0.48 \times 50 + 540 + 100 + 79.7)$ cal = 74.4 kcal

---

8 •• A 50-g piece of aluminum at 20°C is cooled to −196°C by placing it in a large container of liquid nitrogen at that temperature. How much nitrogen is vaporized? (Assume that the specific heat of aluminum is constant and is equal to 0.90 kJ/kg·K.)

$m_N L_{vN} = m_{Al}c_{Al}\Delta T_{Al}$; solve for m_N $\quad m_N = (50 \times 0.9 \times 216/199)$ g = 48.8 g

---

9* •• If 500 g of molten lead at 327°C is poured into a cavity in a large block of ice at 0°C, how much of the ice melts?

$-m_{Pb}(L_{f,Pb} + c_{pb}\Delta T) + m_w L_{f,w} = 0$; solve for m_w $\quad m_w = [500(24.7 + 0.128 \times 327)/333.5]$ g = 99.8 g

---

10 •• A 30-g lead bullet initially at 20°C comes to rest in the block of a ballistic pendulum. Assume that half the initial kinetic energy of the bullet is converted into thermal energy within the bullet. If the speed of the bullet was 420 m/s, what is the temperature of the bullet immediately after coming to rest in the block?

$Q = ½(½mv^2) = mc_{Pb}\Delta T$; $\Delta T = v^2/4c_{Pb}$; c_{Pb} in J/kg $\quad \Delta T = (420^2/4 \times 128)$ K = 344.5 K

---

11 •• A 1400-kg car traveling at 80 km/h is brought to rest by applying the brakes. If the specific heat of steel is 0.11 cal/g·K, what total mass of steel must be contained in the steel brake drums if the temperature of the brake drums is not to rise by more than 120°C?

1. Find $Q = ½mv^2$ $\quad Q = ½ \times 1400 \times 22.2^2$ J = 345.7 kJ = 82.6 kcal
2. $M = Q/c\Delta T$ $\quad M = (82.6/0.11 \times 120)$ kg = 6.26 kg

---

12 • A 200-g piece of lead is heated to 90°C and is then dropped into a calorimeter containing 500 g of water that is initially at 20°C. Neglecting the heat capacity of the container, find the final temperature of the lead and water. In the solution of Problems 12 to 23 (Calorimetry) the fundamental relationship $Q_{out} = Q_{in}$ is used.

$m_{Pb}c_{Pb}(t_{Pb} - t_f) = m_w c_w(t_f - t_w)$; solve for t_f $\quad t_f(500 + 200 \times 0.128)=(500 \times 20 + 200 \times 0.128 \times 90)$;

$t_f = 23.4$ °C

13* • The specific heat of a certain metal can be determined by measuring the temperature change that occurs when a piece of the metal is heated and then placed in an insulated container made of the same material and containing water. Suppose a piece of metal has a mass of 100 g and is initially at 100°C. The container has a mass of 200 g and contains 500 g of water at an initial temperature of 20.0°C. The final temperature is 21.4°C. What is the specific heat of the metal?

$m_1c(t_{1i} - t_f) = m_2c(t_f - t_{2i}) + m_w c_w(t_f - t_{2i})$; find c $\quad 78.6c = 2.8c + 7$; $c = 0.093$ cal/g·K

---

14 •• A 25-g glass tumbler contains 200 mL of water at 24°C. If two 15-g ice cubes each at a temperature of −3°C are dropped into the tumbler, what is the final temperature of the drink? Neglect thermal conduction between the tumbler and the room.

$(m_g c_g + 200)(24 - t_f) = 3m_{ice}c_{ice} + m_{ice}(L_f + t_f)$; find t_f $\quad 205 \times 24 - 205t_f = 44.1 + 30(79.7 + t_f)$; $t_f = 10.6$ °C

---

15 •• A 200-g piece of ice at 0°C is placed in 500 g of water at 20°C. The system is in a container of negligible heat capacity and is insulated from its surroundings. (*a*) What is the final equilibrium temperature of the system? (*b*) How much of the ice melts?

(*a*) To melt the ice requires 200 × 79.7 cal ≈ 16 kcal; reducing the temperature of the water to 0°C releases 10 kcal. Therefore, not all the ice melts, and the final temperature is 0°C.

(*b*) The mass of ice that melts is (10 kcal)/(79.7 cal/g) = 125 g.

---

16 •• A 3.5-kg block of copper at a temperature of 80°C is dropped into a bucket containing a mixture of ice and water whose total mass is 1.2 kg. When thermal equilibrium is reached the temperature of the water is 8°C. How much ice was in the bucket before the copper block was placed in it? (Neglect the heat capacity of the bucket.)

$m_{Cu}c_{Cu}\Delta t_{Cu} = m_w \Delta t_w + m_{ice}L_f$; solve for m_{ice} $\quad 3.5 \times 0.0923 \times 72 = 1.2 \times 8 + 79.7m_{ice}$; $m_{ice} = 0.171$ kg

---

17* •• A well-insulated bucket contains 150 g of ice at 0°C. (*a*) If 20 g of steam at 100°C is injected into the bucket, what is the final equilibrium temperature of the system? (*b*) Is any ice left afterward?

(*a*), (*b*) $m_{st}L_v + m_{st}(100 - t_f) = m_{ice}(L_f + t_f)$; solve for t_f $\quad t_f = 4.97$°C; (*b*) Since $t_f > 0$°C, no ice is left.

---

18 •• A calorimeter of negligible mass contains 1 kg of water at 303 K and 50 g of ice at 273 K. Find the final temperature *T*. Solve the same problem if the mass of ice is 500 g.

1. $m_{ice}L_f + m_{ice}t_f = m_w(30 - t_f)$; solve for t_f, $m_{ice} = 50$ g $\quad 50(79.7 + t_f) = 1000(30 - t_f)$; $t_f = 24.8$°C
2. For $m_{ice} = 500$ g, only 376 g will melt. $\quad t_f = 0$°C

---

19 •• A 200-g aluminum calorimeter contains 500 g of water at 20°C. A 100-g piece of ice cooled to −20°C is placed in the calorimeter. (*a*) Find the final temperature of the system, assuming no heat loss. (Assume that the specific heat of ice is 2.0 kJ/kg·K.) (*b*) A second 200-g piece of ice at −20°C is added. How much ice remains in the system after it reaches equilibrium? (*c*) Would you give a different answer for (*b*) if both pieces of ice were added at the same time?

(*a*) $(m_{Al}c_{Al} + m_w c_w)(20 - t_f) = 20m_{ice}c_{ice} + m_{ice}(L_f + t_f)$ $\quad t_f = 3.0$°C

(*b*) Find Q released to lower Al and H_2O to 0°C $\quad Q = (200 \times 0.215 + 600)3.0 = 1929$ cal

Find Q_{ice} required to raise 200 g of ice to 0°C $\quad Q_{ice} = 200 \times 0.478 \times 20 = 1912$ cal

Find the amount of ice melted by 17 cal $\quad m = 17/79.7$ g $= 0.21$ g; ice remaining = 199.8 g

(*c*) The initial and final conditions are the same $\quad$ No

---

20 •• The specific heat of a 100-g block of material is to be determined. The block is placed in a 25-g copper calorimeter that also holds 60 g of water. The system is initially at 20°C. Then 120 mL of water at 80°C are added to the calorimeter vessel. When thermal equilibrium is attained, the temperature of the water is 54°C. Determine the specific heat of the block.

$(m_B c_B + m_{Cu} c_{Cu} + m_{w1})\Delta t_B = m_{w2}\Delta t_2$; here $\Delta t_B = 34$ K, $\Delta t_2 = 26$ K, $m_{w1} = 60$ g, $m_{w2} = 120$ g. Solve for c_B. $c_B = 0.295$ cal/g·K

21* •• Between innings at his weekly softball game, Stan likes to have a sip or two of beer. He usually consumes about 6 cans, which he prefers at exactly 40°F. His wife Bernice puts a six-pack of 12-ounce aluminum cans of beer (1 ounce has a mass of 28.4 g) originally at 80°F in a well-insulated Styrofoam container and begins adding ice. How many 30-g ice cubes must she add to the container so that the final temperature is 40°F? (Neglect heat losses through the container and the heat removed from the aluminum and assume that the beer is mostly water.)

1. Convert to Celsius degrees. 40 °F = 4.44°C; 80°F = 26.67°C
2. $m_B c_B \Delta t_B = n_{ice} m_{ice}(L_f + 4.44)$; solve for n_{ice} $n_{ice} = (6 \times 12 \times 28.4 \times 22.2/30 \times 84.14) = 18$

---

22 •• A 100-g piece of copper is heated in a furnace to a temperature t. The copper is then inserted into a 150-g copper calorimeter containing 200 g of water. The initial temperature of the water and calorimeter is 16°C, and the final temperature after equilibrium is established is 38°C. When the calorimeter and its contents are weighed, 1.2 g of water are found to have evaporated. What was the temperature t?

1. $1.2L_v + (m_w + m_{cal} c_{Cu})\Delta t_{cal} = m_{Cu} c_{Cu} \Delta t_{Cu}$; find Δt_{Cu} $\Delta t_{Cu} = [1.2 \times 540 + (200 + 150 \times 0.0923) \times 22]/(100 \times 0.0923)$ C°
2. $t = t_f + \Delta t_{Cu}$ $\Delta t_{Cu} = 578$ C°; $t = (578 + 38)$°C = 616°C

---

23 •• A 200-g aluminum calorimeter contains 500 g of water at 20°C. Aluminum shot of mass 300 g is heated to 100°C and is then placed in the calorimeter. (*a*) Using the value of the specific heat of aluminum given in Table 19-1, find the final temperature of the system, assuming that no heat is lost to the surroundings. (*b*) The error due to heat transfer between the system and its surroundings can be minimized if the initial temperature of the water and calorimeter is chosen to be $\frac{1}{2}\Delta t_w$ below room temperature, where Δt_w is the temperature change of the calorimeter and water during the measurement. Then the final temperature is $\frac{1}{2}\Delta t_w$ above room temperature. What should the initial temperature of the water and container be if the room temperature is 20°C?

(*a*) $m_{sh} c_{Al}(100 - t_f) = (m_{cal} c_{al} + m_w)(t_f - 20)$; find t_f $6450 - 64.5t_f = 532.3t_f - 10645$; $t_f = 28.6$°C

(*b*) For the calorimeter, let $t_i = t_r - t_0$ and $t_f = t_r + t_0$, where $t_r = 20$°C; write the calorimetry equation, solve for t_0, t_i. $m_{sh} c_{Al}(100 - 20 - t_0) = (m_{cal} c_{Al} + m_w)(2t_0)$; $64.5 \times 80 - 64.5t_0 = 2 \times 532.3t_0$; $t_0 = 4.57$°C; $t_i = 15.43$°C.

---

24 • Joule's experiment establishing the mechanical equivalence of heat involved the conversion of mechanical energy into internal energy. Give some examples of the internal energy of a system being converted into mechanical energy.

Steam turbine; internal combustion engine; a person performing mechanical work, e.g., climbing a hill.

25* • Can a system absorb heat with no change in its internal energy?

Yes

---

26 • In the equation $Q = \Delta U + W$ (the formal statement of the first law of thermodynamics), the quantities Q and W represent

(*a*) the heat supplied to the system and the work done by the system.

(*b*) the heat supplied to the system and the work done on the system.

(*c*) the heat released by the system and the work done by the system.

(*d*) the heat released by the system and the work done on the system.

(*a*)

---

27 • A diatomic gas does 300 J of work and also absorbs 600 cal of heat. What is the change in internal energy of the gas?

From Equ. 19-10, $\Delta U = Q - W$ $\Delta U = (600 \times 4.184 - 300)\text{ J} = 2210\text{ J}$

---

28 • If 400 kcal is added to a gas that expands and does 800 kJ of work, what is the change in the internal energy of the gas?

From Equ. 19-10, $\Delta U = Q - W$ $\Delta U = (400 \times 4.184 - 800)\text{ kJ} = 874\text{ kJ}$

29* • A lead bullet moving at 200 m/s is stopped in a block of wood. Assuming that all of the energy change goes into heating the bullet, find the final temperature of the bullet if its initial temperature is 20°C.

$Q = \frac{1}{2}mv^2 = mc\Delta t = mc(t_f - t_i)$; $t_f = t_i + v^2/2c$ $t_f = (20 + 200^2/2 \times 128)°\text{C} = 176°\text{C}$

---

30 • (*a*) At Niagara Falls, the water drops 50 m. If the change in potential energy goes into the internal energy of the water, compute the increase in its temperature. (*b*) Do the same for Yosemite Falls, where the water drops 740 m. (These temperature rises are not observed because the water cools by evaporation as it falls.)

(*a*), (*b*) Here $\Delta U = mgh = mc\Delta t$; $\Delta t = gh/c$ (*a*) $\Delta t = 9.81 \times 50/4184 = 0.117$ K; (*b*) $\Delta t = 1.74$ K

---

31 • When 20 cal of heat are absorbed by a gas, the system performs 30 J of work. What is the change in the internal energy of the gas?

From Equ. 19-10, $\Delta U = Q - W$ $\Delta U = (20 \times 4.184 - 30)\text{ J} = 53.7\text{ J}$

---

32 •• A lead bullet initially at 30°C just melts upon striking a target. Assuming that all of the initial kinetic energy of the bullet goes into the internal energy of the bullet to raise its temperature and melt it, calculate the speed of the bullet upon impact.

$Q = \frac{1}{2}mv^2 = mc(T_{MP} - T_i) + mL_f$; solve for v $v = \sqrt{2[c(T_{MP} - T_i) + L_f]} = 354\text{ m/s}$

33* •• A piece of ice is dropped from a height *H*. (*a*) Find the minimum value of *H* such that the ice melts when it makes an inelastic collision with the ground. Assume that all the mechanical energy lost goes into melting the ice. (*b*) Is it reasonable to neglect the variation in the acceleration of gravity in doing this problem? (*c*) Comment on the reasonableness of neglecting air resistance. What effect would air resistance have on your answer?

(*a*) To melt the ice (at $t = 0°\text{C}$), $mgh = mL_f$; $h = L_f/g$ $h = 333.5/9.81\text{ km} = 34\text{ km}$

(*b*) Yes. Since $h << R_E = 6370$ km, one can neglect the variation of g.

(*c*) The piece of ice (depending on its mass and shape) will reach its terminal velocity long before striking the ground, and some of the ice will melt before it reaches the ground. However, the relation $\Delta U = mgh = mL_f$ remains valid, so air resistance does not affect h.

---

34 •• On a cold day you can warm your hands by rubbing them together. (*a*) Assume that the coefficient of friction between your hands is 0.5, that the normal force between your hands is 35 N, and that you rub them together at an average speed of 35 cm/s. What is the rate at which heat is generated? (*b*) Assume further that the mass of each of your hands is approximately 350 g, that the specific heat of your hands is about 4 kJ/kg·K, and that all the heat generated goes into raising the temperature of your hands. How long must you rub your hands together to produce a 5-C° increase in their temperature?

(*a*) $dQ/dt = P = f_f v$. $dQ/dt = 35 \times 0.5 \times 0.35$ J/s $= 6.125$ W

(*b*) $\Delta Q = (dQ/dt)\Delta t = mc\Delta T$; $\Delta t = mc\Delta T/(dQ/dt)$ $\Delta t = 0.35 \times 4 \times 10^3 \times 5/6.125$ s $= 1143$ s $= 19.0$ min

---

35 • A real gas cools during a free expansion, though an ideal gas does not. Explain.

For the ideal gas, U is a function of T only. Since $W = 0$ and $Q = 0$ in free expansion, $\Delta U = 0$ and T is constant. For a real gas, U depends on the density of the gas because the molecules do exert weak attractive forces on each other. In free expansion these forces reduce the average kinetic energy of the molecules and, consequently, the temperature.

---

36 • An ideal gas at one atmosphere pressure and 300 K is confined to half of an insulated container by a thin partition. The partition is then removed and equilibrium is established. At that point, which of the following is correct?

(*a*) The pressure is half an atmosphere and the temperature is 150 K.

(*b*) The pressure is one atmosphere and the temperature is 150 K.

(*c*) The pressure is half an atmosphere and the temperature is 300 K.

(*d*) None of the above.

(*c*)

37* • A certain gas consists of ions that repel each other. The gas undergoes a free expansion with no heat exchange and no work done. How does the temperature of the gas change? Why?

The temperature of the gas increases. The average kinetic energy increases with increasing volume due to the repulsive interaction between the ions.

---

38 • A gas changes its state reversibly from A to C (Figure 19-16). The work done by the gas is

(*a*) greatest for path A →B →C.

(*b*) least for path A →C.

(*c*) greatest for path A →D →C.

(*d*) the same for all three paths.

(*a*)

In Problems 39 through 42, the initial state of 1 mol of an ideal gas is $P_1 = 3$ atm, $V_1 = 1$ L, and $U_1 = 456$ J, and its final state is $P_2 = 2$ atm, $V_2 = 3$ L, and $U_2 = 912$ J.

---

39 • The gas is allowed to expand at constant pressure to a volume of 3 L. It is then cooled at constant volume until its pressure is 2 atm. (*a*) Show this process on a *PV* diagram, and calculate the work done by the gas. (*b*) Find the heat added during this process.

(*a*) The path from the initial state I to the final state F is shown on the *PV* diagram. Here, *P* is in atmospheres and *V* in liters. The work done by the gas is equal to the area under the path.

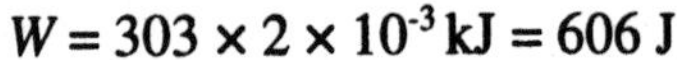

$W = 303 \times 2 \times 10^{-3}$ kJ $= 606$ J

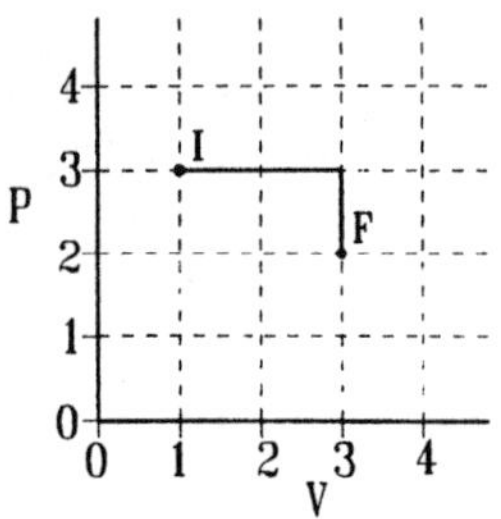

(*b*) $Q = W + \Delta U = W + (U_f - U_i)$ $\qquad Q = (606 + 456)\text{ J} = 1062\text{ J}$

---

40 • The gas is first cooled at constant volume until its pressure is 2 atm. It is then allowed to expand at constant pressure until its volume is 3 L. (*a*) Show this process on a *PV* diagram, and calculate the work done by the gas. (*b*) Find the heat added during this process.

(*a*) The path from the initial state I to the final state F is shown on the *PV* diagram. Here, *P* is in atmospheres and *V* in liters. The work done by the gas is equal to the area under the path.

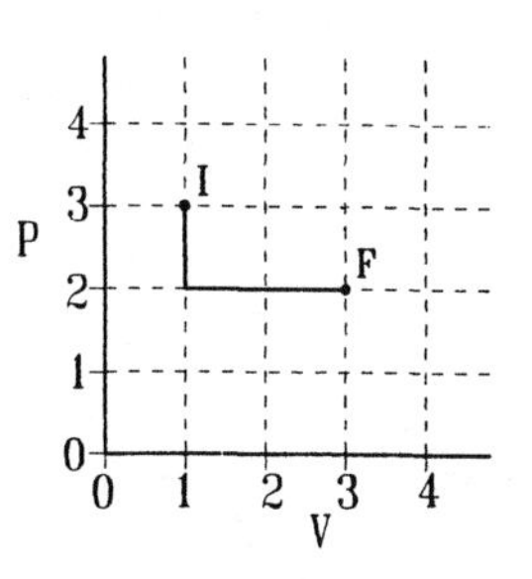

$W = 202 \times 2 \times 10^{-3}\text{ kJ} = 404\text{ J}$

(*b*) $Q = W + \Delta U$ $\qquad Q = (404 + 456)\text{ J} = 860\text{ J}$

41* •• The gas is allowed to expand isothermally until its volume is 3 L and its pressure is 1 atm. It is then heated at constant volume until its pressure is 2 atm. (*a*) Show this process on a *PV* diagram, and calculate the work done by the gas. (*b*) Find the heat added during this process.

(*a*) The path from the initial state I to the final state B is shown on the *PV* diagram. Here, *P* is in atmospheres and *V* in liters. The work done by the gas is equal to the area under the path. Here, the work (area under the curve) is given by Equ. 19-16. We replace nRT_1 by P_1V_1.

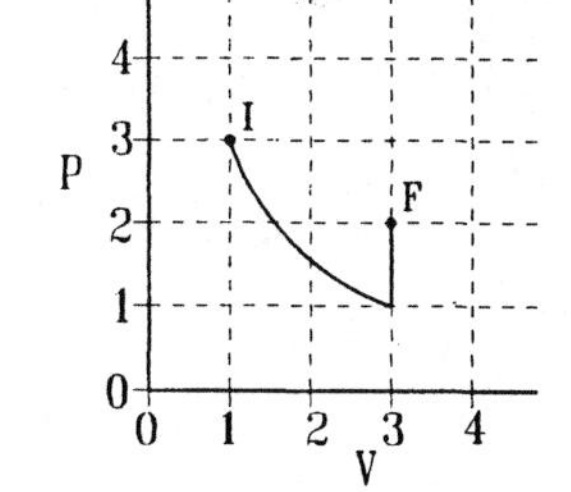

$W = 303 \ln(3)\text{ kJ} = 333\text{ kJ}.$

(*b*) $Q = W + \Delta U$ $\qquad Q = (333 + 456)\text{ J} = 789\text{ J}$

---

42 •• The gas is heated and is allowed to expand such that it follows a straight-line path on a *PV* diagram from its initial state to its final state. (*a*) Show this process on a *PV* diagram, and calculate the work done by the gas. (*b*) Find the heat added during this process.

(*a*) The path from the initial state I to the final state F is shown on the *PV* diagram. Here, *P* is in atmospheres and *V* in liters. The work done by the gas is equal to the area under the path. Here, the work (area under the curve) is

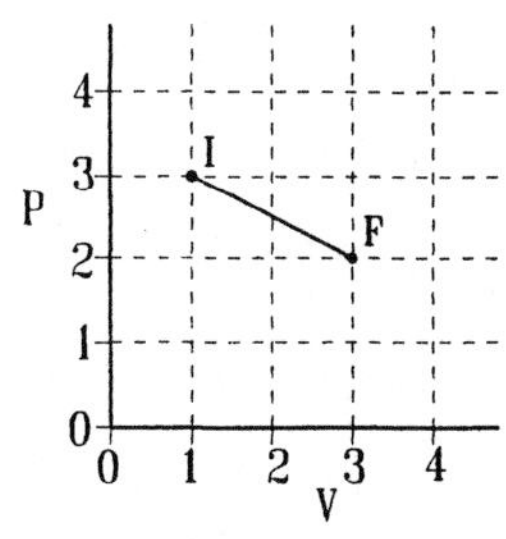

$W = 5\text{ atm}\cdot\text{L} = 505\text{ J}.$

(*b*) $Q = W + \Delta U$ $\qquad Q = (505 + 456)\text{ J} = 961\text{ J}$

---

43 •• One mole of the ideal gas is initially in the state $P_0 = 1$ atm, $V_0 = 25$ L. As the gas is slowly heated, the plot of its state on a *PV* diagram moves in a straight line to the state $P = 3$ atm, $V = 75$ L. Find the work done by the gas.

The path on a *PV* diagram is shown. Here *P* is in atmospheres and *V* in liters. The work done by the gas is the area under the curve in units of atm·L.

$W = 100\text{ atm}\cdot\text{L} = 10.1\text{ kJ}$

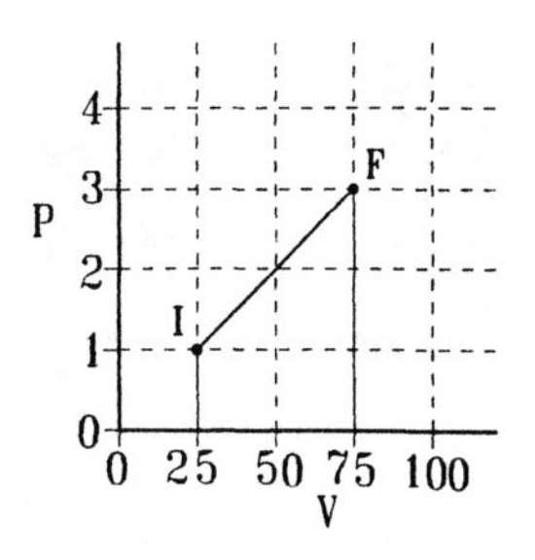

44 •• One mole of the ideal gas is heated so that $T = AP^2$, where A is a constant. The temperature changes from T_0 to $4T_0$. Find the work done by the gas.

1. From the ideal gas law, $PV = RT = RAP^2$. If T changes from T_0 to $4T_0$, the pressure increases from P_0 to $2P_0$. The volume is given by $V = RAP$, so the volume increases linearly with pressure from V_0 to $2V_0$. It now follows that the work done by the gas, the area under the straight line from P_0, V_0 to $2P_0$, $2V_0$ is $W = P_0V_0 + \frac{1}{2}P_0V_0 = 3P_0V_0/2$.

45* •• One mole of an ideal gas initially at a pressure of 1 atm and a temperature of 0°C is compressed isothermally and quasi-statically until its pressure is 2 atm. Find (*a*) the work needed to compress the gas, and (*b*) the heat removed from the gas during the compression.

(*a*) Use Equ. 19-16, and $V_2/V_1 = P_1/P_2$ — W (on the gas) $= 8.314 \times 273 \ln(2) = 1573$ J

(*b*) $\Delta T = 0$, $\Delta U = 0$; $Q = W$ (by the gas) — Q (removed from the gas) $= 1573$ J

46 •• An ideal gas initially at 20°C and 200 kPa has a volume of 4 L. It undergoes a quasi-static, isothermal expansion until its pressure is reduced to 100 kPa. Find (*a*) the work done by the gas, and (*b*) the heat added to the gas during the expansion.

(*a*) Use Equ. 19-16; $nRT = PV$ — $W = 800 \ln 2$ J $= 555$ J

(*b*) For an isothermal process, $\Delta U = 0$, $Q = W$ — $Q = 555$ J

47 • The heat capacity at constant volume of a certain amount of a monatomic gas is 49.8 J/K. (*a*) Find the number of moles of the gas. (*b*) What is the internal energy of the gas at $T = 300$ K? (*c*) What is the heat capacity of the gas at constant pressure?

(*a*) Monatomic gas, $C_v = (3/2)nR$ — $n = 2 \times 49.8/3 \times 8.314 = 4$

(*b*) $U = C_vT$ — $U = 49.8 \times 300$ J $= 14.9$ kJ

(*c*) $C_p = C_v + nR = (5/2)nR$ — $C_p = 49.8 \times 5/3$ J/K $= 83$ J/K

48 • The Dulong–Petit law was originally used to determine the molecular mass of a substance from its measured heat capacity. The specific heat of a certain solid is measured to be 0.447 kJ/kg·K. (*a*) Find the molecular mass of the substance. (*b*) What element is this?

(*a*), (*b*) $C = 3nR$; $3R = CM$ (M = molecular mass) — (*a*) $M = 3 \times 8.314/0.447 = 55.8$; (*b*) Iron

49* •• The specific heat of air at 0°C is listed in a handbook as having the value of 1.00 J/g·K measured at constant pressure. (*a*) Assuming that air is an ideal gas with a molar mass $M = 29.0$ g/mol, what is its specific heat at 0°C and constant volume? (*b*) How much internal energy is there in 1 L of air at 0°C and at 1 atm?

(*a*) For a diatomic gas, $C_v = (5/7)C_p$ — $C_v = 0.714$ J/g·K

(*b*) $\rho_{air} = 1.29$ g/L; $U = C_vT$ — $U = 1.29 \times 0.714 \times 273$ J $= 252$ J

50 •• One mole of an ideal diatomic gas is heated at constant volume from 300 to 600 K. (*a*) Find the increase in internal energy, the work done, and the heat added. (*b*) Find the same quantities if this gas is heated from 300 to 600 K at constant pressure. Use the first law of thermodynamics and your results for (*a*) to calculate the work done. (*c*) Calculate the work done in (*b*) directly from $dW = P\,dV$.

(*a*) $Q = C_v\Delta T$; $W = 0$ for constant volume; $\Delta U = Q$ — $Q = (5/2)8.314 \times 300 = 6.236$ kJ $= \Delta U$; $W = 0$

(*b*) $Q = C_p\Delta T$; $\Delta U = C_v\Delta T$; $W = Q - \Delta U$ — $Q = (7/5) \times 6.236$ kJ $= 8.73$ kJ; $\Delta U = 6.236$ kJ; $W = 2.49$ kJ

(*c*) $W = \int_{V_i}^{V_f} P\,dV = P(V_f - V_i) = R(T_f - T_i)$ — $W = 8.314 \times 300$ J $= 2.49$ kJ

---

51 •• A diatomic gas (molar mass *M*) is confined to a closed container of volume *V* at a pressure P_0. What amount of heat *Q* should be transferred to the gas in order to triple the pressure? (Express your answer in terms of P_0 and *V*.)

1. Use the ideal gas law to find ΔT. — $T_0 = P_0V/nR$; $T_f = 3P_0V/nR = 3T_0$; $\Delta T = 2T_0 = 2P_0V/nR$
2. $Q = C_v\Delta T$ — $Q = (5/2)nR(2P_0V/nR) = 5P_0V$

---

52 •• One mole of air ($c_v = 5R/2$) is confined at atmospheric pressure in a cylinder with a piston at 0°C. The initial volume occupied by gas is *V*. Find the volume of gas *V′* after the equivalent of 13,200 J of heat is transferred to it.

1. Find ΔT and T_f; $C_p\Delta T = Q$; $C_p = 7R/2$ — $\Delta T = 13200 \times 2/7 \times 8.314$ K $= 454$ K; $T_f = 754$ K
2. $V = 22.4$ L; $V' = V(T_f/T_i)$ — $V' = 22.4(754/300)$ L $= 56.3$ L

53* •• The heat capacity of a certain amount of a particular gas at constant pressure is greater than that at constant volume by 29.1 J/K. (*a*) How many moles of the gas are there? (*b*) If the gas is monatomic, what are C_v and C_p? (*c*) If the gas consists of diatomic molecules that rotate but do not vibrate, what are C_v and C_p?

(*a*) $nR = 29.1$ J/K — $n = 29.1/8.314 = 3.5$

(*b*) $C_v = 3nR/2$; $C_p = 5nR/2$ — $C_v = 1.5 \times 29.1$ J/K $= 43.65$ J/K; $C_p = 72.75$ J/K

(*c*) $C_v = 5nR/2$; $C_p = 7nR/2$ — $C_v = 72.75$ J/K; $C_p = 101.85$ J/K

---

54 •• One mole of a monatomic ideal gas is initially at 273 K and 1 atm. (*a*) What is its initial internal energy? (*b*) Find its final internal energy and the work done by the gas when 500 J of heat are added at constant pressure. (*c*) Find the same quantities when 500 J of heat are added at constant volume.

(*a*) $U = C_vT = 3RT/2$ — $U = (3 \times 8.314 \times 273/2)$ J $= 3405$ J

(*b*) $\Delta T = Q/C_p$; $\Delta U = C_v\Delta T = QC_v/C_p$; $W = Q - \Delta U$ — $\Delta U = 0.6 \times 500$ J $= 300$ J; $U = 3705$ J; $W = 200$ J

(*c*) $\Delta U = Q$; $W = 0$ (constant volume) — $\Delta U = 500$ J; $U = 3905$ J; $W = 0$

---

55 •• A certain molecule has vibrational energy levels that are equally spaced by 0.15 eV. Find the critical temperature T_c such that for $T >> T_c$ you would expect the equipartition theorem to hold and for $T << T_c$ you would expect the equipartition theorem to fail.

T_c is approximately that temperture at which $kT_c = 0.15$ eV $= 2.4 \times 10^{-20}$ J; $T_c = 2.4 \times 10^{-20}/1.38 \times 10^{-23}$ K $= 1740$ K

---

56 • When an ideal gas is subjected to an adiabatic process,

(*a*) no work is done by the system.

(*b*) no heat is supplied to the system.

(*c*) the internal energy remains constant.

(*d*) the heat supplied to the system equals the work done by the system.

(*b*)

57* • One mole of an ideal gas ($\gamma = \frac{5}{3}$) expands adiabatically and quasi-statically from a pressure of 10 atm and a temperature of 0°C to a pressure of 2 atm. Find (*a*) the initial and final volumes, (*b*) the final temperature, and (*c*) the work done by the gas.

(*a*) $V_i = 22.4 \times 1/P_i$ L; from Equ. 19-37, $V_f = V_i(P_i/P_f)^{1/\gamma}$ — $V_i = 2.24$ L; $V_f = 2.24(5)^{0.6}$ L $= 5.88$ L

(*b*) $T_f = P_fV_f/R$ — $T_f = (202 \times 5.88/8.314)$ K $= 143$ K

(*c*) $W = Q - \Delta U = 0 - C_v\Delta T = -C_v\Delta T$ — $W = 1.5 \times 8.314 \times 130$ J $= 1.62$ kJ

58 • An ideal gas at a temperature of 20°C is compressed quasi-statically and adiabatically to half its original volume. Find its final temperature if (*a*) $C_v = \frac{3}{2}nR$, and (*b*) $C_v = \frac{5}{2}nR$.

(*a*) Here, $\gamma = 5/3$; from Equ. 19-36, $T_f = T_i(V_i/V_f)^{\gamma-1}$ — $T_f = 293(2)^{0.67}$ K $= 465$ K

(*b*) Here, $\gamma = 7/5$. — $T_f = 293(2)^{0.4}$ K $= 387$ K

59 • Two moles of neon gas initially at 20°C and a pressure of 1 atm are compressed adiabatically to one-fourth of their initial volume. Determine the temperature and pressure following compression.

For Neon, $\gamma = 5/3$. $T_f = T_i(V_i/V_f)^{\gamma-1}$; $P_f = P_i(V_i/V_f)^{\gamma}$ — $T_f = 293(4)^{0.67}$ K $= 738$ K; $P_f = 1(4)^{1.67}$ atm $= 10.1$ atm

60 •• Half a mole of an ideal monatomic gas at a pressure of 400 kPa and a temperature of 300 K expands until the pressure has diminished to 160 kPa. Find the final temperature and volume, the work done, and the heat absorbed by the gas if the expansion is (*a*) isothermal, and (*b*) adiabatic.

(*a*) 1. Find V_i from the ideal gas law. — $V_i = (8.314 \times 300/2 \times 400)$ L $= 3.12$ L

2. For the isothermal case, $V_f = V_i(P_i/P_f)$ — $V_f = 3.12(400/160)$ L $= 7.8$ L

3. $T_f = T_i$ — $T_f = 300$ K

4. Use Equ. 19-16; $W = nRT\ln(V_f/V_i)$ — $W = 0.5 \times 8.314 \times 300 \times \ln(2.5)$ J $= 1.14$ kJ

5. $Q = \Delta U + W$; $\Delta U = 0$ — $Q = 1.14$ kJ

(*b*) 1. From Equ. 19-37, $V_f = V_i(P_i/P_f)^{1/\gamma}$ — $V_i = 3.12$ L; $V_f = 3.12(2.5)^{0.6}$ L $= 5.41$ L

2. $T_f = P_fV_f/nR$ — $T_f = (160 \times 5.41/0.5 \times 8.314)$ K $= 208$ K

3. $W = -\Delta U = -C_v\Delta T$; $Q = 0$ — $W = (0.5 \times 1.5 \times 8.314 \times 92)$ J $= 574$ J; $Q = 0$

61* •• Repeat Problem 60 for a diatomic gas.

(*a*) See Problem 60. — $V_i = 3.12$ L; $V_f = 7.8$ L; $T_f = 300$ K; $W = Q = 1.14$ kJ

(*b*) 1. Here $\gamma = 1.4$; proceed as in Problem 60. — $V_i = 3.12$ L; $V_f = 3.12(2.5)^{0.714} = 6.0$ L

2. $T_f = P_fV_f/nR$ — $T_f = (160 \times 6.0/0.5 \times 8.314)$ K $= 231$ K

3. $\Delta U = C_v\Delta T$; $Q = 0$; $W = -\Delta U$ — $W = (0.5 \times 2.5 \times 8.314 \times 69)$ J $= 717$ J; $Q = 0$

62 •• One-half mole of helium is expanded adiabatically and quasi-statically from an initial pressure of 5 atm and temperature of 500 K to a final pressure of 1 atm. Find (*a*) the final temperature, (*b*) the final volume, (*c*) the work done by the gas, and (*d*) the change in the internal energy of the gas.

(*a*) From Equs. 19-36 and 37, $T_f = T_i(P_f/P_i)^{1-1/\gamma}$ — For He, $\gamma = 5/3$; $T_f = 500(0.2)^{0.4}$ K $= 263$ K (*b*) 1. Find $V_i = nRT_i/P_i$ $V_i = (0.5 \times 8.314 \times 500/505)$ L $= 4.12$ L

2. $V_f = V_i(P_i/P_f)^{1/\gamma}$ $V_f = 4.12(5)^{0.6}$ L $= 10.8$ L

(*c*), (*d*) $\Delta U = C_v\Delta T$; $W = -\Delta U$ $\Delta U = -(0.5 \times 1.5 \times 8.314 \times 237)$ J $= -1.48$ kJ; $W = 1.48$ kJ

---

63 ••• A hand pump is used to inflate a bicycle tire to a gauge pressure of 482 kPa (about 70 lb/in^2). How much work must be done if each stroke of the pump is an adiabatic process? Atmospheric pressure is 1 atm, the air temperature is initially 20°C, and the volume of the air in the tire remains constant at 1 L.

Consider the process to be accomplished in a single compression. The initial pressure is 1 atm = 101 kPa. The final pressure is (101 + 482) kPa = 583 kPa, and the final volume is 1 L. We can now determine the initial volume of the air. Since air is a mixture of diatomic gases, $\gamma_{air} = 1.4$.

1. Find $V_i = V_f(P_f/P_i)^{1/\gamma}$ $V_i = 1(583/101)^{0.714}$ L $= 3.5$ L

2. Use Equ. 19-39, where W is work done by the gas. W (on the gas) $= -(3.5 \times 101 - 583)/0.4$ J $= 574$ J

---

64 ••• An ideal gas at initial volume V_1 and pressure P_1 expands quasi-statically and adiabatically to volume V_2 and pressure P_2. Calculate the work done by the gas directly by integrating $P\,dV$, and show that your result is the same as that given by Equation 19-39.

For the adiabatic process, PV^γ = constant = C.

So, $P = C/V^\gamma$ and $W = \int_{V_1}^{V_2} P\,dV = C\int_{V_1}^{V_2} V^{-\gamma}dV = \frac{C}{1-\gamma}(V_2^{1-\gamma} - V_1^{1-\gamma})$. $CV_2^{1-\gamma} = P_2V_2^\gamma$ and $CV_1^{1-\gamma} = P_1V_1^\gamma$. With these substitutions we obtain the desired result: $W = \frac{P_1V_1 - P_2V_2}{\gamma - 1}$.

65* •• One mole of N_2 ($C_v = \frac{5}{2}R$) gas is originally at room temperature (20°C) and a pressure of 5 atm. It is allowed to expand adiabatically and quasi-statically until its pressure equals the room pressure of 1 atm. It is then heated at constant pressure until its temperature is again 20°C. During this heating, the gas expands. After it reaches room temperature, it is heated at constant volume until its pressure is 5 atm. It is then compressed at constant pressure until it is back to its original state. (*a*) Construct an accurate *PV* diagram showing each process in the cycle. (*b*) From your graph, determine the work done by the gas during the complete cycle. (*c*) How much heat is added or subtracted from the gas during the complete cycle? (*d*) Check your graphical determination of the work done by the gas in (*b*) by calculating the work done during each part of the cycle.

(*a*) 1. Find V at start of cycle, point A, from $V_A = nRT/P$.

$V_A = (8.314 \times 293/505)$ L $= 4.82$ L.

2. Find V_B. $V_B = V_A(P_A/P_B)^{1/\gamma} = 4.82(5)^{0.714}$ L $= 15.2$ L

3. Find $V_C = V_D = (8.314 \times 293/101)$ L $= 24.0$ L

The complete cycle is shown in the diagram. Here P is in atmospheres and V in liters.

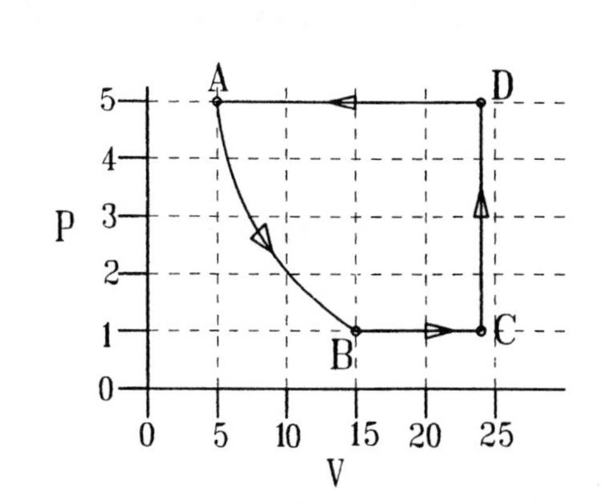

(*b*) Note that for the paths A-B and B-C, W, the work done by the gas, is positive. For the path D-A, W is negative, and greater in magnitude than $W_{A\text{-}C}$. The total work done by the gas is negative and its magnitude is the area enclosed by the cycle. Each rectangle of the dotted lines equals 5 atm·L. Counting these rectangles, the approximate work done by the gas is about -13×5 atm·L = -65 atm·L.

(*c*) Since U is a state function, $\Delta U = 0$ for the complete cycle. Consequently, $Q = W = -65$ atm·L = -6.43 kJ

(*d*) 1. A-B is an adiabatic process. $Q_{A\text{-}B} = 0$

2. B-C, $Q = C_P\Delta T$; $C_P = 7R/2$; $T_B = T_A(V_A/V_B)^{\gamma-1}$ $\quad T_B = 293(4.82/15.2)^{0.4}$ K = 185 K; $Q_{B\text{-}C} = 3.14$ kJ

3. C-D, $Q = C_v\Delta T$; $C_v = 5R/2$; $T_D = P_DV_D/R$ $\quad T_D = (505 \times 24/8.314)$ K = 1458 K; $Q_{C\text{-}D} = 24.2$ kJ

4. D-A, $Q = C_p\Delta T$ $\quad Q_{D\text{-}A} = [7 \times 8.314 \times (-1165)/2]$ J = -33.9 kJ

5. $Q_{tot} = Q_{A\text{-}B} + Q_{B\text{-}C} + Q_{C\text{-}D} + Q_{D\text{-}A}$ $\quad Q_{tot} = (3.14 + 24.2 - 33.9)$ kJ = -6.54 kJ; fair agreement with -6.43 kJ of part (*c*).

---

66 •• Two moles of an ideal monatomic gas have an initial pressure $P_1 = 2$ atm and an initial volume $V_1 = 2$ L. The gas is taken through the following quasi-static cycle: It is expanded isothermally until it has a volume $V_2 = 4$ L. It is then heated at constant volume until it has a pressure $P_3 = 2$ atm. It is then cooled at constant pressure until it is back to its initial state. (*a*) Show this cycle on a *PV* diagram. (*b*) Calculate the heat added and the work done by the gas during each part of the cycle. (*c*) Find the temperatures T_1, T_2, and T_3.

(*a*) The cycle is shown in the diagram. Here P is in atmospheres and V is in liters.

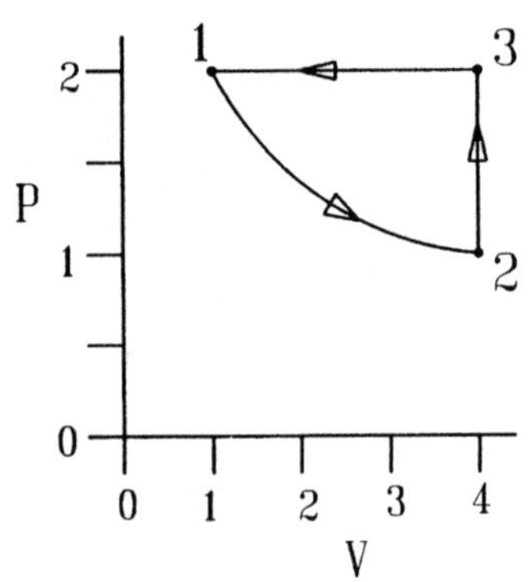

(*c*) 1. Find $T_1 = P_1V_1/nR$ $\quad T_1 = 24.3$ K

2. Find T_2; isothermal expansion, $T_2 = T_1$. $\quad T_2 = 24.3$ K

3. Find $T_3 = P_3V_3/nR$ $\quad T_3 = 48.6$ K

(*b*) 1-2: $Q = W = nRT\ln(V_2/V_1)$ $\quad Q_{1\text{-}2} = 280$ J; $W = 280$ J

2-3: $Q = C_v\Delta T$; $C_v = 3R$ $\quad Q_{2\text{-}3} = 606$ J; $W = 0$

3-1: $Q = C_p\Delta T$; $C_p = 5R$; $W = Q - C_v\Delta T$ $\quad Q_{3\text{-}1} = -1010$ J; $W = -404$ J

---

67 ••• At point D in Figure 19-17 the pressure and temperature of 2 mol of an ideal monatomic gas are 2 atm and 360 K. The volume of the gas at point B on the *PV* diagram is three times that at point D and its pressure is twice that at point C. Paths AB and CD represent isothermal processes. The gas is carried through a complete cycle along the path DABCD. Determine the total amount of work done by the gas and the heat supplied to the gas along each portion of the cycle.

1. Find the volume at D from $V = nRT/P$ $\quad V_D = (2 \times 8.314 \times 360/202)$ L = 29.6 L

2. Find V_B, V_C, P_C, and P_B $\quad V_B = V_C = 3V_D = 88.8$ L; $P_C = (2/3)$ atm, $P_B = (4/3)$ atm

3. Find T_C, T_B, and T_A $\quad T_C = T_D = 360$ K; $T_B = T_A = 2T_D = 720$ K

	A	B	C	D
P (atm)	4	4/3	2/3	2
V (L)	29.6	88.8	88.8	29.6
T (K)	720	720	360	360

4. D-A: $W = 0$; $Q = \Delta U = C_v\Delta T$ — $W_{\text{D-A}} = 0$; $Q_{\text{D-A}} = \Delta U_{\text{D-A}} = 3 \times 8.314 \times 360 \text{ J} = 8.98 \text{ kJ}$

   A-B: $W = nRT\ln(V_B/V_A) = Q$; $\Delta U = 0$ — $W_{\text{A-B}} = 2 \times 8.314 \times 720 \times \ln(3) \text{ J} = 13.16 \text{ kJ}$; $Q_{\text{A-B}} = 13.16 \text{ kJ}$

   B-C: $W = 0$; $Q = \Delta U = C_v\Delta T$ — $W_{\text{B-C}} = 0$; $Q_{\text{B-C}} = -3 \times 8.314 \times 360 \text{ J} = -8.98 \text{ J}$

   C-D: $W = nRT\ln(V_D/V_C) = Q$; $\Delta U = 0$ — $W_{\text{C-D}} = -2 \times 8.314 \times 360 \times \ln(3) \text{ J} = -6.58 \text{ kJ}$; $Q_{\text{C-D}} = -6.58 \text{ kJ}$

5. $W_{\text{tot}} = W_{\text{D-A}} + W_{\text{A-B}} + W_{\text{B-C}} + W_{\text{C-D}}$ — $W_{\text{tot}} = 6.58 \text{ kJ}$

---

68 ••• Repeat Problem 67 with the paths AB and CD representing adiabatic processes.

1. Find P_C; $P_C = P_D(V_D/V_C)^\gamma$ — $P_C = 202 \times (1/3)^{1.67} \text{ kPa} = 32.25 \text{ kPa}$
2. Find T_C; $T_C = P_CV_C/nR$ — $T_C = (32.3 \times 88.8/2 \times 8.314) \text{ K} = 172 \text{ K}$
3. Use the information given to construct the table shown below.

	A	B	C	D
P (kPa)	404	64.5	32.25	202
V (L)	29.6	88.8	88.8	29.6
T (K)	720	344	172	360

4. D-A: $W = 0$; $Q = \Delta U = C_v\Delta T$ — $Q_{\text{D-A}} = 8.98 \text{ kJ}$; $W_{\text{D-A}} = 0$

   A-B: $W = -C_v\Delta T$; $Q = 0$ — $Q_{\text{A-B}} = 0$; $W_{\text{A-B}} = 3 \times 8.314 \times 376 \text{ J} = 9.38 \text{ kJ}$

   B-C: $W = 0$; $Q = \Delta U = C_v\Delta T$ — $Q_{\text{B-C}} = -4.29 \text{ kJ}$; $W_{\text{B-C}} = 0$

   C-D: $W = -C_v\Delta T$; $Q = 0$ — $Q_{\text{C-D}} = 0$; $W_{\text{C-D}} = -3 \times 8.314 \times 188 = -4.69 \text{ kJ}$

5. $W_{\text{tot}} = W_{\text{D-A}} + W_{\text{A-B}} + W_{\text{B-C}} + W_{\text{C-D}}$ — $W_{\text{tot}} = 4.69 \text{ kJ}$

69* ••• Repeat Problem 67 with a diatomic gas.

Proceed as in Problem 67 with $C_v = 5R$. The pressures, volumes, and temperatures are as shown in the table of Problem 67. The results are as follows

D-A: $W_{\text{D-A}} = 0$; $Q_{\text{D-A}} = (5/3) \times 8.98 \text{ kJ} = 15.0 \text{ kJ}$

A-B: $W_{\text{A-B}} = 13.16 \text{ kJ}$; $Q_{\text{A-B}} = 13.16 \text{ kJ}$

B-C: $W_{\text{B-C}} = 0$; $Q_{\text{B-C}} = -15.0 \text{ kJ}$

C-D: $W_{\text{C-D}} = -6.58 \text{ kJ}$; $Q_{\text{C-D}} = -6.58 \text{ kJ}$

$W_{\text{tot}} = 6.58 \text{ kJ}$. Note that the total work done is the same for the diatomic and monatomic gases.

---

70 ••• Repeat Problem 68 with a diatomic gas.

Proceed as in Problem 68 with $C_v = 5R$ and $\gamma = 1.4$. The pressures, volumes, and temperatures are as shown in the table below. W and Q are calculated as in Problem 68, and the results are given below.

	A	B	C	D
P (kPa)	404	86.8	43.4	202
V (L)	29.6	88.8	88.8	29.6
T (K)	720	464	232	360

D-A: $W = 0$; $Q = \Delta U = C_v\Delta T$ $W_{\text{D-A}} = 0$; $Q_{\text{D-A}} = 15.0$ kJ

A-B: $W = -C_v\Delta T$; $Q = 0$ $W_{\text{A-B}} = 10.64$ kJ; $Q_{\text{A-B}} = 0$

B-C: $W = 0$; $Q = C_v\Delta T$ $W_{\text{B-C}} = 0$; $Q_{\text{B-C}} = -9.64$ kJ

C-D: $W = -C_v\Delta T$; $Q = 0$ $W_{\text{C-D}} = -5.32$ kJ; $Q_{\text{C-D}} = 0$

$W_{\text{tot}} = W_{\text{D-A}} + W_{\text{A-B}} + W_{\text{B-C}} + W_{\text{C-D}}$ $W_{\text{tot}} = 5.32$ kJ

---

71 ••• An ideal gas of n mol is initially at pressure P_1, volume V_1, and temperature T_h. It expands isothermally until its pressure and volume are P_2 and V_2. It then expands adiabatically until its temperature is T_c and its pressure and volume are P_3 and V_3. It is then compressed isothermally until it is at a pressure P_4 and a volume V_4, which is related to its initial volume V_1 by $T_cV_4^{\gamma-1} = T_hV_1^{\gamma-1}$. The gas is then compressed adiabatically until it is back in its original state. (*a*) Assuming that each process is quasi-static, plot this cycle on a *PV* diagram. (This cycle is known as the Carnot cycle for an ideal gas.) (*b*) Show that the heat Q_h absorbed during the isothermal expansion at T_h is $Q_h = nRT_h\ln(V_2/V_1)$. (*c*) Show that the heat Q_c given off by the gas during the isothermal compression at T_c is $Q_c = nRT_c\ln(V_3/V_4)$. (*d*) Using the result that $TV^{\gamma-1}$ is constant for an adiabatic expansion, show that $V_2/V_1 = V_3/V_4$. (*e*) The efficiency of a Carnot cycle is defined to be the net work done divided by the heat absorbed Q_h. Using the first law of thermodynamics, show that the efficiency is $1 - Q_c/Q_h$. (*f*) Using your results from the previous parts of this problem, show that $Q_c/Q_h = T_c/T_h$.

(*a*) The cycle is shown on the adjoining *PV* diagram.

(*b*) Since $\Delta U = 0$, and $T_1 = T_h$, $Q_h = W_h = nRT_h\ln(V_2/V_1)$

(*c*) $\Delta U = 0$; $T_3 = T_c$, $Q_{3\text{-}4} = W_{3\text{-}4} = nRT_c\ln(V_4/V_3)$; this is the heat absorbed by the gas. The heat released is $Q_c = nRT_c\ln(V_3/V_4)$.

(*d*) From Equ. 19-36, we have $V_1/V_4 = (T_c/T_h)^{1/(\gamma-1)} = V_2/V_3$. It then follows that $V_2/V_1 = V_3/V_4$.

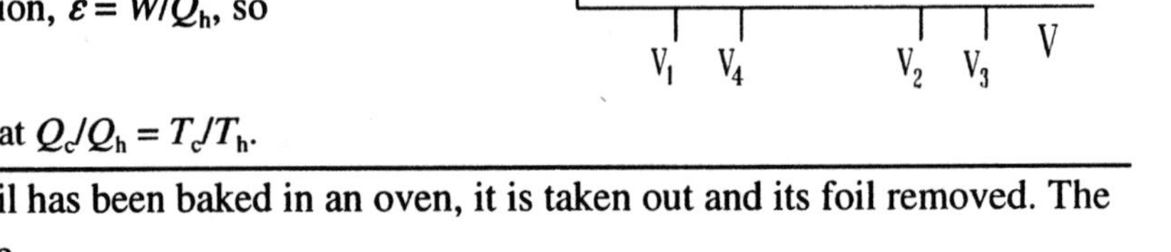

(*e*) Since U is a state function, $\Delta U = 0$ for the complete cycle. Consequently, $W = Q = Q_h - Q_c$. By definition, $\varepsilon = W/Q_h$, so $\varepsilon = 1 - Q_c/Q_h$.

(*f*) Using the result of part (*d*) it follows that $Q_c/Q_h = T_c/T_h$.

---

72 • After a potato wrapped in aluminum foil has been baked in an oven, it is taken out and its foil removed. The foil cools much faster than the potato. Why?

Foil has a smaller mass, a smaller specific heat, and a larger surface area.

73* • True or false:

(*a*) The heat capacity of a body is the amount of heat it can store at a given temperature.

(*b*) When a system goes from state 1 to state 2, the amount of heat added to the system is the same for all processes.

(*c*) When a system goes from state 1 to state 2, the work done on the system is the same for all processes.

(*d*) When a system goes from state 1 to state 2, the change in the internal energy of the system is the same for all processes.

(*e*) The internal energy of a given amount of an ideal gas depends only on its absolute temperature.

(*f*) A quasi-static process is one in which there is no motion.

(*g*) For any material that expands when heated, C_p is greater than C_v.

(*a*) False. (*b*) False. (*c*) False. (*d*) True. (*e*) True. (*f*) False. (*g*) True.

---

74 • If a system's volume remains constant while undergoing changes in temperature and pressure, then

(*a*) the internal energy of the system is unchanged.

(*b*) the system does no work.

(*c*) the system absorbs no heat.

(*d*) the change in internal energy equals the heat absorbed by the system.

(*b*) and (*d*)

---

75 • When an ideal gas is subjected to an isothermal process,

(*a*) no work is done by the system

(*b*) no heat is supplied to the system

(*c*) the heat supplied to the system equals the change in internal energy.

(*d*) the heat supplied to the system equals the work done by the system.

(*d*)

---

76 •• The 1-L fuel tank of a gas grill contains 600 g of propane (C_3H_8) at a pressure of 2MPa. What can you say about the phase state of the propane?

1. For C_3H_8, $M = 44$; find n — $n = 600/44 = 13.6$

2. Find $T = PV/nR$ — $T = (2 \times 10^3/13.6 \times 8.314)$ K $= 17.7$ K

At this low temperature and high pressure, C_3H_8 is a solid. (The melting point of propane at 1 atm pressure is 83 K.)

77* •• An ideal gas undergoes a process during which $P\sqrt{V}$ = constant and the volume of the gas decreases. What happens to the temperature?

$PV = C\sqrt{V} = nRT$. If V decreases, the T decreases.

---

78 • The volume of three moles of a monatomic gas is increased from 50 L to 200 L at constant pressure. The initial temperature of the gas is 300 K. How much heat must be supplied to the gas?

$Q = C_p\Delta T$; $T_f = 4T_i$; $\Delta T = 3T_i$; $C_p = 15R/2$ — $Q = (15 \times 8.314 \times 900/2)$ J $= 56.1$ kJ

---

79 • In the process of compressing n moles of an ideal diatomic gas to one-fifth of its initial volume, 180 kJ of work is done on the gas. If this is accomplished isothermally at room temperature (293 K), how many calories of heat are removed from the gas?

$Q = \Delta U + W$; $\Delta U = 0$; Q (removed) $= W$ (on gas) — Heat removed = 180 kJ

---

80 • What is the number of moles n of the gas in Problem 79?

Equ. 19-16: $W = nRT\ln(V_2/V_1)$; $n = W/[RT\ln(V_2/V_1)]$ — $n = [-180 \times 10^3/(8.314 \times 293 \times \ln 0.2)] = 45.9$

81* • The *PV* diagram in Figure 19-18 represents 3 mol of an ideal monatomic gas. The gas is initially at point *A*. The paths *AD* and *BC* represent isothermal changes. If the system is brought to point *C* along the path *AEC*, find (*a*) the initial and final temperatures, (*b*) the work done by the gas, and (*c*) the heat absorbed by the gas.

Although not required for this problem, we begin by determining pressures, volumes, and temperatures at points A, B, C, D, and E, and then list these as in the table below.

(*a*) Find $T_A = T_D$ and $T_C = T_B$ using $T = PV/nR$ — $T_A = (4 \times 404/3 \times 8.314)$ K $= 64.8$ K; $T_C = 81$ K

Find V_B and V_A using $V = nRT/P$ — $V_B = (81/64.8) \times 4.0 = 5.0$ L; $V_D = 16.0$ L

	A	B	C	D	E
P (atm)	4.0	4.0	1.0	1.0	1.0
V (L)	4.0	5.0	20.0	16.0	4.0
T (K)	64.8	81.0	81.0	64.8	16.2

(*b*) $W_{A\text{-}E} = 0$; $W_{E\text{-}C} = P_E\Delta V$; — $W = 16$ atm·L $= 1.62$ kJ

(*c*) $Q = W + \Delta U$; $\Delta U = C_v\Delta T$; $C_v = 3nR/2$ — $Q = (1.62 + 9 \times 8.314 \times 16.2/2)$ kJ $= 2.23$ kJ

---

82 •• Repeat Problem 81 with the gas following path ABC.

Use the results shown in the table of Problem 81.

(*a*) See table of Problem 81. — $T_A = 64.8$ K; $T_C = 81.0$ K

(*b*) 1. $W_{A\text{-}B} = P_A\Delta V_{A\text{-}B}$; $W_{B\text{-}C} = nRT_B \ln(V_C/V_B)$ — $W_{A\text{-}B} = 0.404$ kJ; $W_{B\text{-}C} = 3 \times 8.314 \times 81 \times \ln(4) = 2.8$ kJ

2. $W = W_{A\text{-}B} + W_{B\text{-}C}$ — $W = 3.20$ kJ

(*c*) $Q = W + \Delta U$; $\Delta U = 0.61$ kJ (see Problem 81) — $Q = 3.81$ kJ

---

83 •• Repeat Problem 82 with the gas following path ADC.

Use the results shown in the table of Problem 81.

(*a*) See the table of Problem 81. — $T_A = 64.8$ K; $T_C = 81.0$ K

(*b*) 1. $W_{A\text{-}D} = nRT_A \ln(V_D/V_A)$; $W_{D\text{-}C} = P_D\Delta V_{D\text{-}C}$ — $W_{A\text{-}D} = 3 \times 8.314 \times 64.8 \times \ln(4)$ J $= 2.24$ kJ; $W_{D\text{-}C} = 0.404$ kJ

2. $W = W_{A\text{-}D} + W_{D\text{-}C}$ — $W = 2.64$ kJ

(*c*) $Q = W + \Delta U$; $\Delta U = 0.61$ kJ (see Problem 81) — $Q = 3.25$ kJ

---

84 •• Suppose that the paths AD and BC represent adiabatic processes. What then are the work done by the gas and the heat absorbed by the gas in following the path ABC?

(*a*) T is a state function. See Problem 81 for T_A, T_C — $T_A = 64.8$ K; $T_C = 81.0$ K

(*b*) 1. Find V_D and V_B using $PV^\gamma =$ constant; $\gamma = 5/3$ — $V_D = 4 \times 4^{0.6}$ L $= 9.19$ L; $V_B = 20/4^{0.6}$ L $= 8.71$ L

2. Find T_D and T_B using ideal gas law. — $T_D = 37.2$ K; $T_B = 141$ K

3. $W_{A\text{-}B} = P_A\Delta V_{A\text{-}B}$; — $W_{A\text{-}B} = 0.404 \times 4.71$ kJ $= 1.90$ kJ;

$W_{B\text{-}C} = -C_v\Delta T_{B\text{-}C}$ — $W_{B\text{-}C} = 9 \times 8.314 \times 60/2$ J $= 2.24$ kJ;

$W = W_{A\text{-}B} + W_{B\text{-}C}$ — $W = 4.14$ kJ

(*c*) $Q = W + \Delta U$; $\Delta U = 0.61$ kJ (see Problem 81) — $Q = 4.75$ kJ

85* •• Repeat Problem 84 for the path *ADC*.

(*a*) See Problem 84. — $T_A = 64.8$ K; $T_C = 81.0$ K

(*b*) See Problem 84 for V_D, T_D; $W_{A\text{-}D} = -C_v\Delta T_{A\text{-}D}$; $W_{D\text{-}C} = P_D\Delta V_{D\text{-}C}$; $W = W_{A\text{-}D} + W_{D\text{-}C}$ — $W_{A\text{-}D} = 9 \times 8.314 \times 27.6/2$ J $= 1.03$ kJ; $W_{D\text{-}C} = 0.101 \times 10.81$ kJ $= 1.09$ kJ; $W = 2.12$ kJ

(*c*) $Q = W + \Delta U$; $\Delta U = 0.61$ kJ — $Q = 2.73$ kJ

---

86 •• At very low temperatures, the specific heat of a metal is given by $c = aT + bT^3$. For the metal copper, $a = 0.0108$ J / kg·K² and $b = 7.62 \times 10^{-4}$ J / kg·K⁴. (*a*) What is the specific heat of copper at 4 K? (*b*) How much heat is required to heat copper from 1 to 3 K?

(*a*) $c = 0.0108T + 7.62 \times 10^{-4}T^3$ J/kg·K $\qquad c(4) = 9.20 \times 10^{-2}$ J/kg·K

(*b*) $Q = \int_{T_i}^{T_f} c(T)dT \qquad Q = \int_1^3 0.0108T\,dT + \int_1^3 7.62\times10^{-4}T^3\,dT = 0.0584$ J/kg

---

87 •• Two moles of a diatomic ideal gas are compressed isothermally from 18 L to 8 L. In the process, 170 calories escape from the system. Determine the amount of work done by the gas, the change in internal energy, and the initial and final temperatures of the gas.

1. $W = nRT\ln(V_f/V_i) = Q + \Delta U;\ \Delta U = 0 \qquad W = -170$ cal $= -711$ J; W (on gas) $= 711$ J
2. $\Delta U = 0$ in an isothermal process. $\qquad \Delta U = 0$
3. $T_i = T_f = W/[nR\ln(V_f/V_i)] \qquad T_i = T_f = -711/[2 \times 8.314 \times \ln(8/18)]$ K $= 52.7$ K

---

88 •• Suppose the two moles of a diatomic ideal gas in Problem 87 are compressed from 18 L to 8 L adiabatically. The work done on the gas is 820 J. Find the initial temperature and the initial and final pressures.

$$W = \frac{P_iV_i - P_fV_f}{\gamma - 1} = \frac{P_i[V_i - V_f(V_i/V_f)^\gamma]}{\gamma - 1};\ \text{find } P_i \qquad P_i = \frac{-0.4\times820}{[18-8(18/8)^{1.4}]\times10^{-3}}\ \text{Pa} = 47.56\ \text{kPa}$$

Find $T_1 = T_i$ from the ideal gas law $\qquad T_i = (47.56 \times 18/2 \times 8.314)$ K $= 51.5$ K

Find $T_2 = T_f$ from $W = -C_v\Delta T \qquad \Delta T = -820/5 \times 8.314$ K $= 19.7$ K; $T_f = 71.2$ K

Find P_f from the ideal gas law $\qquad P_f = 2 \times 8.314 \times 71.2/8$ kPa $= 148$ kPa

---

89* •• Repeat Problem 87 with the diatomic ideal gas replaced by a monatomic ideal gas.

The results are the same for the diatomic gas. See Problem 87.

---

90 •• Repeat Problem 88 with the diatomic ideal gas replaced by a monatomic ideal gas.

Repeat Problem 88 with $\gamma = 1.67$. The results are: $P_i = 42.3$ kPa, $P_f = 164$ kPa; $T_i = 45.8$ K, $T_f = 78.6$ K.

---

91 •• How much work must be done to 30 grams of CO at standard temperature and pressure to compress it to a fifth of its initial volume if the process is (*a*) isothermal; (*b*) adiabatic?

1. Find n; $M = 28 \qquad$ 30 g = 30/28 mol; $n = 1.07$
2. Find V_i; at STP 1 mol = 22.4 L; $V_f = V_i/5 \qquad V_i = 22.4 \times 10.7 = 24.0$ L; $V_f = 4.8$ L

(*a*) $W(\text{on gas}) = -nRT\ln(V_f/V_i) \qquad W(\text{on gas}) = -1.07 \times 8.314 \times 273 \times \ln(0.2) = 3.91$ kJ

(*b*) $W = \frac{P_iV_i - P_fV_f}{\gamma - 1} = \frac{P_i[V_i - V_f(V_i/V_f)^\gamma]}{\gamma - 1};\ \gamma = 1.4 \qquad W(\text{on gas}) = 1(4.8 \times 5^{1.4} - 24)/0.4$ atm·L $= 54.2$ atm·L $= 5.48$ kJ.

---

92 •• Repeat Problem 91 if the gas is CO_2.

From Table 19-5, $c_v = 3.39R$; $c_p = (3.39 + 1.02)R$. Therefore, $\gamma = 1.30$; $M(CO_2) = 44$, so $n = 0.682$.

(*a*) $W(\text{on gas}) = -nRT\ln(V_f/V_i)$; (see Problem 91) $\qquad W(\text{on gas}) = (3.91 \times 0.682/1.07)$ kJ $= 2.49$ kJ (*b*) Find V_i and $V_f = V_i/5$ $\qquad$ $V_i = 0.682 \times 22.4\ \text{L} = 15.3\ \text{L};\ V_f = 3.06\ \text{L}$

$$W = \frac{P_iV_i - P_fV_f}{\gamma - 1} = \frac{P_i[V_i - V_f(V_i/V_f)^\gamma]}{\gamma - 1};\ \gamma = 1.3$$

$W(\text{on gas}) = (3.06 \times 5^{1.3} - 15.3)/0.3\ \text{atm·L} = 3.2\ \text{kJ}$

93* •• Repeat Problem 91 if the gas is argon.

For Ar, $M = 40$, so $n = 0.75$; $\gamma = 1.67$. Following the procedure of the two preceding problems we obtain: $V_i = 16.8$ L and $V_f = 3.36$ L. For (*a*) W(on gas) = 2.74 kJ; for (*b*) W(on gas) = 48.5 atm·L = 4.9 kJ.

---

94 •• A thermally insulated system consists of 1 mol of a diatomic ideal gas at 100 K and 2 mol of a solid at 200 K that are separated by a rigid insulating wall. Find the equilibrium temperature of the system after the insulating wall is removed, assuming that the solid obeys the Dulong-Petit law.

1. Determine $C_{v,\text{solid}}$ and $C_{v,\text{gas}}$. $\qquad$ $C_{v,\text{solid}} = 2 \times 3R = 49.9$ J/K; $C_{v,\text{gas}} = 2.5R = 20.8$ J/K
2. Apply calorimetry equation, $\Delta Q_{\text{tot}} = 0$ $\qquad$ $20.8(T_f - 100) = 49.9(200 - T_f)$
3. Solve for T_f $\qquad$ $T_f = 170.6$ K

---

95 •• When an ideal gas undergoes a temperature change at constant volume, its energy changes by $\Delta U = C_v\Delta T$. (*a*) Explain why this result holds for an ideal gas for any temperature change independent of the process. (*b*) Show explicitly that this result holds for the expansion of an ideal gas at constant pressure by first calculating the work done and showing that it can be written as $W = nR\Delta T$, and then by using $\Delta U = Q - W$, where $Q = C_p\Delta T$.

(*a*) For an ideal gas, the internal energy is the sum of the kinetic energies of the gas molecules, which is proportional to kT. Consequently, U is a function of T only, and $\Delta U = C_v\Delta T$.

(*b*) 1. At constant pressure, $W = P(V_f - V_i) = nR(T_f - T_i) = nR\Delta T$.

2. At constant pressure, $Q = C_p\Delta T$. $\Delta U = Q - W = (C_p - nR)\Delta T = C_v\Delta T$.

---

96 •• One mole of an ideal monatomic gas is heated at constant volume from 300 to 600 K. (*a*) Find the heat added, the work done by the gas, and the change in its internal energy. (*b*) Find these same quantities if the gas is heated from 300 to 600 K at constant pressure.

(*a*) $Q = C_v\Delta T;\ W = \int P dV = 0;\ Q = W + \Delta U$ $\qquad$ $Q = 1.5 \times 8.314 \times 300\ \text{J} = 3.74\ \text{kJ} = \Delta U;\ W = 0$

(*b*) $Q = C_p\Delta T;\ \Delta U = C_v\Delta T;$ $\qquad$ $Q = 2.5 \times 3.74/1.5\ \text{kJ} = 6.23\ \text{kJ};\ \Delta U = 3.74\ \text{kJ};$

$W = Q - \Delta U$ $\qquad$ $W = 2.49$ kJ

97* •• Heat in the amount of 500 J is supplied to 2 mol of an ideal diatomic gas. (*a*) Find the change in temperature if the pressure is kept constant. (*b*) Find the work done by the gas. (*c*) Find the ratio of the final volume of the gas to the initial volume if the initial temperature is 20°C.

(*a*) $\Delta T = Q/C_p;\ C_p = (7/2)nR$ $\qquad$ $\Delta T = (500/7 \times 8.314)\ \text{K} = 8.59\ \text{K}$

(*b*) $W = Q - \Delta U = Q - (5/2)nR\Delta T = nR\Delta T$ $\qquad$ $W = 2 \times 8.314 \times 8.59\ \text{J} = 143\ \text{J}$

(*c*) $V_f/V_i = T_f/T_i = (T_i + \Delta T)/T_i$ $\qquad$ $V_f/V_i = 281.74/273.15 = 1.03$

---

98 •• An insulated cylinder is fitted with a movable piston to maintain constant pressure. The cylinder initially contains 100 g of ice at –10°C. Heat is supplied to the contents at a constant rate by a 100-W heater. Make a graph showing the temperature of the cylinder contents as a function of time starting at $t = 0$, when the temperature is –10°C, and ending when the temperature is 110°C. (Use $c = 2.0$ kJ / kg·K for the average specific heat of ice from –10 to 0°C and of steam from 100 to 110°C.)

1. Find the energy to bring the ice to 0 °C. $Q = 100 \times 2 \times 10$ J $= 2 \times 10^3$ J
2. Find Δt_1, time required at energy input of 100 J/s $\Delta t_1 = 20$ s
3. Time to melt ice $= mL_f/P$, where P is power input $\Delta t_2 = 100 \times 333.5/100$ s $= 333.5$ s
4. Find Δt_3, time to heat water to 100 °C $= mc\Delta T/P$ $\Delta t_3 = 100 \times 4.18 \times 100/100$ s $= 418$ s
5. Find Δt_4, time to vaporize water $= mL_v/P$ $\Delta t_4 = 100 \times 2257/100$ s $= 2257$ s
6. Find Δt_5, time to heat vapor 10 C° $= mc\Delta T/P$ $\Delta t_5 = 100 \times 2 \times 10/100$ s $= 20$ s

The temperature T as a function of time t is shown below. Here T is in Celsius degrees and the times are as follows: $t_1 = 20$ s, $t_2 = 353.5$ s, $t_3 = 771.5$ s, $t_4 = 3028.5$ s, $t_5 = 3048.5$ s.

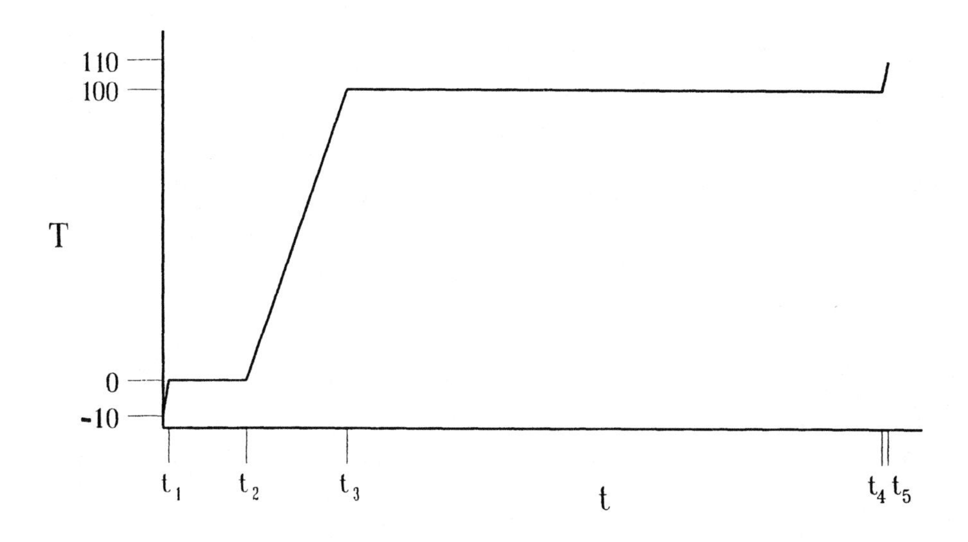

---

99 •• For the winter festival, a sculptor creates a 20-kg statue of a skier made of ice at 0°C. To show off the statue's stability, the sculptor hires someone to repeatedly slide the statue down a plane 8 m long and inclined at 30°. Unfortunately, the sculptor forgot about the thermal energy produced by friction. If the coefficient of sliding friction between the ice and the plane is 0.05, how much ice melts due to the friction on the first run? (Assume that all the mechanical energy lost goes into melting the ice.)

1. Determine the mechanical work, $W = \mu_k F_n L$. $W = 0.05 \times (20 \times 9.81 \times \cos 30°) \times 8$ J $= 68$ J
2. Find mass of ice that melts; $m = W/L_f$ $m = 68/333.5$ g $= 0.2$ g

---

100 •• Two moles of a diatomic ideal gas expand adiabatically. The initial temperature of the gas is 300 K. The work done by the gas during the expansion is 3.5 kJ. What is the final temperature of the gas?

Use Equ. 19-38: $W = -C_v\Delta T$ $\Delta T = -(3.5 \times 10^3/5 \times 8.314)$ K $= -84.2$ K; $T_f = 215.8$ K

101*•• One mole of monatomic gas, initially at temperature T, undergoes a process in which its temperature is quadrupled and its volume is halved. Find the amount of heat Q transferred to the gas. It is known that in this process the pressure was never less than the initial pressure, and the work done on the gas was the minimum possible.

The path for this process is shown on the *PV* diagram. Since $P_fV_f = 4P_iV_i$ and $V_f = V_i/2$, the path for which the work done by the gas is a minimum while the pressure never falls below P_i is shown on the adjacent *PV* diagram. We can now determine W and ΔU in terms of the initial temperature T, initial pressure P_i, and initial volume V_i.

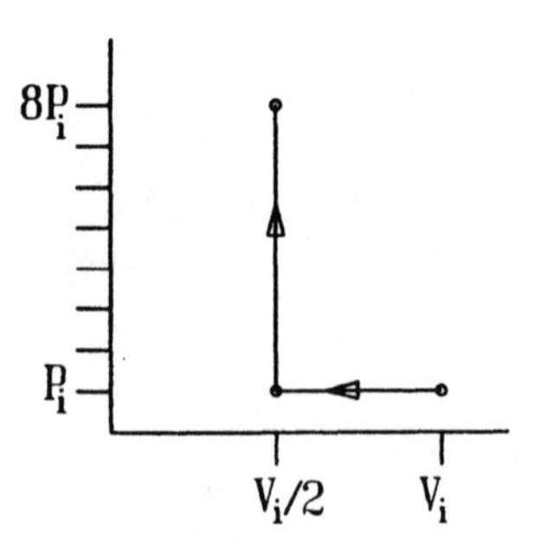

$W = -P_iV_i/2 = -RT/2$. $\Delta U = C_v\Delta T = (3/2)R(3T) = 9RT/2$. $Q = W + \Delta U = 4RT$.

---

102 •• A vertical heat-insulated cylinder is divided into two parts by a movable piston of mass m. Initially the piston is held at rest. The top part is evacuated and the bottom part is filled with 1 mole of diatomic ideal gas at temperature 300 K. After the piston is released and the system comes to equilibrium, the volume, occupied by gas, is halved. Find the final temperature of the gas.

Adiabatic process: $T_2 = T_1(V_1/V_2)^{\gamma-1}$; $\gamma = 1.4$ $\quad$ $T_2 = 300(2)^{0.4}$ K = 396 K

---

103 •• According to the Einstein model of a crystalline solid, the internal energy per mole is given by

$$U = \frac{3N_AkT_E}{e^{T_E/T} - 1}$$

where T_E is a characteristic temperature called the Einstein temperature, and T is the temperature of the solid in kelvins. Evaluate the molar internal energy of diamond (T_E = 1060 K) at 300 K and 600 K, and thereby the increase in internal energy as diamond is heated from 300 K to 600 K.

1. Determine U_{300} and U_{600} $\quad$ $U_{300} = \dfrac{3 \times 1060 \times 8.314}{e^{1060/300} - 1}$ J = 795 J;

$$U_{600} = \frac{3 \times 1060 \times 8.314}{e^{1060/600} - 1} \text{ J} = 5449 \text{ J}$$

2. $\Delta U = U_{600} - U_{300}$ $\quad$ $\Delta U = 4654$ J

---

104 ••• In an isothermal expansion, an ideal gas at an initial pressure P_0 expands until its volume is twice its initial volume. (*a*) Find its pressure after the expansion. (*b*) The gas is then compressed adiabatically and quasi-statically back to its original volume, at which point its pressure is $1.32P_0$. Is the gas monatomic, diatomic, or polyatomic? (*c*) How does the translational kinetic energy of the gas change in these processes?

(*a*) Isothermal process: PV = constant $\quad$ $P_1 = P_0/2$

(*b*) Adiabatic process: PV^γ = constant $\quad$ $P_2 = P_1(V_1/V_0)^\gamma = 1.32P_0 = 0.5P_0(2)^\gamma$; $2.64 = 2^\gamma$

Solve for γ. $\quad$ $\gamma = \ln(2.64)/\ln(2) = 1.4$; gas is diatomic.

(*c*) In an isothermal process, T is constant. $\quad$ Translational kinetic energy is unchanged.

In an adiabatic process, $T_2 = 1.32T_0$ $\quad$ Translational kinetic energy increases by a factor of 1.32

105*••• Prove that the slope of the adiabatic curve passing through a point on the *PV* diagram for an ideal gas is γ times the slope of the isothermal curve passing through the same point.

The slope of the curve on a *PV* diagram is dP/dV. 1. For an isothermal process, PV = constant = C. So,

$P = C/V$, and $dP/dV = -C/V^2 = -P/V$. 2. For an adiabatic process, $PV^\gamma = C$, and $dP/dV = -\gamma P/V$. We see that the slope for the adiabatic process is steeper by the factor γ.

Note: Problems 106 through 109 involve non-quasi-static processes. Nevertheless, assuming that the gases participating in these processes approximate ideal gases, one can calculate the state functions of the end products of the reactions using the first law of thermodynamics and the ideal gas law. For $T > 2000$ K, vibration of the atoms contributes to c_p of H_2O and CO_2 so that c_p of these gases is 7.5R at high temperatures. Also, assume the gases do not dissociate.

---

106 ••• The combustion of benzene is represented by the chemical reaction $2(C_6H_6) + 15(O_2) \rightarrow 12(CO_2) + 6(H_2O)$. The amount of energy released in the combustion of two mol of benzene is 1516 kcal. One mol of benzene and 7.5 mol of oxygen at 300 K are confined in an insulated enclosure at a pressure of 1 atm. (*a*) Find the temperature and volume following combustion if the pressure is maintained at 1 atm. (*b*) If, following combustion, the thermal insulation about the container is removed and the system is cooled to 300 K, what is the final pressure?

For Problems 106–109, the specific heat of the combustion products depends on the temperature. Although c_p increases gradually from $(9/2)R$ per mol to $(15/2)R$ per mol at high temperatures, we shall make the assumption that $c_p = 4.5R$ below $T = 2000$K and $c_p = 7.5R$ above $T = 2000$ K. We shall use $R = 2.0$ cal/mol·K.

(*a*) 1. Find V_i for 8.5 mol of gas at 300 K and 1 atm. — $V_i = (22.4 \times 8.5 \times 300/273)\ \text{L} = 209.2\ \text{L}$

2. Find the total heat released. — $Q = (1516/2)\ \text{kcal} = 758\ \text{kcal}$

3. Find Q needed to form the products at 100°C; there are 3 mol of H_2O, 6 mol of CO_2. — Q to form steam $= 3 \times 18(73 + 540)$ cal $= 33.10$ kcal; Q to heat $CO_2 = 6 \times 4.5 \times 2.0 \times 73$ cal $= 3.94$ kcal

4. Find Q to heat 9 mol of gas to 2000 K — $Q = 9 \times 4.5 \times 2.0 \times 1627\ \text{cal} = 131.79\ \text{kcal}$

5. Find Q available to heat gases above 2000 K — $Q = (758 - 131.79 - 3.94 - 33.10)\ \text{kcal} = 589.2\ \text{kcal}$

6. Find final T of 9 mol of triatomic gases. — $\Delta T = 589.2 \times 10^3/(9 \times 7.5 \times 2.0)\ \text{K} = 4364\ \text{K}$; $T_f = 6364$ K

7. Find V_f from the ideal gas law. — $V_f = (9 \times 8.314 \times 6364/101)\ \text{L} = 4715\ \text{L} = 4.715\ \text{m}^3$

(*b*) Since $T_f = T_i$, $P_f = P_i(n_f/n_i)(V_i/V_f)$ — $P_f = (1 \times (9/8.5)(209.2/4715)\ \text{atm} = 0.047\ \text{atm}$

---

107 ••• Repeat Problem 106, parts (*a*) and (*b*), using as the combustible substance 1 mol of acetylene for which the combustion reaction is $2(C_2H_2) + 5(O_2) \rightarrow 4(CO_2) + 2(H_2O)$. The combustion of 1 mol of acetylene releases 300 kcal.

(*a*) 1. Find V_i of 3.5 mol at 300 K and 1 atm. — $V_i = (22.4 \times 3.5 \times 300/273)\ \text{L} = 86.15\ \text{L}$

2. Find the total heat released. — $Q = 300$ kcal

3. Find Q needed to form products at 100°C; there are 1 mol of H_2O and 2 mol of CO_2. — Q to form steam $= 18(73 + 540)$ cal $= 11.03$ kcal; Q to heat $CO_2 = 2 \times 4.5 \times 2.0 \times 73$ cal $= 1.31$ kcal

4. Find Q to heat 3 mol of gas to 2000 K — $Q = 3 \times 4.5 \times 2.0 \times 1627\ \text{cal} = 43.93\ \text{kcal}$

5. Find Q available to heat gases above 2000 K — $Q = (300 - 43.93 - 1.31 - 11.03)\ \text{kcal} = 243.7\ \text{kcal}$

6. Find T_f of 3 mol of triatomic gases. — $\Delta T = 243.7 \times 10^3/(3 \times 7.5 \times 2.0)\ \text{K} = 5416\ \text{K}$; $T_f = 7418$ K

7. Find V_f using ideal gas law. — $V_f = (3 \times 8.314 \times 7416/101)\ \text{L} = 1831\ \text{L} = 1.831\ \text{m}^3$

(*b*) $P_f = P_i(n_f/n_i)(V_i/V_f)$ — $P_f = 1(3/3.5)(86.15/1831)\ \text{atm} = 0.0403\ \text{atm}$

108 ••• Carbon monoxide and oxygen combine to form carbon dioxide with an energy release of 280 kJ/mol of CO according to the reaction $2(CO) + O_2 \rightarrow 2(CO_2)$. Two mol of CO and one mol of O_2 at 300 K are confined in an 80-L container; the combustion reaction is initiated with a spark. (*a*) What is the pressure in the container prior to the reaction? (*b*) If the reaction proceeds adiabatically, what are the final temperature and pressure? (*c*) If the resulting CO_2 gas is cooled to 0°C, what is the pressure in the container?

(*a*) 1. Find P_i of 3 mol at 300 K in 80 L — $P_i = (3 \times 8.314 \times 300/80)$ kPa = 93.53 kPa

(*b*) 1. Find C_v of combustion product below 2000 K. — $C_v = 2(7/2) \times 8.314 = 58.2$ J/K

2. Find Q to raise 2 mol of CO_2 to 2000 K — $Q = 58.2 \times 1700$ J = 98.94 kJ

3. Find Q to raise CO_2 above 2000 K — $Q = (560 - 98.94)$ kJ = 461.1 kJ

4. Find T_f — $\Delta T = (461.1 \times 10^3/2 \times 6.5 \times 8.314)$ K = 4266 K; $T_f = 6266$ K

5. Find $P_f = P_i(n_f/n_i)(T_f/T_i)$ — $P_f = 93.53(2/3)(6266/300)$ kPa = 1.30 MPa

(*c*) $P_f = P_i(n_f/n_i)(T_f/T_i)$ — $P_f = 93.53(2/3)(273/300)$ kPa = 54.7 kPa

109*•••Suppose that instead of pure oxygen, just enough air is mixed with the two mol of CO in the container of Problem 108 to permit complete combustion. Air is 80% N_2 and 20% O_2 by weight, and the nitrogen does not participate in the reaction. What then are the answers to parts (*a*), (*b*), and (*c*) of Problem 108?

Note that for N_2, c_v at temperatures above 2000 K is $(5/2)R + R$ since there is only one vibrational mode that contributes to the specific heat.

(*a*) 1. Write the reaction for 2 mol of CO — $2(CO) + O_2 + 4N_2 \rightarrow 2(CO_2) + 4N_2$

2. Find P_i of 7 mol at 300 K in 80 L — $P_i = (7 \times 8.314 \times 300/80)$ kPa = 218.2 kPa

(*b*) 1. Find C_v of product gases for $T < 2000$ K — $C_v = [2 \times (7/2) + 4 \times (5/2)]R = 141.3$ J/K

2. Find Q to heat gases to 2000 K — $Q = 1700 \times 141.3$ J = 240.2 kJ

3. Find Q available to raise gases above 2000 K — $Q = (560 - 240.2)$ kJ = 319.8 kJ

4. Find T_f; note that $C_v = 2 \times 6.5$ — $\Delta T = (319.8 \times 10^3/27 \times 8.314)$ K = 1425 K; $T_f = 3425$ K

5. Find $P_f = P_i(n_f/n_i)(T_f/T_i)$ — $P_f = 218.2(6/7)(3425/300)$ kPa = 2.135 MPa

(*c*) $P_f = P_i(n_f/n_i)(T_f/T_i)$ — $P_f = 2135(273/3425)$ kPa = 170.2 kPa

110 ••• Use the expression given in Problem 103 for the internal energy per mole of a solid according to the Einstein model to show that the molar heat capacity at constant volume is given by

$$c_v' = 3R\left(\frac{T_E}{T}\right)^2 \frac{e^{T_E/T}}{\left(e^{T_E/T} - 1\right)^2}$$

$U = \dfrac{3RT_E}{e^{T_E/T} - 1}$. The specific heat at constant volume is dU/dT and performing the indicated operation one obtains $c_v = 3R\left(\dfrac{T_E}{T}\right)^2 \dfrac{e^{T_E/T}}{(e^{T_E/T} - 1)^2}$.

111 ••• (*a*) Use the results of Problem 110 to show that the Dulong–Petit law, $c_v' \approx 3R$, holds for the Einstein model when $T > T_E$. (*b*) For diamond, T_E is approximately 1060 K. Numerically integrate $\Delta U = \int c_v' dT$ to find the increase in the internal energy if 1 mol of diamond is heated from 300 to 600 K. Compare your result to that obtained in Problem 103.

With $T_E = 1060$ K, c_v has the values shown below

T, K	c_v, J/mol·K
300	9.65
400	14.33
500	17.38
600	19.35

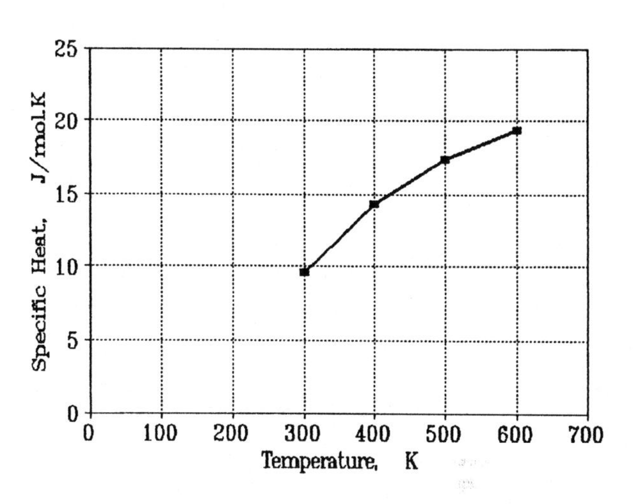

The specific heat as a function of temperature is shown in the Figure. Integrating numerically one obtains $\Delta U = 100[½(9.65) + 14.33 + 17.38) + ½(19.35)]$ = 4621 J in good agreement with the result of Problem 103.

112 ••• A refinement of the Einstein model by Debye resulted in the following expression for the specific heat:

$$c_v' = 9R\left(\frac{T}{T_D}\right)^3 \int_0^y \frac{x^4 e^x}{\left(e^x - 1\right)^2} dx$$

where T_D is called the Debye temperature and $y = T_D / T$. (*a*) Show that when $T >> T_D$, the above expression reduces to the Dulong–Petit result $c_v = 3R$. (Hint: When $T >> T_D$, $y << 1$ and therefore x is always much less than 1. Then $e^x \approx 1 + x$.) (*b*) When $T << T_D$, the integral's upper limit may be approximated by infinity; the definite integral then has the value $4\pi^4 / 15$. Show that at very low temperatures the specific heat is given by $c_v = (12\pi^4/5)R(T/T_D)^3$.

(*a*) When $T >> T_D$, $e^{T_D/T} \approx 1 + (T_D/T)$ and the integral is approximately $\int_0^{T_D/T} \frac{x^4 dx}{x^2} = \int_0^{T_D/T} x^2 dx = \frac{1}{3}\left(\frac{T_D}{T}\right)^3$. The specific heat then reduces to $c_v = 3R$.

(*b*) For $T << T_D$, the upper limit of the integral approaches infinity, and the definite integral has the value $4\pi^4/15$. The specific heat is then as stated, namely $c_v = (12\pi^4/5)R(T/T_D)^3$.

CHAPTER 20

The Second Law of Thermodynamics

1* • Where does the energy come from in an internal-combustion engine? In a steam engine?

Internal combustion engine: From the heat of combustion (see Problems 19-106 to 19-109).

Steam engine: From the burning of fuel to evaporate water and to raise the temperature and pressure of the steam.

2 • How does friction in an engine affect its efficiency?

Friction reduces the efficiency of the engine.

3 • John is house-sitting for a friend who keeps delicate plants in her kitchen. She warns John not to let the room get too warm or the plants will wilt, but John forgets and leaves the oven on all day after his brownies are baked. As the plants begin to droop, John turns off the oven and opens the refrigerator door, intending to use the refrigerator to cool the kitchen. Explain why this doesn't work.

Since a refrigerator exhausts more heat to the room than it extracts from the interior of the refrigerator, the temperature of the room will increase rather than decrease.

4 • Why do power-plant designers try to increase the temperature of the steam fed to engines as much as possible?

Increasing the temperature of the steam increases the Carnot efficiency, and generally increases the efficiency of any heat engine.

5* • An engine with 20% efficiency does 100 J of work in each cycle. (*a*) How much heat is absorbed in each cycle? (*b*) How much heat is rejected in each cycle?

(*a*) From Equ. 20-2, $Q_h = W/\varepsilon$ $\qquad$ $Q_h = 100/0.2\ \text{J} = 500\ \text{J}$

(*b*) $|Q_c| = Q_h(1 - \varepsilon)$ $\qquad$ $|Q_c| = 500 \times 0.8\ \text{J} = 400\ \text{J}$

6 • An engine absorbs 400 J of heat and does 120 J of work in each cycle. (*a*) What is its efficiency? (*b*) How much heat is rejected in each cycle?

(*a*) $\varepsilon = W/Q_h$ $\qquad$ $\varepsilon = 120/400 = 0.3 = 30\%$

(*b*) $|Q_c| = Q_h(1 - \varepsilon)$ $\qquad$ $|Q_c| = 400 \times 0.7\ \text{J} = 280\ \text{J}$

7 • An engine absorbs 100 J and rejects 60 J in each cycle. (*a*) What is its efficiency? (*b*) If each cycle takes 0.5 s, find the power output of this engine in watts.

(*a*) Use Equ. 20-2 $\varepsilon = 1 - 60/100 = 0.4 = 40\%$

(*b*) $P = W/\Delta t = \varepsilon Q_h/\Delta t$ $P = 0.4 \times 100/0.5 \text{ W} = 80 \text{ W}$

---

8 • A refrigerator absorbs 5 kJ of energy from a cold reservoir and rejects 8 kJ to a hot reservoir. (*a*) Find the coefficient of performance of the refrigerator. (*b*) The refrigerator is reversible and is run backward as a heat engine between the same two reservoirs. What is its efficiency?

(*a*) $W = |Q_h| - Q_c$; COP $= Q_c/W$ $W = 3$ kJ; COP $= 5/3 = 1.67$

(*b*) $\varepsilon = W/Q_h$ $\varepsilon = 3/8 = 0.375 = 37.5\%$

9* •• An engine operates with 1 mol of an ideal gas for which $C_v = \frac{3}{2}R$ and $C_p = \frac{5}{2}R$ as its working substance. The cycle begins at $P_1 = 1$ atm and $V_1 = 24.6$ L. The gas is heated at constant volume to $P_2 = 2$ atm. It then expands at constant pressure until $V_2 = 49.2$ L. During these two steps, heat is absorbed by the gas. The gas is then cooled at constant volume until its pressure is again 1 atm. It is then compressed at constant pressure to its original state. During the last two steps, heat is rejected by the gas. All the steps are quasi-static and reversible. (*a*) Show this cycle on a *PV* diagram. Find the work done, the heat added, and the change in the internal energy of the gas for each step of the cycle. (*b*) Find the efficiency of the cycle.

(*a*) The cycle is shown on the right. Here, the pressure *P* is in atm and the volume *V* is in L. To determine the heat added during each step we shall first find the temperatures at points 1, 2, 3, and 4.

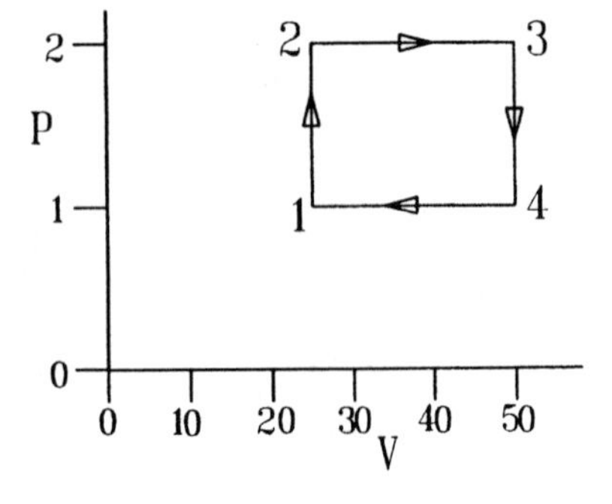

$T_1 = 24.6 \times 273/22.4 \text{ K} = 300 \text{ K}$

$T_2 = 2T_1 = 600 \text{ K}$

$T_3 = 2T_2 = 1200 \text{ K}$

$T_4 = 2T_1 = 600 \text{ K}$

$W_{1\text{-}2} = P\Delta V_{1\text{-}2}$; $Q_{1\text{-}2} = \Delta U_{12} = C_v\Delta T_{1\text{-}2}$ $W_{1\text{-}2} = 0$; $Q_{1\text{-}2} = 1.5 \times 8.314 \times 300 \text{ J} = 3.74 \text{ kJ} = \Delta U_{1\text{-}2}$

$W_{2\text{-}3} = P\Delta V_{2\text{-}3}$; $Q_{2\text{-}3} = C_p\Delta T_{2\text{-}3}$ $W_{2\text{-}3} = 2 \times 24.6 \text{ atm}\cdot\text{L} = 4.97 \text{ kJ}$;

$Q_{2\text{-}3} = 2.5 \times 8.314 \times 600 \text{ J} = 12.47 \text{ kJ}$;

$\Delta U_{2\text{-}3} = (12.47 - 4.97) \text{ kJ} = 7.5 \text{ kJ}$

$W_{3\text{-}4} = P\Delta V_{3\text{-}4}$; $Q_{3\text{-}4} = C_v\Delta T_{3\text{-}4}$ $W_{3\text{-}4} = 0$; $Q_{34} = -1.5 \times 8.314 \times 600 \text{ J} = -7.48 \text{ kJ} = \Delta U_{3\text{-}4}$

$W_{4\text{-}1} = P\Delta V_{4\text{-}1}$; $Q_{4\text{-}1} = C_p\Delta T_{4\text{-}1}$ $W_{4\text{-}1} = -24.6 \text{ atm}\cdot\text{L} = -2.48 \text{ kJ}$; $Q_{4\text{-}1} = -6.24 \text{ kJ}$;

$\Delta U_{4\text{-}1} = -3.76 \text{ kJ}$

(*b*) $\varepsilon = W/Q_{in}$ $W = 2.48$ kJ; $Q_{in} = 16.21$ kJ. $\varepsilon = 0.153 = 15.3\%$

---

10 •• An engine using 1 mol of a diatomic ideal gas performs a cycle consisting of three steps: (1) an adiabatic expansion from an initial pressure of 2.64 atm and an initial volume of 10 L to a pressure of 1 atm and a volume of 20 L, (2) a compression at constant pressure to its original volume of 10 L, and (3) heating at constant volume to its original pressure of 2.64 atm. Find the efficiency of this cycle.

The first process is adiabatic. $Q_1 = 0$

Second step: $Q_2 = C_p\Delta T_2 = 3.5P\Delta V_2$ $Q_2 = -35 \text{ atm}\cdot\text{L}$

Third step: $Q_3 = C_v\Delta T_3 = 2.5V_3\Delta P_3$ — $Q_3 = 25 \times 1.64$ atm·L $= 41$ atm·L

For cycle, $\Delta U = 0$, so $W = Q_1 + Q_2 + Q_3$ — $W = 6$ atm·L

$\varepsilon = W/Q_{in}$ — $\varepsilon = 6/41 = 0.146 = 14.6\%$

---

11 •• An engine using 1 mol of an ideal gas initially at $V_1 = 24.6$ L and $T = 400$ K performs a cycle consisting of four steps: (1) an isothermal expansion at $T = 400$ K to twice its initial volume, (2) cooling at constant volume to $T = 300$ K, (3) an isothermal compression to its original volume, and (4) heating at constant volume to its original temperature of 400 K. Assume that $C_v = 21$ J/K. Sketch the cycle on a *PV* diagram and find its efficiency.

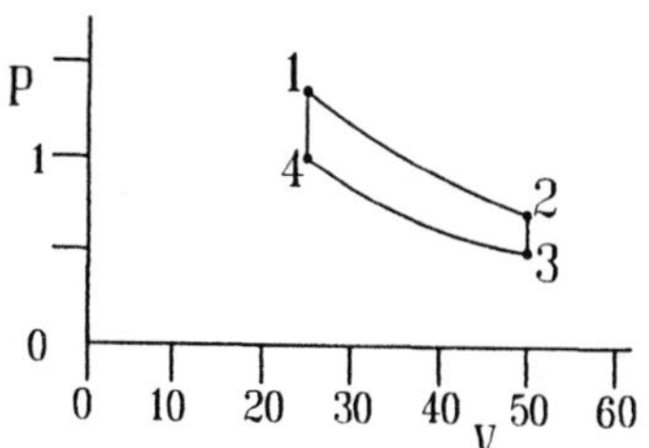

1. The cycle is shown in the figure. Here *P* is in atm and *V* is in L.

2. From the data, $T_1 = T_2 = 400$ K, $T_3 = T_4 = 300$ K, $V_1 = V_4 = 24.6$ L, and $V_2 = V_3 = 49.2$ L.

3. We next determine the work done and the heat absorbed by the gas during each of the four steps.

Step 1-2: $W_{1\text{-}2} = nRT_1\ln(V_2/V_1) = Q_{1\text{-}2}$ — $W_{1\text{-}2} = 8.314 \times 400 \times \ln(2)$ J $= 2.305$ kJ $= Q_{1\text{-}2}$

Step 2-3: $W_{2\text{-}3} = P\Delta V$; $Q_{23} = C_v\Delta T$ — $W_{2\text{-}3} = 0$; $Q_{2\text{-}3} = -21 \times 100$ J $= -2.1$ kJ

Step 3-4: $W_{3\text{-}4} = nRT_3\ln(V_4/V_3) = Q_{3\text{-}4}$ — $W_{3\text{-}4} = 8.314 \times 300 \times \ln(0.5)$ J $= -1.729$ kJ $= Q_{3\text{-}4}$

Step 4-1: $W_{4\text{-}1} = P\Delta V$; $Q_{41} = C_v\Delta T$ — $W_{4\text{-}1} = 0$; $Q_{4\text{-}1} = 21 \times 100$ J $= 2.1$ kJ

Find W and Q_{in} — $W = (2.305 - 1.729)$ kJ $= 0.576$ kJ; $Q_{in} = 4.405$ kJ

Determine $\varepsilon = W/Q_{in}$ — $\varepsilon = 0.576/4.405 = 0.131 = 13.1\%$

---

12 •• One mole of an ideal monatomic gas at an initial volume $V_1 = 25$ L follows the cycle shown in Figure 20-11. All the processes are quasi-static. Find (*a*) the temperature of each state of the cycle, (*b*) the heat flow for each part of the cycle, and (*c*) the efficiency of the cycle.

(*a*) Use $PV = nRT$ — $T_1 = 100 \times 25/8.314$ K $= 300.7$ K; $T_2 = T_3 = 601.4$ K

(*b*) $Q_{1\text{-}2} = C_v\Delta T$ — $Q_{1\text{-}2} = 1.5 \times 8.314 \times 300.7$ J $= 3.75$ kJ

$Q_{2\text{-}3} = W_{2\text{-}3} = nRT_2\ln(V_3/V_2)$ — $Q_{2\text{-}3} = 8.314 \times 601.4 \times \ln(2)$ J $= 3.466$ kJ

$Q_{3\text{-}1} = C_p\Delta T$ — $Q_{3\text{-}1} = -2.5 \times 8.314 \times 300.7$ J $= -6.25$ kJ

(*c*) $W = \Sigma Q$; $Q_{in} = Q_{1\text{-}2} + Q_{2\text{-}3}$; $\varepsilon = W/Q_{in}$ — $W = 0.966$ kJ; $Q_{in} = 7.216$ kJ; $\varepsilon = 0.134 = 13.4\%$

13* •• An ideal gas ($\gamma = 1.4$) follows the cycle shown in Figure 20-12. The temperature of state 1 is 200 K. Find (*a*) the temperatures of the other three states of the cycle and (*b*) the efficiency of the cycle.

(*a*) Use $PV = nRT$; $T_i = T_1(P_iV_i/P_1V_1)$ — $T_1 = 200$ K, $T_2 = 600$ K, $T_3 = 1800$ K, $T_4 = 600$ K

(*b*) Find W = area enclosed by cycle. — $W = 400$ atm·L

Find $Q_{in} = C_v\Delta T_{1\text{-}2} + C_p\Delta T_{2\text{-}3}$ — $Q_{in} = (2.5 \times 200 + 3.5 \times 600)$ atm·L $= 2600$ atm·L

$\varepsilon = W/Q_{in}$ — $\varepsilon = 400/2600 = 0.154 = 15.4\%$

---

14 ••• The *diesel cycle* shown in Figure 20-13 approximates the behavior of a diesel engine. Process *ab* is an adiabatic compression, process *bc* is an expansion at constant pressure, process *cd* is an adiabatic expansion, and process da is cooling at constant volume. Find the efficiency of this cycle in terms of the volumes V_a, V_b, V_c, and V_d.

1. Note that $\varepsilon = 1 - |Q_c|/Q_h$. In the adiabatic portions of the cycle, $Q = 0$. In the portion $b\to c$, $Q = Q_h = C_p(T_c - T_b)$. In the portion $d\to a$, $Q = C_v(T_a - T_d)$ so $|Q_c| = C_v(T_d - T_a)$. Using $\gamma = C_p/C_v$, $\varepsilon = 1 - (T_d - T_a)/\gamma(T_c - T_b)$. The temperatures are related to the volumes by $T_aV_a^{\gamma-1} = T_bV_b^{\gamma-1}$ and $T_cV_c^{\gamma-1} = T_dV_d^{\gamma-1}$. Also, $V_a = V_d$. Using these relations one can now write the efficiency in terms of volume ratios and the ratio T_b/T_c. One obtains

$$\varepsilon = 1 - \frac{(V_c/V_a)^{\gamma-1} - (T_b/T_c)(V_b/V_a)^{\gamma-1}}{\gamma[1 - (T_b/T_c)]}.$$

From the ideal gas law, $T_b/T_c = V_b/V_c$. Simplifying the expression, one obtains $\varepsilon = 1 - \dfrac{(V_c/V_a)^{\gamma} - (V_b/V_a)^{\gamma}}{\gamma[(V_c/V_a) - (V_b/V_a)]}$

---

15 ••• In the *Stirling cycle* shown in Figure 20-14, process ab is an isothermal compression, process bc is heating at constant volume, process cd is an isothermal expansion, and process da is cooling at constant volume. Find the efficiency of the Stirling cycle in terms of the temperatures T_h and T_c and the volumes V_a and V_b.

No work is done in the constant volume segments of the cycle. The work done in the isothermal segments is given by $W_{cd} = nRT_h\ln(V_d/V_c)$ and $W_{ab} = nRT_c\ln(V_b/V_a)$. Since $V_a = V_d$ and $V_b = V_c$, the total work done is $W = nR(T_h - T_c)\ln(V_d/V_c)$.

There is heat input during the segments b→c and c→d. $Q_{bc} = nc_v(T_h - T_c)$; since $\Delta U_{cd} = 0$, $Q_{cd} = nRT_h\ln(V_d/V_c)$. The efficiency is

$$\varepsilon = \frac{W}{Q_h} = \frac{R(T_h - T_c)\ln(V_d/V_c)}{c_v(T_h - T_c) + RT_h\ln(V_d/V_c)}.$$

---

16 ••• The Clausius equation of state is $P(V - bn) = nRT$, where b is a constant. Show that the efficiency of a Carnot cycle is the same for a gas that obeys this equation of state as it is for one that obeys the ideal-gas equation of state, $PV = nRT$.

The Carnot cycle's four segments are: A. An isothermal expansion at $T = T_h$ from V_1 to V_2. B. An adiabatic expansion from V_2 to V_3, at the temperature T_c. C. An isothermal compression from V_3 to V_4. D. An adiabatic compression from V_4 to V_1.

Segment A: $Q_A = W_A = \displaystyle\int_{V_1}^{V_2} P\,dV = nRT_h\int_{V_1}^{V_2}\frac{dV}{V - bn} = nRT_h\ln\left(\frac{V_2 - bn}{V_1 - bn}\right) = Q_h$

Segment C: Following the same procedure as above one obtains $|Q_c| = nRT_c\ln\left(\dfrac{V_3 - bn}{V_4 - bn}\right)$.

For the complete cycle, $\Delta U = 0$, so $W = Q_h - |Q_c|$.

One obtains for the efficiency $\varepsilon = \dfrac{W}{Q_h} = 1 - \dfrac{T_c\ \ln[(V_2 - bn)/(V_1 - bn)]}{T_h\ \ln[(V_3 - bn)/(V_4 - bn)]}$.

The volumes V_1 and V_4, and V_2 and V_3 are related through the adiabatic process for which $dQ = 0 = dU + dW$.

Thus, $C_v dT + PdV = C_v dT + [nRT/(V - bn)]dV = 0$ and $\displaystyle\int\frac{dT}{T} = -\frac{nR}{C_v}\int\frac{dV}{V - bn} = -(\gamma - 1)\int\frac{dV}{V - bn}$. Therefore,

$T(V - bn)^{\gamma-1}$ = constant, and so $T_h(V_2 - bn)^{\gamma-1} = T_c(V_3 - bn)^{\gamma-1}$ and $T_h(V_1 - bn)^{\gamma-1} = T_c(V_4 - bn)^{\gamma-1}$. It follows that

$\frac{V_2 - bn}{V_1 - bn} = \frac{V_3 - bn}{V_4 - bn}$ and the efficiency is $\varepsilon = 1 - T_c/T_h$, the same as for the ideal gas.

17* •• A certain engine running at 30% efficiency draws 200 J of heat from a hot reservoir. Assume that the refrigerator statement of the second law of thermodynamics is false, and show how this engine combined with a perfect refrigerator can violate the heat-engine statement of the second law.

For this engine, $Q_h = 200$ J, $W = 60$ J, and $Q_c = -140$ J. A "perfect" refrigerator would transfer 140 J from the cold reservoir to the hot reservoir with no other effects. Running the heat engine connected to the perfect refrigerator would then have the effect of doing 60 J of work while taking 60 J of heat from the hot reservoir without rejecting any heat, in violation of the heat-engine statement of the second law.

---

18 •• A certain refrigerator takes in 500 J of heat from a cold reservoir and gives off 800 J to a hot reservoir. Assume that the heat-engine statement of the second law of thermodynamics is false, and show how a perfect engine working with this refrigerator can violate the refrigerator statement of the second law.

To remove 500 J from the cold reservoir and reject 800 J to the hot reservoir, 300 J of work must be done on the system. Assuming that the heat-engine statement is false, one could use the 300 J rejected to the hot reservoir to do 300 J of work. Thus, running the refrigerator connected to the "perfect" heat engine would have the effect of transferring 500 J of heat from the cold to the hot reservoir without any work being done, in violation of the refrigerator statement of the second law.

---

19 •• If two adiabatic curves intersect on a *PV* diagram, a cycle could be completed by an isothermal path between the two adiabatic curves shown in Figure 20-15. Show that such a cycle could violate the second law of thermodynamics.

The work done by the system is the area enclosed by the cycle, where we assume that we start with the isothermal expansion. It is only in this expansion that heat is extracted from a reservoir. There is no heat transfer in the adiabatic expansion or compression. Thus we would completely convert heat to mechanical energy, without exhausting any heat to a cold reservoir, in violation of the second law.

---

20 • A Carnot engine works between two heat reservoirs at temperatures $T_h = 300$ K and $T_c = 200$ K. (*a*) What is its efficiency? (*b*) If it absorbs 100 J from the hot reservoir during each cycle, how much work does it do? (*c*) How much heat does it give off during each cycle? (*d*) What is the COP of this engine when it works as a refrigerator between the same two reservoirs?

(*a*) $\varepsilon = 1 - T_c/T_h$ — $\varepsilon = 1 - 2/3 = 1/3 = 0.333 = 33.3\%$

(*b*) $W = \varepsilon Q_h$ — $W = 100/3$ J = 33.3 J

(*c*) $|Q_c| = Q_h - W$ — $|Q_c| = 66.7$ J

(*d*) COP = $|Q_c|/W$ — COP = 2

21* • A refrigerator works between an inside temperature of 0°C and a room temperature of 20°C. (*a*) What is the largest possible coefficient of performance it can have? (*b*) If the inside of the refrigerator is to be cooled to −10°C, what is the largest possible coefficient of performance it can have, assuming the same room temperature of 20°C?

(*a*) Express the COP in terms of T_h and T_c. — COP = $|Q_c|/W = |Q_c|/\varepsilon Q_h = (1 - \varepsilon)/\varepsilon = T_c/(T_h - T_c)$

Evaluate COP — COP = 273/20 = 13.7

(*b*) Evaluate COP — COP = 263/30 = 8.77

---

22 • An engine removes 250 J from a reservoir at 300 K and exhausts 200 J to a reservoir at 200 K. (*a*) What is its efficiency? (*b*) How much more work could be done if the engine were reversible?

(*a*) Use Equ. 20-2 — $\varepsilon = 1 - 200/250 = 0.2 = 20\%$

(*b*) Use Equ. 20-6 for ε_C; find $W = \varepsilon Q_h$ — $\varepsilon = 1 - 200/300 = 0.333$; $W = 83.3$ J

Find W for part (*a*) and ΔW — $W_a = 50$ J; $\Delta W = 33.3$ J

---

23 •• A reversible engine working between two reservoirs at temperatures T_h and T_c has an efficiency of 30%. Working as a heat engine, it gives off 140 J of heat to the cold reservoir. A second engine working between the same two reservoirs also gives off 140 J to the cold reservoir. Show that if the second engine has an efficiency greater than 30%, the two engines working together would violate the heat-engine statement of the second law.
Let the first engine be run as a refrigerator. Then it will remove 140 J from the cold reservoir, deliver 200 J to the hot reservoir, and require 60 J of energy to operate. Now take the second engine and run it between the same reservoirs, and let it eject 140 J into the cold reservoir, thus replacing the heat removed by the refrigerator. If ε_2, the efficiency of this engine, is greater than 30%, then Q_{h2}, the heat removed from the hot reservoir by this engine, is $[140/(1 - \varepsilon_2)]$ J > 200 J, and the work done by this engine is $W = \varepsilon_2 Q_{h2} > 200$ J. The end result of all this is that the second engine can run the refrigerator, replacing the heat taken from the cold reservoir, and do additional mechanical work. The two systems working together then convert heat into mechanical energy without rejecting any heat to a cold reservoir, in violation of the second law.

---

24 •• A reversible engine working between two reservoirs at temperatures T_h and T_c has an efficiency of 20%. Working as a heat engine, it does 100 J of work in each cycle. A second engine working between the same two reservoirs also does 100 J of work in each cycle. Show that if the efficiency of the second engine is greater than 20%, the two engines working together would violate the refrigerator statement of the second law.
If the reversible engine is run as a refrigerator, it will require 100 J of mechanical energy to take 400 J of heat from the cold reservoir and deliver 500 J to the hot reservoir. Now let the second engine, with $\varepsilon_2 > 0.2$, operate between the same two heat reservoirs and use it to drive the refrigerator. Since $\varepsilon_2 > 0.2$, this engine will remove less than 500 J from the hot reservoir in the process of doing 100 J of work. The net result is then that no net work is done by the two systems working together, but a finite amount of heat is transferred from the cold to the hot reservoir, in violation of the refrigerator statement of the second law.

25* •• A Carnot engine works between two heat reservoirs as a refrigerator. It does 50 J of work to remove 100 J from the cold reservoir and gives off 150 J to the hot reservoir during each cycle. Its coefficient of performance COP = Q_c/W = (100 J)/(50 J) = 2. (*a*) What is the efficiency of the Carnot engine when it works as a heat engine between the same two reservoirs? (*b*) Show that no other engine working as a refrigerator between the same two reservoirs can have a COP greater than 2.

(*a*) The efficiency is given by $\varepsilon = W/Q_h$; $\varepsilon = 50/150 = 0.333 = 33.3\%$.

(*b*) If COP > 2, then 50 J of work will remove more than 100 J of heat from the cold reservoir and put more than 150 J of heat into the hot reservoir. So running engine (*a*) to operate the refrigerator with a COP > 2 will result in the transfer of heat from the cold to the hot reservoir without doing any net mechanical work in violation of the second law.

---

26 •• A Carnot engine works between two heat reservoirs at temperatures $T_h = 300$ K and $T_c = 77$ K. (*a*) What is its efficiency? (*b*) If it absorbs 100 J from the hot reservoir during each cycle, how much work does it do? (*c*) How much heat does it give off in each cycle? (*d*) What is the coefficient of performance of this engine when it works as a refrigerator between these two reservoirs?

(*a*) $\varepsilon = 1 - T_c/T_h$	$\varepsilon = 1 - 77/300 = 0.743 = 74.3\%$
(*b*) $W = \varepsilon Q_h$	$W = 100 \times 0.743$ J $= 74.3$ J
(*c*) $\lvert Q_c \rvert = Q_h - W$	$\lvert Q_c \rvert = 25.7$ J
(*d*) COP $= \lvert Q_c \rvert / W$	COP $= 0.345$

---

27 •• In the cycle shown in Figure 20-16, 1 mol of an ideal gas ($\gamma = 1.4$) is initially at a pressure of 1 atm and a temperature of 0°C. The gas is heated at constant volume to $t_2 = 150$°C and is then expanded adiabatically until its pressure is again 1 atm. It is then compressed at constant pressure back to its original state. Find (*a*) the temperature t_3 after the adiabatic expansion, (*b*) the heat entering or leaving the system during each process, (*c*) the efficiency of this cycle, and (*d*) the efficiency of a Carnot cycle operating between the temperature extremes of this cycle.

(*a*) 1. Determine P_2.	$P_2 = 423/273$ atm $= 1.55$ atm
2. Determine $V_3 = V_1(P_2/P_1)^{1/\gamma}$. Note that $V_1 = V$/mol at STP $= 22.4$ L.	$V_3 = 22.4(1.55)^{1/1.4} = 30.6$ L.
3. Find $T_3 = T_1(V_3/V_1)$	$T_3 = 373$ K; $t_3 = 100$°C
(*b*) $Q_{1\text{-}2} = c_v \Delta T$	$Q_{1\text{-}2} = 2.5 \times 8.314 \times 150$ J $= 3.12$ kJ
2→3 is an adiabatic process	$Q_{2\text{-}3} = 0$
$Q_{3\text{-}1} = c_p \Delta T$	$Q_{3\text{-}1} = -3.5 \times 8.314 \times 100 = -2.91$ kJ
(*c*) $W = \Sigma Q$	$W = 0.21$ kJ
$\varepsilon = W/Q_{1\text{-}2}$	$\varepsilon = 0.21/3.12 = 0.0673 = 6.73\%$
(*d*) $\varepsilon_{\text{Carnot}} = 1 - T_c/T_h$	$\varepsilon_{\text{Carnot}} = 1 - T_1/T_2 = 1 - 273/423 = 0.355 = 35.5\%$

---

28 •• A steam engine takes in superheated steam at 270°C and discharges condensed steam from its cylinder at 50°C. Its efficiency is 30%. (*a*) How does this efficiency compare with the maximum possible efficiency for these temperatures? (*b*) If the useful power output of the engine is 200 kW, how much heat does the engine discharge to its surroundings in 1 h?

(*a*) $\varepsilon_{\text{max}} = 1 - T_c/T_h$	$\varepsilon_{\text{max}} = 1 - 323/534 = 40.5\%$
(*b*) $\lvert Q_c \rvert = (1 - \varepsilon)Q_h$; $Q_h = W/\varepsilon = Pt/\varepsilon$	$Q_h = 7.2 \times 10^8/0.3$ J $= 2.4 \times 10^9$ J; $\lvert Q_c \rvert = 1.68$ GJ

29* • A heat pump delivers 20 kW to heat a house. The outside temperature is −10°C and the inside temperature of the hot-air supply for the heating fan is 40°C. (*a*) What is the coefficient of performance of a Carnot heat pump operating between these temperatures? (*b*) What must be the minimum power of the engine needed to run the heat pump? (*c*) If the COP of the heat pump is 60% of the efficiency of an ideal pump, what must the minimum power of the engine be?

(*a*) COP $= T_c/\Delta T$ (see Problem 21)	COP $= 263/50 = 5.26$

(*b*) Use Equ. 20-10; $P = W/t$ $\quad$ $P = [20/(1 + 5.26)]$ kW = 3.19 kW

(*c*) $P' = P/0.6$ $\quad$ $P' = 3.19/0.6$ kW = 5.32 kW

---

30 • Rework Problem 29 for an outside temperature of –20°C.

Follow the procedure of Problem 29 with $T_c = 253$ K, $\Delta T = 60$ K. One obtains (*a*) COP = 4.22; (*b*) $P = 3.83$ kW; (*c*) $P' = 6.39$ kW.

---

31 • A refrigerator is rated at 370 W. (*a*) What is the maximum amount of heat it can remove in 1 min if the inside temperature of the refrigerator is 0°C and it exhausts into a room at 20°C? (*b*) If the COP of the refrigerator is 70% of that of an ideal pump, how much heat can it remove in 1 min?

(*a*) Find COP and $Q_c = (\text{COP}) \times W = (\text{COP}) \times Pt$ $\quad$ COP = 273/20 = 13.65;

$Q_c = 13.65 \times 370 \times 60$ J = 303 kJ

(*b*) $Q_c' = 0.7Q_c$ $\quad$ $Q_c' = 212$ kJ

---

32 • Rework Problem 31 for a room temperature of 35°C.

Follow the same procedure as in Problem 31, with COP = 273/35 = 7.8. One obtains (*a*) $Q_c = 173$ kJ; (*b*) $Q_c' = 121$ kJ.

33* •• On a humid day, water vapor condenses on a cold surface. During condensation, the entropy of the water

(*a*) increases.

(*b*) remains constant.

(*c*) decreases.

(*d*) may decrease or remain unchanged.

(*c*)

---

34 • What is the change in entropy of 1 mol of water at 0°C that freezes?

$\Delta S = \Delta Q/T$ $\quad$ $\Delta S = -(18 \times 333.5/273)$ J/K = −22 J/K

---

35 • Two moles of an ideal gas at $T = 400$ K expand quasi-statically and isothermally from an initial volume of 40 L to a final volume of 80 L. (*a*) What is the entropy change of the gas? (*b*) What is the entropy change of the universe for this process?

(*a*) $\Delta S = \Delta Q/T = nR \ln(V_2/V_1)$ $\quad$ $\Delta S = 2 \times 8.314 \times \ln(2)$ J/K = 11.5 J/K

(*b*) The process is reversible. $\quad$ ΔS of universe = 0; (ΔS of outside = −11.5 J/K)

---

36 • The gas in Problem 35 is taken from the same initial state ($T = 400$ K, $V_1 = 40$ L) to the same final state ($T = 400$ K, $V_2 = 80$ L) by a process that is not quasi-static. (*a*) What is the entropy change of the gas? (*b*) What can be said about the entropy change of the universe?

Since the entropy is a state function, the change in entropy of the gas is as in Problem 35, i.e., 11.5 J/K. In the non-reversible process, the entropy of the universe must increase.

37* • What is the change in entropy of 1.0 kg of water when it changes to steam at 100°C and a pressure of 1 atm?

$\Delta S = \Delta Q/T$ $\quad$ $\Delta S = 2257/373$ kJ/K = 6.05 kJ/K

---

38 • Jay approached his guru in a depressed mood. "I want to change the world, but I feel helpless," he said. The guru turned and pushed a 5-kg rock over a ledge. It hit the ground 6 m below and came to rest. "There," said the guru. "I have changed the world." If the rock, the ground, and the atmosphere are all initially at 300 K, calculate the entropy change of the universe.

$\Delta S = \Delta Q/T$; $\Delta Q = mgh$ $\Delta S = 5 \times 9.81 \times 6/300$ J/K $= 0.981$ J/K

---

39 • What is the change in entropy of 1.0 kg of ice when it changes to water at 0°C and a pressure of 1 atm?

$\Delta S = \Delta Q/T$ $\Delta S = 333.5/273$ kJ/K $= 1.22$ kJ/K

---

40 •• A system absorbs 200 J of heat reversibly from a reservoir at 300 K and gives off 100 J reversibly to a reservoir at 200 K as it moves from state A to state B. During this process, the system does 50 J of work. (*a*) What is the change in the internal energy of the system? (*b*) What is the change in entropy of the system? (*c*) What is the change in entropy of the universe? (*d*) If the system goes from state A to state B by a nonreversible process, how would your answers for parts (*a*), (*b*), and (*c*) differ?

(*a*) $\Delta U = \Delta Q - W$ $\Delta Q = 100$ J; $W = 50$ J; $\Delta U = 50$ J

(*b*) $\Delta S = \Delta S_h - \Delta S_c = Q_h/T_h - Q_c/T_c$ $\Delta S = (200/300 - 100/200)$ J/K $= 0.167$ J/K

(*c*) The process is reversible. $\Delta S_u = 0$

(*d*) S_{system} is a state function; the process is irreversible. (*a*) and (*b*) are the same as before. $\Delta S_u > 0$

41* •• A system absorbs 300 J from a reservoir at 300 K and 200 J from a reservoir at 400 K. It then returns to its original state, doing 100 J of work and rejecting 400 J of heat to a reservoir at a temperature *T*. (*a*) What is the entropy change of the system for the complete cycle? (*b*) If the cycle is reversible, what is the temperature *T*?

(*a*) *S* is a state function of the system. ΔS for complete cycle = 0.

(*b*) $\Delta S = Q_1/T_1 + Q_2/T_2 + Q_3/T_3 = 0$; solve for T_3 1 J/K + 0.5 J/K − (400 J)/$T_3 = 0$; $T_3 = T = 267$ K

---

42 •• Two moles of an ideal gas originally at $T = 400$ K and $V = 40$ L undergo a free expansion to twice their initial volume. What is (*a*) the entropy change of the gas, and (*b*) the entropy change of the universe?

(*a*) See Problem 35; $\Delta S_{gas} = 11.5$ J/K; (*b*) It is an irreversible process; $\Delta S_u > 0$. Since no heat is exchanged, $\Delta S_u = 11.5$ J/K

---

43 •• A 200-kg block of ice at 0°C is placed in a large lake. The temperature of the lake is just slightly higher than 0°C, and the ice melts. (*a*) What is the entropy change of the ice? (*b*) What is the entropy change of the lake? (*c*) What is the entropy change of the universe (the ice plus the lake)?

(*a*) $\Delta S_{ice} = mL_f/T_f$ $\Delta S_{ice} = 200 \times 333.5/273$ kJ/K $= 244.3$ kJ/K

(*b*) $\Delta S_{lake} = -\Delta S_{ice}$ $\Delta S_{lake} = -244.3$ kJ/K

(*c*) $\Delta S_u = 0$. This is only true under the assumption that both the lake and ice are at exactly the same temperature initially. If that were so, then the ice would not melt. Since the temperature of the lake is slightly greater than that of the ice, the magnitude of the entropy change of the lake is less than 244.3 kJ/K and the entropy change of the universe is greater than zero. The melting of the ice is an irreversible process.

---

44 •• A 100-g piece of ice at 0°C is placed in an insulated container with 100 g of water at 100°C. (*a*) When thermal equilibrium is established, what is the final temperature of the water? Ignore the heat capacity of the container. (*b*) Find the entropy change of the universe for this process.

(*a*) Use the calorimetry equation. $100(100 - t) = (100 \text{ g})(79.7 \text{ cal/g}) + 100(t - 0)$

Solve for t — $t = 10.15°C$

(*b*) $\Delta S_{ice} = mL_f/T_f + mc_p \ln(T_f/T_i)$ — $\Delta S_{ice} = [100 \times 333.5/273 + 100 \times 4.184 \times \ln(283.3/273.15)]$ J/K; $\Delta S_{ice} = 137$ J/K

$\Delta S_{water} = mc_p \ln(T_f/T_i)$ — $\Delta S_{water} = 100 \times 4.184 \ln(283/373) = -116$ J/K

$\Delta S_u = \Delta S_{ice} + \Delta S_{water}$ — $\Delta S_u = 21$ J/K; $\Delta S_u > 0$, process is irreversible.

45* •• A 1-kg block of copper at 100°C is placed in a calorimeter of negligible heat capacity containing 4 L of water at 0°C. Find the entropy change of (*a*) the copper block, (*b*) the water, and (*c*) the universe.

(*a*) Use the calorimetry equation to find the final temperature — $1 \times 0.386(100 - t) = 4 \times 4.184(t - 0)$; $t = 2.26°C = 275.4$ K

Find $\Delta S_{Cu} = m_{Cu}c_{Cu} \ln(T_f/T_i)$ — $\Delta S_{Cu} = 0.386 \ln(275.4/373)$ kJ/K $= -117$ J/K

(*b*) $\Delta S_w = m_w c_w \ln(T_f/T_i)$ — $\Delta S_w = 4 \times 4.184 \ln(275.4/273.15)$ kJ/K $= 137$ J/K

(*c*) $\Delta S_u = \Delta S_{Cu} + \Delta S_w$ — $\Delta S_u = 20$ J/K; $\Delta S_u > 0$, the process is irreversible.

---

46 •• If a 2-kg piece of lead at 100°C is dropped into a lake at 10°C, find the entropy change of the universe.

1. Find the heat lost by the lead — $\Delta Q = 2 \times 0.128 \times 90$ kJ $= 23$ kJ
2. Find ΔS_w; T_w remains at 283 K — $\Delta S_w = 23/283$ kJ/K $= 81.4$ J/K
3. Find $\Delta S_{Pb} = m_{Pb}c_{Pb} \ln(T_f/T_i)$ — $\Delta S_{Pb} = 2 \times 0.128 \ln(283/373)$ kJ/K $= -70.7$ J/K
4. Find $\Delta S_u = \Delta S_w + \Delta S_{Pb}$ — $\Delta S_u = 10.7$ J/K

---

47 •• A 1500-kg car traveling at 100 km/h crashes into a concrete wall. If the temperature of the air is 20°C, calculate the entropy change of the universe.

$\Delta S_u = Q/T = \frac{1}{2}mv^2/T$ — $\Delta S_u = \frac{1}{2} \times 1500 \times 27.8^2/293$ J/K $= 1975$ J/K

---

48 •• Find the net change in entropy of the universe when 10 g of steam at 100°C and a pressure of 1 atm are introduced into a calorimeter of negligible heat capacity containing 150 g of water and 150 g of ice at 0°C.

1. Find the heat required to melt 150 g of ice. — $Q_1 = 150 \times 333.5$ J $= 50$ kJ
2. Find the heat released as 10 g of steam at 100°C condense to 0°C. Since $Q_2 < Q_1$, $T_f = 273$ K. — $Q_2 = 10 \times 2257 + 10 \times 4.184 \times 100 = 26.75$ kJ
3. Find m', the mass of ice that melts. — $m' = 26.75/333.5$ kg $= 0.0802$ kg $= 80.2$ g
4. Find $\Delta S_{ice} = Q_2/273$ — $\Delta S_{ice} = 98$ J/K
5. Find $\Delta S_{steam} = -m_s L_v/373 + m_s c_w \ln(273/373)$ — $\Delta S_{steam} = -22750/373 + 41.84 \ln(0.732) = -74$ J/K
6. $\Delta S_u = \Delta S_{ice} + \Delta S_{steam}$ — $\Delta S_u = 24$ J/K

49* •• If 500 J of heat is conducted from a reservoir at 400 K to one at 300 K, (*a*) what is the change in entropy of the universe, and (*b*) how much of the 500 J of heat conducted could have been converted into work using a cold reservoir at 300 K?

(*a*) $\Delta S_u = \Delta S_h + \Delta S_c = -Q/T_h + Q/T_c$ — $\Delta S_u = 500(1/300 - 1/400) = 0.417$ J/K

(*b*) $\varepsilon_{max} = 1 - T_c/T_h$; $W = \varepsilon Q_h$ — $\varepsilon_{max} = 0.25$; $W = 0.25 \times 500 = 125$ J

---

50 •• One mole of an ideal gas first undergoes a free expansion from $V_1 = 12.3$ L and $T_1 = 300$ K to $V_2 = 24.6$ L and $T_2 = 300$ K. It is then compressed isothermally and quasi-statically back to its original state. (*a*) What is the entropy change of the universe for the complete cycle? (*b*) How much work is wasted in this cycle? (*c*) Show that the work wasted is $T\,\Delta S_u$.

(*a*) Although in the adiabatic free expansion no heat is lost by the gas, the process is irreversible and the entropy of the gas increases. In the isothermal reversible process that returns the gas to its original state, the gas releases heat to the surroundings. However, since the process is reversible, the entropy change of the universe is zero. Consequently, the net entropy change is the negative of that of the gas in the isothermal compression.

$\Delta S_{gas} = nR\ln(V_f/V_i) = -\Delta S_u$; $\Delta S_u = nR\ln(V_i/V_f)$ $\Delta S_u = 8.314\ln(2) = 5.76$ J/K

(*b*) If the initial expansion had been isothermal and reversible, no work would have been done in the cycle.

(*c*) The amount of energy that is dissipated is $T\Delta S_u = 1.73$ kJ.

---

51 • In a reversible adiabatic process,

(*a*) the internal energy of the system remains constant.

(*b*) no work is done by the system.

(*c*) the entropy of the system remains constant.

(*d*) the temperature of the system remains constant.

(*c*)

---

52 •• True or false:

(*a*) Work can never be converted completely into heat.

(*b*) Heat can never be converted completely into work.

(*c*) All heat engines have the same efficiency.

(*d*) It is impossible to transfer a given quantity of heat from a cold reservoir to a hot reservoir.

(*e*) The coefficient of performance of a refrigerator cannot be greater than 1.

(*f*) All Carnot engines are reversible.

(*g*) The entropy of a system can never decrease.

(*h*) The entropy of the universe can never decrease.

(*a*) False. (*b*) True. (*c*) False. (*d*) False. (*e*) False. (*f*) True. (*g*) False. (*h*) True.

53* •• An ideal gas is taken reversibly from an initial state P_i, V_i, T_i to the final state P_f, V_f, T_f. Two possible paths are (A) an isothermal expansion followed by an adiabatic compression, and (B) an adiabatic compression followed by an isothermal expansion. For these two paths,

(*a*) $\Delta U_A > \Delta U_B$.

(*b*) $\Delta S_A > \Delta S_B$.

(*c*) $\Delta S_A < \Delta S_B$.

(*d*) none of the above is correct.

(*d*)

---

54 •• Figure 20-17 shows a thermodynamic cycle on an *ST* diagram. Identify this cycle and sketch it on a *PV* diagram.

The processes A-B and C-D are adiabatic; the processes B-C and D-A are isothermal.

The cycle is therefore the Carnot cycle, shown in the adjacent PV diagram.

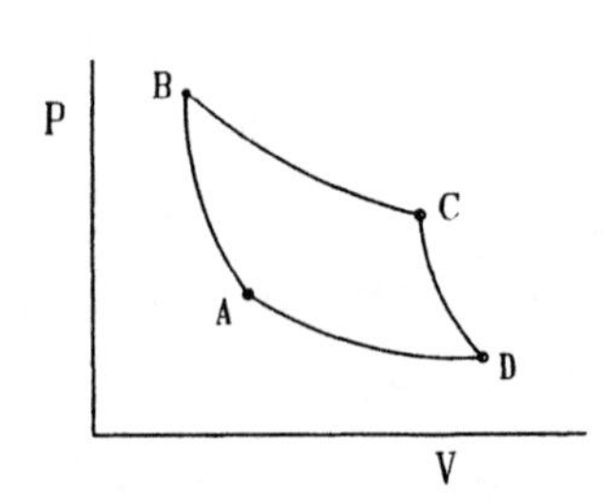

55 •• Figure 20-18 shows a thermodynamic cycle on an SV diagram. Identify the type of engine represented by this diagram.

Note that A-B is an adiabatic expansion. B-C is a constant volume process in which the entropy decreases; therefore heat is released. C-D is an adiabatic compression. D-A is a constant volume process that returns the gas to its original state. The cycle is that of the Otto engine (see Figure 20-3).

56 •• Sketch an ST diagram of the Otto cycle.

Refer to Figure 20-3. Here a-b is an adiabatic compression, so S is constant and T increases. Between b and c, heat is added to the system and both S and T increase. c-d is again isentropic. d-a releases heat and both S and T decrease. The cycle on an ST diagram is sketched in the adjacent figure.

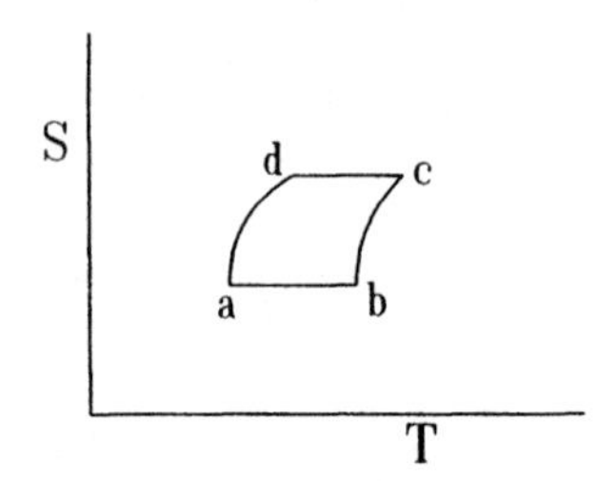

57* •• Sketch an SV diagram of the Carnot cycle.

Referring to Figure 20-8, process 1-2 is an isothermal expansion. In this process heat is added to the system and the entropy and volume increase. Process 2-3 is adiabatic, so S is constant as V increases. Process 3-4 is an isothermal compression in which S decreases and V also decreases. Finally, process 4-1 is adiabatic, i.e., isentropic, and S is constant while V decreases. The cycle is shown in the adjacent SV diagram.

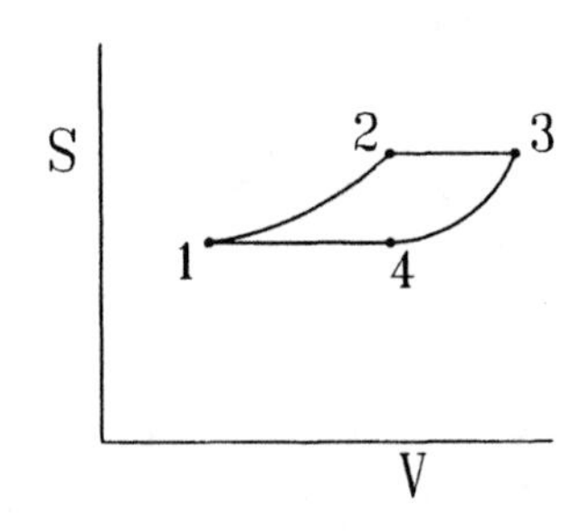

58 •• Sketch an SV diagram of the Otto cycle.

The SV diagram of the Otto cycle is shown in Figure 20-18. (see Problem 55)

59 •• Figure 20-19 shows a thermodynamic cycle on an SP diagram. Make a sketch of this cycle on a PV diagram.

Process A-B is at constant entropy, i.e., an adiabatic process in which the pressure increases. Process B-C is one in which P is constant and S decreases; heat is exhausted from the system and the volume decreases. Process C-D is an adiabatic compression. Process D-A returns the system to its original state at constant pressure. The cycle is shown in the adjacent PV diagram.

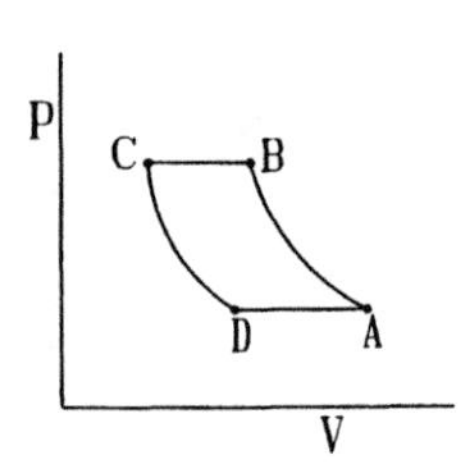

60 • An engine with an output of 200 W has an efficiency of 30%. It works at 10 cycles/s. (*a*) How much work is done in each cycle? (*b*) How much heat is absorbed and how much is given off in each cycle?

(*a*) $W/\text{cycle} = P\Delta t$ $\quad$ $W/\text{cycle} = 200 \times 0.1 \text{ J} = 20 \text{ J}$

(*b*) $Q_h = W/\varepsilon;\ |Q_c| = Q_h - W$ $\quad$ $Q_h = 20/0.3 \text{ J} = 66.7 \text{ J};\ |Q_c| = 46.7 \text{ J}$

61* • Which has a greater effect on increasing the efficiency of a Carnot engine, a 5-K increase in the temperature of the hot reservoir or a 5-K decrease in the temperature of the cold reservoir?

Let ΔT be the change in temperature and $\varepsilon = (T_h - T_c)/T_h$ be the initial efficiency. If T_h is increased by ΔT, ε', the new efficiency is $\varepsilon' = (T_h + \Delta T - T_c)/(T_h + \Delta T)$. If T_c is reduced by ΔT, the efficiency is then $\varepsilon'' = (T_h - T_c + \Delta T)/T_h$. The ratio $\varepsilon''/\varepsilon' = T_h/(T_h + \Delta T) > 1$. Therefore, a reduction in the temperature of the cold reservoir by ΔT increases the efficiency more than an equal increase in the temperature of the hot reservoir.

62 • In each cycle, an engine removes 150 J from a reservoir at 100°C and gives off 125 J to a reservoir at 20°C. (*a*) What is the efficiency of this engine? (*b*) What is the ratio of its efficiency to that of a Carnot engine working between the same reservoirs? (This ratio is called the *second law efficiency.*)

(*a*) $\varepsilon = 1 - |Q_c|/Q_h$ — $\varepsilon = 1 - 125/150 = 0.167 = 16.7\%$

(*b*) $\varepsilon_C = 1 - T_c/T_h$ — $\varepsilon_C = 1 - 293/373 = 0.214 = 21.4\%$; $\varepsilon/\varepsilon_C = 0.777$

63 • An engine removes 200 kJ of heat from a hot reservoir at 500 K in each cycle and exhausts heat to a cold reservoir at 200 K. Its efficiency is 85% of a Carnot engine working between the same reservoirs. (*a*) What is the efficiency of this engine? (*b*) How much work is done in each cycle? (*c*) How much heat is exhausted in each cycle?

(*a*) $\varepsilon = 0.85\varepsilon_C$; $\varepsilon_C = 1 - T_c/T_h$ — $\varepsilon = 0.85(1 - 0.4) = 0.51 = 51\%$

(*b*) $W = \varepsilon Q_h$ — $W = 0.51 \times 200$ kJ $= 102$ kJ

(*c*) $|Q_c| = Q_h - W$ — $|Q_c| = 98$ kJ

64 • To maintain the temperature inside a house at 20°C, the power consumption of the electric baseboard heaters is 30 kW on a day when the outside temperature is −7°C. At what rate does this house contribute to the increase in the entropy of the universe?

$\Delta S/\Delta t = (\Delta Q/T)/\Delta t$ — $\Delta S/\Delta t = 30/266$ kW/K $= 113$ W/K

65* •• The system represented in Figure 20-17 (Problem 54) is 1 mol of an ideal monatomic gas. The temperatures at points A and B are 300 and 750 K, respectively. What is the thermodynamic efficiency of the cyclic process ABCDA?

$\varepsilon = \varepsilon_C$ (see Problem 54); $\varepsilon_C = 1 - T_c/T_h$ — $\varepsilon = 1 - 300/750 = 0.6 = 60\%$

66 •• A sailor is in a tropical ocean on a boat. She has a 2-kg piece of ice at 0°C, and the temperature of the ocean is $T_h = 27$°C. Find the maximum work W that can be done using the fusion of ice.

$\varepsilon_{max} = \varepsilon_C = 1 - T_c/T_h$; $W = \varepsilon_C Q_h = \varepsilon_C m_{ice} L_f$ — $W = (1 - 273/300) \times 2 \times 333.5$ kJ $= 60$ kJ

67 •• (*a*) Which process is more wasteful: (1) a block moving with 500 J of kinetic energy being slowed to rest by friction when the temperature of the atmosphere is 300 K or (2) 1 kJ of heat being conducted from a reservoir at 400 K to one at 300 K? *Hint:* How much of the 1 kJ of heat could be converted into work in an ideal situation? (*b*) What is the change in entropy of the universe for each process?

(*a*) 1. Process (1): All mechanical energy is lost. — Energy loss = 500 J

2. Process (2): Run a Carnot engine. Then $W_{recovered} = (1 - T_c/T_h)Q_h$ — $W_{recovered} = 0.25 \times 1$ kJ $= 250$ J; (1) is more wasteful of *mechanical* energy. (2) is more wasteful of total energy.

(*b*) $\Delta S_1 = \Delta Q/T$; $\Delta S_2 = \Delta Q(1/T_c - 1/T_h)$ $\Delta S_1 = 1.67$ J/K; $\Delta S_2 = 0.833$ J/K; $\Delta S_1 > \Delta S_2$.

68 •• Helium gas ($\gamma = 1.67$) is initially at a pressure of 16 atm, a volume of 1 L, and a temperature of 600 K. It is expanded isothermally until its volume is 4 L and is then compressed at constant pressure until its volume and temperature are such that an adiabatic compression will return the gas to its original state. (*a*) Sketch this cycle on a *PV* diagram. (*b*) Find the volume and temperature after the isobaric compression. (*c*) Find the work done during each cycle. (*d*) Find the efficiency of the cycle.

(*a*) During the isothermal expansion the pressure drops from 16 atm to 4 atm as the volume increases from 1 L to 4 L. The volume at point 3 is determined from the pressure ratio P_1/P_3 and the equation for an adiabatic process, PV^γ = constant; $V_3 = V_1(P_1/P_3)^{1/\gamma} = 1(4)^{0.6}$ L = 2.30 L. The complete cycle is shown on adjacent the *PV* diagram; here *P* is in atm and *V* in L.

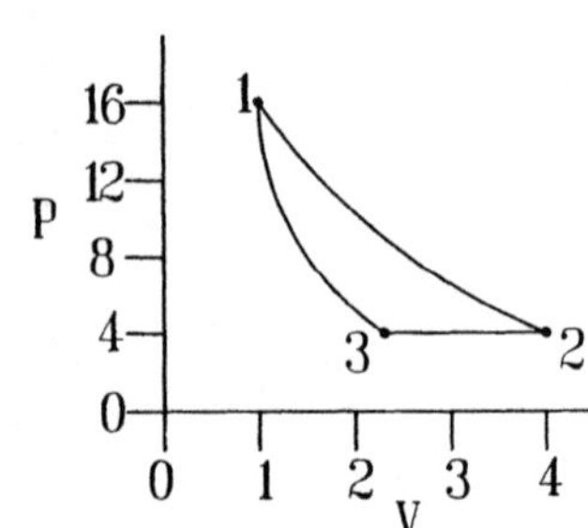

(*b*) The volume at point 3 is $V_3 = 2.30$ L (see above).
$T_3 = T_1(V_1/V_3)^{\gamma-1}$ $T_3 = 600(1/2.3)^{0.667} = 344$ K

(*c*) For process 1-2, $W_{1\text{-}2} = nRT_1 \ln(V_2/V_1) = P_1V_1 \ln(V_2/V_1)$ $W_{1\text{-}2} = 16 \ln(4)$ atm·L = 22.2 atm·L
For process 2-3, $W_{2\text{-}3} = P_2\Delta V$ $W_{2\text{-}3} = 4 \times (2.3 - 4)$ atm·L = −6.8 atm·L
For process 3-1, $W_{3\text{-}1} = -C_v\Delta T = 1.5nR\Delta T = -1.5(P_1V_1 - P_3V_3)$ $W_{3\text{-}1} = -1.5(16 - 9.2)$ atm·L = −10.2 atm·L

(*d*) $Q_{in} = Q_{1\text{-}2} = 22.2$ atm·L; $W = W_{1\text{-}2} + W_{2\text{-}3} + W_{3\text{-}1}$; $\varepsilon = W/Q_{in}$ $\varepsilon = 5.2/22.2 = 0.234 = 23.4\%$

69* •• A heat engine that does the work of blowing up a balloon at a pressure of 1 atm extracts 4 kJ from a hot reservoir at 120°C. The volume of the balloon increases by 4 L, and heat is exhausted to a cold reservoir at a temperature T_c. If the efficiency of the heat engine is 50% of the efficiency of a Carnot engine working between the same reservoirs, find the temperature T_c.

1. Find *W*. $W = 4$ atm·L = 0.404 kJ
2. $\varepsilon = W/Q_h$; $\varepsilon_C = 2\varepsilon = 1 - T_c/T_h$ $\varepsilon_C = 2 \times 0.404/4 = 0.202 = 1 - T_c/393$; $T_c = 313.6$ K
3. $t_c = T_c - 273.15$ $t_c = 40.5$°C

70 •• Show that the COP of a Carnot refrigerator is related to the efficiency of a Carnot engine by COP = $T_c/(\varepsilon_C T_h)$.
By definition, COP = $Q_c/W = (Q_h - W)/W = [1 - (W/Q_h)]/(W/Q_h) = (1 - \varepsilon_C)/\varepsilon_C = (T_c/T_h)/[1 - (T_c/T_h)] = T_c/\varepsilon_C T_h$.

71 •• A freezer has a temperature $T_c = -23$°C. The air in the kitchen has a temperature $T_h = +27$°C. Since the heat insulation is not perfect, some heat flows into the freezer at a rate of 50 W. Find the power of the motor that is needed to maintain the temperature in the freezer.
COP = $T_c/\Delta T$; $P = (dQ_c/dt)/\text{COP} = (dQ_c/dt) \times \Delta T/T_c$ $P = 50 \times 50/250 = 10$ W

72 •• Two moles of a diatomic gas are taken through the cycle ABCA as shown on the *PV* diagram in Figure 20-20. At A the pressure and temperature are 5 atm and 600 K. The volume at B is twice that at A. The segment BC is an adiabatic expansion and the segment CA is an isothermal compression. (*a*) What is the volume of the gas at A? (*b*) What are the volume and temperature of the gas at B? (*c*) What is the temperature of the gas at C?

(*d*) What is the volume of the gas at C? (*e*) How much work is done by the gas in each of the three segments of the cycle? (*f*) How much heat is absorbed by the gas in each segment of the cycle? (*g*) What is the thermodynamic efficiency of this cycle?

(*a*) $V_A = nRT_A/P_A$ $V_A = 2 \times 8.314 \times 600/505$ L = 19.76 L

(*b*) $V_B = 2V_A$; $T_B = 2T_A$ $V_B = 39.52$ L; $T_B = 1200$ K

(*c*) $T_C = T_A$ $T_C = 600$ K

(*d*) $T_B/T_C = (V_C/V_B)^{\gamma-1}$; $V_C/V_B = (T_B/T_C)^{1/(\gamma-1)}$ $V_C = 39.52 \times 2^{2.5} = 224$ L

(*e*) 1. $W_{A\text{-}B} = P_A\Delta V = P_AV_A$ $W_{A\text{-}B} = 5 \times 19.76$ atm·L = 98.8 atm·L = 9.98 kJ

2. $W_{B\text{-}C} = -\Delta U_{B\text{-}C} = -nc_v\Delta T_{B\text{-}C}$ $W_{B\text{-}C} = 2 \times 2.5 \times 8.314 \times 600$ J = 24.9 kJ

3. $W_{C\text{-}A} = nRT\ln(V_A/V_C)$ $W_{C\text{-}A} = 2 \times 8.314 \times 600 \times \ln(19.76/224)$ J = −24.2 kJ

(*f*) 1. $Q_{A\text{-}B} = nc_p\Delta T$ $Q_{A\text{-}B} = 2 \times 3.5 \times 8.314 \times 600 = 34.9$ kJ

2. $Q_{B\text{-}C} = 0$; adiabatic process. $Q_{C\text{-}A} = W_{C\text{-}A}$ $Q_{B\text{-}C} = 0$; $Q_{C\text{-}A} = -24.2$ kJ

(*g*) $\varepsilon = W/Q_{in}$; $W = W_{A\text{-}B} + W_{B\text{-}C} + W_{C\text{-}A}$ $W = 10.68$ kJ, $Q_{in} = 34.9$ kJ; $\varepsilon = 30.6\%$

73* •• Two moles of a diatomic gas are carried through the cycle ABCDA shown in the *PV* diagram in Figure 20-21. The segment AB represents an isothermal expansion, the segment BC an adiabatic expansion. The pressure and temperature at A are 5 atm and 600 K. The volume at B is twice that at A. The pressure at D is 1atm. (*a*) What is the pressure at B? (*b*) What is the temperature at C? (*c*) Find the work done by the gas in one cycle and the thermodynamic efficiency of this cycle.

(*a*) 1. $P_B = P_A(V_A/V_B)$ $P_B = 5/2$ atm = 2.5 atm = 252.5 kPa;

2. Find $V_B = nRT_B/P_B$ $V_B = 2 \times 8.314 \times 600/2.525 \times 10^5$ m^3 = 39.5 L

(*b*) 1. Find $V_C = V_B(P_B/P_C)^{1/\gamma}$; $\gamma = 1.4$ $V_C = 39.5 \times 2.5^{0.714}$ L = 76 L

2. $T_C = T_B(P_CV_C/P_BV_B)$ $T_C = 600(76/98.75)$ K = 462 K

(*c*) 1. $W_{A\text{-}B} = nRT_A\ln(V_B/V_A)$ $W_{A\text{-}B} = 2 \times 8.314 \times 600 \times \ln(2)$ J = 6.915 kJ

2. $W_{B\text{-}C} = -nc_v\Delta T$ $W_{B\text{-}C} = 2 \times 2.5 \times 8.314 \times 138$ J = 5.74 kJ

3. $W_{C\text{-}D} = P_C(V_D - V_C)$; $W_{D\text{-}A} = 0$ $W_{C\text{-}D} = 1(19.75 - 76)$ atm·L = −5.68 kJ; $W_{D\text{-}A} = 0$

4. $W = W_{A\text{-}B} + W_{B\text{-}C} + W_{C\text{-}D} + W_{D\text{-}A}$ $W = 6.975$ kJ

5. $Q_{D\text{-}A} = nc_v(T_A - T_D)$; $T_D = T_A/5$ $Q_{D\text{-}A} = 2 \times 2.5 \times 8.314 \times 480$ J = 20 kJ

6. $Q_{in} = Q_{A\text{-}B} + Q_{D\text{-}A}$; $Q_{A\text{-}B} = W_{A\text{-}B}$; $\varepsilon = W/Q_{in}$ $\varepsilon = 6.975/26.915 = 0.259 = 25.9\%$

---

74 •• Repeat Problem 72 for a monatomic gas.

(*a*), (*b*), and (*c*) are the same as for Problem 72 $V_A = 19.76$ L; $V_B = 39.52$ L; $T_B = 1200$ K; $T_C = 600$ K

(*d*) $V_C = V_B(T_B/T_C)^{1/(\gamma-1)}$; $\gamma = 5/3$ $V_C = 39.52 \times 2^{1.5} = 111.8$ L

(*e*) 1. $W_{A\text{-}B}$ (see Problem 72) $W_{A\text{-}B} = 9.98$ kJ

2. $W_{B\text{-}C} = -nc_v\Delta T_{B\text{-}C}$ $W_{B\text{-}C} = 2 \times 1.5 \times 8.314 \times 600 = 14.97$ kJ

3. $W_{C\text{-}A} = nRT_C\ln(V_A/V_C)$ $W_{C\text{-}A} = 2 \times 8.314 \times 600 \times \ln(19.76/111.8)$ J = −17.29 kJ

(*f*) 1. $Q_{A\text{-}B} = nc_p\Delta T$ $Q_{A\text{-}B} = 2 \times 2.5 \times 8.314 \times 600$ J = 24.94 kJ

2. $Q_{B\text{-}C} = 0$; $Q_{C\text{-}A} = W_{C\text{-}A}$ $Q_{B\text{-}C} = 0$; $Q_{C\text{-}A} = -17.29$ kJ

(*g*) $\varepsilon = W/Q_{in}$; $W = W_{A\text{-}B} + W_{B\text{-}C} + W_{C\text{-}A}$ $W = 7.66$ kJ; $\varepsilon = 7.66/24.94 = 0.307 = 30.7\%$

---

75 •• Repeat Problem 73 for a monatomic gas.

(*a*) Same as for Problem 73. $P_B = 2.5$ atm $= 252.5$ kPa; $V_B = 39.52$ L

(*b*) 1. Find $V_C = V_B(P_B/P_C)^{1/\gamma}$; $\gamma = 5/3$ $V_C = 39.52 \times 2.5^{0.6}$ L $= 68.48$ L

2. $T_C = T_B(P_CV_C/P_BV_B)$ $T_C = 600(68.48/98.8)$ K $= 416$ K

(*c*) 1. $W_{A\text{-}B} = nRT_A \ln(V_B/V_A)$ $W_{A\text{-}B} = 6.915$ kJ (see Problem 73)

2. $W_{B\text{-}C} = -nc_v\Delta T$ $W_{B\text{-}C} = 2 \times 1.5 \times 8.314 \times 184$ J $= 4.59$ kJ

3. $W_{C\text{-}D} = P_C(V_D - V_C)$; $W_{D\text{-}A} = 0$ $W_{C\text{-}D} = 1(19.75 - 68.48)$ atm·L $= -4.92$ kJ; $W_{D\text{-}A} = 0$

4. $W = W_{A\text{-}B} + W_{B\text{-}C} + W_{C\text{-}D} + W_{D\text{-}A}$ $W = 6.585$ kJ

5. $Q_{D\text{-}A} = nc_v(T_A - T_D)$; $T_D = T_A/5$ $Q_{D\text{-}A} = 2 \times 1.5 \times 8.314 \times 480$ J $= 11.97$ kJ

6. $Q_{in} = Q_{A\text{-}B} + Q_{D\text{-}A}$; $Q_{A\text{-}B} = W_{A\text{-}B}$; $\varepsilon = W/Q_{in}$ $\varepsilon = 6.585/18.89 = 0.349 = 34.9\%$

---

76 •• Compare the efficiency of the Otto engine and the Carnot engine operating between the same maximum and minimum temperatures.

The efficiency of the Otto engine is given in Example 20-2: $\varepsilon_O = 1 - \dfrac{T_d - T_a}{T_c - T_b}$, where the subscripts refer to the various points of the cycle as shown in Figure 20-3.

Using the relation $TV^{\gamma-1}$ = constant for the adiabatic process, we have $T_c - T_b = T_d(V_d/V_c)^{\gamma-1} - T_a(V_a/V_b)^{\gamma-1}$. In the Otto cycle, $V_a = V_d$ and $V_b = V_c$; thus, $T_c - T_b = (T_d - T_a)(V_a/V_b)^{\gamma-1}$, and $\varepsilon_O = 1 - (V_b/V_a)^{\gamma-1} = 1 - T_a/T_b$. Note that T_a is the lowest temperature of the cycle, but T_b is not the highest temperature. The high temperature is $T_c = (P_c/P_b)T_b$ which is greater than T_b. A Carnot engine operating between the maximum and minimum temperatures of the Otto will have an efficiency $\varepsilon_C = 1 - T_a/T_c > 1 - T_a/T_b = \varepsilon_O$.

77* •• Compare the efficiency of the Stirling cycle (see Figure 20-14) and the Carnot engine operating between the same maximum and minimum temperatures.

The efficiency of the Sterling cycle, ε_S, is given in Problem 15. Using $T_h - T_c = \varepsilon_C T_h$, that expression can be recast in the form

$$\varepsilon_S = \frac{\varepsilon_C}{1 + \dfrac{c_v\varepsilon_C}{R\ln(V_d/V_c)}}$$

, where V_c and V_d are the volumes indicated in Figure 20-14. Clearly, $\varepsilon_S < \varepsilon_C$.

---

78 ••• Using the equation for the entropy change of an ideal gas when the volume and temperature change and $TV^{\gamma-1}$ is a constant, show explicitly that the entropy change is zero for a quasi-static adiabatic expansion from state (V_1,T_1) to state (V_2,T_2).

In general, $\Delta S = C_v \ln(T_2/T_1) + nR\ln(V_2/V_1)$. For the adiabatic process, $(T_2/T_1) = (V_1/V_2)^{\gamma-1}$. So one has $\Delta S = \ln(V_2/V_1)[nR - C_v(\gamma - 1)]$. But $nR = C_p - C_v$ and $\gamma C_v = C_p$. The expression in the square brackets is therefore equal to zero, and $\Delta S = 0$.

---

79 ••• (*a*) Show that if the refrigerator statement of the second law of thermodynamics were not true, the entropy of the universe could decrease. (*b*) Show that if the heat-engine statement of the second law were not true, the entropy of the universe could decrease. (*c*) An alternative statement of the second law is that the entropy of the universe cannot decrease. Have you just proved that this statement is equivalent to the refrigerator and heat-engine statements?

(*a*) Suppose the refrigerator statement of the second law is violated in the sense that heat Q_c is taken from the cold reservoir and an equal mount of heat is transferred to the hot reservoir and $W = 0$. The entropy change of the universe is then $\Delta S_u = Q_c/T_h - Q_c/T_c$. Since $T_h > T_c$, $\Delta S_u < 0$, i.e., the entropy of the universe would decrease.
(*b*) In this case, is heat Q_h is taken from the hot reservoir and no heat is rejected to the cold reservoir, i.e., $Q_c = 0$, then the entropy change of the universe is $\Delta S_u = -Q_h/T_h + 0$ which is negative. Again, the entropy of the universe would decrease.

(*c*) The heat-engine and refrigerator statements of the second law only state that *some* heat must be rejected to a cold reservoir and *some* work must be done to transfer heat from the cold to the hot reservoir, but these statements do not specify the minimum amount of heat rejected or work that must be done. The statement $\Delta S_u \geq 0$ is more restrictive. The heat-engine and refrigerator statements in conjunction with the Carnot efficiency are equivalent to $\Delta S_u \geq 0$.

---

80 ••• Suppose that two heat engines are connected in series, such that the heat exhaust of the first engine is used as the heat input of the second engine as shown in Figure 20-22. The efficiencies of the engines are ε_1 and ε_2. respectively. Show that the net efficiency of the combination is given by $\varepsilon_{net} = \varepsilon_1 + (1 - \varepsilon_1)\varepsilon_2$

Referring to Figure 20-22, $\varepsilon_1 = W_1/Q_h$ and $\varepsilon_2 = W_2/Q_m$. The overall efficiency is $\varepsilon_{net} = (W_1 + W_2)/Q_h = \varepsilon_1 + (Q_m/Q_h)\varepsilon_2$. Since Q_m is the heat rejected by engine 1, $Q_m/Q_h = 1 - \varepsilon_1$. So $\varepsilon_{net} = \varepsilon_1 + (1 - \varepsilon_1)\varepsilon_2$.

81* ••• Suppose that each engine in Figure 20-22 is an ideal reversible heat engine. Engine 1 operates between temperatures T_h and T_m and Engine 2 operates between T_m and T_c, where $T_h > T_m > T_c$. Show that

$$\varepsilon_{net} = 1 - \frac{T_c}{T_h}$$

This means that two reversible heat engines in series are equivalent to one reversible heat engine operating between the hottest and coldest reservoirs.

Since the engines are ideal reversible engines, $\varepsilon_1 = 1 - T_m/T_h$ and $\varepsilon_2 = 1 - T_c/T_m$. Using the expression derived in Problem 80 one obtains $\varepsilon_{net} = 1 - T_c/T_h$.

---

82 ••• The cooling compartment of a refrigerator and its contents are at 5°C and have an average heat capacity of 84 kJ/K. The refrigerator exhausts heat to the room, which is at 25°C. What minimum power will be required by the the motor that runs the refrigerator if the temperature of the cooling compartment and its contents is to be reduced by 1 C° in 1 min?

1. Determine the COP = $T_c/\Delta T$	COP = 278/20 = 13.9
2. $P = W/t = (\lvert Q_h\rvert/t)/(1 + \text{COP})$; $\lvert Q_h\rvert = C\Delta T = C$	$P = 84 \times 10^3/(60 \times 14.9)$ W = 94 W

---

83 ••• An insulated container is separated into two chambers of equal volume by a thin partition. On one side of the container there are twelve ^{131}Xe atoms, on the other side there are twelve ^{132}Xe atoms. The partition is then removed. Calculate the change in entropy of the system after equilibrium has been established (that is, when the ^{131}Xe and ^{132}Xe atoms are evenly distributed throughout the total volume).

Removing the partition is equivalent to free expansion; i.e., ^{131}Xe and ^{132}Xe each expand freely into the other volume previously unoccupied by these gases. The change in entropy for each gas is given by Equ. 20-18, where here $V_2/V_1 = 2$. We also recall that $R = N_A k$ and $n = N/N_A$, where N_A is Avogadro's number. Thus, $nR = Nk$. Here $N = 12$. We obtain $\Delta S = 2 \times 12 \times 1.38 \times 10^{-23} \ln(2)$ J/K = 2.3×10^{-22} J/K.

CHAPTER 21

Thermal Properties and Processes

1* • Why does the mercury level first decrease slightly when a thermometer is placed in warm water?

The glass bulb warms and expands first, before the mercury warms and expands.

2 • A large sheet of metal has a hole cut in the middle of it. When the sheet is heated, the area of the hole will

(*a*) not change.

(*b*) always increase.

(*c*) always decrease.

(*d*) increase if the hole is not in the exact center of the sheet.

(*e*) decrease only if the hole is in the exact center of the sheet.

(*b*)

3 • A steel ruler has a length of 30 cm at 20°C. What is its length at 100°C?

Apply Equ. 21-2. $\Delta L = (11 \times 10^{-6})(30)(80)$ cm = 0.0264 cm;

$L = 30.0264$ cm

4 • A bridge 100 m long is built of steel. If it is built as a single, continuous structure, how much will its length change from the coldest winter days (–30°C) to the hottest summer days (40°C)?

Apply Equ. 21-2. $\Delta L = (11 \times 10^{-6})(100)(70)$ m = 0.077 m = 7.7 cm

5* •• (*a*) Define a coefficient of area expansion. (*b*) Calculate it for a square and a circle, and show that it is 2 times the coefficient of linear expansion.

(*a*) $\gamma = \dfrac{\Delta A/A}{\Delta T}$. (*b*) For a square, $\Delta A = L^2(1 + \alpha\Delta T)^2 - L^2 = L^2(2\alpha\Delta T + \alpha^2\Delta T^2) = A(2\alpha\Delta T + \alpha^2\Delta T^2)$; in the limit $\Delta T \to 0$, $\Delta A/A = 2\alpha\Delta T$, and $\gamma = 2\alpha$. For the circle, proceed in same way except that now $A = \pi R^2$; again, $\gamma = 2\alpha$.

6 •• The density of aluminum is 2.70×10^3 kg/m³ at 0°C. What is the density of aluminum at 200°C?

Apply Equs. 21-4 and 21-5. $\beta = 3\alpha = 72 \times 10^{-6}\ \mathrm{K}^{-1}$

$\rho = m/V;\ \rho' = m/(V + \Delta V) = \rho/(1 + \beta\Delta T)$ $\rho' = 2.70 \times 10^3/[1+(72 \times 10^{-6})(200)] = 2.66 \times 10^3$ kg/m³

7 •• A copper collar is to fit tightly about a steel shaft whose diameter is 6.0000 cm at 20°C. The inside diameter of the copper collar at that temperature is 5.9800 cm. To what temperature must the copper collar be raised so that it will just slip on the steel shaft, assuming the steel shaft remains at 20°C?

$\alpha\Delta T = 0.02$ cm; use Equ. 21-2 and solve for ΔT — $\Delta T = 0.02/(5.98 \times 17 \times 10^{-6}) = 197$ C°; $T = 217$°C

---

8 •• Repeat Problem 7 when the temperature of both the steel shaft and copper collar are raised simultaneously.

Now $R_{Fe} = R_{Cu}$, and both expand — $6.0000(1 + 11 \times 10^{-6}\Delta T) = 5.9800(1 + 17 \times 10^{-6}\Delta T)$

Solve for ΔT and $T = (20 + \Delta T)$°C — $\Delta T = \dfrac{0.02}{5.98\times17\times10^{-6} - 6.00\times11\times10^{-6}} = 561$ C°; $T = 581$°C

9* •• A container is filled to the brim with 1.4 L of mercury at 20°C. When the temperature of container and mercury is raised to 60°C, 7.5 mL of mercury spill over the brim of the container. Determine the linear expansion coefficient of the container.

1. Express problem statement in terms of V and ΔV — $V_{Hg} = V_c = 1.4$ L; $\Delta V_{Hg} - \Delta V_c = 7.5 \times 10^{-3}$ L
2. Apply Equ. 21-4 and solve for $\beta_{Hg} - \beta_c$ — $\beta_{Hg} - \beta_c = [7.5 \times 10^{-3}/(1.4 \times 40)]\ \mathrm{K}^{-1} = 1.34 \times 10^{-4}\ \mathrm{K}^{-1}$
3. Solve for β_c and apply Equ. 21-5 — $\beta_c = (1.8 - 1.34) \times 10^{-4}\ \mathrm{K}^{-1} = 0.46 \times 10^{-4}\ \mathrm{K}^{-1}$; $\alpha = 15 \times 10^{-6}\ \mathrm{K}^{-1}$

---

10 •• A hole is drilled in an aluminum plate with a steel drill bit whose diameter at 20° C is 6.245 cm. In the process of drilling, the temperature of the drill bit and of the aluminum plate rise to 168°C. What is the diameter of the hole in the aluminum plate when it has cooled to room temperature?

1. Find diameter of the hole (steel drill bit) at 168°C — $d_{Fe} = 6.245(1 + 11 \times 10^{-6} \times 148)$ cm = 6.255 cm
2. Find diameter of the hole in the plate at 20°C — $d_{Al} = 6.255(1 - 24 \times 10^{-6} \times 148)$ cm = 6.233 cm

Note that the diameter of the hole in the plate at 20 °C is less than the diameter of the drill bit at 20°C.

---

11 •• Len sells trees that double in price when they are over 2.00 m high. To make a standard, he cuts an aluminum rod 2.00 m in length, as measured by a steel measuring tape. That day, the temperature of both the rod and the tape is 25°C. What will the tape indicate the length of the rod to be when both the tape and the rod are at (*a*) 0°C and (*b*) 50°C?

1. Apply Equ. 21-2 to the rod and tape — $L_{rod} = 2.00(1 \pm 25\alpha_{Al})$ m; $L_{tape} = 2.00(1 \pm 25\alpha_{Fe})$ m
2. Solve for L_{rod} in terms of L_{tape} — Tape reading $= 2\dfrac{1 \pm 25\times\alpha_{Al}}{1 \pm 25\times\alpha_{Fe}}$ m
3. Use numerical values for α_{Fe} and α_{Al}.
   (*a*) Use negative sign — Tape reading at 0°C = 1.999 m
   (*b*) Use positive sign — Tape reading at 50°C = 2.001 m

---

12 •• A rookie crew was left to put in the final 1 km of rail for a stretch of railroad track. When they finished, the temperature was 20°C, and they headed to town for some refreshments with their coworkers. After an hour or two, one of the old-timers noticed that the temperature had gone up to 25°C, so he said, "I hope you left some gaps to allow for expansion." By the look on their faces, he knew that they had not, and they all rushed back to the work site. The rail had buckled into an isosceles triangle. How high was the buckle?

1. In the figure, L is the length at 20°C, L' the length at 25°C, and h the height of the buckle.

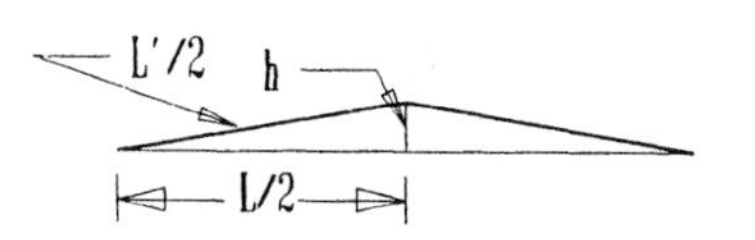

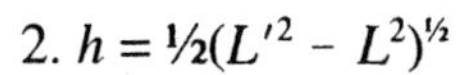

2. $h = \frac{1}{2}(L'^2 - L^2)^{1/2}$
3. $L'^2 = L^2(1 + \alpha\Delta T)^2 \approx L^2(1 + 2\alpha\Delta T)$
4. $h = \frac{1}{2}L(2\alpha\Delta T)^{1/2}$

$h = (500)(2 \times 11 \times 10^{-6} \times 5)^{1/2}$ m = 5.24 m

---

13* •• A car has a 60-L steel gas tank filled to the top with gasoline when the temperature is 10°C. The coefficient of volume expansion of gasoline is $\beta = 0.900 \times 10^{-3}$ K^{-1}. Taking the expansion of the steel tank into account, how much gasoline spills out of the tank when the car is parked in the sun and its temperature rises to 25°C?

Spill = $\Delta V_{gas} - \Delta V_{tank} = V\Delta T(\beta_{gas} - \beta_{tank})$

Spill = $(60)(15)(9 \times 10^{-4} - 3 \times 11 \times 10^{-6})$ L = 0.78 L

---

14 •• A thermometer has an ordinary glass bulb and thin glass tube filled with 1 mL of mercury. A temperature change of 1 C° changes the level of mercury in the thin tube by 3.0 mm. Find the inside diameter of the thin glass tube.

1. Net volume change, $\Delta V = \Delta V_{Hg} - \Delta V_{glass} = A\Delta L$, where A is the area of the capillary.

Note: 1 mL = 10^{-6} m^3

2. $\Delta V = V_0\Delta T(\beta_{Hg} - \beta_{glass})$; solve for A

$$A = \frac{10^{-6}(1.8 \times 10^{-4} - 27 \times 10^{-6})}{3 \times 10^{-3}} \text{ m}^2 = 5.1 \times 10^{-8} \text{ m}^2$$

3. Find d from $A = \pi d^2/4$

$d = \sqrt{4 \times 5.1 \times 10^{-8}/\pi}$ m = 0.255 mm

---

15 •• A mercury thermometer consists of a 0.4-mm capillary tube connected to a glass bulb. The mercury level rises 7.5 cm as the temperature of the thermometer increases from 35°C to 43°C. Find the volume of the thermometer bulb.

1. See Problem 14. $V_0 = A\Delta L/[(\beta_{Hg} - \beta_{glass})\Delta T]$

$$V_0 = \frac{7.5 \times 10^{-2} \times \pi \times (4 \times 10^{-4})^2}{4 \times 8 \times 1.53 \times 10^{-4}} \text{ m}^3 = 7.7 \text{ mL}$$

---

16 ••• A grandfather's clock is calibrated at a temperature of 20°C. (*a*) On a hot day, when the temperature is 30°C, does the clock run fast or slow? (*b*) How much does it gain or lose in a 24-h period? Assume that the pendulum is a thin brass rod of negligible mass with a heavy bob attached to the end.

(*a*) $T_p = 2\pi(L/g)^{1/2}$ is the period of a pendulum; thus as temperature T increases, so does L and T_p, and the clock runs slow.

(*b*) $dT_p/dT = (dT_p/dL)(dL/dT)$

$dT_p/dL = \frac{1}{2}T_p/L$; $dL/dT = \alpha L$.

$\Delta T_p/T_p = \frac{1}{2}\alpha\Delta T$

$\Delta T_p/T_p = \frac{1}{2}(19 \times 10^{-6})(10) = 9.5 \times 10^{-5}$

Loss = $(9.5 \times 10^{-5})(24$ h$)$

$\Delta T = 24 \times 60 \times 60 \times 9.5 \times 10^{-5}$ s = 8.21 s lost in 24 h.

17* ••• A steel tube has an outside diameter of 3.000 cm at room temperature (20°C). A brass tube has an inside diameter of 2.997 cm at the same temperature. To what temperature must the ends of the tubes be heated if the steel tube is to be inserted into the brass tube?

$r_s(1 + \alpha_s\Delta T) = r_b(1 + \alpha_b\Delta T)$; $\Delta T = \dfrac{r_s - r_b}{\alpha_b r_b - \alpha_s r_s}$

$$\Delta T = \frac{3.000 - 2.997}{19 \times 10^{-6} \times 2.997 - 11 \times 10^{-6} \times 3.000} \text{ K} = 125 \text{ K}$$

$T = T_0 + \Delta T$

$T = (20 + 125)$°C = 145°C

---

18 ••• What is the tensile stress in the copper collar of Problem 7 when its temperature returns to 20°C?

Strain = $\Delta L/L$; Stress = $Y(\Delta L/L)$

Stress = $(11 \times 10^{-10})(0.02/5.98)$ N/m^2 = 3.68×10^{8} N/m^2

19 • Mountaineers say that you cannot hard boil an egg on the top of Mount Rainier. This is true because

(*a*) the air is too cold to boil water.

(*b*) the air pressure is too low for stoves to burn.

(*c*) boiling water is not hot enough to hard boil the egg.

(*d*) the oxygen content of the air is too low.

(*e*) eggs always break in their backpacks.

(*c*) (Actually, it can be hard boiled, but it does take quite a bit longer than at sea level.)

20 • Which gases in Figure 21-6 cannot be liquefied by applying pressure at 20°C?

Gases for which $T_c < 293$ K. These are He, A, Ne, H_2, O_2, NO.

21* •• The phase diagram in Figure 21-14 can be interpreted to yield information on how the boiling and melting points of water change with altitude. (*a*) Explain how this information can be obtained. (*b*) How might this information affect cooking procedures in the mountains?

(*a*) With increasing altitude *P* decreases; from curve *OF*, *T* of liquid-gas interphase diminishes, so the boiling temperature decreases. Likewise, from curve *OH*, the melting temperature increases with increasing altitude.

(*b*) Boiling at a lower temperature means that the cooking time will have to be increased.

22 •• For the phase diagram given in Figure 21-14, state what changes (if any) occur for each line segment—AB, BC, CD, and DE—in (*a*) volume and (*b*) phase. (*c*) For what type of substance would *OH* be replaced by *OG*? (*d*) What is the significance of point F?

(*a*) and (*b*). AB: solid sublimates to vapor; volume increases. BC: vapor condenses to liquid; volume decreases. CD: liquid freezes; volume increases. DE: solid changes to liquid; volume decreases.

(*c*) For most materials, the density increases on solidification; for these materials, the phase diagram would have the shape *OG*.

(*d*) F is the critical point.

23 • (*a*) Calculate the volume of 1 mol of steam at 100°C and a pressure of 1 atm, assuming that it is an ideal gas. (*b*) Find the temperature at which the steam will occupy the volume found in part (*a*) if it obeys the van der Waals equation with $a = 0.55$ Pa·m^6/ mol^2 and $b = 30$ cm^3/mol.

(*a*) $V = nRT/P$

$$V = 1 \times 8.314 \times 373/101.3 \times 10^3 \text{ m}^3 = 0.0306 \text{ m}^3 = 30.6 \text{ L}$$

(*b*) Use Equ. 21-6; substitute numerical values

$$T = \frac{\left[1.01\times10^5 + \dfrac{0.55}{(30.6\times10^{-3})^2}\right](30.6\times10^{-3} - 30\times10^{-6})}{8.314}$$

$$T = 374 \text{ K}$$

24 •• From Figure 21-4, find (*a*) the temperature at which water boils on a mountain where the atmospheric pressure is 70 kPa, (*b*) the temperature at which water will boil in a container in which the pressure has been reduced to 0.5 atm, and (*c*) the pressure at which water will boil at 115°C.

(*a*) At 70 kPa, the boiling point is at $T = 90$°C; (*b*) at 0.5 atm, $T_{boil} = 82$ °C; (*c*) for $T_{boil} = 115$°C, $P = 170$ kPa.

25* •• The van der Waals constants for helium are $a = 0.03412\ L^2 \cdot atm/mol^2$ and $b = 0.0237$ L/mol. Use these data to find the volume in cubic centimeters occupied by one helium atom and to estimate the radius of the atom.

In Equ. 21-6, b = volume of 1 mol of molecules — $(0.0237\ L/mol)(1\ mol/6.022 \times 10^{23}\ atoms)(10^3\ cm^3/1\ L)$

For He, 1 molecule = 1 atom — $= 3.94 \times 10^{-23}\ cm^3/atom$.

$V = (4/3)\pi r^3$; solve for r — $r = (3 \times 3.94 \times 10^{-23}/4\pi)^{1/3} = 2.11 \times 10^{-8}$ cm = 0.211 nm

---

26 ••• (*a*) For a van der Waals gas, show that the critical temperature is $8a/27Rb$ and the critical pressure is $a/27b^2$. (*b*) Rewrite the van der Waals equation of state in terms of the reduced variable $V_r = V/V_c$, $P_r = P/P_c$, and $T_r = T/T_c$.

(*a*) At the critical point, $dP/dV = 0$ and $d^2P/dV^2 = 0$. From Equ. 21-6, $dP/dV = -nRT/(V - nb)^2 + 2an^2/V^3 = 0$ (1) and $d^2P/dV^2 = 2nRT/(V - nb)^3 - 6an^2/V^4 = 0$ (2). From (1), $2an^2/V^3 = nRT/(V - nb)^2$ (1*a*); from (2), $6an^2/V^4 = 2nRT/(V - nb)^3$ (2*a*). Dividing (1*a*) by (2*a*) gives $\frac{1}{2}(V - nb) = V/3$; $V_c = 3nb$ (3). Now substitute V_c from (3) into (1*a*): $RT/4nb^2 = 2a/27nb^3$; $T_c = 8a/27Rb$. Now substitute T_c and V_c into Equ. 21 6: $P_c = a/27b^2$.

(*b*) Using the results from (*a*) in Equ. 21-6 one obtains $(P_r + 3/V_r^3)(3V_r - 1) = 8T_r$.

---

27 • A copper bar 2 m long has a circular cross section of radius 1 cm. One end is kept at 100°C and the other end is kept at 0°C. The surface of the bar is insulated so that there is negligible heat loss through it. Find (*a*) the thermal resistance of the bar, (*b*) the thermal current I, (*c*) the temperature gradient $\Delta T/\Delta x$, and (*d*) the temperature of the bar 25 cm from the hot end.

(*a*) $R = \Delta x/kA$ — $R = (2\ m)/[(401 W/m\cdot K)(\pi \times 10^{-4}\ m^2)] = 15.9$ K/W

(*b*) $I = \Delta T/R$ — $I = 100/15.9$ W = 6.3 W

(*c*) Substitute numerical values — $\Delta T/\Delta x = 100/2$ K/m = 50 K/m

(*d*) $T = T_0 + (dT/dx)\Delta x$ — $T = 0°C + 1.75 \times 50°C = 87.5°C$

---

28 • A 20 × 30-ft slab of insulation has an R factor of 11. How much heat (in Btu per hour) is conducted through the slab if the temperature on one side is 68°F and that on the other side is 30°F?

$I = \Delta T/R = A\Delta T/R_f$ — $I = (38)(600)/11$ Btu/h = 2073 Btu/h

---

29* •• Two metal cubes with 3-cm edges, one copper (Cu) and one aluminum (Al), are arranged as shown in Figure 21-15. Find (*a*) the thermal resistance of each cube, (*b*) the thermal resistance of the two-cube system, (*c*) the thermal current I, and (*d*) the temperature at the interface of the two cubes.

(*a*) Use Equ.21-10; substitute numerical values — $R_{Cu} = 1/(0.03 \times 401) = 0.0831$ K/W; $R_{Al} = 0.141$ K/W

(*b*) $R = R_{Cu} + R_{Al}$ — $R = 0.224$ K/W

(*c*) $I = \Delta T/R$ — $I = 80/0.224$ W = 358 W

(*d*) $I_{Cu} = I_{Al} = I$; $\Delta T_{Cu} = I_{Cu}R_{Cu}$ — $\Delta T_{Cu} = 358 \times 0.0831$ K = 29.7 K; $T = 100 - 29.7 = 70.3°C$

---

30 •• The cubes in Problem 29 are rearranged in parallel as shown in Figure 21-16. Find (*a*) the thermal current carried by each cube from one side to the other, (*b*) the total thermal current, and (*c*) the equivalent thermal resistance of the two-cube system.

(*a*) Apply Equ. 21-9 to each cube — $I_{Cu} = 80/0.0831$ W = 963 W; $I_{Al} = 80/0.141$ W = 567 W (*b*) $I = I_{Cu} + I_{Al}$ $I = 1530$ W

(*c*) $R_{equ} = \Delta T/I$ $R = 80/1530$ K/W $= 0.0523$ K/W

---

31 •• A spherical shell of thermal conductivity k has inside radius r_1 and outside radius r_2 (Figure 21-17). The inside of the shell is held at a temperature T_1, and the outside at temperature T_2. In this problem, you are to show that the thermal current through the shell is given by

$$I = \frac{4\pi k r_1 r_2}{r_2 - r_1}(T_2 - T_1)$$

Consider a spherical element of the shell of radius r and thickness dr. (*a*) Why must the thermal current through each such element be the same? (*b*) Write the thermal current I through such a shell element in terms of the area $A = 4\pi r^2$, the thickness dr, and the temperature difference dT across the element. (*c*) Solve for dT in terms of dr and integrate from $r = r_1$ to $r = r_2$. (*d*) Show that when r_1 and r_2 are much larger than $r_2 - r_1$, Equation 21-22 is the same as Equation 21-7.

(*a*) From conservation of energy, the thermal current through each shell must be the same.

(*b*) $I = -kA(dT/dr) = -4\pi kr^2(dT/dr)$; note the minus sign – the heat current is directed opposite to temperature gradient.

(*c*) $dT = -(I/4\pi k)(dr/r^2)$; $\int_{T_1}^{T_2} dT = -\frac{I}{4\pi k}\int_{r_1}^{r_2}\frac{dr}{r^2}$; $T_2 - T_1 = \frac{I}{4\pi k}\left(\frac{1}{r_1} - \frac{1}{r_2}\right)$; $I = \frac{4\pi k r_1 r_2}{r_2 - r_1}(T_2 - T_1)$

(*d*) For $r_2 - r_1 << r_1$, $r_1 \approx r_2 = r$; let $r_2 - r_1 = \Delta r$. Now $I = 4\pi kr^2(\Delta T/\Delta r) = kA(\Delta T/\Delta r)$; Q.E.D.

---

32 •• A group of anthropologists is staying in the high Arctic for a month, and they need accommodation. They are directed to a small company, Inuit Igloos. "How thick do you want the walls?" asks Inuk, the head igloo maker. After some conferring, they reply that it should be 20°C inside when the temperature is −20°C outside. After looking the anthropologists over and poking them a bit, Inuk estimates that they would give off 38 MJ of heat per day. If the inside radius of the hemispherical igloo is to be 2 m, and the thermal conductivity of the compacted snow is 0.209 W / m·K, how thick should the walls be? (As an approximation, assume that the inner surface area of the igloo is equal to the outer surface area.)

1. Use Equ. 21-7; Change MJ/day to J/s — 440 W $= (0.209$ W/m·K$)(2\pi \times 2^2$ m$^2)(40$ K$)/t$
2. Solve for the thickness t. — $t = 0.478$ m $= 47.8$ cm

33* •• For a boiler at a power station, heat must be transferred to boiling water at the rate of 3 GW. The boiling water passes through copper pipes having a wall thickness of 4.0 mm and a surface area of 0.12 m^2 per meter length of pipe. Find the total length of pipe (actually there are many pipes in parallel) that must pass through the furnace if the steam temperature is 225°C and the external temperature of the pipes is 600°C.

1. From Equ. 21-7, $A = I\Delta x/k\Delta T$ — $A = (3 \times 10^9)(4 \times 10^{-3})/[(401)(375)]$ m$^2 = 79.8$ m^2
2. $L = A/(0.12$ m$)$ — $L = 665$ m

---

34 ••• A steam pipe of length L is insulated with a layer of material of thermal conductivity k. Find the rate of heat transfer if the temperature outside the insulation is t_1, the temperature inside is t_2, the outside radius of the insulation is r_1, and the inside radius is r_2.

Proceed as in Problem 31. Consider an element with a cylindrical area of length L, radius r, and thickness dr. The heat current is $I = -2\pi kLr(dT/dr)$. Thus, $dT = -[I/(2\pi kL)]dr/r$. Integrate from T_1 to T_2 and from r_1 to r_2 and solve for the heat current I. $I = 2\pi kL(T_1 - T_2)/\ln(r_1/r_2)$.

Note: If we use the above result in Problem 33 (take 0.12 m^2 to be the outside area per meter of pipe) then $r_1 = 1.91$ cm and $r_2 = 1.51$ cm. Solving for L one obtains $L = 746$ m.

---

35 ••• Brine at -16°C circulating through copper pipes with walls 1.5 mm thick is used to keep a cold room at 0°C. The diameter of each pipe is very large compared to the thickness of its walls. By what fraction is the transfer of heat reduced when the pipes are coated with a 5-mm layer of ice?

1. Use $R_{tot} = R_{Cu} + R_{ice}$; $R = \Delta x/kA$. Also $I = \Delta T/R$ — $I_{tot}/I_{Cu} = R_{Cu}/R = [1 + (\Delta x_{ice}k_{Cu})/(\Delta x_{Cu}k_{ice})]^{-1}$
2. Substitute numerical values. — $I_{tot}/I_{Cu} = 4.43 \times 10^{-4}$; I reduced by a factor of 2260.

---

36 • If the absolute temperature of an object is tripled, the rate at which it radiates thermal energy

(*a*) triples.

(*b*) increases by a factor of 9.

(*c*) increases by a factor of 27.

(*d*) increases by a factor of 81.

(*e*) depends on whether the absolute temperature is above or below zero.

(*d*) The energy radiated is proportional to T^4.

37* • Calculate λ_{max} for a human blackbody radiator, assuming the surface temperature of the skin to be 33°C.

Use Equ. 21-21 and substitute numerical values. — $\lambda_{max} = (2.898 \times 10^{-3})/(273 + 33)$ m $= 9.47$ μm

---

38 • The heating wires of a 1-kW electric heater are red hot at a temperature of 900°C. Assuming that 100% of the heat output is due to radiation and that the wires act as blackbody radiators, what is the effective area of the radiating surface? (Assume a room temperature of 20°C.)

From Equ. 21-20, $A = P_{net}/[e\sigma(T^4 - T_0^4)]$ — $A = 10^3/[1 \times 5.67 \times 10^{-8}(1173^4 - 293^4)]$ m^2 $= 9.35 \times 10^{-3}$ m^2

---

39 •• A blackened, solid copper sphere of radius 4.0 cm hangs in a vacuum in an enclosure whose walls have a temperature of 20°C. If the sphere is initially at 0°C, find the rate at which its temperature changes, assuming that heat is transferred by radiation only.

1. $dQ/dt = mc(dT/dt) = P_{net}$. Find P_{net} — $P_{net} = -4\pi \times 16 \times 10^{-4} \times 5.67 \times 10^{-8}(293^4 - 273^4)$ W $= -2.07$ W
2. Solve for dT/dt; $m = (4/3)\pi\rho r^3$, c = 0.386 kJ/kg·K — $dT/dt = -2.07 \times 3/(4\pi \times 8.96 \times 10^3 \times 64 \times 10^{-6} \times 386)$ K/s $= -2.23 \times 10^{-3}$ K/s

---

40 •• The surface temperature of the filament of an incandescent lamp is 1300°C. If the electric power input is doubled, what will the temperature become? *Hint:* Show that you can neglect the temperature of the surroundings.

1. $P_{net} = e\sigma A(T^4 - T_0^4) = e\sigma AT^4[1 - (T_0/T)^4]$ — $(273/1573)^4 = 9 \times 10^{-4} << 1$; neglect $(T_0/T)^4$.
2. $T \propto P^{1/4}$ — $T = 1573 \times 2^{1/4}$ K $= 1871$ K $= 1598$°C

41* •• Liquid helium is stored at its boiling point (4.2 K) in a spherical can that is separated by a vacuum space from a surrounding shield that is maintained at the temperature of liquid nitrogen (77 K). If the can is 30 cm in diameter and is blackened on the outside so that it acts as a blackbody, how much helium boils away per hour?

1. $dm/dt = P_{net}/L = e\sigma \pi d^2 T^4/L$. Here L is the latent heat of boiling and T_0 can be neglected (see Problem 40)

$dm/dt = (5.67 \times 10^{-8} \times \pi \times 0.3^2 \times 77^4/21 \times 10^3)$ kg/s $= 2.68 \times 10^{-5}$ kg/s $= 9.66 \times 10^{-2}$ kg/h $= 96.6$ g/h.

---

42 • In a cool room, a metal or marble table top feels much colder to the touch than a wood surface does even though they are at the same temperature. Why?

The thermal conductivity of metal and marble is much greater than that of wood; consequently, heat transfer from the hand is more rapid.

---

43 • True or false:

(*a*) During a phase change, the temperature of a substance remains constant.

(*b*) The rate of conduction of thermal energy is proportional to the temperature gradient.

(*c*) The rate at which an object radiates energy is proportional to the square of its absolute temperature.

(*d*) All materials expand when they are heated.

(e) The vapor pressure of a liquid depends on the temperature.

(*a*) True. (*b*) True. (*c*) False. (*d*) False; water contracts on heating between 0°C and 4°C. (*e*) True.

---

44 • Conduction is a method of heat transfer that

(*a*) can proceed in vacuum.

(*b*) involves the transfer of mass.

(*c*) is dominant in solids.

(*d*) depends on the fourth power of the absolute temperature.

(*c*)

45* • The earth loses heat by

(*a*) conduction.

(*b*) convection.

(*c*) radiation.

(*d*) all of the above.

(*c*)

---

46 • Which heat-transfer mechanisms are most important in the warming effect of a fire in a fireplace?

Radiation and convection.

---

47 • Which heat-transfer mechanism is important in the transfer of energy from the sun to the earth?

Radiation is the only mechanism.

---

48 •• Two cylinders made of materials A and B have the same lengths; their diameters are related by $d_A = 2d_B$. When the same temperature difference is maintained between the ends of the cylinders they conduct heat at the same rate. Their thermal conductivities are related by

(*a*) $k_A = k_B/4$

(*b*) $k_A = k_B/2$

(*c*) $k_A = k_B$

(*d*) $k_A = 2k_B$

(*e*) $k_A = 4k_B$

(*e*) From Equ. 21-7, kd^2 must be constant.

49* • A steel tape is placed around the earth at the equator when the temperature is 0°C. What will the clearance between the tape and the ground (assumed to be uniform) be if the temperature of the tape rises to 30°C? Neglect the expansion of the earth.

From Equ. 21-2, $\Delta R = R\alpha\Delta T$

$\Delta R = 6.38 \times 10^6 \times 11 \times 10^{-6} \times 30$ m $= 2.1 \times 10^3$ m $= 2.1$ km

---

50 •• Use the result of Problem 31 (Equation 21-22) to calculate the wall thickness of the hemispherical igloo of Problem 32 without assuming that the inner surface area equals the outer surface area.

Set $r_2 = r_1 + \Delta r$; from Problem 32 (igloo is a hemisphere) $\Delta r = \dfrac{2\pi k r_1^2 \Delta T}{I - 2\pi k r_1 \Delta T}$

For $r_1 = 2$ m, $\Delta T = 40$ K, $I = 440$ W, $k = 0.209$ W/m·K, one obtains $\Delta r = 0.63$ m $= 63$ cm.

---

51 •• Show that change in the density of an isotropic material due to an increase in temperature ΔT is given by $\Delta\rho = -\beta\rho\Delta T$.

$$\rho = \frac{M}{V};\ \frac{d\rho}{dT} = \frac{d\rho}{dV}\frac{dV}{dT} = -\frac{M}{V^2}\beta V = -\rho\beta;\ d\rho = -\beta\rho dT;\ \Delta\rho = -\beta\rho\Delta T$$

---

52 •• The solar constant is the power received from the sun per unit area perpendicular to the sun's rays at the mean distance of the earth from the sun. Its value at the upper atmosphere of the earth is about 1.35 kW/m². Calculate the effective temperature of the sun if it radiates like a blackbody. (The radius of the sun is 6.96×10^8 m.)

1. Determine the total power radiated. $R = 1.5 \times 10^{11}$ m is the sun–earth distance.

   $P = (1.35\ \text{kW/m}^2)(4\pi R^2) = (1.35\ \text{kW/m}^2)4\pi \times (1.5 \times 10^{11}\ \text{m})^2 = 38.2 \times 10^{22}\ \text{kW} = 3.82 \times 10^{26}$ W

2. Use Equ. 21-17; $A = 4\pi R_S^2$, where $R_S = 6.96 \times 10^8$ m

   $T = [3.82 \times 10^{26}/(5.67 \times 10^{-8} \times 4\pi \times 4.84 \times 10^{17})]^{1/4} = 5769$ K

53* •• Lou has patented a cooking timer, which he is marketing as "Nature's Way: Taking You Back To Simpler Times." The timer consists of a 28-cm copper rod having a 5.0-cm diameter. Just as the lower end is placed in boiling water, an ice cube is placed on the top of the rod. When the ice melts completely, the cooking time is up. A special ice cube tray makes cubes of various sizes to correspond to the boiling time required. What is the cooking time when a 30-g ice cube at –5.0°C is used?

1. Find t_1, time to raise the ice cube from -5°C to 0°C.

   $\Delta Q = (0.03\ \text{kg})(2.05 \times 10^3\ \text{J/kg·K})(5\ \text{K}) = 307.5$ J

   $t_1 = \Delta Q/[kA(\Delta T/\Delta x)]$; $\Delta Q = mc \times (5\ \text{K})$

   $t_1 = 307.5/[401 \times (\pi \times 25 \times 10^{-4}/4) \times (102.5/0.28)]$ s $= 1.07$ s

2. t_2, time to melt ice $= mL\Delta x/kA\Delta T$

   $$t_2 = \left(\frac{0.03 \times 333.5 \times 10^3 \times 0.28}{401 \times (25 \times 10^{-4}\pi/4) \times 100}\right)\ \text{s} = 35.6\ \text{s}$$

3. The total time $= t_1 + t_2$

   $t_{tot} = 36.7$ s

54 •• To determine the R value of insulating material that comes in sheets of $\frac{1}{2}$-in thickness, you construct a cubical box of 12 in per side and place a thermometer and a 100-W heater inside the box. After thermal equilibrium has been attained, the temperature inside the box is 90°C when the external temperature is 20°C. Determine the R value of this material.

1. $I = kA(\Delta T/\Delta x)$; $R_f = \Delta x/k = A\Delta T/I$ — $R_f = 6(0.3048\ \text{m})^2(70\ \text{K})/(100\ \text{W}) = 0.39\ \text{K}\cdot\text{m}^2/\text{W}$
2. Convert to U.S. customary units — $R_f = 2.2\ \text{h}\cdot\text{ft}^2\cdot\text{F}^\circ/\text{Btu}$

55 •• A 2-cm-thick copper sheet is pressed against a sheet of aluminum. What should be the thickness of the aluminum sheet so that the temperature of the copper–aluminum interface is $(T_1 + T_2)/2$, where T_1 and T_2 are the temperatures at the copper–air and aluminum–air interfaces?

$\Delta T_{Cu} = \Delta T_{Al} = \frac{1}{2}(T_1 - T_2) = (I/A)(t_{Cu}/k_{Cu})$ $= (I/A)(t_{Al}/k_{Al})$ — $t_{Al} = t_{Cu}(k_{Al}/k_{Cu}) = (2\ \text{cm})(237/401) = 1.18\ \text{cm}$

56 •• At a temperature of 20°C, a steel bar of radius 2.2 cm and length 60 cm is jammed horizontally perpendicular between two vertical concrete walls. With a blowtorch, the temperature of the bar is raised to 60°C. Find the force exerted by the bar on each wall.

$\Delta L = L\alpha\Delta T$; $F = AY\Delta L/L = AY\alpha\Delta T$ — $F = \pi(2.2 \times 10^{-2})^2(2.0 \times 10^{11})(11 \times 10^{-6})(40)\ \text{N}$ $= 1.34 \times 10^5\ \text{N}$

57* •• (*a*) From the definition of β, the coefficient of volume expansion (at constant pressure), show that $\beta = 1/T$ for an ideal gas. (*b*) The experimentally determined value of β for N_2 gas at 0°C is 0.003673 K^{-1}. Compare this value with the theoretical value $\beta = 1/T$, assuming that N_2 is an ideal gas.

(*a*) For an ideal gas, $V = nRT/P$; $\beta = (1/V)(dV/dT) = (P/nRT)(nR/P) = 1/T$. (*b*) $1/273 = 0.003663$ is within 0.3 % of the experimental value.

58 •• One way to construct a device with two points whose separation remains the same in spite of temperature changes is to bolt together one end of two rods having different coefficients of linear expansion as in the arrangement shown in Figure 21-18. (*a*) Show that the distance L will not change with temperature if the lengths L_A and L_B are chosen such that $L_A/L_B = \alpha_B/\alpha_A$. (*b*) If material B is steel, material A is brass, and $L_A = 250$ cm at 0°C, what is the value of L?

(*a*) We want $L = (L_B - L_A) = (L_B + \alpha_B L_B \Delta T) - (L_A + \alpha_A L_A \Delta T)$. Therefore , $\alpha_B L_B = \alpha_A L_A$ or $L_A/L_B = \alpha_B/\alpha_A$.

(*b*) From (*a*) $L_B = L_{Steel} = (250\ \text{cm})(19/11) = 432$ cm, and $L = 182$ cm.

59 •• On the average, the temperature of the earth's crust increases 1.0 C° for every 30 m of depth. The average thermal conductivity of the earth's crust is 0.74 J/m·s·K. What is the heat loss of the earth per second due to conduction from the core? How does this heat loss compare with the average power received from the sun? (The solar constant is about 1.35 kW/m^2.)

1. Heat current/m^2 = $I/\text{m}^2 = k(\Delta T/\Delta x)$ — $I/\text{m}^2 = 0.74/30\ \text{W/m}^2 = 0.0247\ \text{W/m}^2 < 0.002\%$ of the solar constant.
2. $dQ/dt = kA(\Delta T/\Delta x)$ — $dQ/dt = 4\pi(6.38 \times 10^6)^2 \times 0.0247\ \text{W} = 1.26 \times 10^{10}\ \text{kW}$

60 •• A copper-bottomed saucepan containing 0.8 L of boiling water boils dry in 10 min. Assuming that all the heat flows through the flat copper bottom, which has a diameter of 15 cm and a thickness of 3.0 mm, calculate the temperature of the outside of the copper bottom while some water is still in the pan.

1. $I = \Delta Q/\Delta t = kA(\Delta T/\Delta x)$; $\Delta Q = mL_v$; solve for ΔT — $\Delta T = mL_v\Delta x/kA\Delta t$
2. $m = 0.8$ kg, $L_v = 2257$ kJ/kg, $A = .0225\pi/4$ m^2, $\Delta x = 3\times10^{-3}$ m, $k = 401$ W/m·K. — $\Delta T = 4\times0.8\times2257\times10^3\times3\times10^{-3}/401\times0.0225\pi\times600$ $= 1.3$ K; $T_{out} = 100°C + \Delta T = 101.3°C$.

61* •• A hot-water tank of cylindrical shape has an inside diameter of 0.55 m and inside height of 1.2 m. The tank is enclosed with a 5-cm-thick insulating layer of glass wool whose thermal conductivity is 0.035 W/m·K. The metallic interior and exterior walls of the container have thermal conductivities that are much greater than that of the glass wool. How much power must be supplied to this tank to maintain the water temperature at 75°C when the external temperature is 1°C?

We will do this problem twice. First, we shall disregard the fact that the surrounding insulation is cylindrical. We shall then repeat the problem, using the result of Problem 34.

(a) 1. Find the total area — $A_{tot} = [2\times(\pi/4)(0.55)^2 + \pi\times0.55\times1.2]$ m^2 = 2.55 m^2

2. Use Equ. 21-7 — $I = (0.035)(2.55)(74/0.05)$ W = 132 W

(b) 1. Find I through top and bottom surfaces- I_1 — $I_1 = (0.035)[(\pi/2)(0.55)^2](74/0.05)$ W = 24.6 W

2. Find I_c through cylindrical surface (see Problem 34) — $I_c = 2\pi(0.035)(1.2)(74)/\ln(0.65/0.55)$ W = 97.4 W

3. Find the total heat loss $I = I_1 + I_c$ — $I = (24.6 + 97.4)$ W = 122 W

---

62 ••• The diameter of a rod is given by $d = d_0(1 + ax)$, where a is a constant and x is the distance from one end. If the thermal conductivity of the material is k what is the thermal resistance of the rod if its length is L?

$$I = -kA(dT/dx);\ A = (\pi/4)d_0^2(1+ax)^2;\ \int_{T_1}^{T_2} dT = -\frac{4I}{\pi k d_0^2}\int_0^L \frac{dx}{(1+ax)^2};\ T_2 - T_1 = \frac{4IL}{\pi k d_0^2(1+aL)}$$

$$R = \frac{\Delta T}{I} = \frac{4L}{\pi k d_0^2(1+aL)}$$

---

63 ••• A solid disk of radius R and mass M is spinning in a frictionless environment with angular velocity ω_1 at temperature T_1. The temperature of the disk is then changed to T_2. Express the angular velocity ω_2, rotational kinetic energy E_2, and angular momentum L_2 in terms of their values at the temperature T_1 and the linear expansion coefficient α of the disk.

Let $\Delta T = T_2 - T_1$. 1. No torque: $L_2 = L_1$. 2. $I_2 = MR_2^2 = MR_1^2(1 + \alpha\Delta T)^2 = I_1(1 + 2\alpha\Delta T + \alpha^2\Delta T^2) = I_1(1 + 2\alpha\Delta T)$, neglecting the higher order term. $I_1\omega_1 = I_2\omega_2$, $\omega_2 = \omega_1(I_1/I_2) = \omega_1(1 - 2\alpha\Delta T)$. $E_2 = L_2^2/2I_2 = L_1^2/2I_2$; $E_1 = L_1^2/2I_1$; $E_2 = E_1(I_1/I_2) = E_1(1 - 2\alpha\Delta T)$.

---

64 ••• A small pond has a layer of ice 1 cm thick floating on its surface. (*a*) If the air temperature is –10°C, find the rate in centimeters per hour at which ice is added to the bottom of the layer. The density of ice is 0.917 g/cm^3. (*b*) How long does it take for a 20-cm layer to be built up?

(*a*) To freeze m kg of water, $Q = mL_f$. To determine the rate of freezing, $dQ/dt = L_f(dm/dt) = L_f\rho A(dx/dt)$. But we also have $dQ/dt = kA\Delta T/x$, where x is thickness of the ice.

$dx/dt = (k/L_f\rho)\Delta T/x$ $\qquad dx/dt = (0.592/333.5 \times 10^3 \times 917)(10/0.01)$ m/s $= 1.94\ \mu$m/s $= 6.97$ mm/h

(*b*) $$\int_{x_i}^{x_f} x\,dx = \frac{k\Delta T}{\rho L_f}\int_0^t dt;\ \tfrac{1}{2}(x_f^2 - x_i^2) = \frac{k\Delta T}{\rho L_f}t$$ $\qquad t = [(333.5 \times 10^3 \times 917)/(2 \times 0.592 \times 10)](0.04 - 10^{-4})$ s $= 1.03 \times 10^6$ s $\simeq$ 12 days

---

65* ••• A body initially at a temperature T_i cools by convection and radiation in a room where the temperature is T_0. The body obeys Newton's law of cooling, which can be written $dQ/dt = hA(T - T_0)$, where A is the area of the body and h is a constant called the surface coefficient of heat transfer. Show that the temperature T at any time t is given by $T = T_0 + (T_i - T_0)e^{-hAt/mc}$, where m is the mass of the body and c is its specific heat.

1. $dQ = -mcdT$ is heat loss as T diminishes by dT. Thus, $dT = -(1/mc)dQ$ and $dT/dt = -(hA/mc)(T - T_0)$

2. $$\int_{T_i}^{T}\frac{dT}{T - T_0} = -\frac{hA}{mc}\int_0^t dt;\quad \ln\left(\frac{T - T_0}{T_i - T_0}\right) = -\frac{hA}{mc}t,$$ where T_i is the initial temperature.

Take the antilog and solve for T to obtain $T = T_0 + (T_i - T_0)e^{-hAt/mc}$.

---

66 ••• Two 200-g copper containers, each holding 0.7 L of water, are connected by a 10-cm copper rod of cross-sectional area 1.5 cm^2. Initially, one container is at 60°C; the second is maintained at 0°C. (*a*) Show that the temperature t_c of the first container changes over time t according to

$$t_c = t_{c0}e^{-t/RC}$$

where t_{c0} is the initial temperature of the first container, R is the thermal resistance of the rod, and C is the total heat capacity of the container plus the water. (*b*) Evaluate R, C, and the "time constant" RC. (*c*) Show that the total amount of heat Q conducted in time t is

$$Q = Ct_{c0}(1 - e^{-t/RC})$$

(*d*) Find the time it takes for the temperature of the first container to be reduced to 30°C.

(*a*) Heat loss $dQ = -(m_cc_c + m_wc_w)dT_c$, where m_c and c_c are the mass and specific heat of the container and m_w and c_w are the mass and specific heat of the water. Let the sum of those heat capacities be C. Then we can write $dQ/dt = -C(dT_c/dt)$. But we also have, from Equ. 21-7, $dQ/dt = kA(T_c - 0)/\Delta x = T_c/R$. We therefore have

$$\int_{T_{c0}}^{T_c}\frac{dT_c}{T_c} = -\frac{1}{RC}\int_0^t dt;\quad \ln\left(\frac{T_c}{T_{c0}}\right) = -\frac{1}{RC}t;\quad T_c = T_{c0}e^{-t/RC}.$$

(*b*) $R = \Delta x/kA$; $\qquad R = 0.1/(401 \times 1.5 \times 10^{-4})$ K/W $= 1.66$ K/W;

$C = (m_cc_c + m_wc_w)$ $\qquad C = (0.2 \times 386 + 0.7 \times 4180)$ J/K $= 3$ kJ/K; $RC = 4985$ s $= 1.38$ h

(*c*) To determine Q we integrate $dQ = (T_c/R)dt$, where $T_c = T_{c0}e^{-t/RC}$.

$$\int_0^Q dQ = \frac{T_{c0}}{R}\int_0^t e^{-t/RC}\,dt;\quad Q = CT_{c0}(1 - e^{-t/RC})$$

(*d*) $T_c = \frac{1}{2}T_{c0}$; $\exp(-t/RC) = \frac{1}{2}$; $t = RC\ln(2)$ $t = 1.38 \times \ln(2) = 0.96$ h

---

67 ••• Liquid helium is stored in containers fitted with 7-cm-thick "superinsulation" consisting of a large number of layers of very thin aluminized Mylar sheets. The rate of evaporation of liquid in a 200-L container is about 0.7 L per day. Assume the container is spherical and that the external temperature is 20°C. The specific gravity of liquid helium is 0.125 and the latent heat of vaporization is 21 kJ/kg. Estimate the thermal conductivity of superinsulation.

1. Find rate of loss in kg/s. $dm/dt = \rho(dV/dt)$ $dm/dt = (0.125 \times 10^3\text{ kg/m}^3)(0.7 \times 10^{-3}\text{ m}^3)/(86400\text{ s})$ $= 1.01 \times 10^{-6}$ kg/s

2. $I = L_v(dm/dt) = kA(\Delta T/\Delta x)$; $k = \dfrac{L_v\Delta x(dm/dt)}{A\Delta T}$ $A = 4\pi(3 \times 2.0\text{x}10^{-1}/4\pi)^{2/3} = 1.65\text{ m}^2$
$A = 4\pi(3V/4\pi)^{2/3}$ $k = (21\times10^3\times7\times10^{-2}\times1.01\times10^{-6})/(1.65\times288)$ W/m·K
$= 3.1 \times 10^{-6}$ W/m·K
